VOLUME FOUR HUNDRED AND NINETY-SEVEN

METHODS IN ENZYMOLOGY

Synthetic Biology, Part A

Methods for Part/Device Characterization and Chassis Engineering

METHODS IN ENZYMOLOGY

Editors-in-Chief

JOHN N. ABELSON AND MELVIN I. SIMON

Division of Biology
California Institute of Technology
Pasadena, California

Founding Editors

SIDNEY P. COLOWICK AND NATHAN O. KAPLAN

VOLUME FOUR HUNDRED AND NINETY-SEVEN

METHODS IN ENZYMOLOGY

Synthetic Biology, Part A

Methods for Part/Device Characterization and Chassis Engineering

EDITED BY

CHRISTOPHER VOIGT
University of California - San Francisco
San Francisco
California, USA

[redacted]

ELSEVIER

AMSTERDAM • BOSTON • HEIDELBERG • LONDON
NEW YORK • OXFORD • PARIS • SAN DIEGO
SAN FRANCISCO • SINGAPORE • SYDNEY • TOKYO
Academic Press is an imprint of Elsevier

Academic Press is an imprint of Elsevier
525 B Street, Suite 1900, San Diego, CA 92101-4495, USA
225 Wyman Street, Waltham, MA 02451, USA
32 Jamestown Road, London NW1 7BY, UK

First edition 2011

ISBN: 978-0-12-385075-1
ISSN: 0076-6879

Printed and bound in United States of America
11 12 13 14 10 9 8 7 6 5 4 3 2 1

CONTENTS

Section II. Device and System Design, Optimization, and Debugging 157

Contributors

Kirk D. Albrecht
Department of Biology, Colorado State University, Fort Collins, Colorado, USA

Wenlin An[1]
Medical Research Council Laboratory of Molecular Biology, Hills Road, Cambridge, United Kingdom

Rajat Anand
National Centre for Biological Sciences, Tata Institute of Fundamental Research, Bangalore, India

Aseem Z. Ansari
Department of Biochemistry, The Genome Center of Wisconsin, University of Wisconsin-Madison, Madison, Wisconsin, USA

Mauricio S. Antunes
Department of Biology, Colorado State University, Fort Collins, Colorado, USA

J. Thomas Beatty
Department of Microbiology and Immunology, University of British Columbia, Life Sciences Centre, Vancouver, British Columbia, Canada

Yaakov Benenson
Department of Biosystems Science and Engineering, Swiss Federal Institute of Technology (ETH Zurich), Mattenstrasse 26, Basel, Switzerland

Devesh Bhimisaria
Department of Biochemistry, The Genome Center of Wisconsin, University of Wisconsin-Madison, Madison, Wisconsin, USA

Tessa A. Bowen
Department of Biology, Colorado State University, Fort Collins, Colorado, USA

Edward S. Boyden
Synthetic Neurobiology Group, The Media Laboratory and McGovern Institute, Departments of Biological Engineering and Brain and Cognitive Sciences, Massachusetts Institute of Technology, Cambridge, Massachusetts, USA

[1] Current address: MRC Centre for Developmental Neurobiology, New Hunt's House, King's College London, London, United Kingdom

Rainer Breitling
Groningen Bioinformatics Centre, Groningen Biomolecular Sciences and Biotechnology Institute, University of Groningen, Groningen, The Netherlands, and Institute of Molecular, Cell and Systems Biology, College of Medical, Veterinary and Life Sciences, Joseph Black Building, University of Glasgow, Glasgow, United Kingdom

Daniel Camsund
Department of Photochemistry and Molecular Science, Ångström Laboratories, Uppsala University, Uppsala, Sweden

Swapnil Chhabra
VIMSS (Virtual Institute of Microbial Stress and Survival); Physical Biosciences Division, Lawrence Berkeley National Laboratory, Berkeley, and Joint BioEnergy Institute, Emeryville, California, USA

Jason W. Chin
Medical Research Council Laboratory of Molecular Biology, Hills Road, Cambridge, United Kingdom

Brian Y. Chow[2]
Synthetic Neurobiology Group, The Media Laboratory and McGovern Institute, Departments of Biological Engineering and Brain and Cognitive Sciences, Massachusetts Institute of Technology, Cambridge, Massachusetts, USA

Amy S. Chuong
Synthetic Neurobiology Group, The Media Laboratory and McGovern Institute, Departments of Biological Engineering and Brain and Cognitive Sciences, Massachusetts Institute of Technology, Cambridge, Massachusetts, USA

James Dixon
Sharon High School, Sharon, Massachusetts, USA

Leslie J. Donato
Department of Biochemistry, The Genome Center of Wisconsin, University of Wisconsin-Madison, Madison, Wisconsin, USA

John E. Dueber
Department of Bioengineering, and Energy Biosciences Institute, University of California, Berkeley, California, USA

Xavier Duportet
INRIA, Paris-Rocquencourt, 78153 Le Chesnay, France, and Department of Biological Engineering, Massachusetts Institute of Technology, Cambridge, Massachusetts, USA

[2] Future location: Department of Bioengineering, University of Pennsylvania, Philadelphia, Pennsylvania, USA

M. S. Ferry
Department of Bioengineering, University of California, San Diego, California, USA

Kate Franz
Department of Chemistry and Biochemistry, University of California, San Diego, La Jolla, California, USA

Martin Fussenegger
Department of Biosystems Science and Engineering, ETH Zurich, and University of Basel, Faculty of Science, Basel, Switzerland

Justin P. Gallivan
Department of Chemistry and Center for Fundamental and Applied Molecular Evolution, Emory University, Atlanta, Georgia, USA

Delquin Gong
Cardiovascular Research Institute and Department of Biochemistry, University of California San Francisco, San Francisco, California

Carol A. Gross
Department of Microbiology and Immunology, and Department of Cell and Tissue Biology, University of California at San Francisco, San Francisco, California, USA

Saurabh Gupta
Computer Science and Artificial Intelligence Laboratory, and Department of Biological Engineering, Massachusetts Institute of Technology, Cambridge, Massachusetts, USA

Patrick Guye
Computer Science and Artificial Intelligence Laboratory, and Department of Biological Engineering, Massachusetts Institute of Technology, Cambridge, Massachusetts, USA

Klaus M. Hahn
Department of Pharmacology and Lineberger Cancer Center, University of North Carolina, Chapel Hill, North Carolina, USA

J. Hasty
Department of Bioengineering; BioCircuits Institute, and Molecular Biology Section, Division of Biological Sciences, University of California, San Diego, California, USA

Keira L. Havens
Department of Biology, Colorado State University, Fort Collins, Colorado, USA

Li He
Department of Biological Chemistry, Center of Cell Dynamics, Johns Hopkins School of Medicine, Baltimore, Maryland, USA

Thorsten Heidorn
Department of Photochemistry and Molecular Science, Ångström Laboratories, Uppsala University, Uppsala, Sweden

Hsin-Ho Huang
Department of Photochemistry and Molecular Science, Ångström Laboratories, Uppsala University, Uppsala, Sweden

Paul R. Jaschke
Department of Microbiology and Immunology, University of British Columbia, Life Sciences Centre, Vancouver, British Columbia, Canada

Maria Karlsson
Faculty of Biology, Albert-Ludwigs-Universität Freiburg; Centre for Biological Signaling Studies (BIOSS), Albert-Ludwigs-Universität Freiburg, Freiburg, Germany, and Department of Biosystems Science and Engineering, ETH Zurich, Basel, Switzerland

Kimberly L. Keller
University of Missouri, Columbia, Missouri, and VIMSS (Virtual Institute of Microbial Stress and Survival), Berkeley, California, USA

Nathan C. Klapoetke
Synthetic Neurobiology Group, The Media Laboratory and McGovern Institute, Departments of Biological Engineering and Brain and Cognitive Sciences, Massachusetts Institute of Technology, Cambridge, Massachusetts, USA

Natalie Kuldell
MIT, Department of Biological Engineering, Cambridge, Massachusetts, USA

J. C. Liao
Department of Chemical and Biomolecular Engineering, University of California, Los Angeles, Los Angeles, California, USA

Wendell A. Lim
Department of Cellular and Molecular Pharmacology, and Howard Hughes Medical Institute, University of California San Francisco, San Francisco, California

Pia Lindberg
Department of Photochemistry and Molecular Science, Ångström Laboratories, Uppsala University, Uppsala, Sweden

Peter Lindblad
Department of Photochemistry and Molecular Science, Ångström Laboratories, Uppsala University, Uppsala, Sweden

Wenxiu Ma
Department of Computer Science, Stanford University, Stanford, California, USA

Marnix H. Medema
Department of Microbial Physiology, and Groningen Bioinformatics Centre, Groningen Biomolecular Sciences and Biotechnology Institute, University of Groningen, Groningen, The Netherlands

June I. Medford
Department of Biology, Colorado State University, Fort Collins, Colorado, USA

Denise Montell
Department of Biological Chemistry, Center of Cell Dynamics, Johns Hopkins School of Medicine, Baltimore, Maryland, USA

Kevin J. Morey
Department of Biology, Colorado State University, Fort Collins, Colorado, USA

Joseph R. Nevins
Institute for Genome Sciences and Policy, and Department of Molecular Genetics and Microbiology, Duke University, Durham, North Carolina, USA

Elke Nevoigt
School of Engineering and Science, Jacobs University gGmbH, Bremen, Germany

Stephan Noll
Gene Bridges GmbH, Im Neuenheimer Feld 584, Heidelberg, Germany

Paulo Oliveira
Department of Photochemistry and Molecular Science, Ångström Laboratories, Uppsala University, Uppsala, Sweden

Navneet Rai
National Centre for Biological Sciences, Tata Institute of Fundamental Research, Bangalore, and Department of Biosciences and Bioengineering, Indian Institute of Technology, Mumbai, India

Adrian Randall
Computational and Systems Biology Initiative, Massachusetts Institute of Technology, Cambridge, Massachusetts, USA

I. A. Razinkov
Department of Bioengineering, University of California, San Diego, California, USA

Virgil A. Rhodius
Department of Microbiology and Immunology, and Department of Pharmaceutical Chemistry, University of California at San Francisco, San Francisco, California, USA

Gürol Süel
University of Texas Southwestern Medical Center at Dallas, Dallas, Texas, USA

Rafael G. Saer
Department of Microbiology and Immunology, University of British Columbia, Life Sciences Centre, Vancouver, British Columbia, Canada

C. R. Shen
Department of Chemical and Biomolecular Engineering, University of California, Los Angeles, Los Angeles, California, USA

Abhyudai Singh
Department of Chemistry and Biochemistry, University of California, San Diego, La Jolla, California, USA

Joy Sinha
Department of Chemistry and Center for Fundamental and Applied Molecular Evolution, Emory University, Atlanta, Georgia, USA

Karin Stensjö
Department of Photochemistry and Molecular Science, Ångström Laboratories, Uppsala University, Uppsala, Sweden

Gregory Stephanopoulos
Department of Chemical Engineering, Massachusetts Institute of Technology, Cambridge, Massachusetts, USA

Jeffrey J. Tabor
Department of Bioengineering, Rice University, Houston, Texas, USA

Eriko Takano
Department of Microbial Physiology, Groningen Biomolecular Sciences and Biotechnology Institute, University of Groningen, Groningen, The Netherlands

Mukund Thattai
National Centre for Biological Sciences, Tata Institute of Fundamental Research, Bangalore, India

Joshua R. Tietjen
Department of Biochemistry, The Genome Center of Wisconsin, University of Wisconsin-Madison, Madison, Wisconsin, USA

Jared E. Toettcher
Cardiovascular Research Institute and Department of Biochemistry, and Department of Cellular and Molecular Pharmacology, University of California San Francisco, San Francisco, California

Shana Topp
Department of Chemistry and Center for Fundamental and Applied Molecular Evolution, Emory University, Atlanta, Georgia, USA

Jared F. Troupe
Department of Biology, Colorado State University, Fort Collins, Colorado, USA

Keith E. J. Tyo
Department of Chemical and Biological Engineering, Northwestern University, Evanston, Illinois, USA

Judy D. Wall
University of Missouri, Columbia, Missouri, and VIMSS (Virtual Institute of Microbial Stress and Survival), Berkeley, California, USA

Xiaobo Wang
Department of Biological Chemistry, Center of Cell Dynamics, Johns Hopkins School of Medicine, Baltimore, Maryland, USA

Wilfried Weber
Faculty of Biology, Albert-Ludwigs-Universität Freiburg; Centre for Biological Signaling Studies (BIOSS), Albert-Ludwigs-Universität Freiburg, Freiburg, Germany, and Department of Biosystems Science and Engineering, ETH Zurich, Basel, Switzerland

Leor S. Weinberger
Department of Chemistry and Biochemistry, University of California, San Diego, La Jolla, California, USA

Orion D. Weiner
Cardiovascular Research Institute and Department of Biochemistry, University of California San Francisco, San Francisco, California

Ron Weiss
Departments of Biological Engineering and Electrical Engineering and Computer Science, and Computer Science and Artificial Intelligence Laboratory, Massachusetts Institute of Technology, Cambridge,Massachusetts, USA

Weston R. Whitaker
Department of Bioengineering, University of California, Berkeley, California, USA

Jeffrey V. Wong
Department of Biomedical Engineering, and Institute for Genome Sciences and Policy, Duke University, Durham, North Carolina, USA

Wing Hung Wong
Department of Statistics, and Department of Health Research and Policy, Stanford University, Stanford, California, USA

Yi I. Wu[3]
Department of Pharmacology and Lineberger Cancer Center, University of North Carolina, Chapel Hill, North Carolina, USA

[3] Present address: Robert D. Berlin Center for Cell Analysis and Modeling, and Department of Genetics and Developmental Biology, University of Connecticut Health Center, Farmington, Connecticut, USA

Guang Yao[4]
Institute for Genome Sciences and Policy, Duke University, Durham, North Carolina, USA

Lingchong You
Department of Biomedical Engineering; Institute for Genome Sciences and Policy, and Center for Systems Biology, Duke University, Durham, North Carolina, USA

[4] Current address: Department of Molecular and Cellular Biology, University of Arizona, Tucson, Arizona, USA

Preface

This is the first volume of a two-part series in *Methods in Enzymology* on tools and techniques used in synthetic biology. Synthetic biology is an engineering discipline that seeks to construct living systems that do not exist in nature. The field refers to the process by which genetic systems are designed and constructed, as opposed to any particular application. Along these lines, these volumes are organized into two areas. Volume I focuses on the assay techniques and design principles underlying the characterization of genetic parts, their combination into devices and programs, and their integration into various hosts. Volume II focuses on computational tools and biophysical models to aid in the design and organization of genetic programs and modern methods to synthesize and assemble the associated DNA.

The chapters in Volume I have been organized to reflect the hierarchy from simple genetic parts to complete systems in complex hosts. Genetic parts are units of DNA that encode a simple function, such as a promoter- or ribosome-binding site. The first set of chapters focus on new methods to rapidly construct and characterize such parts. This includes methods to rapidly assay transcription (e.g., ChIP-seq) and translational processes, and the application of directed evolution to design promoter libraries. The next set of chapters deal with the design principles for assembling parts into various devices. These devices can encode genetic circuits that function analogously to their electronic counterparts (e.g., logic gates, oscillators, pulse generators). The biochemistry that underlies these circuits varies, and examples are included that rely on proteins and RNA.

Devices often encode complex, dynamic functions, making their characterization a challenge. To this end, a variety of advanced techniques are described, including the application of microscopy and microfluidics to measure the dynamics of single cells. The characterization of a circuit requires its perturbation. Typically, this has been accomplished through the addition (or removal) of an inducer such as IPTG. This process is slow and inaccurate and relies on a transcriptional process. New methods in "optogenetics" allow the perturbation of signaling networks using light. This allows rapid (millisecond) activation and deactivation of a pathway.

The design of metabolic devices is also included. Several chapters are focused on projects that have been designed as teaching modules in undergraduate labs, including a banana-scented bacterium and bacterial photography/edge detection.

The most common host organism in synthetic biology for prokaryotes is *Escherichia coli* and for eukaryotes is *Saccharomyces cerevisiae*. This is due to the availability of simple methods for genetic manipulation. In the natural world, there are innumerable organisms with useful functions and applications. Much progress has been made in establishing genetic tools for model organisms. To this end, chapters have been included that focus on photosynthetic organisms (*Rhodobacter* and *Cyanobacteria*), organisms relevant for fuel/chemical/pharmaceutical production (*Streptomyces* and *Desulphovibrio*), plants, and mammalian viruses.

Christopher Voigt

Methods in Enzymology

Volume XXXV. Lipids (Part B)
Edited by John M. Lowenstein

Volume XXXVI. Hormone Action (Part A: Steroid Hormones)
Edited by Bert W. O'Malley and Joel G. Hardman

Volume XXXVII. Hormone Action (Part B: Peptide Hormones)
Edited by Bert W. O'Malley and Joel G. Hardman

Volume XXXVIII. Hormone Action (Part C: Cyclic Nucleotides)
Edited by Joel G. Hardman and Bert W. O'Malley

Volume XXXIX. Hormone Action (Part D: Isolated Cells, Tissues, and Organ Systems)
Edited by Joel G. Hardman and Bert W. O'Malley

Volume XL. Hormone Action (Part E: Nuclear Structure and Function)
Edited by Bert W. O'Malley and Joel G. Hardman

Volume XLI. Carbohydrate Metabolism (Part B)
Edited by W. A. Wood

Volume XLII. Carbohydrate Metabolism (Part C)
Edited by W. A. Wood

Volume XLIII. Antibiotics
Edited by John H. Hash

Volume XLIV. Immobilized Enzymes
Edited by Klaus Mosbach

Volume XLV. Proteolytic Enzymes (Part B)
Edited by Laszlo Lorand

Volume XLVI. Affinity Labeling
Edited by William B. Jakoby and Meir Wilchek

Volume XLVII. Enzyme Structure (Part E)
Edited by C. H. W. Hirs and Serge N. Timasheff

Volume XLVIII. Enzyme Structure (Part F)
Edited by C. H. W. Hirs and Serge N. Timasheff

Volume XLIX. Enzyme Structure (Part G)
Edited by C. H. W. Hirs and Serge N. Timasheff

Volume L. Complex Carbohydrates (Part C)
Edited by Victor Ginsburg

Volume LI. Purine and Pyrimidine Nucleotide Metabolism
Edited by Patricia A. Hoffee and Mary Ellen Jones

Volume LII. Biomembranes (Part C: Biological Oxidations)
Edited by Sidney Fleischer and Lester Packer

Volume LIII. Biomembranes (Part D: Biological Oxidations)
Edited by Sidney Fleischer and Lester Packer

Volume LIV. Biomembranes (Part E: Biological Oxidations)
Edited by Sidney Fleischer and Lester Packer

Volume LV. Biomembranes (Part F: Bioenergetics)
Edited by Sidney Fleischer and Lester Packer

Volume LVI. Biomembranes (Part G: Bioenergetics)
Edited by Sidney Fleischer and Lester Packer

Volume LVII. Bioluminescence and Chemiluminescence
Edited by Marlene A. DeLuca

Volume LVIII. Cell Culture
Edited by William B. Jakoby and Ira Pastan

Volume LIX. Nucleic Acids and Protein Synthesis (Part G)
Edited by Kivie Moldave and Lawrence Grossman

Volume LX. Nucleic Acids and Protein Synthesis (Part H)
Edited by Kivie Moldave and Lawrence Grossman

Volume 61. Enzyme Structure (Part H)
Edited by C. H. W. Hirs and Serge N. Timasheff

Volume 62. Vitamins and Coenzymes (Part D)
Edited by Donald B. McCormick and Lemuel D. Wright

Volume 63. Enzyme Kinetics and Mechanism (Part A: Initial Rate and Inhibitor Methods)
Edited by Daniel L. Purich

Volume 64. Enzyme Kinetics and Mechanism (Part B: Isotopic Probes and Complex Enzyme Systems)
Edited by Daniel L. Purich

Volume 65. Nucleic Acids (Part I)
Edited by Lawrence Grossman and Kivie Moldave

Volume 66. Vitamins and Coenzymes (Part E)
Edited by Donald B. McCormick and Lemuel D. Wright

Volume 67. Vitamins and Coenzymes (Part F)
Edited by Donald B. McCormick and Lemuel D. Wright

Volume 68. Recombinant DNA
Edited by Ray Wu

Volume 69. Photosynthesis and Nitrogen Fixation (Part C)
Edited by Anthony San Pietro

Volume 70. Immunochemical Techniques (Part A)
Edited by Helen Van Vunakis and John J. Langone

Volume 89. Carbohydrate Metabolism (Part D)
Edited by Willis A. Wood

Volume 90. Carbohydrate Metabolism (Part E)
Edited by Willis A. Wood

Volume 91. Enzyme Structure (Part I)
Edited by C. H. W. Hirs and Serge N. Timasheff

Volume 92. Immunochemical Techniques (Part E: Monoclonal Antibodies and General Immunoassay Methods)
Edited by John J. Langone and Helen Van Vunakis

Volume 93. Immunochemical Techniques (Part F: Conventional Antibodies, Fc Receptors, and Cytotoxicity)
Edited by John J. Langone and Helen Van Vunakis

Volume 94. Polyamines
Edited by Herbert Tabor and Celia White Tabor

Volume 95. Cumulative Subject Index Volumes 61–74, 76–80
Edited by Edward A. Dennis and Martha G. Dennis

Volume 96. Biomembranes [Part J: Membrane Biogenesis: Assembly and Targeting (General Methods; Eukaryotes)]
Edited by Sidney Fleischer and Becca Fleischer

Volume 97. Biomembranes [Part K: Membrane Biogenesis: Assembly and Targeting (Prokaryotes, Mitochondria, and Chloroplasts)]
Edited by Sidney Fleischer and Becca Fleischer

Volume 98. Biomembranes (Part L: Membrane Biogenesis: Processing and Recycling)
Edited by Sidney Fleischer and Becca Fleischer

Volume 99. Hormone Action (Part F: Protein Kinases)
Edited by Jackie D. Corbin and Joel G. Hardman

Volume 100. Recombinant DNA (Part B)
Edited by Ray Wu, Lawrence Grossman, and Kivie Moldave

Volume 101. Recombinant DNA (Part C)
Edited by Ray Wu, Lawrence Grossman, and Kivie Moldave

Volume 102. Hormone Action (Part G: Calmodulin and Calcium-Binding Proteins)
Edited by Anthony R. Means and Bert W. O'Malley

Volume 103. Hormone Action (Part H: Neuroendocrine Peptides)
Edited by P. Michael Conn

Volume 104. Enzyme Purification and Related Techniques (Part C)
Edited by William B. Jakoby

VOLUME 123. Vitamins and Coenzymes (Part H)
Edited by FRANK CHYTIL AND DONALD B. MCCORMICK

VOLUME 124. Hormone Action (Part J: Neuroendocrine Peptides)
Edited by P. MICHAEL CONN

VOLUME 125. Biomembranes (Part M: Transport in Bacteria, Mitochondria, and Chloroplasts: General Approaches and Transport Systems)
Edited by SIDNEY FLEISCHER AND BECCA FLEISCHER

VOLUME 126. Biomembranes (Part N: Transport in Bacteria, Mitochondria, and Chloroplasts: Protonmotive Force)
Edited by SIDNEY FLEISCHER AND BECCA FLEISCHER

VOLUME 127. Biomembranes (Part O: Protons and Water: Structure and Translocation)
Edited by LESTER PACKER

VOLUME 128. Plasma Lipoproteins (Part A: Preparation, Structure, and Molecular Biology)
Edited by JERE P. SEGREST AND JOHN J. ALBERS

VOLUME 129. Plasma Lipoproteins (Part B: Characterization, Cell Biology, and Metabolism)
Edited by JOHN J. ALBERS AND JERE P. SEGREST

VOLUME 130. Enzyme Structure (Part K)
Edited by C. H. W. HIRS AND SERGE N. TIMASHEFF

VOLUME 131. Enzyme Structure (Part L)
Edited by C. H. W. HIRS AND SERGE N. TIMASHEFF

VOLUME 132. Immunochemical Techniques (Part J: Phagocytosis and Cell-Mediated Cytotoxicity)
Edited by GIOVANNI DI SABATO AND JOHANNES EVERSE

VOLUME 133. Bioluminescence and Chemiluminescence (Part B)
Edited by MARLENE DELUCA AND WILLIAM D. MCELROY

VOLUME 134. Structural and Contractile Proteins (Part C: The Contractile Apparatus and the Cytoskeleton)
Edited by RICHARD B. VALLEE

VOLUME 135. Immobilized Enzymes and Cells (Part B)
Edited by KLAUS MOSBACH

VOLUME 136. Immobilized Enzymes and Cells (Part C)
Edited by KLAUS MOSBACH

VOLUME 137. Immobilized Enzymes and Cells (Part D)
Edited by KLAUS MOSBACH

VOLUME 138. Complex Carbohydrates (Part E)
Edited by VICTOR GINSBURG

VOLUME 139. Cellular Regulators (Part A: Calcium- and Calmodulin-Binding Proteins)
Edited by ANTHONY R. MEANS AND P. MICHAEL CONN

VOLUME 140. Cumulative Subject Index Volumes 102–119, 121–134

VOLUME 141. Cellular Regulators (Part B: Calcium and Lipids)
Edited by P. MICHAEL CONN AND ANTHONY R. MEANS

VOLUME 142. Metabolism of Aromatic Amino Acids and Amines
Edited by SEYMOUR KAUFMAN

VOLUME 143. Sulfur and Sulfur Amino Acids
Edited by WILLIAM B. JAKOBY AND OWEN GRIFFITH

VOLUME 144. Structural and Contractile Proteins (Part D: Extracellular Matrix)
Edited by LEON W. CUNNINGHAM

VOLUME 145. Structural and Contractile Proteins (Part E: Extracellular Matrix)
Edited by LEON W. CUNNINGHAM

VOLUME 146. Peptide Growth Factors (Part A)
Edited by DAVID BARNES AND DAVID A. SIRBASKU

VOLUME 147. Peptide Growth Factors (Part B)
Edited by DAVID BARNES AND DAVID A. SIRBASKU

VOLUME 148. Plant Cell Membranes
Edited by LESTER PACKER AND ROLAND DOUCE

VOLUME 149. Drug and Enzyme Targeting (Part B)
Edited by RALPH GREEN AND KENNETH J. WIDDER

VOLUME 150. Immunochemical Techniques (Part K: *In Vitro* Models of B and T Cell Functions and Lymphoid Cell Receptors)
Edited by GIOVANNI DI SABATO

VOLUME 151. Molecular Genetics of Mammalian Cells
Edited by MICHAEL M. GOTTESMAN

VOLUME 152. Guide to Molecular Cloning Techniques
Edited by SHELBY L. BERGER AND ALAN R. KIMMEL

VOLUME 153. Recombinant DNA (Part D)
Edited by RAY WU AND LAWRENCE GROSSMAN

VOLUME 154. Recombinant DNA (Part E)
Edited by RAY WU AND LAWRENCE GROSSMAN

VOLUME 155. Recombinant DNA (Part F)
Edited by RAY WU

VOLUME 156. Biomembranes (Part P: ATP-Driven Pumps and Related Transport: The Na, K-Pump)
Edited by SIDNEY FLEISCHER AND BECCA FLEISCHER

VOLUME 157. Biomembranes (Part Q: ATP-Driven Pumps and Related Transport: Calcium, Proton, and Potassium Pumps)
Edited by SIDNEY FLEISCHER AND BECCA FLEISCHER

VOLUME 158. Metalloproteins (Part A)
Edited by JAMES F. RIORDAN AND BERT L. VALLEE

VOLUME 159. Initiation and Termination of Cyclic Nucleotide Action
Edited by JACKIE D. CORBIN AND ROGER A. JOHNSON

VOLUME 160. Biomass (Part A: Cellulose and Hemicellulose)
Edited by WILLIS A. WOOD AND SCOTT T. KELLOGG

VOLUME 161. Biomass (Part B: Lignin, Pectin, and Chitin)
Edited by WILLIS A. WOOD AND SCOTT T. KELLOGG

VOLUME 162. Immunochemical Techniques (Part L: Chemotaxis and Inflammation)
Edited by GIOVANNI DI SABATO

VOLUME 163. Immunochemical Techniques (Part M: Chemotaxis and Inflammation)
Edited by GIOVANNI DI SABATO

VOLUME 164. Ribosomes
Edited by HARRY F. NOLLER, JR., AND KIVIE MOLDAVE

VOLUME 165. Microbial Toxins: Tools for Enzymology
Edited by SIDNEY HARSHMAN

VOLUME 166. Branched-Chain Amino Acids
Edited by ROBERT HARRIS AND JOHN R. SOKATCH

VOLUME 167. Cyanobacteria
Edited by LESTER PACKER AND ALEXANDER N. GLAZER

VOLUME 168. Hormone Action (Part K: Neuroendocrine Peptides)
Edited by P. MICHAEL CONN

VOLUME 169. Platelets: Receptors, Adhesion, Secretion (Part A)
Edited by JACEK HAWIGER

VOLUME 170. Nucleosomes
Edited by PAUL M. WASSARMAN AND ROGER D. KORNBERG

VOLUME 171. Biomembranes (Part R: Transport Theory: Cells and Model Membranes)
Edited by SIDNEY FLEISCHER AND BECCA FLEISCHER

VOLUME 172. Biomembranes (Part S: Transport: Membrane Isolation and Characterization)
Edited by SIDNEY FLEISCHER AND BECCA FLEISCHER

Volume 190. Retinoids (Part B: Cell Differentiation and Clinical Applications)
Edited by Lester Packer

Volume 191. Biomembranes (Part V: Cellular and Subcellular Transport: Epithelial Cells)
Edited by Sidney Fleischer and Becca Fleischer

Volume 192. Biomembranes (Part W: Cellular and Subcellular Transport: Epithelial Cells)
Edited by Sidney Fleischer and Becca Fleischer

Volume 193. Mass Spectrometry
Edited by James A. McCloskey

Volume 194. Guide to Yeast Genetics and Molecular Biology
Edited by Christine Guthrie and Gerald R. Fink

Volume 195. Adenylyl Cyclase, G Proteins, and Guanylyl Cyclase
Edited by Roger A. Johnson and Jackie D. Corbin

Volume 196. Molecular Motors and the Cytoskeleton
Edited by Richard B. Vallee

Volume 197. Phospholipases
Edited by Edward A. Dennis

Volume 198. Peptide Growth Factors (Part C)
Edited by David Barnes, J. P. Mather, and Gordon H. Sato

Volume 199. Cumulative Subject Index Volumes 168–174, 176–194

Volume 200. Protein Phosphorylation (Part A: Protein Kinases: Assays, Purification, Antibodies, Functional Analysis, Cloning, and Expression)
Edited by Tony Hunter and Bartholomew M. Sefton

Volume 201. Protein Phosphorylation (Part B: Analysis of Protein Phosphorylation, Protein Kinase Inhibitors, and Protein Phosphatases)
Edited by Tony Hunter and Bartholomew M. Sefton

Volume 202. Molecular Design and Modeling: Concepts and Applications (Part A: Proteins, Peptides, and Enzymes)
Edited by John J. Langone

Volume 203. Molecular Design and Modeling: Concepts and Applications (Part B: Antibodies and Antigens, Nucleic Acids, Polysaccharides, and Drugs)
Edited by John J. Langone

Volume 204. Bacterial Genetic Systems
Edited by Jeffrey H. Miller

Volume 205. Metallobiochemistry (Part B: Metallothionein and Related Molecules)
Edited by James F. Riordan and Bert L. Vallee

VOLUME 206. Cytochrome P450
Edited by MICHAEL R. WATERMAN AND ERIC F. JOHNSON

VOLUME 207. Ion Channels
Edited by BERNARDO RUDY AND LINDA E. IVERSON

VOLUME 208. Protein–DNA Interactions
Edited by ROBERT T. SAUER

VOLUME 209. Phospholipid Biosynthesis
Edited by EDWARD A. DENNIS AND DENNIS E. VANCE

VOLUME 210. Numerical Computer Methods
Edited by LUDWIG BRAND AND MICHAEL L. JOHNSON

VOLUME 211. DNA Structures (Part A: Synthesis and Physical Analysis of DNA)
Edited by DAVID M. J. LILLEY AND JAMES E. DAHLBERG

VOLUME 212. DNA Structures (Part B: Chemical and Electrophoretic Analysis of DNA)
Edited by DAVID M. J. LILLEY AND JAMES E. DAHLBERG

VOLUME 213. Carotenoids (Part A: Chemistry, Separation, Quantitation, and Antioxidation)
Edited by LESTER PACKER

VOLUME 214. Carotenoids (Part B: Metabolism, Genetics, and Biosynthesis)
Edited by LESTER PACKER

VOLUME 215. Platelets: Receptors, Adhesion, Secretion (Part B)
Edited by JACEK J. HAWIGER

VOLUME 216. Recombinant DNA (Part G)
Edited by RAY WU

VOLUME 217. Recombinant DNA (Part H)
Edited by RAY WU

VOLUME 218. Recombinant DNA (Part I)
Edited by RAY WU

VOLUME 219. Reconstitution of Intracellular Transport
Edited by JAMES E. ROTHMAN

VOLUME 220. Membrane Fusion Techniques (Part A)
Edited by NEJAT DÜZGÜNEŞ

VOLUME 221. Membrane Fusion Techniques (Part B)
Edited by NEJAT DÜZGÜNEŞ

VOLUME 222. Proteolytic Enzymes in Coagulation, Fibrinolysis, and Complement Activation (Part A: Mammalian Blood Coagulation Factors and Inhibitors)
Edited by LASZLO LORAND AND KENNETH G. MANN

Volume 256. Small GTPases and Their Regulators (Part B: Rho Family)
Edited by W. E. Balch, Channing J. Der, and Alan Hall

Volume 257. Small GTPases and Their Regulators (Part C: Proteins Involved in Transport)
Edited by W. E. Balch, Channing J. Der, and Alan Hall

Volume 258. Redox-Active Amino Acids in Biology
Edited by Judith P. Klinman

Volume 259. Energetics of Biological Macromolecules
Edited by Michael L. Johnson and Gary K. Ackers

Volume 260. Mitochondrial Biogenesis and Genetics (Part A)
Edited by Giuseppe M. Attardi and Anne Chomyn

Volume 261. Nuclear Magnetic Resonance and Nucleic Acids
Edited by Thomas L. James

Volume 262. DNA Replication
Edited by Judith L. Campbell

Volume 263. Plasma Lipoproteins (Part C: Quantitation)
Edited by William A. Bradley, Sandra H. Gianturco, and Jere P. Segrest

Volume 264. Mitochondrial Biogenesis and Genetics (Part B)
Edited by Giuseppe M. Attardi and Anne Chomyn

Volume 265. Cumulative Subject Index Volumes 228, 230–262

Volume 266. Computer Methods for Macromolecular Sequence Analysis
Edited by Russell F. Doolittle

Volume 267. Combinatorial Chemistry
Edited by John N. Abelson

Volume 268. Nitric Oxide (Part A: Sources and Detection of NO; NO Synthase)
Edited by Lester Packer

Volume 269. Nitric Oxide (Part B: Physiological and Pathological Processes)
Edited by Lester Packer

Volume 270. High Resolution Separation and Analysis of Biological Macromolecules (Part A: Fundamentals)
Edited by Barry L. Karger and William S. Hancock

Volume 271. High Resolution Separation and Analysis of Biological Macromolecules (Part B: Applications)
Edited by Barry L. Karger and William S. Hancock

Volume 272. Cytochrome P450 (Part B)
Edited by Eric F. Johnson and Michael R. Waterman

Volume 273. RNA Polymerase and Associated Factors (Part A)
Edited by Sankar Adhya

Volume 293. Ion Channels (Part B)
Edited by P. Michael Conn

Volume 294. Ion Channels (Part C)
Edited by P. Michael Conn

Volume 295. Energetics of Biological Macromolecules (Part B)
Edited by Gary K. Ackers and Michael L. Johnson

Volume 296. Neurotransmitter Transporters
Edited by Susan G. Amara

Volume 297. Photosynthesis: Molecular Biology of Energy Capture
Edited by Lee McIntosh

Volume 298. Molecular Motors and the Cytoskeleton (Part B)
Edited by Richard B. Vallee

Volume 299. Oxidants and Antioxidants (Part A)
Edited by Lester Packer

Volume 300. Oxidants and Antioxidants (Part B)
Edited by Lester Packer

Volume 301. Nitric Oxide: Biological and Antioxidant Activities (Part C)
Edited by Lester Packer

Volume 302. Green Fluorescent Protein
Edited by P. Michael Conn

Volume 303. cDNA Preparation and Display
Edited by Sherman M. Weissman

Volume 304. Chromatin
Edited by Paul M. Wassarman and Alan P. Wolffe

Volume 305. Bioluminescence and Chemiluminescence (Part C)
Edited by Thomas O. Baldwin and Miriam M. Ziegler

Volume 306. Expression of Recombinant Genes in Eukaryotic Systems
Edited by Joseph C. Glorioso and Martin C. Schmidt

Volume 307. Confocal Microscopy
Edited by P. Michael Conn

Volume 308. Enzyme Kinetics and Mechanism (Part E: Energetics of Enzyme Catalysis)
Edited by Daniel L. Purich and Vern L. Schramm

Volume 309. Amyloid, Prions, and Other Protein Aggregates
Edited by Ronald Wetzel

Volume 310. Biofilms
Edited by Ron J. Doyle

Volume 345. G Protein Pathways (Part C: Effector Mechanisms)
Edited by Ravi Iyengar and John D. Hildebrandt

Volume 346. Gene Therapy Methods
Edited by M. Ian Phillips

Volume 347. Protein Sensors and Reactive Oxygen Species (Part A: Selenoproteins and Thioredoxin)
Edited by Helmut Sies and Lester Packer

Volume 348. Protein Sensors and Reactive Oxygen Species (Part B: Thiol Enzymes and Proteins)
Edited by Helmut Sies and Lester Packer

Volume 349. Superoxide Dismutase
Edited by Lester Packer

Volume 350. Guide to Yeast Genetics and Molecular and Cell Biology (Part B)
Edited by Christine Guthrie and Gerald R. Fink

Volume 351. Guide to Yeast Genetics and Molecular and Cell Biology (Part C)
Edited by Christine Guthrie and Gerald R. Fink

Volume 352. Redox Cell Biology and Genetics (Part A)
Edited by Chandan K. Sen and Lester Packer

Volume 353. Redox Cell Biology and Genetics (Part B)
Edited by Chandan K. Sen and Lester Packer

Volume 354. Enzyme Kinetics and Mechanisms (Part F: Detection and Characterization of Enzyme Reaction Intermediates)
Edited by Daniel L. Purich

Volume 355. Cumulative Subject Index Volumes 321–354

Volume 356. Laser Capture Microscopy and Microdissection
Edited by P. Michael Conn

Volume 357. Cytochrome P450, Part C
Edited by Eric F. Johnson and Michael R. Waterman

Volume 358. Bacterial Pathogenesis (Part C: Identification, Regulation, and Function of Virulence Factors)
Edited by Virginia L. Clark and Patrik M. Bavoil

Volume 359. Nitric Oxide (Part D)
Edited by Enrique Cadenas and Lester Packer

Volume 360. Biophotonics (Part A)
Edited by Gerard Marriott and Ian Parker

Volume 361. Biophotonics (Part B)
Edited by Gerard Marriott and Ian Parker

Volume 362. Recognition of Carbohydrates in Biological Systems (Part A)
Edited by Yuan C. Lee and Reiko T. Lee

Volume 363. Recognition of Carbohydrates in Biological Systems (Part B)
Edited by Yuan C. Lee and Reiko T. Lee

Volume 364. Nuclear Receptors
Edited by David W. Russell and David J. Mangelsdorf

Volume 365. Differentiation of Embryonic Stem Cells
Edited by Paul M. Wassauman and Gordon M. Keller

Volume 366. Protein Phosphatases
Edited by Susanne Klumpp and Josef Krieglstein

Volume 367. Liposomes (Part A)
Edited by Nejat Düzgüneş

Volume 368. Macromolecular Crystallography (Part C)
Edited by Charles W. Carter, Jr., and Robert M. Sweet

Volume 369. Combinational Chemistry (Part B)
Edited by Guillermo A. Morales and Barry A. Bunin

Volume 370. RNA Polymerases and Associated Factors (Part C)
Edited by Sankar L. Adhya and Susan Garges

Volume 371. RNA Polymerases and Associated Factors (Part D)
Edited by Sankar L. Adhya and Susan Garges

Volume 372. Liposomes (Part B)
Edited by Nejat Düzgüneş

Volume 373. Liposomes (Part C)
Edited by Nejat Düzgüneş

Volume 374. Macromolecular Crystallography (Part D)
Edited by Charles W. Carter, Jr., and Robert W. Sweet

Volume 375. Chromatin and Chromatin Remodeling Enzymes (Part A)
Edited by C. David Allis and Carl Wu

Volume 376. Chromatin and Chromatin Remodeling Enzymes (Part B)
Edited by C. David Allis and Carl Wu

Volume 377. Chromatin and Chromatin Remodeling Enzymes (Part C)
Edited by C. David Allis and Carl Wu

Volume 378. Quinones and Quinone Enzymes (Part A)
Edited by Helmut Sies and Lester Packer

Volume 379. Energetics of Biological Macromolecules (Part D)
Edited by Jo M. Holt, Michael L. Johnson, and Gary K. Ackers

Volume 380. Energetics of Biological Macromolecules (Part E)
Edited by Jo M. Holt, Michael L. Johnson, and Gary K. Ackers

Volume 381. Oxygen Sensing
Edited by Chandan K. Sen and Gregg L. Semenza

Volume 382. Quinones and Quinone Enzymes (Part B)
Edited by Helmut Sies and Lester Packer

Volume 383. Numerical Computer Methods (Part D)
Edited by Ludwig Brand and Michael L. Johnson

Volume 384. Numerical Computer Methods (Part E)
Edited by Ludwig Brand and Michael L. Johnson

Volume 385. Imaging in Biological Research (Part A)
Edited by P. Michael Conn

Volume 386. Imaging in Biological Research (Part B)
Edited by P. Michael Conn

Volume 387. Liposomes (Part D)
Edited by Nejat Düzgüneş

Volume 388. Protein Engineering
Edited by Dan E. Robertson and Joseph P. Noel

Volume 389. Regulators of G-Protein Signaling (Part A)
Edited by David P. Siderovski

Volume 390. Regulators of G-Protein Signaling (Part B)
Edited by David P. Siderovski

Volume 391. Liposomes (Part E)
Edited by Nejat Düzgüneş

Volume 392. RNA Interference
Edited by Engelke Rossi

Volume 393. Circadian Rhythms
Edited by Michael W. Young

Volume 394. Nuclear Magnetic Resonance of Biological Macromolecules (Part C)
Edited by Thomas L. James

Volume 395. Producing the Biochemical Data (Part B)
Edited by Elizabeth A. Zimmer and Eric H. Roalson

Volume 396. Nitric Oxide (Part E)
Edited by Lester Packer and Enrique Cadenas

Volume 397. Environmental Microbiology
Edited by Jared R. Leadbetter

Volume 398. Ubiquitin and Protein Degradation (Part A)
Edited by Raymond J. Deshaies

Volume 464. Liposomes, Part F
Edited by Nejat Düzgüneş

Volume 465. Liposomes, Part G
Edited by Nejat Düzgüneş

Volume 466. Biothermodynamics, Part B
Edited by Michael L. Johnson, Gary K. Ackers, and Jo M. Holt

Volume 467. Computer Methods Part B
Edited by Michael L. Johnson and Ludwig Brand

Volume 468. Biophysical, Chemical, and Functional Probes of RNA Structure, Interactions and Folding: Part A
Edited by Daniel Herschlag

Volume 469. Biophysical, Chemical, and Functional Probes of RNA Structure, Interactions and Folding: Part B
Edited by Daniel Herschlag

Volume 470. Guide to Yeast Genetics: Functional Genomics, Proteomics, and Other Systems Analysis, 2nd Edition
Edited by Gerald Fink, Jonathan Weissman, and Christine Guthrie

Volume 471. Two-Component Signaling Systems, Part C
Edited by Melvin I. Simon, Brian R. Crane, and Alexandrine Crane

Volume 472. Single Molecule Tools, Part A: Fluorescence Based Approaches
Edited by Nils G. Walter

Volume 473. Thiol Redox Transitions in Cell Signaling, Part A Chemistry and Biochemistry of Low Molecular Weight and Protein Thiols
Edited by Enrique Cadenas and Lester Packer

Volume 474. Thiol Redox Transitions in Cell Signaling, Part B Cellular Localization and Signaling
Edited by Enrique Cadenas and Lester Packer

Volume 475. Single Molecule Tools, Part B: Super-Resolution, Particle Tracking, Multiparameter, and Force Based Methods
Edited by Nils G. Walter

Volume 476. Guide to Techniques in Mouse Development, Part A Mice, Embryos, and Cells, 2nd Edition
Edited by Paul M. Wassarman and Philippe M. Soriano

Volume 477. Guide to Techniques in Mouse Development, Part B Mouse Molecular Genetics, 2nd Edition
Edited by Paul M. Wassarman and Philippe M. Soriano

Volume 478. Glycomics
Edited by Minoru Fukuda

Volume 479. Functional Glycomics
Edited by Minoru Fukuda

Volume 480. Glycobiology
Edited by Minoru Fukuda

Volume 481. Cryo-EM, Part A: Sample Preparation and Data Collection
Edited by Grant J. Jensen

Volume 482. Cryo-EM, Part B: 3-D Reconstruction
Edited by Grant J. Jensen

Volume 483. Cryo-EM, Part C: Analyses, Interpretation, and Case Studies
Edited by Grant J. Jensen

Volume 484. Constitutive Activity in Receptors and Other Proteins, Part A
Edited by P. Michael Conn

Volume 485. Constitutive Activity in Receptors and Other Proteins, Part B
Edited by P. Michael Conn

Volume 486. Research on Nitrification and Related Processes, Part A
Edited by Martin G. Klotz

Volume 487. Computer Methods, Part C
Edited by Michael L. Johnson and Ludwig Brand

Volume 488. Biothermodynamics, Part C
Edited by Michael L. Johnson, Jo M. Holt, and Gary K. Ackers

Volume 489. The Unfolded Protein Response and Cellular Stress, Part A
Edited by P. Michael Conn

Volume 490. The Unfolded Protein Response and Cellular Stress, Part B
Edited by P. Michael Conn

Volume 491. The Unfolded Protein Response and Cellular Stress, Part C
Edited by P. Michael Conn

Volume 492. Biothermodynamics, Part D
Edited by Michael L. Johnson, Jo M. Holt, and Gary K. Ackers

Volume 493. Fragment-Based Drug Design
Tools, Practical Approaches, and Examples
Edited by Lawrence C. Kuo

Volume 494. Methods in Methane Metabolism, Part A
Methanogenesis
Edited by Amy C. Rosenzweig and Stephen W. Ragsdale

Volume 495. Methods in Methane Metabolism, Part B
Edited by Amy C. Rosenzweig and Stephen W. Ragsdale

SECTION ONE

MEASURING AND ENGINEERING CENTRAL DOGMA PROCESSES

CHAPTER ONE

Sequence-Specificity and Energy Landscapes of DNA-Binding Molecules

Joshua R. Tietjen, Leslie J. Donato, Devesh Bhimisaria, *and* Aseem Z. Ansari

Contents

Abstract

A central goal of biology is to understand how transcription factors target and regulate specific genes and networks to control cell fate and function. An equally important goal of synthetic biology, chemical biology, and personalized medicine is to devise molecules that can regulate genes and networks in a programmable manner. To achieve these goals, it is necessary to chart the sequence specificity of natural and engineered DNA-binding molecules. Cognate site identification (CSI) is now achieved via unbiased, high-throughput platforms that interrogate an entire sequence space bound by typical DNA-binding molecules. Analysis of these comprehensive specificity profiles is facilitated through the use of sequence-specificity landscapes (SSLs). SSLs reveal new modes of sequence cognition and overcome the limitations of current approaches that yield amalgamated "consensus" motifs. The landscapes also reveal the impact of nonconserved flanking sequences on binding to cognate sites. SSLs also serve as comprehensive binding energy landscapes that provide insights into the energetic thresholds at which natural and engineered molecules function

Department of Biochemistry, The Genome Center of Wisconsin, University of Wisconsin-Madison, Madison, Wisconsin, USA

Methods in Enzymology, Volume 497
ISSN 0076-6879, DOI: 10.1016/B978-0-12-385075-1.00001-9

within cells. Furthermore, applying the CSI binding data to genomic sequence (genomescapes) provides a powerful tool for identification of potential *in vivo* binding sites of a given DNA ligand, and can provide insight into differential regulation of gene networks. These tools can be directly applied to the design and development of synthetic therapeutic molecules and to expand our knowledge of the basic principles of molecular recognition.

1. Introduction

To understand signal-responsive regulation of gene networks, it is crucial to obtain a comprehensive view of the molecular recognition events that underlie transcription factor–DNA interactions. Similarly, from a bioengineering perspective, designing programmable regulators of gene expression requires a detailed understanding of the molecular recognition events between DNA and the engineered protein or small molecule. However, for the majority of the natural transcription factors, there is limited knowledge of where these proteins bind across the genome to regulate specific genes and networks in response to cell-signaling events. In an effort to understand and control desired gene networks, synthetic and chemical biologists have engineered small molecules and proteins to mimic the sequence specificity and regulatory properties of natural DNA-binding factors (Ansari and Mapp, 2002; Dervan *et al.*, 2005; Gonzalez *et al.*, 2010; Gottesfeld *et al.*, 2000; Lee and Mapp, 2010; Mapp and Ansari, 2007; Rodríguez-Martínez *et al.*, 2010; Sera, 2010; Wolfe *et al.*, 2000). Defining the DNA sequence specificity of these engineered molecules is an important step in targeted regulation of genes and networks. Moreover, defining entire gene networks that coordinately execute a cellular response can reveal key nodes to which such engineered regulators can be targeted.

The importance of understanding the specificity of DNA-binding molecules has stimulated the development of several experimental methods over the years. Most methods for determining the protein–DNA or small molecule–DNA-binding signatures have been quite apt at identifying the highest affinity sequences. These include methods like the systematic evolution of ligands by exponential enrichment (SELEX) and cyclic amplification and selection of targets (CASTing) (Blackwell and Weintraub, 1990; Oliphant *et al.*, 1989; Tuerk and Gold, 1990; Wright *et al.*, 1991). While valuable, such motifs often display limited ability to predict sequences that are bound in a biological context, because physiologically relevant binding sites frequently deviate from the highest affinity sequences (Farnham, 2009; Ward and Bussemaker, 2008). Often, moderate or even low-affinity sites regulate expression in a switch-like manner, and cooperative binding of multiple transcription factors is required (Moretti and Ansari, 2008; Ptashne and

Gann, 2002). In fact, cooperative assembly integrates multiple signals combinatorially to fine-tune the regulation of genes, which outnumber the transcription factors encoded in genomes by an order of magnitude (Lander *et al.*, 2001; Maniatis *et al.*, 1998; Ptashne and Gann, 2002; Venter *et al.*, 2001; Yamamoto *et al.*, 1998). Studies that comprehensively define these binding properties in cells usually correlate with their corresponding binding patterns in cells (Harbison *et al.*, 2004; Horak *et al.*, 2002). However, when they do not correlate, unexpected mechanistic insights can be obtained.

Recently, methods such as the cognate site identification (CSI) have been developed that can reveal the entire spectrum of DNA-binding specificities of a protein or a small molecule binding individually or in cooperative complexes (Carlson *et al.*, 2010; Stormo and Zhao, 2010; Warren *et al.*, 2006). Array-based methods display millions of unique sequences of B-form DNA in spatially resolved microfabricated arrays on glass slides. A DNA-binding factor is applied to the surface and allowed to query the entire sequence space. The factor is either directly fluorescently labeled or can be indirectly detected. After incubation, the unbound factor is washed away and molecules that remain bound to DNA sequences on the array are visualized. Alternative methods, such as fluorescence intercalator displacement (FID), surface plasmon resonance (SPR), or gated microfluidic chambers (MITOMI), have also been used to determine specificity and affinity of binding (Brockman *et al.*, 1999; Campbell and Kim, 2007; Hauschild *et al.*, 2009; Maerkl and Quake, 2007).

In each of the array-based methods, the complexity of the DNA library is limited by the number of sequences that can be displayed, which in turn limits the size of the variable region in the DNA sequences queried. Currently, the available technology limits the number of features on an array to 2.1 (for NimbleGen arrays) or 6 millions (for Affymetrix arrays). This restricts the size of the variable region to every sequence permutation of up to 10–12 contiguous base pairs or the same number of base pairs distributed within a larger DNA duplex. The complexity of the sequences displayed on an array can be expanded through the use of De Bruijn sequences, but this approach decreases the quality of the data obtained. For example, arrays that rely on the De Bruijn approach to display every 10mer only yield optimal data for 8mer binding sites. In addition to loss of information, the approach also increases the complexity of analyzing the resulting data (Philippakis *et al.*, 2008).

To expand beyond the limitations on the sequence space queried by array-based methods, *cognate site identity* can be determined by directly sequencing DNA that binds to a protein of interest (Jolma *et al.*, 2010; Stormo and Zhao, 2010; Zhao *et al.*, 2009; Zykovich *et al.*, 2009). In these experiments, a protein of interest is incubated with a library of DNA sequences that is not subject to the limitations on length or number of features that can be spatially resolved on a surface. The micromole-scale

DNA synthesis permits the examination of sequence space of 14–25mer within a much larger duplex. In these solution-based assays, the protein–DNA complexes are isolated and the DNA is sequenced by next-generation DNA sequencers. The frequency of detecting a certain sequence directly correlates with the affinity of the protein for that sequence. As costs for direct sequencing decrease and the ease of use increases, we anticipate that this approach to defining the sequence specificity of natural or engineered DNA-binding molecules will become more prevalent.

With the magnitude of such comprehensive molecular recognition data comes the challenge of mining that rich vein of information and identifying the full spectrum of binding specificities. The processing and visualization of such multidimensional data can be achieved using sequence-specificity landscapes (SSLs), which are designed to allow user-friendly visualization of DNA-binding characteristics of a protein or a small molecule of interest (Carlson *et al.*, 2010). SSLs provide a convenient method for displaying and analyzing the binding specificity profile for natural or engineered DNA-binding molecules across the entire sequence space mapped by CSI. This displays both high- and low-affinity binding sites, rather than only the best bound sequences. Importantly, since binding intensities on the array correlate with affinities in solution, SSLs also yield comprehensive energy landscapes for DNA-binding molecules (Carlson *et al.*, 2010; Keles *et al.*, 2008; Puckett *et al.*, 2007; Warren *et al.*, 2006). The CSI specificity and energy landscapes can also be used to accurately annotate binding sites across the genome (genomescapes) and provide probabilities of binding to sequences that deviate from the consensus (Carlson *et al.*, 2010).

2. Array-Based Cognate Sequence Identification

2.1. Labeling DNA-binding molecules

To determine DNA-binding specificities of molecules, several high-throughput array-based methods have been developed (Berger and Bulyk, 2009; Liu *et al.*, 2005; Ragoussis *et al.*, 2006; Wang *et al.*, 2005; Warren *et al.*, 2006). One such technique is the CSI approach. CSI is an effective and unbiased method for determining the comprehensive DNA-binding specificities of small molecules or proteins. This technique involves the use of hairpin DNA, consisting of a variable region flanked by two constant regions: a hairpin loop and a complementary sequence. This design places the variable region in the stem of the hairpin. Previous studies confirm that this region of the stem adopts B-form structure under standard buffers that closely approximate physiological conditions (Kuznetsov *et al.*, 2007). The sequences can be synthesized by photolithography on a silanized glass surface similar to standard fabrication procedures for DNA microarrays, and these microarrays can be used to probe the sequence specificity of a DNA-binding molecule. The range of sequences that can be probed is only limited by the length of the DNA and the number of unique features that can be effectively synthesized by the microarray synthesis technology chosen. In other words, approximately half a million features bearing unique sequences are required to display the entire sequence space covered by 10 bp and each additional variable base pair requires approximately fourfold more features. The De Bruijn approach offers to extend the complexity without greatly expanding the number of features. However, this approach fails to capture the full spectrum of binding and it appears that differences between closely related sites are often lost.

A first step in a CSI experiment involves the design of the hairpin DNA probes to be synthesized on the silanized microarray surface. A commonly used design is 5′-GCGC-$N^1N^2N^3N^4N^5N^6N^7N^8N^9N^{10}$-GCGC-**GNA**-GCGC-$N^{10'}N^{9'}N^{8'}N^{7'}N^{6'}N^{5'}N^{4'}N^{3'}N^{2'}N^{1'}$-GCGC-3′, synthesized 3′–5′, where N can be any of the four common bases (A, T, C, G), the GNA trinucleotide forms the hairpin loop, and the GCGC sequences form stabilizing constant regions flanking the variable region of the hairpin (Fig. 1.1). The 5′-GNA-3′ sequence is ideal for loop formation in DNA hairpins (Hirao *et al.*, 1993).

Once the microarray is designed and synthesized, the sequences are induced to form hairpins before performing a CSI experiment:

(1) Incubate the array in 7 *M* urea for 30 min at 65 °C.
(2) Transfer the array to a 1× phosphate-buffered saline (PBS) solution at 65 °C for 15 min to induce the duplex formation.

Next, the surface of the array is blocked or pacified to prevent adventitious binding to the surface from interfering with the detection of the sequence-specific DNA-binding events. This step may not be necessary in all cases, but is especially important if a given DNA-binding molecule has a propensity to bind the glass surface. Blocking the array is performed as follows:

(1) Solubilize powdered milk to a final concentration of 2–3% (20–30 mg/mL) in sterile H_2O.
(2) Apply the chamber to the microarray (i.e., Grace Bio-Labs, Item No. 623503).
(3) Rinse the chamber twice with distilled H_2O, allowing the surface wetting during the second wash to proceed for 1–2 min.
(4) Remove all liquid from the chamber and inject a sufficient volume of the milk solution into the chamber such that approximately three-fourth of the chamber is filled. The remaining one-fourth volume is occupied by an air bubble that rotates as the slide is rotated on a Southern blot hybridization wheel or another similar device. This constant rotation of the air bubble within the chamber mixes the solution evenly across the array.

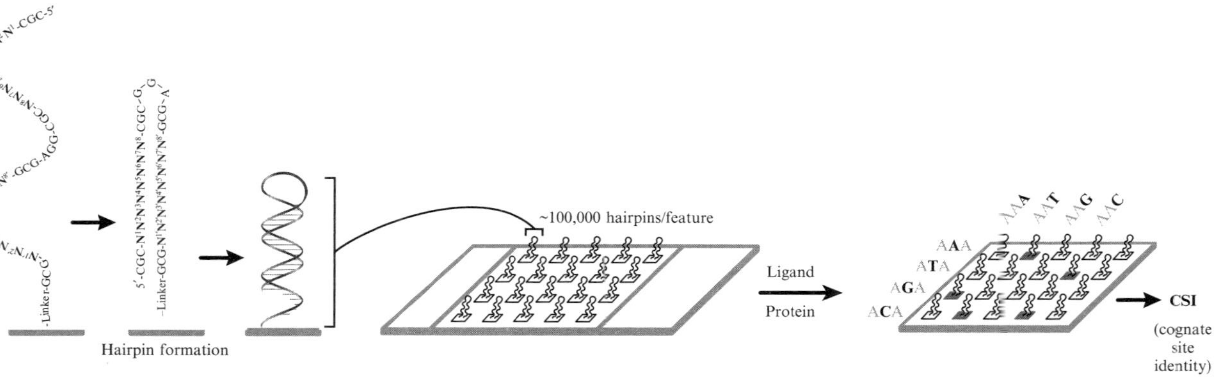

Figure 1.1 *The format of cognate sequence identification (CSI) microarrays*. Self-complementary, single-stranded DNA is synthesized 3′–5′ on a glass slide. Hairpin formation is then induced, which leads to B-form DNA formation. Within each hairpin is a variable region (the Nmer) covering the entire sequence space up to 10 positional variants within a larger DNA duplex. Each "feature" of the microarray contains approximately 100,000 hairpins of a specific sequence (up to 2–6 million different sequences can be displayed currently). Once the DNA-binding protein or small molecule, or a combination of the two, is incubated on the microarray, fluorescence intensity can be used to quantitate the sequence affinity and specificity of the molecule.

(5) Once the samples are injected, seal the chamber as required by the manufacturer (typically done with a removable sticker) and spin the array on a rotor wheel for 1.5 h at 25 °C.
(6) Remove all of the liquid and repeat the rinsing from step (3).

Once the array is blocked, it can be incubated with the protein or small molecule of interest. These experiments can be performed using a fluorescently labeled DNA-binding protein or small molecule, or through indirect detection as discussed below.

2.1.1. Synthetic DNA-binding molecules

Of the many DNA-binding small molecules, polyamides are particularly important, as they can be specifically designed to target a variety of sequences. Polyamides bind DNA in a modular manner through various combinations of *N*-methylpyrrole and *N*-methylimidazole rings, even when the DNA is wrapped around nucleosomes (Dervan and Bürli, 1999; Dervan and Edelson, 2003; Suto *et al.*, 2003; White *et al.*, 1998). Hairpin polyamides have been used to regulate targets involved in many diseases (Burnett *et al.*, 2006; Nickols *et al.*, 2007; Olenyuk *et al.*, 2004). Engineering these molecules to bind specific genomic sites requires a detailed understanding of the molecular recognition rules. To evaluate the DNA-binding specificity of polyamides and other engineered molecules, we developed the CSI approach on a microarray platform. The specific protocol is described below.

Buffer and hybridization mix preparation:

(1) Prepare wash buffer (1.22 *M* MES + 0.89 *M* [Na^+] at pH 6.7–7.2), making sure that the MES stays covered at all times to prevent exposure to light.
(2) Thaw the control oligonucleotide that hybridizes to the control sequences in specific locations on the array. These control features permit data extraction after the hybridization.
(3) Prepare the small molecule/oligo/buffer mix with 5 n*M* control oligonucleotide and an amount of the DNA-binding molecule consistent with its binding affinity in solution.

After the reagents are prepared and the array is ready, the CSI array experiment can begin:

(1) Clean the inside of the chamber + array setup twice with wash buffer. Pipette the solution in and out of the chamber gently several times, allowing the second wash to wet the array surface for several minutes.
(2) Remove the wash buffer from the chamber and then fill the chamber completely with the small molecule/oligonucleotide/buffer mix.
(3) Remove approximately 20% of the solution to create one air bubble within the chamber for mixing.

(4) Seal the chamber as before, attach the microarray to the rotor wheel, and cover it with aluminum foil to protect it from light exposure.
(5) Begin rotating the array and observe the air bubble to ensure it moves smoothly around the array surface to prevent the array from drying out.

The final stage of the process before data analysis consists of washing, drying, and scanning the array:

(1) Prepare a 1× final wash buffer (FWB) solution in a 50-mL falcon tube from the 10× stock (NimbleGen Systems Inc., Cat. No. R10019-2).
(2) Unseal the chamber by removing the adhesive sticker from the holes, and remove all of the liquid. Then wash the array twice with wash buffer, allowing the buffer to sit in the chamber for a few minutes the second time.
(3) Remove the chamber from the microarray surface and immerse the slide in the FWB solution for 30 s.
(4) Remove the slide and dry using a microarray centrifuge.
(5) Scan the array using a standard microarray scanner ($\geq$ 5 μm resolution) with filters at wavelengths that permit the detection of fluorescent markers (i.e., 532 and 635 nm for Cy3 and Cy5 dyes, respectively). For our analysis, we use the Axon GenePix 4000B Microarray Scanner and the GenePix Pro software to scan the arrays.
(6) Extract the array feature intensity data using compatible software, such as the NimbleScan software from Roche-NimbleGen.

2.1.2. Natural or engineered DNA-binding proteins

Similar to small molecule analysis described above, the comprehensive sequence-specificity profiles of proteins can also be determined using high-throughput DNA arrays. Unlike small molecules, each protein requires its own optimal binding conditions, depending on its biochemical properties. These binding conditions should be determined prior to performing the CSI array experiment. One should start with buffer conditions in which the protein of interest will bind to DNA in other experiments such as electrophoretic mobility shift assays (EMSAs), nuclease protection (Footprinting), or fluorescence polarization (FP) (Anderson *et al.*, 2008; Fried and Crothers, 1981; Galas and Schmitz, 1978; Garner and Revzin, 1981; Heyduk and Heyduk, 2002; Heyduk *et al.*, 1996). The binding reaction will contain buffer, salts, small molecules, and other cofactors that are required for the protein to bind DNA.

Additionally, the detection method will need to be determined. Proteins can be detected on the surface of the arrays by a variety of different methods. The protein can be conjugated directly to a fluorescent moiety using common commercially available compounds. If the direct labeling method is employed, the protein should be minimally labeled to limit the effect the label has on its DNA recognition or functional conformation. Thus, proteins that are directly conjugated to a fluorophore require verification that the label does not adversely affect the specificity profiles of the conjugated protein. A less invasive approach for detecting the DNA-binding preferences of a protein involves using a fluorescently labeled antibody that is specific to the protein. In this indirect labeling approach, the antibody specifically recognizes the protein of interest or a peptide tag engineered onto the protein (such as 6–10 histidine residues or Flag, Myc, or GST tags). It is therefore the detection of the fluorescently labeled antibody that identifies the DNA sequences that are bound by the protein of interest. If antibody detection is to be used, it is important to ensure that the antibody does not interfere with the DNA-binding function of the protein of interest.

Here, we describe a protein CSI array experiment as a sample for adaptation to other proteins and conditions. In this example, we measure the binding profile of an unlabeled His-tagged fusion protein, with a binding reaction composed of a buffer (50 m*M* NaCl, 10 m*M* Tris, pH 7.4, 1 m*M* $MgCl_2$, 0.5 m*M* EDTA), the DNA-binding protein (one may use between 1 and 10 n*M* as a starting point depending on the affinity of the protein to DNA), 0.25% milk for protein stabilization, and a 1:500 dilution of a Cy5-labeled anti-His antibody.

Some things to remember when optimizing array–protein interactions:

- The protein concentration that will be optimal on the array (i.e., give the best binding profile with highest specific DNA binding and lowest non-specific or surface binding) will need to be determined for each protein.
- Additives can greatly increase the likelihood of obtaining a good binding profile. Such additives may include bovine serum albumin (BSA) as a molecular crowding agent, milk as a nonspecific competitor, small amounts of detergent (0.002%) to prevent protein aggregation, or dithiothreitol (DTT) to maintain a reducing environment.
- The stability of the protein of interest may be optimized at different temperatures. Therefore, the experiments must be performed at a temperature that is optimal for protein–DNA binding.
- If chemically prelabeling the protein of interest with a fluorophore, the added fluorophore must not interfere with the DNA recognition properties of the protein.
- If detecting the protein–DNA interaction using a fluorescently labeled antibody (such as a Cy5-labeled anti-His antibody to detect a 6-histidine-tagged fusion protein, or a Cy3-labeled anti-Myc antibody to detect a Myc-tagged fusion protein), the antibody must not interfere with the DNA recognition properties of the protein.
- Some commonly used fusion proteins are 6 histidine, GST, Myc, Flag, etcetera. Interestingly, we have found mixed results using a fusion protein of maltose binding protein (MBP).

After blocking of the array with milk, the protein is hybridized to the DNA microarray under optimal binding conditions (in a process similar to the small molecule procedure above). In brief, the steps are:

(1) Clean the inside of the chamber + array setup twice with the predetermined buffer by pipetting the solution in and out of the chamber gently several times, allowing the second wash to wet the array surface for several minutes.
(2) Remove the buffer from the chamber and then fill the chamber completely with the protein/buffer mix.
(3) Remove approximately 20% of the solution to create one air bubble within the chamber for mixing.
(4) Seal the chamber as before, attach the microarray to the rotor wheel, and cover it with aluminum foil to protect it from light exposure.
(5) Begin rotating the array and observe the air bubble to ensure it moves smoothly around the array surface so that none of the array dries out.
(6) Store any remaining buffer until the run is completed (typically 1–2 h).

The final stage of the process before data analysis consists of washing, drying, and scanning the array:

(1) Prepare a 1× FWB solution in a 50-mL falcon tube from the 10× stock (NimbleGen Systems Inc., Cat. No. R10019-2).
(2) Unseal the chamber and remove all of the liquid, then wash twice with buffer, allowing the buffer to remain in the chamber for a few minutes the second time.
(3) Remove the chamber from the microarray surface and immerse the slide in the FWB solution for 30 s.
(4) Remove the slide and dry using a microarray centrifuge.
(5) Scan the array and extract the data as mentioned in the previous small-molecule CSI protocol section (steps 5 and 6).

2.2. Label-free detection

In cases where a DNA-binding molecule cannot be fluorescently labeled, or in situations where this labeling perturbs its DNA-binding characteristics, an alternative application of CSI may be used. This method combines the use of FID with CSI microarrays. FID examines the binding affinity and specificity of small molecules to DNA through the displacement of a small fluorescent DNA intercalator, such as ethidium bromide. This has been used previously as a plate-based assay (Boger *et al.*, 2001), where a small library of DNA hairpin oligonucleotides in individual wells is initially bound by the intercalator and the fluorescence for each sequence is measured. Then, the DNA-binding molecule of interest is applied to the wells and the binding displaces the intercalator from the DNA. Thus, binding is measured through the *decrease* in fluorescence intensity within each well. One major limitation of the FID assay itself is the library complexity that can be achieved in a plate-based system. Thus, the hybrid CSI–FID approach was developed to overcome the library size limitations, allowing for a more complex library—one with a diversity of sequences as much as three to four orders of magnitude larger (Fig. 1.2; Hauschild *et al.*, 2009).

Similar to the CSI array, this assay uses DNA microarrays having DNA hairpin oligonucleotides synthesized on the surface of the array. These arrays also require hairpin formation prior to the first use:

(1) Incubate the array in 7 *M* urea for 30 min at 65 °C.
(2) Transfer the array to a 1× PBS solution at 65 °C for 15 min to induce the duplex formation.

After hairpin formation, the DNA intercalator can be applied to the array. This procedure is performed on two arrays: one as the control for baseline EtBr binding and the other for the intercalation displacement experiment. The control array is necessary to prevent disruption of the intercalation equilibrium, which would occur if only one array were used and if the array were dried and scanned prior to incubation with the DNA-binding molecule:

(1) Seal the arrays with the appropriate chamber (as mentioned above), add a solution of 6 μ*M* EtBr in binding buffer (100 m*M* NaCl + 100 m*M* Tris, pH 8.0), and then seal the chambers.
(2) Incubate the arrays at room temperature for 1 h while on the rotor wheel.
(3) Apply the DNA-binding molecule of interest to one of the arrays (concentration to be determined empirically) and incubate at optimal binding temperature while rotating.
(4) Wash both arrays twice with binding buffer and then dry and scan as described above for CSI experiments. EtBr can be detected using the same array scanner filter as used for the Cy3 dye.

Alternative methods that utilize high-throughput SPR as well as other highly sensitive detection technologies are being developed by several laboratories. In the future, such alternate detection strategies will obviate the need to label DNA-binding molecules directly or indirectly.

3. Solution-Based Cognate Sequence Identification

The application of high-throughput sequencing technology allows for the expansion of the sequence space of the library beyond the restrictions imposed by microarray-based assays. In addition to increasing the library

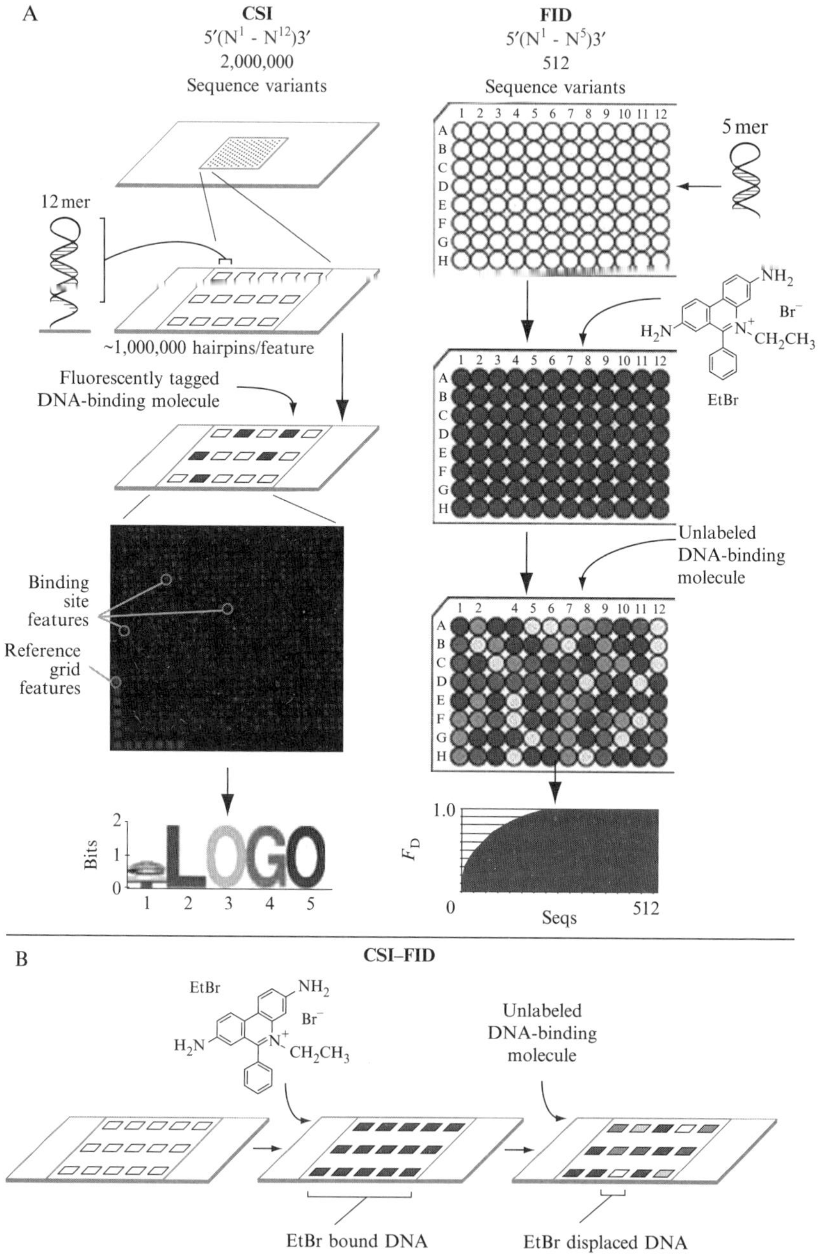

Figure 1.2 *CSI–FID provides an important label-free tool for DNA-binding affinity and specificity determination.* (A) Left column displays a standard CSI microarray experiment (as described in Fig. 1.1) showing the application of a fluorescently tagged DNA-binding molecule to a microarray containing approximately 2 million double-stranded

complexity, new sequencing technologies, such as that used by the Illumina Genome Analyzer, can sequence many samples in parallel. This is accomplished through the use of unique barcode sequences, which are included in the design of the sequences in the library being examined. Multiple binding experiments can be performed in parallel, each with a unique barcode, and the resulting samples can be pooled and sequenced simultaneously (Jolma *et al.*, 2010; Stormo and Zhao, 2010; Zykovich *et al.*, 2009).

As with the microarray-based methods, CSI by sequencing begins with the design of the DNA library to be synthesized. The library design is tailored to the sequencing technology that will be used in the experiment. For Illumina sequencing, the DNA must have a constant primer at the 5′ and 3′ ends of each piece of DNA. In addition, unique barcodes should be designed to identify the sample being sequenced. These barcodes can be as short or long as required to uniquely identify each sample. It is recommended that these barcodes vary by more than a single nucleotide from one another in order to prevent saturation of the detectors during sequencing. Finally, the central region of the DNA sequence should contain the randomized sequences to be bound by the protein or small molecule of interest. These randomized regions are no longer limited to the 10–12mer sequence space used in microarray-based assays. As many as 22 randomized nucleotides have been used to date (Zykovich *et al.*, 2009). However, a balance must be established between the synthetic possibilities and the ability to perform binding in the appropriate concentration regimes. The greater the complexity of the library, the greater the challenges of retaining sufficient molecules of each sequence at high enough concentrations to permit equilibrium binding with the DNA-binding molecules of interest. A particularly thoughtful example is the use of a 14mer library, rather than a larger library, to interrogate the binding by several eukaryotic transcription factors (Jolma, *et al.* 2010).

Following the design and synthesis of the DNA libraries is the binding experiment itself. This can be performed with limited rounds of selection and enrichment to avoid loss of moderate-to-low-affinity binding sequences (i.e., SELEX; Jolma *et al.*, 2010; Zhao *et al.*, 2009) or with a single binding event followed by purification on a column or via EMSA

---

hairpin DNA sequences. The fluorescence intensity readout yields DNA sequence specificity and affinity binding data. Right column describes a standard FID experiment in a 96-well plate format. Unique DNA hairpin sequences are distributed individually across the wells of the plate and then incubated with a fluorescent DNA intercalator (EtBr). The DNA-binding molecule is added to the wells and the decrease in fluorescence intensity is used to quantitate the DNA-binding events. (B) Application of the FID process using CSI microarrays increases the sequence space available for interrogation (up to ~6 million sequences) and allows for label-free detection of the individual binding events, negating the need for directly labeling the DNA-binding molecule or detecting it indirectly.

(Zykovich *et al.*, 2009). Once the DNA–protein or DNA–small molecule complexes are isolated, the DNA can be released and purified for sequencing. One caveat to consider is that it may be necessary at this point to preamplify the bound DNA sequences to have enough material for the sequencing reaction. If this becomes necessary, one must take great care in selecting a method of amplification which will cause the least bias in the resulting DNA samples. It may be possible to avoid amplification if there are a sufficient number of samples being pooled to yield the required amount of DNA for sequencing.

The above protocol can be executed as follows (Fig. 1.3):

(1) Design the library to be used for each experiment and the unique barcode used to identify the sample. As an example, we use a library synthesized by Integrated DNA Technologies (IDT) containing a 15mer randomized library adjacent to a 6-bp barcode and flanked by two adapter sequences required by the Illumina sequencing technology.
(2) Generate dsDNA from the ssDNA library: mix 1 μM of the ssDNA library with 10 μM of the reverse primer in EconoTaq Plus 2× master mix (Lucigen, Cat. No. 30035-1). Run the PCR reaction as follows:
   (a) 94 °C for 2 min 30 s.
   (b) 55 °C for 2 min.
   (c) 72 °C for 30 min.
(3) Perform the binding reaction by incubating the protein of interest with the DNA library: combine 50 mM biotin-labeled protein, 0.01μM dsDNA library, 50 ng/μL polydI•dC in a total volume of 25 μL and incubate at optimal binding conditions for 2 h.
(4) Pull down biotin-labeled protein-bound DNA using Dyna beads M280 (Invitrogen, Cat. No. 112-05D) according to manufacturer's protocol.
(5) Wash the bead-bound complexes six times with 1× PBS + 0.1% BSA solution (100-fold excess buffer).
(6) Transfer the beads to PCR strip tubes and add 20 μM forward and reverse primers with a standard PCR reaction mix to each tube and run the following thermal cycle:
   (a) 94 °C for 2 min.
   (b) 94 °C for 30 s.
   (c) 55 °C for 30 s.
   (d) 72 °C for 30 s.
   (e) Repeat steps (b) through (d) 19 more times for a total of 20 cycles.
(7) Place the samples on a magnetic block and manually remove the supernatant to separate the samples from the beads.
(8) Purify the amplified DNA from the samples using the QIAquick PCR Purification Kit (QIAGEN, Cat. No. 28104).

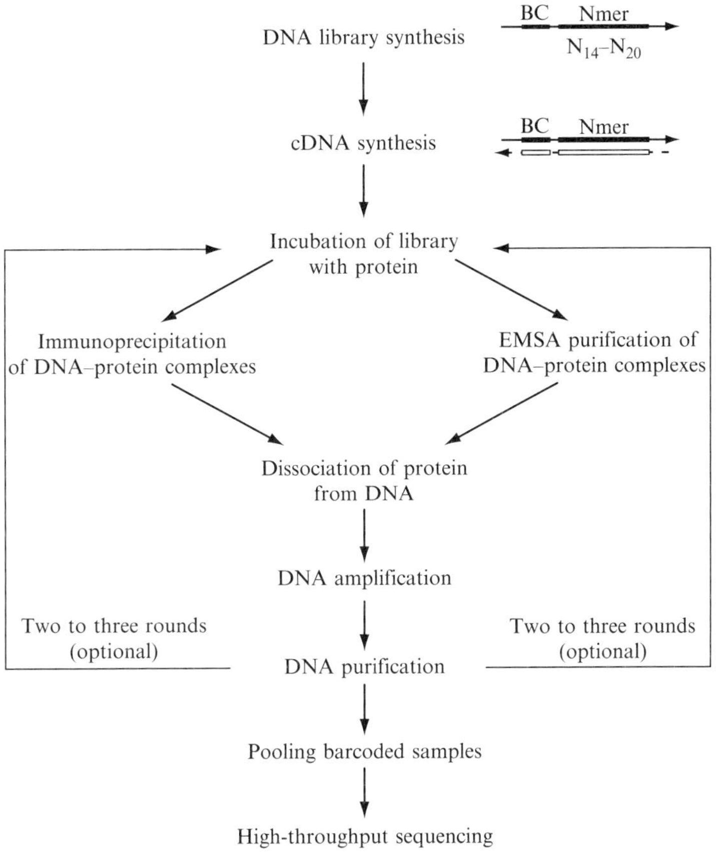

Figure 1.3 *Cognate site identification by sequencing.* Libraries of DNA molecules (Nmer) are incubated with the DNA-binding protein or small molecule and can undergo multiple rounds of selection and amplification to enrich for sequences with preferred binding sites. If each library is barcoded (BC), then several libraries can be pooled and simultaneously sequenced by Illumina, 454, or other next-generation platforms. This reduces the time and cost of data generation in these experiments.

(9) Repeat steps (3) through (8) two more times to select for optimally bound sequences and reduce background binding noise.
(10) Submit the samples for Illumina sequencing.

The binding data obtained by microarrays or by high-throughput sequencing can be analyzed in a similar manner to identify the consensus binding motif as well as the full spectrum of binding affinities of a natural or a synthetic molecule. These high-content data with multidimensional relationships require new computational tools to visualize and comprehend in their entirety. Below, we describe our approach to this challenge.

4. Data Analysis and Visualization of Specificity, Binding Energy, and Genomic-Association Landscapes

4.1. Sequence-specificity landscapes

At present, the sequence preferences of DNA-binding proteins are described by consensus motifs. Typically, a limited set of sequences with high affinity for a particular protein or small molecule are aligned to identify a consensus motif and the frequency of occurrence of a given nucleotide at a specific position is used as a measure of its contribution to the binding event. Such approaches yield a position matrix and several variations have been developed to account for the contributions of adjacent and nonadjacent nucleotides to binding. Collectively, such position weight matrices (PWMs) or position-specific scoring matrices (PSSMs) are then displayed using an intuitive Logo display that scales the size of a nucleotide to the information content at each position (Berg and von Hippel, 1987; Schneider and Stephens, 1990; Stormo and Zhao, 2010; Von Hippel and Berg, 1986).

The highly textured binding data across the entire sequence space from CSI and related methods are not adequately described using a simple PWM (Carlson *et al.*, 2010; Keles *et al.*, 2008; Maerkl and Quake, 2007). PWMs can only display limited information about the binding specificities of DNA-binding molecules. Often, PWMs do not capture the effects of neighboring nucleotides on a given nucleotide's binding probability because they are typically generated with limited binding data and under the assumption that each nucleotide contributes independently to the binding of a protein to its cognate site (Benos *et al.*, 2002). Moreover, the underlying focus on indentifying a "consensus motif" leads to amalgamation of several related motifs into a single consensus. Unrelated motifs that do not fit the consensus are often ignored or obfuscated by the consensus, leading to an inaccurate representation of the binding preferences of a given molecule (Carlson *et al.*, 2010). Finally, the current motif search algorithms do not consider the impact of nonconserved sequences that flank a "consensus" motif on the overall binding affinity for an otherwise optimal site. While PWM/PSSMs are very useful and often suffice in describing the binding preferences of a DNA-binding molecule, they fail to accurately describe binding to suboptimal sites that are often used to regulate biological outputs. Recent efforts to consider di- or trinucleotides in developing PWMs, or to search for more than one motif in the binding data, have resolved some of the limitations of past motif searching algorithms (Stormo and Zhao, 2010). Nevertheless, these matrices and their Logo displays are inherently limited and, in the context of the comprehensive binding data obtained from CSI and related methods, it is important to view the entire specificity landscape to determine the range of activity of a natural or engineered DNA-binding molecule.

Recently, a novel 3D method of displaying this type of data has been developed. SSLs can capture the full binding landscape of an Nmer sequence space for a given binding motif (Carlson *et al.*, 2010). An SSL is generated from a seed motif and represents the data in concentric rings with the z-axis displaying the relative binding affinity—either as the fluorescence intensity on the array or as the number of reads of a given sequence from high-throughput sequencing methods. The center ring contains the data for all the sequences bound which contain the seed motif exactly. Each successive ring moving outward has an increasing number of mismatches to the seed motif (Fig. 1.4). The data within the "zero mismatch" ring are sorted alphabetically by the sequences flanking the seed motif. The mismatch rings are sorted first by the position of the mismatch within the motif, then alphabetically by the mismatched base(s), and finally sorted alphabetically by the sequences flanking the seed motif. Each ring is aligned in the clockwise direction starting with the first (5′) residue of the motif. The impact of changing any single residue becomes apparent in quadrants that represent variations at specific positions. Thus, all changes at the first position of the motif are aligned at the beginning of each mismatch ring. Multidimensional relationships between several positions can be readily visualized in specific quadrants along the rings. In the example, the importance of changing the first two residues of the Nkx2.5 binding site is readily visualized as peaks in circles that are one or two

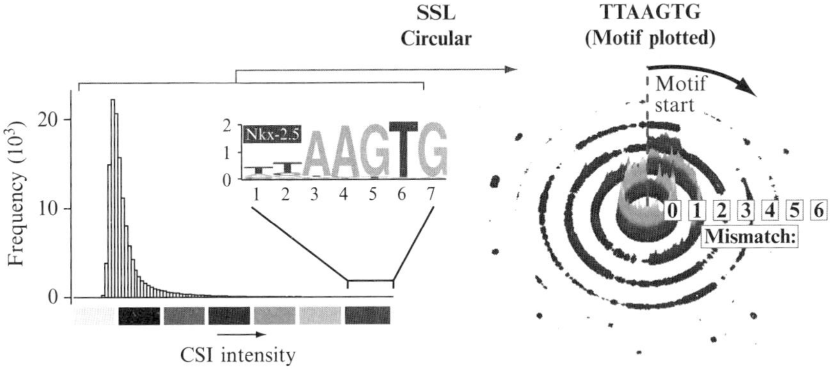

Figure 1.4 *Sequence-specificity landscapes (SSLs).* SSL generated from CSI intensity data for the Nkx-2.5 protein. The best bound sequences were used to generate the seed motif (shown in the LOGO and above the landscape). Color scale indicates normalized fluorescence intensity corresponding to how well a given sequence is bound. The sequences are distributed among the rings as described in the text. The distance between each ring and the next concentric ring outward is equivalent to a Hamming distance of 1. (Adapted from Carlson *et al.*, 2010). (See Color Insert.)

mismatches (Hamming distances) from the seed motif (Fig. 1.4). The data also reveal that the PWM compresses two motifs—TNAAGTG and NTAAGTG into a single consensus wherein the first two T residues are considered to have low information content. This is not the case, as an arginine on an unstructured arm of Nkx2.5 interacts with one T residue either at position 1 or at position 2 and this interaction contributes to high-affinity binding (Carlson *et al.*, 2010; Keles *et al.*, 2008). Similarly, low-to-moderate affinity binding sites with no sequence similarity to the accepted consensus can be deconvoluted by SSL display of the entire data. Using the SSL tool, we found that nearly 40% of the current motifs reported appear to be a forced compression of related motifs (Carlson *et al.*, 2010).

To emphasize the importance of viewing the entire SSL, rather than focusing on the PWM or using the PWM to predict binding properties, we display the CSI data for GATA4 (Fig. 1.5). GATA4 is a nuclear receptor type of zinc-finger transcription factor. It regulates genes involved in embryogenesis and in myocardial differentiation and function, and malfunction of GATA4 has been shown to cause septal defects. Comprehensive GATA4 sequence-specificity profiles were obtained via CSI platform bearing the entire 8mer sequence space. The top 300 binding sequences yielded a PWM and a consensus motif of 5′-GATA-3′. This consensus motif agrees

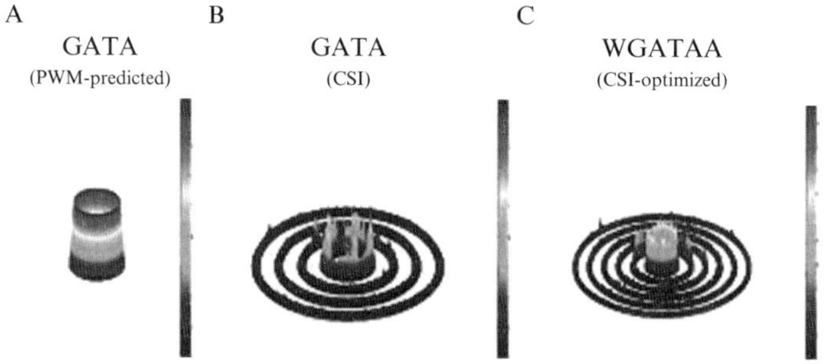

Figure 1.5 *SSLs for GATA4.* Panel (A) displays the specificity landscape predicted from the position weight matrix (PWM) generated from the CSI data. The PWM-based landscape predicts uniform binding wherever the exact GATA sequence is found. Panel (B) shows the actual intensity of every sequence on the array that contains the GATA motif. The data clearly indicate that unlike the prediction from the PWM, the binding is not uniform in every instance where this sequence is found. Panel (C) displays the SSL obtained after iterative refinement of the motif permitted by landscapes. Inspection of binding data in this manner reveals that many of the motifs of DNA-binding proteins and molecules compress the data and do not properly account for the influences of flanking sequences. The color scale reflects the fluorescence intensity of each sequence on the *z*-axis and the seed motif is shown above each landscape. (See Color Insert.)

with the "known" binding motif for this protein. As expected, when the underlying PWM is used to predict protein binding to sequences that vary from the consensus, we see little potential binding to mismatch sites (Fig. 1.5A). Moreover, the PWM-based landscape would predict equivalent binding to the consensus motif irrespective of its flanking sequence context (Fig. 1.5A). On the contrary, when the entire binding data (32,896 sequences) are displayed on a specificity landscape, several unanticipated properties emerge (Fig. 1.5B). First, the central "zero mismatch" ring containing all DNA hairpins with the perfect consensus core motif does not show equivalent binding. Second, the SSL for this consensus motif has very obvious regions of suboptimal binding within the zero mismatch ring, indicating that this motif is not restrictive enough to exclude poorly bound sequences. Third, the outer rings show some low-affinity binding that may have biological relevance. The actual CSI data displayed as a specificity landscape highlights the limitations of the PWMs and Logo displays in describing the full spectrum of molecular recognition properties of a DNA-binding molecule.

The specificity landscapes overcome these limitations and also reveal the significant impact of flanking sequences on the binding potential. It is important to note that in organizing the specificity landscape, we used the consensus motif obtained from the PWM. However, regardless of the source of the seed motif, SSLs can iteratively refine the motif for a given DNA-binding molecule. The quality of the seed motif can be assessed by inspecting where the highest binding regions are in the landscape and adjusting the sequence accordingly to bring all the best binding sequences into the center ring and move poorly bound sequences out. Using this iterative application of SSLs, the optimal binding motif can be refined from the original seed motif. In the case of GATA4, such iterations reveal a larger motif of 5′-WGATAA-3′ as the optimal binding motif (Fig. 1.5C). The data also reveal the contribution of nonconserved flanking sequences on the differential binding even if the optimal motif is present. In the earlier example of Nkx2.5, SSLs revealed the existence of closely related motifs (TNAAGTG and NTAAGTG) with high-information content residues (T at position 1 or 2) that were compressed by standard motif searching algorithms that seek to find a consensus site. In fact, as mentioned earlier, SSL evaluation of the current high-throughput binding data revealed that approximately 40% of reported motifs have two or more related motifs amalgamated into a single consensus (Carlson *et al.*, 2010).

To facilitate the evaluation of binding data by SSL, we have developed a user-friendly software package. The software used to generate these SSLs will be available online at http://www.biochem.wisc.edu/faculty/ansari/ and it can generate SSLs, assist with seed motif identification, and even overlay CSI data on genomic DNA (see genomescape description below; Fig. 1.6). This software requires the user to input the data file containing a

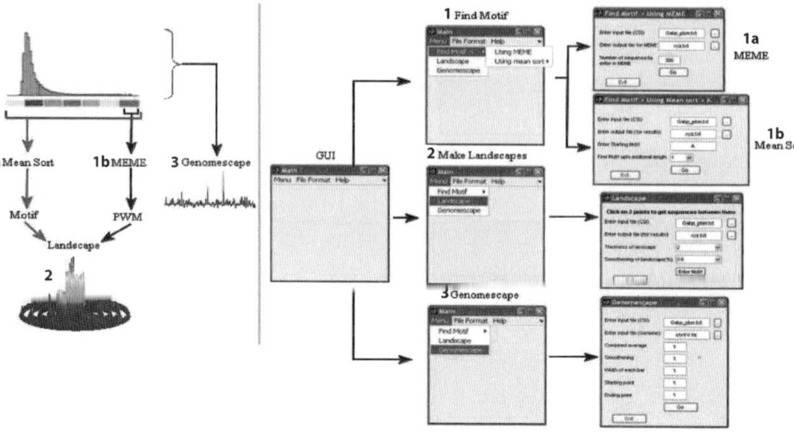

Figure 1.6 *SSL and genomescape software process flow diagram.* The left panel displays the experimental process flow for determining the DNA-binding profile of a given protein or small molecule by CSI. The range of fluorescence intensities of the sequences bound is displayed as a histogram, and subsets of these data are used to find the motif that most accurately represents the binding preferences of the molecule. The motif can be found by the program in two ways, as shown in (1a) and (1b) in both left and right panels, using either the Mean Sort algorithm or MEME. Once a motif is identified, it can be used to seed the SSL shown in (2). Finally, the CSI data can be applied to the genome to generate genomescapes, as shown in (3). On the right are windows displayed in the SSL and genomescape software at each step of the workflow of this process. (See Color Insert.)

forward sequence, the reverse compliment of that sequence, and the data value (intensity, *Z*-score, sequence read count) for that sequence in a tab-delimited text file. In addition, the program requires a seed motif in order to generate the SSLs. There are a number of ways to generate such a motif if the information is not already available from biochemical assays (EMSAs, SELEX, etc.). One such method is provided by the SSL software and uses "mean sorting" to find the optimal motif (1b in Fig. 1.6). This method calculates the mean and standard deviation of all possible sequences of a given length from CSI data and sorts them to show those with the highest mean intensity first. These motifs are then used as seed motifs and are iteratively refined into the optimally bound sequence motif.

Another method for selecting a seed motif involves the use of Multiple EM for Motif Elicitation (MEME) software, which is freely available online at http://meme.nbcr.net/meme4_4_0/intro.html. This program searches for motifs among a set of related DNA or protein sequences and represents the output with Logos and PWMs (1a in Fig. 1.6). The program requires a FASTA formatted list of sequences, weights for the sequences (if applicable), and input parameters such as number of motifs to find and the size of the motifs. One limitation to this method is the limitation on the number and length of sequences that can be included in the input. The program cannot handle all of the sequences from a CSI experiment, and the typical input is the top 300 sequences bound. However, this program can still provide a reasonable starting seed motif from which one can refine an optimal motif through iterative use of SSLs.

Both microarray and sequencing data, bearing millions of different sequences, can be used to obtain a comprehensive view of the molecular recognition properties of DNA-binding molecules. The multidimensional relationships between individual residues within the motif can be intuitively comprehended and used to evaluate the specificity determinants of natural or engineered binding molecules.

4.2. Binding energy landscapes

The specificity landscapes also serve as comprehensive binding energy landscapes. The fluorescence intensities obtained from microarrays are often subject to the caveat that binding at the surface may not reflect true equilibrium due to mass transport problems at surfaces and perturbation of equilibrium during the washing steps. To address this challenge, microfluidic devices, high-throughput SPR, and even fluorescence displacement approaches are being actively developed, but they are still limited to 100- to 1000-fold fewer sequences than could be displayed on a microarray.

Based on systematic solution binding measurements, we find that the fluorescence intensity observed when labeled molecules bind a DNA sequence on the microarray is directly proportional to its binding affinity in solution (Fig. 1.7; Carlson *et al.*, 2010; Puckett *et al.*, 2007; Warren *et al.*, 2006). CSI analysis of a synthetic polyamide and Nkx-2.5, a homeodomain protein, was performed via microarrays. Based on the resulting SSLs, specific sequences were chosen that span the range of fluorescence intensities (shown by the color scale). These intensities were compared to affinities measured by two different solution-based equilibrium binding methods. The binding affinities of the polyamide to these selected sequences were determined via DNaseI nuclease protection, whereas EMSA was used to measure the solution binding of Nkx-2.5. In both cases, we observed strong correlation between binding on the array and binding in solution ($R^2 > 0.99$). Therefore, the binding affinity is directly measurable from the range of fluorescence intensities on the microarray itself. This platform allows one to capture binding affinities of a given DNA-binding molecule or complex across millions of DNA sequences simultaneously. Thus, an SSL also provides a comprehensive binding energy landscape for the entire sequence space. This permits an unprecedented understanding of the energy and specificity thresholds that define the function of natural DNA-binding molecules in living cells. It also provides a framework to engineer artificial molecules that have desired biological potential.

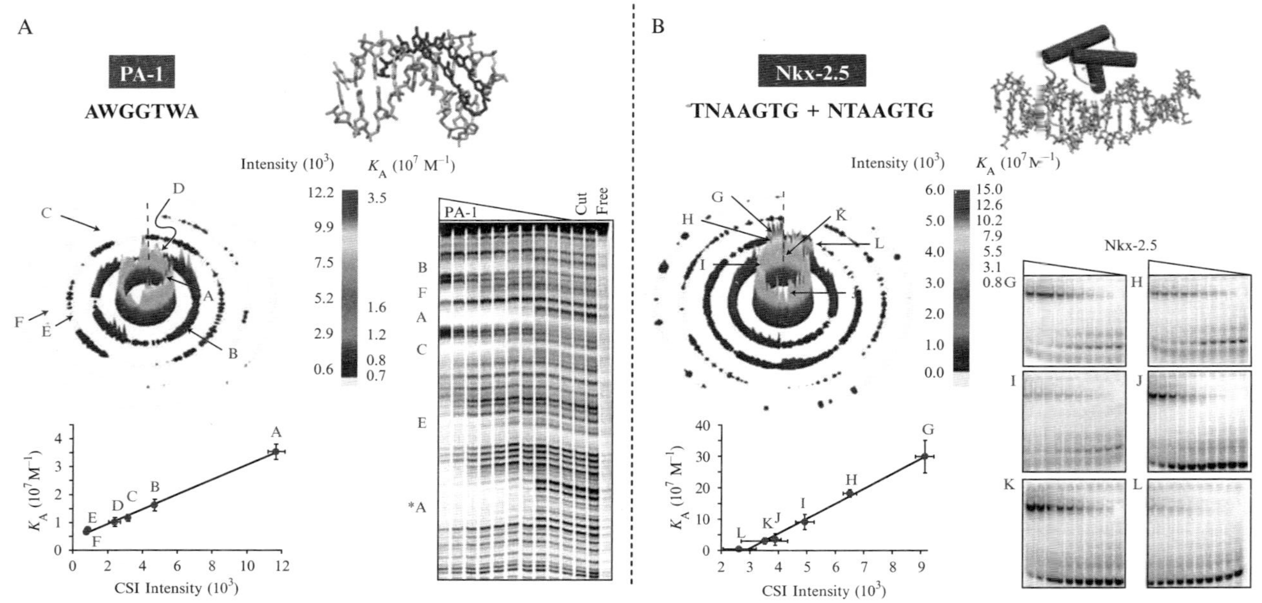

Figure 1.7 *Sequence-specificity landscapes also function as comprehensive binding energy landscapes.* (A) Polyamide PA-1 SSL as determined by CSI. Six sequences with intensities spanning a broad range were chosen (labeled A–F) and nuclease protection was performed to examine the correlation between K_A and CSI intensities. The color scale (center) shows the relationship of SSL–CSI intensities to their corresponding ΔG values ($\Delta G = -RT \ln K_A$). (B) Nkx-2.5 SSL as determined by CSI microarray. Six sequences were also chosen from the Nkx-2.5 CSI data (labeled G–L) and tested using EMSA to measure their K_A. The linear relationship between CSI intensity and DNA-binding characteristics measured in solution is readily detected (adapted from Carlson *et al.*, 2010). (For interpretation of the references to color in this figure legend, the reader is referred to the Web version of this chapter.)

4.3. Genomescapes

Another powerful feature of the SSL software is the ability to overlay CSI data on genomic DNA sequences (Fig. 1.8). The CSI data capture not only the affinity of a given motif but also the contribution of the flanking (contextual) sequences. This entire sequence, core along with flanking residues, can be used to annotate the genome. Thus, a highly textured annotation of potential binding sites across the genome can be readily obtained. This comprehensive and unbiased "genomescape" further reduces the dependence on PWMs to evaluate genomes and identify potential binding sites. One must be cognizant of the fact that the occurrence of a binding site is not sufficient to assume binding of factors to those sites in living cells. This is in part due to the presence of chromatin and multiple epigenetic marks that are superimposed on the underlying sequence. Moreover, it is quite possible that proteins might associate with specific sites in the genome via interactions with other DNA-binding proteins, thus functioning in the absence of discernable cognate sites. These additional layers can be revealed by comparing the CSI-determined genomescape with the newly developed methods for determining transcription factor binding patterns in cells. Thus, when combined with chromatin immunoprecipitation on microarrays (ChIP-chip), sequencing (CSI-seq), or gene expression data, the "genomescape" can reveal new mechanism of regulation (Iyer *et al.*, 2001; Johnson *et al.*, 2007; Ren *et al.*, 2000).

The genomescapes are represented as bar graphs across a given stretch of genomic DNA and can be useful in identifying possible *in vivo* binding sites for a given DNA-binding molecule. The software is capable of generating these genomescapes when provided with the SSL input data file (mentioned above) and a FASTA format text file containing the genomic sequence of interest (up to, and including the full length of a chromosome). The user can even provide boundaries for a region of the chromosome that was provided and expand that region for closer study (Fig. 1.8).

As an example, CSI binding data were obtained for an engineered DNA-binding small molecule (polyamide) and used to generate a genomescape across chromosome 6 of the human genome (Carlson *et al.*, 2010; Puckett *et al.*, 2007). This particular polyamide was designed to target and inhibit the binding of the Hif1a transcription factor to its cognate sites (Olenyuk *et al.*, 2004). HIF1a is known to stimulate the expression of VEGF, a protein closely associated with angiogenesis and cancer (Kim and Kaelin, 2006; Underiner *et al.*, 2004). The engineered polyamide targets the Hif1a binding sites at VEGF and other genes, prevents HIF1 from binding and activating transcription, and counteracts its action in oncogenesis (Kageyama *et al.*, 2006; Nickols *et al.*, 2007; Olenyuk *et al.*, 2004). While the polyamide achieves its function at VEGF, it regulates Hif1a function differently at sites that have similar cognate sites (Olenyuk *et al.*, 2004).

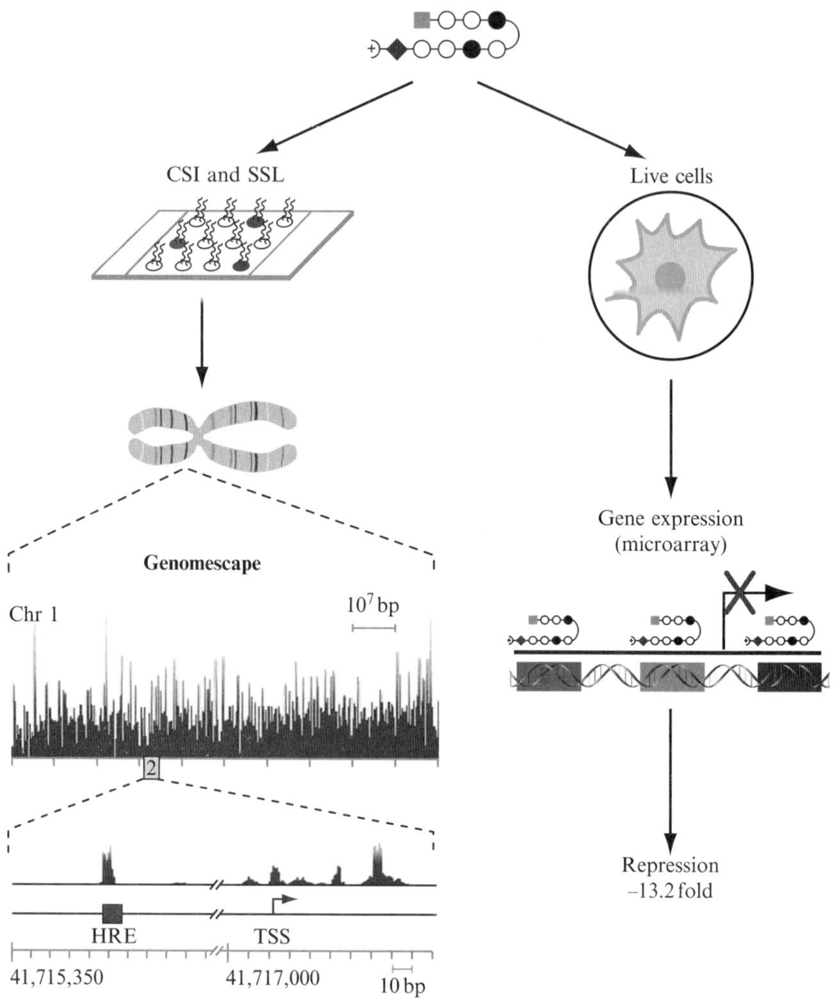

Figure 1.8 *Genomescapes from CSI binding data.* Schematic showing the parallel workflow of determining DNA-binding profiles *in vitro* (via CSI), and in live cells via genome-wide expression profiling, for a polyamide. The purple diamond represents a β-alanine, the orange square represents a chlorothiophene, the open circle represents an *N*-methylpyrrole, and the filled circle represents an *N*-methylimidazole. In genomescapes, the CSI data are used to annotate the genome with potential binding sites for a given DNA-binding molecule. This analysis, when combined with genome-wide expression or binding data, can reveal possible *in vivo* regulatory sites—as reflected by the transcription repression or activation seen in the expression profiling (adapted from Carlson *et al.*, 2010). (For interpretation of the references to color in this figure legend, the reader is referred to the Web version of this chapter.)

This difference cannot be simply explained by differential affinity to different Hif1a binding sites. The genomescapes revealed multiple moderate-to-low binding sites for the polyamide at the endothelin 2 (*ET-2*) gene, where the synthetic molecule was particularly effective in inhibiting gene expression (Fig. 1.8). Thus, genomescapes reveal the cumulative effect of a cluster of weak binding sites that overlap with functional elements in the genome. The application of genomescapes, alongside ChIP-seq or transcriptomics analyses, will provide a far more mechanistic view of gene regulatory patterns.

CSI is a powerful approach for measuring the specificity profile of *any* natural or synthetic DNA-binding molecule, and is not limited to just proteins. Complexes between proteins and engineered small molecules have also been examined by CSI to reveal the changes in sequence specificity of individual molecules due to cooperative assembly on DNA (Warren *et al.*, 2006). The ability to simultaneously interrogate the entire sequence space of a typical DNA-binding site yields a comprehensive view of the specificity spectrum of a DNA-binding molecule, while also uncovering the context effects of neighboring bases on binding affinity. When displayed as SSLs, the data reveal, in unprecedented detail, the different modes of sequence cognition and overcome the limitations of current motif searching algorithms that amalgamate several cognate sites into a single consensus motif. Moreover, the direct relationship observed between binding on microarrays and binding in solution allows SSLs to serve as binding energy landscapes. These specificity-energy landscapes help define the molecular recognition properties of DNA-binding molecules and facilitate the design of synthetic molecules to target genomic sites. Finally, overlaying the CSI data across genomic sequence gives a more accurate annotation of potential binding sites across the genome. Such CSI genomescapes not only complement the current approaches to genome-wide expression and binding studies but also facilitate the precision-tailoring of artificial molecules to target and regulate desired genes and networks. The applications of this methodology range from biophysical explorations of the principles that underlie specificity of molecular interactions, to evolutionary drift of regulatory motifs, to functional genomics, to synthetic regulation of gene networks and cellular fate.

REFERENCES

Anderson, B., Larkin, C., Guja, K., and Schildbach, J. (2008). Using fluorophore-labeled oligonucleotides to measure affinities of protein-DNA interactions. *Methods Enzymol.* **450,** 253–272.

Ansari, A., and Mapp, A. (2002). Modular design of artificial transcription factors. *Curr. Opin. Chem. Biol.* **6,** 765–772.

Benos, P., Bulyk, M., and Stormo, G. (2002). Additivity in protein-DNA interactions: How good an approximation is it? *Nucleic Acids Res.* **30,** 4442–4451.

Berg, O., and von Hippel, P. (1987). Selection of DNA binding sites by regulatory proteins: Statistical-mechanical theory and application to operators and promoters. *J. Mol. Biol.* **193,** 723–750.

Berger, M., and Bulyk, M. (2009). Universal protein-binding microarrays for the comprehensive characterization of the DNA-binding specificities of transcription factors. *Nat. Protoc.* **4,** 393–411.

Blackwell, T., and Weintraub, H. (1990). Differences and similarities in DNA-binding preferences of MyoD and E2A protein complexes revealed by binding site selection. *Science* **250,** 1104–1110.

Boger, D., Fink, B., Brunette, S., Tse, W., and Hedrick, M. (2001). A simple, high-resolution method for establishing DNA binding affinity and sequence selectivity. *J. Am. Chem. Soc.* **123,** 5878–5891.

Brockman, J., Frutos, A., and Corn, R. (1999). A multistep chemical modification procedure to create DNA arrays on gold surfaces for the study of protein-DNA interactions with surface plasmon resonance imaging. *J. Am. Chem. Soc.* **121,** 8044–8051.

Burnett, R., Melander, C., Puckett, J., Son, L., Wells, R., Dervan, P., and Gottesfeld, J. (2006). DNA sequence-specific polyamides alleviate transcription inhibition associated with long GAA•TTC repeats in Friedreich's ataxia. *Proc. Natl. Acad. Sci. USA* **103,** 11497–11502.

Campbell, C., and Kim, G. (2007). SPR microscopy and its applications to high-throughput analyses of biomolecular binding events and their kinetics. *Biomaterials* **28,** 2380–2392.

Carlson, C., Warren, C., Hauschild, K., Ozers, M., Qadir, N., Bhimsaria, D., Lee, Y., Cerrina, F., and Ansari, A. (2010). Specificity landscapes of DNA binding molecules elucidate biological function. *Proc. Natl. Acad. Sci. USA* **107,** 4544–4549.

Dervan, P., and Bürli, R. (1999). Sequence-specific DNA recognition by polyamides. *Curr. Opin. Chem. Biol.* **3,** 688–693.

Dervan, P., and Edelson, B. (2003). Recognition of the DNA minor groove by pyrrole-imidazole polyamides. *Curr. Opin. Struct. Biol.* **13,** 284–299.

Dervan, P., Doss, R., and Marques, M. (2005). Programmable DNA binding oligomers for control of transcription. *Curr. Med. Chem. Anticancer Agents* **5,** 373–387.

Farnham, P. (2009). Insights from genomic profiling of transcription factors. *Nat. Rev. Genet.* **10,** 605–616.

Fried, M., and Crothers, D. (1981). Equilibria and kinetics of lac repressor-operator interactions by polyacrylamide gel electrophoresis. *Nucleic Acids Res.* **9,** 6505–6525.

Galas, D., and Schmitz, A. (1978). DNAase footprinting: A simple method for the detection of protein-DNA binding specificity. *Nucleic Acids Res.* **5,** 3157–3170.

Garner, M., and Revzin, A. (1981). A gel electrophoresis method for quantifying the binding of proteins to specific DNA regions: Application to components of the Escherichia coli lactose operon regulatory system. *Nucleic Acids Res.* **9,** 3047–3060.

Gonzalez, B., Schwimmer, L., Fuller, R., Ye, Y., Asawapornmongkol, L., and Barbas, C. (2010). Modular system for the construction of zinc-finger libraries and proteins. *Nat. Protoc.* **5,** 791–810.

Gottesfeld, J., Turner, J., and Dervan, P. (2000). Chemical approaches to control gene expression. *Gene Expr.* **9,** 77–91.

Harbison, C., Gordon, D., Lee, T., Rinaldi, N., Macisaac, K., Danford, T., Hannett, N., Tagne, J., Reynolds, D., and Yoo, J. (2004). Transcriptional regulatory code of a eukaryotic genome. *Nature* **431,** 99–104.

Hauschild, K., Stover, J., Boger, D., and Ansari, A. (2009). CSI-FID: High throughput label-free detection of DNA binding molecules. *Bioorg. Med. Chem. Lett.* **19,** 3779–3782.

Heyduk, T., and Heyduk, E. (2002). Molecular beacons for detecting DNA binding proteins. *Nat. Biotechnol.* **20,** 171–176.

Heyduk, T., Ma, Y., Tang, H., and Ebright, R. (1996). Fluorescence anisotropy: Rapid, quantitative assay for protein-DNA and protein-protein interaction. *Methods Enzymol.* **274,** 492–503.

Hirao, I., Kawai, G., Yoshizawa, S., Nishimura, Y., Ishido, Y., Watanabe, K., and Miura, K. (1993). Structural features and properties of an extraordinarily stable hairpin-turn structure of d (GCGAAGC). *Nucleic Acids Symp. Ser.* **29,** 205–206.

Horak, C., Mahajan, M., Luscombe, N., Gerstein, M., Weissman, S., and Snyder, M. (2002). GATA-1 binding sites mapped in the beta-globin locus by using mammalian chIp-chip analysis. *Proc. Natl. Acad. Sci. USA* **99,** 2924–2929.

Iyer, V., Horak, C., Scafe, C., Botstein, D., Snyder, M., and Brown, P. (2001). Genomic binding sites of the yeast cell-cycle transcription factors SBF and MBF. *Nature* **409,** 533–538.

Johnson, D., Mortazavi, A., Myers, R., and Wold, B. (2007). Genome-wide mapping of in vivo protein-DNA interactions. *Science* **316,** 1497–1502.

Jolma, A., Kivioja, T., Toivonen, J., Cheng, L., Wei, G., Enge, M., Taipale, M., Vaquerizas, J., Yan, J., and Sillanpää, M. (2010). Multiplexed massively parallel SELEX for characterization of human transcription factor binding specificities. *Genome Res.* **20,** 861–873.

Kageyama, Y., Sugiyama, H., Ayame, H., Iwai, A., Fujii, Y., Eric Huang, L., Kizaka-Kondoh, S., Hiraoka, M., and Kihara, K. (2006). Suppression of VEGF transcription in renal cell carcinoma cells by pyrrole-imidazole hairpin polyamides targeting the hypoxia responsive element. *Acta Oncol.* **45,** 317–324.

Keles, S., Warren, C., Carlson, C., and Ansari, A. (2008). CSI-Tree: A regression tree approach for modeling binding properties of DNA-binding molecules based on cognate site identification (CSI) data. *Nucleic Acids Res.* **36,** 3171–3184.

Kim, W., and Kaelin, W. (2006). Molecular pathways in renal cell carcinoma—Rationale for targeted treatment. *Seminars in Oncology* **33**(5), 588–595.

Kuznetsov, S., Ren, C., Woodson, S., and Ansari, A. (2007). Loop dependence of the stability and dynamics of nucleic acid hairpins. *Nucleic Acids Res.* **36,** 1098–1112.

Lander, E. S., *et al.* (2001). Initial Sequencing and Analysis of the Human Genome. *Nature* **409**(6822), 860–921.

Lee, L., and Mapp, A. (2010). Transcriptional switches: Chemical approaches to gene regulation. *J. Biol. Chem.* **285,** 11033–11038.

Liu, X., Noll, D., Lieb, J., and Clarke, N. (2005). DIP-chip: Rapid and accurate determination of DNA-binding specificity. *Genome Res.* **15,** 421–427.

Maerkl, S., and Quake, S. (2007). A systems approach to measuring the binding energy landscapes of transcription factors. *Science* **315,** 233–237.

Maniatis, T., Falvo, J., Kim, T., Kim, T., Lin, C., Parekh, B., and Wathelet, M. (1998). Structure and Function of the Interferon-beta Enhanceosome. *Cold Spring Harbor Symp. Quant. Biol.* **63,** 609–620.

Mapp, A., and Ansari, A. (2007). A TAD further: Exogenous control of gene activation. *ACS Chem. Biol.* **2,** 62–75.

Moretti, R., and Ansari, A. (2008). Expanding the specificity of DNA targeting by harnessing cooperative assembly. *Biochimie* **90,** 1015–1025.

Nickols, N., Jacobs, C., Farkas, M., and Dervan, P. (2007). Modulating hypoxia-inducible transcription by disrupting the HIF-1–DNA interface. *ACS Chem. Biol.* **2,** 561–571.

Olenyuk, B., Zhang, G., Klco, J., Nickols, N., Kaelin, W., Jr., and Dervan, P. (2004). Inhibition of vascular endothelial growth factor with a sequence-specific hypoxia response element antagonist. *Sci. STKE* **101,** 16768–16773.

Oliphant, A., Brandl, C., and Struhl, K. (1989). Defining the sequence specificity of DNA-binding proteins by selecting binding sites from random-sequence oligonucleotides: Analysis of yeast GCN4 protein. *Mol. Cell. Biol.* **9,** 2944–2949.

Philippakis, A., Qureshi, A., Berger, M., and Bulyk, M. (2008). Design of compact, universal DNA microarrays for protein binding microarray experiments. *J. Comput. Biol.* **15,** 655–665.

Ptashne, M., and Gann, A. (2002). Genes & Signals. CSHL Press, Cold Spring Harbor, NY.

Puckett, J., Muzikar, K., Tietjen, J., Warren, C., Ansari, A., and Dervan, P. (2007). Quantitative microarray profiling of DNA-binding molecules. *J. Am. Chem. Soc.* **129,** 12310–12319.

Ragoussis, J., Field, S., and Udalova, I. (2006). Quantitative profiling of protein-DNA binding on microarrays. *Methods Mol. Biol.* **338,** 261–280.

Ren, B., Robert, F., Wyrick, J., Aparicio, O., Jennings, E., Simon, I., Zeitlinger, J., Schreiber, J., Hannett, N., and Kanin, E. (2000). Genome-wide location and function of DNA binding proteins. *Sci. STKE* **290,** 2306–2309.

Rodríguez-Martínez, J., Peterson Kaufman, K., and Ansari, A. (2010). Small-molecule regulators that mimic transcription factors. *Biochim. Biophys. Acta* **1799**(10–12), 768–774.

Schneider, T., and Stephens, R. (1990). Sequence logos: A new way to display consensus sequences. *Nucleic Acids Res.* **18,** 6097–6100.

Sera, T. (2010). Generation of cell-permeable artificial zinc finger protein variants. *Methods Mol. Biol.* **649,** 91–96.

Stormo, G., and Zhao, Y. (2010). Determining the specificity of protein–DNA interactions. *Nat. Rev. Genet.* **11,** 751–760.

Suto, R., Edayathumangalam, R., White, C., Melander, C., Gottesfeld, J., Dervan, P., and Luger, K. (2003). Crystal structures of nucleosome core particles in complex with minor groove DNA-binding ligands. *J. Mol. Biol.* **326,** 371–380.

Tuerk, C., and Gold, L. (1990). Systematic evolution of ligands by exponential enrichment: RNA ligands to bacteriophage T4 DNA polymerase. *Science* **249,** 505–510.

Underiner, T., Ruggeri, B., and Gingrich, D. (2004). Development of vascular endothelial growth factor receptor (VEGFR) kinase inhibitors as anti-angiogenic agents in cancer therapy. *Curr. Med. Chem.* **11,** 731–745.

Venter, J. C., *et al.* (2001). The Sequence of the Human Genome. *Science* **291**(5507), 1304–1351.

Von Hippel, P., and Berg, O. (1986). On the specificity of DNA-protein interactions. *Proc. Natl. Acad. Sci. USA* **83,** 1608–1612.

Wang, J., Li, T., and Lu, Z. (2005). A method for fabricating uni-dsDNA microarray chip for analyzing DNA-binding proteins. *J. Biochem. Biophys. Methods* **63,** 100–110.

Ward, L., and Bussemaker, H. (2008). Predicting functional transcription factor binding through alignment-free and affinity-based analysis of orthologous promoter sequences. *Bioinformatics* **24,** i165–171.

Warren, C., Kratochvil, N., Hauschild, K., Foister, S., Brezinski, M., Dervan, P., Phillips, G., and Ansari, A. (2006). Defining the sequence-recognition profile of DNA-binding molecules. *Proc. Natl. Acad. Sci. USA* **103,** 867–872.

White, S., Szewczyk, J., Turner, J., Baird, E., and Dervan, P. (1998). Recognition of the four Watson–Crick base pairs in the DNA minor groove by synthetic ligands. *Nature* **391,** 468–471.

Wolfe, S., Nekludova, L., and Pabo, C. (2000). DNA recognition by Cys2His2 zinc finger proteins. *Annu. Rev. Biophys. Biomol. Struct.* **29,** 183–212.

Wright, W., Binder, M., and Funk, W. (1991). Cyclic amplification and selection of targets (CASTing) for the myogenin consensus binding site. *Mol. Cell. Biol.* **11,** 4104–4110.

Yamamoto, K., Darimont, B., Wagner, R., and Iniguez-Lluhi, J. (1998). Building Transcriptional Regulatory Complexes: Signals and Surfaces, Vol. 63, Cold Spring Harbor Laboratory Press, New York, pp. 587–598.

Zhao, Y., Granas, D., and Stormo, G. (2009). Inferring binding energies from selected binding sites. *PLoS Comput. Biol.* **5,** e1000590.

Zykovich, A., Korf, I., and Segal, D. (2009). Bind-n-Seq: High-throughput analysis of in vitro protein-DNA interactions using massively parallel sequencing. *Nucleic Acids Res.* **37,** e151.

CHAPTER TWO

Promoter Reliability in Modular Transcriptional Networks

Rajat Anand,* Navneet Rai,*,† *and* Mukund Thattai*

Contents

Abstract

Synthetic biologists engineer systems with desired properties from simple and well-characterized biological parts. Among the most popular and versatile parts are tunable promoters and the transcription factors (TFs) that regulate them. Individual TFs can transduce physical or chemical signals to regulate gene expression; networks of TFs regulating each other's expression can filter signals, reduce noise, store memories, and oscillate. However, the biochemical parameters that describe TF–promoter interactions are often context dependent, making it challenging to build systems that reliably achieve specific outcomes. Here, we explore this problem using plasmid-borne transcriptional networks in *Escherichia coli*. We demonstrate that the expression properties of a positive-feedback module quantitatively and qualitatively change when this module is embedded within the context of a larger network, where the original TF is used to drive new outputs. A mathematical model suggests this might be due in part to the sequestration of the TF by additional copies of its cognate promoter. The parameters describing

* National Centre for Biological Sciences, Tata Institute of Fundamental Research, Bangalore, India
† Department of Biosciences and Bioengineering, Indian Institute of Technology, Mumbai, India

Methods in Enzymology, Volume 497

ISSN 0076-6879, DOI: 10.1016/B978-0-12-385075-1.00002-0

TF–promoter interactions (the Hill coefficient and half-saturation constant) can vary depending on promoter copy number. This problem is acute for plasmid-borne systems where promoter concentrations exceed the TF–promoter equilibrium constant. In this regime, we advocate the use of operator buffers: passive multimeric stretches of TF-binding sites that insulate promoter properties from context. If such buffers are included in a standard host chassis, promoters once characterized can be reliably integrated into larger networks.

Transcriptional networks are groups of genes that regulate one-another's expression (Kauffman, 1969). Experiments demonstrating the striking capabilities of small engineered networks of transcription factors (TFs) were among the early successes of synthetic biology (Becskei and Serrano, 2000; Elowitz and Leibler, 2000; Gardner *et al.*, 2000). Since then, a variety of transcriptional networks with useful properties have been designed, built, and tested (Andrianantoandro *et al.*, 2006; Hasty *et al.*, 2002): noise reduction systems, threshold devices, signal processors, hysteretic switches, oscillators, multicell communication systems, and more. In parallel with this engineering effort, synthetic networks have been used to explore the function of small modules that are the building blocks of natural prokaryotic and eukaryotic transcriptional networks (Bashor *et al.*, 2010; Lee *et al.*, 2002; Shen-Orr *et al.*, 2002; Sprinzak and Elowitz, 2005). The hierarchical architecture of natural networks mirrors that of electronic circuits (Itzkovitz *et al.*, 2005): basic parts are assembled into functional modules, which are then integrated to form more complex systems. It has been suggested that this modular approach is the key to building synthetic networks with predictable functions (Endy, 2005; Heinemann and Panke, 2006; Purnick and Weiss, 2009).

Transcriptional networks are a good test bed in which to study modular design. We have access to a wide variety of TFs and their cognate promoters (Boyle and Silver, 2009; Voigt, 2006); in many instances, the biochemical parameters related to TF–promoter binding have been quantitatively determined (e.g., von Hippel *et al.*, 1984); and by specifying which promoters drive the expression of which genes, the networks can be easily rewired (Guet *et al.*, 2002). But many practical difficulties must be overcome before we can routinely build complex transcriptional networks through a hierarchy of parts, devices, modules, and systems. Here, we focus on the problem of context-dependent parameters:

> Adding or removing devices will change the module and change any previously estimated parameters. Thus, parameters derived in one context may not apply in another.
>
> Andrianantoandro *et al.* (2006)

One solution is to build systems that are robust to changes in parameter values (Barkai and Leibler, 1997; Stelling *et al.*, 2005); another is to maintain a diverse library of TF–promoter pairs, and use only those parts that are suitable in any given context (Ellis *et al.*, 2009). Here, we discuss a complementary approach: we explore the underlying causes of this context dependence, with the goal of developing molecular systems that improve promoter reliability. Modular engineering might then be achieved by combining reliable parts with robust design.

1. Results

1.1. Measuring the dynamics of an embedded network

We set out to measure whether the properties of a small transcriptional module were modified when it was embedded within a larger network. One of the simplest examples of such an embedding is the well-studied two-gene oscillator, where gene *A* activates the expression of gene *B*, while *B* in turn inhibits the expression of *A* (Atkinson *et al.*, 2003; Danino *et al.*, 2010; Stricker *et al.*, 2008). The net negative feedback tends to damp oscillations unless the system is sufficiently nonlinear; this can be achieved by requiring *A* to activate its own expression. The complete network can thus be regarded as having an embedded positive-feedback module.

We studied the response of such a module using synthetic networks built from the following components (Section 3; Voigt 2006 and references therein): the LacI repressor and P_{Lac} promoter from *Escherichia coli*; the LuxI enzyme, the LuxR transcriptional activator, and the P_R promoter from *Vibrio fischeri*. LacI binds to and represses expression at P_{Lac}, but this repression is relieved in the presence of the inducer isopropyl β-D-thiogalactopyranoside, or IPTG (e.g., Ozbudak *et al.*, 2004). LuxI and LuxR are components of the *V. fischeri* quorum-sensing machinery (Fuqua and Greenberg, 2002; Waters and Bassler, 2005): LuxI generates the signaling molecule acyl homoserine lactone (AHL), which freely diffuses across the cell membrane; the net rate of AHL production is thus proportional to cell density; AHL binds to LuxR, causing it to multimerize (Urbanowski *et al.*, 2004); and AHL-bound LuxR activates expression at the P_R promoter. (This LuxI/LuxR machinery has been demonstrated to function in an *E. coli* background; e.g., Balagadde *et al.*, 2005; Haseltine and Arnold, 2008.)

We constructed a positive-feedback module (*P*) by placing LuxR downstream of its cognate P_R promoter, while expressing LuxI from the P_{Lac} promoter (Fig. 2.1A). We converted this to an embedded positive/negative feedback system (*PN*) by expressing LacI downstream of an additional copy of the P_R promoter (Fig. 2.1B). These constructs were expressed on high copy number plasmids in an *E. coli* background. For both constructs, we tracked the levels of LuxI and LuxR with single-cell-resolved flow cytometry, using polycistronic copies of the cyan fluorescent protein (CFP) and yellow fluorescent protein (YFP), respectively, as proxies (Section 3).

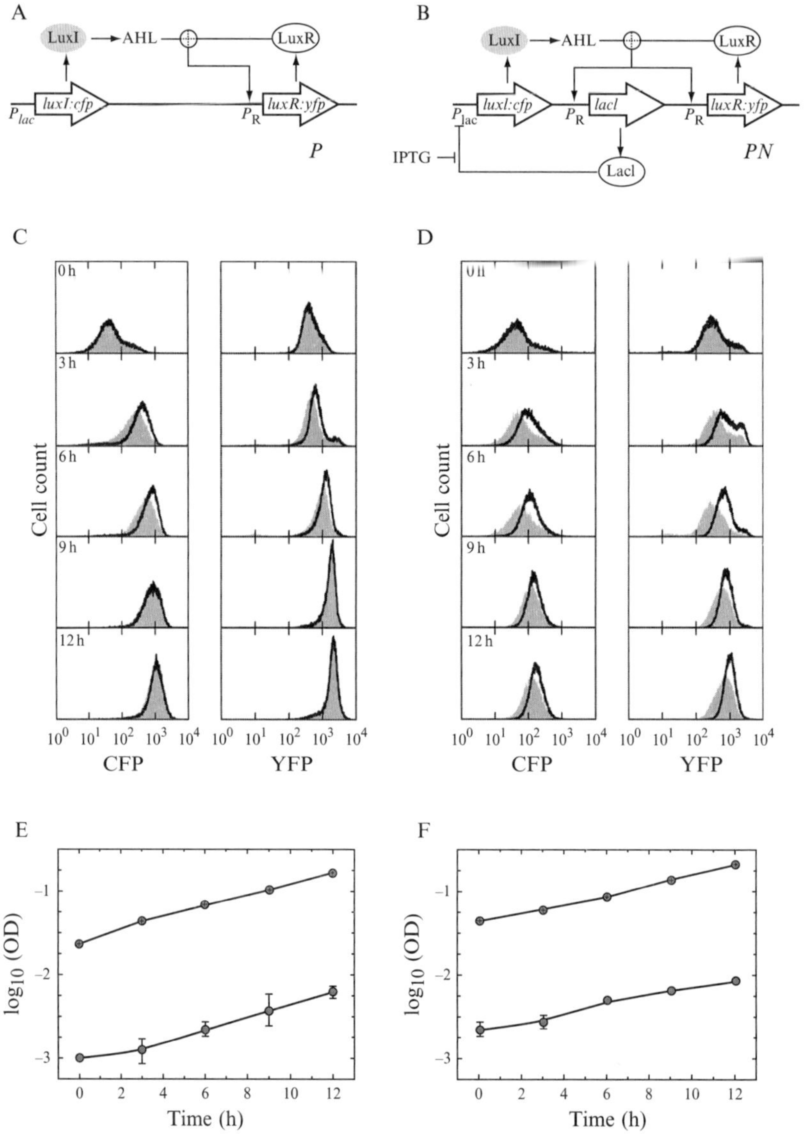

Figure 2.1 Measuring the dynamics of an embedded positive-feedback module. (A and B) Synthetic transcriptional networks *P* (the isolated positive-feedback module) and *PN* (the two-gene oscillator with positive/negative feedback). LuxI synthesizes the diffusible molecule AHL; LuxR, when bound to AHL, activates expression at the P_R promoter; LacI represses expression at the P_{Lac} promoter, but repression is relieved in the presence of IPTG. We track LuxI and LuxR levels by proxy, using polycistronic genes encoding the fluorescent reporters CFP and YFP. In the network *P*, LuxI is expressed constitutively; the network *PN* is identical to *P* except for an extra copy of

The response of two-gene *PN* network, which is predicted to oscillate, is the subject of another study. Our goal here is to examine the response of the positive-feedback module embedded with it, compared to the response of the isolated positive-feedback network *P*. For this we used the following strategy: we carried out experiments in an *E. coli* host strain that expressed high amounts of LacI from a chromosomal gene copy. In this background, the P_{Lac} promoter is completely repressed in the absence of IPTG, regardless of the level of LacI expressed from P_R. We grew cells for 12 h in minimal medium lacking IPTG, during which time CFP (and by proxy, LuxI) reached its basal steady-state level. By growing cells at extremely low density ($OD_{600} < 0.05$ after 12 h of exponential growth), we ensured that any residual AHL accumulation was minimized; this was confirmed by noting that YFP (and by proxy, LuxR) expression was close to background. Next, we transferred cells to minimal medium containing saturating amounts of IPTG (1 m*M*, at which LacI is predominantly in its inactive form), thus driving maximal expression at P_{Lac}. We subsequently tracked CFP and YFP levels in single cells as a function of time over 12 h, for cells in exponential growth with a range of initial cell densities spanning a factor of 25. This experimental protocol was designed to isolate the behavior of the embedded positive-feedback module: in the absence of IPTG, LacI is dominated by expression from the chromosomal copy; in the presence of saturating amounts of IPTG, LacI is predominantly inactive; therefore, in both cases, the negative feedback mediated by LacI is abolished, and the positive-feedback subsystem exists essentially in isolation.

Naively (or optimistically), we would expect LuxR dynamics to be identical for the two networks, *P* and *PN*, under these conditions. In fact, there are several differences. Following the first 12-h growth phase with P_{Lac} repressed, we see that basal CFP distributions are essentially identical for networks *P* and *PN* (Fig. 2.1C and D, top left panels). This indicates that

the promoter P_R that expresses LacI, mediating negative feedback. The experiments discussed here are performed in a host *E. coli* strain expressing LacI in high copy from a chromosomal gene (not shown in this schematic). (C and D) Single-cell-resolved CFP and YFP distributions measured by flow cytometry. At the 0 h timepoint, cells grown in minimal medium in the absence of IPTG are filtered and resuspended in medium containing 1 m*M* IPTG, at varying densities. IPTG induces the synthesis of LuxI and CFP; the resulting AHL accumulation drives the synthesis of LuxR and YFP. The initial state is at 0 h is measured just prior to resuspension; following resuspension, the low-density (solid gray) and high-density (black line) fluorescence distribution trajectories are nearly identical for the network *P*, and broadly overlapping for the embedded network *PN*. (E and F) Cell densities (optical densities measured by absorbance at 600 nm) corresponding to the low- and high-density experiments in (C) and (D). There is a difference of approximately 25 fold between these cases; experiments performed at intermediate densities showed similar results (not shown here).

LacI is present in sufficient amounts to completely repress P_{Lac}, and also that plasmid copy numbers are similar for the two systems. At this timepoint, YFP levels are also close to background (Fig. 2.1C and D, top right panels). Once IPTG is added, CFP levels begin to increase. Over the next 12-h period, AHL is synthesized by the newly expressed LuxI; this AHL is expected to activate LuxR and kick-start the positive-feedback module. We see for both networks that YFP dynamics are not significantly influenced by cell density (Fig. 2.1C and D, overlays; Fig. 2.1E and F), indicating that AHL levels are very quickly in excess of the amounts required to activate LuxR. At the 12-h timepoint, the network *P* achieves much higher CFP and fractionally higher YFP levels than the network *PN* (Fig. 2.1C and D, bottom panels). This could be a consequence of two factors. First, it is possible that the addition of 1 m*M* IPTG is not sufficient to overcome the extra LacI expressed from the P_{R} promoter in the network *PN*, leading to lower CFP expression for that network. However, if AHL levels are indeed in excess, this would not be expected to impact YFP levels. This suggests a second contributing factor that plasmid copy numbers are marginally lower for the network *PN*, possibly because of the additional metabolic burden of LacI expression. Leading up to this 12-h timepoint following IPTG addition, the dynamics of YFP expression show interesting features. For both networks, YFP distributions pass through a transient bimodal stage while shifting to higher fluorescence levels (Fig. 2.1C and D, right panels). Though the precise form and timing of the bimodal distributions are different between the two networks, the occurrence of bimodality is a robust qualitative feature of the dynamics. Our optimism appears to be partly vindicated in this instance: qualitatively at least, the two networks *P* and *PN* show similar behaviors, and embedding appears not to affect the system. However, there is one unavoidable aspect of the embedding process that can, under the right circumstances, affect even the qualitative dynamics of a transcriptional module.

1.2. Modeling an embedded positive-feedback module

To explore the general consequences of embedding, we developed a mathematical model of a positive-feedback module in which an activating TF is placed downstream of its cognate promoter, thus driving its own transcription (Ninfa and Mayo, 2004). This module can be embedded within a larger network by using the same TF to drive additional outputs, or by integrating external inputs that modify TF or promoter properties (Fig. 2.2A). The two dynamical variables in our model are the TF (designated A for activator) and its cognate promoter (designated D for DNA). Let A_T represents the total activator concentration and D_T the total promoter concentration; if D–A binding is in rapid equilibrium, it can be modeled by a Hill equation:

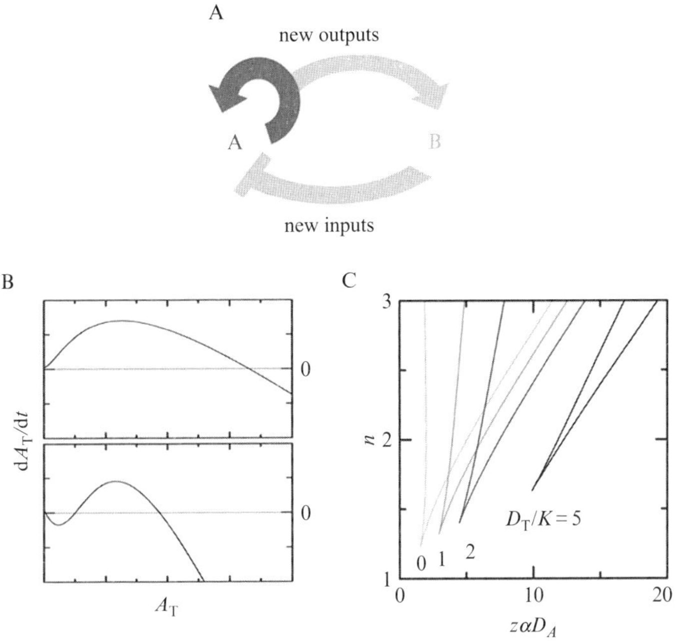

Figure 2.2 Modeling the general consequences of embedding. (A) A positive-feedback module in which a TF (labeled A) activates its own expression (dark gray) can be embedded in a larger network by using the same TF to drive new outputs, or by integrating new inputs (light gray). The network shown here is precisely the two-gene oscillator corresponding to the network PN (Fig. 2.1B). (B) Depending on parameter values in Eq. (2.4), the positive-feedback module can be monostable (top panel: $n = 1.3$, $z\alpha D_A = 4$, $\beta = 0.01$, $K = 1$, $D_T \sim 0$) or bistable (bottom panel: $n = 2$, $z\alpha D_A = 2.4$, $\beta = 0.01$, $K = 1$, $D_T \sim 0$). (C) Parameter regions leading to monostable (outside the cusp) and bistable (within the cusp) behavior. Here, n and $z\alpha D_A$ are varied while holding $\beta = 0.01$. The region of bistability depends on the ratio of the total DNA concentration D_T to the equilibrium constant K: going from light gray to dark gray, D_T/K is varied from approximately 0 to 5.

$$D + nA \overset{K^n}{\leftrightarrow} DA_n, \quad \text{with} \quad D_T = [D] + [DA_n],\ A_T = [A] + n[DA_n], \tag{2.1}$$

$$f([A]) \equiv \frac{[DA_n]}{D_T} = \frac{[A]^n}{K^n + [A]^n}, \tag{2.2}$$

where f represents the fraction of time the promoter is bound to the TF; n is the dimensionless Hill coefficient which captures multimerization or cooperative binding; and K is the equilibrium constant with dimensions of concentration. To model the effect of embedding, assume there are two subsets of promoters: one from which the activator is expressed (D_A) and another which makes new output connections (D_B), both of which contribute to the total promoter concentration:

$$D_T = D_A + D_B. \tag{2.3}$$

If transcription occurs at maximal rate α from the TF-bound promoter and at basal rate $\alpha\beta << \alpha$ from the bare promoter, we have the following dynamics for the TF:

$$\frac{dA_T}{dt} = z\frac{D_A}{D_T}(\alpha\beta[D] + \alpha[DA_n]) - A_T = z\alpha D_A(\beta + (1-\beta)f([A])) - A_T, \tag{2.4}$$

where $0 \leq z \leq 1$ is the fraction of synthesized TF that is functional, representing the influence of external inputs such as inducers. We choose time units so the TF degradation rate is unity. Note that while this differential equation describes the dynamics of total TF (A_T), it is only the free TF which enters into the function $f([A])$.

Steady-state values of A_T are the roots of the algebraic equation that results when the time derivative in Eq. (2.4) is set to zero. A monostable system has a single root, corresponding to a unique steady state. Positive feedback creates the potential for bistability, where the equation has three roots (Fig. 2.2B; Ozbudak *et al*, 2004; Ninfa and Mayo, 2004): an uninduced steady state and a fully induced steady state, separated by an unstable threshold. Figure 2.2C shows parameter regions that result in monostable or bistable outcomes. The key point is that the boundaries between these regions depend on the total promoter concentration ($D_A + D_B$), not only on the concentration of promoters from which the TF is expressed. Embedding can thus influence system response, both quantitatively and qualitatively.

We can connect this model to our experiments as follows: LuxR plays the role of the activator A, and its cognate promoter P_R plays the role of D. In the absence of IPTG, AHL levels are low and LuxR is predominantly in its inactive form; soon after IPTG is added, AHL levels are high and LuxR is predominantly in its active form. This corresponds to suddenly switching from $z = 0$ to $z = 1$, and allowing the system to evolve to its new steady state according to Eq. (2.4). We can get some sense of the intervening dynamics by imagining what would happen if z were varied slowly from 0 to 1, allowing the system to reach steady state for each intermediate value. If the system is in a monostable region, single-cell expression levels will be unimodal; if it passes through a bistable region, expression levels will be bimodal as cells stochastically transition between the two stable states (Ozbudak *et al.*, 2004). In our experiments, the number of copies of the P_R promoter is doubled going from network *P* to network *PN*. As seen in

Fig. 2.2C, this increase in D_T has the potential to shift or even abolish the occurrence of bimodality. We do not see such qualitative changes in our experiments; whether it happens in practice depends not on the relative increase in promoter concentration but on its absolute magnitude.

1.3. Promoter properties depend on copy number

To understand the circumstances under which the positive feedback loop is influenced by promoter concentration, we must further examine Eq. (2.4). The value D_T only enters this equation implicitly, through the function $f([A])$. Applying Eqs. (2.1) and (2.2), we can write

$$A_T = [A] + nD_T \frac{[A]^n}{K^n + [A]^n} = K\left(\frac{f}{1-f}\right)^{1/n} + nD_T f. \tag{2.5}$$

The very process by which one copy of the promoter responds to TF levels—equilibrium binding—also causes it to sequester the protein from other promoters. At high promoter levels, the free TF concentration A can be significantly different from the total A_T. While we cannot in general represent $f([A])$ as a function of A_T in closed form, we can approximate it by a Hill-type function with effective parameters:

$$f([A]) \rightarrow f(A_T) \approx \frac{A_T^{neff}}{Keff^{neff} + A_T^{neff}}, \tag{2.6}$$

where *neff* and *Keff* are the effective Hill coefficient and half-saturation constant, respectively. One route to find the values of these effective parameters is to numerically fit the inverse function in Eq. (2.5) to a Hill-type form using a least-squares score (Fig. 2.3A). Another is by analogy to the true Hill function at half saturation:

$$f(A_T = Keff) \equiv \frac{1}{2} \quad \text{and} \quad neff \equiv 4A_T \frac{\partial f(A_T)}{\partial A_T}\Big|_{A_T = Keff}. \tag{2.7}$$

This gives us the following equations:

$$\begin{aligned} Keff &= K\left(1 + \frac{n}{2} D_T/K\right), \\ neff &= n\frac{4 + 2n(D_T/K)}{4 + n^2(D_T/K)}. \end{aligned} \tag{2.8}$$

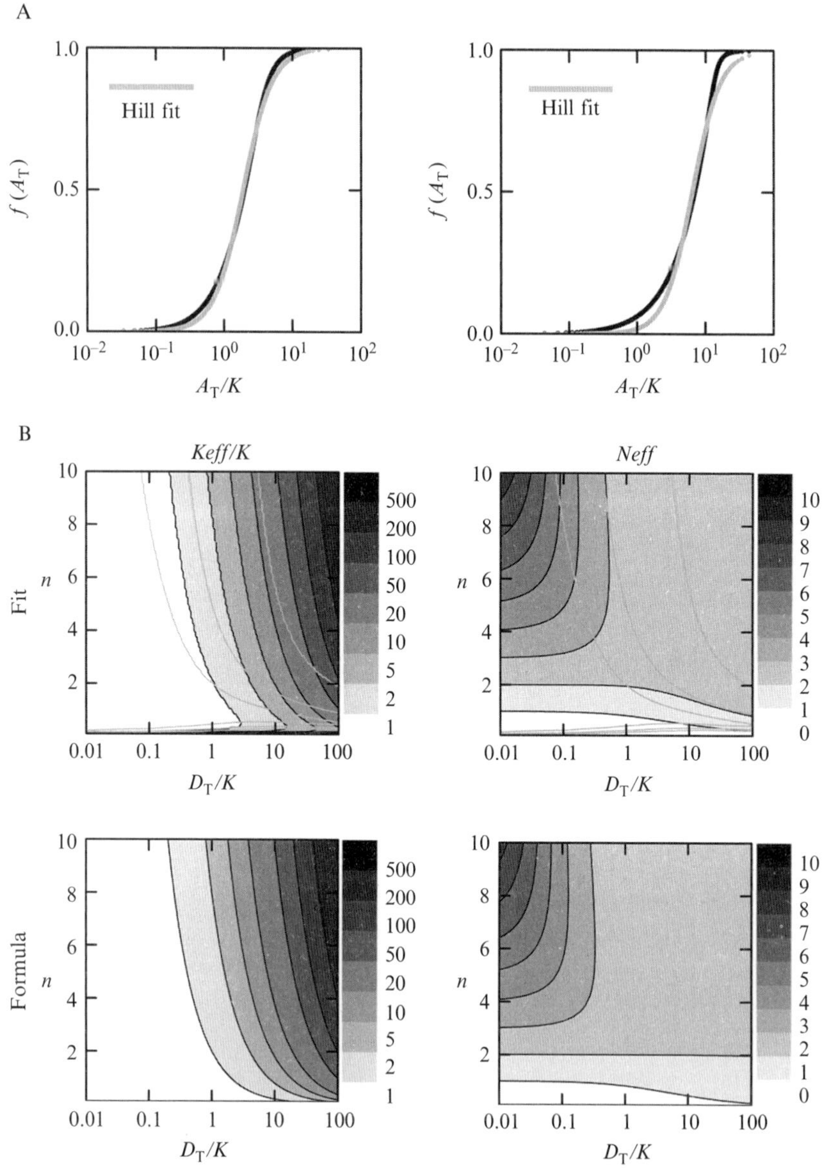

Figure 2.3 Promoter properties depend on copy number. (A) The fraction of time a promoter is bound, $f(A_T)$, plotted as a function of total TF concentration. The black curve shows the exact value and the gray curve shows a least-squares fit to the Hill form of Eq. (2.6) (left panel: $n = 2$, $D_T/K = 1$, $R^2 = 0.995$; right panel: $n = 2$, $D_T/K = 6$, $R^2 = 0.98$). (B) Effective half-saturation concentrations *Keff* (left panels) and effective Hill coefficients *neff* (right panels) obtained either by least-squares fitting (top panels) or from the formulas in Eq. (2.8) (bottom panels). In each plot, we show the values of the

Both these strategies produce similar parameter estimates, as shown in Fig. 2.3B.

As expected, the effective parameters approach their true values for $D_T \ll K$; but they deviate from these values as soon as the promoter concentration becomes comparable to the equilibrium constant ($D_T \approx K$). For very high promoter concentrations ($D_T \gg K$), the effective Hill coefficient approaches $neff \approx 2$ regardless of the true Hill coefficient, while the effective half-saturation constant approaches $Keff \approx \frac{n}{2} D_T$ regardless of the true equilibrium constant. For networks expressed on high copy plasmids (100 plasmids in a μm-scale cell corresponds to a concentration of 0.1 μM, comparable to physiological equilibrium constants) this suggests there are fundamental restrictions on the range of achievable promoter properties.

1.4. Operator buffers insulate promoter properties from context

According to Eq. (2.8), promoter properties depend only on the total promoter concentration. More accurately, they depend only on the total concentration of TF operator sites that bind and sequester the protein. One strategy to ensure that promoters remain reliable even as their numbers increase is to introduce a much larger number of passive operator sites that compete to sequester the TF. In this case, the *relative* change in the number of operator sites will be small as extra promoter copies are added, and variations in free TF will be buffered (Fig. 2.4A). Suppose we interpret the two promoter subpopulations in Eq. (2.3) as corresponding to active promoter copies (D_A) and passive operators in the buffer (D_B). If the active promoters could be ignored, the half-saturation concentration of the TF would be $A_T = K + \frac{n}{2} D_B$. At this concentration, the promoter activity will be given by:

$$f(A_T) \approx \frac{1}{2}\left(1 - \frac{D_\mathrm{A}}{4K/n^2 + D_\mathrm{T}}\right). \tag{2.9}$$

We would consider the promoter to be reliable if this number is close to half. This can be achieved under two different conditions: either (i) negligible binding: $D_A << K$ or (ii) operator buffered: $D_A \ll D_B \approx D_T$. At the half-saturation level, the concentration of free TF is equal to the

effective parameters as a grayscale heat-map (scale bar), as the underlying values of n and D_T/K are varied. For the least-squares fits, we have shown three contours (light gray curves) corresponding to R^2 values of 0.995, 0.98, and 0.965. The fits and the formulas produce similar estimates of effective parameters.

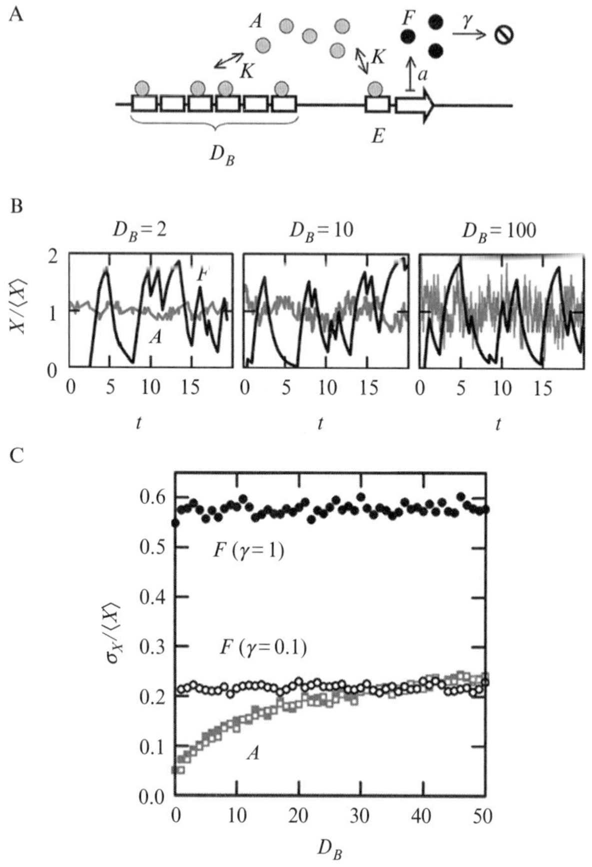

Figure 2.4 Stochastic simulation of operator buffers. (A) The TF can be sequestered by the D_B operator copies in the buffer, leading to fluctuations in free TF number. mRNA F is transcribed when the promoter E is bound to the TF. (B) Trajectories of the free TF number A (gray) and the mRNA number F (black). TF and mRNA numbers are scaled by their mean value. These trajectories were generated using the following parameter values: $k_+ = 0.1$, $k_- = 1$, $K = k_-/k_+ = 10$, $a = 200$, $\gamma = 1$, $E_T = 1$, $A_T = Keff = 0.5(D_B + E_T) + K$. Under these conditions, the free TF number $A = K = 10$ on average so the promoter is at half saturation, and the mean mRNA level is $F = 100$. Simulations of the reaction scheme shown in Eqs. (2.13) and (2.14) were performed using Gillespie's algorithm (Gillespie, 1977); mRNA trajectories were assumed to be noise-free, and generated by direct integration of Eq. (2.14) after computing the stochastic trajectory of the TF-bound promoter EA. The three panels show results for $D_B = 2$, 10, and 100. Although fluctuations in free TF grow as the size of the operator buffer is increased, mRNA trajectories are essentially unchanged. (C) Standard deviations of free TF number (gray squares) and mRNA number (black circles) as a function of D_B, for $\gamma = 1$ (filled symbols) and $\gamma = 0.1$ (empty symbols). Fluctuations in free TF increase approximately as the square-root of D_B, while fluctuations in mRNA number are independent of D_B.

equilibrium constant, $A = K$. Under condition (i), the total TF amount is simply $A_T \approx K$ and the bound fraction is negligible; under condition (ii), the bulk of the TF is bound but the total is precisely that needed to generate the desired free amount.

This discussion of operator buffers has so far ignored the possibility of fluctuations, which can be significant at low molecule number (Kaern *et al.*, 2005; McAdams and Arkin, 1999; Thattai and van Oudenaarden, 2001). Specifically, given a promoter concentration D_B and a total TF concentration $A_T = K + \frac{n}{2} D_B$, the free TF concentration will be $A = K$ *on average*, but could deviate from this level at any instant. To explore the magnitude and impact of these deviations, it is useful to measure concentration in units of molecules per cell. Suppose the buffer consists of D_B operator copies, and that there are A_T TF molecules in a cell. At half saturation, each operator copy will by definition have a probability $f = \frac{1}{2}$ of being bound to n copies of the TF. The actual number of bound operators will be approximately binomially distributed:

$$\langle DA_n \rangle = \frac{1}{2} D_B, \quad \langle \delta DA_n^2 \rangle = \frac{1}{4} D_B, \quad \langle \delta A^2 \rangle = n^2 \langle \delta DA_n^2 \rangle = \frac{1}{4} n^2 D_B, \tag{2.10}$$

where the angular braces represent population averages. Fluctuations in free TF therefore scale as the square-root of the operator copy number:

$$\langle A \rangle = K, \quad \sigma_A \equiv \langle \delta A^2 \rangle^{1/2} = \frac{1}{2} n \sqrt{D_B}. \tag{2.11}$$

Strictly, these results hold under the assumption that the total TF concentration is fixed, that the binding of n TF copies is instantaneous, and that fluctuations in free TF are much smaller than the amount of free TF; but they do capture the essential behavior.

If a single additional promoter copy is placed in this environment, fluctuations in free TF will impact its response. The key, however, is to notice that these fluctuations are rapid compared to the dynamics of a single promoter: over the time that this promoter remains in a TF-free state, there will be on the order of D_B binding or unbinding events in the operator buffer. The time average $\bar{A}$ of free TF over this period—approximately the average of D_B samples of the distribution from Eq. (2.11)—will therefore show much smaller deviations from the average, essentially independent of the number of total operator copies:

$$\sigma_{\bar{A}} \sim \frac{\sigma_A}{\sqrt{D_B}} \sim n. \tag{2.12}$$

The expression of any gene downstream of this additional promoter will therefore be unaffected by the presence of the operator buffer.

We can corroborate these general arguments using exact stochastic simulations. For simplicity, we consider the case when the promoter binds a single TF molecule, so $n = 1$. We expand the equilibrium scheme of Eq. (2.1) into the following kinetic scheme:

$$D + A \underset{k-}{\overset{k_+}{\leftrightarrow}} DA, \quad E + A \underset{k-}{\overset{k_+}{\leftrightarrow}} EA,$$
$$D_B = [D] + [DA], \quad E_T = [E] + [EA] = 1, \quad A_T = [A] + [DA] + [EA], \tag{2.13}$$

where we have explicitly separated operator copies in the buffer (D_B) from the single copy of active promoter ($E_T = 1$). We next model the expression of mRNA (F) from the TF-bound promoter, as well as its decay:

$$EA \overset{a}{\mapsto} F, \quad F \overset{\gamma}{\rightarrow} \phi. \tag{2.14}$$

As shown in Fig. 2.4, although fluctuations in the free TF concentration increase with the size of the operator buffer, fluctuations in mRNA levels remain constant. Thus, the operator buffer can improve reliability with essentially no impact on noise.

2. Conclusion

This study began with a practical motivation, when we noticed that a positive-feedback module which had previously been well characterized displayed some unexpected properties once it was embedded within a larger network. We have argued that variations in promoter copy number provide one possible explanation for this behavior. Electronic circuits are designed so that the connection of new inputs or outputs does not impact the properties of a given module. However, in transcriptional networks, these new connections are typically made by adding new promoter copies. Far from being passive entities, these promoters bind and sequester proteins, and can therefore have pleiotropic effects over the entire network. Specifically, the addition of promoter copies can modify the input–output properties of every other promoter in the system that binds the same TF. To some extent, this influence can be compensated by increasing protein levels; for example, the increase in the half-saturation concentration caused by extra promoter copies requires a corresponding increase in TF concentration in order to maintain promoter activity at original levels. However, other features of the original input–output response, like its steepness or Hill coefficient, cannot be recovered in this way.

If the total promoter concentration is small compared to the TF–promoter equilibrium constant, sequestration effects are negligible and we expect promoter properties to be reliable. Reliability only suffers at high promoter concentrations, such as those that occur in plasmid-borne systems. In the latter case, a potential solution is to actually embrace the sequestration effect, in the form of an operator buffer. This approach does suffer from several limitations: buffering would only be effective for a limited range of active promoter copies; having a high-copy operator buffer requires in turn a high TF expression level; promoter properties in a buffered situation are much more restricted than they would be at low copy; and variations in the buffer copy number, during DNA replication for example, might itself have undesirable consequences. However, it is relatively simple and practical to implement: extra copies of TF-binding operators have already been demonstrated to modulate the availability of free TF (e.g., Ozbudak *et al.*, 2004). It even seems to echo the architecture of natural transcriptional networks (Shen-Orr *et al.*, 2002), where promoters corresponding to any given TF are present only a handful of copies per cell (the negligible binding regime) or in very large numbers in the form of dense regulons (the operator buffered regime). Moreover, electronic circuits themselves use active mechanisms to ensure component reliability. In the final analysis, the benefits of reliability outweigh the costs or complications of the underlying mechanisms which might promote it. Strategies like those we have discussed here will, if successful, allow the community of synthetic biologists to develop, apply, and reuse modules, on the path to ever-more-complex engineered biological systems.

3. Methods

3.1. Strains and media

All experiments were performed in the host *E. coli* strain K12Z1. This is a derivative of the K12 MG1655 strain (Blattner *et al.*, 1997) that contains a chromosomal gene cassette encoding LacI (the Lac repressor, expressed at $\sim$3000 copies per cell), TetR (the tetracycline repressor), and a spectinomycin resistance marker; this cassette was introduced by P1 transduction from the strain DH5αZ1 (Lutz and Bujard, 1997; Master's Thesis, S. Dabholkar, 2007). The host strain was maintained at 4 °C on Luria–Bertani (LB) agar containing 50 μg/ml of spectinomycin; plasmid-transformed cells (see below) were maintained in media containing 100 μg/ml of ampicillin. Prior to each experiment, a fresh bacterial colony was inoculated in LB broth containing the appropriate antibiotic, and grown for 10 h at 37 °C. For subsequent growth and expression experiments, cells were spun down and transferred to 1% succinate–M9 minimal medium (Sambrook *et al.*, 1989) with no antibiotics, and maintained at 37 °C in an incubated shaker.

3.2. Plasmids and constructs

All constructs were built using components from the Registry of Standard Biological Parts (Shetty *et al.*, 2008; partsregistry.org). Constructs were assembled using the standard BioBrick assembly strategy, and maintained in the ampicillin-resistant pSB1A2 plasmid backbone (partsregistry.org/Part: pSB1A2) with a pMB1 origin of replication (copy numbers 100–300). Table 2.1 lists the BioBrick parts we used; Fig. 2.5 shows the schematic assembly strategy.

3.3. Dynamic measurements

Cells were grown in succinate–M9 medium for 12 h at low density (final $OD_{600} < 0.05$), then extracted with a 0.22-μm filter (Millipore) and resuspended in 1 ml warm succinate–M9 medium. These cells were then

Table 2.1 List of BioBrick parts

Part	Description
BBa_R0010	*E. coli* P_{Lac} promoter
BBa_R0062	*V. fischeri* P_R promoter
BBa_B0034	Strong ribosome binding site
BBa_B0015	Transcription terminator
BBa_C0161	*V. fischeri luxI*
BBa_C0062	*V. fischeri luxR*
BBa_E0020	CFP
BBa_E0030	YFP
BBa_C0012	*lacI* with LVA degradation tag

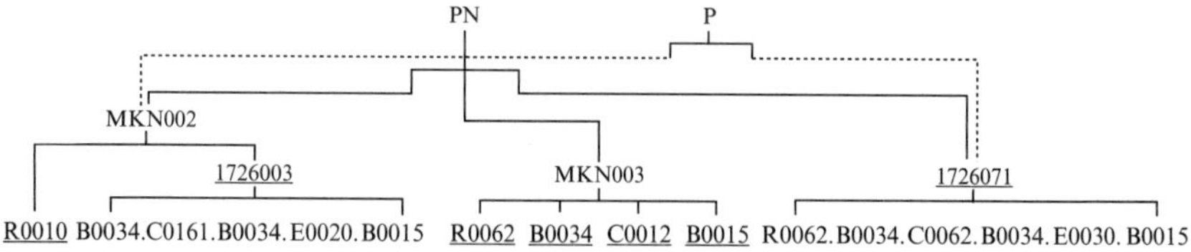

Figure 2.5 Assembly strategy. We generated our *P* and *PN* constructs by standard BioBrick assembly (Shetty *et al.*, 2008), using basic or composite parts (underlined) obtained from the Registry of Standard Biological Parts (partsregistry.org). Basic part descriptions are given in Table 2.1.

transferred at various dilutions into 25 ml succinate–M9 medium containing 1 m*M* IPTG. Every 3 h for the subsequent 12 h, 1 ml of this medium was extracted for cell density and fluorescence measurements. Cell density was measured by optical absorbance at 600 nm using a Thermo spectronic spectrophotometer.

3.4. Flow cytometry

All flow cytometry data were acquired using the CyAn ADP device (Dako). Each sample generated more than 30,000 events. For each event, we recorded forward- and side-scatter, as well as CFP levels (405 nm excitation laser; Violet1 filter set) and YFP levels (488 nm excitation laser; FITC filter set). Cells were gated using tight forward- and side-scatter limits prior to analysis.

ACKNOWLEDGMENTS

We are indebted to Registry of Standard Biological Parts, from which all the parts in this study were sourced. M. T. is partly supported by a WellcomeTrust-DBT India Alliance Fellowship; N. R. is supported by a Council of Scientific and Industrial Research Fellowship.

REFERENCES

Andrianantoandro, E., Basu, S., Karig, D. K., and Weiss, R. (2006). Synthetic biology: New engineering rules for an emerging discipline. *Mol. Syst. Biol.* **2,** 2006.0028.

Atkinson, M. R., Savageau, M. A., Myers, J. T., and Ninfa, A. J. (2003). Development of genetic circuitry exhibiting toggle switch or oscillatory behavior in Escherichia coli. *Cell* **113,** 597–607.

Balagadde, F. K., You, L., Hansen, C. L., Arnold, F. H., and Quake, S. R. (2005). Long-term monitoring of bacteria undergoing programmed population control in a microchemostat. *Science* **309,** 137–140.

Barkai, N., and Leibler, S. (1997). Robustness in simple biochemical networks. *Nature* **387,** 913–917.

Bashor, C. J., Horwitz, A. A., Peisajovich, S. G., and Lim, W. A. (2010). Rewiring cells: Synthetic biology as a tool to interrogate the organizational principles of living systems. *Annu. Rev. Biophys.* **39,** 515–537.

Becskei, A., and Serrano, L. (2000). Engineering stability in gene networks by autoregulation. *Nature* **405,** 590–593.

Blattner, F. R., Plunkett, G., III, Bloch, C. A., Perna, N. T., Burland, V., Riley, M., *et al.* (1997). The complete genome sequence of Escherichia coli K-12. *Science* **277,** 1453–1474.

Boyle, P. M., and Silver, P. A. (2009). Harnessing nature's toolbox: Regulatory elements for synthetic biology. *J. R. Soc. Interface* **6,** S535–S546.

Danino, T., Mondragón-Palomino, O., Tsimring, L., and Hasty, J. (2010). A synchronized quorum of genetic clocks. *Nature* **463,** 326–330.

Ellis, T., Wang, X., and Collins, J. J. (2009). Diversity-based, model-guided construction of synthetic gene networks with predicted functions. *Nat. Biotechnol.* **27,** 465–471.

Elowitz, M. B., and Leibler, S. (2000). A synthetic oscillatory network of transcriptional regulators. *Nature* **403,** 335–338.

Endy, D. (2005). Foundations for engineering biology. *Nature* **438,** 449–453.

Fuqua, C., and Greenberg, E. P. (2002). Listening in on bacteria: Acyl-homoserine lactone signalling. *Nat. Rev. Mol. Cell Biol.* **3,** 685–695.

Gardner, T. S., Cantor, C. R., and Collins, J. J. (2000). Construction of a genetic toggle switch in Escherichia coli. *Nature* **403,** 339–342.

Gillespie, D. T. (1977). Exact stochastic simulation of coupled chemical reactions. *J. Phys. Chem.* **81,** 2340–2361.

Guet, C. C., Elowitz, M. B., Leibler, S., *et al.* (2002). Combinatorial synthesis of genetic networks. *Science* **296,** 1466–1470.

Haseltine, E. L., and Arnold, F. H. (2008). Implications of rewiring bacterial quorum sensing. *Appl. Environ. Microbiol.* **74,** 437–445.

Hasty, J., McMillen, D., and Collins, J. J. (2002). Engineered gene circuits. *Nature* **420,** 224–230.

Heinemann, M., and Panke, S. (2006). Synthetic biology—Putting engineering into biology. *Bioinformatics* **22,** 2790–2799.

Itzkovitz, S., Levitt, R., Kashtan, N., Milo, R., Itzkovitz, M., and Alon, U. (2005). Coarse-graining and self-dissimilarity of complex networks. *Phys. Rev. E* **71,** 016127.

Kaern, M., Elston, T. R., Blake, W. J., and Collins, J. J. (2005). Stochasticity in gene expression. *Nat. Rev. Genet.* **6,** 451–464.

Kauffman, S. (1969). Metabolic stability and epigenesis in randomly constructed genetic nets. *J. Theor. Biol.* **22,** 437–467.

Lee, T. I., Rinaldi, N. J., Robert, F., Odom, D. T., Bar-Joseph, Z., Gerber, G. K., *et al.* (2002). Transcriptional regulatory networks in Saccharomyces cerevisiae. *Science* **298,** 799–804.

Lutz, R., and Bujard, H. (1997). Independent and tight regulation of transcriptional units in Escherichia coli via the LacR/O, the TetR/O and AraC/I1-I2 regulatory elements. *Nucleic Acids Res.* **25,** 1203–1210.

McAdams, H. H., and Arkin, A. (1999). It's a noisy business! Genetic regulation at the nanomolar scale. *Trends Genet.* **15,** 65–69.

Ninfa, A. J., and Mayo, A. E. (2004). Hysteresis vs. graded responses: The connections make all the difference. *Sci. STKE* **232,** pe20.

Ozbudak, E. M., Thattai, M., Lim, H. N., Shraiman, B. I., and van Oudenaarden, A. (2004). Multistability in the lactose utilization network of Escherichia coli. *Nature* **427,** 737–740.

Purnick, P. E. M., and Weiss, R. (2009). The second wave of synthetic biology: From modules to systems. *Nat. Rev. Mol. Cell Biol.* **10,** 410–422.

Sambrook, J., Fritsch, E. F., and Maniatis, T. (1989). Molecular Cloning: A Laboratory Manual. Cold Spring Harbor Laboratory Press, New York.

Shen-Orr, S. S., Milo, R., Mangan, S., and Alon, U. (2002). Network motifs in the transcriptional regulation network of Escherichia coli. *Nat. Genet.* **31,** 64–68.

Shetty, R. P., Endy, D., and Knight, T. F., Jr. (2008). Engineering BioBrick vectors from BioBrick parts. *J. Biol. Eng.* **2,** 5.

Sprinzak, D., and Elowitz, M. B. (2005). Reconstruction of genetic circuits. *Nature* **438,** 443–448.

Stelling, J., Sauer, U., Szallasi, Z., Doyle, F. J., and Doyle, J. (2005). Robustness of cellular functions. *Cell* **118,** 675–685.

Stricker, J., Cookson, S., Bennett, M. R., Mather, W. H., Tsimring, L. S., and Hasty, J. (2008). A fast, robust and tunable synthetic gene oscillator. *Nature* **456,** 516–519.

Thattai, M., and van Oudenaarden, A. (2001). Intrinsic noise in gene regulatory networks. *Proc. Natl. Acad. Sci. USA* **98,** 8614–8619.

Urbanowski, M. L., Lostroh, C. P., and Greenberg, E. P. (2004). Reversible acyl-homoserine lactone binding to purified Vibrio fischeri LuxR protein. *J. Bacteriol.* **186,** 631–637.

Voigt, C. A. (2006). Genetic parts to program bacteria. *Curr. Opin. Biotechnol.* **17,** 548–557.

von Hippel, P. H., Bear, D. G., Morgan, W. D., and McSwiggen, J. A. (1984). Protein-nucleic acid interactions in transcription: A molecular analysis. *Annu. Rev. Biochem.* **53,** 389–446.

Waters, C. M., and Bassler, B. L. (2005). Quorum sensing: Cell-to-cell communication in bacteria. *Annu. Rev. Cell Dev. Biol.* **21,** 319–346.

CHAPTER THREE

The Analysis of ChIP-Seq Data

Wenxiu Ma* *and* Wing Hung Wong[†,‡]

Contents

Abstract

Chromatin immunoprecipitation coupled with ultra-high-throughput parallel DNA sequencing (ChIP-seq) is an effective technology for the investigation of genome-wide protein–DNA interactions. Examples of applications include the studies of RNA polymerases transcription, transcriptional regulation, and histone modifications. The technology provides accurate and high-resolution mapping of the protein–DNA binding loci that are important in the understanding of many processes in development and diseases. Since the introduction of ChIP-seq experiments in 2007, many statistical and computational methods have been developed to support the analysis of the massive datasets from these experiments. However, because of the complex, multistaged analysis workflow, it is still difficult for an experimental investigator to conduct the analysis of his or her own ChIP-seq data. In this chapter, we review the basic

* Department of Computer Science, Stanford University, Stanford, California, USA
† Department of Statistics, Stanford University, Stanford, California, USA
‡ Department of Health Research and Policy, Stanford University, Stanford, California, USA

Methods in Enzymology, Volume 497
ISSN 0076-6879, DOI: 10.1016/B978-0-12-385075-1.00003-2

design of ChIP-seq experiments and provide an in-depth tutorial on how to prepare, to preprocess, and to analyze ChIP-seq datasets. The tutorial is based on a revised version of our software package CisGenome, which was designed to encompass most standard tasks in ChIP-seq data analysis. Relevant statistical and computational issues will be highlighted, discussed, and illustrated by means of real data examples.

1. Introduction

Chromatin Immunoprecipitation (ChIP) coupled with oligonucleotide hybridization genome tiling array (ChIP-chip) (Carroll *et al.*, 2006; Cawley *et al.*, 2004; Kapranov *et al.*, 2002) and with ultra-high-throughput sequencing (ChIP-seq) (Chen *et al.*, 2008; Johnson *et al.*, 2007; Robertson *et al.*, 2007; Wederell *et al.*, 2008) have been widely used to study transcription factor (TF) regulation in the entire genome. In these experiments, cells are treated with formaldehyde to crosslink DNA-associated protein factors such as TFs or histones to the DNA. The DNA molecules are then randomly sheared to sub-kilobase sized double-strand DNA fragments. TF-bound fragments are targeted by a specific antibody and collected during the immunoprecipitation (IP) process. After the crosslinks between TF and DNA have been reversed, the DNA fragments with length falling within a certain range are selected and amplified. In ChIP-chip experiments, which were popular until recently, these fragments are hybridized to a microarray with millions of 25–75mer probes tiling the whole genome. In contrast, in the ChIP-seq protocol, oligonucleotide linkers or adapters are ligated to the both ends of the ChIP fragments to produce the ChIP-seq library which are then sequenced by the next-generation sequencing machine in a massively parallel manner.

Although the ChIP preparation step is essentially the same in both the ChIP-chip and the ChIP-seq platforms, the subsequent steps of the two approaches are quite different. The fluorescence intensity of each probe is captured and digitalized in the ChIP-chip, whereas raw nucleotide short tags (aka *reads*) are sequenced base-by-base in the ChIP-seq. Moreover, the noise sources are distinct: the major noise in microarray experiments results from the probe affinity effect and the cross-hybridization effect, whereas linker/adaptor contamination, background noise, image processing error, and others all contribute to the ChIP-seq error profile.

Several studies have compared results from ChIP-chip experiments and ChIP-seq experiments on the same TF (Euskirchen *et al.*, 2007; Ji *et al.*, 2008; Robertson *et al.*, 2007). Their analysis revealed that higher sensitivity and sharper resolution of protein–DNA bindings are achieved using ChIP-seq. As sequencing cost continues to decrease rapidly, the ChIP-seq technique is expected to become more and more dominant in the study of transcriptional regulatory pathways and networks. However, because ChIP-seq datasets are massive and complex, their analysis requires advanced statistical methods, efficient computational algorithms, and user-friendly software for processing and visualization. After a brief discussion of some design issues related to ChIP-seq experiments, we will examine the pipeline of ChIP-seq data analysis step by step (Fig. 3.1). We will illustrate the analysis by using the software CisGenome (Ji *et al.*, 2008) to analyze two datasets (PolII ChIP-seq and STAT1 ChIP-seq) (Rozowsky *et al.*, 2009) (Table 3.1) that were produced as part of the ENCODE project (The ENCODE Project Consortium, 2007).

2. Planning of ChIP-Seq Experiments

The planning of a ChIP-seq experiment involves the consideration of many practical issues: Which sequencing platform to use? How many short reads need to be sequenced? Is a control sample necessary? How to choose and design the control experiment? How many biological replicates are recommended for each ChIP-seq experiment? It is useful to have a brief discussion of these issues before our treatment of data analysis.

2.1. Choices of sequencing platforms

Several commercial ultra-high-throughout sequencing platforms have emerged on the market since 2005. Popular brand names include Illumina Solexa, ABI SOLiD, and Roche 454. Smith *et al.* have compared the accuracy and efficiency of the above three platforms in the study of a mutant strain of *Pichia stipitis* (Smith *et al.*, 2008). The authors found that all three next-generation sequencing platforms successfully identified nucleotide variations between the reference genome and the mutant strains given sufficient coverage. They concluded all three are suitable for accurate and high-throughput sequencing studies. However, there are differences in their error profiles: the primary sequencing error for Illumia Solexa and ABi SOLiD is base substitution, whereas Roche 454 has difficulty in sequencing stretches of repetitive identical bases (*homopolymers*) (Margulies *et al.*, 2005), thus leading to a higher rate of insertion and deletion (indel) errors. In *de novo* assembly study without a reference sequence, longer read platform such as Roche 454 is favored. For the purpose of ChIP-seq experiment, it is common to use either Illumina Solexa or ABI SOLiD because of their ability to deliver a much higher number of sequence reads in parallel.

Table 3.1 Overview of ChIP-seq datasets

Sample name	Replicate	Number of Illumina lanes	Number of raw reads	Number of mappable reads	Number of nonredundant reads	% of nonredundant reads out of mappable reads
IFN-γ STAT1 ChIP	1	4	48,138,968	10,619,323	9,850,217	92.76
	2	2	31,896,881	17,446,502	16,253,194	93.16
	Total	6	80,035,849	28,065,825	25,818,743	91.99
IFN-γ input control	1	6	50,515,792	24,851,515	22,538,851	90.69
PolII ChIP	1	2	15,768,600	8,699,929	7,769,678	89.62
	2	5	21,301,982	11,628,339	10,171,478	87.47
	3	4	20,789,343	10,601,602	9,221,336	86.98
	Total	11	57,859,925	30,899,870	24,822,736	80.33
Input control	1	13	60,452,858	30,827,818	24,602,505	79.81

The sequencing datasets of STAT1 and PolII are obtained from the public gene expression omnibus (GEO) database (GSE12783 series), which is generated as part of the ENCODE project. One dataset contains STAT1 ChIP-seq data on interferon (IFN)-γ simulated human HeLa cells and the total DNA input control on the IFN-γ human HeLa cells; the other dataset contains PolII ChIP-seq data on the unsimulated human HeLa cells and the total DNA input control of the unsimulated cells. There are two biological replicates for the IFN-γ STAT1 ChIP-seq data and three biological replicates for the PolII data. The first 25 bp of the 27–32 bp raw reads are mapped back to the human genome assembly (hg18/NCBI Build 36) obtained from the UCSC Genome Browser. *Mappable reads* are those that map to a unique location in the genome (with up to two mismatches allowed). *Nonredundant reads* are mappable reads that occur only once in the dataset.

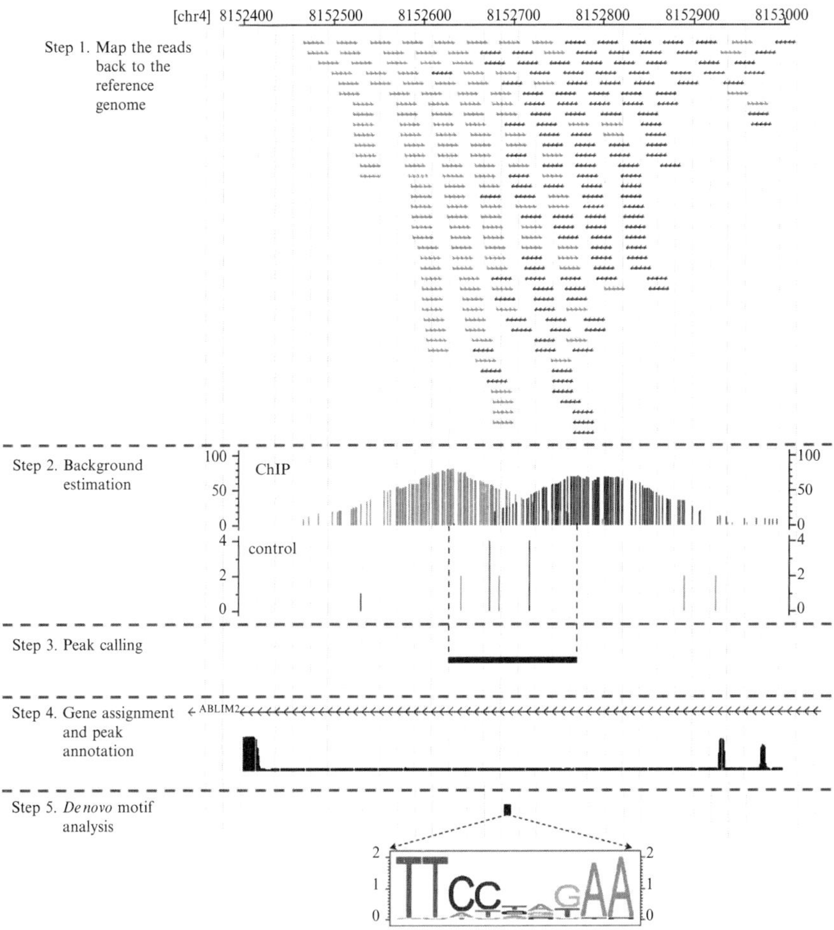

Figure 3.1 Work flow of ChIP-seq data processing and analyzing pipeline.

2.2. Sequencing statistics and quality control

For sequence-specific TF-binding localization and pattern discovery, single-end 25–35 bp reads are commonly used in ChIP-seq studies. We define a *mappable read* as a read that maps (aka aligns) to a unique location in the genome (with up to two mismatches allowed); *nonredundant reads* are the mappable reads that occur only once in the dataset. The goal of a ChIP-seq experiment is to gain an adequate number of mappable reads aggregated at the target regions. At current Solexa Illumina capacity, a single sequencing lane yields tens of millions of short reads and approximately half of them can be uniquely aligned back to the reference genome. In mammalian genomes, a coverage of 10 million reads typically provides clear binding signals at a large fraction of the binding sites. For a more compact genome such as *Drosophila melanogaster*, one can attain higher signal intensity at the same sequencing depth.

Before any large-scale production run of a ChIP-seq experiment, which may consume valuable biological samples and may require large amounts of reagents and machine time, it is useful to first conduct a pilot experiment run on a single lane (a sequencing machine typically has multiple lanes that can be used in parallel in each run). As a very rough guideline, a successful pilot experiment should meet the following four requirements. First, the number of raw sequencing reads should meet the expected depth within the same sequencing batch. Low yield in one particular lane implies the failure of the antibody or the library preparation. Second, the percentage of uniquely mappable reads achieves at least one-third of the total reads for mammalian genomes. Otherwise, contamination of the library should be suspected and investigated. Third, the percentage of nonredundant reads (two reads are *redundant* if they give identical sequence), should be greater than 50% of the total mappable reads. The nonredundant rate is a powerful measurement of the quality of a sequencing experiment as well as an informative estimation of the saturation status. As the sequencing depth increases, the nonredundant rate will decrease. However, at the first pilot lane of the experiment, we do not expect the saturation to occur so soon, unless the antibody failed to pull down desirable number of protein–DNA complexes. Finally, visual examination should reveal instances of clearly defined peaks with the expected form (see Section 3.2). The pilot run will provide valuable data for quality control, and will save the experimenter both time and money, should there be a need to troubleshoot.

2.3. Saturation

If the pilot experiment is satisfactory, the next step is to generate more reads in production runs. The *saturation* point of the sequencing depth is defined as the minimum number of reads which would enable the detection of all true protein–DNA binding loci. We suggest an evaluation procedure similar to the one used in Robertson's paper (Robertson *et al.*, 2007) to test whether a ChIP-seq dataset is saturated. First, we run the peak calling program on the full dataset. Then, we randomly sample one half of the reads and call peaks from one of the half sets using the sample peak calling parameters. The same set of controls will be used, if applicable.

The peak numbers and quality between results from the full set and from the half set are then compared. Peaks are binned at different false discovery rates (FDRs): for instance, 0.01, 0.05, 0.1, and 0.2; three statistics are calculated for each bin of peaks: (1) the motif site enrichment within the peak regions; (2) the peak conservation score; and (3) the conserved motif site enrichment. If these statistics improve when we double the size of the reads, it demonstrates that the data have not yet reached the saturation point. In this case, obtaining more reads will definitely be helpful.

2.4. Negative controls

In addition to sequence reads from positive ChIP samples, it is recommended that sequence reads be generated also from negative control samples. Use of negative controls can significantly increase the sensitivity and specificity of the peak detection (Ji *et al.*, 2008; Rozowsky et al., 2009). Negative controls that are commonly employed in ChIP-seq experiments can be classified into three types. The first type is total DNA input control, where non-IP'ed DNA is sheared, size-selected, and sequenced. The second type is Mock IP control, where we use a nonspecific antibody, for example, the immunoglobulin G (IgG) antibody on the same cells. The third type includes all specially designed controls. For instance, if the ChIP is performed on simulated cells, then using the same antibody on unsimulated cells is a good negative control. Also, in some studies, when the antibody of the TF of interest is not available or the antibody affinity is not strong enough, we might use FLAG or other epitope tags in the IP step. In this case, utilizing the FLAG antibody on un-FLAGed cells will provide a good negative control.

Specifically, negative controls are important for several reasons. First, it provides a background distribution to aid the FDR estimation. In this viewpoint, the input control is thought to be a better control than the mock IP because the reads of the input control will have a more balanced distribution throughout the genome. Whereas the reads of the mock IP control constitute numerous repetitive reads sequenced from the DNA fragments that the antibody pulled off, therefore, leaving fewer randomly distributed reads for background estimation. Furthermore, negative controls provide a means for us to indentify the genomic regions that are expected to have more reads for reasons largely unrelated to the binding of the TF of interest. For example, data from Illumina Solexa sequencer may exhibit a bias toward the GC-rich sequences (Dohm *et al.*, 2008). In addition, Rozowsky *et al.* discovered that input controls have small peaks in transcribed regions especially near transcription start sites (TSSs), because the chromatin tends to be more open at these regions (Rozowsky *et al.*, 2009). Finally, specially designed negative controls will aid in the detection of abnormal associations between the antibody and the DNA sequences. In the situation of the FLAG antibody, the negative control that uses the FLAG antibody on un-FLAGed cells is the best control for detecting potential bindings between the FLAG antibody and the DNA sequences.

2.5. Biological replicates

Although ChIP-seq data are believed to be much less noisy as compared to ChIP-chip data, it is still important to use multiple biological samples whenever it is possible. Having biological replicates helps to reduce sample-specific or sequence-specific biases, which can be caused by a variety of reasons, such as antibody affinities, sonication and amplification variations, library contaminations, and sequencing errors.

If the variability in the replicates is unacceptably large, for example, if the set of peaks detected are largely inconsistent, then one may need to improve the experimental protocol and repeat the experiments to obtain acceptable data. Sometimes, if there are enough replicates and only one of them is inconsistent with the others, then it may be reasonable to proceed with the analysis after removing the outlier sample. In any case, after we have obtained replicate samples with largely consistent results, then we are still faced with the question of how to combine the information in the replicates in the subsequent steps of the analysis.

In this tutorial, we handle replicates by a simple procedure, which calls peaks from individual biological replicate separately and then intersects them to obtain the final common peak regions. We choose this intersecting approach because we found it to be a safe and effective way to detect robust and reproducible binding events when there are noticeable differences among the biological replicates. An alternative and common practice is to pool all reads from multiple biological replicates together and call peak regions from pooled data. For example, Rozowsky *et al.* sampled the same number of reads from each replicate and combined them to proceed to further analysis (Rozowsky *et al.*, 2009). However, the sampling method or the linear scaling method might not be effective in some dataset, because the ratio of signal intensities between biological replicates is not always equal to the ratio of sequencing depths (Fig. 3.2). Thus, a more rigorous statistical method to address the multiple sample normalization and the multiple sample consistency testing problem is desired.

3. Processing and Analyzing ChIP-Seq Datasets

In this section, we demonstrate step by step how to use a revised version of the CisGenome software suite (Ji *et al.*, 2008) to process and analyze two published ChIP-seq datasets (for STAT1 and PolII) (Fig. 3.1). In this revised version, we designed and implemented an improved peak calling procedure based on the use of an iterative background count estimation technique. The new peak caller is described in more details in Section 3.3

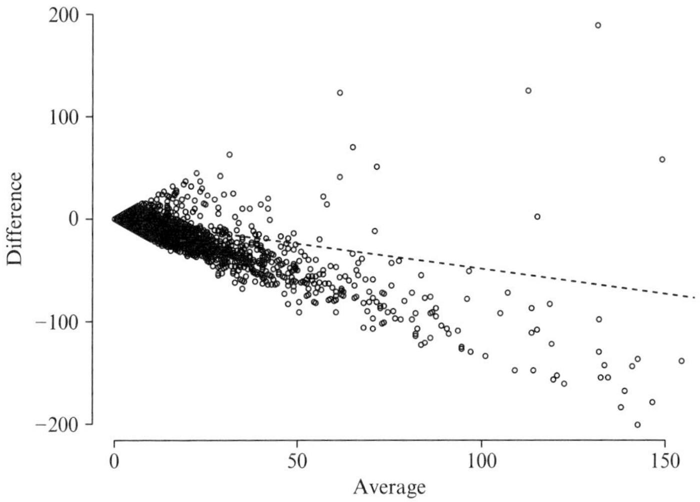

Figure 3.2 *M–A plot of two STAT1 biological replicates.* The entire human genome is divided into 100 bp nonoverlapping windows and the number of reads from the two STAT1 biological replicates in each window is counted separately. The M-A plot is drawn for only the window counts on chromosome 1. *X*-axis: the average number of reads between the two STAT1 biological replicates in each window. *Y*-axis: the difference between the number of reads from the first STAT1 biological replicate and the number of reads from the second STAT1 replicate in each window. The black-dashed line represents the relation between the sequencing depths of the two replicates. This figure demonstrated that the differences between the two biological samples are more significant than the differences between the sequencing depths.

below. Although our discussion is illustrated by the CisGenome system, there are other software tools available which may be used to accomplish similar analyses (Boyle *et al.*, 2008; Fejes *et al.*, 2008; Johnson *et al.*, 2007; Jothi *et al.*, 2008; Kharchenko *et al.*, 2008; Rozowsky *et al.*, 2009; Tuteja *et al.*, 2009; Valouev *et al.*, 2008; Zhang *et al.*, 2008).

3.1. Step 1. Map the reads back to the reference genome

Almost always, the first step in a ChIP-seq data analysis is the mapping of reads to a reference genome. In this step our goal is to identify, for each short read in the dataset, all the locations in a reference genome that show perfect or near perfect (say with no more than two mismatches in a 25-bp read) matches to the read (Fig. 3.1, Step 1).

There are quite a few programs available to map the short reads back to the reference genome (Jiang and Wong, 2008; Langmead *et al.*, 2009;

Li *et al.*, 2008, 2009). Mapping is a straightforward task because there is only one correct result, given the raw reads, the reference genome assembly, and the number of mismatches allowed. Therefore, it does not matter which short-read mapping program is used. The differences among these software tools lay primarily on the algorithm designs and computational efficiencies. A couple of them have add-on features: Bowtie (Langmead *et al.*, 2009) is one of the fastest short-read mapping program; Maq (Li *et al.*, 2008) can leverage on the reads quality scores; and SeqMap (Jiang and Wong, 2008) considers insertions and deletions (indels).

However, there is a tradeoff between the length of the reads to be used in the mapping and the yield of uniquely mappable reads. We usually do not use the full length of the reads in the mapping step, primarily because the sequencing error at each base increases rapidly near the end of the read. For example, in the STAT1 ChIP-seq data, 27–28 bp reads were sequenced. The error rate at positions 26–28 is much higher than that in positions 10–12 in a typical sequencing lane (Fig. 3.3). For ChIP-seq, it is standard to use the first 25 bases of the raw reads to map back to the reference genome assembly. Considering the fact that the human genome is highly repetitive (same for the other mammalian genomes), theoretically about 75% of the human genome can be uniquely mapped using 25 bp reads within two mismatches (McKernan *et al.*, 2009). A mappability profile that counts the redundancy of a read beginning at each nucleotide position on the genome has been calculated to improve peak detection accuracy and specificity (Rozowsky *et al.*, 2009).

The mapping results and statistics of the PolII and STAT1 dataset are listed in Table 3.1. The original reads have varying length 27–35 bp. Here, we remapped the first 25 bases of each read to the human genome assembly (hg18/NCBI Build 36) obtained from the UCSC Genome Browser Database (Rhead *et al.*, 2010; The Genome Sequencing Consortium, 2001) using the SeqMap software (Jiang and Wong, 2008). From the mapping statistics, we can deduce that this dataset is a deeply sequenced, comprehensive, and not highly redundant.

3.2. Step 2. Background estimation

Based on the ChIP-seq protocol and technique, ideally, all reads should be sequenced from the ends of the ChIP fragments that were bound by the target TF. However, in any ChIP-seq datasets, a considerable fraction of the reads may not have originated from these ChIP fragments. For instance, the antibody might target proteins other than the one studied, therefore capturing nonspecific fragments. Other factors that may induce such extraneous reads include library contamination, PCR amplification selection,

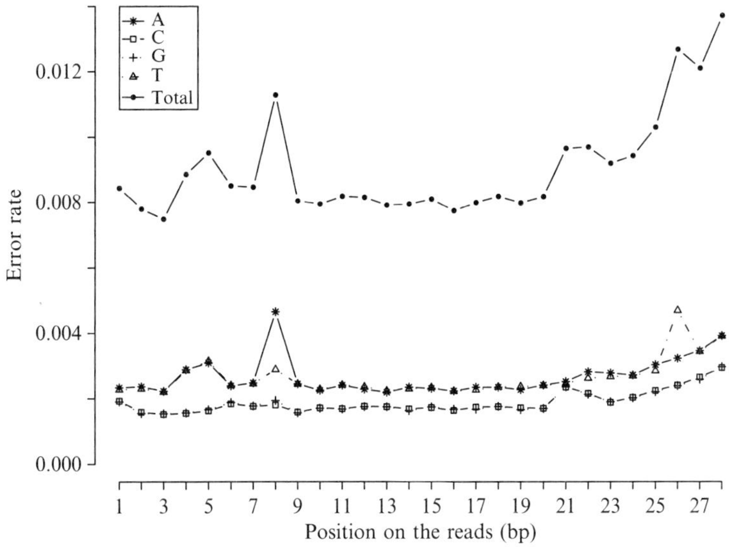

Figure 3.3 *Sequencing error rate comparing to the reference genome.* The 28-bp reads from land B, replicate 1 of the STAT1 ChIP-seq dataset are mapped back to the human genome (hg18/NCBI Build 36). For each uniquely mappable (up to two mismatches) read, we count each position on the read which is different from the reference genome. *X*-axis: nucleotide position on the short reads (1–28 bases); *Y*-axis: the error rate comparing to the total number of uniquely mappable reads in that lane. Asterisks on solid line: the error rate for nucleotide A (the reference genome has A in that position but the sequencing read has C/G/T); boxes on dashed line: C; crosses on dotted line: G; triangles on dot-dashed line: T; and finally, points on solid line: the addition of these four types of errors.

linker/adapter contamination, and image processing errors. What is more, because of sequencing errors, a read that originated in one location of the genome may be uniquely mapped to a different location of the genome that has sequence similarity with the original source.

We can regard all these reads that are unrelated to the binding events of interest, as "background reads" in our ChIP-seq experiment. It is not easy to ascertain whether each read results from a true binding event in the cells or from background. However, we can attempt to estimate the rate of occurrence of the background reads. Knowledge on the background rate is important for the assessment of the statistical significance of the binding regions detected by the peak caller (see Section 3.3). For each ChIP-seq sample, we define the background rate to be the ratio of the number of background reads to the total number of reads in the sample. We call a read a *true signal read* if it falls into the called peak regions (from the actual binding fragments). Otherwise, we call it a *background read.* Since the estimation of the background rate and the detection of the peak regions are dependent on each other, we propose an iterative method to solve this problem (see Section 3.3).

In other words, the peak calling problem is a signal-over-noise detection problem. We can take advantage of classical methods and measuring standards in the signal-over-noise problem. At the global scale, we use the background rate to estimate the ratio of the ChIP signal intensity to the landscape intensity; in each called peak region, we take the fold change of the ChIP signal intensity to the control signal intensity as the local estimate of the signal-to-noise ratio.

3.3. Step 3. Peak calling

In this step, we discuss the most critical task in the ChIP-seq data analysis pipeline. This is to identify the ChIP signal enriched genomic regions. In other words, where did the TF bind? This process is referred to as *peak calling* because the count of reads from the same strand of DNA (Watson or Crick) in a TF-bound fragment should show a peak near the binding location (Fig. 3.1, Step 2).

If a protein factor has a sharply focused binding site, in a successful experiment, one should be able to observe the bi-horned peaks nice bell-shaped peaks will be shaped at both the Watson strand and the Crick strand. This is because a fragment is always sequenced from its ends toward its mid-point. A Watson read represents the 5′-end of a ChIP fragment, whereas a Crick read represents the 3′-end. Therefore, the Watson peaks (as the left red peak in Fig. 3.1) and the Crick peaks (as the right blue peak in Fig. 3.1) are located in the opposite sides of the TF-binding site (TFBS). Thus the two peaks may be used to define a candidate binding region.

In contrast, if the ChIP-seq experiment fails because of the weak affinity of the antibody, extremely high, block-shaped peaks at repetitive regions or generally flat signal across the entire genome would be observed.

As depicted in Fig. 3.1, the mapped reads and the signal profile (defined below) can be visualized in a genome browser. Here, we use the CisGenome Browser (Jiang *et al.*, 2010) that is integrated with the CisGenome software, but other genome browsers, such as the UCSC Genome Browser (Kent *et al.*, 2002) and the Affymetrix Integrated Genome Browser (Nicol *et al.*, 2009) can also be used after we export the suitable files from CisGenome. To obtain the signal profile, we use a fixed window size w and count the number of the Watson and the Crick reads that fall into each nonoverlapping window along the entire genome. Window size $w = 100$ is recommended for sequence-specific TF-binding ChIP-seq data (Ji *et al.*, 2008).

In the signal profile track, we can see that the Watson reads and the Crick reads are clearly separated. Some programs shift the Watson and Crick reads toward their mid-points, and then detect peak regions by combining the shifted reads (Ji *et al.*, 2008; Kharchenko *et al.*, 2008; Valouev *et al.*, 2008). Kharchenko *et al.* used the cross-correlation magnitude to find the optimal shifting distance, while Ji *et al.* used half of the average peak length to shift. An alternative strategy, which we have used to identify the candidate region marked by the black line in the third step of Fig. 3.1, is to call peaks from the Watson strand and from the Crick strand separately, and then regard the region bracketed by the two strand-specific peak locations as the candidate TF-binding region. One advantage of this strategy is that we can easily check the balance between the numbers of reads in the coupled peaks. Moreover, since the length of the ChIP fragments varies greatly among peak regions, using a fixed global shifting distance may not be sufficient.

The first ChIP-seq peak caller, implemented in (Johnson *et al.*, 2007), was an intuitive, *ad hoc* method. It arbitrarily decides a genome-wide cutoff of signal intensity, and defines peaks as the regions above the predetermined cutoff level. A limitation of this method is that it does not provide significance and ranking of the detected peak regions. Similar to Johnson's approach, there are numerous other tag-aggregation methods in which the ranking of the peaks is based solely on the number of tags assembled in each peak region. The rationale for this ranking rule is based on the assumption that the height of a peak is a linear function of the proportion of cells that have the TF bound in the peak locus. Thus, the most significant binding will produce the highest peak. Nevertheless, this assumption is not always true. Certain chromatin regions are open and, therefore, easily fragmented, leading to stronger peaks in or around the transcribed gene neighborhood (Rozowsky *et al.*, 2009). In addition, if the TF binds to multiple locations close to each other, the multiple peak signal strength will add together and shape a broader, stronger, and continuous peak.

To improve the above fixed-cutoff method, a more advanced peak calling program may first estimate the background distribution and then use it to help assess statistical significance of the peaks.

Commonly used distributions for the ChIP-seq background counts include the Poisson distribution (Ji *et al.*, 2008; Zhang *et al.*, 2008) and the negative binomial distribution (Ji *et al.*, 2008). Comparing to the Poisson distribution, the negative binomial distribution is a better fit to the ChIP-seq background reads distribution (Ji *et al.*, 2008).

When a negative control sample is available, the same peak caller can be performed on both the ChIP sample and the control sample. A simple way to filter out false peaks in the control samples is to require a minimum fold change of the ChIP signal to the control signal (Johnson *et al.*, 2007; Valouev *et al.*, 2008). A more statistically rigorous approach to this two-sample problem is implemented in the CisGenome peak caller (Ji *et al.*, 2008). For each read in a given genomic window, we regard it as a *success* if it is from the ChIP sample; a *failure* if from the negative control sample. Thus, given k_{1i} is the number of ChIP reads in the window and k_{2i} is the number of control reads, the number of ChIP reads in that window follows a conditional binomial model, that is, $k_{1i}|n_i \sim \text{binomial}(n_i, p_0)$, where $n_i = k_{1i} + k_{2i}$ and p_0 is the probability of seeing a ChIP read in that window.

In CisGenome, Ji *et al.* divided the entire genome into nonoverlapping windows of w-bp width and counted the number of ChIP reads and the number of negative control reads in each window. Then they took the ratio of the number of windows that contain only one ChIP read and the number of windows that contain only one read (either a ChIP read or a control read) as the estimation of p_0. We have found that this estimation method sometimes failed when a ChIP-seq sample was highly redundant (either because the sample is over-saturated or because the antibody failed) and had very few windows containing only one read, thus leading to a biased estimation of p_0.

To deal with this problem, we have recently developed an improved version of this conditional binomial approach. The main idea is to estimate the expected success rate p_0 in an iterative manner. Assuming that background reads are uniformly distributed, r_0 is the ratio of the probability of seeing a ChIP read to the probability of seeing a control read at any genomic position, that is: r_0 = (number of background reads in a ChIP sample)/(number of background reads in control). Similarly, it is assumed that the number of ChIP reads within a sliding window of w-bp follows a conditional binomial model, that is, $k_{1i}|n_i \sim \text{binomial}(n_i, p_0)$, where $p_0 = r_0/(1 + r_0)$. The program starts with r_0 equals to the total number of reads in the ChIP sample divided by the total number of reads in the control sample (r_0 = total number of ChIP reads/total number of control reads). The program identifies peaks using the initial estimation of r_0, and then filters out the ChIP and control reads that fall into peak regions. Once the peak regions have been filtered out, r_0 is reestimated and iterations continue until r_0 converges.

This peak caller program is applied to the Watson and the Crick reads separately. After all of the peaks have been identified, the Watson and Crick strand peaks are combined. Only those peaks containing a balanced number of Watson and Crick reads are paired. The peak boundaries are set as the modes of the coupled peaks. The fold change between the ChIP signal and the control signal is also calculated for each peak region. An example of output file of the top 20 STAT1 peaks is displayed in Table 3.2.

In addition to acquiring a set of peak regions, we are also interested in the significances of the peak regions. In CisGenome, we calculate the FDR based on the read distribution in both the ChIP and control samples. To be more specific, the FDR of each w-bp window with k_{1i} ChIP reads and n_i total reads, is the ratio of the expected number of windows that have equal to or more than k_{1i} ChIP reads out of n_i total reads given p_0, divided by the observed number of such windows (see Methods in Ji, *et al.*, 2008). The better the background estimation fits the data, the more accurate is the FDR estimation.

In our case study, two-sample peak calling is performed on each biological replicate of the interferon-γ (IFN-γ) STAT1 ChIP versus

Table 3.2 Top 20 STAT1 peaks

Rank	Chromosome	Start	End	Length	Peak height	Number of ChIP reads	Number of control reads	Fold change of ChIP/control
1	chr20	48342552	48342733	182	1328.25	5698	183	31.05
2	chr2	191593263	191593466	204	1204	4606	84	54.84
3	chr14	23700077	23700317	241	1124	5198	102	50.96
4	chr6	30565062	30565269	208	1077.75	3720	65	56.8
5	chr15	42808242	42808406	165	985.25	4447	104	42.76
6	chr5	131854389	131854584	196	956.25	3506	95	36.72
7	chr5	131860571	131860746	176	848.25	6768	220	30.76
8	chr16	55580792	55580988	197	815.25	2897	112	25.75
9	chr16	18845403	18845625	223	771.5	2283	46	49.64
10	chr12	107546406	107546594	189	741.25	3047	87	34.83
11	chr19	10242625	10242835	211	685.25	2505	77	32.33
12	chr17	37794087	37794446	360	662.25	3649	115	31.59
13	chr1	148801113	148801409	297	658.5	3301	109	30.15
14	chr12	47532339	47532527	189	656	3339	121	27.48
15	chr5	43076129	43076301	173	613.75	2995	133	22.44
16	chr16	10830348	10830537	190	608.75	2415	71	33.78
17	chr17	55218675	55218884	210	583.5	2373	49	48.43
18	chr3	126327407	126327607	201	582.25	2258	55	40.68
19	chr7	101493956	101494129	174	565.25	2273	74	30.72
20	chr16	28451190	28451421	232	562.25	2741	58	46.86

The peak regions are ranked by the peak height, which is the average of the maximum numbers of reads in a 100-bp window on the Watson and Crick strands. Alternatively, we can rank the peak regions by the last column, which is the fold change of ChIP signal intensity to the control signal intensity. The peak start is the mode of the Watson peak; the peak end is the mode of the Crick peak. The number of ChIP (or control) reads is the average count in the coupled Watson and Crick peaks.

IFN-γ input control comparison and also each biological replicate of the PolII ChIP versus input control comparison. Peak numbers, the percentage of ChIP reads fell into peak regions, and the ChIP/input signal fold change enrichment in the called peak regions (normalized by the background ratio between the ChIP and the input control samples) at the FDR 0.01 cutoff are provided as output (Table 3.3). For sequence-specific transcriptional factor ChIP-seq data sequenced at a depth of 10 million mappable reads in mammalian genomes, we expect to see at least 3% of ChIP reads originating from the binding peaks (varies from factors and samples) and a ChIP/input fold change above 5.0. After peak detection on each biological replicate, we intersect peak regions of individual replicate to get the common regions as the final list. Finally, we have 3347 reproducible STAT1 peaks and 9087 PolII peaks. Because of the noticeable differences among the biological replicates, using this stringent intersection approach, we get 54–86% fewer peaks than the published peaks in the Rozowsky *et al.*'s paper.

3.4. Step 4. Gene assignment and peak annotation

After we obtain a list of peak coordinates, it is important to study the biological implications of the protein–DNA bindings. Certain questions have always been asked: what are the genomic annotations and the functions of these peak regions?

Because many cis-regulatory elements are close to TSSs of their targets, by default CisGenome associates each peak to its nearest gene, either upstream or downstream. In our example dataset, the PolII peaks are closer

Table 3.3 Summary of peak calling results of STAT1 and PolII

		Replicate 1	Replicate 2
IFN-γ STAT1 versus IFN-γ input control peak calling			
Number of peaks		3,822	14,644
Percentage of ChIP reads fell into peak regions		4.04	11.54
ChIP/input (normalized) signal ratio in peak regions		11.09	9.48
	Replicate 1	**Replicate 2**	**Replicate 3**
PolII versus input control peak calling			
Number of peaks	16,893	21,585	22,296
Percentage of ChIP reads fell into peak regions	27.14	48.71	51.15
ChIP/input (normalized) signal ratio in peak regions	15.51	20.00	22.10

For each called peak list, we calculated the percentage of ChIP sample reads fell into the called peak regions and the ChIP/input signal fold change in the peak regions (normalized by the background read ratios in both samples).

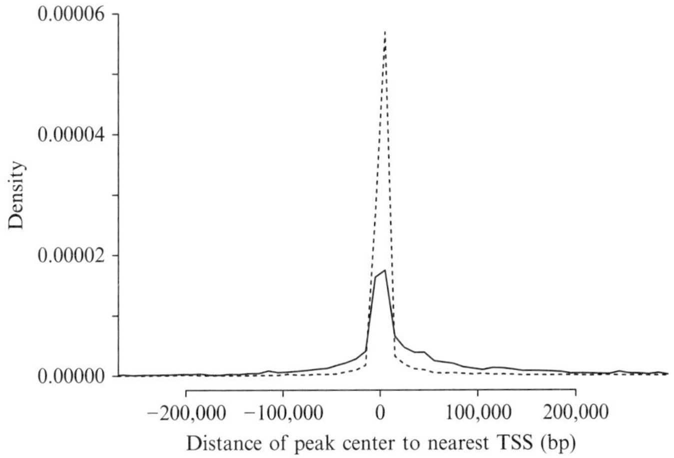

Figure 3.4 *Density of peak locations relative to nearest genes.* *X*-axis: distance from the peak center to the nearest transcription start site (TSS), where 0 is the position of the TSS. Negative numbers: 5′ of TSS. Positive numbers: 3′ of TSS. Only distance from −250 to 250 kbp is displayed in this figure. *Y*-axis: probability of seeing a peak within each bin of 10 kbp (percentage of peaks in each bin/bin size). Solid line: 3347 STAT1 peaks. Dashed line: 9087 PolII peaks.

to the nearest TSSs than the STAT1 peaks (Fig. 3.4). The STAT1 peaks are more enriched in the intron regions and the intergenic regions (Table 3.4), which is consistent with its known cis-regulatory function.

In CisGenome, we can also associate peaks with all genes in their neighborhood. This function is more informative and helpful in screening potential cis-regulatory targets, especially when transcription profiling data under the same cellular condition are available. As a different approach, Ouyang *et al.* (2009) computed a weighted average of peak signals detected near a gene and regard it as a quantitative score for the strength of the association between the TF and the gene. The authors showed that such scores can be used to build good predictive models of the absolute gene expression in mouse embryonic stem cells.

In the CisGenome Browser, the genetic landscape around each peak can be displayed at any resolution. Other genome features such as phylogenetic conservation can also be added to the visualization (Fig. 3.1, Step 4).

3.5. Step 5. *De novo* motif analysis

Another important task in the analysis of the predicted peak regions is *de novo* motif discovery. In some studies, the exact sequence to which the TF binds is known, or even better, a set of validated binding sites is available. However, if this information is not available, we will need to recover the binding motifs from the peak sequences as well as from their

Table 3.4 Genomic location of peak regions comparing to the nearest genes

	3347 STAT1 peaks		9087 PolII peaks	
Peak location	#	%	#	%
IntraGenic	1430	42.72	5505	60.58
5′-UTR	143	4.27	2288	25.18
3′-UTR	32	0.96	167	1.84
CDS	40	1.20	936	10.30
Intron	1215	36.30	2114	23.26
Exon	215	6.42	3391	37.32
InterGenic	1917	57.28	3582	39.42
Upstream	1237	36.96	2849	31.35
Downstream	680	20.32	733	8.07

IntraGenic: transcribed region of a gene; 5′-UTR: 5′-untranslated region; 3′-UTR: 3′-untranslated region; CDS: coding sequence; InterGenic: outside the transcribed gene regions.

orthologous sequences. CisGenome has incorporated a Gibbs sampling module (Lawrence *et al.*, 1993; Liu, 1994) which can recover enriched motifs from the sequences of the peak regions.

In CiGenome Browser, we can visualize the top motif logos. The degree of consistency between the known or published motif and the *de novo* discovered motif can be used to assess the success of the experiment. Motif occupancy and enrichment in peak regions and motif conservation scores offer additional means for assessments.

Three independent runs of Gibbs sampler are performed on the FDR 0.01 STAT1 peak regions. Top enriched *de novo* motifs include the canonical STAT1 motif and the activating protein 1 (AP-1) motif (Table 3.5). There are 2544 (76.01%) of the 3347 STAT1 ChIP-Seq peaks that contained one or more STAT1 *de novo* recovered motif sites within the peak boundaries. The STAT1 motif sites are close to the peak center (Fig. 3.5A). About 33.49% of these STAT1-containing peaks have conserved STAT1 motif sites which are located within the top 10% conserved genomic regions (conservation scores of the 44-vertebrate alignment phastCons scores for the human hg18 genome are obtained from the UCSC Genome Browser Database (Rhead *et al.*, 2010; Siepel *et al.*, 2005)). On average, the conservation scores for the motif sites are significantly higher than their neighborhood regions (Fig. 3.5B).

4. Discussion

In summary, we have provided a systematic discussion of issues related to the analysis of ChIP-seq data. We demonstrated how several key steps, including data exploration and visualization, peak calling, genomic

Table 3.5 Enrichments of *de novo* discovered motifs

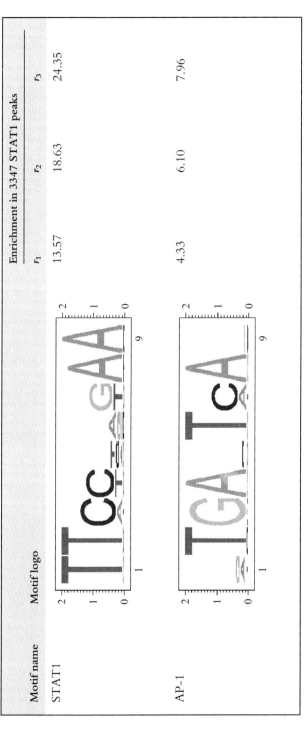

Motif name	Motif logo	Enrichment in 3347 STAT1 peaks		
		r_1	r_2	r_3
STAT1		13.57	18.63	24.35
AP-1	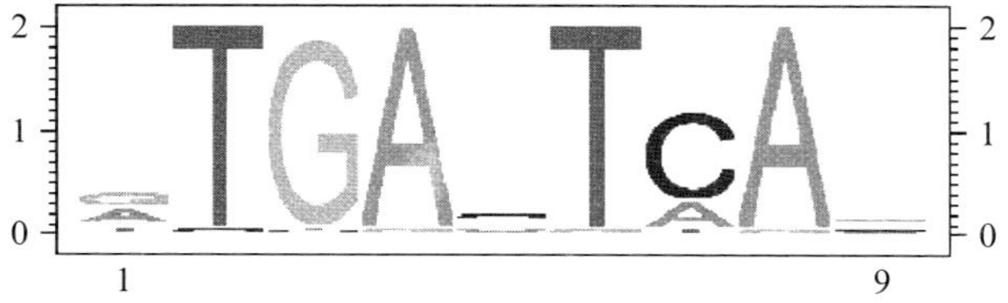	4.33	6.10	7.96

Three motif enrichment ratios are calculated as described in Ji *et al.*'s (2006) paper: r_1 = percentage of peak regions containing the indicated motif(s) versus percentage of matched control regions with the same motif(s); r_2 = percentage of phylogenetically conserved peak regions containing the motif(s) versus percentage of phylogenetically conserved matched control regions with the same motif(s); r_3 = percentage of regions containing the phylogenetically conserved motif(s) versus percentage of matched control regions with the phylogenetically conserved same motif(s). Matched control regions are randomly selected such that the distance between the control region and the closest transcription start site (TSS) has the same probability distribution as the distance between the peaks region and the TSS. For TF ChIP-seq peaks, we expect the r_1, r_2, and r_3 to be simultaneously greater than 5.0 for the primary binding factor, and to be simultaneously greater than 2.0 for other collaborating binding factors.

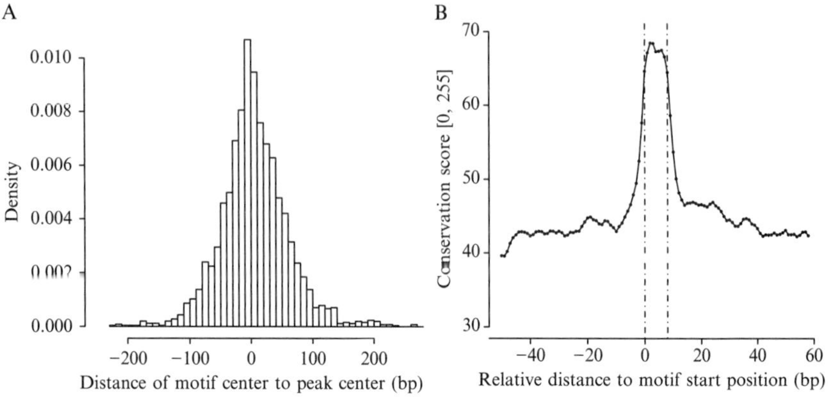

Figure 3.5 *Motif analysis on the STAT1 peak regions.* (A) Histogram of STAT1 *de novo* motif resolution. *X*-axis: distance from the center of STAT1 recovered motifs to the peak center. *Y*-axis: density (percentage of motifs) per bin. Bin size: 10 bp. STAT1 motifs are most often located close to the peak center. (B) Average conservation scores for motif sites and flanking positions. *X*-axis: distance to the motif start position, where 0 is the first base of the motif site. Negative numbers: 5′ of the motif sites. Positive numbers: 3′ of the motif sites. The flanking region of 50 bp on either side is displayed in this figure. *Y*-axis: converted UCSC phastCon scores of 44 vertebrate alignments to the hg18 human genome. The converted conservation score is ranging from 0 (least conserved) to 255 (most conserved). Black solid line: 3218 STAT1 sites within peak regions. Black dashed line: STAT1 motif sites boundaries.

annotation, and downstream motif analyses, can be accomplished by a user-friendly software package CisGenome. We rely on other specialized software kits for the low-level analyses, such as base calling, image processing, error filtering, and so on.

The example datasets include TF ChIP-seq and polymerases transcription ChIP-seq. Finally, our analysis pipeline can be extended to analyze histone modification ChIP-seq dataset. For such applications, some minor modifications on the peak calling algorithms have to be made, including enlarging window size for data exploration and background estimation, and shifting/coupling strategies on the Watson and Crick strands.

The CisGenome software is available at http://www.biostat.jhsph.edu/~hji/cisgenome/.

ACKNOWLEDGMENT

This chapter is based on research supported by NIH Grants R01HG004634 and R01HG003903.

REFERENCES

Boyle, A. P., Guinney, J., Crawford, G. E., and Furey, T. S. (2008). F-Seq: A feature density estimator for high-throughput sequence tags. *Bioinformatics* **24,** 2537–2538.

Carroll, J. S., Meyer, C. A., Song, J., Li, W., Geistlinger, T. R., Eeckhoute, J., Brodsky, A. S., Keeton, E. K., Fertuck, K. C., Hall, G. F., Wang, Q., Bekiranov, S., *et al.* (2006). Genome-wide analysis of estrogen receptor binding sites. *Nat. Genet.* **38,** 1289–1297.

Cawley, S., Bekiranov, S., Ng, H. H., Kapranov, P., Sekinger, E. A., Kampa, D., Piccolboni, A., Sementchenko, V., Cheng, J., Williams, A. J., Wheeler, R., Wong, B., *et al.* (2004). Unbiased mapping of transcription factor binding sites along human chromosomes 21 and 22 points to widespread regulation of noncoding RNAs. *Cell* **116,** 499–509.

Chen, X., Xu, H., Yuan, P., Fang, F., Huss, M., Vega, V. B., Wong, E., Orlov, Y. L., Zhang, W., Jiang, J., Loh, Y. H., Yeo, H. C., *et al.* (2008). Integration of external signaling pathways with the core transcriptional network in embryonic stem cells. *Cell* **133,** 1106–1117.

Dohm, J. C., Lottaz, C., Borodina, T., and Himmelbauer, H. (2008). Substantial biases in ultra-short read data sets from high-throughput DNA sequencing. *Nucleic Acids Res.* **36,** e105.

Euskirchen, G. M., Rozowsky, J. S., Wei, C. L., Lee, W. H., Zhang, Z. D., Hartman, S., Emanuelsson, O., Stolc, V., Weissman, S., Gerstein, M. B., Ruan, Y., and Snyder, M. (2007). Mapping of transcription factor binding regions in mammalian cells by ChIP: Comparison of array- and sequencing-based technologies. *Genome Res.* **17,** 898–909.

Fejes, A. P., Robertson, G., Bilenky, M., Varhol, R., Bainbridge, M., and Jones, S. J. (2008). FindPeaks 3.1: A tool for identifying areas of enrichment from massively parallel short-read sequencing technology. *Bioinformatics* **24,** 1729–1730.

Ji, H., Vokes, S. A., and Wong, W. H. (2006). A comparative analysis of genome-wide chromatin immunoprecipitation data for mammalian transcription factors. *Nucleic Acids Res.* **34**(21), e146.

Ji, H., Jiang, H., Ma, W., Johnson, D. S., Myers, R. M., and Wong, W. H. (2008). An integrated software system for analyzing ChIP-chip and ChIP-seq data. *Nat. Biotechnol.* **26,** 1293–1300.

Jiang, H., and Wong, W. H. (2008). SeqMap: Mapping massive amount of oligonucleotides to the genome. *Bioinformatics* **24,** 2395–2396.

Jiang, H., Wang, F., Dyer, N. P., and Wong, W. H. (2010). CisGenome Browser: A flexible tool for genomic data visualization. *Bioinformatics* **26,** 1781–1782.

Johnson, D. S., Mortazavi, A., Myers, R. M., and Wold, B. (2007). Genome-wide mapping of in vivo protein-DNA interactions. *Science* **316,** 1497–1502.

Jothi, R., Cuddapah, S., Barski, A., Cui, K., and Zhao, K. (2008). Genome-wide identification of in vivo protein-DNA binding sites from ChIP-Seq data. *Nucleic Acids Res.* **36,** 5221–5231.

Kapranov, P., Cawley, S. E., Drenkow, J., Bekiranov, S., Strausberg, R. L., Fodor, S. P., and Gingeras, T. R. (2002). Large-scale transcriptional activity in chromosomes 21 and 22. *Science* **296,** 916–919.

Kent, W. J., Sugnet, C. W., Furey, T. S., Roskin, K. M., Pringle, T. H., Zahler, A. M., and Haussler, D. (2002). The human genome browser at UCSC. *Genome Res.* **12,** 996–1006.

Kharchenko, P. V., Tolstorukov, M. Y., and Park, P. J. (2008). Design and analysis of ChIP-seq experiments for DNA-binding proteins. *Nat. Biotechnol.* **26,** 1351–1359.

Langmead, B., Trapnell, C., Pop, M., and Salzberg, S. L. (2009). Ultrafast and memory-efficient alignment of short DNA sequences to the human genome. *Genome Biol.* **10,** R25.

Lawrence, C. E., Altschul, S. F., Boguski, M. S., Liu, J. S., Neuwald, A. F., and Wootton, J. C. (1993). Detecting subtle sequence signals: A Gibbs sampling strategy for multiple alignment. *Science* **262,** 208–214.

Li, H., Ruan, J., and Durbin, R. (2008). Mapping short DNA sequencing reads and calling variants using mapping quality scores. *Genome Res.* **18,** 1851–1858.

Li, R., Yu, C., Li, Y., Lam, T. W., Yiu, S. M., Kristiansen, K., and Wang, J. (2009). SOAP2: An improved ultrafast tool for short read alignment. *Bioinformatics* **25,** 1966–1967.

Liu, J. S. (1994). The collapsed Gibbs sampler with applications to a gene regulation problem. *J. Am. Stat. Assoc.* **89,** 958–966.

Margulies, M., Egholm, M., Altman, W. E., Attiya, S., Bader, J. S., Bemben, L. A., Berka, J., Braverman, M. S., Chen, Y. J., Chen, Z., Dewell, S. B., Du, L., *et al.* (2005). Genome sequencing in microfabricated high-density picolitre reactors. *Nature* **437,** 376–380.

McKernan, K. J., Peckham, H. E., Costa, G. L., McLaughlin, S. F., Fu, Y., Tsung, E. F., Clouser, C. R., Duncan, C., Ichikawa, J. K., Lee, C. C., Zhang, Z., Ranade, S. S., *et al.* (2009). Sequence and structural variation in a human genome uncovered by short-read, massively parallel ligation sequencing using two-base encoding. *Genome Res.* **19,** 1527–1541.

Nicol, J. W., Helt, G. A., Blanchard, S. G., Jr., Raja, A., and Loraine, A. E. (2009). The Integrated Genome Browser: Free software for distribution and exploration of genome-scale datasets. *Bioinformatics* **25,** 2730–2731.

Ouyang, Z., Zhou, Q., and Wong, W. H. (2009). ChIP-Seq of transcription factors predicts absolute and differential gene expression in embryonic stem cells. *Proc. Natl. Acad. Sci. USA* **106,** 21521–21526.

Rhead, B., Karolchik, D., Kuhn, R. M., Hinrichs, A. S., Zweig, A. S., Fujita, P. A., Diekhans, M., Smith, K. E., Rosenbloom, K. R., Raney, B. J., Pohl, A., Pheasant, M., *et al.* (2010). The UCSC Genome Browser database: Update 2010. *Nucleic Acids Res.* **38,** D613–D619.

Robertson, G., Hirst, M., Bainbridge, M., Bilenky, M., Zhao, Y., Zeng, T., Euskirchen, G., Bernier, B., Varhol, R., Delaney, A., Thiessen, N., Griffith, O. L., *et al.* (2007). Genome-wide profiles of STAT1 DNA association using chromatin immunoprecipitation and massively parallel sequencing. *Nat. Methods* **4,** 651–657.

Rozowsky, J., Euskirchen, G., Auerbach, R. K., Zhang, Z. D., Gibson, T., Bjornson, R., Carriero, N., Snyder, M., and Gerstein, M. B. (2009). PeakSeq enables systematic scoring of ChIP-seq experiments relative to controls. *Nat. Biotechnol.* **27,** 66–75.

Siepel, A., Bejerano, G., Pedersen, J. S., Hinrichs, A. S., Hou, M., Rosenbloom, K., Clawson, H., Spieth, J., Hillier, L. W., Richards, S., Weinstock, G. M., Wilson, R. K., *et al.* (2005). Evolutionarily conserved elements in vertebrate, insect, worm, and yeast genomes. *Genome Res.* **15,** 1034–1050.

Smith, D. R., Quinlan, A. R., Peckham, H. E., Makowsky, K., Tao, W., Woolf, B., Shen, L., Donahue, W. F., Tusneem, N., Stromberg, M. P., Stewart, D. A., Zhang, L., *et al.* (2008). Rapid whole-genome mutational profiling using next-generation sequencing technologies. *Genome Res.* **18,** 1638–1642.

The ENCODE Project Consortium (2007). Identification and analysis of functional elements in 1% of the human genome by the ENCODE pilot project. *Nature* **447,** 799–816.

The Genome Sequencing Consortium (2001). Initial sequencing and analysis of the human genome. *Nature* **409,** 860–921.

Tuteja, G., White, P., Schug, J., and Kaestner, K. H. (2009). Extracting transcription factor targets from ChIP-Seq data. *Nucleic Acids Res.* **37,** e113.

Valouev, A., Johnson, D. S., Sundquist, A., Medina, C., Anton, E., Batzoglou, S., Myers, R. M., and Sidow, A. (2008). Genome-wide analysis of transcription factor binding sites based on ChIP-Seq data. *Nat. Methods* **5,** 829–834.

Wederell, E. D., Bilenky, M., Cullum, R., Thiessen, N., Dagpinar, M., Delaney, A., Varhol, R., Zhao, Y., Zeng, T., Bernier, B., Ingham, M., Hirst, M., *et al.* (2008). Global analysis of in vivo Foxa2-binding sites in mouse adult liver using massively parallel sequencing. *Nucleic Acids Res.* **36,** 4549–4564.

Zhang, Y., Liu, T., Meyer, C. A., Eeckhoute, J., Johnson, D. S., Bernstein, B. E., Nussbaum, C., Myers, R. M., Brown, M., Li, W., and Liu, X. S. (2008). Model-based analysis of ChIP-Seq (MACS). *Genome Biol.* **9,** R137.

CHAPTER FOUR

Using DNA Microarrays to Assay Part Function

Virgil A. Rhodius[*,†] *and* Carol A. Gross[*,‡]

Contents

[*] Department of Microbiology and Immunology, University of California at San Francisco, San Francisco, California, USA

[†] Department of Pharmaceutical Chemistry, University of California at San Francisco, San Francisco, California, USA

[‡] Department of Cell and Tissue Biology, University of California at San Francisco, San Francisco, California, USA

Methods in Enzymology, Volume 497

ISSN 0076-6879, DOI: 10.1016/B978-0-12-385075-1.00004-4

Abstract

In recent years, the capability of synthetic biology to design large genetic circuits has dramatically increased due to rapid advances in DNA synthesis technology and development of tools for large-scale assembly of DNA fragments. Large genetic circuits require more components (parts), especially regulators such as transcription factors, sigma factors, and viral RNA polymerases to provide increased regulatory capability, and also devices such as sensors, receivers, and signaling molecules. All these parts may have a potential impact upon the host that needs to be considered when designing and fabricating circuits. DNA microarrays are a well-established technique for global monitoring of gene expression and therefore are an ideal tool for systematically assessing the impact of expressing parts of genetic circuits in host cells. Knowledge of part impact on the host enables the user to design circuits from libraries of parts taking into account their potential impact and also to possibly modify the host to better tolerate stresses induced by the engineered circuit. In this chapter, we present the complete methodology of performing microarrays from choice of array platform, experimental design, preparing samples for array hybridization, and associated data analysis including preprocessing, normalization, clustering, identifying significantly differentially expressed genes, and interpreting the data based on known biology. With these methodologies, we also include lists of bioinformatic resources and tools for performing data analysis. The aim of this chapter is to provide the reader with the information necessary to be able to systematically catalog the impact of genetic parts on the host and also to optimize the operation of fully engineered genetic circuits.

1. Introduction

The focus of synthetic biology has been the design and implementation of small-scale genetic circuits (Elowitz and Leibler, 2000; Ham *et al.*, 2008; Tabor *et al.*, 2009), including the transplantation and reconstruction of small metabolic pathways in suitable hosts (Lee *et al.*, 2008; Steen *et al.*, 2010). The focus on small systems reflected, in part, the laborious processes of DNA fragment construction and assembly required to optimize designed systems. The rapid expansion of DNA synthesis capacity (Czar *et al.*, 2009; Tian *et al.*, 2009) and the development of simple protocols for large-scale assembly of DNA fragments (Gibson *et al.*, 2008,2009) have broadened the potential focus of synthetic biology. However, larger synthetic circuits require more components (Voigt, 2006), and their reliable operation requires accurate assessments of the impact of each of these components on the host cell processes. When such circuits overburden the host, mutations will rapidly accumulate to relieve the stresses that are introduced. An accurate assessment of the impact of synthetic circuits on host physiology will enable intelligent choice of the circuits chosen for implementation.

The components of synthetic circuits and their impact on the host can be broadly classified into two categories: (1) *Regulatory components* comprised transcription factors, sigma factors, and viral RNA polymerases, which enable controlled expression of individual circuit components. Importantly, the DNA sequence specificities of the regulators may result in aberrant and possibly deleterious gene expression within the host. (2) *Circuit devices* comprised sensors, receivers, signaling molecules, enzymes, etc. These components receive information and then command the cell to perform a task, such as producing chemicals and fuels, secreting proteins, and sending out communication signals. Individual circuit components may be deleterious to the host when overexpressed. Additionally, certain combinations of components may be deleterious even when the individual components have no deleterious effect. Consequently, it is importantly to monitor and catalog the impact of individual and combinations of circuit components on the host in order to facilitate the design process, the choice of components for a particular circuit, and troubleshooting of large synthetic circuits. In addition, a good understanding of the impact of components may facilitate modification of the host to better tolerate the circuit. For example, high level expression of some proteins can result in the accumulation of unfolded products within the cytoplasm, triggering the cytoplasmic heat shock response. This can be relieved by overexpression of cytosolic chaperones.

DNA microarrays provide an easy way to monitor changes in gene expression in the host (Rhodius *et al.*, 2002). They can be used to pinpoint the effects of regulator parts of genetic circuits and provide a useful tool for identifying stress-response pathways that are upregulated in response to circuit devices. In this chapter, we will describe the process of performing microarray experiments and associated data analysis to monitor gene expression. The overall process is illustrated in Fig. 4.1 and involves the following steps: (1) *Experimental setup*: the biological question addressed, experimental design, and performing the microarray experiment(s); (2) *Preprocessing*: data quality control and normalization prior to analysis; and (3) *Analysis*: the statistical tools that identify significantly differentially expressed genes, clustering to identify coregulated genes or similar datasets, and functional annotation to identify common and/or enriched properties of the gene products in the final datasets. We discuss each step and indicate the utility of this technology for synthetic biology.

2. Different Microarray Platforms

The selection of microarray platforms is summarized in Table 4.1. Early microarrays used cDNA libraries, oligonucleotides, or PCR products fabricated by individual laboratories and printed onto polylysine- or

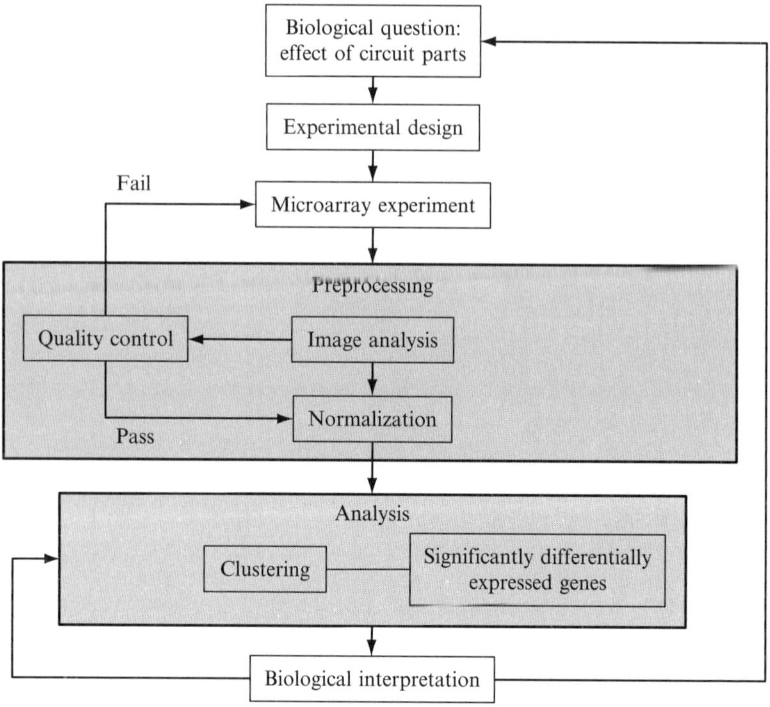

Figure 4.1 Flowchart of the microarray process.

epoxy-treated glass slides. Although inexpensive, the process is laborious, results can be inconsistent and usually limited to a single datapoint for each open reading frame (ORF). Commercially available platforms range from low density arrays with a single printed oligonucleotide probe per ORF, to various high-density platforms with multiple probes per ORF, in which the oligonucleotides are synthesized *in situ*. Additionally, tiled arrays have oligonucleotide probes that anneal to overlapping targets across the genome and some platforms contain probes for intergenic regions, both coding and noncoding strands. Many platforms duplicate their probes across the array surface to reduce hybridization artifacts, thereby improving data reliability. On most platforms, it is possible to perform competitive two-color hybridizations, in which both the reference sample and experimental sample are labeled with different dyes and hybridized to the same array, thereby increasing experimental efficiency. Affymetrix arrays are designed for one-color hybridizations; consequently, the reference and experimental sample are hybridized to two separate arrays. The advantage of one-color hybridizations is their flexibility in comparing samples: only one reference hybridization is required for multiple experimental hybridizations, and all samples can be directly compared with each other when calculating gene expression ratios.

Table 4.1 Different microarray platforms

Probe[a]	Attachment[b]	No. of features/array[c]	Hybridization[d]	Supplier	Notes[e]
≤60 mers	Photodeposition	$1 \times 385K$, $4 \times 72K$ ≤22 probes/ORF	One- or two-color	NimbleGen, www.nimblegen.com	All sequenced bacteria/ custom-reusable × 3
25 mers	Photolithography	1.3×10^6~11 probes/ORF	One-color	Affymetrix (NimbleExpress), www.affymetrix.com	*E. coli*/custom: Requires specialist equipment
35–40 mers	Electrochemical detritylation	$12K$ ~3 probes/ORF	Two-color	CombiMatrix, www.combimatrix.com	Custom-reusable × 3
<60 mers	Ink jet *in situ* synthesis	$4 \times 44K$, $8 \times 15K$	Two-color	Oxford Gene Technology, www.ogt.co.uk	Select bacteria/custom
60 mers	Ink jet *in situ* synthesis	$4 \times 44K$, $8 \times 15K$	Two-color	Agilent, www.agilent.com	No bacteria/custom
50 mers	Self-spot	1 probe/gene	Two-color	Ocimum, www.ocimumbio.com	Select bacteria/custom/ oligo sets
70 mers	Self-spot	1 probe/gene	Two-color	Operon, www.operon.com	Select bacteria/custom/ oligo sets
65 mers	Self-spot	1 probe/gene	Two-color	Sigma-Aldrich, www.sigmaaldrich.com	Select bacteria/custom/ oligo sets

[a] Probe length (nt): mer, oligonucleotide.
[b] Method of probe synthesis and attachment to array surface. Self-spot, user self-spots oligos onto glass slides.
[c] Number of features per array and number of probes per gene: $K = 1000$ (e.g., $72K = 72{,}000$).
[d] Sample hybridization: one-color, one sample is hybridized to the array (often labeled Cy3); two-color, competitive hybridization of two samples labeled Cy3 and Cy5.
[e] Notes on whether arrays are predesigned for specific organisms or can be custom designed. Some arrays can be reused up to three times. Oligo sets are for self-spotting.

The choice of array platform depends upon the type of microarray experiment being performed. The advantage of commercial high-density arrays is that they provide multiple probes often in duplicate for each coding region, which increases data sensitivity and reliability. These probe sets are optimized for signal sensitivity and to reduce cross-hybridization with other homologous genomic sequences. Probes designed toward known ORFs are sufficient when the user is only concerned about the expression status of known annotated genes and transcribed regions. However, several studies demonstrate that there are large number of transcripts from nonannotated regions of the genome, such as intergenic regions and also antisense to known coding regions (Filiatrault *et al.*, 2010; Guell *et al.*, 2009; Selinger *et al.*, 2000; Sharma *et al.*, 2010). The functions of most of these transcripts are unknown, but likely include mRNAs of previously unannotated short coding regions (Hemm *et al.*, 2008,2010), sRNAs that regulate specific target mRNAs (Beisel and Storz, 2010), and regulatory antisense transcripts (Thomason and Storz, 2010). Consequently, high-density platforms that contain probes to both coding and noncoding strands and intergenic regions provide more comprehensive analysis of the transcriptome. It is also possible to custom design arrays to include specific probes of your choice, which is useful if you need to probe specific regions of the genome not covered in the commercial array sets. Finally, many commercial arrays are available in multiplex format in which the array surface is subdivided into sections. Each section contains a complete set of probes, enabling different samples to be separately hybridized to each section in the same experiment, thereby increasing the throughput of experiments.

Several studies have addressed the issue of data quality and reproducibility between commercial array platforms (Bammler *et al.*, 2005; Irizarry *et al.*, 2005; Larkin *et al.*, 2005; Shi *et al.*, 2006,2008). Satisfyingly, these studies find that the results across platforms are remarkably consistent, and the observed fold change in gene expression levels correlates closely with qRT-PCR. However, it is important to standardize protocols for RNA labeling, hybridization, array processing, data acquisition, and normalization. If the array experiments are part of a consortium, agreeing upon a common array platform, strains and growth conditions will greatly facilitate the comparability of datasets produced in different laboratories. The final choice of array platform may also depend on the availability of additional resources in the host institute that are required for microarray experiments such as array hybridization stations and scanners. These are often quite expensive and limited to certain array platforms, and are often shared between multiple laboratories within institutes.

Deep sequencing (RNA-seq) is rapidly becoming a viable alternative for expression analysis (Cho *et al.*, 2009; Filiatrault *et al.*, 2010; Guell *et al.*, 2009; Sharma *et al.*, 2010; Sorek and Cossart, 2010; Wang *et al.*, 2009). In this approach, purified RNA is converted to cDNA flanked with adaptamers that are then sequenced using high-throughput sequencers. The sequence reads are mapped back to the genome and the frequency of reads provides a digital read-out of RNA levels. The main advantage of this approach is that all transcripts are sequenced; consequently, the quantitative information is not limited to existing genomic sequences or restricted to annotations of ORFs. In addition, RNA-seq can provide nucleotide resolution information of transcript boundaries (5′ ends, operon structures, etc.), a greater dynamic range of expression values compared to microarrays and requires small amounts of RNA. However, microarrays currently are cheaper to perform, easier to high-throughput for large experiment sets, ideal for characterizing expression of known RNAs, and both the experimental and data handling issues have been optimized over several years.

3. Experimental Design

Microarray experiments can easily generate large lists of differentially expressed genes making it difficult to unravel the underlying biology. Consequently, careful experimental design is critical for interpretable data. Optimal datasets are where only a few cellular systems are disrupted by a designated perturbation (e.g., induced overexpression of candidate gene). Here, the response usually occurs within a short time frame of the perturbation and therefore can be monitored by following gene expression over a short time course (Nonaka *et al.*, 2006; Rhodius *et al.*, 2006). In contrast, cells exposed to long-term or steady-state differences (e.g., wild-type vs. constitutively expressing mutant strain) can generate complex responses as a result of a cascade of transcriptional effects, making data interpretation difficult.

The gene expression comparisons considered here will be the consequence to the host cell of expressing either a single part or device of a synthetic circuit or a combination of components. First, it is important to establish a standardized expression data set for all circuit parts. This enables cross-comparisons of the effects of different components on host gene expression and therefore can be used as a guide for selecting parts when designing and constructing a circuit. The aim is to catalog the short-term effects of induced overexpression of components within the host under a series of defined growth conditions. Here, gene expression is monitored after induction using a time course in which samples are removed at intervals for up to 1 doubling of culture growth (e.g., at 5, 10, 20, and 40 min after induction). Note that it is essential to standardize the defined growth conditions and the control wild-type host strain used in the transcriptome experiments to enable easy comparison of data between laboratories and scientific communities. It may also be necessary to examine the effects of components under the specific operating conditions of the proposed genetic circuits, especially if these conditions are dramatically different to the standardized conditions or involve long-term expression over several days. This may involve performing much longer time courses of part overexpression (e.g., at 8 h intervals for 48 h), or monitoring the effects of parts or even complete circuits under the specific growth conditions. This will enable fine-tuning of the circuit design and also provide information for whether it is necessary to modify the host in order to optimize circuit performance.

Time-course expression analysis is an effective method for yielding information about the succession of transcriptional changes induced by any component of a circuit. Typically, monitoring expression for 1 doubling after induction is sufficient for identifying direct and any indirect effects, as by that point, the induced protein will reach at least 50% of its maximum induced levels in the cell. Overexpression of a transcription factor part will likely result in a rapid transcriptional response, even if this is as a result of aberrant gene expression; however, overexpression of a device that may exert stresses upon the cell may take longer for transcriptional effects to become apparent as these will likely be indirect. Typically, RNA samples are harvested at select time intervals after induction and are compared on microarrays with samples harvested at equivalent time points from a noninduced control culture. This is best achieved by splitting the starting culture into two aliquots immediately prior to induction: inducing one aliquot and maintaining the other as a control (Fig. 4.2). Note that using completely separate starting cultures for any microarray comparison is inferior as there is increased biological variation in the separate cultures, and hence increased variability in the expression profiles. Harvesting samples at comparable time points from both the control and induced culture will correct for any growth state changes that may occur over the duration of the time course. Sometimes, overexpression of a protein may alter the growth rate of the induced culture. In this case, it is best to harvest samples at comparable culture densities to compensate. However, it is still important to be aware of growth rate differences when interpreting the final gene expression datasets as expression of the ribosomal operons are growth rate dependent and hence will likely be differentially expressed.

4. Experimental Variation

Experimental variation derives either from biological or technical issues. Variability in biology is more difficult to control and can come from issues with the biological sample, the growth conditions of the culture, and alteration in gene expression levels. We discuss each of these in turn and then conclude with the technical issues contributing to variability.

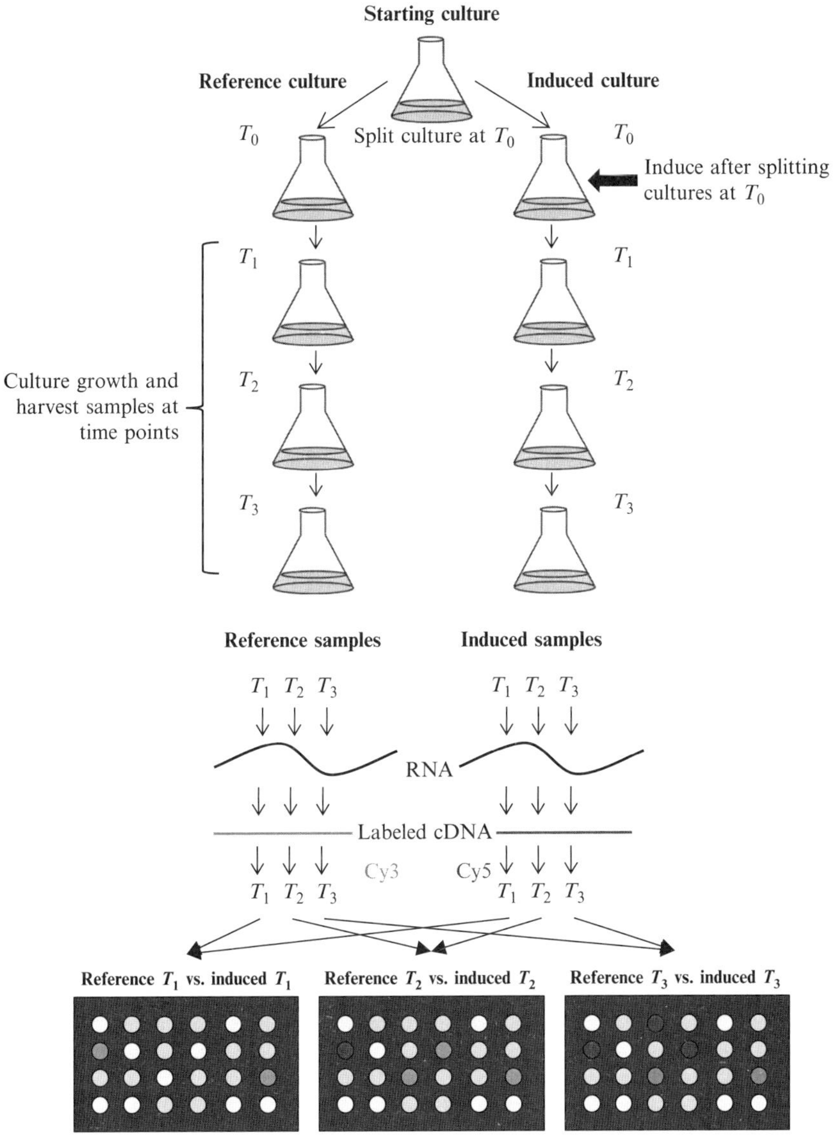

Figure 4.2 Time-course microarray comparison for analyzing "part" effects. Example of a time-course comparison in which the original starting culture is split into two separate cultures ("Reference" and "Induced") at T_0. Immediately afterward, inducer (e.g., IPTG) is added to the "Induced" culture to overexpress the desired circuit part. Both cultures are maintained in identical growth conditions and samples from each culture harvested at the same time points after induction. From each sample, RNA is isolated and cDNA prepared. cDNA from the references samples is labeled with Cy3 and from the induced samples labeled with Cy5. For each time point, Cy3- (reference) and Cy5-labeled (induced) cDNAs are then mixed and hybridized to an array.

Issues related to the biological sample itself are best described by its complexity, quantity, and quality.

1. *Complexity* relates to whether the samples are *simple,* that is, one organism under defined growth conditions (e.g., single homogenous cultures), or *complex,* that is, several organisms or across different growth states (e.g., mixed cultures, including biofilms and host-pathogen models). Sample complexity can dramatically increase biological variation due to poor reproducibility of growth conditions, variability of organism composition of mixed populations, and the isolation of total RNAs from multiple organisms that result in increased cross-hybridization and decreased signal specificity with target probes on microarray. Most biological sample effects are minimized when monitoring part effects under defined growth conditions; however, if the proposed operating conditions of the genetic circuit involve complex environments, then this will dramatically increase the biological variability and consequently the expression profiles measured under these conditions.
2. *Sample quantity* is a factor, especially from complex environments that are not standard cultures and therefore are difficult to grow in large enough quantities. Low sample yield reduces signal strength on microarrays, thereby increasing noise and signal variability.
3. *Sample quality* relates to the purity and integrity of the isolated RNA and labeled cDNA probes prior to hybridization on the microarray. For example, contaminating RNases reduces RNA quality: RNases vary with different host strains, growth conditions, and cellular stress levels.

Growth conditions are another important source of variation, even with defined homogenous cultures. Batch experiments relate to growing cultures in flasks and generally have low or moderate levels of variability. Samples compared from separate starting cultures have higher levels of variability than samples from the same culture. Consequently, it is generally better to split a culture and to induce gene expression in one culture and use the other as a comparative control (see Fig. 4.2). Experiments performed in chemostats tend to have low background variability; however, long chemostat runs can result in the accumulation of mutations, especially in stressed cultures, which can dramatically increase variability from run to run. Defined media, such as M9 (and others) decreases experiment to experiment variability, compared with complex media such as Lennox broth, where the composition of the broth can vary between lot numbers. Finally, the expression levels of weakly expressed genes are inherently more variable due to the sensitivity of probes on the array.

Technical variation comprises sample preparation and labeling, sample hybridization, and microarray slide variability (Yang *et al.*, 2002). As discussed earlier, microarray variability has been shown to be small between most commercial arrays and largest with home-made arrays. Typical sources of variation include differences in array platform and probe design. This latter category includes probe length, specificity, cross-hybridization issues, and different probe sequences for the same target genes on different array platforms. For large-scale studies, these effects can be minimized by using the same array platform, and if possible, arrays manufactured from the same batch. Sample preparation, labeling, and hybridization will be discussed in the next section. With large-scale studies, confounding effects by extraneous factors can be minimized by orthogonalizing variable components, for example, by using equal numbers of arrays from different batches.

5. Sample Preparation

Sample preparation includes RNA harvesting, cDNA synthesis and labeling with fluorescent dyes, and sample hybridization on the microarray. Many array manufacturers recommend protocols for these steps, some of which are specific to the microarray platform (e.g., Affymetrix). In addition, there are many published protocols, for example, Beyhan and Yildiz (2007), Botwell and Sambrook (2003), and Rhodius and Wade (2009). Here, we will discuss important issues of sample preparation and present protocols successfully utilized by many laboratories for two-color hybridizations on most array platforms. Sample preparation involves harvesting the biological material, extracting and purifying the RNA, generating cDNA containing the modified nucleotide amino-allyl dUTP (aa-dUTP) and covalently linking the Cy3 or Cy5 fluorophores to the cDNA samples. Typically in two-color hybridizations to the same array, the reference sample is labeled with Cy3 (scans as red) and the experimental with Cy5 (scans as green). After labeling, both samples are mixed and hybridized to the same array. Reverse labeling in which the dyes are switched (reference is Cy5 and experimental Cy3) is often not necessary, as most dye-dependent biases are removed during data normalization (see later).

5.1. Materials

5.1.1. Solutions

1. All solutions should be RNase free, either by treating with 0.1% DEPC or by purchasing as RNase free from suppliers such as Ambion.
2. Ethanol/phenol stop solution: H_2O-saturated phenol (pH $<$ 7.0) in ethanol (5%, v/v).
3. Lysozyme solution: 500 μg/mL lysozyme in 10 mM Tris (pH 8.0), 1 mM EDTA (pH 8.0). Prepare fresh just before use.
4. Nuclease-free water, Ambion #AM9938.
5. Sodium acetate, pH 5.5, 3 M (100 mL), Ambion #AM9740.

6. 25× aa-dUTP/dNTP mix: 12.5 m*M* dATP/dGTP/dCTP, 5 m*M* dTTP, 7.5 m*M* aa-dUTP (Ambion, #8439). The 3:2 ratio of aa-dUTP:dTTP is optimized for *Escherichia coli* based on the (G + C) content of the template cDNA. Higher GC content organisms may require a lower aa-dUTP to dTTP ratio; lower GC content organisms may require a higher ratio.
7. 1 *M* KPO_4, pH 8.5: 9.5 mL 1 *M* K_2HPO_4, 0.5 mL KH_2PO_4. Prepare daily to ensure optimal pH.
8. Phosphate wash buffer: 5 m*M* KPO_4, 80% EtOH. May be slightly cloudy. Prepare daily.
9. Phosphate elution buffer: 4 m*M* KPO_4, pH 8.5. Prepare daily.
10. 20× SSC (1 L), Ambion #AM9763.
11. SDS, 10% solution (100 mL), Ambion #AM9822.

5.1.2. Supplies

1. All microfuge tubes and pipette tips should be RNase and DNase free.
2. TURBO DNA-free Kit, Ambion #AM1907. Contains TURBO DNase, 10× buffer, and DNase inactivation reagent.
3. Random octamer oligonucleotide, any oligo company.
4. SuperScript III Reverse Transcriptase, Invitrogen #18080-093. Includes 5× first-strand synthesis buffer.
5. MinElute PCR purification kit (50), QIAGEN #28004.
6. CyDye postlabeling reactive dye packs, GE Healthcare #RPN5661.

5.2. Sample harvesting

Careful isolation of RNA from biological samples is essential in order to accurately capture the RNA profile present at time of harvesting. Transcription profiles can rapidly change in response to the cellular stresses of harvesting (media change, centrifugation, temperature change, etc.). Also, RNA is extremely sensitive to degradation by RNases present both in the harvested cells and introduced during the purification procedure. Consequently, it is essential to use RNase-free materials and solutions throughout the purification and cDNA synthesis steps. It is also essential to "freeze" the RNA profile during sample harvesting. This is achieved by mixing the cells with a reagent designed to inactivate transcription and prevent RNA degradation; for example, RNAprotect (QIAGEN); a solution of 40% methanol, 62.5 m*M* HEPES, pH 6.5 at −45 °C (Pieterse *et al.*, 2006); or 5% acid phenol in ethanol stop solution as outlined below:

1. Transfer 10 mL of culture ($0.3 - 1 \times 10^9$ cells/mL) into a 15-mL conical tube containing 1.25 mL of ice-cold ethanol/phenol stop solution.

2. Harvest cells by centrifugation at 6700 × *g* for 2 min at 4 °C. Remove media by aspiration.
3. Rapidly freeze cell pellet in liquid nitrogen. Store at −80 °C until required.

5.3. Total RNA preparation

There are multiple RNA purification methods and kits available. The method of choice will be determined by the properties of your strain (growth conditions, ease of lysis, and level of RNases), the experimental design, and the quantity of available sample versus yield of RNA required for each array experiment. Consequently, it may be necessary to modify protocols to enhance lysis or to cope with high levels of endogenous RNases. It is also important to note that some kit protocols that have affinity purification steps may not give representative yields of small RNAs. Total RNA from bacterial cultures can be isolated using the hot phenol method outlined below, or by using several commercial reagents (e.g., TRIzol, Invitrogen), or kits (e.g., RNeasy, QIAGEN; RiboPure, Ambion). Also, there are protocols and kits available that (1) enrich mRNA from total RNA preparations by using oligos that bind to 16S and 23S rRNA, enabling these RNAs to be subtracted from the RNA prep (e.g., MICROBExpress, Ambion); (2) purify bacterial RNA from complex host-bacterial samples using a similar oligo subtraction method to remove contaminating rRNA and poly-A mRNAs from select eukaryote hosts (e.g., MICROBEnrich, Ambion); and (3) amplify RNA from low yield samples using a linear cDNA amplification step (Gao *et al.*, 2007). We find that for *E. coli* cultures the hot phenol method outlined below yields the best quality RNA preps in terms of RNA integrity, size, purity, and yield.

1. Resuspend the cell pellet (from 10 mL of culture) in 800 μL lysozyme solution. Transfer lysate to 2-mL microfuge tube containing 80 μL of 10% SDS, mix by inversion, and incubate at 64 °C for 2 min.
2. Add 88 μL 1 *M* sodium acetate solution (pH 5.2) and mix by inversion.
3. To the lysate add an equal volume (∼1 mL) of H_2O-saturated phenol (pH < 7.0). Mix by inverting 10 times. Incubate in a 64 °C water bath for 6 min, continuing to mix the tube contents by inverting every 40–60 s.
4. Place tube on ice to chill for 2 min. Afterward, centrifuge at 16,000×*g* for 10 min at 4 °C.
5. Remove the upper aqueous phase into a fresh tube, taking care not to disturb the interface (this is a common point of RNase contamination of preps). Also, perform this step quickly, as the aqueous layer can rapidly become cloudy after centrifugation, making it difficult to separate the layers. If this happens, recentrifuge the sample.

6. Add to the solution an equal volume (~1 mL) of 1:1 mix of H_2O-saturated phenol:chloroform. Invert the tube 6–10 times to mix and centrifuge at 16,000×*g* for 2 min.
7. Carefully remove the upper aqueous phase to a fresh microfuge tube. Repeat the H_2O-saturated phenol:chloroform extractions until the interface is clear (usually ≥ 2–3 times). Some strains may require extensive phenol:chloroform extractions to completely remove contaminating RNases.
8. Divide the final extracted solution equally between two 1.5-mL microfuge tubes. Precipitate by adding 0.1 volume 3 *M* sodium acetate (pH 5.5) and 2.5 volumes 100% cold ethanol. Incubate at −80 °C for 30 min.
9. Recover the RNA by centrifugation at 16,000×*g* for 30 min at 4 °C.
10. Wash the RNA pellet with 1 mL 80% cold ethanol. Centrifuge at 16,000×*g* for 5 min at 4 °C. Carefully remove the ethanol solution by aspiration and dry the RNA pellet in a speed vacuum.
11. Redissolve and pool each pair of pellets in a final volume of 87 μL and place in a fresh 1.5-mL microfuge tube.

5.4. DNase treatment

1. To each RNA preparation (87 μL), add 10 μL 10× TURBO DNase buffer and 3 μL TURBO DNase. Incubate reaction at 37 °C for 30 min.
2. Add an additional 3 μL TURBO DNase and incubate a further 30 min at 37 °C.
3. Add 10 μL DNase inactivation reagent and incubate at room temp, mixing four times.
4. Centrifuge at 10,000×*g* for 1.5 min and transfer the supernatant containing the RNA to a fresh tube. Store at −20 °C until required.

5.5. Assessing RNA quality and yield

1. Determine RNA concentration by measuring the absorbance of a 1:100 dilution in H_2O at 260 nm (concentration, *c* (μg/μL), in a 1-mL quartz cuvette with a 1 cm path length: $c = A_{260} \times f \times 0.04$ μg/μL, where *f* is the dilution factor). Typical yields are 70–300 μg RNA from 10 mL of culture, depending on strain of *E. coli*, growth conditions, and culture density upon harvesting.
2. Check purity by measuring the absorbance ratio of nucleic acid versus protein (A_{260}/A_{280}) of a 1:100 dilution in 10 m*M* Tris–HCl buffer, pH 7.5 (note that the absorbance ratio is sensitive to pH: as RNA is acidic, the ratio must be measured in a low salt neutral buffer). Good RNA preps free from protein contamination give values between 1.8 and 2.1.

3. If required, the integrity of the RNA can be analyzed on a denaturing formaldehyde 1% agarose gel (Ausubel *et al.*, 1998). Upon visualizing the gel, the 23S and 16S ribosomal RNA should be easily observed. For good RNA, the 23S species should be twice as intense as the 16S with little or no smearing between or below these bands.

5.6. cDNA synthesis and RNA hydrolysis

cDNA is synthesized using random octamer primers and a dNTP mix containing aa-dUTP (aa-dUTP).

1. In 0.2-mL PCR tube, mix 15 μg total RNA with 16 μg random octamer primer to give a final volume of 35 μL. Incubate at 70 °C for 10 min and then chill on ice for 10 min.
2. cDNA synthesis reaction: Prepare a cocktail on ice containing for each reaction: 12 μL 5× first-strand synthesis buffer; 2.4 μL 25× aa-dUTP/dNTP mix; 6 μL 0.1 *M* DTT, 2.3 μL SuperScript III RT; 2.3 μL H_2O. Add 25 μL cocktail to annealed RNA sample and incubate at 50 °C for 3 h.
3. RNA is removed from the completed reverse transcription reaction by hydrolysis. To the sample, add 1.2 μL 0.5 *M* EDTA and 6 μL 1 N NaOH and incubate for 10 min at 65 °C.
4. Neutralize the reaction by adding 65 μL 1 *M* HEPES, pH 7, and mix well.

5.7. Sample cleanup

Unincorporated aa-dUTP and competing free amines must be removed from the sample to enable successful coupling of the amino-allyl cDNA with the Cy3/Cy5 dyes. Consequently, Tris-based buffers cannot be used in the following cleanup steps. Each sample is cleaned using the QIAGEN MinElute PCR Purification Kit that has been modified, replacing the QIAGEN wash and elute buffers (PE and EB), which contain free amines, with *phosphate-based* wash and elute buffers (Beyhan and Yildiz, 2007; Hasseman, J., TIGR Aminoallyl Labeling of RNA for Microarrays & TIGR Microarray Labeled Probe Hybridization).

1. Remove the reverse transcription reactions from the PCR tube to a fresh 1.5-mL microfuge tube.
2. Add 500 μL QIAGEN Buffer PB to each sample.
3. Load samples on to MinElute Columns and centrifuge at ≥ 10,000×*g* for 1 min.
4. Discard the flow-through and add 750 μL *phosphate wash buffer* to each column and centrifuge at ≥ 10,000×*g* for 1 min.

5. Discard the flow-through and centrifuge again at $\geq 10{,}000 \times g$ for 1 min.
6. Add 10 μL phosphate elution buffer to the center of each column matrix, incubate for 1 min, and then elute using a fresh collection tube by centrifugation at $\geq 10{,}000 \times g$ for 1 min. Samples can be stored at -20 °C if required.

5.8. Cy3/Cy5 coupling

The Cy dyes are shipped as a desiccate in sealed packs. Note that they are extremely sensitive to light and moisture, therefore each pack is opened and resuspended in DMSO immediately prior to use. Each pack is sufficient for approximately five reactions.

1. To each cDNA sample, add 1 μL 1 *M* sodium bicarbonate, pH 9.0 (note the bicarbonate becomes carbon dioxide with time; therefore, use fresh solution <1 month old).
2. Resuspend each fresh tube of Cy3 or Cy5 in 10 μL DMSO. Keep in dark until ready for use.
3. Add 2 μL of either Cy3 or Cy5 solution to each cDNA sample. Incubate for 2 h at room temperature in the dark.
4. Unincorporated Cy dyes are removed using the QIAGEN MinElute PCR purification kit following the procedure previously described in sample cleanup, steps 2–6. Note that the eluted samples should be colored red for Cy3 and blue for Cy5.
5. cDNA yield and labeling efficiency of the eluted samples can be calculated by using a nanodrop to measure their absorbance at 260, 550, and 650 nm against a water blank (Botwell and Sambrook, 2003). For each sample, calculate

$$\text{pmol nucleotides} = [\text{OD}_{260} \times \text{vol} \times 37\,\text{ng/ml} \times 1000\,\text{pg/ng}]/324.5\,\text{pg/pmol}$$

$$\text{pmol Cy3} = \text{OD}_{550} \times \text{vol}/0.15$$

$$\text{pmol } Cy5 = \text{OD}_{650} \times \text{vol}/0.25$$

$$\text{nucleotide/dye ratio} = \text{pmol cDNA/pmol Cy dye.}$$

Where

$$1\,\text{OD}_{260} = 37\,\text{ng/ml for cDNA;}$$

324.5 pg/pmol average MW of a dNTP
vol = sample volume (μL)

Optimal labeling values for hybridizations are incorporation of >150–200 pmol dye per sample and <20 nucleotides/dye molecule.

6. It is preferable to use the eluted samples the same day for hybridization, storing in the dark until required. For longer storage or if the hybridization volumes are small (<20 μL), combine pairs of Cy3 and Cy5 samples (2 μg each cDNA) to be hybridized together, dry down in a speedvac in the dark. Store at −20 °C.

5.9. Sample hybridization

Hybridization protocols, sample volumes, and quantities vary depending on the microarray platform and hybridization chamber. For some array platforms, the hybridization sample is applied under a lifter-slip placed over the array on the glass slide. The slide is then placed in a small hybridization chamber and incubated between 42 and 65 °C (depending on the length of the array probes) for up to 16 h. In general, these hybridizations require large sample volumes (≥ 40 μL) and are prone to sample drying and uneven hybridization intensities over the array surface, resulting in poor or discarded data. Recently, several high-density array platforms (e.g., Nimblegen) use special hybridization systems (e.g., MAUI hybridization system) that dramatically improve hybridization efficiency. In these cases, mixers are applied over the array surface creating a sealed chamber, enabling the application of small sample volumes that are actively mixed during the hybridization process. This generates an even hybridization, minimizes sample evaporation, increases signal sensitivity, increases reproducibility, and shortens hybridization times. Below, we give a sample hybridization mix used for arrays with lifter-slips (volume = 50 μL): volumes can be scaled accordingly, maintaining the correct final concentration of SSC, HEPES, and SDS. Note water and all solutions must be filtered (e.g., with a 0.2 μm filter) to prevent small particles damaging the surface of the array.

1. For each hybridization, combine Cy3 and Cy5 sample pairs, using 2μg cDNA for each sample, in a 0.2-mL microfuge tube (if combined samples were dried down, resuspend in 10 μL H_2O).
2. To each hybridization reaction, add 7.5 μL 20× SSC, 1.25 μL 1 *M* HEPES, pH 7.0, 1.25 μL 10% SDS and H_2O to a final volume of 50 μL (3× SSC, 25 m*M* HEPES, 0.25% SDS final).
3. Incubate reaction at 99 °C for 2 min and then allow to cool at room temperature for 5 min. Lightly vortex sample to mix and spin down before applying to surface of microarray, following the hybridization instructions for your microarray and hybridization chamber.

5.10. Slide washing and scanning

Prior to scanning, hybridized slides are washed to remove any sample nonspecifically bound to the slide surface. As slide washing protocols will vary according to the manufacturer, we give a washing protocol commonly used for oligo and ORF PCR arrays printed onto polylysine-coated slides. Note that all wash stock solutions should be filtered before using. After washing, the Cy dyes are extremely unstable: Cy5 is rapidly degraded by ozone in minutes (Branham *et al.*, 2007; Fare *et al.*, 2003). Slides should be dried and scanned in a low ozone chamber; alternatively, some companies supply wash solutions that stabilize the dyes (e.g., Agilent).

1. Prepare the following wash solutions in glass slide dishes: two glass slide dishes each containing 500 mL Wash Solution I (897 mL Milli-Q-water, 100 mL 20× SSC, 3 mL 10% SDS). Place an empty slide rack in one of the dishes. If using oligo arrays, Wash Solution I should be preheated to 60 °C and poured into the slide dishes immediately prior to washing the slides. This is essential to remove nonspecific hybridization on oligo arrays; two glass slide dishes each containing 500 mL Wash Solution II (950 mL Milli-Q-water, 50 mL 20× SSC)and one glass slide dish containing 500 mL Wash Solution III (495 mL Milli-Q-water, 5 mL 20× SSC).
2. Carefully remove slide from the hybridization chamber; keeping the array level, submerge into the slide dish containing Wash Solution I with no slide rack.
3. Once submerged, using fine forceps carefully remove the cover slip or mixer-assembly following the manufacturer's instructions, taking care not to scratch the surface of the array.
4. After removing the cover, place the array on the rack in the second slide dish containing Wash Solution I.
5. Repeat steps 2–4 for any other remaining slides. When finished, plunge the rack up and down 10–20 times.
6. Immediately transfer the slide rack to Wash Solution II, and plunge up and down for 60 s.
7. Drain the rack for 5 s, and then place in the second dish containing Wash Solution II and plunge up and down for 60 s.
8. Drain rack for 5 s, and then transfer to Wash Solution III and plunge up and down for 60 s.
9. Dry the arrays by centrifugation at 600 rpm for 2 min in a low ozone chamber.
10. Scan the arrays as soon as possible in a low ozone chamber to reduce degradation of Cy5.

There are several scanners and software available for processing slides and the generated image files; some are specific to certain slide platforms (e.g., Affymetrix). One of the most popular systems that handles several different slide platforms is the GenePix scanner and software from Molecular Devices and also SpotReader from Niles Scientific. Users should scan their slides following the manufacturer's instructions. Finally, it is also possible to reuse slides from some manufacturers (e.g., Nimblegen) for up to three subsequent hybridizations without significant loss of hybridization signal by using a series of slide wash steps that remove the hybridized sample (stripping) but leaves the probes intact. Specific protocols are supplied by the manufacturer.

6. Microarray Preprocessing

For two-color arrays, each slide is excited at two wavelengths, 532 nm for the Cy3 (green)-labeled reference sample and 650 nm for Cy5 (red)-labeled experiment sample, to measure the fluorescence of the hybridized samples to each probe (feature) on the array. Note that during array scanning it is important to individually adjust the scanning voltages at each wavelength, which in turn controls the detected fluorescent intensities. This is required to (1) approximately balance the signal intensities from each channel and (2) to ensure an optimal signal dynamic range for the features on the array, reducing the number of saturated (overexcited) probes while maximizing the detection of weakly fluorescing probes. Scanning generates two high-resolution 16-bit tiff images (one for each channel) that contain the fluorescence intensities for each feature. Image files are also generated when scanning one-color Affymetrix arrays: here, two slides are scanned to generate separate image files for the reference and experimental samples. Preprocessing involves several steps that analyze the image files to generate a single expression ratio for each gene represented on the array. This involves

1) *Image analysis and quality control.* The fluorescent intensities of every feature are determined by overlaying a grid on each image file to map the location and identity of each feature, and to quantify their specific signal intensity and surrounding background signal. This generates separate data files for each channel that lists all features and their associated specific and background fluorescent intensities. From this preliminary analysis diagnostic reports can be generated on the quality of the hybridization. These measure the quality of grid alignment with the features, calculate average specific versus background intensity ratios for all features, the number of saturated features and features with signals above threshold, and assess the uniformity of the background and specific signals to determine if there was any bias in the hybridization across the array surface.
2) *Probe set summarization.* High-density arrays contain multiple probes for each gene. Summarization collates the fluorescent values of probe sets to

generate a single intensity for each gene, and also discards or reduces the influence of any probes within the set that have unusual fluorescence values.

3) *Within and between array normalization.* Prior to calculating gene expression ratios between the reference and experiment data sets, it is important to correct for any systematic errors in fluorescence measurements and sample quantification. Within array normalization corrects for systematic errors between the two compared samples; between array normalization corrects for systematic errors across multiple arrays (experiments), enabling more accurate comparisons of expression ratios across multiple experiments. Normalization is discussed in more detail in the next section.

Most companies provide associated software for array preprocessing, and additional software is freely available that enables more advanced preprocessing and array normalizations (Table 4.2). Note that Bioconductor is an open source project that provides many high quality tools and documentation for microarray data analysis (Table 4.2). These tools run in the programming language, *R*, and whilst the learning curve is steep, they are highly recommended for the serious microarray analyst! Here, we focus on tools and procedures that are commonly available across freely available software. First, we present common diagnostic plots used to assess general features of the expression ratios and subsequent data normalization.

6.1. Diagnostic plots of gene expression ratios

Plots of the gene expression ratios provide very useful information on the quality of array experiments and facilitate comparisons between experimental repeats and across experiment sets. Note that in all subsequent data analysis, the gene expression ratios are log transformed: $\log_2 (R/G)$, where R = Cy5 intensities (experimental) and G = Cy3 intensities (reference). Generally, the Cy3 and Cy5 values are background subtracted, which is the default setting on most software programs. However, this can decrease the accuracy of determining expression ratios for weakly expressed genes with signals close to background. For this reason, it is common to filter datasets removing features that have signal intensities less than 2–3 standard deviations above the mean background.

1) *Histograms.* The simplest gene expression plot is a histogram of $\log_2 (R/G)$ expression ratios for all genes on the array (Fig. 4.3A). Most gene expression experiments only alter a small fraction of genes. Consequently, good histograms should contain a single symmetrical peak with no shoulders, and have a narrow distribution with only a small number of genes with large expression ratios in the distribution tails.

Table 4.2 Free tools and resources

Software	Functions	URL/Reference
R	Open source statistical computing environment	http://www.r-project.org/
Bioconductor	Open source software for bioinformatics that runs in *R*. Extensive tools for preprocessing, normalization, cluster, statistical and gene set enrichment analysis	http://www.bioconductor.org/
TM4 Microarray Software Suite	Four major software packages. Image analysis, normalization, clustering, statistical and gene set enrichment analysis	http://www.tm4.org/ (Saeed *et al.*, 2006)
GenePattern	Genomic analysis platform. Clustering, statistical and gene set enrichment analysis	http://www.broadinstitute.org/cancer/software/genepattern/ (Reich *et al.*, 2006)
limmaGUI	Linear Models for microarrays Graphical User Interface. Normalization, diagnostic plots, differential expression. Also available in Bioconductor (R/limma)	http://bioinf.wehi.edu.au/limmaGUI/ (Wettenhall and Smyth, 2004)
RMAExpress	Robust Multichip Average. Background adjustment, quantile normalization and probe summarization for Affymetrix Genechip data	http://rmaexpress.bmbolstad.com/ (Bolstad *et al.*, 2003; Irizarry *et al.*, 2003a,b)
SAM	Significance Analysis of Microarrays. Statistical technique for differential expression; gene set analysis. Available as Excel macro, Bioconductor (R/sam) and in TM4 suite (MeV)	http://www-stat.stanford.edu/~tibs/SAM/ (Storey, 2002; Taylor *et al.*, 2005; Tusher *et al.*, 2001)
Cyber-T	Statistics program for differential expression. Also available for Bioconductor (R/hdarray; R/bayesreg; R/bayesAnova)	http://cybert.microarray.ics.uci.edu/ (Baldi and Long, 2001)
Cluster 3.0 + Java TreeView	Clustering algorithms and visualization tools	http://bonsai.hgc.jp/~mdehoon/software/cluster/

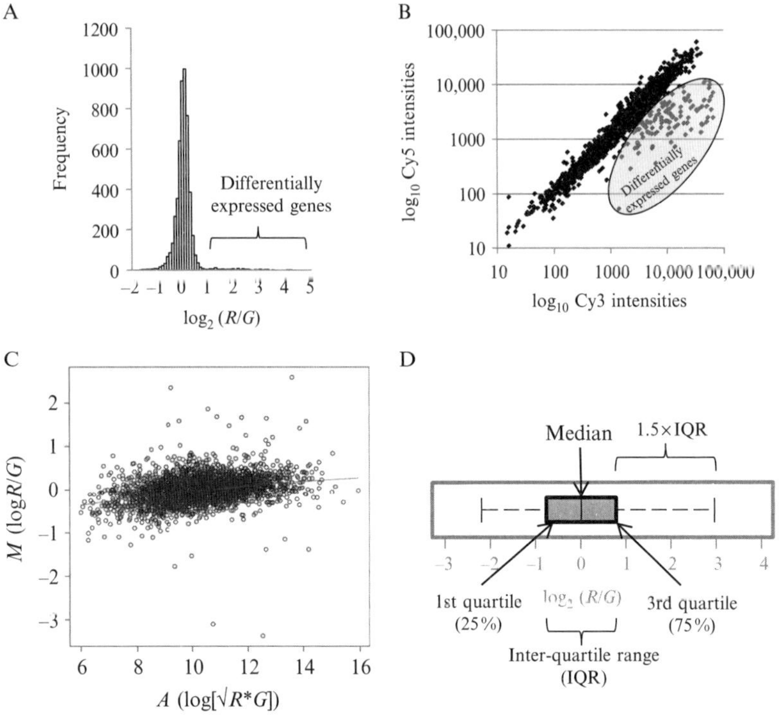

Figure 4.3 Common diagnostic plots for analyzing microarray expression data. (A) Histogram of log expression ratios ($\log_2 [R/G]$) from one array. Histogram should be symmetrical with no shoulders. Tails represent differentially expressed genes. (B) Scatter plot of log Cy5 versus log Cy3 fluorescent intensities (background subtracted) of features on an array. Scatter plot should be linear and evenly distributed. Differentially expressed genes are indicated. (C) MA plot of log gene expression ratios ($M = \log_2 [R/G]$) versus average intensity ($A = \log_2 \sqrt{[R \times G]}$) for one array. Plot should be linear with good dynamic range (range of A values) (D) Box plot of log gene expression ratios ($M = \log_2 [R/G]$) for one array, illustrating the quartile distribution of M values.

2) *Scatter plots*. These are plots of log Cy3 versus log Cy5 fluorescent intensities for every gene, and hence are more informative as they provide a visual overview of Cy3 and Cy5 signals for every gene (Fig. 4.3B). These plots should give a linear distribution with a good dynamic range and little scatter from the diagonal.
3) *MA plots*. MA plots enable visualization of variation of gene expression ratios ($M = \log_2 [R/G]$) as a function of average signal intensity ($A = \log_2 \sqrt{[R \times G]}$; Yang *et al.*, 2002). Visually, MA plots are similar to scatter plots rotated clockwise by 45° (Fig. 4.3C). MA plots are good for detecting an artifact that arises when the labeling reaction is nonoptimal; this is apparent as a "skew" in the plot, such that low intensity ratios tend

to be more negative ($G > R$) than high intensity ratios. This can be corrected by intensity-dependent normalization (see later).

4) *Box plots*. These are useful for comparing the spread and median of gene expression ratios either between sectors on an array or between arrays (Fig. 4.3D).

6.2. Data normalization

Normalization scales the red and green intensities within an array or across arrays by a common factor: $\log_2 R/G \rightarrow \log_2 R/(kG)$. Normalization attempts to correct systematic technical differences between the two channels (reference and experiment) or across experiments (multiple arrays). These include unequal RNA quantities, labeling efficiency, biases in measured expression levels, scanner settings, and array batch variations. It is also important that the normalization process maintains the original biological variation in the signal and that the assumptions of the normalization procedure are understood. Normalization requires common references between the two samples that remain unchanged. Possible approaches to normalization are:

1) *Normalization to housekeeping genes*. This is not considered to be an acceptable method as there is no evidence for a class of genes that has constant expression under different conditions.
2) *Normalization to reference RNA*. In these cases, each sample is spiked with RNA standards from another organism and accompanying target probes are present on the array. The expression values of each channel are then normalized to the standards. This is useful if there are large-scale changes in gene expression and hence the assumptions of global scaling are not valid (described next). However, the disadvantage of this approach is that errors are easily introduced from the accuracy of quantifying the references, application of references to the samples, and their signal intensity measurements from the array.
3) *Normalization by global scaling*. The most common form of normalization is by global scaling, which assumes that the total amount of mRNA remains constant under the various experimental conditions and only a small subset of genes change expression. Note, if this assumption is true, then the histogram of expression ratios remains "balanced": that is, symmetrical with small tails. Simple forms of global scaling sum total mRNA intensities from each channel and scale one to the other. More advanced global normalizations discussed below correct for biases in expression data that are a function of signal intensity or have altered distributions of expression ratios that can be affected by array batches and labeling efficiencies.
4) *Intensity-dependent normalization (loess smoothing)*. Gene expression ratios visualized using MA plots often display a skew in the median intensities

that is a function of signal intensity and is due to differences in dye stability, efficiency of dye incorporation, and scanner settings (Fig. 4.4A and B). This can be corrected using loess smoothing, in which a robust locally weighted regression curve (loess) is fitted to the data using overlapping windows of signal intensity (typically ~30% of the data). The data is then normalized to the curve such that the distribution of log gene expression ratios is centered on zero across the range of signal intensities (Yang *et al.*, 2002).

5) *Scale normalization*. This adjusts the spread of gene expression ratios so that they have similar distributions between different arrays, and hence normalizes for differences in dye stability, dye incorporation, scanner settings, and array batch variations (Yang *et al.*, 2002). Note that the

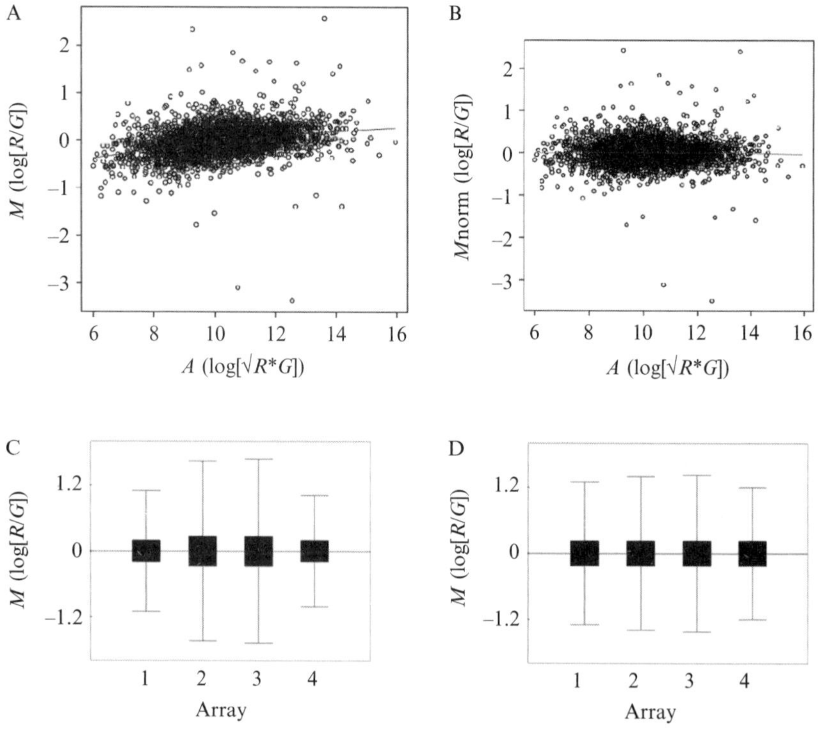

Figure 4.4 MA and Box plots illustrating normalization of array data. (A) MA plot of prenormalized expression data from one array illustrating the loess curve. (B) MA plot of the same expression data in which the *M* values have been normalized to the loess curve (loess smoothing). (C) Box plots of four replicate microarray experiments, illustrating differences in distribution of *M* values. (D) Box plots of same four microarray data in which the *M* values have been normalized by standard deviation regularization.

assumption here is that the biological distribution of mRNA expression is similar across multiple experiments. This is applicable across replicates, but can reduce biological information if comparing different conditions that result in dramatically different variations in gene expression distributions; for example, time-course experiments. Simple scale normalization can be visualized using box-plots and involves regularizing the variance of log gene expression ratios across multiple experiments (Fig. 4.4C and D). Quantile normalization regularizes the distribution of probe intensities such that they are same across multiple arrays (Bolstad *et al.*, 2003).

A variety of programs are freely available that perform array preprocessing tasks; some are listed in Table 4.2.

7. Clustering

Clustering is an exploratory data analysis process for datasets containing multiple array experiments. It is used to: discover patterns in the data; group "similar" patterns together either by clustering genes (rows) with similar expression profiles or by clustering arrays/experiments (columns) with similar profiles, or both genes and experiments; reduce the complexity of the data into several distinct patterns; and provide a method to order and organize the data (reviewed in Boutros and Okey, 2005; D'Haeseleer, 2005; Quackenbush, 2001). Consequently, clustering is extremely useful for data visualization, hypothesis generation, and selection of genes for further consideration. Clustering is an extremely useful tool for characterizing the transcriptional effects of circuit parts on the host. First, clustering expression experiments of different parts will identify parts that have similar effects on the host. This is useful for classifying parts into different categories based on their effects on the host, thereby aiding the design of large circuits with multiple parts. For example, the user may decide to select parts that affect different systems in the host to avoid "overstressing" any one particular system. Second, for a specific part, clustering genes across multiple experiments that measure the effect of the part under different growth conditions or across an induction time-course will identify genes with similar expression profiles and therefore aid detection of the cellular systems affected by the part across the different conditions. This is extremely useful, as it is well-documented that genes with similar expression profiles are involved in similar cellular processes and are often transcriptionally coregulated by regulators.

Most microarray data clustering is unsupervised; that is, no prior information (e.g., gene functional categories, operon structure, etc.) is used to guide the clustering. The goal of unsupervised clustering is to discover patterns from the data; however, even if the data is random such clustering assumes that there is still an underlying pattern. Consequently, clustering always works! Therefore it is important to cluster with caution, use different algorithms, and where possible, filter the data so that only genes with the most varied expression profiles are used. When clustering by array/experiment, the aim is to identify conditions that generate similar expression profiles; consequently, it is typical to use all the expression data. However, when clustering by genes to identify coregulated genes it is important to filter the dataset prior to clustering. This serves both to reduce the dataset to a meaningful size and, importantly, to remove genes with expression profiles that have little variance across the dataset, thereby reducing "noise" in the clusters. Commonly used filters are: (1) genes with expression ratios present in a given fraction of experiments; (2) number of genes with expression ratios above a certain threshold in a given fraction of experiments; (3) number of genes with variance or standard deviation of their expression ratios across the experiments exceeding a specified value. Finally, when preparing data for clustering, it is important to use normalized data that has been log transformed.

7.1. Measures of similarity between genes and distance between clusters

The basic principles of clustering will be discussed in the context of clustering only genes; however, these principles also apply to clustering by array/experiment. Clustering measures the *similarity* in expression profiles between genes and also the *distance* between clusters of genes with similar profiles. The aim of clustering algorithms is to identify clusters that maximize the similarity of gene expression profiles within clusters whilst maximizing the distance between clusters to obtain distinct clusters. It is also important to be able to identify outliers that contain expression profiles that do not easily fit into any particular cluster; however, only some algorithms do this.

There are two main measures of similarity between genes: correlation coefficients that are scale-invariant; and distance metrics that are scale-dependent. Scale-invariant measures identify similar patterns of "ups" and "downs" in gene expression ratios, while scale-dependent measures also consider the magnitude of the expression patterns. Commonly used correlations and distance metrics are listed in Table 4.3A. Euclidean distance is the most commonly used metric and considers both the patterns of ups and downs and also the magnitude of the expression ratios, thereby preserving more information about the data. There are three commonly used measures of distance between clusters listed in Table 4.3B. The average linkage is the least sensitive to outliers; however, complete distance often generates the most discrete clusters, while single linkage often performs very poorly.

Table 4.3 Commonly used measures of similarity between genes and distance between clusters

Measure	Comments
(A) *Measures of similarity between genes*	
Correlation (Pearson)	Identifies similar patterns of ups and downs
Uncentered correlation (cosine-angle)	Identifies similar patterns of ups and downs and also magnitude
Absolute correlation	Identifies similar patterns of ups or downs
City Block (Manhatten) distance	Corresponds to the sum of differences across dimensions. Yields diamond shaped clusters and is less sensitive to outliers
Euclidean distance	Corresponds to the geometric distance in multidimensional space. Identifies similar patterns of ups and downs and also the magnitude of the patterns. Yields spherical shaped clusters
(B) *Measures of distance between clusters*	
Single (minimum) linkage	Shortest distance between members of clusters. Yields elongated clusters and is sensitive to outliers
Compact (maximum) linkage	Largest (outside) distance between members of clusters. Yields compact clusters and is sensitive to outliers
Average (Mean) linkage	Average distance between members of clusters. Generates "in between" sized clusters and is insensitive to outliers

7.2. Clustering algorithms

Clustering algorithms can be divided into two types: (1) hierarchical, in which genes with similar patterns are joined together by a dendrogram, for example, hierarchical clustering; and (2) partitioning, in which the data is divided into groups or clusters with similar patterns, for example, Self-Organizing Maps (SOMs) and *k*-means. Some freely available clustering software is listed in Table 4.2 and simple clustering algorithms outlined below.

1) *Hierarchical clustering*. This is an agglomerative process that starts with every gene considered as its own cluster (Eisen *et al.*, 1998). The most similar pair of clusters are joined together to form a parent cluster, then the next most similar pair of clusters are joined together, and the process repeated until there is just one large cluster. During this process, all genes are scored using the similarity metrics, and the distance between clusters measured using the distance metrics. The merging of the clusters is illustrated using a dendrogram connecting each gene and the expression profiles illustrated using a heatmap (Fig. 4.5A). The dendrogram and associated heatmap provides a good overall visual guide of patterns in the

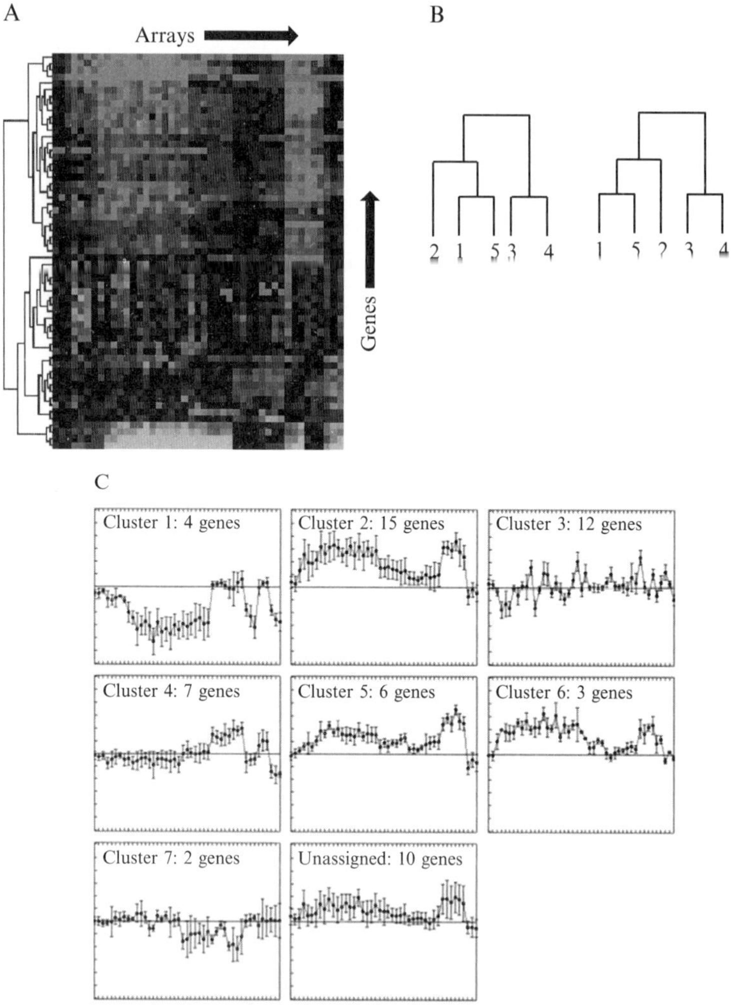

Figure 4.5 Clustering algorithms. (A) Hierarchical clustering of top 1.5% most variable genes (59/4249 genes) across 45 experimental conditions (arrays). Color coded heat map indicates $\log_2$ (R/G) expression ratio of genes (red = $R > G$; green = $G > R$; black = no change). (B) Dendrograms linking four genes illustrating different arrangement of nodes. (C) *k*-means clustering of same dataset in (A) specifying seven (*k*) clusters, 10 *k*-means runs and requirement for genes to cocluster in at least eight runs. Note presence of 10 unassigned genes that did not meet cocluster threshold.

data; however, there are several caveats. First, the order of the nodes in the dendrogram is arbitrary, placing genes adjacent to each other that may be not that similar. Some software programs flip the dendrogram nodes to optimize the ordering of genes based on their similarity (Fig. 4.5B). Second, as the agglomerative process of connecting gene expression profiles is rigid and relies on joining the most similar clusters first, any poor clusters generated early in the process affects the quality of clusters later on. Also, unrelated gene expression profiles are eventually joined. Third, it is difficult to define discrete clusters using hierarchical clustering.

2) *Self-organizing maps (SOMs).* SOMs is a partitioning algorithm that requires the user to input the desired number of clusters (centroids; Tamayo *et al.*, 1999). The centroids take the form of a grid of dimensions x times dimensions y (e.g., 2×3) that is overlain on the data in n dimensional expression space. Next, an initial gene is chosen at random and the closest centroid moved toward that gene and the process repeated in turn for all genes, completing one cycle or iteration. Multiple iterations are performed until the program is terminated. As the initial order of gene selection is random, slightly different clusters are generated with each run; consequently, it is important to perform multiple runs in order to identify stable clusters. The disadvantage of SOMs is that all genes are forced into clusters; consequently, outliers can distort optimal clusters.
3) *k-means.* This is also a partitioning algorithm in which the user inputs the desired number of clusters (k) (Tavazoie *et al.*, 1999). The algorithm then chooses k centroids at random that represent random gene expression profiles. Each gene is assigned to the closest centroid until all genes are assigned, then the centroids are reset to the average of their assigned genes (cluster). All genes are then reassigned to the new closest centroid and the process repeated for a defined number of iterations or until no more genes change cluster. Similar to SOMs, k-means generates different clusters in multiple runs depending on the initial centroid positions. Some software programs (e.g., k-means support in MEV; see Table 4.2) take advantage of the random initiation by deliberately rerunning the algorithm multiple times to identify genes that frequently cocluster (i.e., form "consensus" clusters). Here, a user-defined threshold is applied such that gene members are required to cocluster in X% of k-means runs. The advantage of this approach is that robust clusters are generated, giving the user a feel for the significance of clusters. Also, genes that do not commonly cocluster (i.e., outliers) remain unassigned and therefore do not distort existing clusters (see Fig. 4.5C).
4) *Figures of Merit (FOM).* The disadvantage of both SOMs and k-means is that the user is required to estimate the optimal number of clusters to describe the data. The FOM algorithm estimates the predictive power of a clustering algorithm and therefore can be used as a guide for determining the optimal number of clusters (Yeung *et al.*, 2001). For example, FOM can be applied by running k-means repeatedly over a range of cluster numbers. This generates a curve of the predictive power of the clustering algorithm versus the number of clusters. The predictive power increases with the number of clusters; hence the curve can be used to select the optimal number of clusters for running k-means.

In summary, when clustering gene expression profiles, it is very important to use the most variable profiles for clustering and to "explore" the data using several different clustering algorithms in order to obtain robust clusters. The aim is to identify clusters that describe distinct expression patterns and to identify/remove outliers that can detract from the quality of the obtained clusters.

8. Differential Expression Analysis

Specialized statistical methods are required to identify significantly differentially expressed genes from microarray data (Allison *et al.*, 2006; Dudoit *et al.*, 2003). Use of these methods is essential to reliably identify genes that are perturbed by expression of parts in a host. Data from a single microarray experiment without further experimental validation is insufficient to reliably identify differentially expressed genes. This is because application of a fold cutoff does not take into account the uncertainty in measuring gene expression ratios introduced from biological sample variability and from technical issues. For example, expression ratios of weakly expressed genes are often inherently unreliable as their fluorescence measurements have low signal to noise ratios. Consequently, it is essential to perform replicate experiments from *separate biological cultures* (i.e., *biological* rather than *technical* replicates) to enable the construction of test statistics that incorporate variability estimates for each gene. This provides assessment of the statistical significance (e.g., p-value) in the differential expression values for each gene, enables a cutoff to be applied to identify significantly differentially expressed genes, and also provides an estimate of the type-I (false positive) and type-II (false negative) error rates at each applied cutoff. In addition, prior to statistical analysis, all microarray expression data should be normalized and $\log_2$ transformed.

Microarray datasets present unique problems for identifying significantly differentially expressed genes. Statistical analysis of expression measurements of a single gene in condition A versus condition B is relatively straightforward. Here, simple statistics such as the Student's t-test can be used to derive p-values of whether the mean of the replicated log ratios differ from a null hypothesis of 0, in which there is no change in expression in condition A versus condition B. However, this approach is problematic when scaled up to large datasets such as microarrays that contain measurements of 1000s of genes with few experimental replicates. Here, additional methods are used including modified t-statistics and controls for the increased type-I error rate due to multiple testing.

1) *Modified t-statistic.* The t-statistic is derived from the difference in means divided by the sample variance. Due to the few replicates in microarray

experiments, gene-specific variance estimates are imprecise, which can result in highly variable *t*-statistics when applied across 1000s of genes. Specialized microarray statistical methods such as SAM (Tusher *et al.*, 2001), LIMMA (Smyth, 2004), and Cyber-T (Baldi and Long, 2001; see Table 4.2) calculate a modified *t*-statistic in which the estimated gene-specific variance is combined with a predicted variance derived from all genes on the microarray. This improves the estimate of variance for each gene, thereby increasing the power of the *t*-statistic. However, this approach assumes that the null distribution of the test statistics is the same across all transcripts and that all transcripts are independent, which is not necessarily true.

2) *Controlling the type-I error rate.* Applying a cutoff of $p < 0.05$ for replicates of a single gene experiment predicts 1 false positive from 20 independent trials. However, microarrays involve multiple statistical testing of 1000s of genes, which dramatically increases the type-I error (false positive) rate for a given α (p-value). For example, applying a cutoff of $p < 0.05$ for a typical microbial genome of 4500 genes would generate $4500 \times 0.05 = 225$ genes as false positives even if there is no significant differential gene expression. Consequently, specialized statistical methods are applied to microarray data to correct for multiple testing and thereby control for type-I errors. The Bonferroni correction is a simple method for controlling Family-Wise Error Rate (FWER; probability of making type-I errors) at level α, where $\tilde{p}_g \leq \alpha$. Here, the p-value for each gene (p_g) is adjusted: $\tilde{p}_g = Np_g$, where $N =$ number of genes. However, the Bonferroni correction is a very stringent adjustment as it decreases FWER to 0, which can result in missing many true positives (i.e., false negatives, or type-II error rate), and also assumes independence amongst genes. There are multiple methods for controlling type-I error. The Šidák procedure, min P and max T (Westfall and Young, 1993) adjust p-values to control FWER (see also Dudoit *et al.*, 2003). However, often it is more useful to have an estimate of the false detection rate (FDR): that is, for a given threshold, what fraction are false positives (Benjamini and Hochberg, 1995). Methods include controlling FDR below a certain level by adjusting p-values based on their ranking (Benjamini and Hochberg, 1995), and mixture-models that treat genes as either as differentially or not differentially expressed (Allison *et al.*, 2002; Datta, 2005; Do *et al.*, 2005; Pounds and Morris, 2003). One popular microarray statistical tool, SAM (Statistical Analysis of Microarrays; Tusher *et al.*, 2001), provides a method for estimating FDR for a chosen cutoff value of test statistic (Storey, 2002). This is achieved by permuting the datasets to determine if the expression of any of the genes is significantly related to the response. Any test statistic of the permuted dataset exceeding the cutoff is counted as a "false positive." The appealing feature of SAM is that the user chooses what cutoff to apply based on an FDR they are

comfortable in dealing with in their significant data set. Recent versions of SAM have also incorporated a "Miss rate" table that also estimates the false negative rate of genes that do not make the cutoff (Taylor *et al.*, 2005). In addition, SAM generates a *q*-value for each gene, which describes the lowest FDR rate at which the gene is called significant (Storey, 2002).

Most microarray statistical analysis is *one-class* where only one response variable is tested; for example, condition *A* versus condition *B*. The one-class problem tests whether the mean log expression ratios differ from the null hypothesis of 0. It is also possible to perform *two-class* comparisons; for example, condition *A* versus condition *B* (expt 1) compared against condition *A* versus condition *C* (expt 2), where the mean log ratios of expt 1 are compared for significant difference against the mean log ratios of expt 2. This is useful to compare if there are significant difference between different parts when compared to a common control. It is also possible to apply statistical analysis to time-course data to identify significant differentially expressed genes using either a one-class or two-class comparison based on the consistent increase or decrease in gene expression over time (e.g., SAM; Table 4.2).

A common question for statistical analysis of microarray data is: How many replicates are sufficient? Most algorithms require at least four biological replicates to obtain reasonable statistics. However, more replicates increase the statistical power: that is, maximizing the detection of true positives and minimizing FDR. Several approaches are available for estimating sample size (number of replicates) based on the observed variability of gene expression ratios in pilot experiments in order to achieve a desired statistical power (Lee and Whitmore, 2002; Li *et al.*, 2005; Tibshirani, 2006; software SAM, R/size; Table 4.2).

Finally, it is common in many microarray experiments for statistical analysis to yield long lists of significant genes. Consequently, some users to employ both a statistical and a fold cutoff to reduce their candidate list and also employing the reasoning that genes with high-expression ratios are more likely to be directly regulated and therefore easier to biologically interpret. Note that, for published reports, it is necessary to describe the statistics and expected number of false positives within the significant dataset.

9. Data Analysis: Understanding the Perturbation

Biological interpretation of candidate gene lists identified through clustering and by significant differential expression is essential in order to identity the biological processes or systems perturbed by expression of parts. Expression of regulators may result in general aberrant ectopic gene expression due to recognition of miscellaneous sites throughout the genome.

However, expression of circuit devices may target specific cellular processes that will likely be reflected in the expression patterns. Several approaches can be used to identify candidate cellular processes (see Table 4.4).

1) *Metabolic pathways.* Expression data can be overlain on metabolic maps in order to identify genes in reactions or pathways that are differentially regulated.

Table 4.4 Resources for biological interpretation

Resource	Function	URL
Metabolic pathways		
KEGG	Kyoto Encyclopedia of Genes and Genomes database providing a metabolic pathway viewer for overlaying gene expression data	http://www.genome.jp/kegg/
EcoCyc	Model organism database for *E. coli* providing a metabolic pathway viewer for overlaying gene expression data	http://biocyc.org/ecocyc/
Functional categories		
GO	Gene Ontology website provides gene functional classification and tools to analyze overrepresentation among lists of genes	http://www.geneontology.org/
Protein–protein interactions		
EcoCyc	Model organism database for *E. coli* providing experimentally verified protein–protein interaction networks	http://biocyc.org/ecocyc/
Transcriptional networks and motif-finding algorithms		
EcoCyc	Model organism database for *E. coli* providing experimentally verified transcription networks	http://biocyc.org/ecocyc/
BioProspector	Motif-finding algorithm suited for two-block motifs	http://ai.stanford.edu/~xsliu/BioProspector/
Consensus	Motif-finding algorithm	http://bifrost.wustl.edu/consensus/
Gibbs Motif Sampler	Motif-finding algorithm	http://bayesweb.wadsworth.org/gibbs/gibbs.html
MEME	Motif-finding algorithm	http://meme.sdsc.edu/meme/intro.html

2) *Functional categories.* A common query is whether particular functional groups are enriched within datasets. A useful classifier for this purpose is Gene Ontology (GO) that categorizes genes according to their associated biological processes, cellular components and molecular functions. A Fisher's exact or chi-square test can be performed to determine if a particular GO term is overrepresented within a set of differentially expressed genes compared to the whole genome (additional tools are listed in Table 4.4). There are several limitations to this approach: an arbitrary cutoff is required to identify differentially expressed genes; many differentially expressed genes are required to provide robust statistics; nondifferentially expressed genes are not used; the level of gene expression is not incorporated. A better alternative is to ask whether the genes associated with a GO term are "differentially expressed" within the dataset. This is termed Gene Set Enrichment Analysis (GSEA) and can be tested using Wilcoxon rank sum test, modified Kolmogorov-Smirnov statistic (Subramanian *et al.*, 2005) or using tools available in various software packages (see Table 4.4).
3) *Protein interactions.* High-throughput studies of protein–protein interactions have revealed that proteins involved in the same process often interact (Arifuzzaman *et al.*, 2006; Butland *et al.*, 2005). These interaction maps can be used to determine if any protein networks are overrepresented within the expression data; however, these datasets may have many false positives as there is little overlap between them. Alternatively, the EcoCyc database has a collection of low throughput (experimentally verified) data (Table 4.4).
4) *Transcriptional networks.* Coregulated genes are often controlled by common transcription factors. Consequently, it is often useful to search for overrepresented motifs within the promoter regions of genes that cocluster or are differentially expressed (see algorithms listed in Table 4.4).

10. Closing Remarks

The key to successful microarray experiments are careful experiment design that enables the user to capture the direct effects of the introduced perturbation, in this case, the effect of expressing parts within a host. Systematic analysis of multiple parts requires the use of carefully defined growth conditions and control samples to enable cross-comparison of expression data, both within and between laboratories. Equally important is careful data analysis in order to maximize interpretation of the data. This requires knowledge of the assumptions behind data normalization, clustering, identification of significantly differentially expressed genes to select gene lists, and also biological interpretation to identify known cellular systems that are being modulated. The goal is to be able to identify and understand the cellular stress circuits that are being triggered by expression of parts of circuits. This enables the selection of parts that have orthogonal effects on the host to minimize the impact of the fully engineered circuit, and also to facilitate modifying the host to better tolerate the genetic circuit. Finally, it is important to accurately document the microarray experiment; both as a requirement for publication and also to enable others to repeat, modify, or critically evaluate your work. Sadly, this is a critical problem in the field; a recent study was unable to reproduce 10/18 published microarray experiments (Ioannidis *et al.*, 2009). Most journals require MIAME compliance (Minimum information about a microarray experiment) for publication of microarray data (Ball *et al.*, 2004; Brazma *et al.*, 2001). This requires uploading array data onto public repositories such as GEO (http://www.ncbi.nlm.nih.gov/geo/), ArrayExpress (http://www.ebi.ac.uk/microarray-as/ae/), and CIBEX (http://cibex.nig.ac.jp/). All these databases require detailed documentation of experimental design, samples and their preparation, hybridization, array design, and data measurement and analysis.

REFERENCES

Allison, D. B., Gadbury, G. L., Heo, M. S., Fernandez, J. R., Lee, C. K., Prolla, T. A., and Weindruch, R. (2002). A mixture model approach for the analysis of microarray gene expression data. *Comput. Stat. Data Anal.* **39,** 1–20.

Allison, D. B., Cui, X., Page, G. P., and Sabripour, M. (2006). Microarray data analysis: From disarray to consolidation and consensus. *Nat. Rev. Genet.* **7,** 55–65.

Arifuzzaman, M., Maeda, M., Itoh, A., Nishikata, K., Takita, C., Saito, R., Ara, T., Nakahigashi, K., Huang, H. C., Hirai, A., Tsuzuki, K., Nakamura, S., *et al.* (2006). Large-scale identification of protein-protein interaction of *Escherichia coli* K-12. *Genome Res.* **16,** 686–691.

Ausubel, F. M., Brent, R., Kingston, R. E., Moore, D. D., Seidman, J. G., and Struhl, K. (1998). Current Protocols in Molecular Biology. John Wiley & Sons, Inc, Hoboken, NJ, USA.

Baldi, P., and Long, A. D. (2001). A Bayesian framework for the analysis of microarray expression data: Regularized t -test and statistical inferences of gene changes. *Bioinformatics* **17,** 509–519.

Ball, C. A., Brazma, A., Causton, H., Chervitz, S., Edgar, R., Hingamp, P., Matese, J. C., Parkinson, H., Quackenbush, J., Ringwald, M., Sansone, S. A., Sherlock, G., *et al.* (2004). Submission of microarray data to public repositories. *PLoS Biol.* **2,** E317.

Bammler, T., Beyer, R. P., Bhattacharya, S., Boorman, G. A., Boyles, A., Bradford, B. U., Bumgarner, R. E., Bushel, P. R., Chaturvedi, K., Choi, D., Cunningham, M. L., Deng, S., *et al.* (2005). Standardizing global gene expression analysis between laboratories and across platforms. *Nat. Methods* **2,** 351–356.

Beisel, C. L., and Storz, G. (2010). Base pairing small RNAs and their roles in global regulatory networks. *FEMS Microbiol. Rev.* **34,** 866–882.

Benjamini, Y., and Hochberg, Y. (1995). Controlling the false discovery rate—A practical and powerful approach to multiple testing. *J. R. Stat. Soc. B Methodol.* **57,** 289–300.

Beyhan, S., and Yildiz, F. (2007). Bacterial gene expression analysis using microarrays. *J. Vis. Exp.* **4,** 206.

Bolstad, B. M., Irizarry, R. A., Astrand, M., and Speed, T. P. (2003). A comparison of normalization methods for high density oligonucleotide array data based on variance and bias. *Bioinformatics* **19,** 185–193.

Botwell, D., and Sambrook, J. (2003). DNA Microarrays: A Molecular Cloning Manual. Cold Spring Harbor Press, New York.

Boutros, P. C., and Okey, A. B. (2005). Unsupervised pattern recognition: An introduction to the whys and wherefores of clustering microarray data. *Brief. Bioinform.* **6,** 331–343.

Branham, W. S., Melvin, C. D., Han, T., Desai, V. G., Moland, C. L., Scully, A. T., and Fuscoe, J. C. (2007). Elimination of laboratory ozone leads to a dramatic improvement in the reproducibility of microarray gene expression measurements. *BMC Biotechnol.* **7,** 8.

Brazma, A., Hingamp, P., Quackenbush, J., Sherlock, G., Spellman, P., Stoeckert, C., Aach, J., Ansorge, W., Ball, C. A., Causton, H. C., Gaasterland, T., Glenisson, P., *et al.* (2001). Minimum information about a microarray experiment (MIAME)-toward standards for microarray data. *Nat. Genet.* **29,** 365–371.

Butland, G., Peregrin-Alvarez, J. M., Li, J., Yang, W., Yang, X., Canadien, V., Starostine, A., Richards, D., Beattie, B., Krogan, N., Davey, M., Parkinson, J., *et al.* (2005). Interaction network containing conserved and essential protein complexes in Escherichia coli. *Nature* **433,** 531–537.

Cho, B. K., Zengler, K., Qiu, Y., Park, Y. S., Knight, E. M., Barrett, C. L., Gao, Y., and Palsson, B. O. (2009). The transcription unit architecture of the *Escherichia coli* genome. *Nat. Biotechnol.* **27,** 1043–1049.

Czar, M. J., Anderson, J. C., Bader, J. S., and Peccoud, J. (2009). Gene synthesis demystified. *Trends Biotechnol.* **27,** 63–72.

Datta, S. (2005). Empirical Bayes screening of many p-values with applications to microarray studies. *Bioinformatics* **21,** 1987–1994.

D'Haeseleer, P. (2005). How does gene expression clustering work? *Nat. Biotechnol.* **23,** 1499–1501.

Do, K. A., Muller, P., and Tang, F. (2005). A Bayesian mixture model for differential gene expression. *J. R. Stat. Soc. C Appl. Stat.* **54,** 627–644.

Dudoit, S., Shaffer, J. P., and Boldrick, J. C. (2003). Multiple hypothesis testing in microarray experiments. *Stat. Sci.* **18,** 71–103.

Eisen, M. B., Spellman, P. T., Brown, P. O., and Botstein, D. (1998). Cluster analysis and display of genome-wide expression patterns. *Proc. Natl. Acad. Sci. USA* **95,** 14863–14868.

Elowitz, M. B., and Leibler, S. (2000). A synthetic oscillatory network of transcriptional regulators. *Nature* **403,** 335–338.

Fare, T. L., Coffey, E. M., Dai, H., He, Y. D., Kessler, D. A., Kilian, K. A., Koch, J. E., LeProust, E., Marton, M. J., Meyer, M. R., Stoughton, R. B., Tokiwa, G. Y., *et al.* (2003). Effects of atmospheric ozone on microarray data quality. *Anal. Chem.* **75,** 4672–4675.

Filiatrault, M. J., Stodghill, P. V., Bronstein, P. A., Moll, S., Lindeberg, M., Grills, G., Schweitzer, P., Wang, W., Schroth, G. P., Luo, S., Khrebtukova, I., Yang, Y., *et al.* (2010). Transcriptome analysis of *Pseudomonas syringae* identifies new genes, noncoding RNAs, and antisense activity. *J. Bacteriol.* **192,** 2359–2372.

Gao, H., Yang, Z. K., Gentry, T. J., Wu, L., Schadt, C. W., and Zhou, J. (2007). Microarray-based analysis of microbial community RNAs by whole-community RNA amplification. *Appl. Environ. Microbiol.* **73,** 563–571.

Gibson, D. G., Benders, G. A., Axelrod, K. C., Zaveri, J., Algire, M. A., Moodie, M., Montague, M. G., Venter, J. C., Smith, H. O., and Hutchison, C. A., 3rd (2008). One-step assembly in yeast of 25 overlapping DNA fragments to form a complete synthetic *Mycoplasma genitalium* genome. *Proc. Natl. Acad. Sci. USA* **105,** 20404–20409.

Gibson, D. G., Young, L., Chuang, R. Y., Venter, J. C., Hutchison, C. A., 3rd, and Smith, H. O. (2009). Enzymatic assembly of DNA molecules up to several hundred kilobases. *Nat. Methods* **6,** 343–345.

Guell, M., van Noort, V., Yus, E., Chen, W. H., Leigh-Bell, J., Michalodimitrakis, K., Yamada, T., Arumugam, M., Doerks, T., Kuhner, S., Rode, M., Suyama, M., *et al.* (2009). Transcriptome complexity in a genome-reduced bacterium. *Science* **326,** 1268–1271.

Ham, T. S., Lee, S. K., Keasling, J. D., and Arkin, A. P. (2008). Design and construction of a double inversion recombination switch for heritable sequential genetic memory. *PLoS ONE* **3,** e2815.

Hemm, M. R., Paul, B. J., Schneider, T. D., Storz, G., and Rudd, K. E. (2008). Small membrane proteins found by comparative genomics and ribosome binding site models. *Mol. Microbiol.* **70,** 1487–1501.

Hemm, M. R., Paul, B. J., Miranda-Rios, J., Zhang, A., Soltanzad, N., and Storz, G. (2010). Small stress response proteins in *Escherichia coli*: Proteins missed by classical proteomic studies. *J. Bacteriol.* **192,** 46–58.

Ioannidis, J. P., Allison, D. B., Ball, C. A., Coulibaly, I., Cui, X., Culhane, A. C., Falchi, M., Furlanello, C., Game, L., Jurman, G., Mangion, J., Mehta, T., *et al.* (2009). Repeatability of published microarray gene expression analyses. *Nat. Genet.* **41,** 149–155.

Irizarry, R. A., Bolstad, B. M., Collin, F., Cope, L. M., Hobbs, B., and Speed, T. P. (2003a). Summaries of Affymetrix GeneChip probe level data. *Nucleic Acids Res.* **31,** e15.

Irizarry, R. A., Hobbs, B., Collin, F., Beazer-Barclay, Y. D., Antonellis, K. J., Scherf, U., and Speed, T. P. (2003b). Exploration, normalization, and summaries of high density oligonucleotide array probe level data. *Biostatistics* **4,** 249–264.

Irizarry, R. A., Warren, D., Spencer, F., Kim, I. F., Biswal, S., Frank, B. C., Gabrielson, E., Garcia, J. G., Geoghegan, J., Germino, G., Griffin, C., Hilmer, S. C., *et al.* (2005). Multiple-laboratory comparison of microarray platforms. *Nat. Methods* **2,** 345–350.

Larkin, J. E., Frank, B. C., Gavras, H., Sultana, R., and Quackenbush, J. (2005). Independence and reproducibility across microarray platforms. *Nat. Methods* **2,** 337–344.

Lee, M. L., and Whitmore, G. A. (2002). Power and sample size for DNA microarray studies. *Stat. Med.* **21,** 3543–3570.

Lee, S. K., Chou, H., Ham, T. S., Lee, T. S., and Keasling, J. D. (2008). Metabolic engineering of microorganisms for biofuels production: From bugs to synthetic biology to fuels. *Curr. Opin. Biotechnol.* **19,** 556–563.

Li, S. S., Bigler, J., Lampe, J. W., Potter, J. D., and Feng, Z. (2005). FDR-controlling testing procedures and sample size determination for microarrays. *Stat. Med.* **24,** 2267–2280.

Nonaka, G., Blankschien, M., Herman, C., Gross, C. A., and Rhodius, V. A. (2006). Regulon and promoter analysis of the *E. coli* heat-shock factor, sigma32, reveals a multifaceted cellular response to heat stress. *Genes Dev.* **20,** 1776–1789.

Pieterse, B., Jellema, R. H., and van der Werf, M. J. (2006). Quenching of microbial samples for increased reliability of microarray data. *J. Microbiol. Methods* **64,** 207–216.

Pounds, S., and Morris, S. W. (2003). Estimating the occurrence of false positives and false negatives in microarray studies by approximating and partitioning the empirical distribution of p-values. *Bioinformatics* **19,** 1236–1242.

Quackenbush, J. (2001). Computational analysis of microarray data. *Nat. Rev. Genet.* **2,** 418–427.

Reich, M., Liefeld, T., Gould, J., Lerner, J., Tamayo, P., and Mesirov, J. P. (2006). GenePattern 2.0. *Nat. Genet.* **38,** 500–501.

Rhodius, V. A., and Wade, J. T. (2009). Technical considerations in using DNA microarrays to define regulons. *Methods* **47,** 63–72.

Rhodius, V., Van Dyk, T. K., Gross, C., and LaRossa, R. A. (2002). Impact of genomic technologies on studies of bacterial gene expression. *Annu. Rev. Microbiol.* **56,** 599–624.

Rhodius, V. A., Suh, W. C., Nonaka, G., West, J., and Gross, C. A. (2006). Conserved and variable functions of the sigma(E) stress response in related genomes. *PLoS Biol.* **4,** 43–59.

Saeed, A. I., Bhagabati, N. K., Braisted, J. C., Liang, W., Sharov, V., Howe, E. A., Li, J., Thiagarajan, M., White, J. A., and Quackenbush, J. (2006). TM4 microarray software suite. *Methods Enzymol.* **411,** 134–193.

Selinger, D. W., Cheung, K. J., Mei, R., Johansson, E. M., Richmond, C. S., Blattner, F. R., Lockhart, D. J., and Church, G. M. (2000). RNA expression analysis using a 30 base pair resolution *Escherichia coli* genome array. *Nat. Biotechnol.* **18,** 1262–1268.

Sharma, C. M., Hoffmann, S., Darfeuille, F., Reignier, J., Findeiss, S., Sittka, A., Chabas, S., Reiche, K., Hackermuller, J., Reinhardt, R., Stadler, P. F., and Vogel, J. (2010). The primary transcriptome of the major human pathogen Helicobacter pylori. *Nature* **464,** 250–255.

Shi, L., Reid, L. H., Jones, W. D., Shippy, R., Warrington, J. A., Baker, S. C., Collins, P. J., de Longueville, F., Kawasaki, E. S., Lee, K. Y., Luo, Y., Sun, Y. A., *et al.* (2006). The MicroArray Quality Control (MAQC) project shows inter- and intraplatform reproducibility of gene expression measurements. *Nat. Biotechnol.* **24,** 1151–1161.

Shi, L., Perkins, R. G., Fang, H., and Tong, W. (2008). Reproducible and reliable microarray results through quality control: Good laboratory proficiency and appropriate data analysis practices are essential. *Curr. Opin. Biotechnol.* **19,** 10–18.

Smyth, G. K. (2004). Linear models and empirical bayes methods for assessing differential expression in microarray experiments. *Stat. Appl. Genet. Mol. Biol.* **3,** Article3.

Sorek, R., and Cossart, P. (2010). Prokaryotic transcriptomics: A new view on regulation, physiology and pathogenicity. *Nat. Rev. Genet.* **11,** 9–16.

Steen, E. J., Kang, Y., Bokinsky, G., Hu, Z., Schirmer, A., McClure, A., Del Cardayre, S. B., and Keasling, J. D. (2010). Microbial production of fatty-acid-derived fuels and chemicals from plant biomass. *Nature* **463,** 559–562.

Storey, J. D. (2002). A direct approach to false discovery rates. *J. R. Stat. Soc. B Stat. Methodol.* **64,** 479–498.

Subramanian, A., Tamayo, P., Mootha, V. K., Mukherjee, S., Ebert, B. L., Gillette, M. A., Paulovich, A., Pomeroy, S. L., Golub, T. R., Lander, E. S., and Mesirov, J. P. (2005). Gene set enrichment analysis: A knowledge-based approach for interpreting genome-wide expression profiles. *Proc. Natl. Acad. Sci. USA* **102,** 15545–15550.

Tabor, J. J., Salis, H. M., Simpson, Z. B., Chevalier, A. A., Levskaya, A., Marcotte, E. M., Voigt, C. A., and Ellington, A. D. (2009). A synthetic genetic edge detection program. *Cell* **137,** 1272–1281.

Tamayo, P., Slonim, D., Mesirov, J., Zhu, Q., Kitareewan, S., Dmitrovsky, E., Lander, E. S., and Golub, T. R. (1999). Interpreting patterns of gene expression with self-organizing maps: Methods and application to hematopoietic differentiation. *Proc. Natl. Acad. Sci. USA* **96,** 2907–2912.

Tavazoie, S., Hughes, J. D., Campbell, M. J., Cho, R. J., and Church, G. M. (1999). Systematic determination of genetic network architecture. *Nat. Genet.* **22,** 281–285.

Taylor, J., Tibshirani, R., and Efron, B. (2005). The "miss rate" for the analysis of gene expression data. *Biostatistics* **6,** 111–117.

Thomason, M. K., and Storz, G. (2010). Bacterial antisense RNAs: How many are there, and what are they doing? *Annu. Rev. Genet.* **44,** 167–188.

Tian, J., Ma, K., and Saaem, I. (2009). Advancing high-throughput gene synthesis technology. *Mol. Biosyst.* **5,** 714–722.

Tibshirani, R. (2006). A simple method for assessing sample sizes in microarray experiments. *BMC Bioinform.* **7,** 106.

Tusher, V. G., Tibshirani, R., and Chu, G. (2001). Significance analysis of microarrays applied to the ionizing radiation response. *Proc. Natl. Acad. Sci. USA* **98,** 5116–5121.

Voigt, C. A. (2006). Genetic parts to program bacteria. *Curr. Opin. Biotechnol.* **17,** 548–557.
Wang, Z., Gerstein, M., and Snyder, M. (2009). RNA-Seq: A revolutionary tool for transcriptomics. *Nat. Rev. Genet.* **10,** 57–63.
Westfall, P. H., and Young, S. S. (1993). Resampling-Based Multiple Testing: Examples and Methods for p-Value Adjustment. Wiley, New York.
Wettenhall, J. M., and Smyth, G. K. (2004). limmaGUI: A graphical user interface for linear modeling of microarray data. *Bioinformatics* **20,** 3705–3706.
Yang, Y. H., Dudoit, S., Luu, P., Lin, D. M., Peng, V., Ngai, J., and Speed, T. P. (2002). Normalization for cDNA microarray data: A robust composite method addressing single and multiple slide systematic variation. *Nucleic Acids Res.* **30,** e15.
Yeung, K. Y., Haynor, D. R., and Ruzzo, W. L. (2001). Validating clustering for gene expression data. *Bioinformatics* **17,** 309–318.

CHAPTER FIVE

Orthogonal Gene Expression in *Escherichia coli*

Wenlin An[1] *and* Jason W. Chin

Contents

Abstract

Here, we describe a route orthogonal gene expression which combines orthogonal transcription and translation using library-based selections. We show how orthogonal gene expression can be used to create a minimal orthogonal ribosome and describe how to create orthogonal transcription–translation feed forward loops that introduce tailored information processing delays into gene expression.

Medical Research Council Laboratory of Molecular Biology, Hills Road, Cambridge, United Kingdom
[1] Current address: MRC Centre for Developmental Neurobiology, New Hunt's House, King's College London, London, United Kingdom

Methods in Enzymology, Volume 497
ISSN 0076-6879, DOI: 10.1016/B978-0-12-385075-1.00005-6

1. Introduction

Synthetic biology aims to embed engineered systems into cells to perform new and potentially useful functions (Chin, 2006; de Lorenzo and Danchin, 2008; Endy, 2005; Greber and Fussenegger, 2007; Hasty *et al.*, 2001,2002; Isaacs *et al.*, 2006; Kaern *et al.*, 2003; Kampf and Weber, 2010; Kohanski and Collins, 2008; Serrano, 2007; Sprinzak and Elowitz, 2005; Young and Alper, 2010). A central challenge in synthetic biology is to engineer and embed synthetic systems in cells that take full advantage of the host cell's abilities, but are not limited by the host cell's regulatory networks or evolutionary history. Nowhere is this challenge more acute than in the fundamental process of gene expression, in which genetic information is copied and decoded to produce the networks of molecules that mediate biological function. Selective abstraction of gene expression in synthetic networks from cellular gene expression and its associated regulatory processes would release biology for more effective engineering. The construction of orthogonal gene expression systems, in which the operation of converting the information in DNA into proteins is executed entirely by components that are functionally insulated from the endogenous gene expression machinery, would provide a compact solution to the selective abstraction of gene expression. Gene expression in bacteria relies on the coupled transcription of a DNA template by a DNA-dependent RNA polymerase (RNAP) and translation of the transcript by ribosomes. We realized that orthogonal gene expression might be achieved by the discovery, invention, and integration of components for orthogonal transcription and translation.

1.1. Discovery of orthogonal ribosome–orthogonal mRNA pairs

We have described the evolution and characterization of orthogonal ribosome (O-ribosome)–orthogonal mRNA (O-mRNA) pairs in *Escherichia coli* (Rackham and Chin, 2005a). In these pairs, the O-ribosome efficiently and specifically translates its cognate O-mRNA, which is not a substrate for the endogenous wild-type ribosome (wt-ribosome; Fig. 5.1). The specificity of O-ribosomes for the translation of an O-mRNA arises from the altered 16S rRNA in the O-ribosome that recognizes an altered Shine–Dalgarno sequence in the leader sequence of the O-mRNA in the initiation phase of translation (Ramakrishnan, 2002). In previous work, O-ribosomes have been evolved that decode genetic information in O-mRNAs in new ways. In combination with orthogonal aminoacyl–tRNA synthetases and tRNAs that recognize unnatural amino acids, the evolved O-ribosomes have

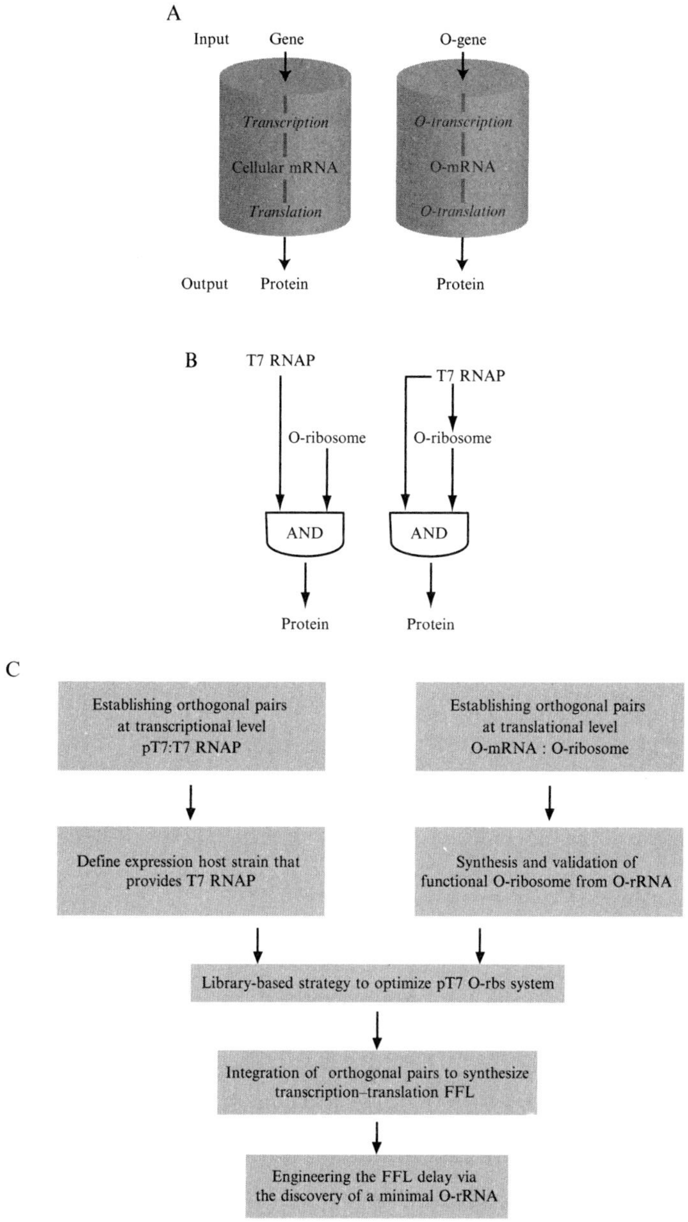

Figure 5.1 Overview of the strategy for synthesis orthogonal transcription–translation networks. (A) Orthogonal gene expression is insulated from cellular gene expression, and operates in parallel with, but independent of endogenous transcription–translation system. (B) The orthogonal AND function (left) and an orthogonal transcription–translation feed forward loop (FFL) (right), both display Boolean AND logic (Bottom). (C) From orthogonal components to programmed delays in gene expression.

allowed us to begin to undo the "frozen accident" of the natural genetic code and direct the efficient incorporation of multiple unnatural amino acids into proteins encoded on O-mRNAs (Neumann *et al.*, 2010; Wang *et al.*, 2007; Xie and Schultz, 2006). O-ribosomes have been also used to create new translational Boolean logic functions that would not be possible to create by using the essential cellular ribosome (Rackham and Chin, 2005b) and to define functionally important nucleotides in the structurally defined interface between the two subunits of the ribosome (Rackham *et al.*, 2006).

1.2. Discovery of orthogonal T7 RNAP-T7 promoter pairs (T7 RNAP: pT7)

T7 RNAP is a small (99 kDa) DNA-dependent RNAP derived from bacteriophage T7 (Chamberlin and Ring, 1973; Golomb and Chamberlin, 1974; Steitz, 2004). The polymerase efficiently and specifically transcribes genes bearing a T7 promoter (PT7). In the absence of T7 RNAP, the promoter does not direct transcription by endogenous polymerases in *E. coli* (Basu and Maitra, 1986). T7 RNAP and its cognate promoters are therefore a natural orthogonal polymerase–promoter pair for transcription in *E. coli* (Fig. 5.1). We realized that it might be possible to direct the transcription and translation of a gene by using a T7 promoter and an O-ribosome binding site (O-rbs) to create an orthogonal transcription–translation system. The resulting gene would be transcribed only to its corresponding mRNA in the presence of T7 RNAP and would be translated only to its encoded protein product in the presence of the O-ribosome, creating an orthogonal gene expression pathway in the cell (Fig. 5.1) that relies on AND logic (Fig. 5.1). Because T7 RNAP and the O-ribosome, unlike the cell's endogenous RNAPs and ribosome, are not responsible for the synthesis of the cell's proteome from its genome, the orthogonal gene expression pathway opens the possibility of inventing and exploring new modes of gene regulation.

1.3. Integration of orthogonal transcription–translation pairs into gene expression networks

Studies on transcriptional gene regulatory networks and other information-processing networks, including the worldwide web, electronic circuits, and the neuronal network of *Caenorhabditis elegans*, indicate that the type 1 coherent feed-forward loop (FFL) is an important module in these networks (Milo *et al.*, 2002). The FFL consists of three components (X, Y, and Z) in which X directly activates Y, and both X and Y are required to activate Z (Mangan and Alon, 2003; Milo *et al.*, 2002). Given the importance of transcriptional FFLs in controlling the timing of gene expression, an important property in almost all aspects of natural and synthetic biology from cell cycle control, developmental control, and circadian control to synthetic dynamic circuits, switches, and oscillators (Kaern *et al.*, 2003), we became interested in synthesizing and characterizing orthogonal transcription–translation FFLs. We define an orthogonal transcription–translation FFL (An and Chin, 2009; Fig. 5.1), as a network in which an orthogonal RNAP (T7 RNAP) transcribes the orthogonal rRNA necessary for the production of O-ribosomes and transcribes a mRNA bearing an O-rbs. The O-ribosome then translates the orthogonal message to produce the output protein. Unlike natural transcriptional FFLs, in which two transcription factors act to produce a single transcript (Mangan and Alon, 2003), the orthogonal transcription–translation FFL is activated at sequential, but coupled steps in gene expression, leading to a short cascade.

Here, we describe the implementation of an orthogonal gene expression pathway in *E. coli* (An and Chin, 2009). We integrate the orthogonal building blocks (transcription–translation pairs) to create orthogonal gene expression networks, including transcription–translation FFLs, and examine their dynamic properties (An and Chin, 2009; Kwok, 2010). We used library-based strategies to screen for compatible control elements that operate in a coordinated manner (An and Chin, 2009). We go on to demonstrate that the transcription–translation networks, which could not be created by using host polymerases or endogenous ribosomes, allow the introduction of distinct delays into gene expression that have not been demonstrated in natural systems (An and Chin, 2009). In the process of creating these networks, we refactor (Chan *et al.*, 2005) the rRNA operon (rrnB) to uncouple O-16S rRNA synthesis and processing from the synthesis and processing of the rest of the rrnB and define a minimal module for O-ribosome production in cells (An and Chin, 2009). The minimal O-ribosome allows us to rationally alter the delay in gene expression (An and Chin, 2009).

2. High-Throughput Screening for Orthogonal T7 Promoter O-rbs System

To create orthogonal gene expression modules, we required an upstream genetic element that would respond specifically and efficiently to T7 RNA polymerase (T7 RNAP) and an O-ribosome. We use Bl21 (DE3) competent cells (Novagen) as host cell strain that contain an IPTG inducible T7 RNAP gene in chromosome, which is under control of the *lac*UV5 promoter in chromosome. In these experiments, O-rRNA for the O-ribosome is produced from pSC101*O-rDNA under control of constitutive promoter P1P2. Initial experiments indicated that a linear combination of the pET vectors leader sequence (that normally directs transcription by T7 RNAP) and an O-rbs sequence does not lead to high-level GFP expression in cells containing both T7 RNAP and the O-ribosome. Because both the T7 promoter and the O-rbs are active in several other contexts, and transcription and translation are coupled processes in bacteria, it seemed reasonable that the sequence between the promoters and ribosome-binding sites might be important for efficient gene expression. We therefore decided to combinatorially optimize the sequence between the T7 promoter and the O-rbs for T7 RNAP-dependent and O-ribosome-dependent gene expression.

To optimize the pT7 O-rbs construct we created a 10^9 member library, in which a 15-nt stretch 3′ to the T7 promoter (Basu and Maitra, 1986) and 5′ to the O-rbs is randomized to all possible combinations (Fig. 5.2). The resulting library (T7n15GFPlib) was screened for O-ribosome-dependent expression from the T7 promoter by three rounds of FACS (see Fig. 5.2). In a first round of positive FACS sorting, we screened for expression of GFP in the presence of O-ribosomes and T7 RNAP. To achieve this, we transformed the T7n15GFP library into cells that constitutively produce

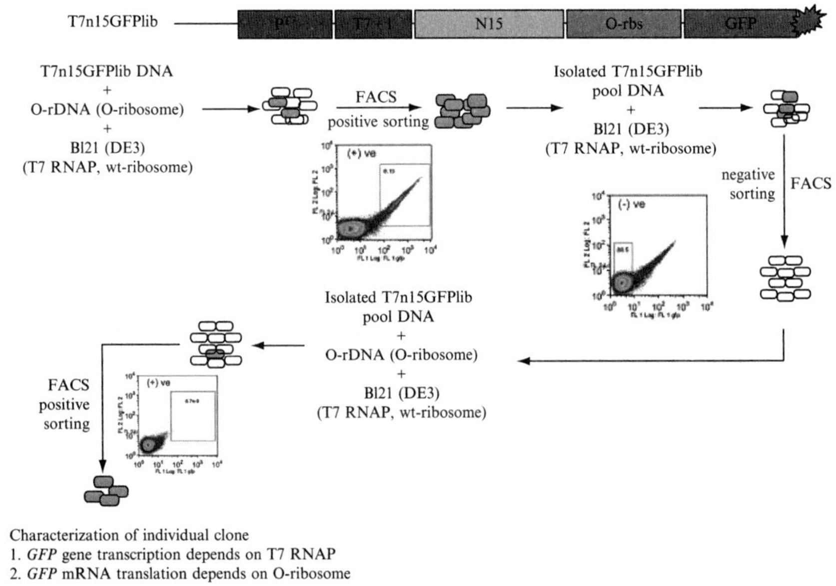

Figure 5.2 Flowchart of high-throughput screening for a T7 promoter O-rbs system. FACS charts shows the patterns of three rounds of screening for a module that specifically and efficiently directs transcription by T7 RNAP and translation by the O-ribosome of a target gene.

O-ribosomes (by virtue of expressing O-rRNA from pSC101*O-ribosome) and that express T7 RNAP. We collected fluorescent cells and isolated the pool of T7n15GFP variants. To remove T7n15GFP variants from the pool that direct expression of GFP by the wt-ribosome, we performed a second round of negative FACS sorting. We transformed the pool of T7n15GFP library members from the positive FACS sort into cells that do not produce O-ribosomes, but do express T7 RNAP and wt-ribosomes and collected cells that do not express GFP, and are therefore not fluorescent. We isolated the resulting T7n15GFP clones and performed a third round of positive FACS sorting. The fluorescence of 96 T7n15GFP clones surviving all three rounds of sorting was examined in cells containing T7 RNAP and the O-ribosome.

The eight clones exhibiting the greatest fluorescence were examined further. The expression of GFP from all eight clones was strongly O-ribosome-dependent and T7 RNAP-dependent. Sequencing the T7n15GFP variants revealed five distinct sequences. In all of the selected sequences, the randomized region was very rich in A and T. The selected sequences may minimize spurious RNA–RNA interactions with the O-rbs-containing sequence. However, because transcription and translation are coupled in bacteria, we cannot rule out more sophisticated explanations in which these sequences modulate coupling. Although all of the selected sequences retain the −9 to +1 sequence most important for transcriptional efficiency *in vitro* (Imburgio *et al.*, 2000), four of these sequences contain deletions in the −11 to−17 region of the promoter where most point mutations have a modest effect on the efficiency of the T7 promoter *in vitro* (Imburgio *et al.*, 2000). Indeed, Northern blots of the T7 RNAP-dependent GFP transcript demonstrate that comparable transcript accumulates with each of the selected T7N15lib sequences. A single selected sequence (T71504) had a wild-type T7 promoter and displayed excellent O-ribosome dependence, so we decided to characterize this sequence in more detail.

To begin to demonstrate the portability of the selected T7 promoter O-rbs combination and confirm that the system shows Boolean AND logic we replaced the *GFP* gene with a *GST–GFP* fusion (creating pT7 O-rbs–GST–GFP). The GST–GFP fusion protein was produced only in the presence of both O-ribosomes and T7 RNAP, as demonstrated by both the level of GFP fluorescence and the purification of GST–GFP from cells that contain the O-ribosome and T7 RNAP, but not from cells containing any other combination of O-ribosomes and T7 RNAP. In addition, the GST–GFP mRNA was produced only in the presence of T7 RNAP. These experiments demonstrate that we have created a genetic element that is heritable in, but unreadable by, the host cell. This genetic element is efficiently transcribed and translated by the orthogonal polymerase and ribosome.

3. Integration of Orthogonal Pairs to Synthesize Transcription–Translation FFL

To construct a transcription–translation FFL, we required an O-16S rRNA that is transcribed from a T7 promoter and assembled into O-ribosomes. Because the mutations in the 3′ end of the O-16S sequence differentiate the translational initiation sites of natural and O-ribosomes, production of the O-ribosome requires the synthesis, processing, and incorporation of O-16S rRNA into 70S ribosomes (Brosius *et al.*, 1981; Srivastava and Schlessinger, 1990). With a single characterized exception (Hartmann *et al.*, 1987a,b, 1981), ribosomal RNA is produced in a single transcript (Brosius *et al.*, 1981). This transcript generally contains a 5′ leader sequence, the 16S rDNA, a spacer that may contain a tDNA, the 23S rDNA, and the 5S rDNA with or without additional tDNAs. The primary transcript is cleaved by a number of ribonucleases, some of which (most notably RNase III) can act cotranscriptionally, and is processed and assembled into ribosomes (Srivastava and Schlessinger, 1990). Processing enzymes are known to be required to different extents for processing different parts of the transcript. For example, RNase III is required for correct end processing of 23S rRNA, but it is dispensable for correct end processing of 16S rRNA (Srivastava and Schlessinger, 1990). To test whether a version of rrnB-producing O-16S rRNA can produce O-ribosomes for the synthesis of an orthogonal transcription–translation FFL, we cloned rrnB containing the O-16S sequence onto a T7 promoter in an RSF vector (creating pT7 RSF O-ribosome). We followed the production of GST–GFP in cells containing T7 RNAP, pT7 O-rbs–GST–GFP, and pT7 RSF O-ribosome over 60 h and the production of GST–GFP in control cells that did not contain T7 RNAP or pT7 RSF O-ribosome.

We found that the orthogonal transcription–translation FFL leads to the expression of GST–GFP that strictly depends on both T7 RNAP and pT7 RSF O-ribosome. In the presence of T7 RNAP, pT7 RSF O-ribosome, and pT7 O-rbs–GST–GFP, GST–GFP was purified in good yield. However, in the absence of T7 RNAP or pT7, RSF O-ribosome no detectable GST–GFP was purified. The GFP fluorescence of cells confirmed the accumulation of GST–GFP is pT7 RSF O-ribosome and T7 RNAP-dependent, demonstrating that we have synthesized an orthogonal transcription–translation FFL that displays AND logic.

One information-processing feature of certain transcriptional FFLs is their capacity to mediate delays in response to input signals (Mangan and Alon, 2003). To investigate whether orthogonal transcription–translation FFLs mediate delays in orthogonal gene expression, we compared the kinetics of gene expression for cells containing T7 RNAP, pT7 O-rbs–GST–GFP, and pT7 RSF O-ribosome with cells in which only the production of T7 RNAP is inducible [pT7 RSF O-ribosome is replaced with a plasmid that constitutively produces O-rrnB (pSC101* O-ribosome)] or cells in which both O-rrnB and T7 RNAP are produced from inducible, T7 RNAP-independent promoters (pT7 RSF O-ribosome is replaced with pTrc RSF O-ribosome, which produces O-rrnB from a Trc promoter using the host RNAP).

We found that orthogonal gene expression is fastest for cells that constitutively produce O-ribosomes (time taken to reach half-maximal expression, $t_{1/2} = 220$ min). These cells are poised to translate the O-rbs–GST–GFP transcript, and the accumulation of GST–GFP protein is therefore only limited by production of T7 RNAP and its transcription of pT7 O-rbs–GST–GFP. Cells containing pTrc RSF O-ribosome show a long delay in gene expression ($t_{1/2} = 580$ min, delay $= 360$ min) relative to cells that constitutively produce O-ribosomes. Upon induction of these cells, O-*rrnB* must be transcribed and processed, and the resulting O-rRNA must be assembled into O-ribosomes. These steps account for the delay observed. Because the Trc promoter is not as strong as the P1P2 promoter on constitutively produced O-rRNA the maximal expression of the O–GST–GFP is $\approx 50\%$ of that realized when O-rRNA is constitutively produced on the P1P2 promoter.

Cells containing pT7 RSF O-ribosome also show a delay in gene expression of 360 min relative to cells that constitutively produce O-ribosomes ($t_{1/2} = 580$ min, delay $= 360$ min). The orthogonal transcription–translation FFL and the simple inducible AND system both show the same $t_{1/2}$, and the same long delay relative to the system in which only transcription is inducible. This finding indicates that, in contrast to previously characterized transcription factor FFLs that introduce distinct delays with respect to the corresponding simple AND function (Mangan and Alon, 2003), the rate of gene expression in the orthogonal transcription–translation system is independent of whether the O-ribosome is in series (FFL) or parallel (AND) with T7 RNAP, although we cannot rule out that there are smaller differences in the timing of gene expression at early time points after induction. We envisioned two limiting scenarios that might lead to the identical kinetics of the simple inducible AND system and the FFL: (i) the production of O-ribosomes from O-rrnB may be much slower than any other step in GST–GFP production, leading to GST–GFP expression kinetics that are insensitive to whether O-ribosome production is in series (FFL) or in parallel (simple AND) with T7 RNAP; and (ii) the transcription of O-rRNA is rate-determining for O-ribosome formation and using the faster T7 RNAP for transcription of O-rrnB cancels out the delay introduced by requiring T7 RNAP synthesis and accumulation before O-rrnB synthesis.

4. Engineering the FFL Delay via the Discovery of a Minimal O-rRNA

We realized that because *rRNA* is produced on a long primary transcript (Brosius *et al.*, 1981; Fig. 5.3), but O-ribosomes only require the production of O-16S rRNA, it might be possible to minimize the transcript. A correctly

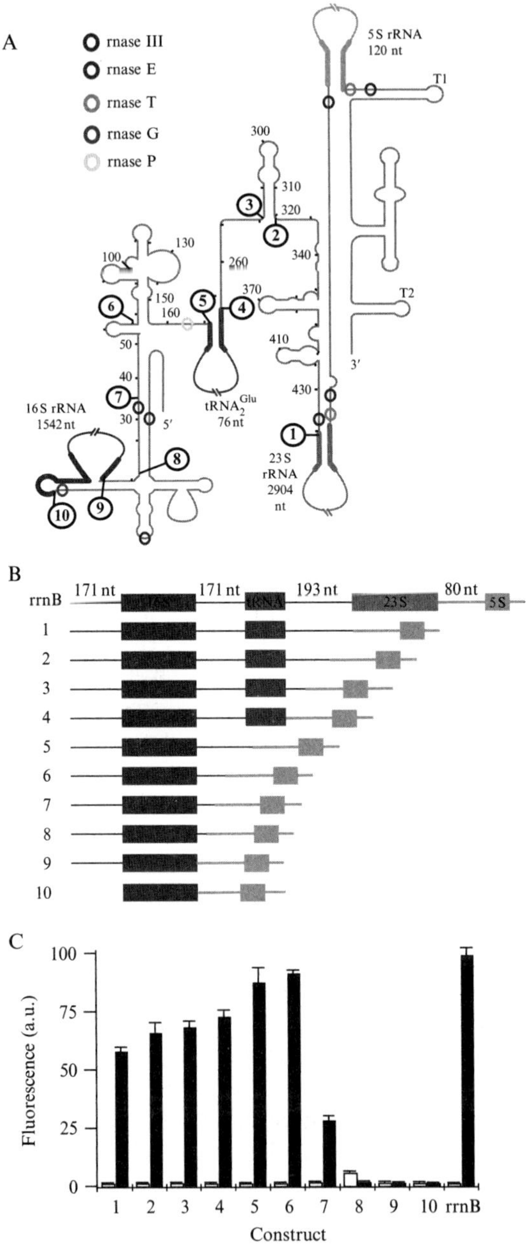

Figure 5.3 Creation of a minimal O-ribosome. (A) The proposed secondary structure of the rrnB primary transcript. T1 and T2 are terminator sequences. Numbered circles indicate the sites of truncation. Colored circles are sites that being recognized and processed via indicated Ribonuclease (rrnase). (B) Schematic of the truncations examined for the production of active O-ribosomes. (C) The activity of O-ribosomes

processed O-16S rRNA transcript would assemble into functional ribosomes with genomically encoded 23S rRNA, 5S rRNA, and ribosomal proteins. A minimal O-16S transcript would be shorter and therefore be transcribed more quickly than the full-length O-rrnB transcript. Moreover, a minimal O-16S transcript would decrease the number of processing steps required to produce O-16S rRNA for O-ribosomes and would remove the requirement for processing steps in other parts of the operon. In particular, 23S processing steps might be limiting for the release of O-16S from O-rrnB transcribed by T7 RNAP. We realized that because *rRNA* is produced on a long primary transcript (Brosius *et al.*, 1981; Fig. 5.3), but O-ribosomes only require the production of O-16S rRNA, it might be possible to minimize the transcript. A correctly processed O-16S rRNA transcript would assemble into functional ribosomes with genomically encoded 23S rRNA, 5S rRNA, and ribosomal proteins. A minimal O-16S transcript would be shorter and therefore be transcribed more quickly than the full-length O-rrnB transcript. Moreover, a minimal O-16S transcript would decrease the number of processing steps required to produce O-16S rRNA for O-ribosomes and would remove the requirement for processing steps in other parts of the operon. In particular, 23S processing steps might be limiting for the release of O-16S from O-rrnB transcribed by T7 RNAP, because it has been reported that 23S rRNA produced by T7 RNAP [which is five times faster than host polymerases (Chamberlin and Ring, 1973; Golomb and Chamberlin, 1974)] is not efficiently cotranscriptionally processed and is incorporated into nonfunctional 50S subunits (Lewicki *et al.*, 1993). If either transcription or processing of O-16S RNA from pT7 RSF O-ribosome is rate limiting, then a minimal O-16S transcript might decrease the observed delay in the orthogonal transcription–translation FFL.

To create a minimal O-16S expression construct, we prepared a series of deletion mutants (Fig. 5.3) in pTrc O-ribosome [a version of rrnB that is transcribed from the IPTG-inducible pTrc promoter and contains the O-16S sequence in the rrnB operon (Rackham and Chin, 2005a)]. We assayed the function of these deletion mutants by their ability to form O-ribosomes and produce GFP from a gene with a constitutive promoter and an O-rbs (pR22).

Deletion of the 23S rRNA from pTrc O-ribosome led to a decrease in GFP fluorescence to half that of the full-length operon. However, further deletion of the spacer and tRNA led to rescue of the GFP fluorescence to levels close to that observed for the full-length operon. The maximally active truncated operons (Fig. 5.3, constructs 5 and 6) contain the 5′ leader sequence of 16S rRNA and the

---

produced from each truncation construct (constructs 1–10, filled bars) compared with the full-length operon (O-rrnB). Fluorescence was measured in cells containing pXR1 (a tetracycline-resistant p15A plasmid that directs GFP expression from a constitutive promoter and O-rbs). The empty bars show the expression of GFP produced when pXR1 is combined with wild-type ribosomes in the absence of O-ribosomes. (See Color Insert.)

region of the spacer immediately 3′ to 16S rRNA that is believed to form a base-paired helix with the 5′ leader sequence in the primary transcript (Fig. 5.3; Brosius *et al.*, 1981). This helix contains RNase III sites that are cleaved to release a 16S rRNA-containing fragment from the primary transcript in rrnB. Deletion of sequences close to these RNase III sites leads to a loss of functional O-ribosome production, as judged by the drastic decrease in GFP signal (Fig. 5.3, constructs 7–10). These experiments refactor the O-rrnB operon and define a minimal O-16S expression construct (Fig. 5.3, construct 6). The minimal O-16S expression construct reduces the transcript required to produce O-ribosomes to 50% of its original length, from 5486 to 2247 nt. The minimal O-ribosome construct will take less time to transcribe than rrnB, moreover it uncouples O-16S rRNA synthesis from 23S rRNA synthesis and tRNA synthesis, and therefore uncouples O-16S rRNA processing from processing steps that may limit the production of O-16S rRNA from O-rrnB. The decreased transcription time and the potentially decreased processing time could act together to decrease the time required to produce a functional O-ribosome.

To investigate whether the minimal O-16S leads to altered gene expression kinetics, we assembled an orthogonal transcription–translation FFL in which the minimal O-16S sequence was transcribed from a T7 promoter in the presence of T7 RNAP and assembled into functional O-ribosomes (Fig. 5.4). T7 RNAP transcribes pT7 O-rbs–GST–GFP, and the resulting mRNA is translated by the O-ribosome. Expression of GST–GFP strictly depends on T7 polymerase and the plasmid encoding O-16S rRNA from the T7 promoter (pT7 RSF O-16S), as judged by GFP fluorescence. These data confirm that the construction of the FFL depends on both inputs.

We measured the kinetics of gene expression in the FFL by inducing the production of T7 RNAP with IPTG and after the increase in fluorescence as a function of time (Fig. 5.4). Comparison of the FFL to a system in which O-ribosomes are constitutively produced and that is simply regulated by transcription demonstrates that the orthogonal transcription–translation FFL introduces a delay of 170 min relative to the simple transcription regulation case ($t_{1/2} = 390$ min, delay $= 170$ min). This delay is approximately half the length of that observed with the full-length rrnB on a T7 promoter to produce O-16S rRNA, suggesting that processing outside of 16S sequence or transcription of rRNA determines the kinetics of gene expression. These experiments demonstrate that the minimal O-ribosome construct alters the delay in gene expression.

5. Discussion

We have combined orthogonal transcription by T7 RNAP and orthogonal translation by O-ribosomes to create an orthogonal gene expression pathway in the cell. This pathway specifically directs

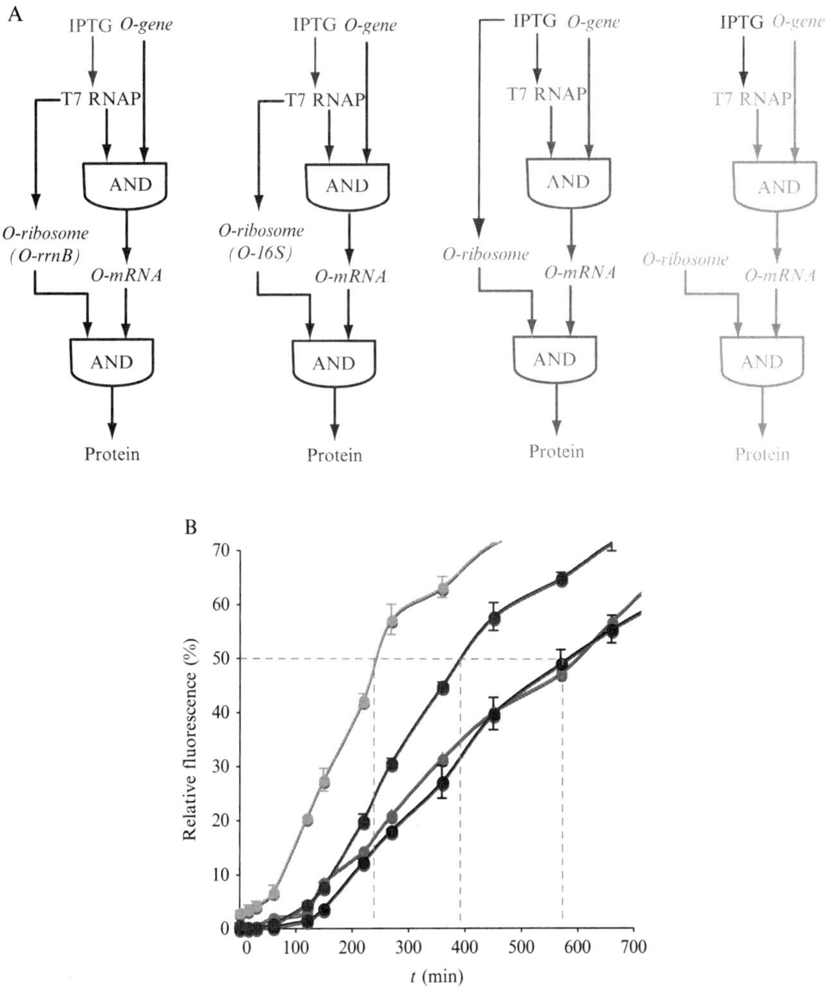

Figure 5.4 The FFLs with the minimal O-ribosome and the progenitor O-ribosome have identical topologies but mediate distinct delays. (A) The FFL using the progenitor O-ribosome (maroon) and the minimal O-ribosome (orange). (B) The delays in gene expression created by the orthogonal transcription–translation networks. Orange solid circles, BL21 (DE3), pT7 RSF O-16S, pT7 O-rbs GST-GFP. The time taken to reach 50% of maximal expression, used to quantify the delay (Mangan and Alon, 2003), is indicated. (See Color Insert.)

the decoding of genetic information from an orthogonal gene that is heritable in the host, but is unreadable by the host. Orthogonal genes might form one basis for creating nontransmissible, safe synthetic genetic circuits.

Because both O-ribosomes and T7 RNAP are nonessential and directed specifically to orthogonal genes, the expression of orthogonal genes can be regulated and altered in ways not possible with the cells' essential polymerases and ribosome that have been evolutionarily trapped by the requirement to synthesize the proteome and ensure cell survival. We have demonstrated that the modular combination of an orthogonal RNAP and an O-ribosome allows the construction of compact transcription–translation networks with predictable properties. These networks, which could not be created by using host polymerases or endogenous ribosomes, allow the introduction of translational delays into gene expression that are not possible in natural systems. We have created four orthogonal gene expression networks with different expression kinetics that control the timing of gene expression in another way. The fastest orthogonal gene expression system ($t_{1/2} = 3.5$ h) uses constitutively produced O-ribosomes and requires induction of T7 RNAP. The slowest systems ($t_{1/2} = 10$ h) require induction of full-length rrnB, either in series (FFL) or in parallel (simple AND) with T7 RNAP. In the process of creating these networks, we refactored the rrnB to uncouple O-16S rRNA synthesis and processing from the synthesis and processing of the rest of the operon, and we defined a minimal module for O-ribosome production in cells. This minimal O-ribosome allowed the creation of a FFL with an intermediate delay ($t_{1/2} = 6.5$ h).

Overall, this work creates an orthogonal gene expression system that has the properties of a parallel operating system embedded in the cell. The orthogonal gene expression system is more flexible and amenable to abstraction and engineering than natural transcription and translation (that has the evolutionarily inherited burden of decoding the genome and synthesizing the proteome), yet the orthogonal system is bootstrapped to the natural system and takes advantage of the natural system's capabilities, including the cells' nonorthogonal components for genetic encoding, and replication.

It will be interesting to investigate the properties of other compact orthogonal gene expression circuits, including orthogonal transcription–translation FFLs in which an O-ribosome, rather than an orthogonal polymerase, is the master regulator. Moreover, using mutually O-ribosomes (Chubiz and Rao, 2008; Rackham and Chin, 2005a), other translational control strategies (Anderson *et al.*, 2007), and mutually orthogonal polymerases, novel sigma factors, and other transcriptional control strategies it will be possible to assemble combinatorially more complex systems, allowing us to simultaneously boot up multiple, mutually orthogonal parallel operating systems within the cell. It may be possible to regulate the timing of host gene expression by using the orthogonal systems to investigate the effects of altering timing on the phenotypic or cellular decision outcome of signaling. Finally, because the orthogonal gene expression system is composed of nonessential components, it may be possible to use genetic selections to discover systems that display tailored delays in gene expression or other interesting and useful dynamic properties.

6. Material and Methods

6.1. Construction of T7-O-rbs libraries by EIPCR

To create the library pT7 N15lib, we performed enzymatic inverse PCR (Rackham and Chin, 2005a; Stemmer and Morris, 1992) with primers T7n15O-rbsf (GAACCGAGATCTCGATCCCGCGAAATTAATACG ACTCACTATAGGGAGANNNNNNNNNNNNNNNTTTCATAT CCCTCCGCAAATGCGTAAAGGAG) and T7R (ATCGAGATCTC GGGCAGCGTTGGGTCCTGGC) on plasmid C. The resulting enzymatic inverse PCR products (20 μg) were digested with *Bgl*II (10 h, 37 °C), digested with *Dpn*I (2 h, 37 °C), and ligated (T4 DNA ligase, 16 h, 16 °C). The library DNA was ethanol-precipitated and transformed into Mega X DH10B (Invitrogen).

To construct a plasmid for constitutive expression of *GFP* from an O-rbs, the sequence ATA in lac operator of pGFPmut3.1 was replaced by CTCGAG. The entire flanking sequence between lac operator and *GFP* was replaced by the conserved 18-nt (TTTCATATCCCTCCGCAA), producing the vector pR22. The *GFP* gene, O-rbs, flanking sequence, and terminator were amplified from R22 by using the primers xr1GFPnotIf (ATATGCGGCCGCAACCGTATTACCGCCTTTGA) and xr1GF Pbglr (TGACAGATCTACATTTCCCCGAAAAGTGC). The PCR product was digested with *Bgl*II and *Not*I. A fragment containing the tetracycline resistance gene and the p15A origin was amplified from pO-CAT with the PCR primers pcatbglf (TATAGCGGCCGC-CAAAGCC GTTTTTCCATAGG) and pcatNotIr (CAGTA-GATCTTCCGCG TTTCCAGACTTTAC), and digested with *Bgl*II and *Not*I. The pR22 fragment and the pO-CAT fragments were ligated (T4 DNA ligase, 16 h, 16 °C) to yield pXR1.

6.2. Selection of an optimized T7 promoter/O-rbs system

An optimized T7 promoter/O-rbs system was selected by three rounds of FACS screening (positive, negative, and positive; Fig. 5.2). In the positive rounds of screening BL21 (DE3; Novagen) containing pSC101*O-ribosome were transformed with pT7n15GFPlib and grown overnight (37 °C, 12 h, 250 rpm) in 100 mL of LB-AK (LB media containing 25 μg·mL^{-1} ampicillin and 12.5 μg·mL^{-1} kanamycin). Five milliliters of overnight culture was diluted 1:20 in fresh LB-AK and incubated (1.5–2 h, 250 rpm, 37 °C). At $OD_{600} \approx 0.5$–0.8, IPTG (1 m*M*) was added, and the cells were incubated (1.5–2 h, 250 rpm, 37 °C). The cultures were filtered through 70 μm BD Falcon Cell Strainers (BD Biosciences) to remove cell debris and diluted 1:100 in PBS. The samples were subjected to FACS by MoFlo (MoFlo Cytomation), with a flow rate of 10,000 events per second using a 100-μm nozzle. A total of 3.7×10^7 cells were sorted and 3×10^6 cells were collected. The collected GFP-positive cells were amplified in LB-2AKG (LB media containing 25 $\mu g \cdot mL^{-1}$ ampicillin, 25 $\mu g \cdot mL^{-1}$ kanamycin, and 2% glucose; 37 °C, 250 rpm, 16 h). Total plasmid DNA was isolated from cells and pT7n15lib DNA was separated from pSC101*O-ribosome DNA by 1% agarose gel electrophoresis. The pT7n15GFPlib DNA was extracted from the gel for use in the next round of screening or for characterization of individual clones.

For negative FACS sorting pT7n15lib DNA surviving the positive sort was transformed into BL21 (DE3) containing pSC101*BD (this vector produces rrnB from the native ribosomal P1P2 promoter). Cultures were prepared for FACS sorting as described for the positive FACS sort. In the negative FACS sort, 10^8 cells were sorted and 88.5% of the cells were collected, as they had a level of fluorescence comparable to negative controls. The collected negative clones were amplified and their pT7n15lib DNA was resolved and extracted, as described above for positive sort clones. In the final positive sort, 10^8 cells were sorted and 6×10^3 cells with strong fluorescence were collected.

6.3. Characterizing pT7 O-rbs GFP expression constructs

Individual pT7n15lib clones were transformed into BL21 (DE3) containing either the wild-type (pTrc RSF wt ribosome or pSC101*BD) or the O-ribosome (pTrc RSF-O-rDNA or pSC101*O-ribosome). Transformed cells were grown overnight (37 °C, 12 h, 250 rpm) in 10 mL of LB-AK. Overnight culture (0.5 mL) was diluted 1:20 into 10 mL of fresh LB-AK and incubated (1.5–2 h, 250 rpm, 37 °C). At $OD_{600}\ \mu \approx 0.5$–0.8, IPTG (1 m*M*) was added, and the cells were incubated (12 h, 250 rpm, 37 °C). Fluorescence was quantified by using a fluorescent plate reader (Tecan safire II plate reader). The excitation wavelength was 488 nm and the emission was measured at 515 nm with a 10-nm bandpass. The GFP values were normalized by OD_{600} values. Clones from the selection showing good O-ribosome-dependent fluorescence were sequenced.

To further characterize clone pT71504 resulting from the selection, we replaced *GFP* in pT71504 by a *GSTsfGFP* fusion to create the pT7 O-rbs–*GST–GFP* [sfGFP is superfolding green fluorescent protein (Pedelacq *et al.*, 2006)]. A *GSTsfGFP* containing fragment was amplified by using the primers 1504G9GfGFPf (TGCCCGAGATCTCGATCCCGCGAA ATTAATACGACTCACTATAGGGAGACTATATCTGTTATTTTT TCATATCCCTCCGCAAATGTCC) and sfGFPHindr (CAACTAAGC

TTATTAATGGTGATGATGATGGTGGCTGCCTTTATACAGTTC ATCCATACC) and ligated between the *Bgl*II and *Hind*III sites in pT71504 to generate pT7 O-rbs–*GST–GFP*.

To demonstrate that pT7 O-rbs–*GST–GFP* displays Boolean AND logic, we transformed BL21 (T1R; Sigma/Aldrich) and BL21 (DE3) with pT7 O-rbs–*GST–GFP* and either pSC101*O-ribosome or pSC101*BD. We expressed and purified the resulting GST–GFP protein and examined the protein made by SDS/PAGE. Briefly, the transformed cells were cultured in LB-2AKG media (37 °C, 250 rpm, 16 h). Overnight cultures were inoculated (1:100) into 100 mL of LB-AK and incubated (37 °C, 3 h, 250 rpm). At $OD_{600} \approx 0.7$–0.9, IPTG (1 m*M*) was added and the cells, which were incubated for a further 3–5 h (37 °C, 3–5 h, 250 rpm). Fifty milliliters of cells was harvested by centrifugation (4000×*g*, 10 min), and the pellets were washed once with 1 mL of ice-cold PBS. BugBuster Protein Extraction Reagent (1 mL; Novagen) containing complete protease inhibitor mixture (Roche) and 1 m*M* PMSF (Sigma/Aldrich) were added to cell pellets and incubated at 25 °C for 30 min. The supernatant was collected after centrifugation (16,000×*g*, 10 min, 4 °C). Glutathione Sepharose 4B beads (40 μL; GE Healthcare BioscienceAB) were added to the supernatant and incubated at 4 °C (1 h) with rotating. The beads were washed four times with ice-cold PBS. Proteins were eluted from the beads by the addition of 60 μL of NuPAGE SDS sample buffer (Invitrogen). The mixture was boiled for 5 min at 95 °C and the beads were pelleted by centrifugation. Samples of the supernatant (15 L) were subjected to SDS/PAGE on 4–12% gel (400 mA, 2 h). Proteins were visualized by InstantBlue staining (www.expedeon.com).

6.4. Characterization of minimal O-rRNA for O-ribosomes

To compare the activity of O-ribosomes produced by each O-rrnB truncation, each construct (Fig. 5.3, constructs 1–10) was cotransformed with pXR1 into Genehog *E. coli* (Invitrogen). Transformed cells were grown overnight (37 °C, 12 h, 250 rpm) in 10 mL of LB-AK. Overnight culture (0.5 mL) was diluted 1:20 into fresh LB-AK and incubated (1.5–2 h, 250 rpm, 37 °C). At $OD_{600} \approx 0.5$–0.8, IPTG (1 m*M*) was added, and the cells were incubated (12 h, 250 rpm, 37 °C). Fluorescence was quantified by using a fluorescent plate reader (Tecan safire II plate reader). The excitation wavelength was 488 nm and the emission was measured at 515 nm with a 10-nm bandpass. The GFP values were normalized by OD_{600} values.

6.5. Characterization of orthogonal gene expression kinetics

To characterize the orthogonal gene expression kinetics (Fig. 5.4), we used BL21 (DE3) containing pT7 O-rbs–GST–GFP with pRSF O-ribosome, pSC101*O-ribosome, pT7 RSF O-ribosome, or pT7 RSF O-16S.

Overnight culture (0.5 mL) grown in LB-AKG was used to innoculate 10 mL of LB-AK. These cultures were incubated (37 °C, 250 rpm, 2 h to $OD_{600} \approx 0.5$) and then induced with IPTG (1 m*M*) and incubated a further 56 h. Fluorescence was quantified by using a fluorescent plate reader (Tecan safire II plate reader). The excitation wavelength was 488 nm and the emission was measured at 515 nm with a 10-nm bandpass. To determine the effect of T7 RNAP on gene expression, the experiments were carried out using BL21-TIR instead of BL21 (DE3). To determine the effect of the O-ribosome on gene expression, the experiments were carried out using the wt-ribosome equivalent of the O-rRNA vectors described above.

ACKNOWLEDGMENTS

We are grateful to the European Research Council and the Medical Research Council for financial support.

REFERENCES

An, W., and Chin, J. W. (2009). Synthesis of orthogonal transcription-translation networks. *Proc. Natl. Acad. Sci. USA* **106,** 8477–8482.

Anderson, J. C., Voigt, C. A., and Arkin, A. P. (2007). Environmental signal integration by a modular AND gate. *Mol. Syst. Biol.* **3,** 133.

Basu, S., and Maitra, U. (1986). Specific binding of monomeric bacteriophage T3 and T7 RNA polymerases to their respective cognate promoters requires the initiating ribonucleoside triphosphate (GTP). *J. Mol. Biol.* **190,** 425–437.

Brosius, J., Dull, T. J., Sleeter, D. D., and Noller, H. F. (1981). Gene organization and primary structure of a ribosomal RNA operon from Escherichia coli. *J. Mol. Biol.* **148,** 107–127.

Chamberlin, M., and Ring, J. (1973). Characterization of T7-specific ribonucleic acid polymerase. 1. General properties of the enzymatic reaction and the template specificity of the enzyme. *J. Biol. Chem.* **248,** 2235–2244.

Chan, L. Y., Kosuri, S., and Endy, D. (2005). Refactoring bacteriophage T7. *Mol. Syst. Biol.* **1**(2005), 0018.

Chin, J. W. (2006). Modular approaches to expanding the functions of living matter. *Nat. Chem. Biol.* **2,** 304–311.

Chubiz, L. M., and Rao, C. V. (2008). Computational design of orthogonal ribosomes. *Nucleic Acids Res.* **36,** 4038–4046.

de Lorenzo, V., and Danchin, A. (2008). Synthetic biology: Discovering new worlds and new words. *EMBO Rep.* **9,** 822–827.

Endy, D. (2005). Foundations for engineering biology. *Nature* **438,** 449–453.

Golomb, M., and Chamberlin, M. (1974). Characterization of T7-specific ribonucleic acid polymerase. IV. Resolution of the major in vitro transcripts by gel electrophoresis. *J. Biol. Chem.* **249,** 2858–2863.

Greber, D., and Fussenegger, M. (2007). Mammalian synthetic biology: Engineering of sophisticated gene networks. *J. Biotechnol.* **130,** 329–345.

Hartmann, R. K., Ulbrich, N., and Erdmann, V. A. (1987a). An unusual rRNA operon constellation: In Thermus thermophilus HB8 the 23S/5S rRNA operon is a separate entity from the 16S rRNA operon. *Biochimie* **69,** 1097–1104.

Hartmann, R. K., Ulbrich, N., and Erdmann, V. A. (1987b). Sequences implicated in the processing of Thermus thermophilus HB8 23S rRNA. *Nucleic Acids Res.* **15,** 7735–7747.

Hartmann, R. K., Toschka, H. Y., and Erdmann, V. A. (1991). Processing and termination of 23S rRNA-5S rRNA-tRNA(Gly) primary transcripts in Thermus thermophilus HB8. *J. Bacteriol.* **173,** 2681–2690.

Hasty, J., Isaacs, F., Dolnik, M., McMillen, D., and Collins, J. J. (2001). Designer gene networks: Towards fundamental cellular control. *Chaos* **11,** 207–220.

Hasty, J., McMillen, D., and Collins, J. J. (2002). Engineered gene circuits. *Nature* **420,** 224–230.

Imburgio, D., Rong, M., Ma, K., and McAllister, W. T. (2000). Studies of promoter recognition and start site selection by T7 RNA polymerase using a comprehensive collection of promoter variants. *Biochemistry* **39,** 10419–10430.

Isaacs, F. J., Dwyer, D. J., and Collins, J. J. (2006). RNA synthetic biology. *Nat. Biotechnol.* **24,** 545–554.

Kaern, M., Blake, W. J., and Collins, J. J. (2003). The engineering of gene regulatory networks. *Annu. Rev. Biomed. Eng.* **5,** 179–206.

Kampf, M. M., and Weber, W. (2010). Synthetic biology in the analysis and engineering of signaling processes. *Integr. Biol. Camb.* **2,** 12–24.

Kohanski, M. A., and Collins, J. J. (2008). Rewiring bacteria, two components at a time. *Cell* **133,** 947–948.

Kwok, R. (2010). Five hard truths for synthetic biology. *Nature* **463,** 288–290.

Lewicki, B. T., Margus, T., Remme, J., and Nierhaus, K. H. (1993). Coupling of rRNA transcription and ribosomal assembly *in vivo*. Formation of active ribosomal subunits in *Escherichia coli* requires transcription of rRNA genes by host RNA polymerase which cannot be replaced by bacteriophage T7 RNA polymerase. *J. Mol. Biol.* **231,** 581–593.

Mangan, S., and Alon, U. (2003). Structure and function of the feed-forward loop network motif. *Proc. Natl. Acad. Sci. USA* **100,** 11980–11985.

Milo, R., Shen-Orr, S., Itzkovitz, S., Kashtan, N., Chklovskii, D., and Alon, U. (2002). Network motifs: Simple building blocks of complex networks. *Science* **298,** 824–827.

Neumann, H., Wang, K., Davis, L., Garcia-Alai, M., and Chin, J. W. (2010). Encoding multiple unnatural amino acids via evolution of a quadruplet-decoding ribosome. *Nature* **464,** 441–444.

Pedelacq, J. D., Cabantous, S., Tran, T., Terwilliger, T. C., and Waldo, G. S. (2006). Engineering and characterization of a superfolder green fluorescent protein. *Nat. Biotechnol.* **24,** 79–88.

Rackham, O., and Chin, J. W. (2005a). A network of orthogonal ribosome × mRNA pairs. *Nat. Chem. Biol.* **1,** 159–166.

Rackham, O., and Chin, J. W. (2005b). Cellular logic with orthogonal ribosomes. *J. Am. Chem. Soc.* **127,** 17584–17585.

Rackham, O., Wang, K., and Chin, J. W. (2006). Functional epitopes at the ribosome subunit interface. *Nat. Chem. Biol.* **2,** 254–258.

Ramakrishnan, V. (2002). Ribosome structure and the mechanism of translation. *Cell* **108,** 557–572.

Serrano, L. (2007). Synthetic biology: Promises and challenges. *Mol. Syst. Biol.* **3,** 158.

Sprinzak, D., and Elowitz, M. B. (2005). Reconstruction of genetic circuits. *Nature* **438,** 443–448.

Srivastava, A. K., and Schlessinger, D. (1990). Mechanism and regulation of bacterial ribosomal RNA processing. *Annu. Rev. Microbiol.* **44,** 105–129.

Steitz, T. A. (2004). The structural basis of the transition from initiation to elongation phases of transcription, as well as translocation and strand separation, by T7 RNA polymerase. *Curr. Opin. Struct. Biol.* **14,** 4–9.

Stemmer, W. P., and Morris, S. K. (1992). Enzymatic inverse PCR: A restriction site independent, single-fragment method for high-efficiency, site-directed mutagenesis. *Biotechniques* **13,** 214–220.

Wang, K., Neumann, H., Peak-Chew, S. Y., and Chin, J. W. (2007). Evolved orthogonal ribosomes enhance the efficiency of synthetic genetic code expansion. *Nat. Biotechnol.* **25,** 770–777.

Xie, J., and Schultz, P. G. (2006). A chemical toolkit for proteins–An expanded genetic code. *Nat. Rev. Mol. Cell Biol.* **7,** 775–782.

Young, E., and Alper, H. (2010). Synthetic biology: Tools to design, build, and optimize cellular processes. *J. Biomed. Biotechnol.* **2010,** 130781.

CHAPTER SIX

Directed Evolution of Promoters and Tandem Gene Arrays for Customizing RNA Synthesis Rates and Regulation

Keith E. J. Tyo,* Elke Nevoigt,† *and* Gregory Stephanopoulos‡

Contents

Abstract

Manipulating RNA synthesis rates is a primary method the cell uses to adjust its physiological state. Therefore to design synthetic genetic networks and circuits, precise control of RNA synthesis rates is of the utmost importance. Often, however, a native promoter does not exist that has the precise characteristics required for a given application. Here, we describe two methods to change the rates and regulation of RNA synthesis in cells to create RNA synthesis of a desired specification. First, error-prone PCR is discussed for diversifying the properties of native promoters, that is, changing the rate of synthesis in constitutive promoters and the induction properties for an inducible promoter. Specifically, we describe techniques for generating diversified promoter libraries of the constitutive promoters $P_L$tetO-1 in *Escherichia coli* and *TEF1* in

* Department of Chemical and Biological Engineering, Northwestern University, Evanston, Illinois, USA
† School of Engineering and Science, Jacobs University gGmbH, Bremen, Germany
‡ Department of Chemical Engineering, Massachusetts Institute of Technology, Cambridge, Massachusetts, USA

Methods in Enzymology, Volume 497
ISSN 0076-6879, DOI: 10.1016/B978-0-12-385075-1.00006-8

Saccharomyces cerevisiae as well as the inducible, oxygen-repressed promoter *DAN1* in *S. cerevisiae*. Beyond generating promoter libraries, we discuss techniques to quantify the parameters of each new promoter. Promoter characteristics for each promoter in hand, the designer can then pick and choose the promoters needed for the specific genetic circuit described *in silico*. Second, Chemically Induced Chromosomal Evolution (CIChE) is presented as an alternative method to finely adjust RNA synthesis rates in *E. coli* by variation of gene cassette copy numbers in tandem gene arrays. Both techniques result in precisely defined RNA synthesis and should be of great utility in synthetic biology.

1. Introduction

Transcription is one of the two basic processes in the central dogma of molecular biology and changing RNA synthesis rates is a central mechanism by which a cell reconfigures its state in response to extra- and intracellular stimuli. It is obvious that tools for altering RNA synthesis rates and regulation are essential in synthetic biology. A key tenet of synthetic biology is to develop biological networks and circuits using computer aided design (CAD) similar to paradigms in electrical, civil, and chemical engineering (Tyo *et al.*, 2010). By this, cellular engineering is first performed *in silico* using mathematical models. Once a desired genetic circuit has been specified *in silico*, DNA must be constructed that will exhibit the parameters that were specified from the simulation. Such an approach saves significant time and resources.

To adjust the rate and/or regulation of RNA synthesis according to the model predictions, appropriate tools for DNA modifications are necessary in order to generate DNA sequences which have the required specifications. The specifications may include (i) a defined RNA synthesis rate or (ii) dynamic changes in RNA synthesis rate (for regulated transcription). The precise adjustment of maximal RNA synthesis rates is indeed important in the design of synthetic genetic circuits as recently demonstrated (Ellis *et al.*, 2009). Even when engineering a simple linear pathway as depicted in Fig. 6.1, precise control is important. In this example, enzyme B must be expressed to a specific level that is in balance with enzymes A and C. If B expression is too low, the pathway will be bottlenecked at B, and α may accumulate to toxic levels. If B expression is too high, cellular resources are wasted on excess enzyme, and the intermediate following B, β, may accumulate to toxic levels.

Beyond steady state considerations, dynamic changes of RNA synthesis rates in response to an input signal are of great importance in synthetic biology. Different genetic networks such as feed-forward loop or genetic timer networks require precisely regulated promoters. In industrial practice,

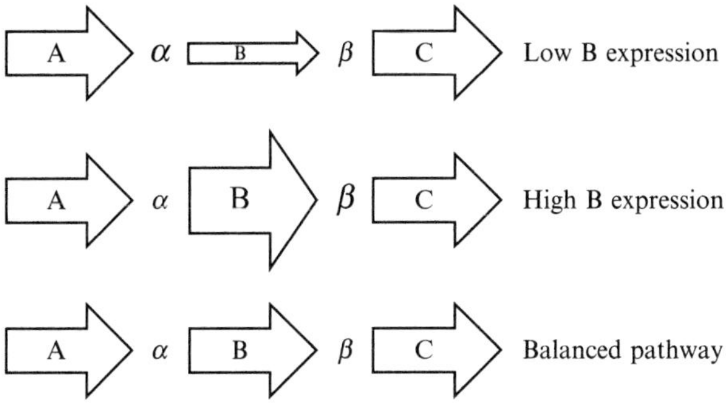

Figure 6.1 Fine-tuned RNA synthesis rates are required for balanced pathways and gene regulatory networks. Arrows A, B, and C are enzymatic reactions in a metabolic pathway. α and β are metabolite intermediates in the pathway. Size of arrow/letter denotes concentration. See Section 1 for a discussion of this example.

it may be necessary to allow a cell-growth phase that is separated from a production phase. In order to manage this, RNA synthesis rates of the product forming pathway enzymes have to respond to a cell concentration stimulus. As native promoters may not have the correct induction properties, tools for customizing the sensitivity in response to the stimulus are essential.

Here, we discuss different possibilities for customizing the maximal rate and the dynamic changes of RNA synthesis at the level of DNA. One method is the directed evolution of promoters which can be applied in order to customize the maximal rate and regulatory properties of RNA synthesis. The second method is the generation of tandem gene arrays which can be alternatively applied in order to adjust the maximal synthesis rate of a target (m)RNA.

In detail, directed evolution of promoters to alter RNA synthesis rates and regulation in *Escherichia coli* and *Saccharomyces cerevisiae* will be accomplished by generating promoter sequence diversity using error-prone polymerase chain reaction (PCR). This includes altering strength of constitutive promoters and affecting transfer functions in inducible promoters. Specifically, the *E. coli* promoter $P_L$tetO-1 is mutagenized to obtain a range of constitutive promoters. Likewise in yeast, the *TEF1* promoter is mutagenized to create a range of constitutive promoters. Also in yeast, the oxygen-repressed *DAN1* promoter is mutagenized to change to effective concentration of 50% activation (EC50) of oxygen for this promoter. Tandem gene arrays in *E. coli* will be described using an approach called Chemically Induced Chromosomal Evolution (CIChE). This approach is useful for increasing RNA synthesis rates above a level achieved with one copy of a very strong promoter and has improved stability properties compared to plasmids.

2. Promoter Modification by Error-Prone PCR

To exploit CAD-based engineering as described in Section 1, a registry of promoters is needed that allows the designer to pick a DNA sequence that will give the desired RNA synthesis properties. One way to achieve this is by developing a library of different promoters with well-characterized properties that span the range of useful expression levels or regulatory properties. Then, after completing the design phase, the appropriate promoters for each gene could be selected from the library and used. A recent example of this strategy has been described (Ellis *et al.*, 2009). A good library should contain (a) quantitatively characterized promoters whose RNA synthesis parameters are clearly established, (b) a wide dynamic range of expression, that is, from very low to very high, (c) small increments, or fine-tuned gradations along the expression range, and (d) in the case of inducible promoters, a variety of transfer functions or induction curves (Fig. 6.2). A comprehensive open-source library is being developed in the Standard Biological Parts repository (http://partsregistry.org).

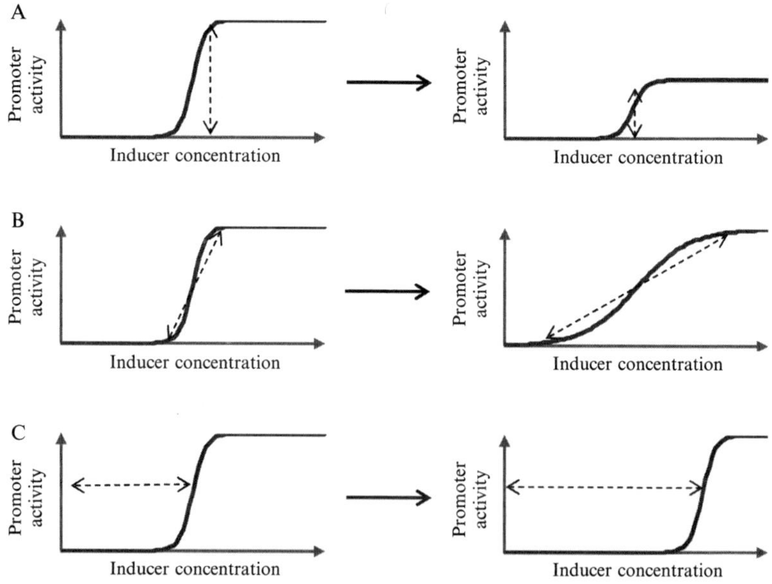

Figure 6.2 Altering transfer functions for inducible promoters. In an inducible promoter, the transfer function that describes how RNA synthesis rate changes with inducer concentration has several parameters. (A) Dynamic Range—the ratio of promoter strengths from "on" to "off" state. (B) Sensitivity—the rate at which induction increases with inducer. (C) EC50—the level of inducer that results in 50% of the RNA synthesis rate increase. For a given genetic circuit, each of these is important to determine the dynamics of the system.

One avenue to generate promoter libraries with the above-mentioned specifications for CAD-based engineering would be to take a native promoter and introduce mutations to generate many versions of the promoter (Alper *et al.*, 2005; Nevoigt *et al.*, 2006, 2007). In general, error-prone PCR can be used in order to generate random nucleotide changes in DNA sequences. Promoter regions are located upstream of the coding sequence and bind the RNA polymerase (RNAP) or associated transcription factors (TFs) during transcription initiation. Mutations in the promoter sequence can alter the binding affinity of the promoter sequence between RNAP and TFs thereby changing the transcription initiation rates. In the case of inducible promoters, that is, those whose RNA synthesis rate is regulated by a stimulus, mutations will result in changes to the dynamic range, sensitivity, and EC50 of the promoter. Here, we will first describe how to modify a native promoter by error-prone PCR to generate a library of promoters with mutated DNA sequences causing a range of RNA synthesis rates (for constitutive promoters) and transfer functions (for inducible promoters).

2.1. Promoter selection

The first step in generating a range of promoter strengths is to identify a suitable native promoter that will be diversified. While there are thousands of promoters to choose from for a given organism, it can be helpful to use a well-studied promoter to best understand the possible caveats in using it. The promoter should have an expression level near the desired maximum level of the dynamic range since it is unlikely to obtain higher RNA synthesis rates after error-prone PCR while it is easy to generate low promoter activities by this method even when starting with a strong promoter.

Importantly, for these RNA synthesis parts to be useful in synthetic genetic systems, the promoters must also be robust. By this, a promoter must have a RNA synthesis rate that has a small variance and is stable over time. Cell-to-cell variation can often be significant, as well as constitutive and inducible-specific problems discuss below. Thorough characterization of a new promoter at the single-cell level under different growth conditions is therefore important for quality control.

2.1.1. Constitutive promoter considerations

Constant expression level is called for in many synthetic biology applications. Constitutive promoters can be defined in the literature as either a constant expression rate *or* always "on," although the transcription rate may vary significantly. Promoters which are considered to be constitutive according to the second definition could cause problems in genetic circuit design. In general, choosing exogenous promoters from phage or other sources have the advantage of being largely free of host regulation and exhibit some of the highest expression levels. As mentioned above, thorough characterization of the promoter under different conditions is best to determine any effects due to cellular regulation. In *E. coli*, phage promoters, such as P_{T5}, P_{L}tetO-1 (lacking tetR), and P_{T7}, are desirable in this context because they are minimally influenced by cellular signals. P_{T7} relies on the T7 RNAP, avoiding effects of changing native RNAP levels. P_{L}tetO-1 is repressed by tetR, but in its absence binds RNAP efficiently (Lutz and Bujard, 1997). The ribosomal promoters, rrnB P1 and rrnB P2, have high expression levels, but are growth rate dependent (i.e., rrnB is upregulated while growing but downregulated in stationary phase; Murray *et al.*, 2003).

In yeast, there are a number of commonly used promoters considered to be constitutive and strong. Seven of these promoters have been recently thoroughly compared using the same reporter gene and growth conditions (Partow *et al.*, 2010). This study revealed that most of these promoters significantly varied in their activity during the course of a batch culture with five of the seven decreasing. In terms of both strength and robustness, the *TEF1* promoter was the best under all conditions tested. The *TEF1* gene encodes the transcriptional elongation factor EF-1a. For moderate expression in yeast, the promoter of the *CYC1* gene (encoding cytochrome *c*) can be used (Mumberg *et al.*, 1995).

2.1.2. Inducible promoter considerations

An inducible promoter should be (i) tightly regulated, that is, does not have a significant level when repressed (S_{min}), (ii) a high or adjustable level after induction (S_{max}), (iii) a fast and complete switch between off and on state in all cells, (iv) an inducer (triggering molecule or condition) which does not cause pleiotropic effects, and (v) a convenient induction method which does not require cumbersome medium exchanges. In addition to these requirements, industrial applications, particularly the production of pharmaceutical proteins call for triggering molecule or conditions which are nontoxic and inexpensive in order to be compatible with downstream processing regulations and process economics (Weber and Fussenegger, 2007). As well, inducible promoters may have undesirable bistability. In a bistable situation, the population-averaged promoter activity may have a smooth increase; however this is due to an increase in the percentage of the population going from an "off" state to an "on" state, rather than a smooth increase in promoter activity of each cell.

Native promoters do not always fulfill the exact requirements in terms of their regulatory properties including dynamic range, sensitivity, and EC50. One possibility to customize these properties is to use a native promoter, apply error-prone PCR plus an appropriate selection procedure in order to identify mutated promoter versions with the required specifications.

2.2. Generating divergent promoters by error-prone PCR

2.2.1. Error-prone PCR of promoter

1. PCR amplify the promoter region and clone into the reporter plasmid. For *E. coli*, $P_L$tetO-1 promoter from pZE21 was used (Alper *et al.*, 2005; Lutz and Bujard, 1997). pZE21 already contained a gfp(ASV) fluorescent reporter. In yeast, the *TEF1* promoter (constitutive) or *DAN1* promoter (inducible) were used in a CEN-ARS plasmid (p416-TEF) driving yECitrine reporter protein (Mumberg *et al.*, 1995; Nevoigt *et al.*, 2006, 2007; Sheff and Thorn, 2004). The *TEF1* promoter was already present in the p416-TEF plasmid (Mumberg *et al.*, 1995) and we clone the yECitrine reporter downstream of the *TEF1* promoter. To clone the *DAN1* promoter, a fragment of 551 bp upstream of the *DAN1* open reading frame was PCR amplified from *S. cerevisiae* genomic DNA and cloned into p416-TEF-yECitrine, replacing the *TEF1* promoter.
2. Generate diversity in the promoter using error-prone PCR. There are several techniques and commercial kits for error-prone PCR, here we used the method using 8-oxo-2′-deoxyguanosine and 6-(2-deoxy-β-D-ribofuranosyl)-3,4-dihydro-8Hpyrimido-[4,5-c][1,2]oxazin-7-one nucleotide analogues (Zaccolo and Gherardi, 1999). The primers used in the PCR reaction are designed to facilitate cloning the mutagenized PCR product. For *E. coli,* the primers contained *Kpn*I and *Mlu*I sites on the sense and antisense primer, respectively. In yeast, the primers overlapped with the promoter flanking regions in the host plasmid. In both cases, the host plasmid was the same as the template plasmid used for error-prone PCR. Table 6.1 shows a suggested PCR reaction mixture

Table 6.1 PCR reaction mixture

Final concentration	Volume (μL)	Component	Stock concentration
	36.25	Nuclease-free water	
	5	10 × PCR reaction buffer	
200 μ*M* each	1	dNTP mix	10 m*M*
20 μ*M*	2.5	8-oxodGTP[a]	10 m*M*
20 μ*M*	2.5	dPTP[b]	10 m*M*
0.5 μ*M*	0.5	Sense primer	50 μ*M*
0.5 μ*M*	0.5	Antisense primer	50 μ*M*
2 μ*M*	1.25	Plasmid template (~2 kb)	100 ng/μL [~80 n*M*]
0.05 U/μL	0.5	Taq DNA polymerase	5 U/μL
	50	Total	

[a] 8-oxo-2′-Deoxyguanosine.
[b] 6-(2-Deoxy-β-D-ribofuranosyl)-3,4-dihydro-8Hpyrimido-[4,5-c][1,2]oxazin-7-one.

Table 6.2 PCR reaction protocol

Cycles	Time	Temperature (°C)	Action
1	30 s	95	Initial melt
10, 20, and 30[a]	10 s	95	Melt
	30 s	$T_{m}{}^{b}-5$	Anneal
	1 min/kb	72	Extend
1	5 min	72	Final extend

[a] Fewer cycles result in a lower mutation frequency.
[b] Lower melting temperature of two primers.

and Table 6.2 shows a standard mutagenesis PCR. The number of cycles in the PCR determines the mutation frequency. To access a range of different mutation rates, 10, 20, and 30 cycles were performed and the PCR products mixed together afterward at equal amounts.

3. Purify PCR products using a spin kit such as the QIAquick PCR Purification Kit (Qiagen, Valencia, CA, USA). If DNA yield is very low (<1 μg), a traditional PCR can be used to amplify the product.

2.2.2. *E. coli* promoter library cloning

1. Digest ~3 μg of the pZE21 reporter plasmid, and gel purify to remove the native $P_L$tetO-1 promoter.
2. Digest and ligate the mutagenized promoter library into the cut version of the reporter plasmid.
3. Transform into library efficiency *E. coli*, such as Library Efficiency DH5α Competent Cells (Invitrogen, Carlsbad, CA, USA). To have a large amount of diversity, we generated 30,000 colonies spread over multiple large (150 mm) agar plates. Multiple transformations may be necessary to reach 30,000 colonies. Plate at a density that prevents colonies from affecting neighbors growth.

2.2.3. *S. cerevisiae* promoter library cloning

1. Digest and gel purify an appropriate amount of p416-TEF1-yECitrine plasmid with *Sac*I and *Xba*1 in order to remove the unmutated *TEF1* promoter.
2. *S. cerevisiae* competent cells prepared with the "FROZEN-EZ-YEAST Transformation" kit according to the protocol provided by the supplier (Zymo Research, Orange, CA, USA) were cotransformed with the *Sac*I/*Xba*I cut plasmid backbone and the mutagenized promoter versions (PCR products; Section 2.2.1) according to the recombinatorial cloning method described (Raymond *et al.*, 1999). In one transformation,

equimolar amounts of vector and products from error-prone PCR (together making up about 200 ng DNA) were transferred into 50 μL competent yeast cells. The library sizes were 13,500 for the *TEF1* promoter and 12,000 for the *DAN1* promoter.

2.3. Characterizing promoter mutants

2.3.1. Initial selection and quality control

An initial screen is necessary to ensure the mutants span a desirable range, are monodispersed in the population, and are not identical promoter sequences.

1. Visual inspection of the fluorescent colonies on a plate should be sufficient to check if a desirable range of promoter strengths is represented in the library. Colony fluorescence can be determined using a transilluminator at the excitation wavelength, such as the Dark Reader (Clare Chemical Research, Dolores, CO, USA) with glasses that filter the excitation wavelength. If all colonies appear to have the same fluorescence, the mutation rates may not have been high enough to generate adequate diversity.
2. Pick colonies with a range of fluorescence and a total number of colonies that is twice the desired resolution of promoter strengths (~40 colonies were chosen commonly). This can be accomplished by either visual inspection or fluorescence-activated cell sorting (FACS). For FACS method, combine all colonies of the library after transformation, inoculate an appropriate volume of medium, grow the mixture up to the exponential growth phase, and subject the cells to a FACS selection. Using FACS, sort single cells from different fluorescence ranges into different wells of a microtiter plate to obtain a collection of finely graded promoter strengths. This procedure was used for selecting the mutagenized yeast promoters while *E. coli* promoters were selected by visual inspection of the transformants on plate.
3. Purify plasmids from the selected clones and transform them into new cells to remove any spontaneous mutations that may have occurred in the host strain.
4. Determine if each mutant is monodispersed. Grow each mutant into exponential phase. Measure fluorescence distribution by FACS. If distribution is monodispersed with a coefficient of variance approximately equal to the WT promoter, include promoter in library. Discard the promoter, if polymodal distribution or exceptionally wide distribution.
5. Sequence all promoters. Remove those with identical sequences.

2.3.2. Reporter considerations

To be useful in designing genetic circuits, the specific transcriptional parameters for each promoter in the library must be measured experimentally. Here we discuss how to measure promoter activity. In an effort to standardize these measurements, the concept of "polymerases per second" (PoPs), or the number of RNAP that pass through a segment of DNA each second, has been put forward, although no method to directly measure this has been described (Kelly *et al.*, 2009). Instead indirect measures are used to estimate PoPs.

In vivo transcriptional rates are commonly evaluated by (a) direct measurement of mRNA concentrations by quantitative RT-PCR, (b) the concentration of a reporter protein via fluorescence (e.g., gfp), or (c) the enzymatic activity of a reporter protein (β-galactosidase, luciferase activity). In general, all these measurements have two major limitations: they (i) only record the actual concentration of transcript or protein in a cell and (ii) they usually provide average values of the entire cell population. Fluorescent proteins offer the useful option of quantifying on a FACS, allowing individual cells to be measured. This gives an estimate of the cell-to-cell variation. For reporter proteins, translational and posttranslational effects can confound the characterization promoter activity. In particular, changes in mRNA or protein half-life of the reporter may affect the concentration in unexpected ways.

2.3.3. *E. coli* constitutive promoter characterization

All *E. coli* characterization was carried out in defined M9G/CAA media (Table 6.3). M9 media was prepared as described (Maniatis *et al.*, 1982).

Table 6.3 M9G/CAA components

Per 1 L	Component
12.8 g	$Na_2HPO_4 \cdot 7H_2O$
3 g	KH_2PO_4
2.5 g	NaCl
1 g	NH_4Cl
After autoclave, added sterile	
2 mL	$MgSO_4$ (1 *M*)
100 μL	$CaCl_2$ (1 *M*)
5 g	D-glucose
1 g	Casamino acids

2.3.3.1. mRNA measurement

1. 50 mL cultures of M9G/CAA were grown at 37 °C and 225 rpm in 250-mL Erlenmeyer flasks and inoculated with 500 μL of an overnight LB preculture.
2. After 3 h, total RNA was extracted from a 1.5 mL sample using RNEasy Mini Kit (Qiagen).
3. RNA concentrations were normalized to 20 μg/mL and stored at −20 °C until analysis.
4. One-step RT-PCR was carried out using iScript One-Step RT-PCR kit with SYBR-Green (Bio-Rad, PLACE) measured using a iCycler (Bio-Rad). Primers were designed for an amplicon in the gfp transcript and were used at 100 n*M*. Template RNA was added at 20 ng/50 μL reaction and was analyzed using iScript protocol for the thermal cycler. Threshold cycles and transcript concentration were calculated using the Bio-Rad software. Comparing expression to a control gene can significantly reduce the error in estimating RNA concentrations (Kelly *et al.*, 2009).

2.3.3.2. Growth-phase yECitrine fluorescence

1. 50 mL cultures of M9G/CAA were grown at 37 °C and 225 rpm in 250 mL Erlenmeyer flasks and inoculated with 500 μL of an overnight LB preculture.
2. Optical density (A_{600}) and fluorescence (EX = 488 nm/EM = 575 nm) were measured throughout the exponential phase. Fluorescence was plotted as a function of optical density and the best-fit slope, representing the exponential phase steady-state fluorophore concentration, f_{SS}, was calculated (Fig. 6.3).
3. Transcriptional rate was estimated using a dynamic model that accounted for gfp maturation, protease degradation, and growth (Leveau and Lindow, 2001). The rate of promoter driven production, *P*, can be calculated by Eq. (6.1).

$$P = f_{ss}\left(\mu\left(1+\frac{\mu}{m}\right) + D\left(2+\frac{\mu}{m}\right)\right) \tag{6.1}$$

In Eq. (6.1), μ is growth rate, *m* is the maturation constant for oxygen-dependent fluorophore activation of gfp (1.5 h^{-1} by Andersen *et al.*, 1998), and *D* is the first-order rate constant for protease-mediated degradation set to 0.23 h^{-1} according to Cormack *et al.* (1996). Kelly *et al.* (2009) has given an alternative formulation to calculate RNA synthesis rates.

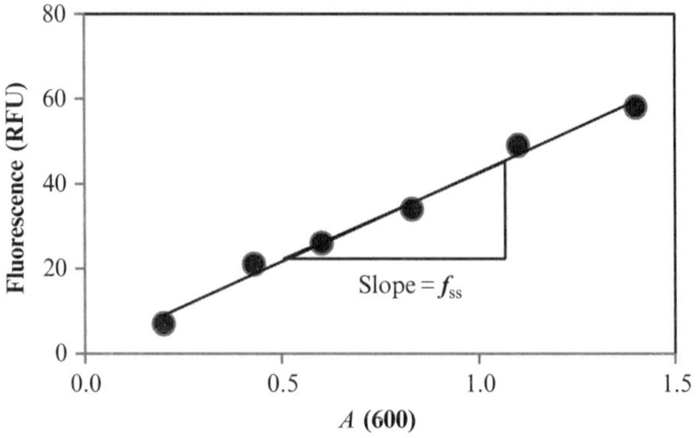

Figure 6.3 Estimating steady state fluorescence.

2.3.3.3. Chloramphenicol Minimum Inhibitory Concentration Another reporter was incorporated into the study to characterize the promoters. Resistance to the antibiotic chloramphenicol is dependent on the concentration of chloramphenicol acetyl transferase (*cat*) in the cell. Therefore, the concentration of chloramphenicol that first inhibits cell growth is a proxy for *cat* concentration and by extension the RNA synthesis rate of *cat*.

1. The promoter constructs were digested with *Kpn*I/*Mlu*I to remove gfp (ASV) and replaced with *cat*, conferring resistance to the antibiotic chloramphenicol, such that increased promoter strength should yield increased resistance to the antibiotic. The *cat* open reading frame was amplified from pACYC184 using primers that introduced *Kpn*I/*Mlu*I sites for cloning into the promoter vectors.
2. Cells were grown to exponential phase in LB containing kanamycin, the antibiotic used to maintain the plasmid.
3. Each strain was streaked onto agar plates containing chloramphenicol concentrations of 0, 17, 34, 68, 225, 282, 338, 372, 425, and 450 μg/mL and incubated overnight at 37 °C.
4. The lowest concentration that inhibited growth was recorded.

2.3.4. *S. cerevisiae* constitutive promoter characterization

To characterize yeast transformants using *URA3* as a selectable marker, we used a yeast synthetic complete medium lacking uracil (YSC-Ura$^-$; Table 6.4). For the growth experiments using respiratory carbon sources, 2% ethanol and 2% glycerol were added to the medium instead of 2% glucose.

Table 6.4 YSC-Ura$^-$components

Per 950 mL	Component
6.7 g	Yeast nitrogen base without amino acids (Difco)
0.77 g	CSM-URA (Qbiogene, Irvine, CA)
After autoclave, added sterile	
50 mL	40% glucose
	or
	40% ethanol/40% glycerol

2.3.4.1. mRNA measurement

1. 20 mL cultures of YSC-Ura$^-$medium were inoculated with 1 mL of an overnight preculture and grown at 30 °C and 200 rpm in 250 mL Erlyenmeyer flasks.
2. After 6 h, total RNA was extracted from a 1.0 mL sample using the RiboPureTM-Yeast Kit (Ambion, Austin, TX).
3. 100 μL of the eluate was mixed with 100 μL deionized H_2O. The concentrations were determined for all samples (between 70 and 95 ng/μL) and the samples were stored at −20 °C until use. Shortly before RT-PCR, 100 μL of a 10 ng/μL total RNA solution was prepared for each sample.
4. RT-PCR results were obtained out using the same kit, PCR cycler and software described above for the *E. coli* promoter characterization. RT-PCR primers were designed to amplify a 160 bp fragment within the coding sequence of yECitrine. Primer concentration used in the RT-PCR was 300 n*M*. Template RNA was added at 100 ng (10 μL of 10 ng/μL solution) in a 50 μL PCR reaction. We also amplified the actin mRNA for normalization. However, the results were similar to those obtained by simply using exactly the same total RNA amount in the RT-PCR reactions.

2.3.4.2. Growth-phase yECitrine fluorescence

1. 20 mL cultures of CSM-Ura$^-$medium, containing either (a) 2% glucose or (b) 2% ethanol and 2% glycerol were inoculated with 1 mL of an overnight preculture and grown at 30 °C and 200 rpm in 250 mL Erlenmeyer flasks. Different carbon sources were used to evaluate the promoter under fermentative and nonfermentative conditions.
2. Measurements of specific fluorescence were performed using cells harvested from the logarithmic phase ($A_{600} \sim 1$). The fluorescence of yECitrine was measured in diluted cultures (optical density at 600 nm [A_{600}] was alwaysbetween 0.1 and 0.3) in cuvettes using a fluorescence

spectrometer (HITACHI F-2500) (EX = 502 nm/EM = 532 nm). The specific fluorescence is referred to as the ratio of the fluorescence level measured and the A_{600} measured in the same cuvette.

2.4. Using iterative diversification/characterization to generate specific inducibility properties

To demonstrate the feasibility of directed evolution for customizing promoter regulation, the oxygen sensitivity of the *DAN1* promoter of *S. cerevisiae* was taken as an example. This promoter is inactive under aerobic conditions but highly active under anaerobic ones (Kwast *et al.*, 2002; Piper *et al.*, 2002; ter Linde *et al.*, 1999). The *DAN1* promoter requires fastidiously anaerobic conditions for induction, which requires sparging cultures with nitrogen (Cohen *et al.*, 2001; Sertil *et al.*, 2003). A more practical and convenient induction method would involve simple elimination or reduction of aeration. This could be accomplished by increasing the oxygen concentration by which the promoter transitions from active to repressed, essentially the EC50. To demonstrate the applicability of error-prone PCR for modifications of transfer functions of inducible promoters, we evolved the oxygen-repressed *S. cerevisiae DAN1* promoter for a changing EC50 value in terms of the degree of oxygen limitation which is required for induction.

2.4.1. Varying oxygen availability

In order to achieve different oxygen availabilities in small scale experiments, the following conditions were used. Erlenmeyer flasks (20 mL medium in 250 mL flasks) were used to provide aerobic conditions. To achieve microaerobiosis, the culture was grown in closed, air-tight vials. The latter conditions do not completely deplete dissolved oxygen at the beginning of the experiment but sharply lower oxygen availability during growth of the cells when residual dissolved oxygen is consumed. Fastidious anaerobiosis was obtained by sparging the culture in the air-tight vials with pure nitrogen.

2.4.2. Evolving promoter to increase oxygen EC50

1. The mutant library from Section 2.2.3 (transformants were scraped of the plates and mixed) and the control strain (containing the reporter plasmid with the wild-type *DAN1* promoter) were pregrown overnight in 20 mL cultures using YSC-Ura$^-$ medium at 30 °C and 200 rpm in 250 mL Erlenmeyer flasks.
2. 20 mL YSC-Ura$^-$medium was inoculated with the overnight preculture adjusting $A_{600} \sim 0.2$ and grown at the above conditions for 6 h.

3. For flow cytometry/FACS, 1.5 mL of cells were harvested from aerobic culture, centrifuged at 600×*g* and resuspended in 2 mL of ice-cold deionized water.
4. Mutants that were repressed were selected by FACS. The histogram of the WT *DAN1* promoter on aerobic conditions can be used in order to set the FACS selection conditions for the first round. By this, all promoter mutants in the library which showed a specific fluorescence under aerobic conditions (repressed state) above the yeast autofluorescence (control sample) were removed.
5. In order to prepare the second FACS round, we needed three cultures: (i) the selected cells from the first FACS round (Step 4) under microaerobic conditions, (ii) the control strain under microaerobiosis, and (iii) the control strain under fastidious anaerobiosis. An appropriate volume of YSC-Ura^-medium containing 10 mg/L ergosterol and 420 mg/L Tween 80 was inoculated with overnight precultures of the required strains adjusting $A_{600} \sim 0.2$ and grown aerobically (as in Steps 1 and 2) for 2 h.
6. Afterward, the cultures were transferred to microaerobic growth and fastidious anaerobic conditions. These cultures were stored in an incubator at 30 °C without mixing. After 4 h, 3 mL of the micoaerobic and fastidiously anaerobic cultures were transferred into Erlenmeyer flasks and shaken for 45 min at 200 rpm and 30 °C in order to allow oxygen-dependent maturation of yECitrine (Tsien, 1998). The incubation time of 45 min was previously determined to be a good compromise between converting about ~ 75% of the generated yECitrine into the mature/active form and preventing any falsification by newly synthesized yECitrine during the 45 min maturation period. In fact, the addition of 300 μg/mL cycloheximide, a translation inhibitor, during this phase did not change the results at all.
7. For FACS, 1.5 mL of the "maturation cultures" (Step 6) were harvested (microaerobic and anaerobic), centrifuged at 600×*g* and resuspended in 2 mL of ice-cold deionized water.
8. In the second FACS round, mutants that were activated in microaerobic growth were selected. The selection criteria were set to only isolate single cells which had a very high expression/fluorescence. In order to evaluate what range of fluorescence could be considered as a high expression in terms of the induced *DAN1* promoter, we used the control strain (WT *DAN1*) grown under fastidious anaerobiosis conditions. During the second selection round, the cells with high-specific fluorescence were directly sorted into single wells of a microtiter plate containing 100 μL YSC-Ura^-medium. The resulting clones were used to isolate the plasmids and the latter were retransferred into fresh yeast before detailed characterization.

2.4.3. Inducible promoter characterization

Measurements of mRNA levels and specific fluorescence were performed as described in Sections 2.3.4.1 and 2.3.4.2, respectively with a few modifications.

1. The preevaluation of the FACS selected clones was carried out by measuring the specific fluorescence after incubation for 4 h under microaerobic conditions plus maturation period as described in Step 6 of Section 2.4.2 and comparing them with the fluorescence of the control strain under the same conditions. Strains with increased microaerobic fluorescence compared to the control were selected.
2. In order to characterize the dynamics of the two best-performing mutated promoters (after preevaluation) in comparison to the unmodified native *DAN1* promoter we characterized them in small scale under aerobic and microaerobic conditions as described above (Section 2.4.1). A_{600} and specific fluorescence was hourly recorded during a time period of 5 h after inoculation. Total RNA for RT-PCR was isolated from the same time points. Media used for microaerobic or anaerobic cultivation were also supplemented with 10 mg/L ergosterol and 420 mg/L Tween 80.

3. Generating Stable Tandem Gene Arrays for Controlling RNA Synthesis Rate

A second way to change RNA synthesis rates is to change the copy number of the promoter/gene/terminator construct (expression cassette) in the cell, thereby altering the number of locations that transcription initiation can occur simultaneously. We note that changing copy number can be used in combination with mutagenized promoters discusses in Section 2. Copy number can be changed by modifying the origin of replication of a plasmid or by changing the number of copies on the genome. The plasmid based-approach can be achieved in *E. coli* by placing the cassette on either low copy (pSC101), medium copy (pBR322), or high copy (pUC18) plasmids. In yeast, one can choose between yeast CEN/ARS plasmids for single copy or the 2 μm replicon for high-copy gene expression.

Because plasmids have unfavorable stability properties (Tyo *et al.*, 2009), genomic integration is desirable, particularly in industrial applications. However, multiple integration of an expression cassette into different loci is cumbersome. Here, tandem gene arrays are a useful alternative. Tandem gene arrays refer to many genomic copies of an expression cassette that are linked head-to-tail to each other. By being physically linked, tandem gene arrays have increased stability properties (Tyo *et al.*, 2009).

Here we describe a method for creating high-copy tandem gene arrays in *E. coli* called CIChE. The method works by integrating one copy of an expression cassette alongside an antibiotic resistance marker, flanked on either side by homologous repeats. Copy number amplification through tandem gene duplication is achieved by gradually increasing antibiotic concentration. The increased cellular requirements for antibiotic resistance create a selection pressure for generation of multicopy tandem gene arrays. In this protocol, we will rely on the pTGD plasmid (Tyo *et al.*, 2009) which contains the *cat* gene conferring resistance to chloramphenicol, a multiple cloning site close to the resistance marker where the expression cassette can be inserted and two flanking homologous regions. This method is currently *E. coli* specific.

A common caveat in using multicopy constructs is inadequate cellular transcription factor concentration. If this occurs, increases in copy number will not correlate with increases in mRNA synthesis. Therefore, it is necessary to check the promoter properties have changed as expected when shifting to higher copy number. This problem can be mitigated by overexpressing associated transcription factors, but the pleiotropic effects of such a change must be carefully examined in a strain.

1. Clone the expression cassette of interest into the pTGD vector (Fig. 6.4) by standard methods. The pTGD plasmid contains unique *Sph*I, *Sac*I, *Xho*I, *Pst*I, *Kpn*I, and *Mlu*I site in the multicloning site.
2. pTGD can be integrated into the *E. coli* genome directly using the Lambda InCh system (Boyd *et al.*, 2000). A detailed protocol is available at the Lambda InCh website (http://beck2.med.harvard.edu/resources/InCh/download/Lambda_InCh_Manual.pdf). Alternatively, the plasmid can be used as a template for PCR amplification of the integration cassette used by the Lambda Red-based integration method (Datsenko and Wanner, 2000). For Lambda Red integration, the construct, including tandem repeats (from *Bam*HI to *Xba*I sites on Fig. 6.3) should be PCR amplified. PCR primers should have 5′ homology with the target location in the genome. By either method, the promoter/gene construct, flanked by the homologous region on either side is delivered to the genome.
3. Once a successful integration is completed, the strain copy number is increased by ramping up the chloramphenicol concentration in the culture. 5 mL cultures in 14 mL culture tubes of LB + 13.6 μg/mL chloramphenicol were inoculated with the CIChE-integrated strain. Cells were grown at 37 °C and 225 rpm to stationary phase (about 12 h). 50 μL of this culture was added to a new 14-mL tube containing the same media, but with 27.2 μg/mL chloramphenicol. Subculturing was continued and the chloramphenicol concentration was increased each time (i.e., 54.4, 108, 170, 340, 515, 680, and 1360 μg/mL).

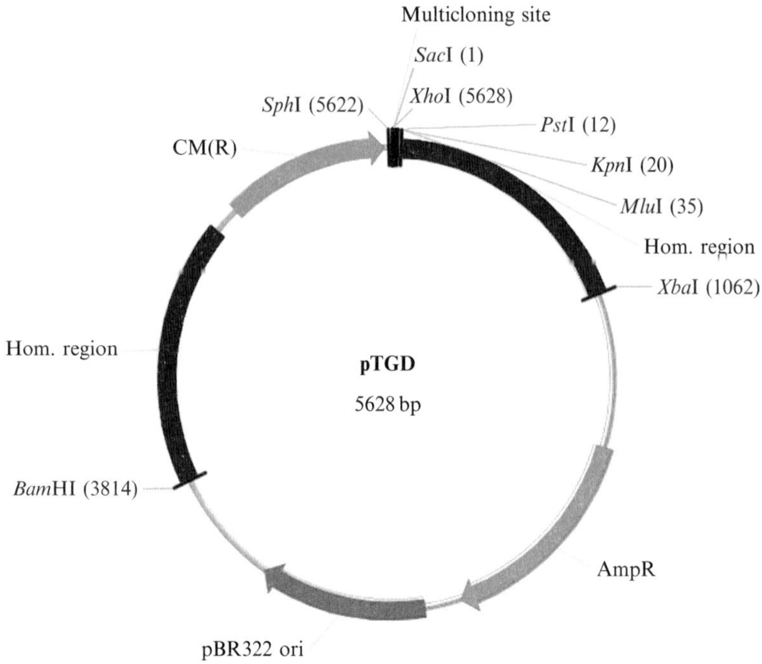

Figure 6.4 pTGD plasmid map. AmpR—betalactamase, pBR322 ori—pBR322 origin of replication, Hom. Region—1 kilobase DNA fragment from *Synechocystis* PCC 6803 chlB necessary for the tandem duplication, CM(R)—chloramphenicol acetyl transferase. Gene(s)-of-interest can be cloned into the multicloning site.

Copy number of the tandem array increases with increasing chloramphenicol concentrations.

4. At the desired chloramphenicol concentration, *recA* is deleted directly. P1 phage transduction is used to deliver a *recA*::*kan* allele from BW26,547 to the evolved strain using the Sauer protocol (http://openwetware.org/wiki/Sauer:P1vir_phage_transduction). *recA* deletion is conveniently verified by UV sensitivity at 3000 μJ using a Stratalinker UV cross-linker (Stratagene, La Jolla, California).

   Note: During Steps 3 and 4, mutations can occur in the expression cassette and copy number can decrease if cells are not grown on the target chloramphenicol concentration. For this reason, it is advisable to minimize (if possible) the number of cell doublings until *recA* is deleted.
5. At this point, both gene copy number and mRNA concentration should be measured. mRNA concentration can be measured as above (2.3.3.1). Gene copy number can be measured by qPCR on the extracted genomic DNA.

4. Concluding Remarks

In this chapter, we describe methods to modify the rate or regulation of RNA synthesis. The rate of transcription (promoter strength) can be changed either by altering promoter-binding interactions through DNA mutations or by varying the copy number of expression cassette. Altering DNA-binding interactions can also be used to modify the way RNA synthesis responds to a stimulus (inducer). Error-prone PCR-based promoter engineering is a desirable approach to create precisely specified promoters. This approach can yield additional insights into the mechanisms of promoter control by comparative sequence analysis of different mutants. Moreover, computational methods can be used in order to link causative mutations with the phenotype (expression level/ regulatory properties). However, computer aided analysis would require the sequencing and phenotyping of a huge number of mutated promoter versions.

An alternate method to alter RNA synthesis rates is to change the amino acid sequence of RNAP or associated transcription factors. Global transcription machinery engineering (gTME) uses this strategy. By mutagenizing a single native transcription factors, the activity of many promoters is simultaneously changed (Alper *et al.*, 2006; Lam *et al.*, 2010). This is a useful approach for addressing multigenic phenotypic traits.

As more libraries are made, the deposition of promoter information (DNA sequence, activity information, etc.) to the Registry of Standard Biological Parts will significantly increase the value of this resource. As a community effort, a library of mapping of DNA sequences to promoter activity can quickly be compiled, especially for constitutive promoters. We would recommend users to deposit characterized promoters to this repository. If you are in the process of designing genetic circuits for a purpose, checking this repository for available promoters may save the time of generating and characterizing promoters yourself.

There are still challenges in characterizing promoters. Most significantly there is no convenient way to directly measure RNA synthesis rates or PoPs. Instead, this parameter is estimated using models that account for cell growth and protein degradation. As well, inducible promoters are more difficult to fully characterize as approximately five inducer concentrations are necessary to estimate the transfer function, much more than one measurement for constitutive promoters. Technologies that can facilitate this data acquisition would be very helpful.

REFERENCES

Alper, H., Fischer, C., Nevoigt, E., and Stephanopoulos, G. (2005). Tuning genetic control through promoter engineering. *Proc. Natl. Acad. Sci. USA* **102,** 12678–12683.

Alper, H., Moxley, J., Nevoigt, E., Fink, G. R., and Stephanopoulos, G. (2006). Engineering yeast transcription machinery for improved ethanol tolerance and production. *Science* **314,** 1565–1568.

Andersen, J., Sternberg, C., Poulsen, L., Bjorn, S., Givskov, M., and Molin, S. (1998). New unstable variants of green fluorescent protein for studies of transient gene expression in bacteria. *Appl. Environ. Microbiol.* **64,** 2240–2246.

Boyd, D., Weiss, D. S., Chen, J. C., and Beckwith, J. (2000). Towards single-copy gene expression systems making gene cloning physiologically relevant: Lambda InCh, a simple Escherichia coli plasmid-chromosome shuttle system. *J. Bacteriol.* **182,** 842–847.

Cohen, B. D., Sertil, O., Abramova, N. E., Davies, K. J. A., and Lowry, C. V. (2001). Induction and repression of DAN1 and the family of anaerobic mannoprotein genes in *Saccharomyces cerevisiae* occurs through a complex array of regulatory sites. *Nucleic Acids Res.* **29,** 799–808.

Cormack, B. P., Valdivia, R. H., and Falkow, S. (1996). FACS-optimized mutants of the green fluorescent protein (GFP). *Gene* **173,** 33–38.

Datsenko, K. A., and Wanner, B. L. (2000). One-step inactivation of chromosomal genes in *Escherichia coli* K-12 using PCR products. *Proc. Natl. Acad. Sci. USA* **97,** 6640–6645.

Ellis, T., Wang, X., and Collins, J. J. (2009). Diversity-based, model-guided construction of synthetic gene networks with predicted functions. *Nat. Biotechnol.* **27,** 465–471.

Kelly, J., Rubin, A., Davis, J., Ajo-Franklin, C., Cumbers, J., Czar, M., de Mora, K., Glieberman, A., Monie, D., and Endy, D. (2009). Measuring the activity of BioBrick promoters using an in vivo reference standard. *J. Biol. Eng.* **3,** 4.

Kwast, K. E., Lai, L. C., Menda, N., James, D. T., Aref, S., and Burke, P. V. (2002). Genomic analyses of anaerobically induced genes in *Saccharomyces cerevisiae*: Functional roles of Rox1 and other factors in mediating the anoxic response. *J. Bacteriol.* **184,** 250–265.

Lam, F. H., Hartner, F. S., Fink, G. R., and Stephanopoulos, G. (2010). Enhancing stress resistance and production phenotypes through transcriptome engineering. *Methods Enzymol.* **470,** 509–532, Vol. 470: Guide to Yeast Genetics.

Leveau, J. H. J., and Lindow, S. E. (2001). Predictive and interpretive simulation of green fluorescent protein expression in reporter bacteria. *J. Bacteriol.* **183,** 6752–6762.

Lutz, R., and Bujard, H. (1997). Independent and tight regulation of transcriptional units in *Escherichia coli* via the LacR/O, the TetR/O and AraC/I1-I2 regulatory elements. *Nucleic Acids Res.* **25,** 1203–1210.

Maniatis, T., Fritsch, E., and Sambrook, J. (1982). Molecular Cloning: A Laboratory Manual. Cold Spring Harbor Laboratory Press, Plainville, NY.

Mumberg, D., Muller, R., and Funk, M. (1995). Yeast vectors for the controlled expression of heterologous proteins in different genetic backgrounds. *Gene* **156,** 119–122.

Murray, H. D., Appleman, J. A., and Gourse, R. L. (2003). Regulation of the *Escherichia coli* rrnB P2 Promoter. *J. Bacteriol.* **185,** 28–34.

Nevoigt, E., Kohnke, J., Fischer, C. R., Alper, H., Stahl, U., and Stephanopoulos, G. (2006). Engineering of promoter replacement cassettes for fine-tuning of gene expression in *Saccharomyces cerevisiae*. *Appl. Environ. Microbiol.* **72,** 5266–5273.

Nevoigt, E., Fischer, C., Mucha, O., Matthaus, F., Stahl, U., and Stephanopoulos, G. (2007). Engineering promoter regulation. *Biotechnol. Bioeng.* **96,** 550–558.

Partow, S., Siewers, V., Bjorn, S., Nielsen, J., and Maury, J. (2010). Characterization of different promoters for designing a new expression vector in Saccharomyces cerevisiae. *Yeast* **27**(11), 955–964. Pub Med ID: 20625983.

Piper, M. D. W., Daran-Lapujade, P., Bro, C., Regenberg, B., Knudsen, S., Nielsen, J., and Pronk, J. T. (2002). Reproducibility of oligonucleotide microarray transcriptome analyses—An interlaboratory comparison using chemostat cultures of *Saccharomyces cerevisiae*. *J. Biol. Chem.* **277,** 37001–37008.

Raymond, C. K., Pownder, T. A., and Sexson, S. L. (1999). General method for plasmid construction using homologous recombination. *BioTechniques* **26,** 134.

Sertil, O., Kapoor, R., Cohen, B. D., Abramova, N., and Lowry, C. V. (2003). Synergistic repression of anaerobic genes by Mot3 and Rox1 in *Saccharomyces cerevisiae*. *Nucleic Acids Res.* **31,** 5831–5837.

Sheff, M. A., and Thorn, K. S. (2004). Optimized cassettes for fluorescent protein tagging in *Saccharomyces cerevisiae*. *Yeast* **21,** 661–670.

ter Linde, J. J. M., Liang, H., Davis, R. W., Steensma, H. Y., van Dijken, J. P., and Pronk, J. T. (1999). Genome-wide transcriptional analysis of aerobic and anaerobic chemostat cultures of *Saccharomyces cerevisiae*. *J. Bacteriol.* **181,** 7409–7413.

Tsien, R. Y. (1998). The green fluorescent protein. *Annu. Rev. Biochem.* **67,** 509–544.

Tyo, K. E., Ajikumar, P. K., and Stephanopoulos, G. (2009). Stabilized gene duplication enables long-term selection-free heterologous pathway expression. *Nat. Biotechnol.* **27,** 760–765.

Tyo, K. E. J., Kocharin, K., and Nielsen, J. (2010). Toward design-based engineering of industrial microbes. *Curr. Opin. Microbiol.* **13,** 255–262.

Weber, W., and Fussenegger, M. (2007). Inducible product gene expression technology tailored to bioprocess engineering. *Curr. Opin. Biotechnol.* **18,** 399–410.

Zaccolo, M., and Gherardi, E. (1999). The effect of high-frequency random mutagenesis on in vitro protein evolution: A study on TEM-1 beta-lactamase. *J. Mol. Biol.* **285,** 775–783.

SECTION TWO

DEVICE AND SYSTEM DESIGN, OPTIMIZATION, AND DEBUGGING

CHAPTER SEVEN

Design and Connection of Robust Genetic Circuits

Adrian Randall,* Patrick Guye,†,‖ Saurabh Gupta,†,‖ Xavier Duportet,‡,‖ *and* Ron Weiss§,¶

Contents

Abstract

Phenotypic robustness is a highly sought after goal for synthetic biology. There are many well-studied examples of robust systems in biology, and for the advancement of synthetic biology, particularly in performance-critical applications, fundamental understanding of how robustness is both achieved and maintained is very important. A synthetic circuit may fail to behave as expected for a multitude of reasons, and since many of these failures are difficult to

* Computational and Systems Biology Initiative, Massachusetts Institute of Technology, Cambridge, Massachusetts, USA
† Computer Science and Artificial Intelligence Laboratory, Cambridge, Massachusetts, USA
‡ INRIA, Paris-Rocquencourt, 78153 Le Chesnay, France
§ Departments of Biological Engineering and Electrical Engineering and Computer Science, Cambridge, Massachusetts, USA
‖ Department of Biological Engineering, Massachusetts Institute of Technology, Cambridge, Massachusetts, USA
¶ Computer Science and Artificial Intelligence Laboratory, Massachusetts Institute of Technology, Cambridge, Massachusetts, USA

Methods in Enzymology, Volume 497
ISSN 0076-6879, DOI: 10.1016/B978-0-12-385075-1.00007-X

predict a *priori*, a better understanding of a circuit's behavior as well as its possible failures are needed. In this chapter, we outline work that has been done in developing design principles for robust synthetic circuits, as well as sharing our experiences designing and constructing gene circuits.

1. Introduction

Robustness is a hallmark of natural biological systems and is an essential goal for engineered systems. Robust biological systems maintain function in unpredictable environments using unreliable components despite external and internal perturbations (Kitano, 2007). Biological robustness is not of a new topic of study (e.g., Waddington, 1942), but recent advances in molecular and network-level understanding of biological systems have sparked a renewed interest in this field. It is inherently difficult to define a generally applicable definition of robustness (Morohashi *et al.*, 2002), as particular behavioral specifications can drastically change the properties and analysis of robustness for a given system (Stelling *et al.*, 2004). Several different notions, such as canalization, stability, parameter insensitivity, buffering, capacitance, and homeostasis, have been used to describe and quantify particular robustness properties in a variety of different contexts (Goulian, 2004).

In comparison to natural systems, it is arguably more tractable to define and evaluate robustness for synthetic biology. Most, if not all efforts in synthetic biology focus on obtaining specific behaviors. As such, we can define robustness as the continuation of a desired engineered behavior despite perturbations and designate any phenomenon that causes the organism to deviate from performing the originally prescribed function (e.g., by gaining fitness at the cost of performing the engineered function) to be detrimental and less robust. One can view natural evolution as a destructive force for synthetic biology, a force that can act counter to engineered constructs in an attempt to rid the organism of any costly unessential processes (Dougherty and Arnold, 2009).

In addition, the study of robustness in natural biological systems will have a different meaning between steady performance in the short-term and evolution and adaptation in the long-term. This issue of timescales is also a problem to consider in the evaluation of robustness in synthetic versus natural systems. While it is inevitable that engineered organisms will eventually evolve away from the initial design, this may not preclude their use as long as they are stable for a sufficiently long duration. For instance, in industrial settings, when engineered organisms fail or evolve, a reasonable action is to "reboot" the system and start with fresh cells. For a variety of medical applications, for example, the therapeutic agent should function for as long as it is clinically relevant, but does not necessarily need to persist longer. Another approach for improving robustness of the engineered systems is to implement a "fail-safe" or "kill switch" that is activated when an error is detected or used when otherwise needed. Whichever mechanisms are employed to improve robustness, it is clear that for critical applications of synthetic biology that robustness is thoroughly evaluated and failures understood and mitigated as much as possible.

Despite the different focuses and goals, recent efforts to understand and quantify robustness in natural systems have been beneficial for engineering applications. Understanding the sources of failure in natural systems and mechanisms that biology uses to cope with such potential problems is essential for the creation of robust synthetic systems. For example, useful design practices can be gleaned from studies of how particular network motifs are able to overcome potentially detrimental phenomenon such as gene expression noise. At the same time, the construction and careful analysis of synthetic circuits can shed light on the robust functioning of natural systems. In this chapter, we review studies relevant to understanding robustness in natural systems, describe efforts in synthetic biology to engineer robust systems and outline how lessons learned from these efforts can be applied to the construction of future synthetic systems.

2. Sources of Failure

In order to design robust systems, one must first understand and account for potential sources of failure. Engineered biological systems can cease to perform the originally designated function for a number of reasons. For example, genetic mutations can affect critical components of a circuit, or circuits can fail due to design shortcomings. With respect to the negative impact of mutations on genetically engineered systems, little quantitative information exists in the synthetic biology literature regarding the exact magnitude of this problem. Anecdotal evidence suggests that depending on a variety of factors, including the DNA sequence, DNA delivery, the maintenance mechanism used, and the metabolic burden placed on the organism, synthetic circuits can fail in a matter of hours or conversely persist for weeks even in the presence of heavy selective pressure against the synthetic constructs (e.g., Balagadde *et al.*, 2005). Mutation rates vary significantly across different species. For example, bacteria have been reported to have mutation rates as low as 1 error per 10^8–10^9 base pairs replicated. As a general rule, high-copy number plasmids with repeat sequences and heavy metabolic load placed on a fast replicating bacterial host are unlikely to persist fully intact for long durations. However, at the opposite end of the spectrum, chromosomal integration of synthetic constructs with light metabolic load and stable DNA sequences in a slowly replicating cell with high-fidelity DNA replication mechanisms is likely to result in DNA constructs that persist for longer periods of time.

Even in the absence of genetic mutations, achieving long-term robust operation by synthetic systems is a formidable task, and one of the major culprits is gene expression noise, an inherent property of biological systems. Noise and randomness serve a variety of important functions in biology, for example, by generating errors in DNA replication and hence enabling evolution, by producing heterogeneity in cell populations, and by creating divergence in cell fates (Rao *et al.*, 2002). Recent studies have elucidated some of the biochemical mechanisms and the genetic parameters that contribute to gene expression noise. For example, high transcription rates and low translation rates reduce fluctuations in protein concentrations (Ozbudak *et al.*, 2002), thereby mitigating the effects of noise. In their study, Ozbudak and colleagues analyzed several inefficiently translated regulatory genes, and suggested that even though efficient translation is energetically more favorable, the importance of maintaining stable concentrations of these proteins resulted in selection of genetic parameters with low-noise characteristics but higher metabolic load. One example is the *cya* gene of *Escherichia coli*, required for production of (Fig. 7.1) cyclic AMP (cAMP), since fluctuations in cAMP levels can be toxic for the cell (Ozbudak *et al.*, 2002). Other recent studies found that genes essential for cellular function, such as genes involved in protein synthesis and

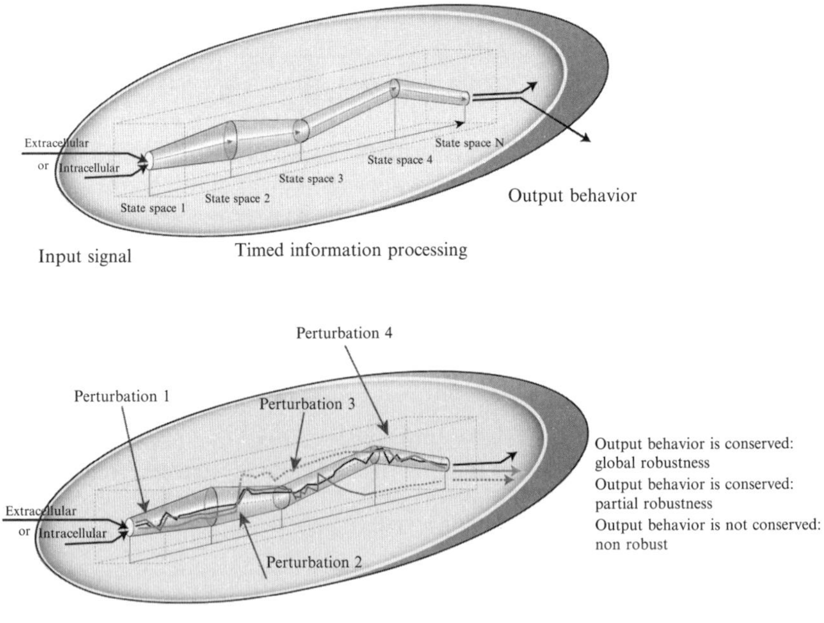

Figure 7.1 A diagram representing how fluctuations can cause a circuit to fail to behave robustly.

degradation, often had less noisy expression levels, while nonessential genes, such as stress-response genes, tend to be more noisy (Bar-Even *et al.*, 2006; Blake *et al.*, 2006; Newman *et al.*, 2006). As a testament to the negative implications of noise, aging has also been correlated with increased fluctuations in gene expression. For example, in individual murine cardiac myocytes and other muscle tissue, several housekeeping and cell type-specific genes exhibit increased noise as a function of organism age (Bahar *et al.*, 2006; Newlands *et al.*, 1998). Perhaps, not surprisingly, the capabilities of some of the initial synthetic circuits such as the toggle switch and repressilator were limited by gene expression noise (Ozbudak *et al.*, 2002). As will be discussed later in this chapter, several synthetic circuits have been constructed since then to quantify gene expression noise both in single promoter and in various simple network topologies. Ultimately, however, a general theory and predictive computational models of noise propagation in gene networks is needed in order to help synthetic biologists mitigate, or exploit, the effects of gene expression noise.

A source of failure that is more difficult to quantify is the robustness of the system in the face of unknown and varying perturbations and changes in environmental conditions. One approach is to employ parameter sensitivity analysis to reveal how potential changes in rate constants affect system performance. However, our incomplete understanding of the biological cell places limits on our ability to guarantee the performance of synthetic systems. For example, undesired and unexpected crosstalk and imperfect insulation from endogenous elements may harm the performance of a synthetic system, often times in situations and under conditions that were not previously tested in a laboratory setting. Designing systems that function well in unexpected and unanticipated conditions remains beyond the scope of most (if not all) existing synthetic biology efforts. However, there are many promising application areas for synthetic biology where the environmental conditions and relevant cellular machinery have been characterized to the point that engineering efforts can focus on (and solve) issues related to obtaining robust system performance under known conditions.

3. Robustness Principles and Examples in Natural Systems

Biology has evolved many mechanisms from the molecular to the population level that help maintain robust behavior in the presence of environmental stresses and perturbations. Just like engineered systems, natural systems have specific operating regimes and are only robust within specific parameters. However, what sets apart biology from most engineered systems is the ability to evolve and adapt. That feature has enabled organisms to thrive in new environments with different temperatures, food sources, predators, and many other factors. It is important to examine what features from the molecular to the population level promote robustness, and incorporate these principles when appropriate into the design of synthetic circuits (Fig. 7.1).

At the molecular level, biology has evolved many mechanisms to ensure fidelity and robustness. For example, the handling of genetic material from replication to transcription and expression are high fidelity processes in many organisms. Mutation rates for DNA replication in prokaryotes and eukaryotes have been estimated to be on the order of 10^{-8} per base pair per division (Drake *et al.*, 1998); there are many mechanisms for DNA repair and proofreading that allow for this level of fidelity. More subtly, degeneracy in the amino acid code leads to synonymous mutations that do not affect protein function (although may affect protein expression; Carlini and Stephan, 2003). In eukaryotic cells, two chromosomes can help rescue a cell from DNA damage and mutations by using a second correct sequence as a template for repair (Wyman *et al.*, 2004). As another example, in yeast it has been shown that a large fraction of knockout mutations (up to 50%) do not affect survival as mutated functions are readily compensated for by other genes (Thatcher *et al.*, 1998).

Developmental biology is a rich source for engineers to understand the inherent robustness of biological systems. A developing embryo is tasked with the problem of growing from a single cell into a patterned, multicellular organism. For example, in *Drosophila melanogaster*, despite variations of initial concentrations and kinetic parameters of constituents in the polarity gradient formed during embryogenesis, the same number of repetitive stripes of differential gene expression are consistently produced (von Dassow *et al.*, 2000). In addition, while the posterior–anterior *bicoid* gradient exhibits significant variability, the morphogen *hunchback* that it regulates has a precise expression profile through a noise filtering mechanism of noise associated with *bicoid* (Houchmandzadeh *et al.*, 2002). Later studies showed that filtering may be possible through an as-of-yet undiscovered third protein that counter-regulates *hunchback* (Hardway *et al.*, 2008). Thus, even though biological circuits contain inherent noise and variation, critical functions like development are able to mitigate the effects of inherent noise and variation and achieve significant precision and order.

Disease and pathogens are also able to demonstrate significant robustness toward hosts that aim to mitigate them. For example, cancer can be regarded as a very robust disease because of its ability to propagate uncontrollably, even in the presence of drugs, the body's immune response and other factors aimed at limiting its growth (Kitano, 2003, 2004b). As highlighted by Kitano (2007), tumors have at least two specific behaviors that enhance their survival in the human body: functional redundancy where heterogeneity in a cancer enables it to maintain population robustness, and feedback-control systems which allow a tumor to counteract environmental stresses like hypoxia and cytotoxins.

HIV is another classic example of a robust pathogen against its host stemming, in part, from its ability to evade the immune system and failure in innate pathways designed to facilitate its destruction. During the acute phase of HIV infection when the immune response begins to detect and respond to the virus, initial selective pressure is placed on the virus, which causes it to mutate rapidly in order to evade the adaptive and innate immune responses (Arnott *et al.*, 2010). As an example of population-level robustness, HIV's reverse transcriptase exhibits a high level of mutation that allows individual virions to evolve and adapt rapidly (Preston *et al.*, 1988). In addition, integration of its genome followed by latency allows the virus to persist for many years relatively unencumbered (Trono *et al.*, 2010), as the genome is maintained in a quiescent state inside of cells.

4. Methods for Obtaining Robust Synthetic Circuits

In order to maintain a robust, engineered phenotype, it is critical that the underlying genotype is maintained or buffered well. Biology has evolved many mechanisms by which phenotypic stability is maintained, such as high-fidelity DNA replication, DNA damage repair, multiple copies of genes, and redundancy. Through these processes, cells are able to reasonably maintain their phenotype through many generations and in the presence of many environmental stresses. Here, we outline what an engineer would need to take into consideration during the design of a robust synthetic circuit.

4.1. Improved genetic stability

When introducing synthetic circuits into cells, in order to maintain the constancy of desired phenotypes, the genetic sequence of the circuit as well as its ability to be expressed need to be maintained. DNA replication is not error-free, and cells can obtain mutations that disrupt behavior of a non-beneficial synthetic genetic circuit will likely be selected for and eventually predominate. Given this selection process, significant care and consideration must be taken in order to ensure the genetic stability of a synthetic circuit during construction, testing, and deployment. Natural systems contain many ways to compensate for the loss of knocked out genes; however, artificial circuits may not contain the same benefits. Therefore, for a robust phenotype, the underlying genotype must be ensured.

Synthetic gene networks are normally either stably integrated into the chromosome, or maintained as mobile genetic elements (plasmids, transposons). Mobile elements are, in general, subject to rapid loss. Synthetic circuits built on these mobile platforms usually carry a selection mechanism to prevent loss in the system, since in the absence of selective pressure to maintain the mobile element, cells not containing the element will be selected for quickly (Smith and Bidochka, 1998). In general, though, mobile elements are easier to build and introduce into target cells, as stable integration and selection may take days to weeks to properly screen and test for circuit fidelity. However, recent improvements in recombination and genetic manipulation may allow broad adaptation of stable, efficient, rapid, and site-specific integration mechanisms of synthetic circuits within the chromosome.

Each method and approach to introducing recombinant DNA has particular benefits and drawbacks that must be considered depending on the application for the circuit. Integration of constructs into yeast and bacteria generally takes advantage of homologous recombination; sites of homology to the genomic locus one would like to integrate into can be added to 5′ and 3′ to the genetic circuit and combined with appropriate selection to allow for clones with stably integrated circuits (Kopecko and Cohen, 1975; Schiestl and Petes, 1991). In mammalian cells if the synthetic circuit is to be maintained for many generations, transfections of DNA without a means for replication and selection of the recombinant DNA are inappropriate, as transiently transfected DNA will be diluted quickly. Episomal vectors have been designed with origins of replication compatible with mammalian cells that allow for stable expression of a construct for longer time period; however, insertional mutagenesis due to spontaneous recombination is possible (Ehrhardt *et al.*, 2008).

Recombination can also act as a valuable tool for DNA integration and rearrangement in cells because of the possibility to integrate large circuits into a specific locus in the genome. There are two major classes of recombination systems that are widely used to engineer cells: site-specific recombination (SSR) and homologous recombination. SSR systems, such as bacteriophage CRE-*lox*, *Lambda*, *HK022*, and yeast FLP-*frt*, can mediate both integration and excision of a genetic circuit (Keravala *et al.*, 2006; Thomson and Ow, 2006). Many of these systems have been optimized to mediate SSR in a broad spectrum of organisms, and efforts are ongoing to acheive higher efficiency and higher specificity for genomic integration.

Using a viral strategy to integrate synthetic circuits in mammalian genomes allows integration and expression of the target construct, as the number of integration events is proportional to the number of viral particles used to infect the cell. In addition, techniques like SSR and homologous recombination have been used frequently for DNA integration and rearrangements in cells. For example, researchers have used zinc-finger nucleases (ZFNs) to cause double-stranded breaks in the genome at specific loci, whereby genetic constructs can be inserted through homologous recombination

(Moehle *et al.*, 2007). However, while integrating a synthetic system into the chromosome has the benefit of increasing the stability of the system and fixing the gene copy number there are disadvantages to this approach. One of the major drawbacks of genomic integration is the decrease in protein expression levels due to the relatively low gene copy number. In addition, if integration is not site-specific, it may disrupt the function of other genes in the chromosome. In mammalian cells, gene silencing is a major issue that often occurs due to epigenetic modifications and chromatin remodeling (Turker, 2002). Therefore one must take care to add selectable markers, integrate into sites known not to be silenced, and/or incorporate genetic elements that counteract silencing. Viral vectors tend to insert themselves into actively transcribing regions of the genome (Matrai *et al.*, 2010), which can knockout the function of endogenous genes. However the use of ZFNs and homologous recombination allows insertion of a circuit into a predefined target. While these technologies allow for site-specificity, there are size limits to integration and ZFNs can cut in off-target locations causing mutations and cytotoxicity (Moehle *et al.*, 2007; Orlando *et al.*, 2010).

Homologous recombination involves nucleotide sequence exchange between two similar or identical molecules of DNA, where natural systems, homologous recombination is vital to repair fatal DNA damage during DNA replication. However, homologous recombination between synthetic vectors and the cell chromosome can also occur spontaneously. In addition to working with strains that have diminished recombination activity, for example, *E. coli* cells that are recA-, designing circuits with minimum internal and chromosomal homology and the chromosome will reduce the likelihood of aberrant recombination events.

4.2. Basic network motifs

For a complex circuit with many components, how parts interact and provide feedback to one another are not only important considerations for desired phenotype but also robustness. For example, it has been shown in *D. melanogaster* segment polarity formation that positive autoregulatory feedback of *wingless* and *engrained* is responsible for the robustness of pattern formation during embryogenesis (Ingolia, 2004). With regards to synthetic biology, a helpful abstraction for understanding feedback has been the application of principles from other engineering disciplines, including control theory and digital/analog circuit design (Purnick and Weiss, 2009). These abstractions with modeling can help a researcher gain insight into the *dynamics* of the system as it functions over time. As cells and organisms are dynamic with respect to time and levels of molecules in a system, understanding how constituents regulate each other and provide feedback is critical for the rational design of robust synthetic circuits.

Naturally occurring network motifs and patterns are often used in new ways in the design of synthetic circuits. A good starting point or understanding and designing robust synthetic circuits is through the analysis of how robust natural circuits employ feedback in useful ways (reviewed in Alon, 2007). For example, network topologies of natural circadian clocks have been used as templates for creating synthetic oscillators (Fig. 7.2). At the heart of these systems is a set of transcriptionally regulated activators and repressors employing feedback to allow for a robust dynamic behavior. Another example of feedback that has been used for creating a robust circuit is through a toggle switch using mutual transcriptional repression (Gardner *et al.*, 2000; Greber *et al.*, 2008). Thus, from understanding natural examples of feedback, engineers can use these same principles in the design of new synthetic circuits.

Noise is a common characteristic to natural genetic circuits, and cells have evolved many mechanisms to both deal with and take advantage of noise. For the robust behavior of a synthetic circuit, one would need to both understand and modulate the noise characteristics as needed. Biological noise stems from a gene's own fluctuations, upstream noise, as well as global noise in a system (Pedraza and van Oudenaarden, 2005). For example, synthetic transcriptional cascades have been used to attenuate noise as well as modulate signal levels and time delays in bacteria (Hooshangi *et al.*, 2005). It has been suggested that there may be an optimal cascade length for minimizing noise (Thattai and van Oudenaarden, 2002), as there appears to be a tradeoff between attenuation of input noise and the noise added through each level of a signaling cascade. Positive and negative feedback can help modulate the gain of a signal in order to regulate the noise of a synthetic circuit; however, the most simple and widely seen mechanism for attenuating noise is through negative feedback (Rao *et al.*, 2002). For example, Serrano and colleagues showed that negative feedback using tetracycline repressor can greatly diminish the variability of GFP expression in *E. coli* (Becskei and Serrano, 2000), however follow-up studies by

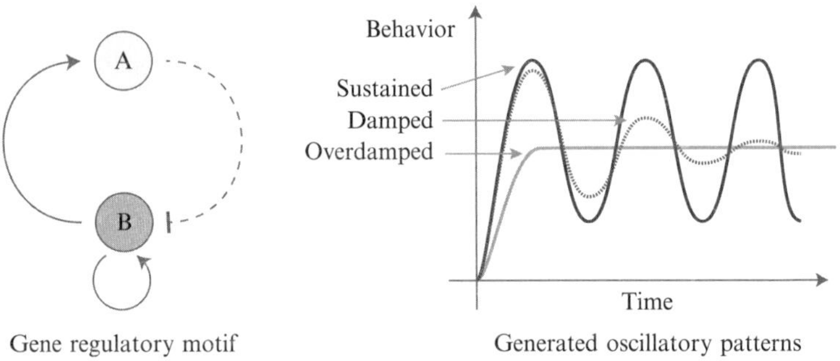

Figure 7.2 Negative feedback loop can generates oscillations.

Hooshangi and colleagues have shown that cascades of negative feedback can amplify noise (Hooshangi *et al.*, 2005).

An important consideration for synthetic circuits, especially when connecting functional modules, is threshold or signal level matching of the components (Purnick and Weiss, 2009). In general, this can be done by either modulation of upstream components through feedback, or through filtering circuits. For example, filters can be used to select for ranges of signals upon which to actuate. Simple negative feedback results in a low-pass filter that prevents short-lived perturbations in inputs from affecting downstream components. In addition, varying repression levels coupled to feed-forward loops can implement a band detect, demonstrating robust, non-monotonic, multicellular behaviors in response to upstream signals. (Basu *et al.*, 2005).

Ideally, a circuit should function properly despite kinetic perturbations, and to achieve this behavior may require sophisticated regulatory motifs to obtain. Integral feedback for obtaining adaptation to perturbation is a commonly used motif, as this form of feedback amplifies intermediate frequencies while attenuating high and low frequencies (Yi *et al.*, 2000). This type of feedback is used in signaling pathways to attenuate themselves even in the continued presence of signal input, an important trait for improving sensitivity to external stimuli. Due to density-dependent kinetics and signals that quickly saturate, this type of feedback may be challenging to implement in a synthetic circuit. A recent theoretical study proposes a three-promoter network to construct a perfectly adapting circuit (Ang *et al.*, 2010), but this has not been proven experimentally yet.

The oscillator provides an insightful case study demonstrating the increased ability to design robust synthetic circuits. Following Liebler and colleagues's repressilator in 2000, there have been several other oscillators that have obtained more robust oscillatory behavior, demonstrating our increased understanding of synthetic circuit construction. The original repressilator comprised three repressors in a cycle of repression (Elowitz and Leibler, 2000). More recent oscillator designs, such as Hasty's fast tunable oscillator and quorum of clocks, use a combination of positive and negative feedback to attain more robust oscillations (Danino *et al.*, 2010; Stricker *et al.*, 2008). In those circuits, computational modeling and experimental data show that the main factor for robustness in oscillators is positive feedback followed by delayed negative feedback, as they found positive feedback allows for synchronization of oscillations as well as tunability in the period.

As circuits become more and more complex, there is an increasing need for *a priori* modeling that can provide accurate predictions about the behavior of a designed circuit. While the behavior of individual modules can be intuitively understood and experimentally validated, the interaction of components within a context of a network or a whole cell makes understanding behaviors like noise much more challenging, highlighting the need for computational models (Rao *et al.*, 2002). In addition, tools like directed evolution combined with rational design may play a role in designing more complicated synthetic circuits, where circuit "templates" can be constructed and further optimized to the desired behavior. (Yokobayashi *et al.*, 2002).

4.3. Modularity

A fundamental aspect of engineered large-scale systems is hierarchical structures, which allows a system to be divided and subdivided into smaller functional modules. A module is a compartmentalized set of devices that performs a task or a set of tasks, where devices within a module are usually highly connected and interdependent. However, modules enjoy certain insulation from external perturbations (and other modules). In keeping with this framework, natural systems can be analyzed in terms of functional modules (Barabási and Oltvai, 2004). For example, at the molecular level, biological systems can be divided into multiple layers (DNA, transcription, mRNA/ncRNA's, translation, protein domains, proteins, organelles, cells, intercellular milieu) through which information flows from module to module. In addition, a cell can be treated as a "module" where a meaningful collection of cells constitutes an organ, which can be considered as a module as well in the context of a complex organism.

In a synthetic circuit, modularity allows for isolation of perturbations from the rest of the system (Kitano, 2007; Stelling *et al.*, 2004), thereby minimizing their effect on the whole system (Kitano, 2004a,b). A robust system can hence be designed by relying on the encapsulation of its components where one can guarantee the confinement of local perturbations to separable parts. Independence and insulation of the different components provided by encapsulation allow them to be easily rewired, where rearrangement of encapsulated components may still allow for correct function.

Based on this idea, synthetic biologists are focusing on modular construction of genetic circuits. To this end, the Registry of Standard Biological Parts was established at MIT as, a collection of characterized pieces of DNA called Bio Bricks that are amenable to modular DNA assembly techniques. These parts are freely distributed to registered scientists, and in turn, scientists can submit their own parts to the registry for others to use. An effective proof of concept of the modular assembly of parts to obtain a complex within a working synthetic circuit is the genetic edge detector (Tabor *et al.*, 2009). In this work, modules with specific functions (sensing, amplifier, cell–cell communication, read-out, AND-gate) were combined to create a circuit with a desired overall function. Each module was characterized using transfer functions and the combination of characterized modules allowed for a more complex a circuit with predicted overall behavior.

A highly valuable aspect of modularity is to abstract and deconvolve complexity. In attempting to understand biological hierarchy and organization, significant emphasis has been placed on treating a biological system as a set of functional "modules," thereby relating biology to other disciplines of engineering where one can describe a system consists of modules with inputs, outputs, and state variables (Milo *et al.*, 2002). This concept is very useful while designing integrated circuits in electrical engineering (Alexander *et al.*, 2009). Downstream circuits can retroactively affect the upstream circuit by diverting electrons. However, one can overcome this problem by including insulation devices or amplifiers in their circuits as necessary. An interesting developing paradigm in studying biological modularity is the concept of retroactivity and insularity (Del Vecchio *et al.*, 2008). When combining multiple genetic modules together, it is imperative to understand how linking of modules affects the overall behavior of the system and incorporate mechanisms to overcome downstream components affected upstream ones. Modeling has been used to study how the addition of upstream or downstream components can affect system behavior (Del Vecchio *et al.*, 2008). For example, if one wants to build an oscillatory circuit that drives the expression of a downstream gene, it is often the case that the downstream circuit is not completely independent of the oscillator, and will thus have possibly unintended effects on the upstream behavior. Frameworks to theoretically analyze these problems and to minimize retroactivity between various modules in a synthetic biology circuit have been recently developed and are beginning to be explored experimentally (Del Vecchio *et al.*, 2008; Saez-Rodriguez *et al.*, 2008).

Hierarchical modularity is a type of organizational structure in which independent modules with discrete functions are combined in a meaningful way (Bashor *et al.*, 2010). Protocols are rules that allow system designers to form proper abstractions at various levels for interactions between modules without omitting necessary details. In the organization of genetic regulatory networks, one node may consist of a regulatory region (promoter) and protein coding regions. Such nodes can interact with other nodes downstream to form regulatory interactions in which the upstream node is an input and the downstream node is an output. These regulatory interactions are often linked together in defined patterns to form a motif (Alon, 2007). Similar to nodes, each motif also has inputs and outputs and performs an information processing function that is one level higher in the network hierarchy (Milo *et al.*, 2002). In this way, a higher-order complex genetic network interaction can be regarded as multiple motifs interacting with each other (Milo *et al.*, 2004). Protocols are the rules by which nodes or motifs are connected to achieve desired functionality. Hence, protocols determine the flow of information from one module to another and therefore play a critical role in the robustness of any genetic circuit.

Engineering gene-regulatory networks to be scale-free is argued to provide robustness for the global system. Scale-free networks have the property of common vertices or "hubs" that allows for fault-tolerant behavior. The disruption of a particular node is unlikely to perturb other regions of the network as the probability for this node to be a hub for an adjacent node is very small (Goulian, 2004). Researchers have also demonstrated that dynamics can be rendered robust when the scale-free topology is used in a circuit, preventing the network from behaving chaotically after small perturbations have been applied (Aldana and Cluzel, 2003). Robustness can also be achieved by combining multiple levels of regulation in the circuit. This hierarchy could rely on the superposition of layers such as controlled transcription, translation, posttranslational modification, and degradation (Stelling *et al.*, 2004). Jim Collins and colleagues have postulated that biosensors as robust as those of natural organisms could be designed using these hybrid solutions, harnessing the desired characteristics of each layer of regulation (Khalil and Collins, 2010).

Division into hierarchies is a common characteristic of engineering systems and is very useful in the abstraction of large complex systems into simpler, more basic systems. This type of abstraction helps hide tedious details and helps the architect focus on overall functionality. This is also beneficial for managing organization since engineers can work in a coordinated fashion and somewhat independently at each level of hierarchy. (Heinemann and Panke, 2006). A complex biological system shares some important properties to a complex dynamic engineering system, and hence it is argued that decomposing it into simple hierarchical modules aids in understanding its overall behavior. Based on this approach, synthetic biologists have successfully engineered simple regulatory modules which have laid the foundation for building more intricate designs. Since these modules have well-defined interfaces, they can be systematically connected in multiple alternate topologies. As an example, this kind of analysis has been applied to evolution and how optimal networks are chosen for a particular function (Boyle and Silver, 2009).

4.4. Decoupling/orthogonality

In order to minimize crosstalk between an engineered circuit and a host's machinery, synthetic biologists often aim to create circuits using biological parts that are decoupled from their host cell machinery as much as possible. Being able to characterize the biological function of these basic parts well and provide a "plug and play" feature is a goal for synthetic biology. However, in order to characterize and validate parts, the effects a host plays on their behavior should be either minimized, or where applicable, quantified. In the design and use of synthetic circuits, the concepts of decoupling and orthogonality from engineering have been applied toward understanding the need to minimize host effects.

Decoupling signifies the desire to isolate low-level variation from high-level functionalities and "buffer" different components. Natural mobile regulatory circuits like integrons, phages, transposons present very good examples of orthogonal systems that are evolutionarily selected to be largely context-independent (Frost *et al.*, 2005; Mazel, 2006). For example, T7 phage polymerase is able to transcribe genes from only the T7 promoter sequence in most hosts. A similar example of a synthetic orthogonal translation system based on engineered ribosomes is described by Wang *et al.* (2007), where orthogonal ribosome-mRNA pairs are used to process information in parallel and independent of the wild-type counterparts. In addition, by designing synthetic systems which are translated and transcribed by completely orthogonal gene expression machinery can act as another way of taking advantage of a host cell without interfering with its components (An and Chin, 2009). The authors put forth an idea that interference from the endogenous cellular networks can be reduced if a synthetic system has its own set of ribosomes and polymerases. With the creation of orthogonal regulatory elements like previously described, minimal synthetic gene networks can function more independently of the cellular context and thereby allow for better orthogonality with higher predictability and robustness.

It is often difficult to achieve true modularity in genetic circuits because of noise and interference from other parts of the host cell. As described earlier, careful decoupling of individual modules should improve predictability and repeatability for synthetic systems (Khalil and Collins, 2010). As an example, Kobayashi and colleagues have shown a way to decouple synthetic circuits from the interference presented by the chassis of the system through the use of a bistable toggle to control another functional module (Kobayashi *et al.*, 2002). The authors were able to make the response of synthetic circuit more robust and digital-like by minimizing the effects of intrinsic noise through the binary "on/off" behavior of their toggle.

Usually, synthetic circuits are affected by random fluctuations in concentrations of its components or by signals that temporarily change system parameters because of stochastic effects due, for instance, to plasmid copy number. This results in a mixed population of cells with different expression states. A study by Murphy *et al.* (2010) demonstrated that one of the ways to decouple the engineered gene networks from noise is by coupling it to the regulatory circuitry of the cell. The authors replicated this concept found in natural systems and illustrated how simple changes can be employed to control and decouple noise in synthetic eukaryotic gene expression. In their study, a tet-regulated GAL promoter controlling the expression of TetR repressor was modified with mutations in the TATA box to finely tune and control gene expression. This incorporation of TATA box mutations within an upstream regulatory promoter is still able to maintain a similar dynamic range of gene expression as the parent circuit, however, they highlight how promoter mutations can be used to mitigate noise without hampering circuit function.

Undesired evolution is a significant cause for a synthetic circuit to lose its function. One of the ways to mitigate it is to use endogenous DNA-repair systems to correct any mismatches that might lead the host to not express the circuits from implanted DNA (de Lorenzo and Danchin, 2008). In this publication, the authors argue that developing orthogonal parts is one end of the spectrum of solutions. However, evolution can be used to develop evolvable circuits, which can aid in maintaining a circuit in a mutating host (Silva-Rocha and de Lorenzo, 2008). Ultimately, the question is whether the synthetic circuits can be made resistant to mutations even though they are under continuous evolutionary pressure to mutate.

4.5. Redundant pathways/alternative (or fail-safe) mechanisms

A common mechanism in engineered systems for improving robustness is the implementation of *N*-modular redundancy, where *N* identical units together regulate the operation of a system. The separate units typically receive the same inputs, each computes votes for a particular action to carry out, and the action with the highest number of votes is the one that is pursued. In these systems, a single module failure does not result in system malfunction (or *m* failures, where $m < N$ based on the actual voting scheme used). These alternative, or fail-safe, mechanisms provide higher tolerance against the failure of individual components through redundancy.

It is interesting to note that while gene duplication is observed in natural system and certain network motifs recur frequently, there have been no discoveries yet of precisely duplicated circuits (Kitano, 2004a,b). Such precise gene circuit duplication was deemed to be evolutionary unstable, and in situations where duplications were found, one of the duplicates was ultimately silenced (Kafri *et al.*, 2005). Apparently, maintenance of a duplicate circuit requires a change of function, either toward a more specialized role or a completely new role (Kafri *et al.*, 2005). An important question is then why severe mutations often do not give rise to observably abnormal phenotypes (Fig. 7.3). The answer probably lies with redundant paralog circuits, which are expressed in a dissimilar fashion in most growth conditions (Kafri *et al.*, 2005). In the case of a mutation, the intact paralog is able, at least partially, to provide backup and maintain the function originally provided by the now-mutant circuit.

For synthetic circuits, implementing *N*-modular redundancy or similar mechanisms may be nonetheless useful since the engineered circuits do not necessarily need to function over long evolutionary time scales. One can certainly envision the creation of completely redundant pathways that are able to significantly improve the mean-time-to-failure of a synthetic system, and perhaps integrating these with a mechanism for aborting the host cell as soon as a single failure is detected.

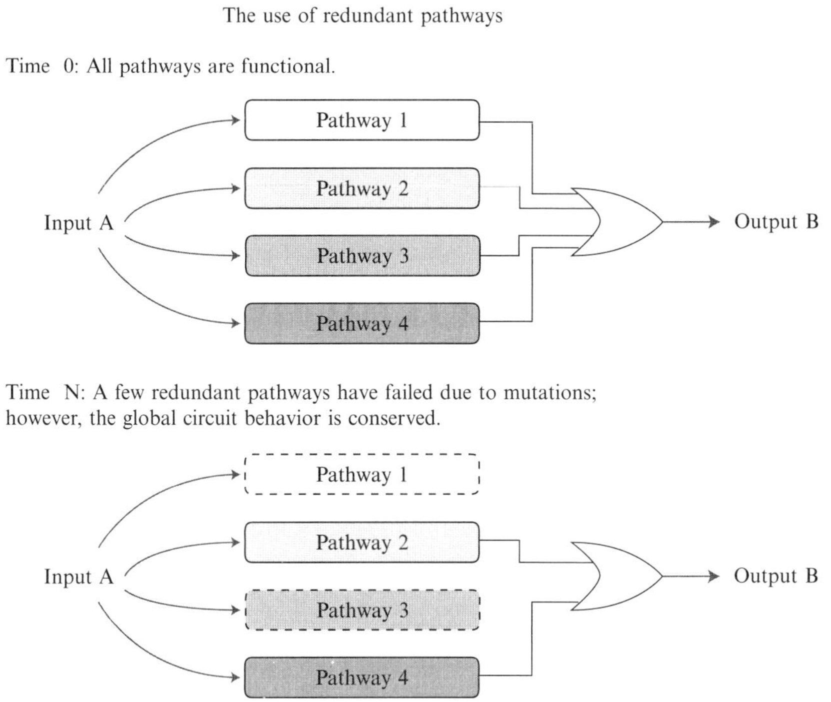

Figure 7.3 Some synthetic gene circuits may take advantage of redundant pathways, where a failure in one pathway does not disrupt overall behavior due to other pathways able to compensate for the failure.

4.6. Multicellularity (redundancy and heterogeneity)

There are limitations to what single cells can do on their own. Multicellular and multispecies systems allow a population of cells to synchronize, coordinate, and compute information relevant to other cells in this population, which can ultimately lead to very complex behaviors. Some applications, such as tissue engineering or artificial ecosystems explicitly depend on using heterogeneous multicellular systems. This section focuses on the most prominent characteristics of communities, multicellular-, and multispecies systems: heterogeneity/specialization, redundancy, intercellular communication, and homeostasis, and how they can lead to overall improved system robustness.

4.6.1. Intercellular communication

One of the most challenging aspects of engineering artificial multicellular systems is intercellular communication, the exchange of information between cells. In bacteria, acyl-homoserine lactones (AHL) and naturally secreted metabolites have been employed for cell-cell signalling (Basu *et al.*,

2005; Bulter *et al.*, 2004; Danino *et al.*, 2010; Ninfa, 2010), but there is a need for additional communication channels. Tabor *et al.* (2009) used AHL-based cell–cell communication combined with an intricate cellular processing in bacteria to detect edges. In addition to AHL, You *et al.* designed two *E. coli* populations to act as a synthetic predator–prey system (Balagadde *et al.*, 2008). A distributed quorum of clocks in *E. coli* was engineered recently using AHL (Danino *et al.*, 2010). Using a plant hormone and its receptors in yeast has been shown to allow for artificial cell–cell communication using parts from multiple kingdoms (Chen and Weiss, 2005). Distributed computing through intercellular communication in yeast was demonstrated by Regot *et al.* (2010). Combining yeast cells responding to various extracellular stimuli and a variable wiring molecule, the authors engineered circuits with a wide range of logic functions (XOR, NAND, multiplexer, 1-bit adder). In mammalian cells, there are currently no published orthogonal intercellular communication devices. Using nonorthogonal pathways, cell–cell communication in mammalian cells has been demonstrated using acetaldehyde or nitric oxide synthase (Wang *et al.*, 2008; Weber and Fussenegger, 2010). While systems that use endogenous metabolic compounds allow us to quickly test concepts and achieve certain goals, communication systems that do not interfere with endogenous signaling pathways will provide important benefits for a variety of future applications. Many initiatives are under way in the community, for example, in the International Engineered Genetic Machines (iGEM) competition, and one can hope that soon there will be such systems available.

4.6.2. Heterogeneity

Heterogeneity may be phenotypic, genetic, or both. Coordinated phenotypic heterogeneity may be achieved by intercellular communication, allowing an individual cell to change its phenotype depending on its spatial position and/or the behavior of neighboring cells. Genetic heterogeneity is possible with cells that are engineered to have different genetic circuits (Basu *et al.*, 2005; Balagadde *et al.*, 2008). Multispecies communities are inherently genetically heterogeneous, but they can still contain the same synthetic genetic circuit. This could, for example, be used for synchronizing related species with a quorum sensing circuit. Heterogeneity can enhance robustness, as diverse cells or gene networks are less prone to fail simultaneously, and can have overlapping functions (Kitano, 2003; Liu *et al.*, 2007). Engineering gene circuits for heterogeneous populations on the other hand adds a layer of complexity to the design and implementation.

4.6.3. Redundancy

Multicellular systems can normally cope with the loss of a certain number of cells. Most life forms are able to regenerate damaged organs, tissues, and extremities. Autoregeneration, together with the ability to reproduce is one of the most salient properties of biological systems (Brockes and Kumar, 2005; Orlic *et al.*, 2001). Spreading functionality over a population allows for robust behavior despite some physical logical damage or malfunctioning by averaging out or voting on an outcome (Tamsir *et al.*, 2010).

4.6.4. Homeostasis

With the aid of intercellular interactions, homeostasis regulates the amount and types of individual cells within a multicellular system. In multicellular organisms, the robustness of an organism has priority over the robustness of the single cell. Humans shed an enormous amount of cells every day (skin cells, gut epithelium) in a controlled fashion. During embryogenesis and tissue growth, many cells die in a programmed fashion either because they fulfilled their function or simply because they are at the wrong place in a tissue (Meier *et al.*, 2000). Natural safeguard mechanisms, such as the tumor suppressor protein p53 remove defective or aberrant cells from a population, filtering therefore for integrity. Cancer is a disease where the individual cell gains robustness and performance in an uncontrolled fashion, endangering the entire organism. Using AHL as a quorum sensing molecule, synthetic cell population homeostasis has been demonstrated in a suspension culture of *E. coli* (You *et al.*, 2004). In this system, a population of bacteria contains two plasmids. One expresses LuxI and LuxR, therefore synthesizing AHL (diffusing into the extracellular milieu) and through the LuxR-AHL interaction, allowing for activation of the pLux promoter on the second plasmid to induce expression of a cytotoxin (ccdB). As soon as the population level reaches a certain density, sufficient concentrations of AHL are present to initiate cell death. In this system, noise allowed for better overall population robustness, as cells responded differently to similar AHL levels, thereby leading to some cells dying while others survived.

4.6.5. Noise reduction in multicellular systems

Noise is often detrimental to the reliable functioning of a genetic circuit, and researchers have used multicellular strategies to filter out or reduce noise and achieve better robustness. For example, Pai *et al.* (2009) demonstrate that by using an AHL-based quorum sensing mechanism and specifically destabilizing the receiver protein (LuxR), one can achieve a reduction of noise at the population level. Another possible approach to average out noise in a population was applied to multicellular computing using genetically engineered NOR gates and chemical wires (Tamsir *et al.*, 2010). A colony (spatially restricted part of a population) of bacteria containing the same NOR gate is wired through quorum signals to neighboring bacterial colonies containing another type of NOR gate. Despite the single cell (microscopic level) variability, the colony of bacteria (macroscopic level) behaves in a robust way, effectively absorbing and buffering the variations of its single members. The use of quorum sensing and population averaging represents a common design rule for achieving computational operations robust enough to overcome the stochastic limitations of circuits in individual cells.

4.6.6. Synchronization

Without checkpoints or clocks, multicellular computing is inherently asynchronous. Inputs can be triggered before a pathway has finished and results in incorrect outputs because of mismatched delays in the circuits. A potential mechanism to address this is to implement an external clock by spatially separating the computing clusters (spotting colonies) on a solid phase (Tamsir *et al.*, 2010) or to provide an intercellular signalling molecule to synchronize the population (Danino *et al.*, 2010). This layering of computation forces synchronized and therefore timed processing of the information. Alternatively, one can design circuits that function properly despite the asynchrony, perhaps by drawing inspiration from a synchronous electronic systems.

4.6.7. Exploiting noise in multicellular systems

There are cases where a noisy input may be beneficial for a genetic circuit. For example, in natural systems, stochastic fluctuations of a single protein's concentration can be used for segregating a population into two states which improves population robustness by selecting states that allow for cell survival depending on the conditions. ComK, a key transcription factor for determining competency in *Bacillus subtilis*, positively regulates itself. Two stable states are possible: low or high levels of ComK. Low levels of ComK cannot activate expression of comK, and *B. subtilis* remains in its vegetative state. A small fraction of *B. subtilis* cells stochastically express larger amounts of ComK, positively regulating the comK expression (feedforward) and further boosting ComK levels. By doing so, this bacterial subpopulation switch to their competent state (Mettetal and van Oudenaarden, 2007; Suel *et al.*, 2006). This system ensures that a fraction of the *B. subtilis* population is competent at any point in time, which can be beneficial to allow cells to incorporate possibly beneficial exogenous DNA.

The decision by the lambda phage to undergo lysis or lysogeny in infected *E. coli* has been attributed for a long time to noise, while a correlation with the cell size suggested that some of the cell's inherent properties play an additional role. Recently, it has been demonstrated (Zeng *et al.*, 2010) that for a cell with a multiplicity of infection >1, each phage casts a vote for or against lysogeny. Only if all phages agree on lysogeny this path is pursued. Otherwise, the lytic fate is chosen. This example illustrates nicely that even though the result of a complex process in a multicellular system might look as it was the result of noise, a closer observation might show that it is indeed the result of a precisely defined computational process within the population.

Combining a clock, quorum sensing, and antagonistic self-regulation may allow for generating in a robust way complex developmental patterns or oscillations within a population of cells (Rodrigo *et al.*, 2010). This publication demonstrates how *in silico* designing can lead to circuits that exhibit a complex behavior. A simple operon encompassing pLac-lux, *lacI*, *luxR*, and *luxI* lies at the core of this system and oscillates depending on the spatial location and time. Perturbations are robustly compensated through an autoregulatory feedback loop and an AHL-based quorum sensing mechanism allowing synchronization between the cells.

Multicellular systems offer new possibilities and present new design challenges in comparison to single cellular ones. Connecting intercellular signaling to an intercellular information exchange is one of the critical steps in achieving a robust artificial population. While various bacterial quorum sensing systems like AHL are relatively well characterized and widely adopted as an orthogonal carrier of intercellular information, there is clearly a need for additional signaling molecules or other motifs, especially for mammalian synthetic biology.

4.7. Circuit construction strategies to improve robustness

There are many concepts and guidelines that can help design a robust synthetic circuit; however, the strategies described above which are "rational" engineering approaches, might not always enable the synthetic biologist to create experimentally an optimized robust circuit. There are additional ways to address and improve robustness of a synthetic circuit during the experimental conception of the circuit. One strategy is to consider combinatorial synthesis/assembly of various elements in the circuit. By exchanging and reordering the various components of the circuits, unexpected behaviors leading to a more robust system may be observed. Michael Elowitz and colleagues (Guet *et al.*, 2002) produced diverse phenotypes in *E. coli* through changes in network connectivity. The devices were genes encoding transcriptional regulators, and the library of networks generated encoded or exhibited behaviors resembling several logic functions. The synthetic integron developed by Bikard *et al.* (2010) is also an interesting proof of concept that the combinatorial approach can generate a surprisingly large diversity of complex behaviors and phenotypes. In this example, automated intracellular shuffling of the different genes of the tryptophan operon, using a synthetic integron, generated phenotypes with different ratios of tryptophan production versus generation time. However, it has become clear that because of the absence of detailed mechanistic understanding of metabolic pathways, there is a real need for functional diversification and optimization.

Directed evolution has proved to be another powerful complement to rational genetic engineering approaches (Dougherty and Arnold, 2009).

Directed evolution techniques allow for the creation of more robust parts and circuits, where iterations of mutagenesis and artificial selection or screening are used to generate more robusts circuits. Many parts necessary for a robust system cannot be easily found in nature or at least they are yet to be discovered and are currently not readily available. It is also important to note that many of the available parts are not well characterized or standardized to operate with other parts. By tuning properties of an existing characterized part, it is possible to engineer it to exhibit the desired new properties. A very elegant proof of concept is shown by Ellis *et al.* (2009), who constructed feed-forward networks with different and predictable input–output characteristics, using directed evolution to generate a new library of promoters. Another example is the improvement of the compatibility between the two widely used AraC-PBAD and LacI-Plac systems. Directed evolution was used to increase the AraC system's sensitivity to arabinose 10-fold and to reduce its sensitivity to inhibition by IPTG, thereby dramatically reducing crosstalk between the two systems (Lee *et al.*, 2007). Random mutagenesis of specific residues of the binding pocket of AraC followed by FACS-based dual screening resulted in alteration of the binding specificity of AraC and allowed the identification of variants that respond to D-arabinose and not the effector L-arabinose (Tang *et al.*, 2008). An automated method to generate genetic variants with useful, altered phenotypes was developed by Wang *et al.* (2009) termed multiplex automated genome engineering (MAGE). This method was used to optimize the 1-deoxy-D-xylulose-5-phosphate biosynthesis pathway in *E. coli.* The technology is based on the simultaneous targeting of different locations on the chromosome with synthetic oligonucleotides to generate rapid genetic changes (mismatches, insertions, or deletions). MAGE combined with other tools may allow for the directed evolution of entire synthetic circuits to enhance overall circuit behavior and robustness.

However, directed evolution can potentially lead to unwanted and unpredicted interactions with existing host components, which can also result in significant perturbations and decrease in global robustness. Such interactions have been confirmed by comparing the transcriptomes of cells expressing a recombinant gene versus wild-type *E. coli* cells, which exhibited hundreds of unanticipated changes in expression levels (Haddadin and Harcum, 2005). This problem can be addressed at least to some extent by reducing the complexity of the host strains (Dietz and Panke, 2010). Current projects to minimize genomes would allow a given cell to grow under a defined, favorable set of conditions, and would likely reduce the likelihood of undesired interactions (Glass *et al.*, 2006). Some experiments focused on streamlining genomes of model bacteria revealed that genome reduction led to unanticipated beneficial properties, such as high electroporation efficiency and accurate propagation of recombinant genes and plasmids that were unstable in other strains.

5. Robustness Trade-Offs

Robustness comes with a price. Rendering a system more robust to a certain range of perturbations often leads to higher fragility, compromised performance, and an increased need for resources (Kitano, 2007). The balance between robustness, fragility, performance, and consumed resources needs to be carefully assessed when designing a genetic circuit. While robustness makes the behavior of systems more reliable, stable, and resistant to external perturbations, there are cases where the opposite is actually desired. Bacteria, viruses, and large populations of rapidly growing cells seem to prefer reduced robust modes of operation. The individual cell is expendable and robustness on the population level is of greater importance. It is notable that reduced robustness at the cellular level (e.g., induction of apoptosis in response to genomic damage) may benefit, and therefore enhance the robustness of the organism or system as a whole (Fig. 7.4).

Using a high degree of modularity and encapsulation may enhance robustness by avoiding unwanted crosstalk. This is commonly done by avoiding reusing parts and common motifs. However, there are still only a rather limited number of characterized parts currently available to the synthetic biologist, and in turn, engineering robustness can also limit the expandability of the system as a whole. A more complex circuit also consumes greater resources (metabolic load) in the host cell. A careful study of the *E. coli* heat-shock response system uncovered many modules specifically counter acting the loss of response speed or yield of folded proteins due to robust implementations of the core system (Kurata *et al.*, 2006). The increased complexity usually associated with

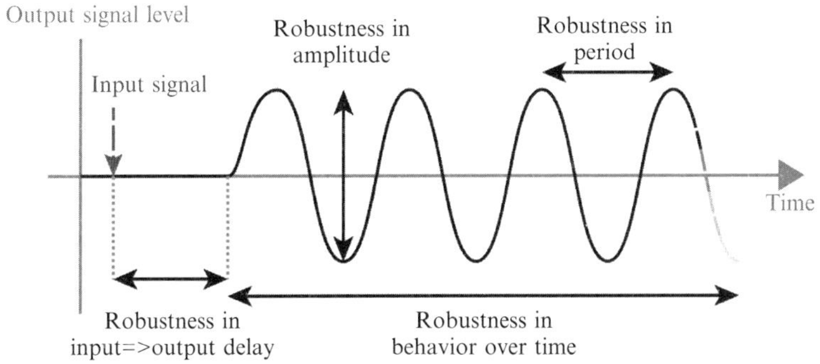

Figure 7.4 Different types of robustness can be defined for each of the properties of the designed synthetic circuit. Depending on the importance of the achievement of these properties for a system to have the final expected behavior, some properties might have to be robust, while others are critical. In the specific case of the need for a sustained oscillation, the amplitude of the signal might not have to be as robust as the period of the oscillation. Hence, one has to take into account where the synthetic circuit has to be robust to generate the expected behavior.

robustness in genetic networks must be carefully balanced (more components, greater need for resources, leading to a slower response and adaptation at the population level) depending on the application. While the design principles for building robust systems are being developed, there is less agreement on where it is essential to allocate resources to increase robustness. Therefore, in each case, a close analysis of the genetic network to be built is necessary to assess the need and potential for robustness.

Systems specifically optimized to resist certain perturbations are often much more sensitive and fragile to unexpected perturbations (Kitano, 2007). The high optimized tolerance (HOT) theory (Zhou *et al.*, 2002) postulates that this enhanced sensitivity of complex systems to unexpected perturbations is unavoidable if one wants enhanced robustness to other specific perturbations. Interestingly, while HOT systems are often robust to fail-off malfunctions (a component or node ceasing to work), they appear to be more fragile against fail-on perturbations (a component or node submitting erroneous information). This effect is known as the Byzantine generals problem, where an army with multiple generals exchanging information in the field is drastically affected by traitors submitting erroneous information.

While robustness is a much sought after property of synthetic genetic circuits, one should critically examine the need for robustness depending on the application. The overhead and increased complexity associated with engineering robust circuit can consume significant development resources. As applications of synthetic biology begin to appear, a framework for understanding and quantifying robustness, as well as objectively evaluating the need for it, will need to be established.

6. Conclusion

Here, we present paradigms for the design, construction, and validation of robust gene circuits for synthetic biology. This chapter is by no means comprehensive but it does reflect experiences designing circuits. While much of the past 10 years has been spent demonstrating functionality, as industrial, therapeutic, and environmental begin to appear using cells engineered with complex synthetic gene circuits, robustness and predictability will become critical design considerations for engineers. Looking forward, we believe that the principles described in this chapter will become of greater importance to the field of synthetic biology.

REFERENCES

Aldana, M., and Cluzel, P. (2003). A natural class of robust networks. *Proc. Natl. Acad. Sci. USA* **100**(15), 8710–8714.

Alexander, R. P., Kim, P. M., *et al.* (2009). Understanding modularity in molecular networks requires dynamics. *Sci. Signal.* **2**(81), pe44.

Alon, U. (2007). Network motifs: Theory and experimental approaches. *Nat. Rev. Genet.* **8**(6), 450–461.
An, W., and Chin, J. W. (2009). Synthesis of orthogonal transcription-translation networks. *Proc. Natl. Acad. Sci. USA* **106**(21), 8477–8482.
Ang, J., Bagh, S., *et al.* (2010). Considerations for using integral feedback control to construct a perfectly adapting synthetic gene network. *J. Theor. Biol.* **266**(4), 723–738.
Arnott, A., Jardine, D., *et al.* (2010). High viral fitness during acute HIV-1 infection. *PLoS ONE* **5**(9), 1–12.
Bahar, R., Hartmann, C. H., *et al.* (2006). Increased cell-to-cell variation in gene expression in ageing mouse heart. *Nature* **441**(7096), 1011–1014.
Balagadde, F. K., You, L., *et al.* (2005). Long-term monitoring of bacteria undergoing programmed population control in a microchemostat. *Science* **309**(5731), 137–140.
Balagadde, F. K., Song, H., *et al.* (2008). A synthetic *Escherichia coli* predator-prey ecosystem. *Mol. Syst. Biol.* **4,** 187.
Bar-Even, A., Paulsson, J., *et al.* (2006). Noise in protein expression scales with natural protein abundance. *Nat. Genet.* **38**(6), 636–643.
Barabási, A. L., and Oltvai, Z. N. (2004). Network biology: understanding the cell's functional organization. *Nat. Rev. Genet.* **5,** 101–113.
Bashor, C. J., Horwitz, A. A., *et al.* (2010). Rewiring cells: Synthetic biology as a tool to interrogate the organizational principles of living systems. *Annu. Rev. Biophys.* **39,** 515–537.
Basu, S., Gerchman, Y., *et al.* (2005). A synthetic multicellular system for programmed pattern formation. *Nature* **434**(7037), 1130–1134.
Becskei, A., and Serrano, L. (2000). Engineering stability in gene networks by autoregulation. *Nature* **405**(6786), 590–593.
Bikard, D., Julie-Galau, S., *et al.* (2010). The synthetic integron: An in vivo genetic shuffling device. *Nucleic Acids Res.* **38**(15), e153.
Blake, W. J., Balazsi, G., *et al.* (2006). Phenotypic consequences of promoter-mediated transcriptional noise. *Mol. Cell* **24**(6), 853–865.
Boyle, P. M., and Silver, P. A. (2009). Harnessing nature's toolbox: Regulatory elements for synthetic biology. *J. R. Soc. Interface* **6**(Suppl. 4), S535–S546.
Brockes, J. P., and Kumar, A. (2005). Appendage regeneration in adult vertebrates and implications for regenerative medicine. *Science* **310**(5756), 1919–1923.
Bulter, T., Lee, S. G., *et al.* (2004). Design of artificial cell-cell communication using gene and metabolic networks. *Proc. Natl. Acad. Sci. USA* **101**(8), 2299–2304.
Carlini, D. B., and Stephan, W. (2003). In vivo introduction of unpreferred synonymous codons into the Drosophila Adh gene results in reduced levels of ADH protein. *Genetics* **163**(1), 239–243.
Chen, M. T., and Weiss, R. (2005). Artificial cell-cell communication in yeast *Saccharomyces cerevisiae* using signaling elements from *Arabidopsis thaliana. Nat. Biotechnol.* **23**(12), 1551–1555.
Danino, T., Mondragon-Palomino, O., *et al.* (2010). A synchronized quorum of genetic clocks. *Nature* **463**(7279), 326–330.
de Lorenzo, V., and Danchin, A. (2008). Synthetic biology: Discovering new worlds and new words. *EMBO Rep.* **9**(9), 822–827.
Del Vecchio, D., Ninfa, A. J., *et al.* (2008). Modular cell biology: Retroactivity and insulation. *Mol. Syst. Biol.* **4,** 161.
Dietz, S., and Panke, S. (2010). Microbial systems engineering: First successes and the way ahead. *Bioessays* **32**(4), 356–362.
Dougherty, M. J., and Arnold, F. H. (2009). Directed evolution: New parts and optimized function. *Curr. Opin. Biotechnol.* **20**(4), 486–491.
Drake, J. W., Charlesworth, B., *et al.* (1998). Rates of spontaneous mutation. *Genetics* **148** (4), 1667–1686.

Ehrhardt, A., Haase, R., *et al.* (2008). Episomal vectors for gene therapy. *Curr. Gene Ther.* **8**(3), 147–161.

Ellis, T., Wang, X., *et al.* (2009). Diversity-based, model-guided construction of synthetic gene networks with predicted functions. *Nat. Biotechnol.* **27**(5), 465–471.

Elowitz, M. B., and Leibler, S. (2000). A synthetic oscillatory network of transcriptional regulators. *Nature* **403**(6767), 335–338.

Frost, L. S., Leplae, R., *et al.* (2005). Mobile genetic elements: The agents of open source evolution. *Nat. Rev. Microbiol.* **3**(9), 722–732.

Gardner, T. S., Cantor, C. R., *et al.* (2000). Construction of a genetic toggle switch in *Escherichia coli*. *Nature* **403**(6767), 339–342.

Glass, J. I., Assad-Garcia, N., *et al.* (2006). Essential genes of a minimal bacterium. *Proc. Natl. Acad. Sci. USA* **103**(2), 425–430.

Goulian, M. (2004). Robust control in bacterial regulatory circuits. *Curr. Opin. Microbiol.* **7**(2), 198–202.

Greber, D., El-Baba, M. D., *et al.* (2008). Intronically encoded siRNAs improve dynamic range of mammalian gene regulation systems and toggle switch. *Nucleic Acids Res.* **36**(16), e101.

Guet, C. C., Elowitz, M. B., *et al.* (2002). Combinatorial synthesis of genetic networks. *Science* **296**(5572), 1466–1470.

Haddadin, F. T., and Harcum, S. W. (2005). Transcriptome profiles for high-cell-density recombinant and wild-type *Escherichia coli*. *Biotechnol. Bioeng.* **90**(2), 127–153.

Hardway, H., Mukhopadhyay, B., *et al.* (2008). Modeling the precision and robustness of Hunchback border during *Drosophila* embryonic development. *J. Theor. Biol.* **254**(2), 390–399.

Heinemann, M., and Panke, S. (2006). Synthetic biology–putting engineering into biology. *Bioinformatics* **22**(22), 2790–2799.

Hooshangi, S., Thiberge, S., *et al.* (2005). Ultrasensitivity and noise propagation in a synthetic transcriptional cascade. *Proc. Natl. Acad. Sci. USA* **102**(10), 3581–3586.

Houchmandzadeh, B., Wieschaus, E., *et al.* (2002). Establishment of developmental precision and proportions in the early *Drosophila* embryo. *Nature* **415**(6873), 798–802.

Ingolia, N. T. (2004). Topology and robustness in the *Drosophila* segment polarity network. *PLoS Biol.* **2**(6), e123.

Kafri, R., Bar-Even, A., *et al.* (2005). Transcription control reprogramming in genetic backup circuits. *Nat. Genet.* **37**(3), 295–299.

Keravala, A., Groth, A. C., *et al.* (2006). A diversity of serine phage integrases mediate site-specific recombination in mammalian cells. *Mol. Genet. Genomics* **276**(2), 135–146.

Khalil, A. S., and Collins, J. J. (2010). Synthetic biology: Applications come of age. *Nat. Rev. Genet.* **11**(5), 367–379.

Kitano, H. (2003). Cancer robustness: Tumour tactics. *Nature* **426**(6963), 125.

Kitano, H. (2004a). Biological robustness. *Nat. Rev. Genet.* **5**(11), 826–837.

Kitano, H. (2004b). Cancer as a robust system: Implications for anticancer therapy. *Nat. Rev. Cancer* **4,** 227–235.

Kitano, H. (2007). Biological robustness in complex host-pathogen systems. *Prog. Drug Res.* **64**(239), 241–263.

Kobayashi, H., *et al.* (2004). Programmable cells: interfacing natural and engineered gene networks. *Proc. Natl. Acad. Sci. USA* **101,** 8414–8419.

Kopecko, D. J., and Cohen, S. N. (1975). Site specific recA–independent recombination between bacterial plasmids: Involvement of palindromes at the recombinational loci. *Proc. Natl. Acad. Sci. USA* **72**(4), 1373–1377.

Kurata, H., El-Samad, H., *et al.* (2006). Module-based analysis of robustness tradeoffs in the heat shock response system. *PLoS Comput. Biol.* **2**(7), e59.

Lee, S. K., Chou, H. H., *et al.* (2007). Directed evolution of AraC for improved compatibility of arabinose- and lactose-inducible promoters. *Appl. Environ. Microbiol.* **73**(18), 5711–5715.

Liu, A. C., Welsh, D. K., *et al.* (2007). Intercellular coupling confers robustness against mutations in the SCN circadian clock network. *Cell* **129**(3), 605–616.

Matrai, J., Chuah, M. K., *et al.* (2010). Recent advances in lentiviral vector development and applications. *Mol. Ther.* **18**(3), 477–490.

Mazel, D. (2006). Integrons: Agents of bacterial evolution. *Nat. Rev. Microbiol.* **4**(8), 608–620.

Meier, P., Finch, A., *et al.* (2000). Apoptosis in development. *Nature* **407**(6805), 796–801.

Mettetal, J. T., and van Oudenaarden, A. (2007). Microbiology. Necessary noise. *Science* **317** (5837), 463–464.

Milo, R., Shen-Orr, S., *et al.* (2002). Network motifs: Simple building blocks of complex networks. *Science* **298**(5594), 824–827.

Milo, R., Itzkovitz, S., *et al.* (2004). Superfamilies of evolved and designed networks. *Science* **303**(5663), 1538–1542.

Moehle, E. A., Rock, J. M., *et al.* (2007). Targeted gene addition into a specified location in the human genome using designed zinc finger nucleases. *Proc. Natl. Acad. Sci. USA* **104** (9), 3055–3060.

Morohashi, M., Winn, A. E., *et al.* (2002). Robustness as a measure of plausibility in models of biochemical networks. *J. Theor. Biol.* **216**(1), 19–30.

Murphy, K. F., Adams, R. M., *et al.* (2010). Tuning and controlling gene expression noise in synthetic gene networks. *Nucleic Acids Res.* **38**(8), 2712–2726.

Newlands, S., Levitt, L. K., *et al.* (1998). Transcription occurs in pulses in muscle fibers. *Genes Dev.* **12**(17), 2748–2758.

Newman, J. R., Ghaemmaghami, S., *et al.* (2006). Single-cell proteomic analysis of *S. cerevisiae* reveals the architecture of biological noise. *Nature* **441**(7095), 840–846.

Ninfa, A. J. (2010). Use of two-component signal transduction systems in the construction of synthetic genetic networks. *Curr. Opin. Microbiol.* **13**(2), 240–245.

Orlando, S. J., Santiago, Y., *et al.* (2010). Zinc-finger nuclease-driven targeted integration into mammalian genomes using donors with limited chromosomal homology. *Nucleic Acids Res.* **38**(15), e152.

Orlic, D., Kajstura, J., *et al.* (2001). Bone marrow cells regenerate infarcted myocardium. *Nature* **410**(6829), 701–705.

Ozbudak, E. M., Thattai, M., *et al.* (2002). Regulation of noise in the expression of a single gene. *Nat. Genet.* **31**(1), 69–73.

Pai, A., Tanouchi, Y., *et al.* (2009). Engineering multicellular systems by cell-cell communication. *Curr. Opin. Biotechnol.* **20**(4), 461–470.

Pedraza, J. M., and van Oudenaarden, A. (2005). Noise propagation in gene networks. *Science* **307,** 1965–1969.

Preston, B. D., Poiesz, B. J., *et al.* (1988). Fidelity of HIV-1 reverse transcriptase. *Science* **242**(4882), 1168–1171.

Purnick, P. E., and Weiss, R. (2009). The second wave of synthetic biology: From modules to systems. *Nat. Rev. Mol. Cell Biol.* **10**(6), 410–422.

Rao, C. V., Wolf, D. M., *et al.* (2002). Control, exploitation and tolerance of intracellular noise. *Nature* **420**(6912), 231–237.

Regot, S., Macia, J., *et al.* (2010). Distributed biological computation with multicellular engineered networks. *Nature* **269,** 207–211.

Rodrigo, G., Carrera, J., *et al.* (2010). Robust dynamical pattern formation from a multifunctional minimal genetic circuit. *BMC Syst. Biol.* **4,** 48.

Saez-Rodriguez, J., Gayer, S., *et al.* (2008). Automatic decomposition of kinetic models of signaling networks minimizing the retroactivity among modules. *Bioinformatics* **24**(16), i213–i219.

Schiestl, R. H., and Petes, T. D. (1991). Integration of DNA fragments by illegitimate recombination in *Saccharomyces cerevisiae*. *Proc. Natl. Acad. Sci. USA* **88**(17), 7585–7589.

Silva-Rocha, R., and de Lorenzo, V. (2008). Mining logic gates in prokaryotic transcriptional regulation networks. *FEBS Lett.* **582**(8), 1237–1244.

Smith, M. A., and Bidochka, M. J. (1998). Bacterial fitness and plasmid loss: The importance of culture conditions and plasmid size. *Can. J. Microbiol.* **44**(4), 351–355.

Stelling, J., Sauer, U., *et al.* (2004). Robustness of cellular functions. *Cell* **118**(6), 675–685.

Stricker, J., Cookson, S., *et al.* (2008). A fast, robust and tunable synthetic gene oscillator. *Nature* **456**(7221), 516–519.

Suel, G. M., Garcia-Ojalvo, J., *et al.* (2006). An excitable gene regulatory circuit induces transient cellular differentiation. *Nature* **440**(7083), 545–550.

Tabor, J. J., Salis, H. M., *et al.* (2009). A synthetic genetic edge detection program. *Cell* **137** (7), 1272–1281.

Tamsir, A., Tabor, J. J., *et al.* (2010). Robust multicellular computing using genetically encoded NOR gates and chemical 'wires. *Nature* **269,** 212–215.

Tang, S. Y., Fazelinia, H., *et al.* (2008). AraC regulatory protein mutants with altered effector specificity. *J. Am. Chem. Soc.* **130**(15), 5267–5271.

Thatcher, J. W., Shaw, J. M., *et al.* (1998). Marginal fitness contributions of nonessential genes in yeast. *Proc. Natl. Acad. Sci. USA* **95**(1), 253–257.

Thattai, M., and van Oudenaarden, A. (2002). Attenuation of noise in ultrasensitive signaling cascades. *Biophys. J.* **82**(6), 2943–2950.

Thomson, J. G., and Ow, D. W. (2006). Site-specific recombination systems for the genetic manipulation of eukaryotic genomes. *Genesis* **44**(10), 465–476.

Trono, D., Van Lint, C., Rouzioux, C., Verdin, E., Barré-Sinoussi, F., Chun, T. W., and Chomont, N. (2010). HIV persistence and the prospect of long-term drug-free remissions for HIV-infected individuals. *Science* **329,** 174–180.

Turker, M. S. (2002). Gene silencing in mammalian cells and the spread of DNA methylation. *Oncogene* **21**(35), 5388–5393.

von Dassow, G., Meir, E., *et al.* (2000). The segment polarity network is a robust developmental module. *Nature* **406**(6792), 188–192.

Waddington, C. H. (1942). Canalization of development and the inheritance of acquired characters. *Nature* **150,** 563–565.

Wang, K., Neumann, H., *et al.* (2007). Evolved orthogonal ribosomes enhance the efficiency of synthetic genetic code expansion. *Nat. Biotechnol.* **25**(7), 770–777.

Wang, W. D., Chen, Z. T., *et al.* (2008). Construction of an artificial intercellular communication network using the nitric oxide signaling elements in mammalian cells. *Exp. Cell Res.* **314**(4), 699–706.

Wang, H. H., Isaacs, F. J., *et al.* (2009). Programming cells by multiplex genome engineering and accelerated evolution. *Nature* **460**(7257), 894–898.

Weber, W., and Fussenegger, M. (2010). Cell-to-cell communication for cell density-controlled bioprocesses. *Cells and Culture* **4,** 407–412.

Wyman, C., Ristic, D., *et al.* (2004). Homologous recombination-mediated double-strand break repair. *DNA Repair* **3**(8–9), 827–833.

Yi, T. M., Huang, Y., *et al.* (2000). Robust perfect adaptation in bacterial chemotaxis through integral feedback control. *Proc. Natl. Acad. Sci. USA* **97**(9), 4649–4653.

Yokobayashi, Y., Weiss, R., *et al.* (2002). Directed evolution of a genetic circuit. *Proc. Natl. Acad. Sci. USA* **99**(26), 16587–16591.

You, L., Cox, R. S., 3rd, *et al.* (2004). Programmed population control by cell-cell communication and regulated killing. *Nature* **428**(6985), 868–871.

Zeng, L., Skinner, S. O., *et al.* (2010). Decision making at a subcellular level determines the outcome of bacteriophage infection. *Cell* **141**(4), 682–691.

Zhou, T., Carlson, J. M., *et al.* (2002). Mutation, specialization, and hypersensitivity in highly optimized tolerance. *Proc. Natl. Acad. Sci. USA* **99**(4), 2049–2054.

CHAPTER EIGHT

Engineering RNAi Circuits

Yaakov Benenson

Contents

Abstract

Engineered "computing" biological networks are a generalization of endogenous regulatory pathways. They are intended to generate novel biological responses based on preprogrammed processing of multiple molecular signals. We have recently introduced an approach to constructing complex signal processing networks in mammalian cells using RNA interference (RNAi) as the underlying regulatory mechanism. The approach is modular and the circuits contain sensory, computational, and actuation modules. In the sensory module, various molecular signals are transduced into RNAi-compatible effectors such as small interfering RNA, or microRNA. In the computational module, multiple

Department of Biosystems Science and Engineering, Swiss Federal Institute of Technology (ETH Zurich), Mattenstrasse 26, Basel, Switzerland

Methods in Enzymology, Volume 497
ISSN 0076-6879, DOI: 10.1016/B978-0-12-385075-1.00008-1

small RNA (sRNA) effectors converge on the small number of output constructs. Here, we describe experimental methods utilized in circuit construction with the focus on the computational module. We emphasize the steps involved in the design of large sRNA sets required for such circuits.

1. Introduction

Biological regulatory pathways "compute" their response to environmental stimuli using complex interaction networks. A biological computation can be quite convoluted and even measuring it experimentally is a major challenge. Some of these computations are well approximated by Boolean abstraction, and theoretical study of regulatory networks as Boolean circuits has a long history (Kauffman *et al.*, 2003). Accordingly, it was surmised (Weiss *et al.*, 1999) that bottom-up rational construction of Boolean molecular circuits could be a method of choice to generate novel, nonnative biological functions to specification for both basic science and applications. One such application could be a "medical biocomputer" that would control the amount of activated therapeutic agent in each individual cell in a patient's body based on preprogrammed analysis of multiple intracellular molecular disease markers (Shapiro and Benenson, 2006).

Biological computers are different from their silicon counterparts in almost all aspects. While a silicon computer has a fixed hardware that is reprogrammed by uploading sets of instructions in a machine language, biological computers are reprogrammed by physical alteration of interactions in a molecular network. Therefore, each new biocomputer program will necessarily require a new physical object (i.e., a new network) to implement it. Yet, if the transition between the program and the corresponding network is sufficiently streamlined, it may be delegated to an automated "compiler." In a futuristic scenario, the compiler will not only specify the network but also operate a DNA synthesizer that will make the physical circuit. Whether fully automated design of biological computing networks based on high-level specifications is possible is an open question, and steps have been taken to bring this vision close to reality.

Among many alternatives, RNA and RNA-based regulation has emerged as the substrate that may enable semiautomated circuit design (Benenson *et al.*, 2004; Faulhammer *et al.*, 2000; Seelig *et al.*, 2006; Stojanovic and Stefanovic, 2003) In mammalian systems, RNA-based regulation includes alternative splicing, RNA interference (RNAi), and antisense RNA to name a few. RNAi is an inhibitory interaction elicited by a short (19–27 nt) RNA strand against mRNA that contains either a fully or partially complementary sequence to this strand in its coding region or 3′-UTR. Short RNAs can be chemically synthesized as small interfering

RNA (siRNA) and delivered exogenously, or transcribed in cells from plasmids as short hairpin RNA (shRNA) or microRNA (miRNA). All these species can elicit very efficient repression if they meet certain structural requirements and are fully complementary to their target. We proposed that multiple RNAi-based interactions can be integrated in synthetic circuits such that certain logic formulae known as disjunctive normal forms (DNF, Fig. 8.1A) or conjunctive normal forms (CNF, Fig. 8.1B) can be computed (Rinaudo *et al.*, 2007). Using RNAi alone leads to DNF circuits and adding an extra transcriptional layer enables CNF circuits. We showed integration of up to five siRNA inputs in a DNF circuit and up to two inputs in a CNF circuit (Rinaudo *et al.*, 2007). We also showed circuits that use synthetic miRNA genes regulated by transcription factors to compute complex transcriptional regulatory functions (Leisner *et al.*, 2010). This chapter focuses on the "logic core" of these circuits where a large number of small RNA (sRNA) species regulate a small number of target transcripts. Choosing and testing the sets presents a number of unique experimental challenges described here. The next step that has not been fully addressed yet is the design of sensory mechanisms that transduce various signals to sRNA. First steps in this direction were taken by researchers who showed that small molecules can affect sRNA activity via aptamer binding (An *et al.*, 2006; Beisel *et al.*, 2008; Tuleuova *et al.*, 2008).

2. Constructing a Computational Logic Core for the RNAi-Based DNF Circuit

The key to successful circuit construction lies in the proper design of sRNA species that can be integrated in a modular fashion in the 3′-UTR of the desired output gene.

A number of requirements need to be fulfilled:

(1) Each sRNA in the set has to be highly active against its intended target yet show minimal interference with endogenous processes
(2) sRNAs in the set need to be mutually orthogonal and not engage in cross talk
(3) sRNA targets need to function independently when placed next to each other.

We propose to follow a stepwise procedure below to address these issues.

2.1. Step 1: Design candidate sRNA sequences

Table 1 in a recent report (Reynolds *et al.*, 2004) contains a number of statistical rules that predict sRNA activity. A number of web tools offer "good" siRNA design, for example, http://www.ambion.com/techlib/

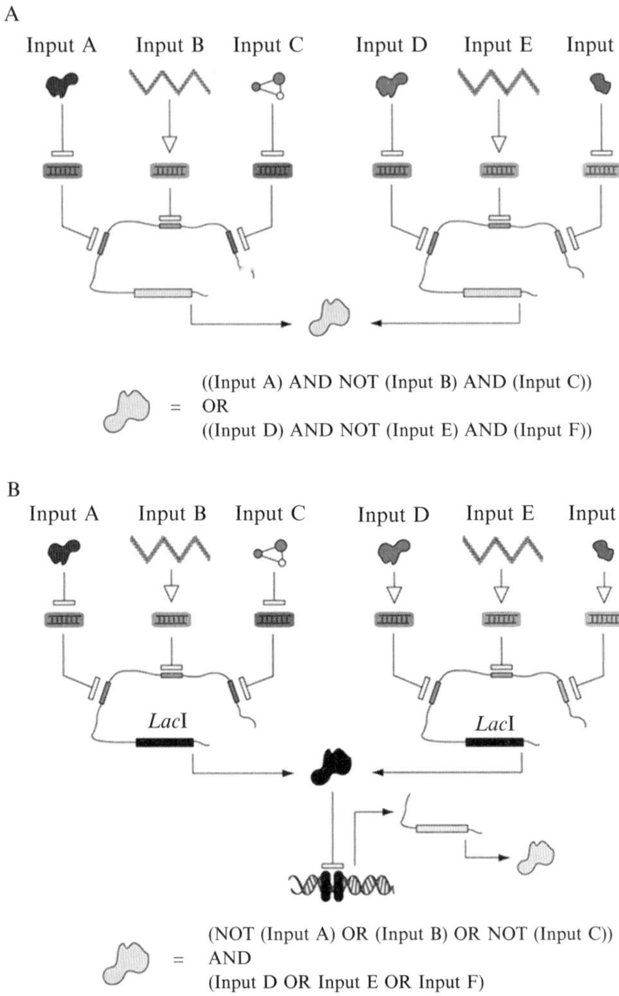

Figure 8.1 Layout of DNF and CNF circuits. (A) A layout of a DNF circuit. DNF is a disjunctive normal form of logic equations. It connects logic variables that may or may not be negated by NOT operators with AND operators, and then connects those groups with OR operators. The OR logic is achieved by using multiple transcripts coding for the same output protein; the AND logic is implemented via multiple sRNA target sites in the output 3′-UTR; and the NOT logic is implemented by the activating interaction between an input and an sRNA molecule. Blunt arrows denote repression and pointed arrows denote activation. (B) A layout of a CNF circuit. CNF is a conjunctive normal form of logic equations where occasionally negated variables are connected by OR operators and their groups are connected in turn by AND operators. The AND logic is achieved by using multiple repressor-producing mRNA to generate a repressor protein. The OR logic is achieved by placing multiple siRNA targets in the repressor mRNA 3′-UTR. The NOT logic is implemented by an inhibitory interaction between an input and its corresponding sRNA.

misc/siRNA_finder.html or http://www.dharmacon.com/designcenter/designcenterpage.aspx. These web sites are geared toward siRNA for endogenous gene knockdown. One way to use them for the purpose of this work is to provide a target gene sequence that is not present in the host cell (e.g., a nonhuman genomic sequence will be treated as a human gene by making the proper selections on the web page, such as minimizing seed region in *Homo sapiens* genome and BLAST against that same genome). Web interface would then generate a ranked list of candidates that include such important information as the number of mismatches with potential side targets in the host transcriptome. It is important to manually curate potential sequence candidates by cross-checking with the list of criteria from Reynolds *et al.* We found that perhaps the most important requirement that ensures siRNA efficiency is the asymmetry (Schwarz *et al.*, 2003) in the siRNA duplex: the 3′-end (the one that is supposed to be unwound by the RISC complex) should be AT-rich while the 5′-end should be CG-rich. It is also important to manually BLAST the top candidates and find out what are the potential side targets in the host organism. It will be helpful to research the expression pattern of the gene in question. Instead of taking a gene or genome sequence from another organism, purely random sequences of about 1000–2000 nt can be generated computationally. It is important to double-check by using BLAST that those sequences are not homologous to any transcribed region in the human genome. The siRNAs should also be blasted against the transcript that generates the circuit's output, especially if the output is exogenous, such as a fluorescent protein.

Another approach that we used in our work is to rely on published siRNA sequences known to target nonhuman genes, or use sequences that are employed as "negative controls." However this method is inherently limited.

2.2. Step 2: Design a set of siRNAs

The above search process will render one or more siRNA candidates with a potential for high efficacy and low off-target activity, but it does not guarantee that those siRNA will not cross talk with each other. A simple way to check a given set is to calculate Hammond distance between different sRNA strands (antisense strands need to be used because they bind to the target). Hammond distance is the number of nucleotide mismatches when the sequences are aligned by their respective termini without frameshift. For N sequences, there are $N(N-1)/2$ pairwise distances and none of them should be higher than some low cut-off value, for example, 50%.

In a more rigorous approach, one should check for potential off-site hybridization events between an siRNA and its unintended targets. It is important to realize that in large circuits, different targets will be placed next to each other or placed repeatedly. Concatenation of different individual targets generates new target sequences that have not been envisioned by the original design. In principle, it is possible to generate all possible sequences that reflect different scenarios of target concatenation, including order permutations as well as using tandem repeats. siRNA sequences from the candidate set should then be checked for off-site hybridization to those combined target sequences. Ideally, they should not have significant homology to any of the unintended sites, in particular not in the seed region. If such homology is found, siRNA set should be modified. Of course, this will also change the set of potential target sequences so the whole process has to be repeated. Unfortunately, there is no off-the-shelf software solution to this problem but it resembles the design problem of DNA "words" in DNA computing experiments (Frutos *et al.*, 1997).

The last requirement from the siRNA set is the compatibility of its targets. We and others found that an important prerequisite of efficient RNAi is the lack of strong mRNA secondary structure at the RISC complex binding site, that is, around the RNAi target. Complex target sequences constructed in the previous step should be folded using mFold (Zuker, 2003) or similar software and examined for extensive base-paired regions. Strong secondary structures can fold not only when different targets are placed to each other, but also when the same target is amplified to generate tandem repeats. While such amplification can in general lead to better repression (Sullivan and Ganem, 2005), sometimes the repression in fact gets poorer due to enhanced secondary structure. We also found that when placing multiple targets next to each other, certain arrangements result in fewer base pairs compared to others. If strong secondary structures are observed for many different arrangements, the siRNA set can be modified and retested. Another possible solution is to introduce low-complexity spacers between the target sites, rather than placing them next to each other. The downside of this solution is more complex cloning and in some cases unintentional insertion of RNA endonuclease targets or RNA protein binding sites.

Overall, the process of designing a good siRNA set for logic circuit construction is quite complex, especially for larger set. No computational tool that addresses all the above aspects has been built yet and this would be an interesting direction to pursue.

2.3. Step 3: Cloning and experimental characterization of "computational" molecules and the siRNA

A stereotypical output construct that we also call a "computational molecule" includes a constitutive promoter, a coding frame of the desired output protein, a sequence of sRNA targets, and transcription termination and polyadenylation signals. As a starting point, we recommend using commercial vectors that express fluorescent proteins with 3′-multiple cloning sites.

The presence of 3′-MCS will ensure a number of unique restriction sites that can be used for insertion of sRNA targets. Most of these constructs are used to prepare C-terminal fusion proteins and the stop codon is placed downstream of the MCS. In this case, a new stop codon needs to be introduced in front of the cloned sRNA targets, to make sure those targets are not translated. A constitutive promoter such as CMV should drive the reporter. The choice of the promoter is not trivial. At the least, the promoter should be active in the cells in which the circuits are intended to work. They may also have an effect of siRNA efficiency (Leisner *et al.*, 2010). Accordingly, the following protocols can be repeated for different output-driving promoters if the necessary resources are available.

For a new set of siRNAs, we commence circuit construction by building a set of constructs that contain one sRNA target each to examine a possibility of cross talk. If the intention is to use tandem target repeats, these individual constructs should contain a number of repeats projected for the ultimate circuit design. To test the cross talk in the set, the constructs are cotransfected with a transfection marker, typically a different fluorescent protein. Good pairs of marker–reporter include CFP-DsRed, GFP-DsRed, or YFP-DsRed. Pairs of CFP/YFP or comparable proteins may cross talk and require compensation. For a set of N siRNAs, the number of measurements is $(N+1)^2$, with the addition of a control construct that has a "mock" target instead of the intended one. For nine targets, this requires 100 experiments, yet the step is a crucial prerequisite for any future work. On the upside, good sets of siRNA can be reused for other circuits so the tests need not be repeated.

An additional test one may want to perform is to measure off-target effects of the siRNAs on the endogenous transcripts. This can be done by microarrays. The caveat in these measurements is that different human cell types express different genes. If the analysis is limited to one or two cell types, there is no guarantee that the off-target effect will not happen in other cell types.

Once the lack of cross talk and off-target effect are confirmed, the next step is to validate the inhibitory effects in the context of complex 3′-UTR sequences involving multiple targets. While all the possible target arrangements can be tested computationally, their physical cloning is infeasible. Moreover, constructs with multiple targets are likely to be the final components of the logic circuits. In order to choose the best arrangements, one should pick among alternatives the one with the lowest secondary structure.

2.3.1. Cloning protocol for individual siRNA targets (CP1)

CP1.1. Identify two adjacent unique restriction sites in the 3′-UTR of the fluorescent reporter (use a commercial construct intended for creating C-terminal fusions). The 3′-site should be as close as possible to the last codon of that protein or even reside inside the coding sequence. For example, in a plasmid pZsYellow-C1 (Clontech) these sites are *Xho*I and *Bam*HI.

CP1.2. Design a pair of DNA oligomers that introduce a stop codon in frame with the fluorescent protein and as close as possible to the last codon, followed by an siRNA target site. Introduce a spacer of 10 nt between the new stop codon and the target site. The double-stranded annealing product should contain sticky ends compatible with the double digestion product of the vector (Fig. 8.2A). Design a negative control insert with a "mock" target sequence and use in to create a "negative control" reporter.

CP1.3. Prepare the dsDNA insert. PAGE purification of raw oligomers is an advantage, but the method often works with the raw oligomers. Anneal oligonucleotides at 25 or 12.5 μ*M* in TE buffer (10 m*M* Tris–HCl and 1 m*M* EDTA, pH 8.0) supplemented with 50 m*M* NaCl in a PCR machine block by heating to 95 °C and cooling down to 10 °C for 50 min. Phosphorylate double-stranded inserts at 3 μ*M* concentration in 50 μl of Polynucleotide Kinase (PNK) reaction buffer (New England Biolabs) by 15 units of PNK and 1 m*M* ATP (Invitrogen) for 30 min at 37 °C. Alternatively, order gel-purified and phosphorylated oligos and use directly after annealing.

CP1.4. Subclone the insert. Ligate the inserts into 25 ng of a digested vector at ~2:1 insert to vector molar ratio in 10 μl of T4 DNA Ligase buffer (New England Biolabs) using 400 units of T4 DNA Ligase (New England Biolabs) for 2.5 h at 15 °C. Transform the reaction mixture into 50 μl of Max Efficiency DH5α *Escherichia coli* cells (Invitrogen, Cat # 18258-012), outgrow for 60 min at 37 °C in SOC medium (Invitrogen) shaken at 300 rpm, and plate on selective plates overnight. Analyze colonies by colony PCR using a pair of primers flanking the insert region. Expand positive colonies in selective medium overnight, isolate plasmid DNA using MiniPrep kit (Qiagen), and verify the insert integrity by sequencing. Note that due to synthetic source of the inserts mutations will be frequent. Sequence at least 5 minipreps.

2.3.2. Testing protocol for individual siRNAs (TP1)

TP1.1. Order siRNA molecules. siRNA can be ordered from a large number of suppliers. We recommend ordering the sense and antisense oligomers separately, as it gives more flexibility in potential structure modifications.

TP1.2. If necessary, reconstitute siRNA by annealing the sense and antisense oligos. Resuspend in water at 25 μ*M* final concentration and anneal by slow cooling down similarly to DNA annealing. Store concentrated stock at −80 °C.

TP1.3. Find the minimal concentration of siRNA that causes efficient gene expression knockdown and quantify off-target effects.

TP1.3.1. Calibrate the reporter and the transfection marker. Transfect increasing amounts of marker–reporter mixtures. For transfections in 12-well plates, try values between 50 and 500 ng of each plasmid.

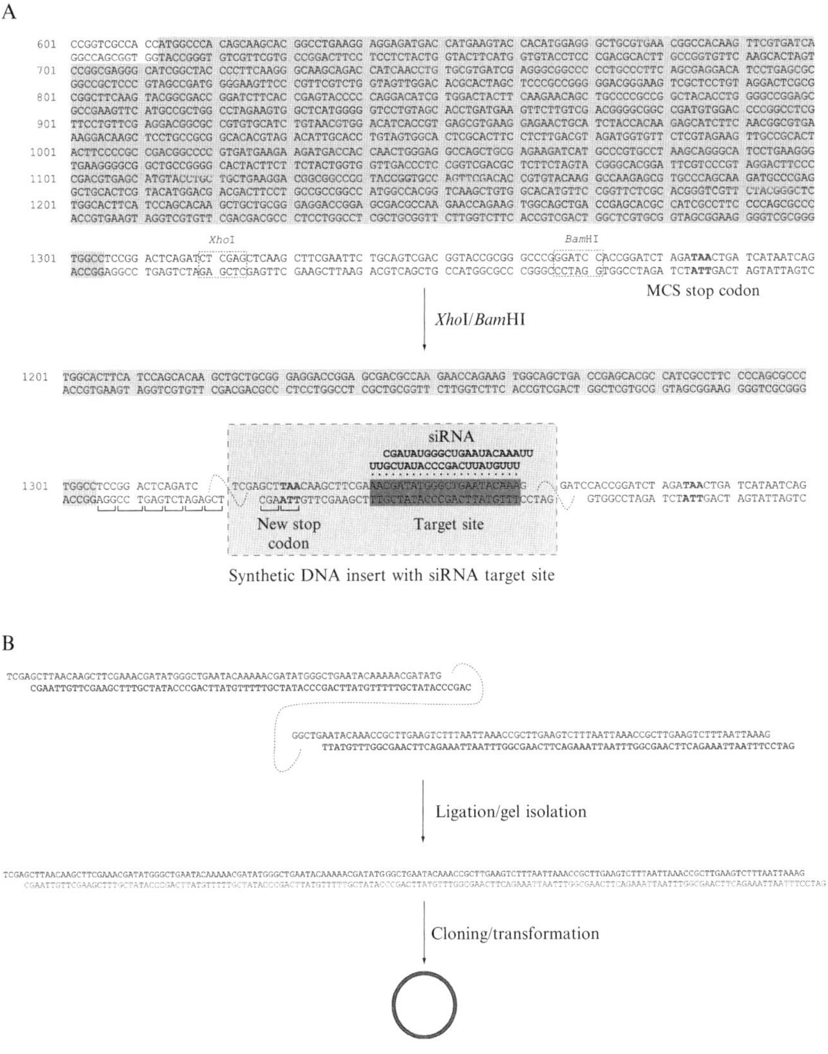

Figure 8.2 Examples of cloning approaches. (A) Subcloning an siRNA target in 3′-UTR of a fluorescent reporter. A plasmid pZsYellow-C1 (Clontech) is used for illustration. Coding sequence of ZsYellow protein is shaded. The insert is designed to contain a preformed sticky end, a stop codon, a spacer, and an sRNA target site/s. (B) Subcloning a long target sequence. The desired insert containing triple repeats of two different targets is made from two separate dsDNA building blocks by test-tube ligation. The resulting insert is subcloned in the output UTR as in (A). (See Color Insert.)

For example, pAmCyan-C1 plasmid (Clontech) or pDsRed-Express-C1 can be used as a marker with pZsYellow-C1 siRNA-targeted derivatives. Use transfection protocol appropriate for the cell line employed in your study.

As a sample protocol for HEK293 cells, consider the following: plate 90–120 thousand cells in 1 ml of complete medium (DMEM (Invitrogen) supplemented with 0.1 m*M* of MEM nonessential amino acids (Invitrogen), 0.045 units/ml of penicillin, and 0.045 g/ml streptomycin (penicillin–streptomycin liquid, Invitrogen), and 10% FBS (Invitrogen)) into each well of 12-well uncoated glass-bottom (MatTek) or plastic (Falcon) plates and grown for 24 h. Shortly before transfection, replace the medium with 1 ml DMEM without supplements with a single medium wash step. Prepare transfection mixtures by mixing all nucleic acids into 40 μl of DMEM. Add 2.4 μl of the Plus reagent (Invitrogen) to the final mix and incubate for 20 min at 24 °C. In parallel, mix 1.6 μl lipofectamine (Invitrogen) with 40 μl DMEM. Mix Plus- and lipofectamine-containing solutions and incubated for 20 more min at 24 °C before application to the cells. Apply the transfection mixture (typically 90 μl) to the wells and mix with the medium by gentle shaking. Three hours after transfection, add 120 μl FBS to the wells and incubate the cells for up to 48 h before the analysis.

Measure fluorescence after 48 h using microscopy and flow cytometry. Use images for qualitative examination and flow cytometry for quantitative characterization. The criteria that guide the choice of the optimal concentration may vary and include transfection efficiency (%Marker$^+$ cells), average intensity per transfected cells (mean(Marker)|Marker$^+$ cells), or integrated signal from the transfected well (mean(Marker)|Marker$^+$ cells) × (%Marker$^+$ cells). As a rule of thumb, one should use the lowest possible plasmid amount that gives a reasonably strong signal. Transfection efficiency should be at least 20–30% for robust signal generation.

TP1.3.2. Use plasmid amounts found in the previous steps and titrate siRNA concentrations. For each siRNA, use a mixture of the appropriate reporter plasmid and the transfection marker. Cotransfect with varying siRNA concentrations from 0 to 60 pmol/well in a 12-well plate. Use small concentration increases at the low end of the range, because often amounts as low as 1 pmol can result in substantial repression. Other values we recommend are 2.5 and 5 pmol. Use transfection reagent that has been shown to work with both DNA and siRNA. The example in TP1.3.1. works for both. Analyze gene expression after 48 h. The quantification method of choice is the ratio of the integrated reporter signal to integrated marker signals, divided by the same ratio in the control experiment where a "mock" siRNA target is used instead of the intended target. Note that this reference ratio may also change as a result of nonspecific siRNA effects. It is common for siRNA effect to reach saturation at less than 100% repression.

The tested siRNAs can then be ranked based on the saturation knockdown efficiency and the minimal siRNA level that gives near-saturation knockdown. The higher the efficiency and the lower this concentration, the better. In general, siRNA that do not elicit their full effect at

2.5–5.0 pmol and/or do not elicit at least 95% knockdown, should not be used in circuit construction. However, they can be reevaluated by inserting multiple tandem repeats of the target. This may lead to increased efficiency and sensitivity of the response element. If this fails for 3–4 tandem repeats, the siRNA should be discarded from the set.

TP1.3.3. Repeat the measurements for all siRNAs in the set. Triplicate measurements are advisable. For each siRNA, determine the lowest concentration that elicits efficient knockdown. If the values are similar, the highest value among them can be used with all the siRNA for convenience. Next, perform the cross talk measurements. For each siRNA in a N-member set {siRNA-1, siRNA-2, ..., siRNA-N}, prepare $N + 1$ wells. In well i, transfect a mixture of the reporter for siRNA-i with the transfection marker. In the well $N + 1$, transfect a mixture of the "mock" reporter with the marker (Fig. 8.3A). Cotransfect the tested siRNA in all the wells using previously found amount. Perform a similar set of measurements for the siRNA/s that you plan to use as a "negative control" (this would normally be one of the commercially available molecules, and it should not be active toward the "mock" target). Since negative control siRNA will be used to replace active siRNAs in the circuits, you should plan for the worst-case scenario when all the siRNAs are replaced by the control. For example, if the set contains four siRNAs and each has to be used at 5 pmol, test 20 pmol of the control siRNA. No effect on the reporters should be observed; if it is present, test an alternative negative control siRNA.

In general, one expects knockdown only when the siRNA is applied to its intended target. Therefore among the $(N + 1)^2$ measurements, only N should give values close to 100% and all the rest should ideally be zero. In reality some off-target effects may be observed. While individual effects could be small, in large circuits they may add up to considerable effect. One could consider a worst-case scenario under the assumption that the effects are additive; in this case it is not advisable to use sets that might generate more than 50% cumulative nonspecific knockdown. Normally only one siRNA would contribute the most, in which case it should be discarded from the set.

2.3.3. Cloning protocol for multitarget constructs (CP2)

CP2.1. Design DNA oligomers that contain multiple siRNA target sites, as required by the desired circuit architecture. If the expected total length of the target region is low (e.g., three to four targets, one repeat each, 60–80 bp), use protocol CP1. If the total length is more than that, either due to larger number of targets or to using target repeats, use a stepwise assembly strategy as follows.

CP2.2. Design the desired final dsDNA insert sequence with built-in terminal ligation-ready sticky ends. Introduce breakpoints in the sense and antisense strands such that the distance between the breakpoints is between

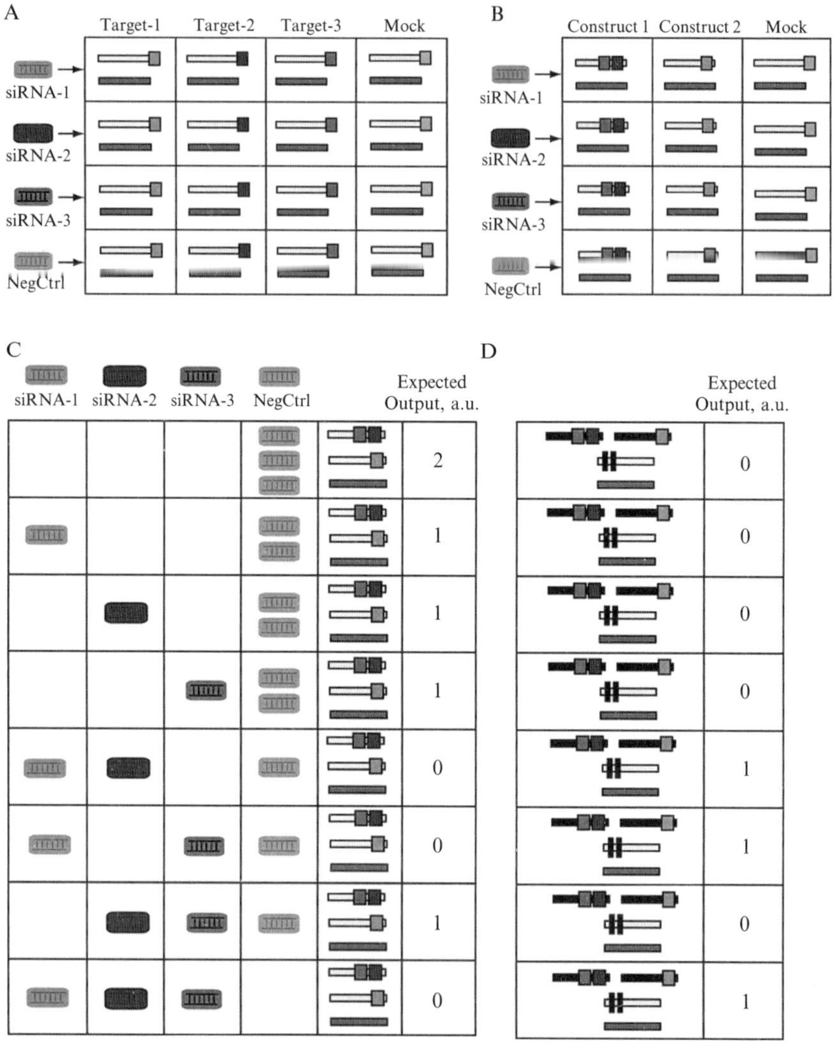

Figure 8.3 Schematics of testing matrices. (A) Testing matrix to measure siRNA cross talk. Each cell in a table represents one transfected well. Each siRNA is cotransfected with all the target constructs. (B) Testing matrix for the "computational" molecules. Each output construct is tested with each siRNA. (C), (D) Testing matrix for the logic DNF (C) and CNF (D) circuits. siRNA molecules cotransfected in each evaluation are shown, together with the anticipated outcome.

80 and 90 nt, and the adjacent breakpoints on the opposing strands are located 5–6 nt from each other. Order, PAGE-purify and phosphorylate the oligos comprising the subdivided sense and antisense strands. Reanneal complementary oligos, generating a number of partially dsDNA fragments

of 80–90 bp that have complementary sticky ends of 5–6 nt long. Verify that the sticky ends are unique and nonpalindromic.

CP2.3. Ligate the fragments to each other in a one-pot ligation mixture. Use 1 μ*M* concentration of each fragment in 300 μl of T4 DNA Ligase buffer (New England Biolabs) in the presence of 40 u/μl of T4 DNA Ligase (New England Biolabs) for 1 h at 20 °C. Purify full-length ligation product from agarose gel. Use the purified fragment for cloning into the output-coding vector (Fig. 8.2B).

CP2.4. This approach should work with up to four fragments or about 320–350 bp total insert length. If longer target regions are required, hierarchical assembly or sequential cloning may be needed. We note that PCR amplification of these inserts is very difficult if each target is present as tandem repeats.

2.3.4. Testing protocol for multitarget constructs (TP2)

TP2.1. The protocol is very similar to TP1. The experimental matrix will contain the different multitarget constructs and the "mock" target construct. Each siRNA is applied to each of the constructs at a previously determined concentration (TP1). A negative control siRNA should also be applied to all the constructs at a concentration equaling the sum of individual siRNA levels (Fig. 8.3B).

TP2.2. Multitarget constructs should be knocked down by the siRNA whose targets are present in their 3′-UTR, but not by the others. Ideally, both the knock-down efficiencies and the off-target effects should be consistent with the ones measured with single-target constructs. If this is not the case, try increasing the siRNA concentration first. If the knockdown remains insufficient, or if the required concentration is too high and is incompatible with full circuit operation, reexamine the 3′-UTR secondary structure and try to rationalize decreased siRNA efficiency. If the same target appears in different constructs, it is instructive to compare the efficiencies. If they are construct-dependent (e.g., siRNA X elicits 95% knockdown when acting against the construct Output-TX-TY but only 80% when acting against the construct Output-TX-TZ-TW), this points toward context-dependent effects. The solution could be repeated computational analysis of the complex target sequence, examination of alternative target arrangements and/or of adding spacer sequence between adjacent targets in order to reduce secondary structure. This should be followed by construction and testing of the alternative target arrangements.

2.4. Step 4: Characterization of the logic core

Once individual target sets have been validates, it is possible to proceed with the full circuit testing. The main difference between the full circuit testing and the preliminary steps is concurrent application of multiple siRNA species in the same mixture. This may create additional side effects not observed previously. The measurements are straightforward. All the required output-coding constructs (two or more for a gate with one or more OR operators) are combined with a transfection marker and subsets of the complete siRNA set used in this circuit. For N siRNAs, this amounts to 2^N measurements. Each siRNA should be added at an optimal concentration found previously with single and multiple target constructs. Omitted siRNA should be substituted by an equivalent amount of the negative control siRNA (Fig. 8.3C). Output level should be measured using flow cytometry and processed as described in TP1.3.2. Normalized output levels should be then compared to expectation, which is determined by the logic computed in the circuit. The levels corresponding to anticipated False output should be significantly lower than those corresponding to True outputs. Building histograms is helpful to analyze circuit performance. The histograms should be bimodal because most computations give 0 or 1. For multi-OR circuits, some measurements are expected to give double or triple levels of expression, resulting in multimodal distribution. Since we consider any outcome that is higher than "1" as true, the focus should still remain on the group of "0" and "1" outcomes. Two useful measures are (1) the worst-case scenario, that is, the ratio between the lowest "1" measurement and the highest "0" measurement; and (2) the ratio of the mean "1" and mean "0" measurement. Comparing standard deviations may be misleading because a single outlier will not affect it by a large factor, and yet such an outlier will invalidate the entire computational core or at least will have to be treated specially (e.g., the logic formula will have to be modified to be consistent with a single error, and the implication of this change will have to be considered).

The outliers normally come from the "0" group, that is, the measured output is too high. Sometimes simply using higher concentration of one of the siRNAs can alleviate this. Indeed, in our experience the outliers would normally be due to a single malfunctioning element or a single nonspecific interaction that leads to decrease of siRNA activity compared to the preliminary measurements.

If, however, the distributions are satisfactory, this completes the construction of the logic core. The siRNA sequences can be used as a basis for miRNA construction (see below), or the basis for other molecular sensing mechanisms that transduce external signals into those sequences.

3. Constructing a Computational Logic Core for the RNAi-Based CNF Circuit

CNF circuits are more complex than DNF due to the extra repression layer. In a sense, the repressor-producing component is equivalent in structure to a DNF circuit apart from the fact that the output is a repressor protein that cannot be visualized (unless fused to a fluorescent reporter). Therefore, testing a DNF equivalent of this subnetwork is an important step and it is a good proxy of what to expect with the full CNF circuit (see below). However, the requirements are more stringent because even low residual levels of the repressor will result in significant downregulation of the output-driving promoter and lead to low output levels even when the repression is supposed to be turned off based on siRNA combination. Accordingly, every effort should be made to design a set of potent siRNAs or to improve knockdown by using multiple target repeats.

An important extra step is the selection of the repressor and the responsive promoter. Many well-studied repressor–promoter combinations are available commercially, but those should be recharacterized prior to making the final decision using a desired model system (e.g., performance may differ between different cell lines). Ideally, the repressor itself should be constitutive, such as *LacI* or tTA. Another degree of freedom is the choice of a constitutive promoter driving the repressor. For consistency, a constitutive part of the repressible promoter can be used to drive the repressor, but at times it is advantageous to use different promoters. They should also be calibrated to determine the right repressor–promoter ratio in transient transfections.

3.1. Testing protocol for repressor–output pair (TP3)

TP3.1. Choose one or more repressor–promoter pairs.

TP3.2. If necessary, replace a constitutive promoter driving a fluorescent reporter of choice with the repressible promoter of choice using standard cloning techniques.

TP3.3. Find the optimal output amount using the approach in TP1.3.1.

TP3.4. Fix the above output/transfection marker amount. Prepare and transfect output/marker mixtures with increasing amount of the repressor-expressing plasmid. One can expect a saturating concentration dependency. If the saturating repression is incomplete (i.e., there is expression leakage), one needs to consider the consequences, for example, the fact that the "Off" output levels will never go below the observed leakage levels. If the repression is efficient at saturation, consider determining the minimal repressor level that generates such an effect. For example, if 50 ng repressor generate 70% repression while 100 ng generate 95% repression, test repressor levels between these values to determine the inflection point of the response curve. Fix this optimal repressor plasmid concentration.

3.2. Testing protocol for siRNA set for a CNF circuit (TP4)

TP4.1. When reusing previously optimized siRNA set, measure their efficacy with the fluorescent output constructs that use the same promoter that drives the repressor gene and keep the concentration of *each output construct* equal to the optimal repressor concentration found above. Measurements can be performed directly with the configuration anticipated in the complete CNF circuit (where a repressor is substituted by the fluorescent reporter, and the repressible output is omitted). Note that in this case, the repression must be as efficient as possible. Consider increasing standard siRNA levels, improving target efficacy by increasing the number of tandem repeats, or testing additional siRNAs if the initial measurements are not satisfactory.

3.3. Cloning protocol for repressor constructs with siRNA targets (CP3)

CP3.1. The simplest way to clone the constructs is to take the reporter plasmids from TP4 and replace the fluorescent coding frame with the repressor-coding frame. Use SII-type enzymes if there are no available unique restriction sites at the right location.

3.4. Testing protocol for the CNF circuit (TP5)

TP5.1. Once all the constructs are ready, perform exhaustive measurements with 2^N siRNA combinations similar to DNF characterization.

TP5.2. The output should be inverted with respect to the testing results in TP4. Note however that the levels will range strictly between "0" and "1," that is, between the maximally repressed output level and the unrepressed expression from the repressible promoter (Fig. 8.3D).

4. Transition from siRNA to miRNA

While siRNA could be the product of a sensory process, we have recently described engineered miRNA-expressing genes as transcription factor sensors. In general, any signal that alters the transcription of an miRNA gene can be then integrated in the logic circuits. This chapter describes the adaptation of siRNA circuits to miRNA circuits.

In general, even if a circuit is designed to use miRNA, it is important to test the siRNA-based version of the same circuit. In our experience, any problem encountered with siRNA is aggravated when using an equivalent miRNA sequence. Therefore, it is unwise to jump straight to the miRNA network because the construction of miRNA expression cassettes is much more labor-intensive. In addition, we note that until now we have only shown DNF circuits with miRNA.

4.1. Cloning protocol for miRNA (CP4)

CP4.1. Use a pPRIME approach (Stegmeier *et al.*, 2005) to design a DNA template for an miRNA hairpin that will generate the same active strand as its equivalent siRNA.

CP4.2. Order the oligomers with built-in *Xho*I- and *Eco*RI sticky ends; in parallel obtain a recipient plasmid, such as pPRIME-CMV-Neo-FF3 or pPRIME-CMV-Neo-recipient (AddGene). These plasmids contain an endogenous miR-30 backbone; following the insertion of a synthetic active miR this leads to a functional miR gene. Process DNA oligos according to CP1.3. Use *Xho*I and *Eco*RI to subclone the synthetic DNA insert.

CP4.3. The resulting Neo-miR fusion is in itself a functional miRNA gene. In many cases, however, cloning the miRNA as an intron is desired. The fragment that should be placed in the intron is the miR-30 backbone with subcloned active miR insert. Design a cloning strategy to place this sequence, together with 3′- and 5′-splicing signals, into a coding sequence of a protein of interest, or simply into an mRNA-coding sequence if concomitant protein expression is not required. Another factor to consider in the cloning strategy is the promoter that drives the expression of an intron-encoded miRNA, and the polyA signal.

4.2. Testing protocol for miRNA circuits (TP6)

TP6.1. The main difference between the siRNA and the miRNA set testing is that the former contains a small number of plasmids (output constructs) and the siRNAs are made chemically; in the latter case all the components are encoded on plasmids. This may place additional strain on the cells due to large DNA amounts, as well as lead to potential issues with cotransfections.

TP6.2. Use approach similar to that described in TP1.3.2. to assess miRNA efficiency and determine their working concentrations. This time, vary nanogram or mole amounts of the miRNA-encoding plasmid. If the miRNA promoter is inducible, make sure that it is fully induced in these experiments.

TP6.3. If you have already measured cross talk and off-target effects of the siRNA molecules that correspond to your miRNA constructs and found them satisfactory, the miRNA set will most likely exhibit similar properties. If time is of the essence, one can skip equivalent measurements with miRNA plasmids. However, there is no guarantee that new effects will not appear due to the transition from siRNA to miRNA, and thus it is advisable to measure cross talk at least once.

TP6.4. Logic gate output should be reevaluated with miRNA constructs even if it has been measured with the corresponding siRNA. More often than not, miRNA constructs are less efficient that the siRNA in these assays and the digital gate behavior may deteriorate.

TP6.5. If the logic gate behavior is satisfactory when miRNA plasmids are added (and fully induced, if needed) or withheld, one can proceed to test the entire circuit when the input signals controlling miRNA expression. In such tests, the entire set of miRNA-coding plasmids and the output constructs is transfected into the cells and the signals that control miRNA expression are turned On and Off in an exhaustive fashion (e.g., by adding appropriate ligands to the cell culture medium). Prior to these measurements, it is important to confirm that the signals indeed efficiently regulate miRNA expression under the specific experimental conditions used (such as specific miRNA plasmid amounts). This can be done by fusing miR introns into fluorescent protein coding genes and measuring fluorescence as a proxy for miRNA levels in cells. For an activating signal, it is crucial that no expression is taking place in the absence of signal; for a repressing signal, it is crucial to achieve full repression when the signal is On.

ACKNOWLEDGMENTS

I would like to thank my lab members at Harvard FAS Center for Systems Biology who took part in the development of these methods, including Keller Rinaudo, Leonidas Bleris, Zhen Xie, Madeleine Leisner, Jason Lohmueller, and David Glass, as well as our collaborators Ron Weiss and Sairam Subramanian. The work was supported by the Bauer Fellows Program and NIGMS grant GM068763 to the Centers of Systems Biology.

REFERENCES

An, C. I., Trinh, V. B., and Yokobayashi, Y. (2006). Artificial control of gene expression in mammalian cells by modulating RNA interference through aptamer-small molecule interaction. *RNA—A Publication of the RNA Society* **12,** 710–716.

Beisel, C. L., Bayer, T. S., Hoff, K. G., and Smolke, C. D. (2008). Model-guided design of ligand-regulated RNAi for programmable control of gene expression. *Mol. Syst. Biol.* **4,** 224.

Benenson, Y., Gil, B., Ben-Dor, U., Adar, R., and Shapiro, E. (2004). An autonomous molecular computer for logical control of gene expression. *Nature* **429,** 423–429.

Faulhammer, D., Cukras, A. R., Lipton, R. J., and Landweber, L. F. (2000). Molecular computation: RNA solutions to chess problems. *Proc. Natl. Acad. Sci. USA* **97,** 1385–1389.

Frutos, A. G., Liu, Q. H., Thiel, A. J., Sanner, A. M. W., Condon, A. E., Smith, L. M., and Corn, R. M. (1997). Demonstration of a word design strategy for DNA computing on surfaces. *Nucleic Acids Res.* **25,** 4748–4757.

Kauffman, S., Peterson, C., Samuelsson, B., and Troein, C. (2003). Random Boolean network models and the yeast transcriptional network. *Proc. Natl. Acad. Sci. USA* **100,** 14796–14799.

Leisner, M., Bleris, L., Lohmueller, J., Xie, Z., and Benenson, Y. (2010). Rationally designed logic integration of regulatory signals in mammalian cells. *Nat. Nanotechnol.* **5,** 666–670.

Reynolds, A., Leake, D., Boese, Q., Scaringe, S., Marshall, W. S., and Khvorova, A. (2004). Rational siRNA design for RNA interference. *Nat. Biotechnol.* **22,** 326–330.

Rinaudo, K., Bleris, L., Maddamsetti, R., Subramanian, S., Weiss, R., and Benenson, Y. (2007). A universal RNAi-based logic evaluator that operates in mammalian cells. *Nat. Biotechnol.* **25,** 795–801.

Schwarz, D. S., Hutvagner, G., Du, T., Xu, Z. S., Aronin, N., and Zamore, P. D. (2003). Asymmetry in the assembly of the RNAi enzyme complex. *Cell* **115,** 199–208.

Seelig, G., Soloveichik, D., Zhang, D. Y., and Winfree, E. (2006). Enzyme-free nucleic acid logic circuits. *Science* **314,** 1585–1588.

Shapiro, E., and Benenson, Y. (2006). Bringing DNA computers to life. *Sci. Am.* **295,** 44–51.

Stegmeier, F., Hu, G., Rickles, R. J., Hannon, G. J., and Elledge, S. J. (2005). A lentiviral microRNA-based system for single-copy polymerase II-regulated RNA interference in mammalian cells. *Proc. Natl. Acad. Sci. USA* **102,** 13212–13217.

Stojanovic, M. N., and Stefanovic, D. (2003). A deoxyribozyme-based molecular automaton. *Nat. Biotechnol.* **21,** 1069–1074.

Sullivan, C. S., and Ganem, D. (2005). A virus-encoded inhibitor that blocks RNA interference in mammalian cells. *J. Virol.* **79,** 7371–7379.

Tuleuova, N., An, C.-I., Ramanculov, E., Revzin, A., and Yokobayashi, Y. (2008). Modulating endogenous gene expression of mammalian cells via RNA-small molecule interaction. *Biochem. Biophys. Res. Commun.* **376,** 169–173.

Weiss, R., Homsy, G. E., and Knight, T. F. (1999). Toward *in vivo* digital circuits. *In* "Evolution as Computation: DIMACS Workshop," (F. Landweber and E. Winfree, eds.), pp. 275–295. Springer, Berlin, Germany.

Zuker, M. (2003). Mfold web server for nucleic acid folding and hybridization prediction. *Nucleic Acids Res.* **31,** 3406–3415.

CHAPTER NINE

From SELEX to Cell: Dual Selections for Synthetic Riboswitches

Joy Sinha, Shana Topp, *and* Justin P. Gallivan

Contents

Abstract

Synthetic riboswitches have emerged as useful tools for controlling gene expression to reprogram cellular behavior. However, advancing beyond proof-of-principle experiments requires the ability to quickly generate new synthetic riboswitches from RNA libraries. In this chapter, we provide a step-by-step overview of the process of obtaining synthetic riboswitches for use in *Escherichia coli*, starting from a randomized RNA library.

1. Introduction

In the past several years, our group and several others have shown that starting with a known aptamer, it is possible to create a variety of synthetic riboswitches, which are RNA sequences that regulate gene expression in a ligand-dependent fashion without the need for protein cofactors (Desai and Gallivan, 2004; Duchardt-Ferner *et al.*, 2010; Hanson *et al.*, 2003, 2005; Lynch and Gallivan, 2009; Lynch *et al.*, 2007; Muranaka *et al.*, 2009a,b; Sinha *et al.*,

Department of Chemistry and Center for Fundamental and Applied Molecular Evolution, Emory University, Atlanta, Georgia, USA

Methods in Enzymology, Volume 497
ISSN 0076-6879, DOI: 10.1016/B978-0-12-385075-1.00009-3

2010; Suess *et al.*, 2004; Topp and Gallivan, 2008a,b; Weigand and Suess, 2007; Weigand *et al.*, 2008; Werstuck and Green, 1998; Win and Smolke, 2007). These synthetic riboswitches can be used to create cells with useful phenotypes, including designer auxotrophs that depend on the presence of a nonmetabolite for survival (Desai and Gallivan, 2004), as well as cells that can selectively localize to high concentrations of an exogenous ligand that they would not otherwise detect (Sinha *et al.*, 2010; Topp and Gallivan, 2007). While taking a known aptamer as a starting point for creating a synthetic riboswitch has proven fruitful, it is not clear that such an approach represents the optimal solution, as not all aptamers lead to successful riboswitches (Sinha *et al.*, 2010; Weigand *et al.*, 2008). For example, an aptamer that requires a highly structured conformation before binding its ligand may not be capable of undergoing the ligand-dependent structural modulation required to function as a riboswitch. Since the molecular details that distinguish effective aptamers from ineffective aptamers are not yet known, there is considerable interest in developing methods that simplify the process of obtaining synthetic riboswitches. We have proposed that a combination of *in vitro* and *in vivo* selection may provide greater opportunity to achieve this goal by optimizing the interplay between binding and switching parameters (Fig. 9.1).

Here, we present a combination of *in vitro* and *in vivo* selection techniques that were used in our lab to create a synthetic riboswitch that activates protein translation in response to the herbicide atrazine.

2. General Precautions

It is critical that the following *in vitro* selection procedures are carried out in an RNase-free environment including the use of sterile, RNase-free reagents, and materials. Frequent glove changes should be employed to protect the RNA sample from RNases present on the hands or in the environment. RNA should always be kept on ice when preparing reactions to reduce unwanted spontaneous cleavage. When using radioactive isotopes, appropriate precautions should be taken to avoid contamination of the user and the surroundings. Additionally, care should be taken while handling unpolymerized acrylamide.

3. *In Vitro* Selection

3.1. Materials

- Template DNA: Template DNA contains a randomized region flanked by two defined regions. The defined regions are used for primer annealing for cDNA synthesis and PCR amplification. The 5′-defined region

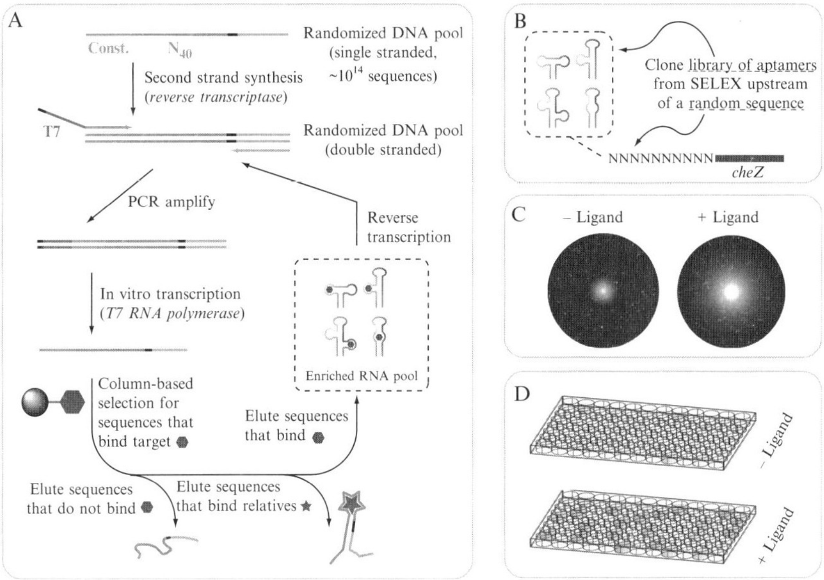

Figure 9.1 Selection strategy to identify synthetic riboswitches using a combination of *in vitro* and *in vivo* selections. (A) Aptamer selection scheme. RNA aptamers are isolated from large pools of randomized sequences using *in vitro* selection. The pool is enriched for sequences that bind an immobilized ligand and are eluted with free ligand. To obtain high-affinity binders, the selected sequences are reverse transcribed, amplified, and subjected to further rounds of selection. (B) Library construction. To select for synthetic riboswitches, a library of aptamers may be cloned upstream of a randomized sequence in the 5′-UTR of *cheZ*, which controls *E. coli* motility. The randomized region can become an expression platform that couples ligand binding and gene expression. (C) Riboswitch enrichment. The pool of *E. coli* cells that moves only in the presence of ligand is picked. (D) Riboswitch verification. Individual riboswitch clones can be confirmed using reporter gene (e.g., *lacZ*) assays. Such assays can also be used to quantify the gene expression of the riboswitch in the presence and absence of ligand.

also contains a T7 RNA polymerase promoter for *in vitro* transcription. The randomized region may vary from 20 to 100 nucleotides (nt). We typically use a 40-nt randomized sequence for *in vitro* selection.

- Ligand of choice: Since the ligand will also be used for *in vivo* selection, it must be cell permeable and nontoxic. Ideally, the ligand will have functional groups capable of interacting with RNA (e.g., hydrogen-bond donor and acceptor groups) and a site for attachment to a solid support so that the ligand can be immobilized.
- Resin for immobilizing the ligand: A wide range of activated agarose and sepharose resins are commercially available and can accommodate a variety of functional groups. The choice of resin depends on the functional groups present on the ligand; ideally the coupling reaction will

occur only at a single site. For example, for selecting atrazine aptamers, an atrazine derivative with a hydroxyl linker was reacted under mild conditions with epoxy-activated Sepharose 6B (GE Healthcare).

- RNA (either radiolabeled or unlabeled) transcribed from template DNA: Radiolabeled RNA is usually obtained by transcribing in the presence of α-^{32}P-ATP (3000 Ci/mmol; total of 250 μCi per transcription reaction).
- Binding Buffer: Any buffer can be chosen as long as two simple parameters are followed: The buffer should not facilitate dissociation of RNA from the ligand, and it should not degrade RNA. Typical binding buffers contain 5–20 m*M* of compounds that can maintain pH near neutrality (such as Tris or Hepes) and contain monovalent (Na^+ or K^+) and divalent cations (Mg^{2+}). We typically use 20 m*M* Tris–HCl (pH 7.4), 5 m*M* $MgCl_2$, 250 m*M* NaCl for RNA buffer.
- AmpliScribe transcription kit (Epicenter)
- Reagents and apparatus for PAGE
- Denaturing 10% polyacrylamide gel (1.5 mm thick)
- Gel loading buffer (2×): 18 *M* urea, 20% (w/v) sucrose, 0.05% (w/v) bromophenol blue, 0.05% (w/v) xylene cyanol, 2× TBE buffer
- Qiagen MinElute kit
- DNase I (we use Epicenter but other suppliers are also acceptable)
- Scintillation counter
- Hand-held UV light, shortwave (254 nm)

3.2. Procedures

The following procedures provide an overview of a typical *in vitro* selection experiment using SELEX as an example. There are several detailed reviews available which provide a more in-depth coverage of SELEX, and the reader is encouraged to use them (Codrea *et al.*, 2010; Conrad *et al.*, 1995; Klug and Famulok, 1994).

3.2.1. RNA transcription and purification

The RNA is transcribed from ~1 to 2 μg of template DNA using the AmpliScribe transcription kit, according to the manufacturer's instructions. Transcribed RNA is purified on a 10% denaturing polyacrylamide gel and visualized by UV shadowing. The bands corresponding to the full length RNA are excised. To elute the RNA, the gel slices are crushed and suspended in RNase-free water (1 mL) and incubated at 4 °C overnight. The resulting solution is passed through a 0.2-μm filter (VWR) to remove polyacrylamide gel particles.

To recover RNA, one-tenth volume of 3 *M* sodium acetate (0.3 *M* final) and 2.5 volumes of cold 100% ethanol are added to the eluted RNA solution, which is incubated at −80 °C for 15 min. RNA is precipitated by centrifuging at 14,000×*g* for 20–30 min at 4 °C. The pellet is rinsed with 70% ethanol, centrifuged again at 14,000×*g* for 15 min, and the supernatant is discarded. The pellet is then dried under vacuum and resuspended in 1× binding buffer (20 m*M* Tris–HCl, 5 m*M* $MgCl_2$, 250 m*M* NaCl, pH 7.4). The quantity of RNA is determined by measuring the absorbance at 260 nm. The yield of RNA can vary, but the above procedure typically yields 10–15 μg of RNA, enough to provide at least 10^{14} molecules for each round of selection.

3.2.2. Column preparation

Two columns are prepared: a preselection column that contains blocked resin and a selection column that contains ligand-coupled resin (atrazine in this case). Each column is packed with resin to obtain a 250-μL bed volume. Each column is rinsed five times with 1 mL of binding buffer to remove any undesired compounds (e.g., preservatives such as sodium azide, often found in commercially available resins, or salts, organics or leaving groups that remain from coupling the target to the resin). Columns are used only once and are suitably disposed after each round of selection. Therefore, the above process is repeated prior to every selection cycle.

3.2.3. *In vitro* selection

RNA (300 pmol; ∼10^{14} sequences) in 1× binding buffer is denatured by incubating at 70 °C for 2 min and then allowed to slow cool to room temperature over 5 min. This will ensure that each species of RNA in the pool folds into its most stable conformation. To remove RNA transcripts that bind to the resin, the radiolabeled RNA pool (300 pmol) is loaded on a preselection column containing 250 μL of blocked sepharose without atrazine and incubated for 30 min at room temperature. The column is washed with 600 μL of binding buffer and RNA that passes through the column is collected. This RNA pool, which does not bind blocked resin, is then loaded on a selection column containing 250 μL of atrazine-sepharose. The column is incubated at room temperature for 30 min and then washed with 1.2 mL of binding buffer. Bound transcripts are eluted with 1 mL of 5 m*M* of atrazine solution. Note that the volume of wash buffer directly affects the stringency of the selection. To favor the accumulation of binding species, it is highly recommended that the wash volume be low during early rounds. The wash volume can then be increased as the selection progresses and if greater stringency is desired.

To the eluted RNA solution, one-tenth volume of 3 *M* sodium acetate (0.3 *M* final), 2.5 volumes of cold 100% ethanol, and 1–2 μg of glycogen are added. The solution is then incubated at −80 °C for 15 min, RNA is precipitated by centrifuging at 14000×*g* for 30 min at 4 °C. The pellet is rinsed with 70% ethanol, centrifuged again at 14000×*g* for 15 min. The supernatant is then discarded and then the pellet dried under vacuum.

Finally, the pellet is redissolved in 30 μL of water. Half of the pool is used for amplification via reverse transcription and PCR. The other half is saved for archival purposes. If a subsequent amplification or selection process fails, the procedure can be easily restarted from this archived pool. DNA from the PCR reaction is purified using the Qiagen MinElute kit, according to the manufacturer's protocol. Approximately 1 μg of the purified amplification reaction is used as a template to transcribe RNA for subsequent *in vitro* selection cycles. For each subsequent selection cycle, RNA is transcribed, gel purified, and ethanol precipitated as described above. RNA is then loaded on a column containing 250 μL of sepharose-atrazine. The selection conditions are as described above.

Depending upon the requirements, negative selection can be performed intermittently. Additionally, counter selection is also used where the pool is exposed to an additional molecule which is structurally very similar to the ligand. This ensures that the final aptamers only recognize the desired ligand and not related structures.

To track the progress of selection, aliquots of the solutions are recovered both before loading onto the selection column and after the elution, and the radioactivity is measured using a scintillation counter. These values are used to calculate percentage of RNA bound to the column.

4. *In Vivo* Selection

4.1. Material for motility-based selection

- T4 DNA Ligase (New England Biolabs)
- 10× T4 DNA Ligase buffer (New England Biolabs)
- 1-butanol (Sigma-Aldrich)
- Ethanol (100% and 70%)
- Luria-Bertani media (EMD Biosciences, Gibbstown, NJ)
- Ampicillin (Fisher, Pittsburgh, PA), final concentration 50 μg/mL of media for all procedures
- Tryptone broth: 10 g/L Bacto Tryptone (Difco) and 5 g/L NaCl
- Tryptone motility agar, prepared as Tryptone broth with 2.5 g/L Bacto Agar (Difco)
- Ligand of interest
- TOP10F′, electrocompetent *Escherichia coli* (Invitrogen, Carlsbad, CA)
- JW1870, electrocompetent *E. coli* (Δ*cheZ*, Keio collection; Baba *et al.*, 2006)
- Sterile Petri dishes, 85 mm (Fisher)
- Bioassay tray, 241 mm × 241 mm (Nalgene, Rochester, NY)
- QIAprep Spin Miniprep Kit (Qiagen, Valencia, CA)

4.2. Material for verification of *in vivo* selection

- Bioassay trays, 241 mm × 241 mm (Nalgene)
- 96-well microtiter plates (Costar)
- Lysis solution: 10:1 mixture of Pop Culture® (Novagen, Madison, WI) and lysozyme (4 U/mL)
- Z-buffer: 60 m*M* Na_2HPO_4, 40 m*M* NaH_2PO_4, 10 m*M* KCl, 1 m*M* $MgSO_4$, 50 m*M* β-mercaptoethanol, pH 7.0
- 2-nitrophenyl-β-D-galactopyranoside (ONPG) (Sigma-Aldrich) solution: 4 mg/mL in 100 m*M* NaH_2PO_4, pH 7
- 1 *M* Na_2CO_3
- Chloroform
- 0.1% SDS
- X-Gal solution (U.S. Biological, Swampscott, MA): 6.25 mg/mL in dimethyl formamide (DMF)
- Microplate reader (BioTek)
- Multichannel pipettor

4.3. Procedures

The following procedures assume that the user has already identified an aptamer (or a pool of potential aptamers) that binds to a target ligand that is cell permeable and nontoxic. Here, we present a library and selection scheme that enables the isolation of riboswitches that upregulate gene expression at the posttranscriptional level, but it is also possible to develop a framework amenable to isolating riboswitches that downregulate gene expression or that act at the level of transcription. The *in vivo* selection procedures described here will focus on using a motility-based selection method, as the resulting clones will be used directly for applications involving cell motility. Motility-based selections are performed with the *cheZ* reporter gene. It is advisable to use a constitutive promoter, although the optimal promoter strength depends upon the strength of the ribosome binding site (RBS). Verification of the riboswitch function is performed with the *lacZ* reporter gene.

4.3.1. Generation of library for *in vivo* selection

To generate a riboswitch from an aptamer, an expression platform is generally added after the *in vitro* selection step. However, adding or deleting nucleotides from an aptamer can adversely affect its binding ability. Additionally, rational designing of the aptamer-expression platform combination is cumbersome. Thus, to develop effective riboswitches, it is best to allow the *in vivo* selection experiment to optimize the expression platform. This step will also allow the identification of a functional RBS. Therefore, a fully randomized region of 10–12 bases is introduced between the aptamer and a region of 4–6 constant bases located immediately before the start codon. This constant region is important to provide suitable spacing between the start codon and the RBS that will be selected as part of the expression platform. Although the sequence of these constant bases is not critical, an example is shown in Fig. 9.2.

The aptamer-expression platform library described above should be cloned upstream of the start codon of a reporter gene within a high-copy number plasmid that confers an antibiotic resistance. In our lab, the vector

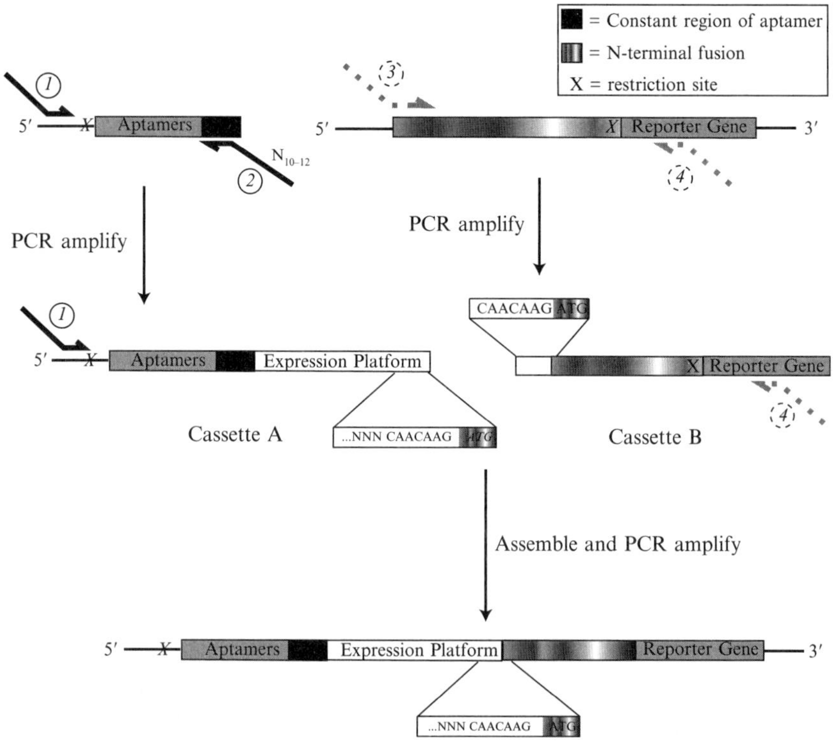

Figure 9.2 A general strategy for constructing a riboswitch library using oligonucleotide-based cassette mutagenesis. Cassette A is generated using forward primer 1 and reverse primer 2. Primer 1 anneals 5′ to the aptamer and reverse primer 2 contains the 10–12 base randomized region. The randomized region is flanked by a constant region complementary to the aptamer and a constant region that includes 4–6 constant bases and the start codon. Cassette B is created with forward primer 3 (the reverse complement of primer 2) and reverse primer 4, which anneals to a region located within the reporter gene. Following amplification and gel purification of each cassette, A and B can then be mixed together and assembled using outer primers 1 and 4 to yield PCR product C. Using the restriction sites at the 5′ and 3′ ends of the PCR product C, the product can then be cloned into a vector containing the desired reporter gene.

of choice is pUC18, which confers resistance to ampicillin. Since both *cheZ* and *lacZ* readily accept N-terminal protein fusions, all of our reporter gene constructs contain a short N-terminal peptide fusion. This strategy has two advantages. First, it facilitates DNA manipulation. Cloning steps are simplified by introducing a unique restriction site within the translational fusion at the N-terminus of either reporter gene. Second, riboswitch expression platforms can potentially interact with the aptamer or with ribonucleotides that encode for amino acids of the reporter protein. Because the reporter protein is changed between selection and verification steps, and different sequences can adversely affect the gene regulation ability of the riboswitch, the incorporation of a short peptide fusion ensures that the sequence immediately 3′ to the expression platform is constant throughout the selection and verification steps.

Figure 9.2 illustrates a general strategy for constructing a riboswitch library using oligonucleotide-based cassette mutagenesis. The assembled riboswitch library (PCR product C) is then digested with the appropriate restriction enzymes, gel purified, and cloned into the initial vector digested with the same enzymes. To prevent recircularization of the vector, the digested plasmid is dephosphorylated with CIP prior to ligation.

Ligation Reaction:

Digested vector	× μL (50 ng total)
Digested PCR product (contains library)	× μL (50–150 ng total)
T4 DNA Ligase buffer	2 μL
T4 DNA Ligase	1 μL
Water to 20 μL final	

To achieve high transformation efficiency in *E. coli*, the ligation reaction is precipitated with 1-butanol. In a 1.5-mL centrifuge tube, 10 volumes of 1-butanol is added to the ligation reaction. The solution is vortexed, and the DNA is pelleted by centrifugation at 10,000×*g* for 10 min at 4 °C. The supernatant is removed carefully and the pellet is washed with 70% ethanol (15 volumes). The pellet is dried under vacuum and redissolved in 10 μL sterile water. One microliter of the ligated product is used to transform electrocompetent cells to generate the library.

4.3.2. Motility selection

To select for riboswitches using motility as a read out, we use *cheZ* as a reporter gene. In the *E. coli* chemotaxis system, the CheZ protein dephosphorylates the CheY-P protein, which binds to the flagellar motor and causes swimming cells to tumble (Kuo and Koshland, 1987). Note that optimum levels of CheZ expression are required for the *E. coli* cells to migrate on a motility agar plate. If too little CheZ is present, the level of CheY-P will increase, and the cells will tumble incessantly and not migrate

(Wolfe *et al.*, 1987). If cells have excess CheZ, they will swim very smoothly and rarely tumble. Because cells that swim extremely smoothly can become embedded in the semisolid media, they cannot migrate (Wolfe and Berg, 1989). Thus it is critical to ensure that CheZ is not overexpressed in these assays. Because the strength of the promoter will ultimately dictate the maximum expression level of the *cheZ* gene, we advise the use of two different promoters: the "weak" *IS10* promoter (Jain and Kleckner, 1993) and the stronger *tac* promoter (de Boer *et al.*, 1983). By using different promoters, the motility selections can readily reveal which promoter provides the appropriate CheZ expression level. In our case, the weaker promoter provided ideal expression levels (Sinha *et al.*, 2010; Topp and Gallivan, 2008a).

4.3.2.1. General procedure for motility Electrocompetent TOP10F′ *E. coli* cells are transformed with 1 μL of the precipitated ligation described above. We typically let cells recover for 1 h at 37 °C with shaking (250 rpm). To determine the library size, 1 μL of recovered cells are plated on a Petri dish (85 mm) with LB/agar, supplemented with ampicillin. The remaining cells are plated on a large (241 mm × 241 mm) bioassay tray containing 300 mL of LB/agar, supplemented with ampicillin. The plates are incubated at 37 °C until colonies become visible (8–10 h). To calculate the total library size, the number of colonies on the small plate is first counted. The colonies from the large bioassay tray are scraped and the cells are suspended in 5 mL of LB supplemented with ampicillin. The culture is incubated for approximately 2 h at 37 °C with shaking (250 rpm) and plasmid DNA is extracted from 3 mL of culture with the QIAprep Spin Miniprep Kit.

Electrocompetent JW1870 *E. coli* cells (Δ*cheZ*) are then transformed with the plasmid library and the cells are allowed to recover for 1 h at 37 °C with shaking (250 rpm). The recovered cells are plated on a large (241 mm × 241 mm) bioassay tray containing 300 mL of LB/agar, supplemented with ampicillin. The plate is incubated at 37 °C until colonies become visible (8–10 h). The cells are scraped off the bioassay tray and resuspended in 5 mL of tryptone broth supplemented with ampicillin. The culture is incubated for approximately 2 h at 37 °C with shaking (250 rpm). An aliquot of this culture (100 μL) is used to inoculate 5 mL of tryptone broth supplemented with ampicillin and grown at 37 °C with shaking for approximately 3 h to an OD_{600} between 0.5 and 0.7. The culture is then diluted using tryptone broth to an OD_{600} of 0.2 (~200,000 cells/μL). Three microliters of diluted culture (~600,000 cells) are then applied at the center of motility agar plates supplemented with ampicillin and either containing or lacking ligand. The cells are allowed to dry in air for 10 min and the plates are incubated (media-side down) at 30 °C for 12–18 h (Fig. 9.3).

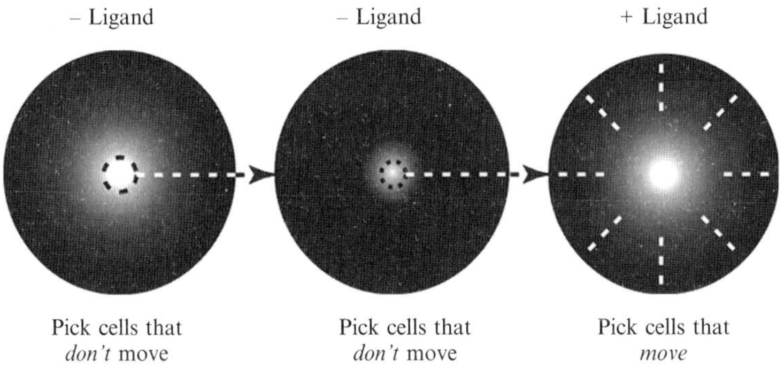

Figure 9.3 Selection strategy to identify synthetic riboswitches using bacterial motility.

4.3.2.2. Procedure for motility selection To select for a riboswitch that upregulates gene expression, the cells are applied to a motility agar plate (supplemented with antibiotic) that contains no ligand. After an overnight incubation, the population of cells remaining at the center of the plate (where the culture was applied) is collected by pipet and is used to innoculate 5 mL of tryptone broth supplemented with ampicillin. The culture is grown to an OD_{600} of 0.5 and then diluted to an OD_{600} of 0.2, as described in the previous section. The cells are applied to a second motility plate containing no ligand. Once again, the cells that remain nonmotile are collected and used to inoculate 5 mL of tryptone broth, as described above. The diluted cell suspension is now applied to plates that contain ligand. In a parallel experiment, the diluted cell suspension is also applied to plates that lack ligand. Ligand concentration can vary; for example, we used 0.75 m*M* ligand for selecting atrazine-responsive riboswitches (Sinha *et al.*, 2010) and 1 m*M* ligand for selecting theophylline-dependent riboswitches (Topp and Gallivan, 2008a). After an overnight incubation, the diameter of the outermost ring for each motility plate is measured to compare the migration distances of the cell population in the presence and absence of ligand. After the third round of selection, the enriched population should show a distinct increase in migration distance on the motility plate containing ligand compared to the plate lacking the ligand.

To identify library members for which gene expression is "on" in the presence of ligand, the cells are pipetted from the outside of the plate containing ligand. Usually, cells are pipetted from just beyond the visible migration edge. Cells are inoculated overnight in 5 mL LB containing ampicillin. Using the QIAprep Spin Miniprep Kit, plasmids from the overnight culture are isolated. These plasmids can now be used to subclone the enriched pool of riboswitches upstream of the *lacZ* reporter gene. The enriched pool of sequences is amplified by PCR, the PCR product is digested with appropriate restriction enzymes, and the gel purified pool is ligated upstream of the *lacZ* reporter gene.

4.3.3. Enzymatic (β-galactosidase) screen for functional riboswitches

To quantify and validate gene expression for the enriched pool from the motility-based selections, *E. coli* TOP10F′ cells are transformed with the ligated library of enriched riboswitches that were cloned upstream of *lacZ*. The plates are incubated at 37 °C for 14 h, and 96 random colonies are picked and inoculated in a 96-well microtiter plate containing 200 μL of LB supplemented with ampicillin. The 96-well microtiter plates are covered with parafilm and incubated at 37 °C with shaking (180 rpm). The following day, a multichannel pipettor is used to inoculate four 96-well plates of media (two sets of two) with 2 μL of the overnight culture. The first set of plates contains 200 μL of LB supplemented with ampicillin. The second set of plates contains 200 μL of LB supplemented with both ampicillin and the ligand (0.5–1 m*M*). Plates are incubated for ~2.5–3 h at 37 °C with shaking (210 rpm) to an OD_{600} of 0.085–0.14 as determined by a microplate reader. This reading corresponds to an OD_{600} of 0.3–0.5 with a 1-cm path length cuvette. When the appropriate level of growth is reached, the OD_{600} of each well is recorded. Twenty-one microliters of lysis solution are added to each well and mixed by gently pipetting up and down. The cultures are allowed to lyse for 5 min at room temperature. Fifteen microliters of the lysed culture is then added to 132 μL of Z-buffer and incubated at 30 °C for 10 min. To each well, 29 μL of ONPG solution is added. ONPG is allowed to hydrolyze for approximately 20 min (or until yellow color is visible), and the reaction is then quenched with 75 μL of Na_2CO_3. The time of hydrolysis is noted and OD_{420} of each well is determined. The β-galactosidase activity of each well is expressed in Miller units, using the following equation:

$$\text{Miller units} = OD_{420}/(OD_{600} \times \text{hydrolysis time} \times [\text{volume of cell lysate}/\text{total volume}]).$$

To identify functioning riboswitches, the ratio of Miller units observed for cultures grown in the presence of the ligand to those grown in the absence are determined. Functioning riboswitches should show consistent results across the two plates, while displaying normal growth rates and significant increases in β-galactosidase expression in the presence of the ligand (Lynch *et al.*, 2007). If these criteria are met, Miller unit ratios greater than 2 should indicate the presence of a functioning switch. To validate the function of identified switches, clones can be subcultured and assayed on a larger scale using a previously described protocol of Jain and Belasco (2000).

REFERENCES

Baba, T., Ara, T., Hasegawa, M., Takai, Y., Okumura, Y., Baba, M., Datsenko, K. A., Tomita, M., Wanner, B. L., and Mori, H. (2006). Construction of *Escherichia coli* K-12 in-frame, single-gene knockout mutants: The Keio collection. *Mol. Syst. Biol.* **2**(2006), 2006. 0008.

Codrea, V., Hayner, M., Hall, B., Jhaveri, S., and Ellington, A. (2010). In vitro selection of RNA aptamers to a small molecule target. *Curr. Protoc. Nucleic Acid Chem.* 1–23, (Chapter 9, Unit 9.5).

Conrad, R. C., Baskerville, S., and Ellington, A. D. (1995). In vitro selection methodologies to probe RNA function and structure. *Mol. Divers.* **1,** 69–78.

de Boer, H. A., Comstock, L. J., and Vasser, M. (1983). The tac promoter: A functional hybrid derived from the trp and lac promoters. *Proc. Natl. Acad. Sci. USA* **80,** 21–25.

Desai, S. K., and Gallivan, J. P. (2004). Genetic screens and selections for small molecules based on a synthetic riboswitch that activates protein translation. *J. Am. Chem. Soc.* **126,** 13247–13254.

Duchardt-Ferner, E., Weigand, J. E., Ohlenschlager, O., Schmidtke, S. R., Suess, B., and Wohnert, J. (2010). Highly modular structure and ligand binding by conformational capture in a minimalistic riboswitch. *Angew. Chem. Int. Ed. Engl.* **49,** 6216–6219.

Hanson, S., Berthelot, K., Fink, B., McCarthy, J. E., and Suess, B. (2003). Tetracycline-aptamer-mediated translational regulation in yeast. *Mol. Microbiol.* **49,** 1627–1637.

Hanson, S., Bauer, G., Fink, B., and Suess, B. (2005). Molecular analysis of a synthetic tetracycline-binding riboswitch. *RNA* **11,** 503–511.

Jain, C., and Belasco, J. G. (2000). Rapid genetic analysis of RNA-protein interactions by translational repression in *Escherichia coli*. *Methods Enzymol.* **318,** 309–332.

Jain, C., and Kleckner, N. (1993). IS10 mRNA stability and steady state levels in *Escherichia coli*: Indirect effects of translation and role of rne function. *Mol. Microbiol.* **9,** 233–247.

Klug, S. J., and Famulok, M. (1994). All you wanted to know about SELEX. *Mol. Biol. Rep.* **20,** 97–107.

Kuo, S. C., and Koshland, D. E., Jr. (1987). Roles of cheY and cheZ gene products in controlling flagellar rotation in bacterial chemotaxis of *Escherichia coli*. *J. Bacteriol.* **169,** 1307–1314.

Lynch, S. A., and Gallivan, J. P. (2009). A flow cytometry-based screen for synthetic riboswitches. *Nucleic Acids Res.* **37,** 184–192.

Lynch, S. A., Desai, S. K., Sajja, H. K., and Gallivan, J. P. (2007). A high-throughput screen for synthetic riboswitches reveals mechanistic insights into their function. *Chem. Biol.* **14,** 173–184.

Muranaka, N., Abe, K., and Yokobayashi, Y. (2009a). Mechanism-guided library design and dual genetic selection of synthetic OFF riboswitches. *Chembiochem* **10,** 2375–2381.

Muranaka, N., Sharma, V., Nomura, Y., and Yokobayashi, Y. (2009b). An efficient platform for genetic selection and screening of gene switches in *Escherichia coli*. *Nucleic Acids Res.* **37,** e39.

Sinha, J., Reyes, S. J., and Gallivan, J. P. (2010). Reprogramming bacteria to seek and destroy an herbicide. *Nat. Chem. Biol.* **6,** 464–470.

Suess, B., Fink, B., Berens, C., Stentz, R., and Hillen, W. (2004). A theophylline responsive riboswitch based on helix slipping controls gene expression in vivo. *Nucleic Acids Res.* **32,** 1610–1614.

Topp, S., and Gallivan, J. P. (2007). Guiding bacteria with small molecules and RNA. *J. Am. Chem. Soc.* **129,** 6807–6811.

Topp, S., and Gallivan, J. P. (2008a). Random walks to synthetic riboswitches—A high-throughput selection based on cell motility. *Chembiochem* **9,** 210–213.

Topp, S., and Gallivan, J. P. (2008b). Riboswitches in unexpected places—A synthetic riboswitch in a protein coding region. *RNA* **14,** 2498–2503.

Weigand, J. E., and Suess, B. (2007). A designed RNA shuts down transcription. *Chem. Biol.* **14,** 9–11.

Weigand, J. E., Sanchez, M., Gunnesch, E. B., Zeiher, S., Schroeder, R., and Suess, B. (2008). Screening for engineered neomycin riboswitches that control translation initiation. *RNA* **14,** 89–97.

Werstuck, G., and Green, M. R. (1998). Controlling gene expression in living cells through small molecule-RNA interactions. *Science* **282,** 296–298.

Win, M. N., and Smolke, C. D. (2007). A modular and extensible RNA-based gene-regulatory platform for engineering cellular function. *Proc. Natl. Acad. Sci. USA* **104,** 14283–14288.

Wolfe, A. J., and Berg, H. C. (1989). Migration of bacteria in semisolid agar. *Proc. Natl. Acad. Sci. USA* **86,** 6973–6977.

Wolfe, A. J., Conley, M. P., Kramer, T. J., and Berg, H. C. (1987). Reconstitution of signaling in bacterial chemotaxis. *J. Bacteriol.* **169,** 1878–1885.

CHAPTER TEN

Using Noisy Gene Expression Mediated by Engineered Adenovirus to Probe Signaling Dynamics in Mammalian Cells

Jeffrey V. Wong,[*,†] Guang Yao,[†,1] Joseph R. Nevins,[†,‡] *and* Lingchong You[*,†,§]

Contents

Abstract

Perturbations from environmental, genetic, and pharmacological sources can generate heterogeneous biological responses, even in genetically identical cells. Although these differences have important consequences on cell physiology and survival, they are often subsumed in measurements that average over the population. Here, we describe in detail how variability in adenoviral-mediated gene expression provides an effective means to map dose responses of signaling pathways. Cell–cell variability is inherent in gene delivery methods used in cell biology, which makes this approach adaptable to many existing experimental systems. We also discuss strategies to quantify biologically relevant inputs and outputs.

* Department of Biomedical Engineering, Duke University, Durham, North Carolina, USA
† Institute for Genome Sciences and Policy, Duke University, Durham, North Carolina, USA
‡ Department of Molecular Genetics and Microbiology, Duke University, Durham, North Carolina, USA
§ Center for Systems Biology, Duke University, Durham, North Carolina, USA
1 Current address: Department of Molecular and Cellular Biology, University of Arizona, Tucson, Arizona, USA

Methods in Enzymology, Volume 497
ISSN 0076-6879, DOI: 10.1016/B978-0-12-385075-1.00010-X

1. Introduction

Bulk measurements of cell properties can belie the rich heterogeneity within isogenic populations (Raser and O'Shea, 2005). Such nongenetic variability has its origins in the probabilistic nature of cellular processes (e.g., transcription and translation) involving small numbers of molecules and in the heterogeneous environment (Elowitz *et al.*, 2002; Taniguchi *et al.*, 2010). Heterogeneity in cellular signaling can propagate through gene regulatory networks (Blake *et al.*, 2003), with resultant effects (e.g., differences in protein levels) persisting over several generations (Sigal *et al.*, 2006). Such nongenetic individuality can greatly impact the interpretation and response to external signals, and the fitness and survival of individual cells (Kussell and Leibler, 2005). Indeed, cell–cell heterogeneity plays a role in determining the fate of hematopoietic progenitors in response to growth factors (Chang *et al.*, 2008) and the differential susceptibility to apoptotic stimuli (Spencer *et al.*, 2009) or anticancer drugs (Singh *et al.*, 2010).

On one hand, cell–cell variability limits the precision of cellular processes and thwarts efforts to predict and manipulate cell behavior (Rueger *et al.*, 2005). On the other hand, it has been used by cells to coordinate gene expression and enable transient differentiation (Eldar and Elowitz, 2010). Variability in gene expression has also been used to infer the topology and the strength of interactions in gene circuits (Dunlop *et al.*, 2008; Geva-Zatorsky *et al.*, 2010). In this sense, it provides a valuable window into processes that are difficult to observe otherwise. In this chapter, we describe a simple but powerful application of variability in viral-mediated gene expression to probe signaling dynamics in mammalian cells (Fig. 10.1).

Wild-type human adenoviruses (Ad) comprise over 52 serotypes that are responsible for various respiratory, ocular, and gastroenterological illnesses. These viruses consist of nonenveloped, icosahedral units that contain a single-stranded genome of about 36 kilobases (kb) (Reddy *et al.*, 2010). Replication-defective, recombinant Ad vectors are routinely used to generate high levels of gene expression in cultured mammalian cells (Campos and Barry, 2007). Their adoption in research is attributed to the fact that they can be amplified to relatively high titers, can transduce a wide range of dividing and nondividing cell types, and do not integrate into the host genome. Strains used in the laboratory can accommodate exogenous DNA payloads of up to 8 kb in place of genes required for viral replication (Bett *et al.*, 1994).

Ad-mediated gene delivery consists of two steps: (1) cell attachment and (2) internalization (Smith and Helenius, 2004). Cell attachment involves the Coxsackie and Adenovirus Receptor (CAR), a cell surface glycoprotein that mediates homotypic cell–cell interactions (Cohen *et al.*, 2001). CAR is the

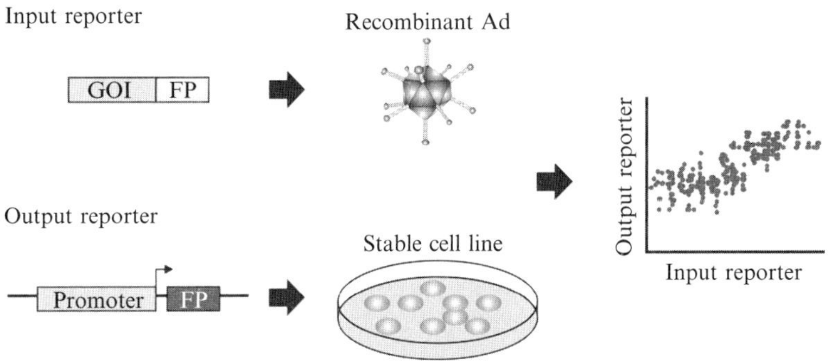

Figure 10.1 Using viral-mediated expression variability to probe gene expression dose response. System input consists of a gene of interest (GOI) along with fluorescent protein (FP) expressed from a recombinant adenovirus (Ad); output is a stably integrated reporter under the control of a gene regulatory sequence of interest (promoter). Cell–cell variability in Ad-mediated input expression provides a convenient way to quantify output dose response in individual cells (green data points). (For interpretation of the references to color in this figure legend, the reader is referred to the Web version of this chapter.)

attachment target of Ad fibers that protrude from the virion capsule (Bergelson *et al.*, 1997; Tomko *et al.*, 1997), and is correlated with Ad-mediated gene expression in different cell lines (Li *et al.*, 1999). The ectopic expression of CAR boosts Ad binding (Bergelson *et al.*, 1997) and subsequent gene expression (Leon *et al.*, 1998). Internalization is mediated through interactions between αv integrins on the cell surface and penton proteins from which Ad fibers emerge (Stewart *et al.*, 1997; Wickham *et al.*, 1993). Engagement of integrins on the cell surface triggers a cascade of intracellular signaling events reminiscent of binding to the extracellular matrix, culminating in actin cytoskeleton rearrangements (Li *et al.*, 1998a,b), receptor-mediated endocytosis of virions (Meier and Greber, 2004), and nuclear delivery of the viral genome (Greber *et al.*, 1997).

Cell-to-cell differences in Ad-mediated gene expression can often span several orders of magnitude in an isogenic cell population (Hitt *et al.*, 2000; Leon *et al.*, 1998). The nature of this broad variability is not well understood. Environmental factors such as "population context" (i.e., size, density, and location within the cell population) can account for a large proportion of the variation in infection (Snijder *et al.*, 2009). Variability in Ad-mediated gene expression may also arise from levels of cell–surface receptors for viral attachment, intracellular signaling pathways that modulate viral translocation to nucleus, and stochastic gene expression.

Regardless of the mechanism, variability in Ad-mediated gene delivery can be utilized as a convenient means to probe signaling dynamics in mammalian cells. As an illustration, our recent work uses this approach to elucidate cellular response to MYC-stimulation in a high-throughput manner. MYC is a transcription factor often amplified in human tumors (Meyer and Penn, 2008). Its downstream target, E2f1, plays an important role in cell cycle regulation (Johnson, 2000). We have combined the variability in Ad-mediated gene delivery with fluorescent protein (FP) reporters to measure the effects of MYC on E2f1 output in individual cells. This approach revealed a biphasic E2f1 response that was previously unknown: low levels of MYC can activate E2f1 whereas elevated MYC suppresses it (Fig. 10.2). This biphasic effect reconciles the seemingly contradictory responses of E2f1 and other genes to MYC observed by a number of groups (Wong *et al.*, 2011). It also reveals an intrinsic safeguard mechanism to curtail potentially oncogenic growth stimulation.

A major focus of synthetic biology has been the engineering of gene circuits with increasing complexity to program cellular behavior in a predictable manner (Basu *et al.*, 2005; Gardner *et al.*, 2000; Tabor *et al.*, 2009). Recent work, however, suggests the potential to use small-scale circuits to probe host cell physiology (Marguet *et al.*, 2010; Tan *et al.*, 2009). Here, we detail the design and construction of adenoviral inputs and transcriptional reporters that both can be tracked with FPs in the context of dose response interrogation for single mammalian cells. We also describe how this approach may be applied to observe other input–output relationships.

2. Design and Construction

2.1. Inputs

This section describes strategies to couple an input gene—the variable "perturbation" of interest—with an FP in the context of an Ad genome. Detailed review of the availability and characteristics of FPs can be found elsewhere (Shaner *et al.*, 2005). Overall, we wish to achieve a strong correlation between the expression level of input genes and the intensity of the coupled FP. The simplest approach involves cloning the expression cassette of the input gene into an Ad vector that carries a built-in FP expression cassette (He *et al.*, 1998). This method ensures that inputs and FP coding sequences are delivered to cells in equal amounts; however, this strategy does not guarantee good agreement between their expression levels.

An alternative approach is to create a transcriptional fusion of an input gene and FP (i.e., cotranscribed as part of the same mRNA), by driving both coding sequences from the same promoter. In this case, translation of the 3′ expression cassette is directed by an internal ribosomal entry site (IRES; Pelletier and Sonenberg, 1988). Here, expression correlation between the input gene and coupled FP depends largely on their relative efficiency of translation. This correlation may break down when an input gene and FP

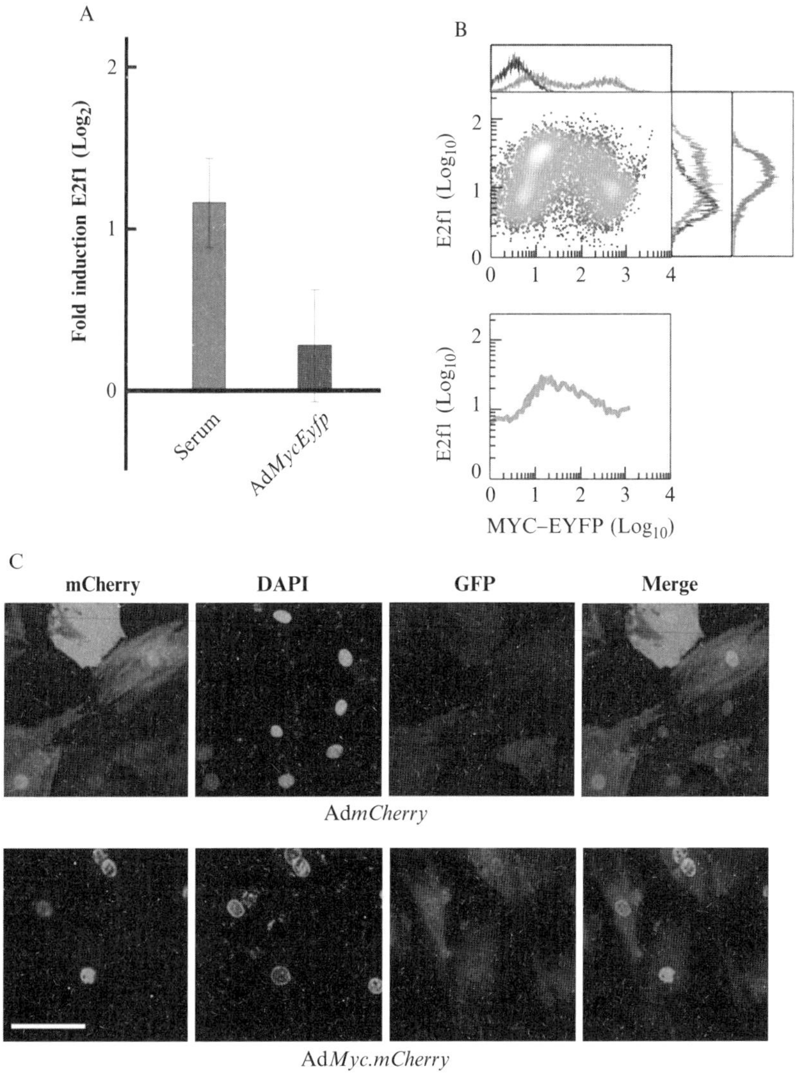

Figure 10.2 Cell–cell variability in MYC input levels reveals biphasic E2f1 response. (A) Real-time PCR results for E2f1 responses. Rat embryonic fibroblasts (REF52) cultured in 0.02% bovine growth serum (BGS) were either switched to 10% BGS (Serum; blue bar) or infected with an Ad vector expressing MYC fused to EYFP (AdMycEyfp; red bar) at an MOI of 1000 for 36 h. Endogenous E2f1 mRNA levels in each sample are expressed relative to cells that were starved or starved and infected with AdEyfp control virus, respectively. (B) Typical flow cytometry results. Data from serum starved REF52 cells harboring integrated GFP reporter under control of E2f1 promoter. (Top) Scatter plot shows E2f1 reporter activity (green fluorescence) in individual cells as function of MYC–EYFP input (yellow fluorescence). Histograms summarize fluorescence output for each respective channel in response to 10% serum

exhibit distinct translation efficiency and/or if gene products exhibit differences in subcellular localization (e.g., nuclear vs. cytoplasmic).

A third approach is to physically link the input gene with an FP. Such "translational fusion" entails cloning an FP expression cassette either upstream or downstream of the input gene in the same reading frame. FP cassettes can also be sandwiched between exons of the input gene by placing it in the context of splice acceptor and donor sequences (Cohen *et al.*, 2008). We have found that a translation fusion of MYC N-terminal to an enhanced yellow FP (MYC–EYFP) maintains the nuclear localization of native MYC rather than the cytoplasmic localization associated with native EYFP (Fig. 10.3A and B). Further, we confirmed that a linear correlation exists between the EYFP fluorescence intensity and MYC protein levels (Fig. 10.3C). Thus, translation fusion can provide accurate assessment of the concentration and subcellular localization of the expressed input gene.

A critical consideration is the potential impact of protein fusion on function. For example, although linking the estrogen receptor hormone-binding domain to either the N-terminus or the C-terminus of MYC results in a protein possessing equivalent MYC activity (Eilers *et al.*, 1989), C-terminal linkage of an FP to granulysin (a lytic protein secreted by lymphocytes) appears to disrupt its intracellular localization (Hanson and Ziegler, 2004). We have observed that linking EYFP to either MYC or E2F reduces its fluorescence intensity. Interference posed by fusion can sometimes be mitigated by separating each species with an intermediate protein linker to reduce steric hindrance (Arai *et al.*, 2001). In all cases, fusion protein activity should be compared with the native protein in order to assess localization, stability, and function.

Several tradeoffs exist when employing FP to track inputs (Table 10.1). For example, native protein stability can be drastically altered by translational fusion. Native MYC has a half-life of between 10 and 60 min in rat fibroblasts (Sears *et al.*, 1999); the typical half-life of FP variants is ~26 h (Corish and Tyler-Smith, 1999). In our experience, fusion of MYC with EYFP results in an intermediate half-life between that of native EYFP and MYC (unpublished observation). Altering protein stability can also drastically affect temporal dynamics of gene expression. Specifically, while the extended half-life of FPs affords a stronger fluorescence signal, it might

(blue); control infection with Ad expressing β-galactosidase (black); and infection with AdMycEyfp (red). (Bottom) Moving median values from scatter plot in (B). (C) Fluorescence microscopy. Starved REF52 cells harboring E2f1 reporter (GFP) were infected with Ad vectors expressing mCherry or MYC-mCherry fusion (500 MOI) for 36 h. DNA was subsequently stained with DAPI. Merge represents overlay of mCherry, DAPI, and GFP signals. Scale bar: 100 μm. (See Color Insert.)

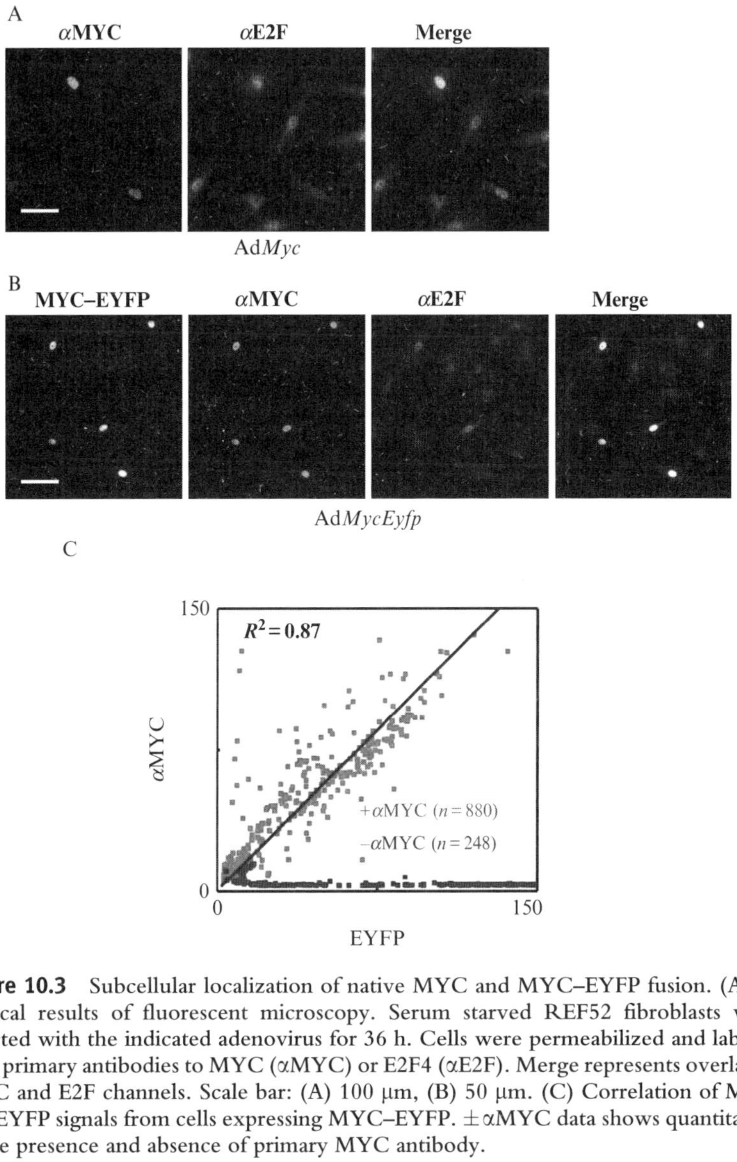

Figure 10.3 Subcellular localization of native MYC and MYC–EYFP fusion. (A, B) Typical results of fluorescent microscopy. Serum starved REF52 fibroblasts were infected with the indicated adenovirus for 36 h. Cells were permeabilized and labeled with primary antibodies to MYC (αMYC) or E2F4 (αE2F). Merge represents overlay of MYC and E2F channels. Scale bar: (A) 100 μm, (B) 50 μm. (C) Correlation of MYC and EYFP signals from cells expressing MYC–EYFP. ±αMYC data shows quantitation in the presence and absence of primary MYC antibody.

poorly reflect the dynamics of short-lived species. For a constitutively expressed gene, the response time (that period required to reach half-maximal expression) is proportional to protein half-life. Conversely, it is often necessary to bring expression of a reporter to low basal levels prior to stimulation (~5 half-lives), which in the case of native FPs, is impractical.

Table 10.1 Trade-offs on fluorescent protein tracking

Method	Advantages	Disadvantages
Independent promoters	Strong FP fluorescence signal	Poorest indicator of target levels
	Native protein activity retained	No spatial information
Transcriptional fusion	Strong FP fluorescence signal	Existence of IRES itself (alternative splicing)
	Native protein activity retained	Baranick *et al.* (2008)
	Shared mRNA levels	No spatial information
Translational fusion	Direct readout of target concentration	Disruption of target function
	Spatial information of target	Alteration of target stability
		Reduced FP fluorescence signal

2.2. Adenoviral construction

Recombinant Ad can be generated using commercially available adenoviral expression systems. We use the ViraPower^{TM} system (Invitrogen, Cat. No. K4930-00). Briefly, the input gene of interest is subcloned into a pENTR Gateway Vector (Invitrogen). Translation initiation efficiency of the input gene can be modified through the introduction or modification of the Kozak consensus (the ribosome binding site in eukaryotes; Kozak, 1987). The input sequence cannot contain a *Pac*I restriction site, which is reserved to linearize the viral genome for replication in cells. Subsequently, the input gene is transferred via *in vitro* recombination (instead of time-consuming ligation procedure) to a destination vector containing the Ad genomic backbone. The destination vector can either direct constitutive expression of the input gene from within the Ad genome (e.g., pAd/CMV/V5-DEST; Invitrogen, Cat. No. V493-20) or can be promoterless (e.g., pAd/PL-DEST; Invitrogen, Cat. No. V49420) if constitutive expression from the strong CMV promoter is not desirable. To generate viral particles from the recombined viral genome, DNA is *Pac*I digested, purified by ethanol precipitation, and transfected into cells (e.g., 293 cells) permissive for virion replication in 6-well dishes using Lipofectamine 2000 (Invitrogen, Cat. No. 11668-019). Forty-eight hours after transfection, cultured cells are transferred to 10 cm dishes to await the emergence of cytopathic effect (CPE) that is a consequence of viral replication. CPE typically occurs within 7–10 days posttransfection but it may take more than 2 weeks for genes with cytotoxic effects such as c-Myc and E2f1.

We do not purify individual plaques from virions produced through DNA transfection. Plaque purification is recommended for large-scale amplification initiated from previously generated Ad preparations to avoid propagation of mutants. We use standard approaches for large-scale growth and maintenance of adenoviral vector stocks (Nevins, 1980). To assess viral concentration or "titer," several methods have been developed previously. Ad particle numbers can be directly counted via electron microscopy (Mittereder *et al.*, 1996), whereas the simplest method is to physically disrupt virions and assess OD260 to estimate viral DNA copy number (Maizel *et al.*, 1968). Assessment of virion number, however, cannot distinguish noninfectious virions (Walters *et al.*, 2002). Thus, a more popular approach is to determine the number "infectious units" in replication permissive 293 cells through evaluation of plaque formation at limiting dilution (Nyberg-Hoffman *et al.*, 1997). Using an assay that identifies the presence of Ad hexon coat protein production in 293 cells (Adeno-X Rapid Titer Kit; Clontech; Cat. No. 632250), we have found that Ad infectious units represent ~25% of the number of virions observed using electron microscopy (unpublished observations). We typically achieve titers on the order of 10^8–10^9 μL^{-1} using the Adeno-X kit. This is our basis to define the multiplicity of infection (MOI)—the number of Ad units per cell.

2.3. Outputs

This section describes design of stable cells harboring fluorescent reporters to probe the cellular responses to input genes introduced by adenoviral vectors. Choice of FP for output reporters involves similar considerations mentioned for inputs, especially with regards to FP stability. In addition, it is important to minimize spectral overlap between input and output FPs. In particular, Ad-mediated expression of input FPs is likely to occur at high levels thereby increasing the potential for "bleed over" into the detection spectrums of output FPs. We recommend utilizing the most spectrally distinct FPs for input and output to minimize complications arising from spectral overlap. Using mCherry and GFP as the input and output coupled FPs, respectively, we have successfully detected the nonmonotonic response of E2f1 to MYC using fluorescence microscopy (Fig. 10.2C), which would be difficult with a GFP/YFP pairing.

Stable cell lines harboring integrated reporter constructs can be generated by "stable transfection" or by using retroviral or lentiviral vectors. We have generated cell lines harboring GFP reporters under the control of promoters from the E2f1 and Cyclin D1 genes by using the Clontech's Retro-X Q vector system (Cat. No. 613515) (Yao *et al.*, 2008). These nonreplicating, self-inactivating retroviruses carry three flavors of drug resistance genes permitting the incorporation of as many reporters. Note that the introduced promoter and reporter sequences should be cloned downstream of the extended packaging signal (ψ+) but upstream of the immediate early CMV promoter (PCMV IE) in the Retro-X Q vectors, to avoid interference between the CMV and introduce promoters. In these retroviral vectors, one must omit a poly adenylation (poly A) signal in reporter sequences, as this would interfere with the complete transcription of the full-length viral genome (Coffin and Varmus, 1996). As with any retroviral system, integration site and copy number can vary among infected cells. It may be desirable to generate populations derived from single cell clones to reduce this genetic heterogeneity.

An alternative method involves targeted integration of the reporter construct into the adenoassociated virus site 1 (AAVS1) of human chromosome 19 (CompoZr Targeted Integration Kit—AAVS1; Sigma, Cat. No. CTI1). Briefly, a plasmid donor carrying the user-defined payload is flanked by regions homologous to the AAVS1 chromosomal site. The integration of the donor plasmid is stimulated by a pair of engineered zinc finger nucleases through a process known as homology directed repair (Urnov *et al.*, 2005). The primary advantage of targeted integration is the elimination of variability introduced by genomic context and copy number (Hockemeyer *et al.*, 2009). Nevertheless, constructed reporter cell lines should be rigorously characterized for integration site, copy number, and functionality.

3. Measurement

3.1. Flow cytometry

Flow cytometry provides a sensitive and efficient method to quantify multiple fluorescent signals at the single cell level. Using this method, we have routinely generated dose response plots for inputs that span several orders of magnitude from $\sim 10^5$ cells. A typical flow cytometry protocol is described in the following sections.

3.1.1. Cell plating

Cells are split into 6-well dishes. Depending on the treatment conditions, we aim to have cells less than 60% confluent at the time of infection, as Ad uptake is severely reduced in dense cultures. For REF52 fibroblasts, we culture 2–4 $\times$ 10^5 cells in each well in 0.02% BGS for 48 h to bring them to quiescence.

3.1.2. Infection

Different cell lines uptake Ad virions with different efficiency (in part due to expression of cell surface receptors). Thus, an optimal MOI range must be determined for each cell type. Also, while the extent of Ad gene expression is roughly correlated with MOI, the actual range of infection across a population fluctuates to a small degree, likely a cause of subtle differences in cell number, density, and measurement timing. To ensure broad coverage of the input domain, we infect over a range of MOI (e.g., between 1 and 1000) and pool samples just prior to measurement.

Dilute Ad into 250 μL serum-free media supplemented with 25 m*M* HEPES. For REF52 cells, we would cover a MOI range up to 1000 in increments of 100. Viral mixture is added drop-wise onto cells and incubated for 90 min at 37 °C with rocking every 15 min to redistribute virus. Subsequently, supplement with the appropriate culture media. Fluorescent signals can be detected starting from 6 to 8 h postinfection under microscopy (see below).

3.1.3. Harvest

Detach cells using trypsin and resuspend in media with 10% serum. Cells at different MOI can be pooled at this time. Pellet cells in centrifuge (2000×*g* for 5 min). Fix cells in phosphate buffered saline (PBS) supplemented with 3.7% formaldehyde (10% stock, methanol free; Polysciences, Cat. No. 04018) to achieve a concentration in the range of 1–2 × 10^6 cells/mL in total volume of less than 500 μL. We have used real-time PCR to quantify mRNA and microRNA levels in cell subpopulations sorted on the basis of fluorescence intensity (Wong *et al.*, 2011). In this case, the cells should not be fixed. Instead, resuspend cell pellets in PBS with 1% bovine serum albumin and keep on ice as much as possible to reduce signal degradation and cell aggregation. Suspensions may also be passed through a CellTrics 30 μ*M* disposable filter to remove cell clumps (Partec, Cat. No. 04-004-2326) prior to measurement/sorting.

3.1.4. Analysis

Spectral overlap between FPs must always be taken into account. The act of subtracting the spurious signals from one channel into another is termed "spectral compensation" (Herzenberg *et al.*, 2006), which requires control cell populations expressing each FP/color alone at levels comparable to those that would be observed in samples. We have simultaneously measured a strong viral-mediated MYC–EYFP input with a weaker EGFP reporter of E2f1 transcription on flow cytometers with specialized optical filters (Table 10.2). For compensation in this case, we used a control virus expressing EYFP (AdEyfp) to correct for the high levels of MYC–EYFP, and correspondingly, the EGFP channel control was obtained by serum stimulation of an E2f1-EGFP reporter. Similarly, we have analyzed up to three colors (mCherry, GFP, and YFP) simultaneously on BD FACVantage with DiVA and BD FACSTAR Plus cytometers (Becton Dickinson, NJ). Typically, we aim to achieve anywhere between 10^4 and 10^5 events.

Table 10.2 Excitation and emission employed for flow cytometry

Protein	Excitation	Emission	Notes
EGFP	488 nm	510/21 nm bandpass	Emission split with 525 nm
EYFP		545/35 nm bandpass	shortpass dichroic
mCherry	600 nm dye laser	630/22 nm bandpass	

3.2. Antibody labeling and fluorescent microscopy

Here, we describe a modified protocol using immunolabeling. Signals from cytoplasmic FPs generally do not survive permeabilization required for intracellular introduction of antibodies. However, we have observed translational fusion with MYC does in fact preserve FP signals, perhaps due to protection afforded by nuclear localization (Fig. 10.3B). In aiding the identification of cells for image processing, a constitutive color or stain (e.g., DAPI) may be included to mark cell location. Alternatively, one may use the total background fluorescence present in all channels to identify cells (see Section 3.2.5) if the compatible stains or channels are unavailable.

3.2.1. Plating

We use glass coverslips affixed to silicone as a cell substratum (SecureSlip™; Sigma, Cat. No. S1815). The quality of glass should be suitable for the microscope (e.g., Number 1.5, 170 μm thickness). Prior to plating, coverslips are placed in 12-well dish and coated with a 0.01% gelatin solution to enhance adhesion. Coverslips are then rinsed twice with PBS. Cells are plated at a density such that they will be subconfluent at time of Ad infection. For REF52 cells, we generally plate about $\sim 10^4$ cells per coverslip.

3.2.2. Infection

Perform as described in Section 3.1. We generally infect cells cultured on a coverslip with a viral mixture of 200 μL/well in a 12-well plate.

3.2.3. Immunolabeling

Cells are washed twice with PBS and fixed with ice-cold methanol for 10 min at −20 °C. Cells are then washed three times with PBS and permeabilized with PBS/0.25% Triton X-100/1% BSA for 10 min. Background binding is blocked by incubation with PBS/3% BSA/0.02% Tween-20 for 30 min. Cells are then incubated with primary antibodies for at least 1 h at 1:100–1:500 in PBS/1% BSA/0.02% Tween-20. Note that for detection of more than one protein simultaneously, the species of origin of each primary antibody must be unique so that secondary antibodies do not cross-react. One may prefer preadsorbed antibodies, which remove antibodies that cross-react with serum from another species. Cells are subsequently washed three times with PBS/1% BSA/0.02% Tween-20 and incubated with secondary antibody in PBS/1% BSA/0.02% Tween-20 for 1 h: for blue channel, we use AlexaFluor405 (Invitrogen); for green channel, AlexaFluor488; for far-red, AlexaFluor594 or AlexaFluor633. Cells are washed three times with PBS/1% BSA/0.02% Tween-20 before microscopy. Coverslips are mounted with SlowFade Gold Antifade with DAPI solution (Invitrogen, Cat. No. S36938).

3.2.4. Imaging

We have used a Zeiss LSM 510 inverted confocal microscope. This instrument is equipped with lasers at 405 nm (AlexaFluor405/DAPI), 488 nm (EGFP/EYFP/AlexaFluor488/FITC), and 594 nm (AlexaFluor594/AlexaFluor633/mCherry).

3.2.5. Analysis

We use a custom Matlab (The Mathworks) script to extract fluorescence data from multicolor TIF images (written by Quanli Wang, Department of Statistical Science, Duke University). First, cell boundaries are identified by a mask. Ideally, this would be performed using the signals from a constitutive cell stain like DAPI (DNA in nuclei). We have also successfully used the sum of the background fluorescence over all the channels of an image to create a mask. In either case, the image should be globally thresholded in order to create the initial mask to define regions from which signals will be extracted. A balance must be struck in this process: setting the threshold too high will run the risk of unnecessarily removing regions and at the same time inflating the mean signal value across a population; if too low, fragmented cell materials and other "junk" will be included and bias signals toward lower values. A rule of thumb is to set the threshold high enough such that the reduction in the number of spots identified reaches a plateau, as has been described previously (Raj and Tyagi, 2010). In general, it is important to ensure by eye that raw images and masks have good agreement. Next, any small but bright particles from the mask are filtered out by setting a minimum number of pixels that will define a cell region. At this point, we are able to extract each respective signal from the bounded regions in the image. We find that the mean and median signals suffice to give an accurate readout of spot intensity.

4. Broader Applications

The preceding describes the simplicity and power in using heterogeneous stimuli to quantify dose responses. Although we have focused on the use of Ad-mediated input gene delivery and fluorescent reporters or immunostaining as outputs, our approach can be generalized to examine other cellular properties where variable input and outputs can be quantified. We envision that time lapse fluorescence microscopy as an ideal way to examine how variable inputs impact different aspects of cells in four dimensions (Muzzey and van Oudenaarden, 2009). Recent advances in microfluidic devices portend that single cell, high-throughput, quantitative interrogation of other aspects of cell physiology will become the norm (Bennett and Hasty, 2009; Le Gac and van den Berg, 2010). For example, single-cell real-time PCR (Taniguchi *et al.*, 2009) could be used to quantify mRNA levels of many genes in response to variable Ad-mediated input.

Are there other ways to deliver variable inputs to cells? Our observations are consistent with the notion that the cell–cell differences in retroviral-mediated FP expression are similar to that mediated by Ad vectors (Wong *et al.*, 2011). Likewise, simple transient transfection can generate very broad expression (Ducrest *et al.*, 2002). This suggests that a large degree of gene expression heterogeneity is inherent in common molecular and cell biology approaches. For this reason, we feel our method may be used extensively in attempts to interrogate the quantitative basis of cellular responses.

References

Arai, R., Ueda, H., Kitayama, A., Kamiya, N., and Nagamune, T. (2001). Design of the linkers which effectively separate domains of a bifunctional fusion protein. *Protein Eng.* **14,** 529–532.

Baranick, B. T., Lemp, N. A., Nagashima, J., Hiraoka, K., Kasahara, N., and Logg, C. R. (2008). Splicing mediates the activity of four putative cellular internal ribosome entry sites. *Proc. Natl. Acad. Sci. USA* **105,** 4733–4738.

Basu, S., Gerchman, Y., Collins, C. H., Arnold, F. H., and Weiss, R. (2005). A synthetic multicellular system for programmed pattern formation. *Nature* **434,** 1130–1134.

Bennett, M. R., and Hasty, J. (2009). Microfluidic devices for measuring gene network dynamics in single cells. *Nat. Rev. Genet.* **10,** 628–638.

Bergelson, J. M., Cunningham, J. A., Droguett, G., Kurt-Jones, E. A., Krithivas, A., Hong, J. S., Horwitz, M. S., Crowell, R. L., and Finberg, R. W. (1997). Isolation of a common receptor for Coxsackie B viruses and adenoviruses 2 and 5. *Science* **275,** 1320–1323.

Bett, A. J., Haddara, W., Prevec, L., and Graham, F. L. (1994). An efficient and flexible system for construction of adenovirus vectors with insertions or deletions in early regions 1 and 3. *Proc. Natl. Acad. Sci. USA* **91,** 8802–8806.

Blake, W. J., Kaern, M., Cantor, C. R., and Collins, J. J. (2003). Noise in eukaryotic gene expression. *Nature* **422,** 633–637.

Campos, S. K., and Barry, M. A. (2007). Current advances and future challenges in Adenoviral vector biology and targeting. *Curr. Gene Ther.* **7,** 189–204.

Chang, H. H., Hemberg, M., Barahona, M., Ingber, D. E., and Huang, S. (2008). Transcriptome-wide noise controls lineage choice in mammalian progenitor cells. *Nature* **453,** 544–547.

Cohen, C. J., Shieh, J. T., Pickles, R. J., Okegawa, T., Hsieh, J. T., and Bergelson, J. M. (2001). The coxsackievirus and adenovirus receptor is a transmembrane component of the tight junction. *Proc. Natl. Acad. Sci. USA* **98,** 15191–15196.

Coffin, J. M., and Varmus, H. E. (eds.), (1996). Retroviruses, Cold Spring Harbor Laboratory Press, New York.

Cohen, A. A., Geva-Zatorsky, N., Eden, E., Frenkel-Morgenstern, M., Issaeva, I., Sigal, A., Milo, R., Cohen-Saidon, C., Liron, Y., Kam, Z., *et al.* (2008). Dynamic proteomics of individual cancer cells in response to a drug. *Science* **322,** 1511–1516.

Corish, P., and Tyler-Smith, C. (1999). Attenuation of green fluorescent protein half-life in mammalian cells. *Protein Eng.* **12,** 1035–1040.

Ducrest, A. L., Amacker, M., Lingner, J., and Nabholz, M. (2002). Detection of promoter activity by flow cytometric analysis of GFP reporter expression. *Nucleic Acids Res.* **30,** e65.

Dunlop, M. J., Cox, R. S., 3rd, Levine, J. H., Murray, R. M., and Elowitz, M. B. (2008). Regulatory activity revealed by dynamic correlations in gene expression noise. *Nat. Genet.* **40,** 1493–1498.

Eilers, M., Picard, D., Yamamoto, K. R., and Bishop, J. M. (1989). Chimaeras of myc oncoprotein and steroid receptors cause hormone-dependent transformation of cells. *Nature* **340,** 66–68.

Eldar, A., and Elowitz, M. B. (2010). Functional roles for noise in genetic circuits. *Nature* **467,** 167–173.

Elowitz, M. B., Levine, A. J., Siggia, E. D., and Swain, P. S. (2002). Stochastic gene expression in a single cell. *Science* **297,** 1183–1186.

Gardner, T. S., Cantor, C. R., and Collins, J. J. (2000). Construction of a genetic toggle switch in *Escherichia coli*. *Nature* **403,** 339–342.

Geva-Zatorsky, N., Dekel, E., Batchelor, E., Lahav, G., and Alon, U. (2010). Fourier analysis and systems identification of the p53 feedback loop. *Proc. Natl. Acad. Sci. USA* **107,** 13550–13555.

Greber, U. F., Suomalainen, M., Stidwill, R. P., Boucke, K., Ebersold, M. W., and Helenius, A. (1997). The role of the nuclear pore complex in adenovirus DNA entry. *EMBO J.* **16,** 5998–6007.

Hanson, D. A., and Ziegler, S. F. (2004). Fusion of green fluorescent protein to the C-terminus of granulysin alters its intracellular localization in comparison to the native molecule. *J. Negat. Results Biomed.* **3,** 2.

He, T. C., Zhou, S., da Costa, L. T., Yu, J., Kinzler, K. W., and Vogelstein, B. (1998). A simplified system for generating recombinant adenoviruses. *Proc. Natl. Acad. Sci. USA* **95,** 2509–2514.

Herzenberg, L. A., Tung, J., Moore, W. A., and Parks, D. R. (2006). Interpreting flow cytometry data: A guide for the perplexed. *Nat. Immunol.* **7,** 681–685.

Hitt, D. C., Booth, J. L., Dandapani, V., Pennington, L. R., Gimble, J. M., and Metcalf, J. (2000). A flow cytometric protocol for titering recombinant adenoviral vectors containing the green fluorescent protein. *Mol. Biotechnol.* **14,** 197–203.

Hockemeyer, D., Soldner, F., Beard, C., Gao, Q., Mitalipova, M., DeKelver, R. C., Katibah, G. E., Amora, R., Boydston, E. A., Zeitler, B., *et al.* (2009). Efficient targeting of expressed and silent genes in human ESCs and iPSCs using zinc-finger nucleases. *Nat. Biotechnol.* **27,** 851–857.

Johnson, D. G. (2000). The paradox of E2F1: Oncogene and tumor suppressor gene. *Mol. Carcinog.* **27,** 151–157.

Kozak, M. (1987). An analysis of 5′-noncoding sequences from 699 vertebrate messenger RNAs. *Nucleic Acids Res.* **15,** 8125–8148.

Kussell, E., and Leibler, S. (2005). Phenotypic diversity, population growth, and information in fluctuating environments. *Science* **309,** 2075–2078.

Le Gac, S., and van den Berg, A. (2010). Single cells as experimentation units in lab-on-a-chip devices. *Trends Biotechnol.* **28,** 55–62.

Leon, R. P., Hedlund, T., Meech, S. J., Li, S., Schaack, J., Hunger, S. P., Duke, R. C., and DeGregori, J. (1998). Adenoviral-mediated gene transfer in lymphocytes. *Proc. Natl. Acad. Sci. USA* **95,** 13159–13164.

Li, E., Stupack, D., Bokoch, G. M., and Nemerow, G. R. (1998a). Adenovirus endocytosis requires actin cytoskeleton reorganization mediated by Rho family GTPases. *J. Virol.* **72,** 8806–8812.

Li, E., Stupack, D., Klemke, R., Cheresh, D. A., and Nemerow, G. R. (1998b). Adenovirus endocytosis via alpha(v) integrins requires phosphoinositide-3-OH kinase. *J. Virol.* **72,** 2055–2061.

Li, D., Duan, L., Freimuth, P., and O'Malley, B. W., Jr. (1999). Variability of adenovirus receptor density influences gene transfer efficiency and therapeutic response in head and neck cancer. *Clin. Cancer Res.* **5,** 4175–4181.

Maizel, J. V., Jr., White, D. O., and Scharff, M. D. (1968). The polypeptides of adenovirus. II. Soluble proteins, cores, top components and the structure of the virion. *Virology* **36,** 126–136.

Marguet, P., Tanouchi, Y., Spitz, E., Smith, C., and You, L. (2010). Oscillations by minimal bacterial suicide circuits reveal hidden facets of host-circuit physiology. *PLoS ONE* **5,** e11909.

Meier, O., and Greber, U. F. (2004). Adenovirus endocytosis. *J. Gene Med.* **6**(Suppl. 1), S152–S163.

Meyer, N., and Penn, L. Z. (2008). Reflecting on 25 years with MYC. *Nat. Rev. Cancer* **8,** 976–990.

Mittereder, N., March, K. L., and Trapnell, B. C. (1996). Evaluation of the concentration and bioactivity of adenovirus vectors for gene therapy. *J. Virol.* **70,** 7498–7509.

Muzzey, D., and van Oudenaarden, A. (2009). Quantitative time-lapse fluorescence microscopy in single cells. *Annu. Rev. Cell Dev. Biol.* **25,** 301–327.

Nevins, J. R. (1980). Definition and mapping of adenovirus 2 nuclear transcription. *Meth. Enzymol.* **65,** 768–785.

Nyberg-Hoffman, C., Shabram, P., Li, W., Giroux, D., and Aguilar-Cordova, E. (1997). Sensitivity and reproducibility in adenoviral infectious titer determination. *Nat. Med.* **3,** 808–811.

Pelletier, J., and Sonenberg, N. (1988). Internal initiation of translation of eukaryotic mRNA directed by a sequence derived from poliovirus RNA. *Nature* **334,** 320–325.

Raj, A., and Tyagi, S. (2010). Detection of individual endogenous RNA transcripts *in situ* using multiple singly labeled probes. *Meth. Enzymol.* **472,** 365–386.

Raser, J. M., and O'Shea, E. K. (2005). Noise in gene expression: Origins, consequences, and control. *Science* **309,** 2010–2013.

Reddy, V. S., Natchiar, S. K., Stewart, P. L., and Nemerow, G. R. (2010). Crystal structure of human adenovirus at 3.5 A resolution. *Science* **329,** 1071–1075.

Rueger, M. A., Winkeler, A., Miletic, H., Kaestle, C., Richter, R., Schneider, G., Hilker, R., Heneka, M. T., Ernestus, R. I., Hampl, J. A., *et al.* (2005). Variability in infectivity of primary cell cultures of human brain tumors with HSV-1 amplicon vectors. *Gene Ther.* **12,** 588–596.

Sears, R., Leone, G., DeGregori, J., and Nevins, J. R. (1999). Ras enhances Myc protein stability. *Mol. Cell* **3,** 169–179.

Shaner, N. C., Steinbach, P. A., and Tsien, R. Y. (2005). A guide to choosing fluorescent proteins. *Nat. Meth.* **2,** 905–909.

Sigal, A., Milo, R., Cohen, A., Geva-Zatorsky, N., Klein, Y., Liron, Y., Rosenfeld, N., Danon, T., Perzov, N., and Alon, U. (2006). Variability and memory of protein levels in human cells. *Nature* **444,** 643–646.

Singh, D. K., Ku, C. J., Wichaidit, C., Steininger, R. J., 3rd, Wu, L. F., and Altschuler, S. J. (2010). Patterns of basal signaling heterogeneity can distinguish cellular populations with different drug sensitivities. *Mol. Syst. Biol.* **6,** 369.

Smith, A. E., and Helenius, A. (2004). How viruses enter animal cells. *Science* **304,** 237–242.

Snijder, B., Sacher, R., Ramo, P., Damm, E. M., Liberali, P., and Pelkmans, L. (2009). Population context determines cell-to-cell variability in endocytosis and virus infection. *Nature* **461,** 520–523.

Spencer, S. L., Gaudet, S., Albeck, J. G., Burke, J. M., and Sorger, P. K. (2009). Non-genetic origins of cell-to-cell variability in TRAIL-induced apoptosis. *Nature* **459,** 428–432.

Stewart, P. L., Chiu, C. Y., Huang, S., Muir, T., Zhao, Y., Chait, B., Mathias, P., and Nemerow, G. R. (1997). Cryo-EM visualization of an exposed RGD epitope on adenovirus that escapes antibody neutralization. *EMBO J.* **16,** 1189–1198.

Tabor, J. J., Salis, H. M., Simpson, Z. B., Chevalier, A. A., Levskaya, A., Marcotte, E. M., Voigt, C. A., and Ellington, A. D. (2009). A synthetic genetic edge detection program. *Cell* **137,** 1272–1281.

Tan, C., Marguet, P., and You, L. (2009). Emergent bistability by a growth-modulating positive feedback circuit. *Nat. Chem. Biol.* **5,** 842–848.

Taniguchi, K., Kajiyama, T., and Kambara, H. (2009). Quantitative analysis of gene expression in a single cell by qPCR. *Nat. Meth.* **6,** 503–506.

Taniguchi, Y., Choi, P. J., Li, G. W., Chen, H., Babu, M., Hearn, J., Emili, A., and Xie, X. S. (2010). Quantifying *E. coli* proteome and transcriptome with single-molecule sensitivity in single cells. *Science* **329,** 533–538.

Tomko, R. P., Xu, R., and Philipson, L. (1997). HCAR and MCAR: The human and mouse cellular receptors for subgroup C adenoviruses and group B coxsackieviruses. *Proc. Natl. Acad. Sci. USA* **94,** 3352–3356.

Urnov, F. D., Miller, J. C., Lee, Y. L., Beausejour, C. M., Rock, J. M., Augustus, S., Jamieson, A. C., Porteus, M. H., Gregory, P. D., and Holmes, M. C. (2005). Highly efficient endogenous human gene correction using designed zinc-finger nucleases. *Nature* **435,** 646–651.

Walters, R. W., Freimuth, P., Moninger, T. O., Ganske, I., Zabner, J., and Welsh, M. J. (2002). Adenovirus fiber disrupts CAR-mediated intercellular adhesion allowing virus escape. *Cell* **110,** 789–799.

Wickham, T. J., Mathias, P., Cheresh, D. A., and Nemerow, G. R. (1993). Integrins alpha v beta 3 and alpha v beta 5 promote adenovirus internalization but not virus attachment. *Cell* **73,** 309–319.

Wong, J. V., Yao, G., Nevins, J. R., and You, L. (2011). Viral-mediated noisy gene expression reveals biphasic E2f1 response to MYC. *Mol. Cell.* **41,** 275–285.

Yao, G., Lee, T. J., Mori, S., Nevins, J. R., and You, L. (2008). A bistable Rb-E2F switch underlies the restriction point. *Nat. Cell Biol.* **10,** 476–482.

CHAPTER ELEVEN

De novo Design and Construction of an Inducible Gene Expression System in Mammalian Cells

Maria Karlsson,*,†,‡ Wilfried Weber,*,†,‡ *and* Martin Fussenegger‡,§

Contents

Abstract

Inducible expression systems represent the founding technology for the emergence of synthetic biology in mammalian cells. The core molecules in these systems are bacterial regulator proteins that bind to or dissociate from a cognate DNA operator sequence in response to an exogenous stimulus like a small-molecule inducer. In this chapter, we describe a generic protocol of how bacterial regulator proteins can be applied to the design, construction, and optimization of an inducible expression system in mammalian cells. By choosing regulator

* Faculty of Biology, Albert-Ludwigs-Universität Freiburg, Freiburg, Germany
† Centre for Biological Signaling Studies (BIOSS), Albert-Ludwigs-Universität Freiburg, Freiburg, Germany
‡ Department of Biosystems Science and Engineering, ETH Zurich, Basel, Switzerland
§ University of Basel, Faculty of Science, Basel, Switzerland

Methods in Enzymology, Volume 497
ISSN 0076-6879, DOI: 10.1016/B978-0-12-385075-1.00011-1

proteins with an appropriate small-molecule inducer, this protocol provides a straightforward approach for establishing biosensors, cell-to-cell communication systems, or tools to control gene expression *in vivo*.

1. Introduction

Inducible systems for exogenously controlling transgene expression represent the core technology that enabled the emergence of synthetic biology (Weber and Fussenegger, 2010). While inducible expression systems were abundantly available in prokaryotes, thus enabling the design and construction of the first synthetic gene networks (Elowitz and Leibler, 2000; Gardner *et al.*, 2000), it took a few more years until a sufficiently large palette of inducible expression systems had been developed for mammalian cells that could be applied for the establishment of complex genetic networks in cell cultures and in mice. Examples of such networks include the design and construction of bi-stable (Kramer *et al.*, 2004b), time-delayed (Weber *et al.*, 2007b), and hysteretic genetic switches (Kramer and Fussenegger, 2005), as well as of networks performing Boolean logic (Kramer *et al.*, 2004a) or exhibiting tuneable oscillating transgene expression (Tigges *et al.*, 2009).

The molecular basis of inducible expression systems in mammalian cells is the DNA-binding proteins that bind to or dissociate from a cognate operator sequence in response to the concentration of a small-molecule inducer. Those proteins originally evolved to control bacterial metabolism in response to small-molecule environmental cues such as nutrients, quorum-sensing signals, or toxic insults (Ramos *et al.*, 2005). To adapt these prokaryotic regulator proteins for controlling gene expression in mammalian cells, two principal designs were developed, one based on transcriptional activation and one on repression (Fig 11.1). In the activation-based system (Fig. 11.1A), the prokaryotic regulator protein is fused to a transactivation (TA) domain while the regulator's cognate operator sequence O (or multiple repeats thereof, O_n) is fused to a minimal promoter (P_{min}) devoid of enhancer sequences. In this configuration, binding of the regulator protein to its operator will enable the assembly of the transcription–initiation complex in close proximity of the minimal promoter which will then trigger transcription. Modulating the binding affinity of the regulator protein to its operator via the addition or removal of the small-molecule inducer therefore enables the expression strength to be adjusted to desired levels. In this configuration, commonly used transactivation domains include human NF-κB-derived p65, *Herpes simplex* VP16, or minimal domains thereof (Weber *et al.*, 2002b). Minimal promoters are commonly derived from the human cytomegalovirus immediate-early

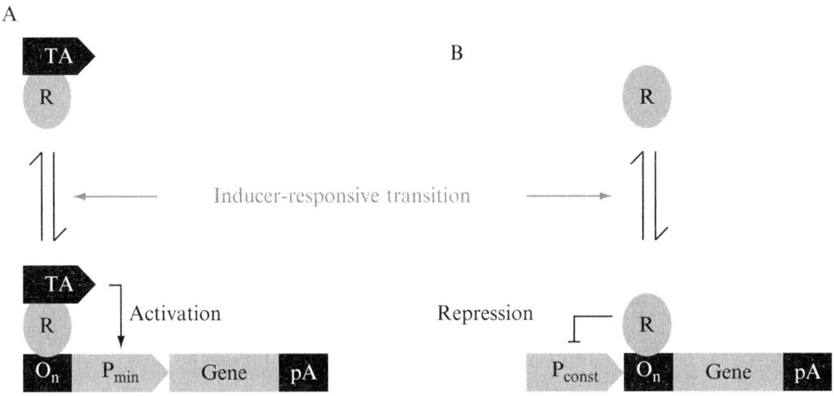

Figure 11.1 General design principle for inducible expression systems. (A) Activation-based regulation of expression. The bacterial regulator protein (R) is fused to a transactivation (TA) domain. Upon the binding of R to its cognate operator sequence (O_n), TA triggers the induction of the minimal promoter P_{min}, thus resulting in expression of the target gene. By modulating the interaction between R and O_n via the concentration of the small-molecule inducer, gene expression can be adjusted to desired levels. (B) Repression-based regulation of gene expression. Binding of the bacterial regulator protein (R) to repeats of its cognate operator (O_n) sterically prevents transcription from the constitutive promoter P_{const}. Upon dissociation of R from O_n via modulating the concentration of the small-molecule inducer, gene expression is adjusted to desired levels.

promoter (Gossen and Bujard, 1992) or from the *Drosophila* heat-shock promoter P_{HSP70} (Fussenegger *et al.*, 2000).

In the repression-based configuration (Fig. 11.1B), several operator repeats are cloned downstream of a constitutive promoter P_{const}, like the simian virus 40 or cytomegalovirus immediate-early promoters. Binding of the regulator protein to its operator will therefore sterically block transcription while dissociating the regulator protein from its operator by modulating the concentration of the small-molecule inducer will relieve repression and result in the initiation of transcription. The repressing effect of regulator proteins can further be increased via their fusion to a transsilencer domain such as KRAB of human *kox1* (Fussenegger *et al.*, 2000).

In the past decade, these design schemes were applied to several regulator proteins (Table 11.1), resulting in a large palette of mutually compatible inducible expression systems in mammalian cells. While first generation inducible expression systems relying on antibiotics (Fussenegger *et al.*, 2000; Gossen and Bujard, 1992; Weber *et al.*, 2002a) were controlled by the exogenous addition of the regulating compound, second generation inducible systems were responsive to metabolites with a concentration that could be modulated by metabolically engineered cells to construct synthetic cell-to-cell communication systems (Weber *et al.*, 2007a,b).

Table 11.1 Bacterial regulator proteins that have been used in the construction of inducible expression systems in mammalian cells

Regulator protein	Origin	Inducer	Interaction with operator in response to inducer	Reference
AlcR [a]	*A. nidulans*	Acetaldehyde	Association	(Weber *et al.*, 2004)
ArgR	*C. pneumoniae*	L-Arginine	Association	(Hartenbach *et al.*, 2007)
BirA	*E. coli*	Biotinyl-AMP	Association	(Weber *et al.*, 2009a)
CymR	*P. putida*	Cumate	Dissociation	(Mullick *et al.*, 2006)
EthR	*M. tuberculosis*	2-Phenylethyl butyrate	Dissociation	(Weber *et al.*, 2008)
HdnoR	*A. nicotinovorans*	6-Hydroxy nicotine	Dissociation	(Malphettes *et al.*, 2005)
HucR	*D. radiodurans*	Uric acid	Dissociation	(Kemmer *et al.*, 2010)
MphR(A)	*E. coli*	Macrolides	Dissociation	(Weber *et al.*, 2002a)
PIP	*S. pristinaespiralis*	Streptogramins	Dissociation	(Fussenegger *et al.*, 2000)
Rex	*S. coelicolor*	NADH	Dissociation	(Weber *et al.*, 2006)
RheA	*S. albus*	Heat	Dissociation	(Weber *et al.*, 2003a)
rTetR	Variant of TetR	Doxycycline	Association	(Gossen *et al.*, 1995)
ScbR	*S. coelicolor*	SCB1	Dissociation	(Weber *et al.*, 2003b)
TetR	*Transposon Tn10*	Tetracyclines	Dissociation	(Gossen and Bujard, 1992)
TraR	*A. tumefaciens*	3-oxo-C8-HSL	Association	(Neddermann *et al.*, 2003)
TtgR	*P. putida*	Phloretin	Dissociation	(Gitzinger *et al.*, 2009)

[a] AlcR is of fungal origin and already contains a transactivation domain, but its remaining molecular configuration is similar to bacterial regulator-based systems. It is not fully understood whether acetaldehyde induces operator binding or whether transcriptional activation is induced by other means.

This previous work on the design and construction of inducible expression systems impressively demonstrates the potential of bacterial regulator proteins for synthetic biology in mammalian cells. The key advantage using this approach is the huge availability of bacterial regulator proteins responsive to a multitude of small-molecule compounds (as reviewed in Ramos *et al.*, 2005). To facilitate the full exploitation of this natural reservoir for mammalian synthetic biology, we describe in this chapter a generic protocol of how a small-molecule-responsive bacterial regulator protein can be used to construct an inducible gene expression system in mammalian cells.

2. Selection of a Conditional DNA-Binding Protein

Bacterial regulator proteins that bind to a cognate DNA operator sequence in response to the concentration of a small-molecule inducer are abundantly available (Ramos *et al.*, 2005). When setting out to select a regulator protein for the design of a mammalian inducible expression system, the following criteria can be applied:

1. Choice of the regulator with regard to its cognate small-molecule inducer. The type of the small-molecule inducer is a key selection criterion for its subsequent application in a synthetic gene network. For example, drug- or toxin-responsive regulators could be used for the construction of biosensors (Aubel *et al.*, 2001), pharmacologically responsive regulators could be used for controlling gene expression *in vivo* (Gossen and Bujard, 2002), and regulators responsive to an inducer that can be enzymatically synthesized or degraded could be used for the construction of cell-to-cell communication networks (Weber *et al.*, 2007a,b).
2. Choice of the regulator with regard to its biochemical properties. Principally, two types of small-molecule-responsive regulator proteins are available, those that bind the operator in the presence of the small-molecule inducer (typically regulator proteins controlling anabolic pathways such as ArgR or BirA, Table 11.1) and those that dissociate from the operator upon inducer addition (typically regulators involved in catabolic or defense pathways such as CymR, PIP, or TetR, Table 11.1). The binding characteristics are best demonstrated by electrophoretic mobility shift assays (EMSA), which also give a first estimation of the required inducer concentrations to modulate the interaction between the regulator protein and its operator.

3. Establishment of the Inducible Expression System

In this paragraph, we describe the example of the *Escherichia coli* biotin repressor BirA (Genbank Accession No. ACX41617.1) and how an inducible expression system can be established in mammalian cells. The *E. coli* biotin repressor BirA binds its cognate operator *birO* in the presence of the biotin-metabolite biotinyl 5′ AMP, the formation of which is also catalyzed by BirA (Lin *et al.*, 1991). The implementation of the inducible expression system comprises the construction of the biotin-responsive transactivator and the BirA-responsive promoter. The cloning strategy is only presented here as an overview (Fig. 11.2) since the cloning techniques are described in detail elsewhere (Sambrook *et al.*, 2000). However, a detailed protocol is provided below for the characterization of the system in mammalian cells.

3.1. Construction of the biotin-responsive transactivator

The biotin-responsive transactivator can be constructed by fusing BirA to a transactivation domain like *Herpes simplex* VP16 (Triezenberg *et al.*, 1988) (Fig. 11.2A). According to experience with previous inducible expression systems, the transactivation domain should be fused to the terminus which is furthest away from the DNA-binding domain to prevent interference with operator binding (the DNA-binding domain can be identified at http://expasy.org/prosite/ or via published crystal structures). As the BirA DNA-binding domain is contained at the N-terminus, the VP16 domain (amino acids Ala 363-Gly 490) was fused to the BirA C-terminus. The 5′ region of the *birA-vp16* fusion gene was further linked to a mammalian cell-compatible ribosome-binding sequence (Kozak sequence, CCACC) and cloned into a standard mammalian expression vector such as pcDNA3.1 (Invitrogen, Carlsbad, CA, cat. no. V790-20).

3.2. Construction of the biotin-responsive promoter

For the construction of the biotin-responsive promoter, the *birO* operator sites were fused upstream of a minimal promoter (Fig. 11.2B), like the minimal human cytomegalovirus promoter $P_{hCMVmin}$ (nucleotides 1093–1212, Accession No. M60321; Gossen and Bujard, 1992). As the expression strength commonly correlates with the number of operator modules, it is advisable to construct plasmids harboring different operator copy numbers and then test for the optimum configuration. Such repeats can easily be constructed by self-ligating 5′-phosphorylated operator-containing

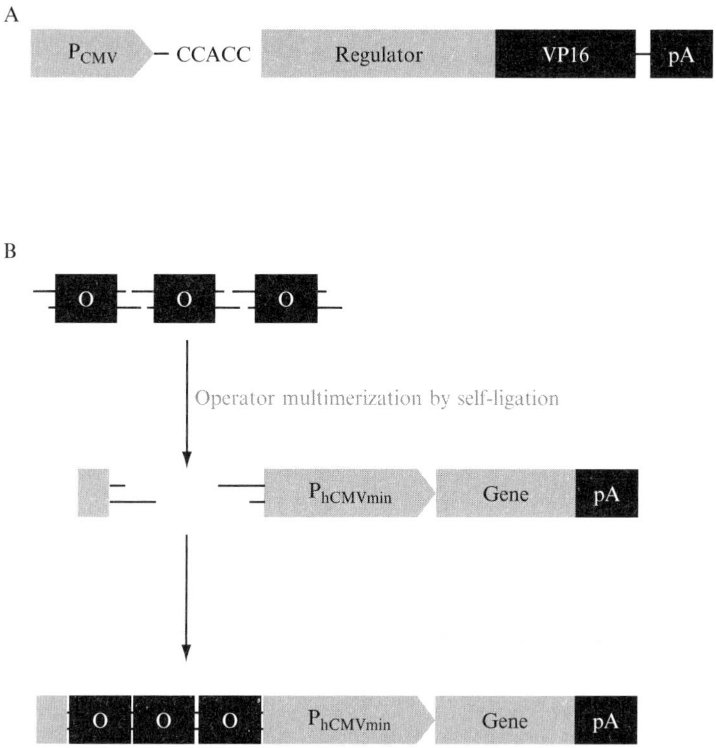

Figure 11.2 Design of an activation-based regulated gene expression system. (A) Design of the transactivator expression plasmid. A constitutive promoter (e.g., the cytomegalovirus promoter P_{CMV}) drives expression of a fusion gene consisting of the genes encoding the regulator protein and the transactivator VP16. The bacterial ribosome-binding sequence was replaced by the Kozak motif CCACC for efficient translation in mammalian cells. (B) Design of the reporter plasmid. The minimal promoter (e.g., the minimal human cytomegalovirus promoter $P_{hCMVmin}$) was fused to a target gene. Several repeats of the operator sequence O were fused upstream of the minimal promoter. This was efficiently done by annealing 5′ phosphorylated operator-encoding double-stranded oligonucleotides in such a way that they self-ligated via compatible sticky ends.

annealed oligos and directly cloning the fused operator repeats to the 5′ region of the minimal promoter (Fig. 11.2B). For monitoring transcription, the chimeric promoter is fused to a reporter gene such as SEAP, human placental secreted alkaline phosphatase, which can easily be quantified in the cell culture supernatant.

When constructing the plasmid carrying the biotin-responsive promoter, care should be taken to avoid the presence of other constitutively active promoters (e.g., for expression of resistance genes) close to the minimal promoter. Enhancer sequences contained in the constitutive promoter might cross-activate the minimal promoter and thus significantly increase leaky expression.

3.3. Initial testing of the system in cell culture

For the initial testing in cell culture, the following setup is performed:

- Transfection of the promoter construct alone to assess leaky expression.
- Transfection of the promoter and transactivator constructs in the presence of increasing biotin concentrations.
- Transfection of a constitutive expression vector for the same reporter gene in the presence of increasing biotin concentrations to quantify possible side effects of the inducer substance on reporter gene expression (e.g., the constitutive SEAP expression vector pSEAP2-control, Clontech, Palo Alto, CA).

As host cell, the biotechnologically important Chinese hamster ovary cell line (CHO-K1) was proven to be a good choice in previous studies. Therefore, the present protocol was adapted to this cell line. Similar protocols can be found in the literature for other cell lines (Pfeier *et al.*, 2010). It is recommended that each condition is transfected at least in triplicate, although six replicas per condition are advisable. The following protocol was scaled to one well of a 24-well cell culture plate (Greiner Bio-One GmbH, Frickenhausen, Germany, cat. no 662 160).

The DNA for transfection should be purified using anion-exchange-based kits such as the Genomed Jetstar 2.0 Midi (Genomed AG, Bad Oeynhausen, Germany, cat. no. 210 050). For the cell culture medium, HTS medium (Cell Culture Technologies, Gravesano, Switzerland) supplemented with 5% FCS (PAN Biotech GmbH, Aidenbach, Germany, cat. no. P281803) and 1% penicillin/streptomycin (PAN Biotech GmbH, Aidenbach, Germany, cat. no. P06-07100) was used (designated as HTS complete medium below). For characterization of the BirA-based expression system, biotin-free HTS medium was used together with dialyzed FCS (PAN Biotech GmbH, Aidenbach, Germany, cat. no. P30-2100).

The following steps were performed:

1. Cell seeding.
   a. Detach cells from a confluent Petri-dish by removing the cell culture medium, incubating the cells in 3 ml trypsin solution (PAN Biotech GmbH, Aidenbach, Germany, cat. no. P10-023500) for 5–10 min in the incubator and removing the cells by gently tapping the plate.
   b. Subsequently mix the cell suspension with 7 ml HTS complete medium, centrifuge for 3 min at 450×*g*, and remove the trypsin-containing supernatant.

c. Resuspend the cells in 5–7 ml HTS complete medium, count cells (either by a haemocytometer or a cell-counting device such as CASY® Model DT (Innovatis AG, Bielefeld, Germany)), and dilute to 80,000 cells ml^{-1} in HTS complete medium. Seed 0.5 ml cell suspension per well.

2. Cell transfection. The following steps were performed 12–18 h after seeding:
   a. Prepare transfection mixture containing 1.2 μg total DNA (e.g., 0.6 μg promoter plasmid and 0.6 μg transactivator plasmid or 0.6 μg promoter plasmid and 0.6 μg empty vector since transfection efficiency depends on the total amount of DNA used) and 500 m*M* $CaCl_2$ in a total volume of 12 μl.
   b. Prepare 12 μl PO_4 solution (50 m*M* HEPES, 280 m*M* NaCl, 1.5 m*M* Na_2HPO_4, pH 7.05).
   c. Add DNA solution from (a) whilst vortexing, wait 25 s to allow for DNA–calcium phosphate precipitate formation, then add 0.4 ml medium (HTS medium, Cell Culture Technologies) containing 2% FCS.
   d. Aspirate the medium from one CHO-K1-containing well.
   e. Add the $CaPO_4$–DNA mix from step (c) to the well while gently shaking.
   f. Centrifuge the cell culture plate at 900×*g* for 5 min to sediment DNA complexes onto the cells and subsequently place in an incubator for 1.5 h (without centrifugation, cells should be incubated for 5 h to allow settling of the DNA complexes).
   g. Aspirate the medium and add 0.4 ml HTS medium supplemented with 2% FCS and 15% glycerol (the glycerol shock facilitates DNA uptake). Aspirate after 30 s and wash cells once with HTS complete medium.
   h. Add 0.4 ml HTS complete medium, increasing the concentrations of the small-molecule inducer, and place in the incubator for 2 days. If the small-molecule inducer is hardly soluble in water, DMSO is often a viable alternative provided that final DMSO concentration in the cell culture remains below 1% to avoid toxic side effects.
3. Analysis of reporter gene expression. SEAP analysis can either be performed using a luminescence-based kit (Roche Applied Science, Rotkreuz, Switzerland, cat. no. 11 779 842 001) or by using the following cheap and fast colorimetric method (Schlatter *et al.*, 2002):
   a. Take 200 μl supernatant, heat in an Eppendorf tube in a thermo-shaker for 15 min at 65 °C (to inactive endogenous phosphatases), and centrifuge for 1 min at 14,000×*g*.
   b. Put 100 μl 2×SEAP buffer (20 m*M* homoarginine, 1 m*M* $MgCl_2$, 21% (v/v) diethanolamine, pH 9.8) into one well of a transparent 96-well plate and add 80 μl of the sample.

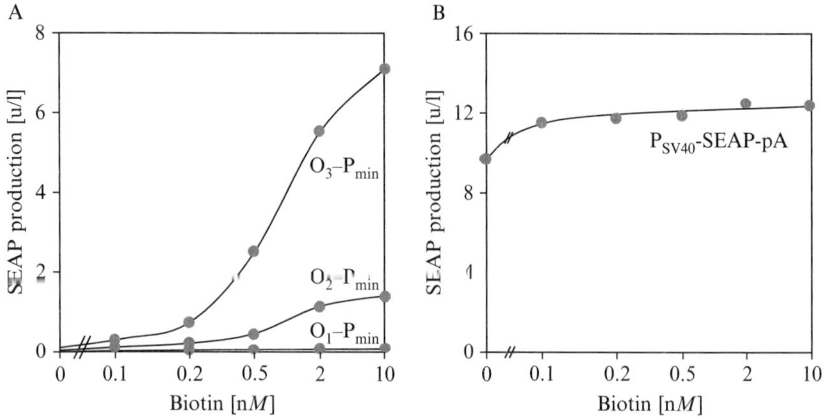

Figure 11.3 Representative performance of the biotin-inducible expression system. (A) Chinese hamster ovary cells (CHO-K1) were transfected with a plasmid encoding the BirA-VP16 transactivator and three variants of the biotin-inducible promoter comprising one, two, or three *birO* operator sites upstream of the minimal promoter P_{min} ($O_1 - P_{min}$, $O_2 - P_{min}$, $O_3 - P_{min}$, respectively). The cells were cultivated in the presence of increasing biotin concentrations for 48 h prior to quantifying SEAP production in the supernatant. (B) Impact of biotin concentration on constitutive gene expression. As the control, a constitutive SEAP expression plasmid based on the simian virus 40 promoter (P_{SV40}) was transfected into CHO-K1 cells. Following cultivation for 48 h in the presence of increasing biotin concentrations, SEAP concentration was profiled in the supernatant.

c. Add 20 μl pNPP solution (120 m*M* *para*-nitrophenyl phosphate in 2×SEAP buffer) and measure the increase in absorbance at 405 nm for approximately 60 min. Calculate the slope (in OD/min) and convert to SEAP activity using Lambert–Beer's law $E = \varepsilon \times c \times d$ where $\varepsilon = 18{,}600\ M^{-1}\ \text{cm}^{-1}$, c = increase of *para*-nitrophenolate per minute [$M\,\text{min}^{-1}$], and d = length of the light path in the sample [cm] (typically 0.5 cm in this configuration). E is the increase in optical density as determined from the slope above. The activity in units per liter is calculated as follows: Activity [U/l] $= c \times 10^6 \times 200/80$ (dilution factor of sample in buffer).

4. Figure 11.3 shows the results of the BirA-based inducible expression system using the protocol described in this chapter.

4. Optimization of the Expression System

Once the system is established and initially characterized, several generic options are available to optimize the regulation performance or to optimally integrate the system into the desired background.

The implementation and testing of the optimized system follows the generic detailed protocol described above.

4.1. Optimization of regulation performance

The following generic approaches have successfully been applied to optimize the regulation performance of the inducible expression systems.

4.1.1. Optimizing the transactivator

Most commonly, the VP16 transactivation domain is fused to the regulator proteins, in general yielding good regulation characteristics. However, using tandem repeats of minimal VP16 domains or exchanging VP16 for the human transactivation domains, p65 or E2F4 was shown to result in graded regulation characteristics enabling expression levels to be tuned in to the desired window (Baron *et al.*, 1997; Weber *et al.*, 2002b).

4.1.2. Optimizing promoter performance

1. The minimal human cytomegalovirus promoter $P_{hCMVmin}$ can be replaced by alternative minimal promoters such as the *Drosophila* heat-shock 70 promoter ($P_{hsp70min}$). In previous settings, $P_{hsp70min}$ resulted not only in lower maximum expression levels but also in lower leaky expression, thus making it the promoter of choice when less leaky expression was critical for a system's performance (Weber *et al.*, 2002b).
2. The spacing between the operator and the minimal promoter (and thus the torsion angle between the transactivator binding site and the minimal promoter) can be modified in order to achieve optimum regulation performance, as exemplified by the stepwise introduction of 2 bp increments (Weber *et al.*, 2002b).

4.1.3. Engineering the cell for superior regulation

In some cases, the regulation performance might be limited as the inducer molecule cannot freely diffuse into the cell. In these cases, it was shown that the expression of a transmembrane transporter for importing the inducer significantly improved regulated expression characteristics (Kemmer *et al.*, 2010).

4.2. Integration of the inducible expression system into the desired network

The following generic modifications enable a streamlined integration of the newly established inducible expression system into the desired synthetic gene network.

1. For long-term expression, the transgenes must be stably integrated into the host genome. A detailed protocol of how to integrate an inducible expression system into CHO-K1 cells was described previously (Weber and Fussenegger, 2004).
2. If several genes are to be expressed under the control of one inducible promoter, multicistronic expression vectors or bi-directional promoters can be applied. In multicistronic vectors, the gene in the first cistron is translated in a classical cap-dependent manner, while the translation of genes in the second or third cistron is mediated by internal ribosome entry sites (IRES). Multicistronic expression vectors have been successfully applied in several inducible expression systems (Fussenegger *et al.*, 1998; Weber *et al.*, 2002c). When using bi-directional promoters, the operator sequence is flanked by two divergently orientated minimal promoters, thus enabling the simultaneous transcription of two transgenes (Baron *et al.*, 1995).
3. For implementing complex expression scenarios such as processing several input signals according to Boolean algebra, promoters can be constructed that harbor several different operator sequences for making them responsive to different transactivators or repressors. Examples of such multi-input promoters have been described previously (Kramer *et al.*, 2004a).
4. For installing inducible expression systems in difficult-to-transfect cells, such as primary cells or tissues, viral vectors are usually the method of choice. During the past few years, lentiviral or adenoviral vectors have prevailed. Detailed protocols of how to apply them and how to use them in combination with inducible expression systems are available elsewhere (Gonzalez-Nicolini and Fussenegger, 2005; Mitta *et al.*, 2004).

5. Summary

During recent years, an extensive platform of inducible expression systems for mammalian cells has evolved, enabling the design and construction of complex generic networks in both cell cultures and mice. At present, the most widely adopted gene regulation systems are those based on a binary arrangement with small-molecule-regulated interactions between a bacterial response regulator and its cognate operator sequence. In this chapter, generic methods and strategies for the development of such systems are provided; the compiled set of protocols facilitates a straightforward approach to the design, implementation, and verification of inducible expression systems. Additionally, techniques for regulation performance optimization and system integration are given. Thus, the collection of methods and strategies presented here serves as a comprehensive aid in the realization of advanced synthetic gene networks in mammalian cells tailored for targeted applications.

ACKNOWLEDGMENTS

The work of MK was supported by the Swiss National Science Foundation (Grant No. CR32I3_125426), the work of WW was supported by the Excellence Initiative of the German Federal and State Governments (EXC 294), and the work of MF was supported by the Swiss National Science Foundation (grant no. 31003A-126022) and in part by the EC Framework 7 (Persist).

REFERENCES

Aubel, D., Morris, R., Lennon, B., Rimann, M., Kaufmann, H., Folcher, M., Bailey, J. E., Thompson, C. J., and Fussenegger, M. (2001). Design of a novel mammalian screening system for the detection of bioavailable, non-cytotoxic streptogramin antibiotics. *J. Antibiot. (Tokyo)* **54,** 44–55.

Baron, U., Freundlieb, S., Gossen, M., and Bujard, H. (1995). Co-regulation of two gene activities by tetracycline via a bidirectional promoter. *Nucleic Acids Res.* **23,** 3605–3606.

Baron, U., Gossen, M., and Bujard, H. (1997). Tetracycline-controlled transcription in eukaryotes: Novel transactivators with graded transactivation potential. *Nucleic Acids Res.* **25,** 2723–2729.

Elowitz, M. B., and Leibler, S. (2000). A synthetic oscillatory network of transcriptional regulators. *Nature* **403,** 335–338.

Fussenegger, M., Mazur, X., and Bailey, J. E. (1998). pTRIDENT, a novel vector family for tricistronic gene expression in mammalian cells. *Biotechnol. Bioeng.* **57,** 1–10.

Fussenegger, M., Morris, R. P., Fux, C., Rimann, M., von Stockar, B., Thompson, C. J., and Bailey, J. E. (2000). Streptogramin-based gene regulation systems for mammalian cells. *Nat. Biotechnol.* **18,** 1203–1208.

Gardner, T. S., Cantor, C. R., and Collins, J. J. (2000). Construction of a genetic toggle switch in Escherichia coli. *Nature* **403,** 339–342.

Gitzinger, M., Kemmer, C., Daoud-El Baba, M., Weber, W., and Fussenegger, M. (2009). Controlling transgene expression in subcutaneous implants using a skin lotion containing the apple metabolite phloretin. *Proc. Natl. Acad. Sci. USA* **106,** 10638–10643.

Gonzalez-Nicolini, V., and Fussenegger, M. (2005). A novel binary adenovirus-based dual-regulated expression system for independent transcription control of two different transgenes. *J. Gene Med.* **7,** 1573–1585.

Gossen, M., and Bujard, H. (1992). Tight control of gene expression in mammalian cells by tetracycline-responsive promoters. *Proc. Natl. Acad. Sci. USA* **89,** 5547–5551.

Gossen, M., and Bujard, H. (2002). Studying gene function in eukaryotes by conditional gene inactivation. *Annu. Rev. Genet.* **36,** 153–173.

Gossen, M., Freundlieb, S., Bender, G., Muller, G., Hillen, W., and Bujard, H. (1995). Transcriptional activation by tetracyclines in mammalian cells. *Science* **268,** 1766–1769.

Hartenbach, S., Daoud-El Baba, M., Weber, W., and Fussenegger, M. (2007). An engineered L-arginine sensor of Chlamydia pneumoniae enables arginine-adjustable transcription control in mammalian cells and mice. *Nucleic Acids Res.* **35,** e136.

Kemmer, C., Gitzinger, M., Daoud-El Baba, M., Djonov, V., Stelling, J., and Fussenegger, M. (2010). Self-sufficient control of urate homeostasis in mice by a synthetic circuit. *Nat. Biotechnol.* **28,** 355–360.

Kramer, B. P., and Fussenegger, M. (2005). Hysteresis in a synthetic mammalian gene network. *Proc. Natl. Acad. Sci. USA* **102,** 9517–9522.

Kramer, B. P., Fischer, C., and Fussenegger, M. (2004a). BioLogic gates enable logical transcription control in mammalian cells. *Biotechnol. Bioeng.* **87,** 478–484.

Kramer, B. P., Viretta, A. U., Daoud-El-Baba, M., Aubel, D., Weber, W., and Fussenegger, M. (2004b). An engineered epigenetic transgene switch in mammalian cells. *Nat. Biotechnol.* **22,** 867–870.

Lin, K. C., Campbell, A., and Shiuan, D. (1991). Binding characteristics of Escherichia coli biotin repressor-operator complex. *Biochim. Biophys. Acta* **1090,** 317–325.

Malphettes, L., Weber, C. C., El-Baba, M. D., Schoenmakers, R. G., Aubel, D., Weber, W., and Fussenegger, M. (2005). A novel mammalian expression system derived from components coordinating nicotine degradation in arthrobacter nicotinovorans pAO1. *Nucleic Acids Res.* **33,** e107.

Mitta, B., Weber, C. C., Rimann, M., and Fussenegger, M. (2004). Design and in vivo characterization of self inactivating human and non-human lentiviral expression vectors engineered for streptogramin-adjustable transgene expression. *Nucleic Acids Res.* **32,** e106.

Mullick, A., Xu, Y., Warren, R., Koutroumanis, M., Guilbault, C., Broussau, S., Malenfant, F., Bourget, L., Lamoureux, L., Lo, R., Caron, A. W., Pilotte, A., *et al.* (2006). The cumate gene-switch: A system for regulated expression in mammalian cells. *BMC Biotechnol.* **6,** 43.

Neddermann, P., Gargioli, C., Muraglia, E., Sambucini, S., Bonelli, F., De Francesco, R., and Cortese, R. (2003). A novel, inducible, eukaryotic gene expression system based on the quorum-sensing transcription factor TraR. *EMBO Rep.* **4,** 159–165.

Pfeier, A., Lim, T., and Zimmermann, K. (2010). Lentivirus transgenesis. *Methods Enzymol.* **477,** 3–15.

Ramos, J. L., Martinez-Bueno, M., Molina-Henares, A. J., Teran, W., Watanabe, K., Zhang, X., Gallegos, M. T., Brennan, R., and Tobes, R. (2005). The TetR family of transcriptional repressors. *Microbiol. Mol. Biol. Rev.* **69,** 326–356.

Sambrook, J., Fritsch, E. F., and Maniatis, T. (2000). Molecular Cloning: A Laboratory Manual. Cold Spring Harbor Laboratory Press, Cold Spring Harbor, NY.

Schlatter, S., Rimann, M., Kelm, J., and Fussenegger, M. (2002). SAMY, a novel mammalian reporter gene derived from Bacillus stearothermophilus alpha-amylase. *Gene* **282,** 19–31.

Tigges, M., Marquez-Lago, T. T., Stelling, J., and Fussenegger, M. (2009). A tunable synthetic mammalian oscillator. *Nature* **457,** 309–312.

Triezenberg, S. J., Kingsbury, R. C., and McKnight, S. L. (1988). Functional dissection of VP16, the trans-activator of herpes simplex virus immediate early gene expression. *Genes Dev.* **2,** 718–729.

Weber, W., and Fussenegger, M. (2004). Inducible gene expression in mammalian cells and mice. *Methods Mol. Biol.* **267,** 451–466.

Weber, W., and Fussenegger, M. (2010). Synthetic gene networks in mammalian cells. *Curr. Opin. Biotechnol.* **21,** 690–696.

Weber, W., Fux, C., Daoud-el Baba, M., Keller, B., Weber, C. C., Kramer, B. P., Heinzen, C., Aubel, D., Bailey, J. E., and Fussenegger, M. (2002a). Macrolide-based transgene control in mammalian cells and mice. *Nat. Biotechnol.* **20,** 901–907.

Weber, W., Kramer, B. P., Fux, C., Keller, B., and Fussenegger, M. (2002b). Novel promoter/transactivator configurations for macrolide- and streptogramin-responsive transgene expression in mammalian cells. *J. Gene Med.* **4,** 676–686.

Weber, W., Marty, R. R., Keller, B., Rimann, M., Kramer, B. P., and Fussenegger, M. (2002c). Versatile macrolide-responsive mammalian expression vectors for multiregulated multigene metabolic engineering. *Biotechnol. Bioeng.* **80,** 691–705.

Weber, W., Marty, R. R., Link, N., Ehrbar, M., Keller, B., Weber, C. C., Zisch, A. H., Heinzen, C., Djonov, V., and Fussenegger, M. (2003a). Conditional human VEGF-mediated vascularization in chicken embryos using a novel temperature-inducible gene regulation (TIGR) system. *Nucleic Acids Res.* **31,** e69.

Weber, W., Schoenmakers, R., Spielmann, M., El-Baba, M. D., Folcher, M., Keller, B., Weber, C. C., Link, N., van de Wetering, P., Heinzen, C., Jolivet, B., Sequin, U., *et al.* (2003b). Streptomyces-derived quorum-sensing systems engineered for adjustable transgene expression in mammalian cells and mice. *Nucleic Acids Res.* **31,** e71.

Weber, W., Rimann, M., Spielmann, M., Keller, B., Daoud-El Baba, M., Aubel, D., Weber, C. C., and Fussenegger, M. (2004). Gas-inducible transgene expression in mammalian cells and mice. *Nat. Biotechnol.* **22,** 1440–1444.

Weber, W., Link, N., and Fussenegger, M. (2006). A genetic redox sensor for mammalian cells. *Metab. Eng.* **8,** 273–280.

Weber, W., Daoud-El Baba, M., and Fussenegger, M. (2007a). Synthetic ecosystems based on airborne inter- and intrakingdom communication. *Proc. Natl. Acad. Sci. USA* **104,** 10435–10440.

Weber, W., Stelling, J., Rimann, M., Keller, B., Daoud-El Baba, M., Weber, C. C., Aubel, D., and Fussenegger, M. (2007b). A synthetic time-delay circuit in mammalian cells and mice. *Proc. Natl. Acad. Sci. USA* **104,** 2643–2648.

Weber, W., Schoenmakers, R., Keller, B., Gitzinger, M., Grau, T., Daoud-El Baba, M., Sander, P., and Fussenegger, M. (2008). A synthetic mammalian gene circuit reveals antituberculosis compounds. *Proc. Natl. Acad. Sci. USA* **105,** 9994–9998.

Weber, W., Lienhart, C., Daoud El-Baba, M., and Fussenegger, M. (2009a). A biotin-triggered genetic switch in mammalian cells and mice. *Metab. Eng.* **11,** 117–124.

Weber, W., Schuetz, M., Denervaud, N., and Fussenegger, M. (2009b). A synthetic metabolite-based mammalian inter-cell signaling system. *Mol. Biosyst.* **5,** 757–763.

CHAPTER TWELVE

BioBuilding: Using Banana-Scented Bacteria to Teach Synthetic Biology

James Dixon* *and* Natalie Kuldell†

Contents

Abstract

Student interest in synthetic biology is detectable and growing. Each year teenagers from around the world participate in iGEM, a summer long synthetic biology competition. As part of their iGEM experience, undergraduates design and construct novel living systems using standardized biological parts. One engineering feat was accomplished by the 2006 MIT iGEM team, who modified the normally putrid smell of bacteria so that the cells generated pleasant scents, such as wintergreen and banana. We have taken advantage of their project as well as other iGEM successes to develop a teaching curriculum for high schools and colleges. The curriculum includes four hands-on activities and two classroom assignments. We envision these activities either complementing existing instruction, for example in an advanced placement biology lab, or replacing some outdated, cookbook lab classes that are often used as gateways to undergraduate research opportunities. The activities we have developed also introduce engineering and technology concepts that are often overlooked in the

* Sharon High School, Sharon, Massachusetts, USA
† MIT, Department of Biological Engineering, Cambridge, Massachusetts, USA

Methods in Enzymology, Volume 497
ISSN 0076-6879, DOI: 10.1016/B978-0-12-385075-1.00012-3

already over-stuffed high school and college curricula. To ease their adoption, the activities include teacher materials, such as annotated instructions, grading rubrics, and animated resources. Here, we detail the student and teacher materials for performing the banana-scented bacteria lab, called "Eau that Smell." Other free teaching materials similar to the content here can be accessed through BioBuilder.org.

1. Introduction

Having observed the profoundly successful learning experiences that high school and college students have had through the international Genetically Machines Competition (iGEM; Mitchell *et al.*, 2010), we have turned to synthetic biology to reinvigorate and reengage students in classes at the high school and college level. In particular, we have recast or extended student iGEM projects so as to align them with existing teaching frameworks, or to meet educational needs in biological engineering laboratory classes (Kuldell, 2007). The content we offer includes a lab activity inspired by the iGEM project from the 2006 MIT team, namely Eau d'coli. These bacteria have been genetically engineered to smell like mint during the exponential ("log") phase of bacterial growth and like bananas during the stationary phase (MIT 2006 iGEM team: http://openwetware.org/wiki/IGEM:MIT/2006).

In transforming the Eau d'coli project and other iGEM successes into hands-on activities at BioBuilder.org that could be widely adopted, we considered generally accepted hallmarks of good curricula as well as more pedestrian logistical questions. The activities had to explicitly address engineering- and technology-teaching standards (National Academies Standards: http://www.nap.edu/catalog.php?record_id=4962, National Education Standards: http://www7.nationalacademies.org/bose/Standards_Framework_Homepage.html) using investigative frameworks (National Research Council, 2000, Wiggins and McTighe, 1998), and had to be provided through an open digital platform (e.g., Khan Academy: http://www.khanacademy.org/). Additionally, since we are both teachers, we are sensitive to implementation issues for these activities. Instructions had to provide real, "rubber meets the road" guidance. For example, the resources provided to teachers had to include guidelines for fitting the activities into short or long lab periods, rubrics for grading the work, and digital forums for posing practical questions and sharing clever solutions.

One illustrative example of the "BioBuilder" curriculum grew from a project called "eChromi" which was the 2009 iGEM Grand Prize winner carried out by students at the University of Cambridge (University of Cambridge 2009 iGEM team: http://www.echromi.com/). As part of their work, students generated a bacterial palette, genetically reprogrammed strains to appear shades of red, green, and purple. Unexpectedly, the performance of these color-generating genetic programs depended on the genetic background of the bacterial chassis. Though the iGEM team did not see this variable behavior as much more than odd, we took advantage of the observation as the point of departure for an investigative teaching lab that could replace the standard transformation lab commonly taught in high school advanced placement (AP) biology classes (Pearson AP Biology Labs: http://www.phschool.com/science/biology_place/labbench/lab6/concepts1.html). In our "BioBuilder" lab 4, students transform a B-type and a K-type bacterial host with the purple and the green color-generating plasmids. As with the traditional lab series, students are able to observe that DNA can be transferred to confer phenotypes and that selectable markers can be used to identify cells with plasmids. In the context of the BioBuilder lab, however, there is motivation for performing this DNA transformation. Students can directly examine the presumed equivalence of host strains and can ask why DNA might be expressed differently. They can identify future experiments and applications to exploit their observations. In other words, they become practitioners of science and engineering, as opposed to technicians.

In this chapter, we detail a teaching laboratory that extends the 2006 MIT iGEM team's "Eau d'coli" project (Fig. 12.1). Students grow four strains of genetically engineered cells. One of the strains serves as the zero-smell standard. Another strain is designed to generate a banana-flavored smell when the cells are growing in stationary phase. Students are asked to investigate the performance of the remaining two strains. These strains use distinct genetic programs to generate the banana smell during the exponential ("log") phase of growth. Observing that equivalent designs can perform differently in real life is an important lesson for budding engineers. The lab activity further offers a chance for students to learn important microbiological techniques and behaviors, in a charismatic and interesting context.

2. Eau d'coli

The banana-flavored smell from the genetically engineered Eau d'coli cells arises from the conversion of isoamyl alcohol to isoamyl acetate. Isoamyl acetate has a banana smell (Fig. 12.2).

Isoamyl alcohol (also called 3-methyl-1-butanol, isopentyl alcohol, or isobutylcarbinol) can be added to the bacterial growth media, where it is efficiently imported into the cells. There, it is converted to isoamyl acetate by the product of *ATFI*, a gene from *Saccharomyces cerevisiae* that was cloned between a bacterial ribosome binding site and a transcriptional terminator to make a three part "banana-odor generator."

To regulate the production of the banana smell, promoters that are active during distinct stages of the bacterial growth cycle were used. These promoters

Figure 12.1 MIT 2006 iGEM team. Members of the summer 2006 team are shown in Drew Endy's research lab at MIT wearing their iGEM team T-shirts. From the left are shown: Stephen Payne, Boyuan Zhu, Tom Knight, Reshma Shetty, Andre Green, Samantha Sutton, Veena Venkatachalam, Jason Kelly, Austin Che, Barry Canton, Kate Broadbent. Source: Heather A. Thomson.

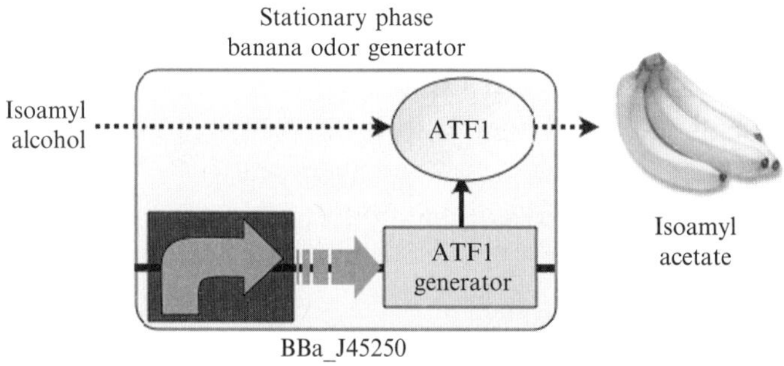

Figure 12.2 Eau d'coli. Diagram illustrating workings of the Eau d'coli system. Cells expressing the *ATF1* gene from *S. cerevisiae* can convert isoamyl alcohol in the growth media to isoamyl acetate, a compound that smells like bananas. A bacterial promoter that is primarily active during stationary phase controls the expression of *ATF1*. Source: 2006 MIT iGEM team.

differ in their affinity for sigma-factors that can associate with the RNA polymerase core. For cells designed to smell like bananas during stationary phase, a sigma-38 regulated promoter, pOsmY (Hengge-Aronis *et al.*, 1993),

Table 12.1 Registry of standard biological parts for "Eau That Smell" experiment

BBa_J45999	BBa_J45199	BBa_J45250	BBa_J45200	BBa_J45990
Indole-free chassis	Banana-odor generator	Sigma-38 controlling banana smell	Sigma-70 controlling banana smell	Sigma-38 promoter plus 4 part genetic inverter generating banana smell

was cloned upstream of the device. For cells designed to smell like bananas during log phase growth, either a sigma-70 regulated promoter, pTetR (Lutz and Bujard, 1997), was cloned upstream of banana-odor generator, or a four part genetic inverter was added to the sigma-38 based construct (BBa_J45990: http://partsregistry.org/Part:BBa_J45990). To best detect the banana smell, the cellular chassis bore a mutation in the *tnaA* gene, inhibiting indole production and effectively eliminating the putrid smell that characterizes most *Escherichia coli* (YYC912 strain reference: http://cgsc.biology.yale.edu/Strain.php?ID=64826). The genetic devices associated with this synthetic system were all entered into the Registry of Standard Biological Parts (Registry Homepage: http://partsregistry.org/Main_Page) and are tabulated here (Table 12.1).

A combination of techniques was employed by the MIT iGEM team to characterize the growth and behavior of their system. To assess the effect of the synthetic devices on cellular growth rate, the team measured changes in turbidity of the cultures over time. Growth curves were then correlated with expression from the devices by fusing the promoters to GFP instead of the banana-odor generator. In this way, the fluorescence of the cells over time could be measured and used as a predictor of timed scent production in that system (Fig. 12.3). Finally, gas chromatography and a "sniff" test were used on the complete systems to look for isoamyl acetate generated by the cells as they grew (Fig. 12.4).

3. "Eau That Smell" Teaching Lab Using the MIT iGEM Team's Eau d'coli Cells

The cells that were engineered by the MIT 2006 iGEM team smell distinctively of bananas, almost like a banana smoothie, by the time they reach fully saturated growth. However, the completed system leaves only a few teachable questions and experimental manipulations open to teachers. We extended the existing behavior of the Eau d'coli project so students

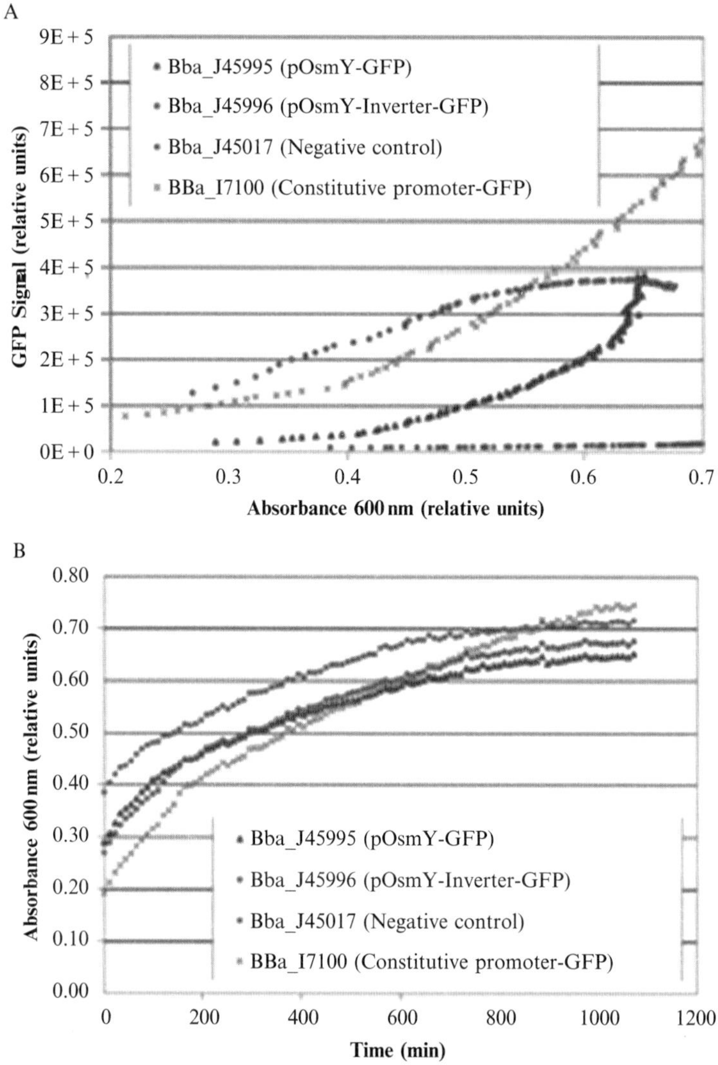

Figure 12.3 Characterization of Eau d'coli growth and cell population control. Portions of the Eau d'coli system were characterized by examining the growth rate (panel A) and timing of gene expression (panel B). In both cases, the relevant promoters were fused to GFP and changes in turbidity and fluorescence were measured over time. The data suggest Eau d'coli's output can be controlled by natural changes in cell population. Source: http://openwetware.org/wiki/IGEM:MIT/2006/osmY_Results.

could actively engage with it. Drawing from the National Science Standards (National Academy Standards website http://www.nap.edu/openbook.php?record_id=4962), we first identified a design opportunity, namely

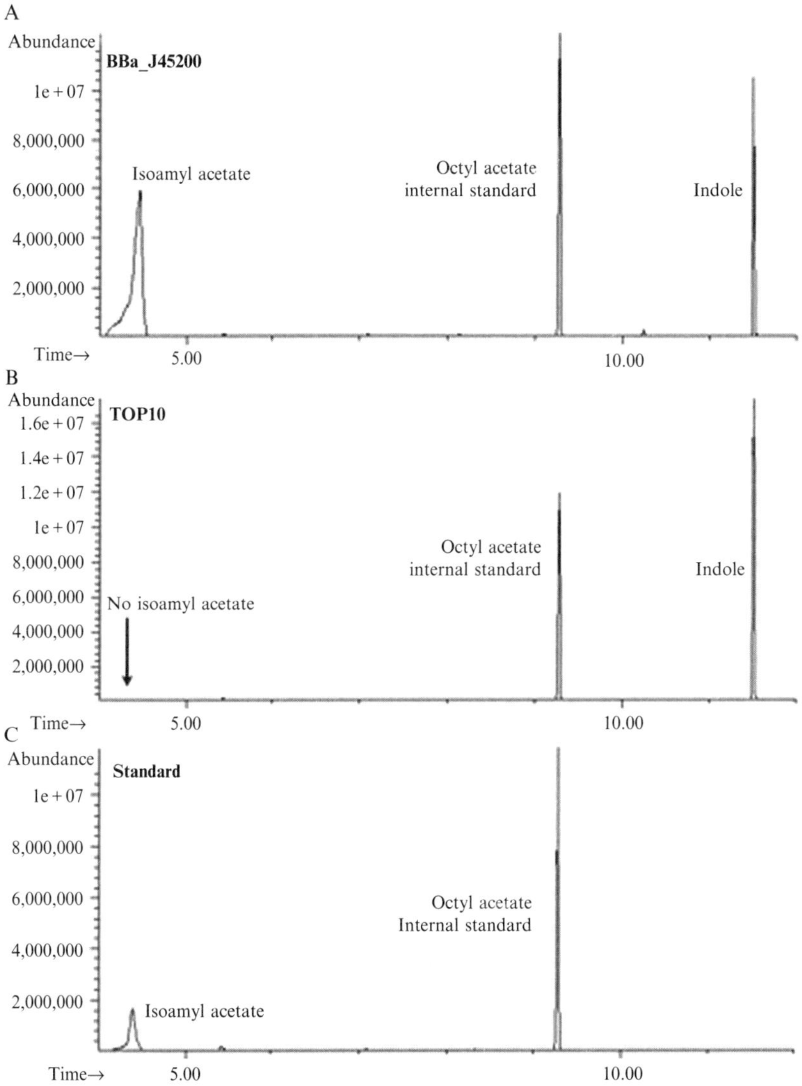

Figure 12.4 Isoamyl acetate output from Eau d'coli. Isoamyl acetate appears as a distinct and detectable peak on a gas chromatography when Eau d'coli cells are grown in the presence of isoamyl acetate (panel A), but only when cells bear the banana-odor generator, BBa_J45200 (panel B). Purified isoamyl acetate was used to confirm the retention rate of the compound (panel C). The pronounced peak for indole reflects the fact that the strains expressing these devices were wild-type for tnaA. Source: http://partsregistry.org/Part:BBa_J45200:Experience.

alternative genetic circuits that both met the stated goal of making log-phase banana smell. Next, we allowed the students to experimentally evaluate and then choose the better performing solution. The teacher's materials that we provide include some guidance for helpful assessment of the student's work, rubrics to guide the student's communication of experimental limitations and questions to motivate future directions.

In our hands, the banana-smell is less intense when directed by a sigma-70 promoter than when controlled by a sigma-38 promoter and a four part genetic inverter. However, the former design does a better job of expressing the device during log-phase only. This experimental result presents an interesting "choice" for the students to weigh (Fig. 12.5).

3.1. Growing starter cultures for the students

To begin this experiment, a high school or college teacher would request a kit from us. We could also send kits to summer iGEM teams who need preliminary training. These kits include four bacterial strains, three that should smell like bananas when grown in the presence of isoamyl alcohol and one negative control strain. The kits will also include, as needed, growth media, banana-scent standards, and turbidity standards. The strains will be sent in the form of a "stab" or "slant," a test tube with a small amount of bacteria on a slanted media. To continue the experiment, teachers and their students will have to further culture the bacteria by streaking out the stabs onto LB (Luria Broth) + ampicillin plates, as instructed below, and then further growing liquid starter cultures the following day. Note that since the host strain is chloramphenicol resistant, the selection can equally well be carried out in LB + ampicillin + chloramphenicol. If used, chloramphenicol stock solutions are made as 34 mg/ml ethanol and are used at a 1:1000 dilution in plates and in liquid culture.

Day 1

1. Using a sterile toothpick or inoculating loop, gather a small amount of bacteria from the stab and transfer it to a petri dish containing LB agar (1% tryptone, 0.5% yeast extract, 1% NaCl, 2% agar) + ampicillin (200 μg/ml final concentration).
2. Repeat with the remaining stab samples, streaking out each onto a different petri dish.
3. Place these cultures in a 37 °C incubator overnight.

Day 2

1. Using a sterile inoculating loop, transfer a bacterial colony from one of the petri dishes to a large sterile culture tube containing 5 ml of LB (recipe is identical to that for petri dishes but the agar is omitted) and 5 μl of ampicillin (stock solution = 100 mg/ml sterile water).

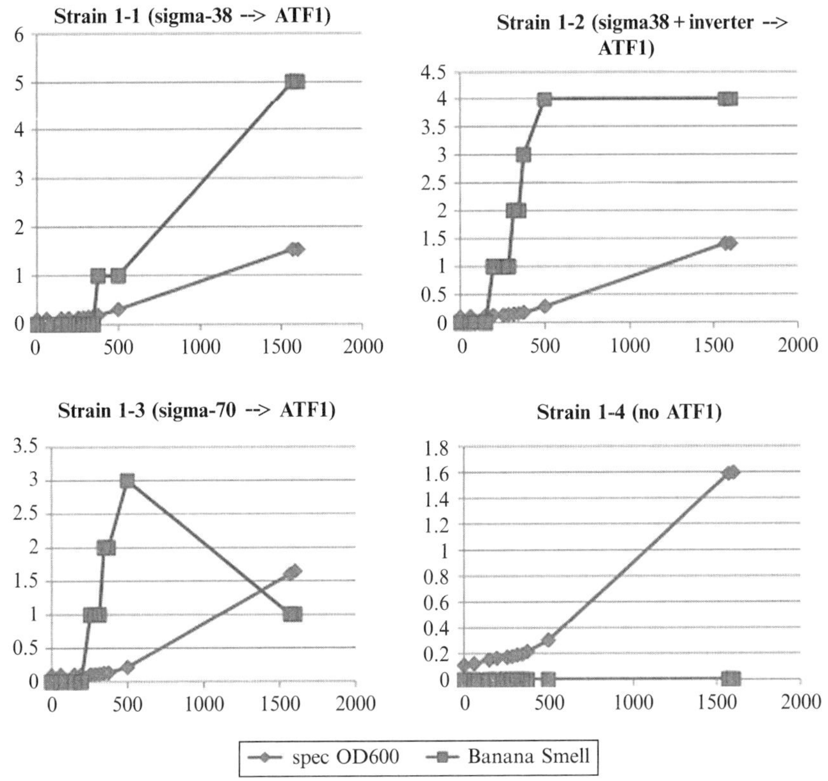

Figure 12.5 Behavior of "Eau that Smell" experimental and control strains. Growth and banana-smell generated by each experimental strain was assessed. Growth curves, shown with the blue lines, are measured as changes in turbidity over time. Banana smell, shown with red lines, was calibrated to the smell standards and is plotted for each time point on the growth curve. Growth time, in minutes, is shown on the x-axis. Unexpectedly, the log-phase promoter (strain 1-3) generates a less pronounced banana-flavored smell but is more tightly controlled to express only during the log-phase of growth. Source: data collected by J. Dixon. (See Color Insert.)

2. Repeat for each strain you will inoculate.
3. Place the culture tubes in the roller wheel in the incubator at 37 °C overnight. Be sure to balance the tubes across from each other to minimize stress on the roller wheel.

3.2. Bacterial growth and scent curves

Once the liquid starter cultures for each strain have been grown, students can inoculate larger volumes of cells to be examined for growth and banana-smell over time. Cells are grown in flasks in sufficiently large volumes so the smell associated with each strain is pronounced, and the volume can support the removal of multiple aliquots for turbidity measurements. We have described the lab with 50 ml of each culture growing in 100 ml Erlenmeyer flasks on stir plates at room temperature. Instructors will have to scale the materials according to the number of students and the availability of equipment.

Depending on the time available for teaching, the experiment can be run in different ways. A longer time period such as a 3 or 4 h college lab would allow the students to follow much of the growth curve, ideally from the end of lag phase through exponential growth and into stationary phase, all in one day. In cases when students are in the lab for an hour or less, the experiment can be carried out over several days, with an instructor placing the growing cells into a refrigerator between days to slow down their growth. In this case, the instructor overseeing the lab should warm the cells back to room temperature in advance of the students' time in lab and should note the time the cultures spend at room temperature so students can accurately determine changes in turbidity over time. It is important that the growing cultures remain covered with foil or plugs while they cool and warm so the smell associated with each strain can be determined. Additionally, students should be discouraged from smelling the cultures too often since the smell will dissipate and, if smelled too often, it can give some students a headache.

Data collected should be plotted with time as the common x-axis, and with both turbidity and smell sharing the y-axis. Turbidity can be measured with optical density at 600 nm using a spectrophotometer or by comparison to the MacFarland Turbidity Scale, described below. The intensity of the banana smell can be described with the following standards, prepared as indicated.

3.2.1. Banana-scent standard

An arbitrary scale has been established to associate numerical values to the intensity of the banana smell. This scale is based on the smell of a series of banana extract dilutions. The banana extract is an oil and will not dissolve in water. However, the concentrations are low and as long as the standard is given a shake before smelling, a suspension is sufficient. We have used banana extract made by "Frontier Natural Flavors" but suspect other brands would work equally well. The dilutions are prepared as indicated in Table 12.2, storing the solutions in plastic 50 ml conical tubes at room temperature until needed.

3.2.2. Measuring banana-smell and cell growth

1. Prepare a stock growth solution with 300 ml LB (1% tryptone, 0.5% yeast extract, 1% NaCl), 300 μl ampicillin (stock solution = 100 mg/ ml sterile water), 250 μl isoamyl alcohol. Note that isoamyl alcohol

Table 12.2 Arbitrary standards for banana-smell and their preparation

Standard	Concentration (%)	Extract in H_2O (final volume 25 ml)
0	0	0
1	0.1	25 μl
2	0.25	62.5 μl
3	0.5	125 μl
4	1	250 μl
5	2.5	625 μl
6	5	1.25 ml

stock should be used in a chemical fume hood since the smell can irritate the eyes, nose, and lungs. Once diluted, the isoamyl alcohol gives off a sweet and not pungent smell, though students and teachers who spend too much time smelling even the dilution may develop a headache or a stomach ache.

2. Mix this stock growth solution, by swirling the bottle or vortexing gently.
3. Set aside 2 ml of this mixture for each student group into a cuvette. This aliquot will serve as the blank for the spectrophotometer.
4. Move 50 ml of the broth solution to a 100 ml sterile Erlenmeyer flask and add 2 ml of bacteria from one of the overnight cultures, for example, strain 1-1.
5. Repeat the addition of 2 ml of bacteria to 50 ml of broth in an Erlenmeyer flask for each of the overnight cultures.
6. Cover the flasks with foil or a cotton plug, and swirl them gently.
7. Remove 2 ml from each sample to read the starting density of each. If you are testing all four samples you should now have five cuvettes, four with bacterial dilutions, and one blank.
8. Prepare the spectrophotometer by setting it to OD_{600}.
9. Note the time and take an "initial" density reading for the bacterial samples. This time should be noted as "T_0." Discard all samples except the blank.
10. Add a stir bar to each culture flask and place the flasks onto stir plates. Stir slowly. Cover the flasks with foil.
11. After 20 min, remove 1–2 ml from each sample and place in a cuvette.
12. Read the uninnoculated sample (blank) and set the % absorbance of this sample to zero.
13. Read each sample and record the % absorbance.
14. Sniff the Erlenmeyer flasks for any evidence of a banana smell, comparing the smell with the banana extract standards. Be sure to shake the standards and swirl the cultures before sniffing. Record your data.

15. At 20-min intervals repeat steps 11–14.
16. Between time points, you can calculate the bacterial population using the approximation of 1 OD_{600} unit = 1×10^9 bacteria.

3.2.3. Notes to teachers

A teacher presenting this lab has an opportunity to teach microbiology techniques, population growth dynamics, molecular genetics, and basic synthetic biology concepts in a meaningful, real world way. Given that engineering practice in general and synthetic biology in particular are not commonly taught, we provide lesson materials to introduce the lab activities found on the "BioBuilding" wiki. For example, the "Eau that Smell" lab can be framed with a one page "bioprimer" in which two characters discuss the relevant merits of the designs that will be compared. Terms that are unfamiliar to the teacher or students can be learned through the glossary link on the BioBuilding site, or through short animations that further extend the narrative between the characters. The animations and activities are collected on BioBuilder.org (Fig. 12.6).

3.3. Assessment

To show their understanding of the system, students can be asked to discuss:

- How well were they able to measure the population growth?
- How well were they able to smell bananas?
- Did each strain and their associated devices produce the same results?
- Did the genetic systems affect the growth curve of the bacteria?

Students performing this lab also have a chance to do meaningful error analysis and examine the difference between quantitative and qualitative results. When the students analyze their data, they might consider:

- How does each part of the experiment add to the conclusion?
- What errors or experimental parameters might lead to data variability?
- How confidently can they state their results?
- Are they equally confident in both the growth data and the smell data?
- Is using smell to measure the isoamyl acetate valid?
- What methods did they use to try to increase their confidence in the results?
- How might they change this system to better quantify the banana smell?
- Is there another kind input or output they would suggest including?
- If they could construct a different genetic system, what might they construct?

This final question could easily be used as the bridge to other "BioBuilding" activities or to a summer iGEM project, depending on the teaching context.

Figure 12.6 BioPrimer #1. To introduce students to the scientific and engineering underpinnings of the "Eau that Smell" lab activity, a one page "BioPrimer" can be circulated and discussed. When accessed from the internet, the BioPrimer has active links to terms that are unfamiliar, allowing the concepts to be learned through animations or additional readings. Multiple BioPrimers could serve as the framework for a semester long biological engineering class or a biotechnology course. Source: Animated Storyboards and BioBuilder.org.

4. Teaching Labs Modified for Resource-Stretched Settings

For teaching settings that do not have a reliable spectrophotometer or a sufficient number of instruments to use in a class setting, we offer a modification to the standard protocol. To assess turbidity, the bacterial solutions can be compared to the MacFarland Turbidity standards. The turbidity standards are prepared as suspensions of $BaCl_2$ in H_2SO_4 and are visually similar to suspensions of growing *E. coli*.

Other activities available through BioBuilder.org offer similar instructions for modifications when equipment is unreliable or limiting. For example, the "iTune Device" lab generates a yellow colored product that the students can compare to a Benjamin Moore Paint Chip to approximate intensity readings.

4.1. Preparation of MacFarland Turbidity standards

Dilutions of 1% $BaCl_2$ in 1% H_2SO_4 are prepared according to Table 12.3. The standards are stable at room temperature for at least a month, more likely longer, and can be aliquoted by the students themselves or in advance by the instructors.

4.2. Comparison of growing cultures to Turbidity Standards

1. Turbidity standards should be aliquoted into small clear test tubes. The tubes should contain enough of each standard to fill the tube to a height of about 1 in. (2.5 cm). Tubes must be properly labeled with its turbidity standard number.
2. The aliquots of the turbidity standards should be placed in a test tube rack that allows the liquid to be viewed from the side.
3. On a blank index card or piece of folded printer paper, two thick black lines should be drawn with a marker. These lines should be placed on the card or paper to fall within the height of the standards.
4. Place the card with the lines into the test tube rack behind the standards (see Fig. 12.7).
5. To compare the bacterial cultures to the turbidity standards, an aliquot of the growing cultures should be moved into a test tube of the same size as the standards. A volume approximately equal to that of the standards should be used and the identity of the samples should be included on the tube's label.

Table 12.3 Standards for turbidity and their preparation

Turbidity scale	OD_{600}	1% $BaCl_2$/1% H_2SO_4 (ml)
0	0	0.0/10
1	0.1	0.05/9.95
2	0.2	0.1/9.9
3	0.4	0.2/9.8
4	0.5	0.3/9.7
5	0.65	0.4/9.6
6	0.85	0.5/9.5
7	1.0	0.6/9.4

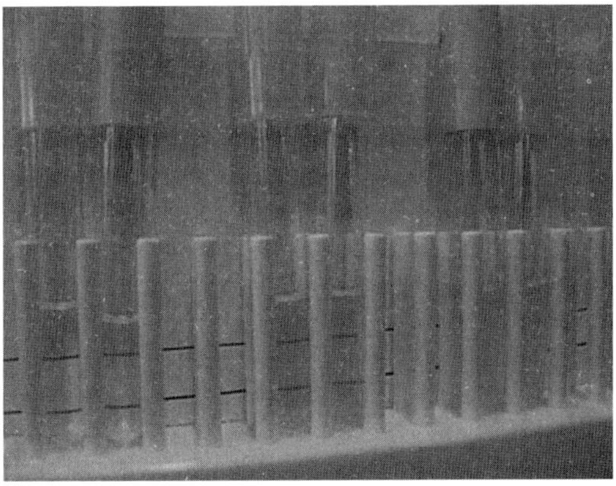

Figure 12.7 Example of turbidity comparison. Samples of bacteria (leftmost member of each pair) are compared to MacFarland Turbidity Standards. Samples are deemed equivalent when the black lines behind the tubes are obscured to the same extent. Source: J. Dixon.

6. The sample tubes can be placed next to the turbidity standards to be compared side-by-side. The standard that best represents the turbidity of the samples will be the one that obscures to the same extent the black lines drawn on the card.
7. Table 12.3 can be used to determine the OD_{600} value using each turbidity standard.
8. If the number of cells is to be calculated, then 1 OD_{600} unit can be approximated as equal to 1×10^9 cells/ml.

5. Summary

Most students at the high school and college level today could hardly imagine a world that did not include digital resources. Indeed, when students in Michael Wesch's "Digital Enthography" class at Kansas State University tackled the question: "what is it like being a student today," they documented their findings as a video ("Vision of Students Today": http://www.youtube.com/watch?v=dGCJ46vyR9o) and posted it to YouTube. In their video, the students present the results of their class surveys by holding 8½ × 11 in. signs or by showing their computer screens to the camera. Some of their messages, like "my average class size is 115" and "I buy hundred dollar textbooks that I never open" are not new issues for students at the university level. However, some aspects they present do seem different, for example, "I facebook through most of my classes" and "My neighbor paid for class but never comes."

How is a teacher to respond to the digital distractions that are a hallmark of this era? One (not unappealing) reply is to say something like, "stop goofing off and study." Indeed, experience says that students who put more into any class are the ones who get more out of it. So when students offer only a portion of their energy and attention in class, they are only short-changing themselves. Nonetheless, most teachers are hopeful that the subject they teach will excite others, and so most teachers look for ways to engage the students.

Perhaps, a more constructive reply to the issue of digital distractions is to meet the students on their technology playing field. If students really do "spend 3.5 h a day online" then perhaps a winning strategy is to teach through technology and online tools. Digital quiz-lets, discussion forums, and internet games can encourage students to interact more often and more thoughtfully with the class content. But when poorly implemented, cyber-oriented teaching can fall short, seeming more like "chocolate-covered broccoli" than any novel and engaging tool for interaction (Gorsky and Blau, 2009). Additionally, the recent failure of Google's "Wave" learning management system is a sobering reminder that technological advances must not be too complicated for average users, and must integrate well with the way people actually teach (Young, 2010).

Synthetic biology offers a mechanism for training students in an engaging and novel way. iGEM continues to attract summer students worldwide (iGEM: http://2010.igem.org/Main_Page), and the Adventures in Synthetic Biology comic is "given out like candy" (DIY-bio: http://diybio4beginners.blogspot.com/2009/02/adventures-in-synthetic-biology-comic.html). These two educational successes are leveraged at the BioBuilder website (http://www. biobuilder.org/) which expands iGEM projects like the banana-smelling bacteria, into digitally accessible materials for teaching and learning.

ACKNOWLEDGMENTS

We thank the 2006 MIT iGEM team for their hard and thoughtful work to establish the Eau d'coli system. We further thank Ginkgo Bioworks for DNA construction. These teaching materials were developed over two summers with the support of SynBERC, an NSF-funded Engineering Research Center, and the support of MIT's Department of Biological Engineering.

REFERENCES

Gorsky, P., and Blau, I. (2009). Online teaching effectiveness: A tale of two instructors. *Int. Rev. Res. Open Dist. Learn.* **10**(3), 1–27.

Hengge-Aronis, R., Lange, R., Henneberg, N., and Fischer, D. (1993). Osmotic regulation of rpoS-dependent genes in *Escherichia coli*. *J. Bacteriol.* **175**(1), 259–265, PMID:8416901.

Kuldell, N. (2007). Authentic teaching and learning through synthetic biology. *J. Biol. Eng.* **1,** 8.

Lutz, R., and Bujard, H. (1997). Independent and tight regulation of transcriptional units in *Escherichia coli* via the LacR/O, the TetR/O and AraC/I1-I2 regulatory elements. *Nucleic Acids Res.* **25**(6), 1203–1210.

Mitchell, R., Dori, Y. J., and Kuldell, N. (2010). Experiential engineering through iGEM—An undergraduate summer competition in synthetic biology. *J. Sci. Educ. Technol.* 10.1007/210956-010-9242-7.

National Research Council (2000). *How People Learn* National Academies Press, Washington, DC0-309-07036-8.

Wiggins, G., and McTighe, J. (1998). Understanding by Design. Merrill Prentice Hall, Columbus, OH0-13-093058-X.

Young, J. (2010). Google Wave, Embraced by many on Campuses, to get Wiped Out. *Chronicle of Higher Education.* http://chronicle.com/blogPost/blogPost-content/26039/ August 5.

SECTION THREE

DEVICE MEASUREMENT, OPTIMIZATION, AND DEBUGGING

CHAPTER THIRTEEN

Use of Fluorescence Microscopy to Analyze Genetic Circuit Dynamics

Gürol Süel

Contents

University of Texas Southwestern Medical Center at Dallas, Dallas, Texas, USA

Methods in Enzymology, Volume 497
ISSN 0076-6879, DOI: 10.1016/B978-0-12-385075-1.00013-5

Abstract

The physiological processes and programs of cells are not typically determined by single genes, but are governed by the patterns of interactions between genes and proteins [Alon, U. (2007). An Introduction To Systems Biology: Design Principles of Biological Circuits. Chapman & Hall/CRC, Boca Raton.]. These interactions are commonly referred to as genetic circuits, and the pattern of these interactions is called the circuit's architecture [Sprinzak, D. and Elowitz, M. B. (2005). Reconstruction of genetic circuits. *Nature* **438**(7067), 443–448.]. Genetic circuits control diverse cellular processes, and each process requires specific dynamic behaviors to properly function. Biochemical evidence aids in the identification of interactions between genes and proteins, but the spatio-temporal dynamics of these interactions are more difficult to probe using conventional techniques. Fluorescence time-lapse microscopy is a powerful tool in the study of genetic circuit dynamics, allowing the measurement of circuit dynamics in single cells [Suel, G.M., *et al.* (2007). Tunability and noise dependence in differentiation dynamics. *Science* **315**(5819), 1716–1719.]. Uncovering the dynamics of genetic circuits allows verification of mathematical models of genetic circuits and aids in the design of forward experiments. By enabling the study of relationships between circuit architecture and dynamic behavior, fluorescence time-lapse microscopy opens new frontiers in synthetic biology, allowing for the alteration of genetic circuits to achieve novel behaviors [Cagatay, T., *et al.* (2009). Architecture-dependent noise discriminates functionally analogous differentiation circuits. *Cell* **139**(3), 512–522.], and even the generation of completely synthetic, purpose built genetic circuits [Elowitz, M. B. and Leibler, S. (2000). A synthetic oscillatory network of transcriptional regulators. *Nature* **403**(6767), 335–338.]. Perhaps more importantly, determination of genetic circuit dynamics can reveal the concepts and principles underlying the biological functions they regulate.

1. Fluorescent Reporters

1.1. Fluorescent proteins

The discovery and cloning of green fluorescent protein (GFP) from the jellyfish *Aequorea victoria* marks the beginning of a revolution in cellular biology (Tsien, 1998). Fluorescent proteins are naturally occurring chromophores, capable of absorbing energy from a photon and returning to a lower energy state by emitting a photon of a different wavelength (Tsien, 1998). Today, various natural and engineered fluorescent proteins span the spectrum of visible light from far red to near violet (Heim and Tsien, 1996; Shagin *et al.*, 2004; Zhang *et al.*, 2002). Fluorescent proteins are genetically encoded, allowing the creation of stable cell lines that express chromophores. Genetic expression of fluorescent molecules holds an advantage over the use of fluorescent dyes, which must be added each time the experiment is performed. However, fluorescent proteins are larger, have limited photostability and brightness, and are subject to proteolytic degradation in the cell. Furthermore, fluorescent proteins require time to be expressed and mature within the cell (chromophore maturation). Derivatives of GFP undergo an autocatalytic chemical reaction during chromophore maturation, but various factors such as pH inside the host cell affects the maturation time of fluorescent proteins (Miyawaki *et al.*, 2003). As examples, in *Escherichia coli* and *Bacillus subtilis* maturation times are on the order of 10 min for GFP derivatives (Nagai *et al.*, 2002). Recent advances utilize point mutations that shorten the maturation time and alter the stability of many new fluorescent proteins. Fluorescent protein stability is typically greater than native proteins of the host cell, as host cell proteases are unlikely to have specificity for exogenous fluorescent proteins. The relative stability of fluorescent proteins favors a high signal to noise ratio. Further information about fluorescent proteins is available from Dr. Roger Tsien in this book.

1.2. Fluorescent dyes

Fluorescent dyes, also known as reactive dyes or fluorophores, have been used by biologists for decades. Fluorescent dyes offer higher photostability and brightness compared to fluorescent proteins and do not require a maturation time. However, fluorescent dyes are usually targeted to proteins of interest by antibody conjugates or peptide tags. This requires fixation of cells, which renders measurement of genetic circuit dynamics impossible. Several fluorescent dyes can be used in living cells, but in many cases their applicability is still limited. The remainder of this chapter will focus on the use of fluorescent proteins as the reporter of choice, but many discussion points also apply to the use of fluorescent dyes.

2. Constructing and Using Genetic Fluorescent Reporters

2.1. Transcriptional reporters

The construction of transcriptional reporters are a common use for fluorescent proteins. The goal is to measure promoter activity, allowing analysis of the dynamics of gene expression. To accomplish this, the fluorescent protein coding sequence is placed downstream of the promoter of interest. In most cases transcriptional reporters do not interfere with any biological functions of the cell, excluding side effects such as phototoxicity and overexpression of fluorescent proteins. The concentration of any protein in a cell is governed by two opposing rates, production and degradation. When analyzing the signal of a fluorescent protein expressed from a promoter of interest, a direct measure of the concentration of the native protein is not possible. While the proteins should theoretically be produced at similar rates, in most cases the increased stability of fluorescent proteins leads to a greater concentration of chromophore than of the native protein. However, fast fluctuations in gene expression may be masked by a highly stable fluorescent protein. Transcriptional reporters often provide information about the dynamics of genetic circuits; however, their use still poses challenging concerns. These concerns are not unique to the use of fluorescent reporters, as many of these issues are also commonly addressed in the field of crystallography, which requires the expression of large quantities of protein for structure determination. For example, in the use of transcriptional reporters, a common problem is a low signal to noise ratio. Due to the inherent cost of protein production, many natural promoters exhibit very low protein expression rates. To increase signal over noise, many strategies have been developed to increase the efficiency of fluorescent protein production without altering the native promoter. Listed below are a few of these important design strategies intended to increase the signal to noise ration of fluorescent measurements in cells:

(a) *Promoter sequence*: Determine and include all important upstream sequences to avoid omitting important regulatory binding sites for transcription factors. However, if upstream sequences may include other genes or promoters, great care must be taken in identifying the minimal sequence that still contains all relevant regulatory binding sites.
(b) *Ribosomal binding site (RBS)*: Different RBSs can affect the efficiency of fluorescent protein translation and thus the signal to noise ratio. Determine the native RBS, and determine if the RBS of the desired promoter is optimal for efficient expression in the cells of interest. The native RBS can be replaced by introducing an optimized RBS during the construction of a transcriptional reporter, and this step may be necessary to increase translational efficiency (Salis *et al.*, 2009).
(c) *RBS to start codon spacing*: Spacing between the RBS and the start codon is critical for proper translational efficiency (Chen *et al.*, 1994). In different organisms, the optimal distance may vary, so the optimal spacing in the organism of choice must be considered during construct design.
(d) *Codon optimization*: Across species, the relative frequency of codon usage varies, and may affect translational efficiency of fluorescent reporters. For example, GFP is derived from the jellyfish *A. Victoria,* and this sequence is not optimized for expression in different kingdoms or phyla. This concern may be addressed through the use of optimized codon sequences, designed for the species of interest. This may be

accomplished through synthesis of the entire coding sequence or site directed mutagenesis of specific codon sequences (Sastalla *et al.*, 2009).

(e) *Termination*: Inclusion of a strong transcriptional terminator is recommended at both the 5′ and 3′ end of a genetic construct. Beginning a construct with a terminator helps prevent cross talk between neighboring promoters that may mask the precise dynamics of the promoter of interest (Nishihara *et al.*, 1994). Incorporating a strong terminator at the end of a construct can improve the efficiency of gene expression (Kim *et al.*, 2003). This strategy can be employed in the construction of transcriptional reporters by adding appropriate terminators at the end of constructs.

2.2. Translational reporters

The above section details methods for measuring the activity of promoters using fluorescent reporters. While this approach provides temporal information about the expression of a protein from a specific promoter, information about the localization or distribution of the gene product is unavailable. Intermediate steps between mRNA production and protein synthesis are often key regulatory events that must be considered to understand genetic circuit behavior. Additionally, some proteins have little or no functionality unless concentrated or localized within the cell. Translational reporters commonly referred to as fusion proteins are used to probe this information. Fluorescent fusion proteins combine the coding sequences of two proteins, the protein of interest (minus stop codon) and a fluorescent reporter on one mRNA molecule. The two sequences are then translated together into one protein molecule. The main advantage of translational reporters is that they provide both temporal and spatial information. In other words, translation reporter constructs allow for measurement of both protein concentration and localization. One problem encountered with translational reporters is low amplitude and diffuse fluorescence signal, which makes accurate measurements difficult. However, localization of translational fusion proteins often results in subcellular regions of high concentration, markedly improving the signal to noise ratio (Feilmeir *et al.*, 2000).

For instance, localization of a protein typically occurs after translation, and fusion proteins allow visualization of the subcellular localization over time. As a result, very fast events and fluctuations in the spatiotemporal dynamics of proteins can be determined. Translational reporters have been utilized to determine the localization (Tsien, 1998), binding and complex assembly (Lakowicz, 2006), and even posttranslational state of proteins (Waldo *et al.*, 1999). These and other applications of fluorescent reporters are discussed further in Chapters 5 and 6 of this edition.

Fusion of a fluorescent protein (~240 amino acids long) can interfere with the activity of the protein it is fused to. In particular, many studies show proteins of interest that are unable to perform catalytic activities or localize when fused to fluorescent reporters. These issues are often due to steric hindrance. To address these problems with fluorescent reporter fusion design, certain considerations must be taken into account. Fluorescent proteins are most often fused to the N- or C-terminal of the protein coding sequence. When designing the fusion reporter, protein function in the cell must be considered to prevent the fused fluorescent protein from interfering. For example, the N- or C-terminal of the protein may be involved in establishing interactions with other proteins in the cell. In some cases, both N- and C-termini can be of functional importance. In these instances researchers have introduced fluorescent protein sequences inside the coding sequence of their protein. Specifically, with structural knowledge of the protein of interest, we can identify functionally noncritical loops into which we can introduce fluorescent reporters. This approach will not affect secondary structural elements such as alpha helices or beta sheets that may be critical to the stability and function of our protein.

2.3. Introducing genetic reporters into cells

Once cloned, fluorescent reporter constructs must be expressed in cells. Plasmid reporter systems and chromosomal integration are the most common methods for expressing reporter constructs in cells. Both methods typically rely on the ability to select for successful introduction of constructs, typically through the use of antibiotic resistance selection. With multiple selection markers, it is often possible to introduce multiple constructs into the same cell. Plasmid reporter systems offer the advantage of increased signal to noise ratio, due to the presence of multiple copies of the reporter in cells. However, because the copy number of plasmid may vary greatly between cells, plasmid based reporter systems often exhibit high variation in signal between cells. Chromosomal integration of constructs utilizes recombination to introduce a single construct copy per cell. With a single copy, the issue of noise due to copy number variability is eliminated, but signal intensity may be lower than from multicopy plasmid systems. One of the important advantages of chromosomal integration is that it offers a very stable introduction of constructs, which are reliably passed to daughter cells. Plasmid systems may remain very stable when under antibiotic selection, but can cure from cells when not under continuous antibiotic selection.

3. Fluorescent Time-Lapse Microscopy

3.1. A simple method for setting up a movie with bacteria

Time-lapse microscopy is one of the most suitable techniques for measuring the dynamics of genetic circuits in living cells. Below is a short protocol for the preparation of cells for time-lapse microscopy that has been optimized for bacterial cells and yeast, but the basic principles of the method are also applicable to cells of higher organisms such as mammalian cells.

Streak bacteria on an agar plate and grow overnight. Start a liquid culture from a single colony and grow to high optical density. Harvest cells by centrifugation. To examine, for example, stress response, wash and resuspend cells in liquid stress media. Allow cells to acclimate to new media conditions by incubating for 1 h. Dilute cells 10- to 1000-fold and pipette 3 μl onto an agarose pad.

Make the agarose pad while cells are growing by adding 1.5% weight by volume of low melting point agarose to the liquid stress media. Microwave the solution to dissolve the agarose, but avoid boiling. To create a smooth, flat surface for easy imaging, pour 5 ml of the melted media onto a glass cover slip. Sandwich the media by placing another cover slip on top. Allow the media to solidify before removing the top glass. Cut the agarose into small square pads of desired size.

After placing cells onto pad, place pad onto heat block until the drop has dried. The dried drop will show up as an opaque halo on the pad surface. Flip the pad onto the glass cover slip bottom of a Wilco dish for fluorescence microscopy. The sample is now ready for imaging.

3.2. Minimal requirements for fluorescence microscopy

Basic system requirements for measuring the dynamics of genetic circuits are as follows:

Microscope: Genetic circuit dynamics can be measured using time-lapse fluorescence microscopy in an automated microscope system. Automation is especially critical for precise image acquisition as a function of time and will be discussed below. All microscope manufacturers offer models with various capabilities specifically designed for fluorescence measurement over time. As a rule one should always demo at least two systems before investing is such a setup.

Illumination: Light illumination is required to excite the chromophore of our reporter. Artifact from nonhomogenous illumination can lead to artificial variation in cell–cell fluorescence intensity. To correct for this problem, identical fluorescent beads are imaged in the same field of view, allowing for the calculation of a parabolic plane of excitation intensity. This data can be used to correct measurements of fluorescence amplitudes in single cells.

Wavelength selection for specificity: Optical filters or dichroic mirrors allow for illumination with specific light wavelengths to selectively excite fluorescent reporter chromophores (described in Section 1 (Fluorescent Reporters)). Another set of filters or mirrors allows for selective passage of the emission wavelength of the reporter. These components are in particular necessary when the objective is to simultaneously measure multiple spectrally distinct fluorescent reporters.

Camera: Electron Multiplying Charge-Coupled Devices (CCDs) are recommended as cameras for the improved detection of low amplitude signals. Electron Multiplying CCD's amplify signal without significantly increasing background noise. Camera choice will depend on many factors, such as the acquisition and transfer speed, or the spatial resolution required for the experiment.

Automation of image acquisition: Automation enables accurate and reproducible measurements over extended periods of time. Even with only a single fluorescent reporter in cells, at minimum automated control of light exposure and image acquisition is recommended. To measure multiple spectrally distinct fluorescent reporters, automated control of excitation and emission filters is necessary. Automation of stage position allows measurement of multiple positions for increased data acquisition per experiment.

Automated focus: One requirement of using time-lapse microscopy for extended measurements of genetic circuit dynamics is maintenance of proper focus. Drift in focus results in dramatic decreases in signal to noise ratio, prohibiting accurate measurement. Several companies offer software capable of taking a stack of images along the *z*-axis and applying contrast and edge detection algorithms to identify the image which is in focus. However, software based auto focusing is typically slow and computationally intensive. Recently, infrared autofocusing systems have also become available for use in microscopes that do not rely on software. Rather, these systems utilize an infrared laser to optically measure the distance between the objective and sample and maintain focus accordingly.

Environmental control: Constant temperature or atmosphere may be critical to experimental success. The microscope may be enclosed in an environmentally controlled chamber, or set up in an atmosphere controlled room. These requirements are determined by the particular cellular system in which measurements will be performed.

3.3. Image analysis

Time-lapse microscopy generates large amounts of image data files. Therefore, automated and quantitative image analysis is a critical component of measuring genetic circuit dynamics when using fluorescence time-lapse microscopy. Image analysis is a very rapidly developing field, and only the basics are discussed here. The goal of the analysis is to utilize differences in intensities of fluorescence or light images to discriminate cells from background and from each other. For example, in recorded images of bacterial cells with phase contrast optics, cells will appear darker than the background. Edge detection software and histograms of image intensities, together with constraints on cell size and shape, allow identification of individual bacterial cells in a process called segmentation. Once the cells have been identified, numerous properties of single cells can be measured, such as the mean or maximum of fluorescence signal. Determination of the spatial localization of translational fusions can also be accomplished. By repeating this process for images of different time points, spatiotemporal dynamics of fluorescence signal can be determined, demonstrating the activity of a component within the genetic circuit of interest.

3.4. Tracking lineages

An important advantage of fluorescence time-lapse microscopy compared to other single cell measurement techniques, such as flow cytometry, is the ability to track individual cells over time and through cell division. Automated cell tracking is possible through software that traces the locations of a cell and its daughters across a series of images. This allows compilation of lineage information such that we can follow individual descendents over many consecutive generations. Lineage tracking information also allows measurement of conditional probabilities. For example, a small percentage ($\sim$5%) of *B. subtilis* differentiates into a state of competence for uptake of extracellular DNA in stress conditions (Cagatay *et al.*, 2009). The hypothesis that the initiation of competence is stochastic was tested by measuring the conditional probability of cells to become competent given that their sister cells were competent. Sister cells shared the same cytoplasm prior to cell division and are typically located adjacent to each other. Therefore, if competence was stochastically initiated, sister cell's fates should be independent. However, if local differences in the microenvironment or lineage history contribute to the probability of competence, sister cells would exhibit an interdependent conditional probability of becoming competent. Lineage analysis revealed that sister cell's probability to become competent was independent, consistent with the hypothesis that competence is triggered in a cell autonomous and stochastic manner. These results were then later confirmed with more direct measurements.

Lineage analysis was also utilized to determine the gene regulation function at the single cell level. These studies revealed how the concentration of a given transcription factor is related to the production rate of protein from its target promoter (Elowitz *et al.*, 2002). Lineage analysis also revealed the autocorrelation time of fluctuations in gene regulation, uncovering the limited accuracy of signal propagation (Pedraza and van Oudenaarden, 2005). Lineage analysis is especially necessary when the dynamics of a genetic circuit are spread over many cell divisions. Therefore, lineage analysis is a very powerful technique that utilizes the advantages provided by time-lapse microscopy.

4. Measuring and Interpreting Dynamics

The previous sections discuss the technical aspects of designing and measuring fluorescent reporter constructs. This chapter describes how fluorescence time-lapse microscopy can be used as a tool to quantitatively measure and interpret the dynamics of genetic circuits in living cells.

4.1. Gene expression

Control of gene expression from promoters by specific transcription factors is an important interaction in genetic circuits. The rate of promoter expression can be measured by using a transcriptional fluorescent reporter as described in Section 4.1. The derivative of the fluorescence time trace defines the change in fluorescent protein concentration over time, and this in turn corresponds to the rate of gene expression. These measurements can provide critical understanding of genetic circuit dynamics.

4.2. Protein concentration and localization

Translation reporter constructs can be used to measure the concentration and localization dynamics of proteins that comprise genetic circuits. One problem encountered with translational reporters is difficultly in accurately measuring low amplitude and diffuse signal. However, localization of translational fusion proteins often results in subcellular regions of high concentration, markedly improving the signal to noise ratio. For example, the targeting of translational reporters to membranes has been utilized to measure small protein concentrations in bacterial cells (Choi *et al.*, 2010). Subcellular localization of proteins often occurs following posttranslational modification. Indirect measurement of modification state can be made by monitoring fluorescence localization. As an example, the posttranslational modifications of transcription factors often dictate their intracellular localization (Whitmarsh and Davis, 2000). Many transcription factors only bind promoter sequences with high affinity when phosphorylated, such that the phosphorylation state of transcription factor protein fusions can be inferred by the presence of localized fluorescent signal upon binding to promoter sequences.

4.3. Combining multiple fluorescent reporters to measure circuit interactions

Single fluorescent reporters alone cannot demonstrate interactions that define genetic circuits. As illustrated in the above examples, simultaneous measurement of multiple fluorescent reporters allows characterization of the dynamics of interactions that comprise genetic circuits.

While the number of spectrally distinct fluorescent reporters that can be functionally measured in a single cell is limited, the number of genetic circuit interactions (Alon, 2007) that can be measured using a small number of reporters is not. As an example, consider a genetic circuit consisting of four genes, A, B, C, and D. While the use of four separate reporters may not be feasible, the creative use of two fluorescent reporters can allow for determination of the relative dynamics of A, B, C, and D. Multiple cell lines can be constructed, each containing a transcriptional fusion to the A promoter, and a transcriptional fusion of one of the other genes. To measure the dynamics of all four components, pair-wise comparisons of A to B, A to C, and A to D activity can be made. Then the time traces from the B, C, and D data sets can be aligned to the common A reporter, creating a complete picture of the A, B, C, and D dynamics within a cell type.

4.4. Binding interactions FRET, etc.

While transcriptional reporters reveal information about the interaction between transcription factors and promoters, other techniques are needed to determine the dynamics of protein–protein interactions. One such technique is Förster (Fluorescence) Resonance Energy Transfer (FRET). This technique utilizes a dipole–dipole coupling mechanism to transfer the energy from an excited donor chromophore to an unexcited recipient. Spectrally distinct translational fusions of each protein are generated to measure the dynamics of interaction (binding). Upon excitation at a specific wavelength the first chromophore is excited, and through resonance energy transfer donates its energy to a second chromophore. In turn, the second chromophore is excited and releases a photon to return to a lower energy state, which is measurable as emitted light (Lakowicz, 2006). FRET is only observed when the two chromophores are in very close proximity, typically only occurring following direct binding of the proteins of interest that are each fused to one the FRET components. Observation of FRET signal over time allows for measurement of protein–protein (binding) interaction dynamics that are critical for determining dynamics of genetic circuits.

5. Applications for Measurement of Circuit Dynamics

Information describing interactions among genes and proteins is accumulating rapidly; however, understanding of the operational principles of genetic circuits has not increased proportionally. Extracting simple concepts from vast amounts of available data is becoming a major challenge for biomedical researchers. Since measurement of dynamics provides critical information that can reveal the design principles of genetic circuits, there are many applications for fluorescence time-lapse microscopy. Design principles can be determined by utilizing genetic circuit dynamics to test mathematical models of circuit behavior and obtain greater conceptual understanding of circuit function.

5.1. Simplifying interactions comprising cellular differentiation circuits

Reducing the complexity of a genetic circuit often uncovers basic principles of biological function (Alon, 2007). For example, measurement of circuit dynamics using fluorescence time-lapse microscopy can be used to differentiate between critical and noncritical circuit interactions.

In the model organism *B. subtilis,* a genetic circuit regulates differentiation into a state called competence (Suel *et al.*, 2006). In stress conditions, a small fraction of *B. subtilis* cells differentiate into competence, allowing the uptake and integration of extracellular DNA. Alternate models explain cellular differentiation into competence. One model suggests that the decision is a result of particular cell–cell signaling molecules in the microenvironment, stimulating cells to initiate competence. Alternatively, the decision to become competent may be made cell autonomously. If competence is triggered cell autonomously, interactions within the genetic circuit of competence that relate to cell–cell signaling may be ignored to simplify our circuit.

Quorum sensing signals affect the initiation of competence. As cells increase in concentration, quorum sensing peptides excreted from *B. subtilis* cells accumulate. While high concentrations of quorum sensing signals are necessary to trigger competence, they may not be sufficient. One promoter activated by high concentrations of quorum sensing signals is the *ComS* promoter. ComS is a peptide that competitively interferes with degradation of ComK, a transcription factor and master regulator of competence. ComK is necessary and sufficient for competence, and activates over 140 competence genes. Reduction of ComK degradation by *ComS* promoter activation should therefore allow for the accumulation of ComK, triggering competence. This suggests that in cells which differentiate into competence, ComS expression will be higher than in cells which do not (Suel *et al.*, 2006). This hypothesis, that ComS expression is responsible for triggering competence, was tested by fluorescence time-lapse microscopy. Spectrally distinct *ComS* and *ComK* transcriptional fluorescent reporters were chromosomally integrated in *B. subtilis*. ComS and ComK expression dynamics were measured in the same cell to determine if ComK expression is activated in cells with higher ComS expression. The expression amplitude of ComS was not found to predict which cells would differentiate into competence. As a result, cell–cell signaling, while important, was not found to dictate cellular entry into competence. Thus, competence was found to be triggered cell autonomously, and the competence genetic circuit can be simplified by ignoring genetic circuit interactions that couple to quorum sensing signals. Thereby the number of interactions which must be considered to understand the genetic circuit that governs competence initiation is reduced.

5.2. Simplifying regulatory inputs into promoters

Fluorescence time-lapse microscopy can also be used to simplify understanding of promoter regulation. As an example, imagine two distinct promoters called A and B in a genetic circuit. Promoter A is exclusively regulated by transcription factor T_1, while B has binding sites for transcription factors T_1, T_2, and T_3. Whether the contributions of T_2 and T_3 in the regulation of B must be considered, or can be simplified to only consider T_1, can be answered using fluorescence time-lapse microscopy. By constructing chromosomal integration of spectrally distinct transcriptional reporters for the A and B promoters, their respective activities may be simultaneously measured. If the temporal dynamics of promoters A and B are identical, both promoters can be assumed to be predominantly regulated by T_1. As a result, we can ignore the interactions between T_2 and T_3 on the B promoter. This approach was applied to simplify the competence circuit of *B. subtilis*, allowing for a simplified model of the genetic circuit, promoting intuitive understanding of the underlying control of competence (Suel *et al.*, 2006).

5.3. Uncovering novel interactions in genetic circuits

Modeling of genetic circuits can be used to form predictions about their dynamic behavior. By measuring the actual dynamic behavior in cells, refinement of models can be accomplished, and new predictions made.

Differences between observations and predictions may hint at novel biologically important interactions. For example, observations revealed the excitable dynamics of the competence genetic circuit in *B. subtilis* (Suel *et al.*, 2006). How this genetic circuit promotes both entry and exit from competence was unknown, but mathematical models suggested that competing positive and negative feedback loops could account for the transient differentiation of cells into competence. While modeling suggested a negative feedback loop, this interaction was uncharacterized. Literature review revealed that the ComS is a competitive inhibitor of ComK degradation. However, no link between ComK and expression of ComS was known. By creating transcriptional reporters for the *ComK* and *ComS* promoters, the levels of ComS and ComK production were found to be inversely correlated in cells, suggesting that ComK negatively regulates ComS expression, leading to increased degradation of ComK (Suel *et al.*, 2006). This confirmed the hypothesis that a negative feedback loop control over ComK was established by ComS. Forward experiments tested and confirmed the existence of the ComS mediated negative feedback loop and its role in the exit from competence. These data established that the genetic circuit governing competence was an excitable system with coupled positive and negative feedback loops.

5.4. Supplementing traditional biochemical promoter analysis

Single point mutations can alter the binding affinity of a transcription factor to its target site. Traditional biochemical or biophysical methods for determining the effect of mutations within promoters are capable of measuring differences in the binding affinities of transcription factors. For example, Isothermal Titration Calorimetry (ITC) measures the heat released or absorbed as a result of a bimolecular binding event. This technique can accurately determine the affinity between a promoter and its transcription factor. However, techniques such as this do not describe how the dynamics of gene expression in cells are affected by changes in transcription factor binding sites. Fluorescence time-lapse microscopy can be utilized to perform single cell experiments to address this question. By creating spectrally distinct transcriptional fluorescent reporters for both wild type and mutant promoters within the same cell, simultaneous measurement of reporter expression can determine if point mutations have an effect on the timing of transcription factor binding. Deviation in correlation between the wild type and mutant promoter reporters reveals how the dynamics of gene expression are altered as a result of the introduced mutation. These measurements can therefore show how single point mutations in promoter sequences can alter the dynamics of gene expression in cells.

5.5. Measuring the noise in genetic circuit dynamics

Two-color quantitative fluorescence time-lapse microscopy enabled the measurement and characterization of stochastic fluctuations (noise) in gene expression. Two identical promoters individually fused to spectrally distinct fluorescent reporters allow measurement of fluctuations in gene expression intrinsic to each promoter (Elowitz *et al.*, 2002). Correlated fluctuations in gene expression are due to fluctuations in components upstream of the gene, such as transcription factor concentration. However, uncorrelated fluctuations in gene expression between the two identical promoters reveal stochastic fluctuations inherent to gene expression from individual promoters. The propagation of noise through a genetic circuit and persistence of noise as a function of time is also measurable using two-color fluorescence time-lapse microscopy. The determination of noise dynamics with a single reporter would require precise quantitative measurements, because often the amplitude of fluctuations is very small. However, the use of two colors allows determination of noise from the amount of correlation between two independent signals, rather than measuring absolute amplitudes in fluctuation of a single reporter (Suel, 2007). In addition, the measurement of fluctuations across a cell lineage allows determination of the persistence of noise over time.

5.6. Engineering novel dynamic behaviors in cells

The above sections utilize reporters to understand the dynamics of natural genetic circuits. However, a powerful tool in uncovering possible selection pressures that may have driven the evolution of genetic circuits is the response of these circuits to artificial perturbations (Cagatay, 2009; Suel, 2007; Zhuravel *et al.*, 2010).

5.6.1. Bifurcation of dynamics through parameter perturbation

Excitable systems are defined by having a single stable resting state, and a second state(s) that is unstable. Stimulation of an excitable system, by for example, noise, can allow the system to escape the attraction of the stable state and progress toward the unstable state. Since the system cannot reside at the unstable state, the system must ultimately return back to its only stable resting state (Strogatz, 1994). Therefore excitable systems generate transient dynamics in the form of pulses that take the system transiently away from its stable resting state. One property of excitable systems is that they are typically able to access a regime of oscillatory dynamics when parameters are altered such that the stable state is eliminated. Such transitions between dynamical regimes are referred to as a bifurcation and are discussed in this edition in much more detail by Dr. Steven Strogatz.

As an example of this concept, we take a look at the competence circuit that has been shown to be a system with excitable dynamics. Using information obtained from fluorescence time-lapse microscopy, mathematical models were generated to aid in predicting ways to induce a bifurcation *in vivo* such that the competence circuit would generate oscillatory dynamics. According to these mathematical models, increasing the basal expression rate of the competence circuit master regulator ComK would eliminate the stable resting state of the system, leaving only one unstable state (Suel *et al.*, 2006). As a result, the system would oscillate around the unstable state of the system. Thus, artificially increasing the basal expression of ComK in the competence circuit should eliminate the stability of the resting state and thus induce oscillatory behavior around the unstable state. To test this prediction, an inducible promoter was used to drive ectopic expression of ComK. As predicted, increase in the basal expression rate of ComK was sufficient to transition the competence circuit from excitable to oscillatory dynamics in single cells.

5.6.2. Rewiring of genetic circuit interactions to generate novel dynamics

The dynamics of a genetic circuit can also be altered by changing its pattern of interactions (Sprinzak and Elowitz, 2005; Toettcher, 2011). For example, the excitable dynamics of the competence circuit are generated by the competition of a positive and negative feedback loop. Similar to above, mathematical modeling suggested that elimination of the negative feedback loop in the competence circuit, which controls exit from competence, would stabilize the unstable fixed point of the excitable system and convert it into a bistable system. As a result, the unstable and thus transient state of competence would become stable making the system a bistable system with two stable states. To test this prediction, competence dynamics were observed in cells in which the negative feedback loop was bypassed. These cells entered competence similarly to unaltered cells, but were unable to exit this differentiated state. Therefore, cells resided in a stable manner either in the uninduced/undifferentiated resting state, or in the competence state. Together, these experiments show that measurement and analysis of genetic circuit dynamics can be utilized to predict the effect of synthetic perturbations in the competence circuit, allowing for the engineering of novel dynamic behaviors in living cells.

5.7. Relationship among genetic circuit architecture, dynamics and biological function

Measurement of genetic circuit dynamics can reveal the relationship among the wiring pattern of interactions (architecture) that constitutes the genetic circuit, its dynamics and biological function. This relationship may reveal why genetic circuits have particular patterns of interactions and how critical the architecture is for the biological function. This question can be approached by identifying equivalent alternative genetic circuit architectures that are mathematically predicted to reconstitute the dynamics of their natural counterparts (Elowitz and Leibler, 2000). If the dominant role of genetic circuit architecture is to generate appropriate dynamics, then genetic circuits that have alternative architectures, but are capable of generating the same dynamic behavior should be able to fully reconstitute the biological function of their native counterpart.

A recent study tackled this problem by substituting the native genetic circuit that regulates competence in *B. subtilis* with a synthetic circuit (SynEx) that differed in architecture (Cagatay *et al.*, 2009). Specifically, both the native and SynEx circuits were comprised of competing positive and negative feedback loops. However, while in the native circuit the negative feedback is comprised by the repression of an activator, in the SynEx the negative feedback loop is comprised by the activation of a repressor. Therefore, the order of consecutive activation and repression reactions is switched in the negative feedback loops of the two circuits. However, the positive feedback loop is the same in both circuits serving as an anchor for direct comparison. By quantitatively comparing the physiological function of the native and SynEx circuits it is possible to determine the role of circuit architecture in terms of biological function.

Comparative analysis using fluorescence time-lapse microcopy revealed that the SynEx circuit reconstructed the mean dynamics and physiological function of the native circuit. Cells with either circuit were able to generate similar excitable dynamics at the single cell level where cells transiently differentiation into the competence state. Furthermore, both circuits allowed cells to take up extracellular DNA and incorporation it into their chromosome with the same efficiency. However, while the mean duration of transient competence events was the same between the two circuits, the variation around the mean was distinct. The native circuit exhibited a higher variation in competence duration times compared to the SynEx circuit. Further experiments showed that the higher variability in competence durations generated by the native circuit provided a biological advantage. Variability in competence durations allowed the population of cells with the native circuit to remain responsive over a broader range of DNA concentrations in the environment. Cells with the SynEx circuit generated less variability and thus had a narrower response range in terms of DNA concentrations in the environment.

Variability in competence durations therefore appeared to determine the range of variability in extracellular DNA concentrations over which the competence circuits functioned biologically. Mathematical modeling predicted that the particular architecture of the negative feedback loop was the source for the variability in competence durations. This prediction was tested and verified experimentally through perturbations of parameter values and measurement of genetic circuit dynamics. Together these results suggest that biological function may select for specific genetic circuit architectures to cope with environmental constraints. In other words, the relationship between circuit architecture and biological function appears to be mediated by the dynamics and variability generated by specific genetic circuit architectures.

In closing, the purpose of this chapter was to describe how fluorescence microscopy can be used to measure the dynamics of genetic circuits and to demonstrate how this technique can be applied to tackle various problems in biology. In particular, there are two critical points we have to keep in mind. (1) All biological systems are comprised of interactions among various components. Only through understanding of interactions can we understand the operational principles of biological systems. (2) All biological systems are subject to change as a function of time. Therefore measurement of dynamics is absolutely essential to expand and deepen our understanding of biological systems. Fluorescence time-lapse microscopy is by no means the only technique, but it is certainly one of the most suitable and perhaps most elegant approaches that literally provides us with a picture of how biological systems such as genetic circuits function in cells.

REFERENCES

Alon, U. (2007). An Introduction to Systems Biology: Design Principles of Biological Circuits. Chapman & Hall/CRC, Boca Raton.

Feilmeir, Bradley, J., *et al.* (2000). Green fluorescent protein function as a reporter for protein localization in *Escherichia coli*. *J. Bact.* **182**(14), 4068–4076.

Cagatay, T., *et al.* (2009). Architecture-dependent noise discriminates functionally analogous differentiation circuits. *Cell* **139**(3), 512–522.

Chen, H. Y., *et al.* (1994). Determination of the optimal aligned spacing between the Shine–Dalgarno Sequence and the translation initiation codon of *Escherichia coli* Messenger-Rnas. *Nucleic Acids Res.* **22**(23), 4953–4957.

Choi, P. J., Xie, X. S., and Shakhnovich, E. I. (2010). Stochastic switching in gene networks can occur by a single-molecule event or many molecular steps. *J. Mol. Biol.* **396**(1), 230–244.

Elowitz, M. B., and Leibler, S. (2000). A synthetic oscillatory network of transcriptional regulators. *Nature* **403**(6767), 335–338.

Elowitz, M. B., *et al.* (2002). Stochastic gene expression in a single cell. *Science* **297**(5584), 1183–1186.

Heim, R., and Tsien, R. Y. (1996). Engineering green fluorescent protein for improved brightness, longer wavelengths and fluorescence resonance energy transfer. *Curr. Biol.* **6** (2), 178–182.

Kim, D., *et al.* (2003). Improved mammalian expression systems by manipulating transcriptional termination regions. *Biotechnol. Prog.* **19**(5), 1620–1622.

Lakowicz, J. R. (2006). Principles of Fluorescence Spectroscopy. Springer Science & Business Media, Baltimore.

Miyawaki, A., Nagai, T., and Mizuno, H. (2003). Mechanisms of protein fluorophore formation and engineering. *Curr. Opin. Chem. Biol.* **7**(5), 557–562.

Nagai, T., *et al.* (2002). A variant of yellow fluorescent protein with fast and efficient maturation for cell-biological applications. *Nat. Biotechnol.* **20**(1), 87–90.

Nishihara, T., Iwabuchi, T., and Nohno, T. (1994). A T7 promoter vector with a transcriptional terminator for stringent expression of foreign genes. *Gene* **145**(1), 145–146.

Pedraza, J. M., and van Oudenaarden, A. (2005). Noise propagation in gene networks. *Science* **307**(5717), 1965–1969.

Salis, H. M., Mirsky, E. A., and Voigt, C. A. (2009). Automated design of synthetic ribosome binding sites to control protein expression. *Nat. Biotechnol.* **27**(10), 946–950.

Sastalla, I., *et al.* (2009). Codon-optimized fluorescent proteins designed for expression in low-GC gram-positive bacteria. *Appl. Environ. Microbiol.* **75**(7), 2099–2110.

Shagin, D. A., *et al.* (2004). GFP-like proteins as ubiquitous metazoan superfamily: Evolution of functional features and structural complexity. *Mol. Biol. Evol.* **21**(5), 841–850.

Sprinzak, D., and Elowitz, M. B. (2005). Reconstruction of genetic circuits. *Nature* **438** (7067), 443–448.

Strogatz, S. H. (1994). Nonlinear Dynamics and Chaos: With Appliations to Physics, Biology, Chemistry, and Engineering. Perseus Books Publishing, LLC, Cambridge.

Suel, G. M., *et al.* (2006). An excitable gene regulatory circuit induces transient cellular differentiation. *Nature* **440**(7083), 545–550.

Suel, G. M., *et al.* (2007). Tunability and noise dependence in differentiation dynamics. *Science* **315**(5819), 1716–1719.

Toettcher, Jared, E., *et al.* (2011). The promise of optogenetics in cell biology: interrogating molecular circuits in space and time. *Nature Methods.* **8,** 35–38.

Tsien, R. Y. (1998). The green fluorescent protein. *Annu. Rev. Biochem.* **67,** 509–544.

Waldo, G. S., *et al.* (1999). Rapid protein-folding assay using green fluorescent protein. *Nat. Biotechnol.* **17**(7), 691–695.

Whitmarsh, A. J., and Davis, R. J. (2000). Regulation of transcription factor function by phosphorylation. *Cell. Mol. Life Sci.* **57**(8–9), 1172–1183.

Zhang, J., *et al.* (2002). Creating new fluorescent probes for cell biology. *Nat. Rev. Mol. Cell Biol.* **3**(12), 906–918.

Zhuravel, D., *et al.* (2010). Phenotypic impact of regulatory noise in cellular stress-response pathways. *Syst. Synth. Biol.* **4,** 105–106.

CHAPTER FOURTEEN

Microfluidics for Synthetic Biology: From Design to Execution

M. S. Ferry,*[,1] I. A. Razinkov,*[,1] *and* J. Hasty*[,†,‡]

Contents

Abstract

With the expanding interest in cellular responses to dynamic environments, microfluidic devices have become important experimental platforms for biological research. Microfluidic "microchemostat" devices enable precise environmental control while capturing high quality, single-cell gene expression data. For studies of population heterogeneity and gene expression noise, these abilities are crucial. Here, we describe the necessary steps for experimental microfluidics using devices created in our lab as examples. First, we discuss the rational design

* Department of Bioengineering, University of California, San Diego, California, USA
† BioCircuits Institute, University of California, San Diego, California, USA
‡ Molecular Biology Section, Division of Biological Sciences, University of California, San Diego, California, USA
1 These authors contributed equally to this work.

Methods in Enzymology, Volume 497
ISSN 0076-6879, DOI: 10.1016/B978-0-12-385075-1.00014-7

of microchemostats and the tools available to predict their performance. We carefully analyze the critical parts of an example device, focusing on the most important part of any microchemostat: the cell trap. Next, we present a method for generating on-chip dynamic environments using an integrated fluidic junction coupled to linear actuators. Our system relies on the simple modulation of hydrostatic pressure to alter the mixing ratio between two source reservoirs and we detail the software and hardware behind it. To expand the throughput of microchemostat experiments, we describe how to build larger, parallel versions of simpler devices. To analyze the large amounts of data, we discuss methods for automated cell tracking, focusing on the special problems presented by *Saccharomyces cerevisiae* cells. The manufacturing of microchemostats is described in complete detail: from the photolithographic processing of the wafer to the final bonding of the PDMS chip to glass coverslip. Finally, the procedures for conducting *Escherichia coli* and *S. cerevisiae* microchemostat experiments are addressed.

1. Part I: Introduction

Microfluidic technology has enjoyed considerable success and interest in recent years. Microfluidic devices have been used for everything from miniaturization of molecular biology reactions to platforms for cell growth and analysis (Bennett *et al.*, 2008; Cookson *et al.*, 2005; Danino *et al.*, 2010; Hersen *et al.*, 2008; Hong *et al.*, 2004; Kurth *et al.*, 2008; Lee *et al.*, 2008; Rowat *et al.*, 2009; Taylor *et al.*, 2009; Thorsen *et al.*, 2002). A driving factor for increased use of microfluidics is the potential for more productive experiments, that is, accomplishing the same or more using fewer resources (primarily less reagents, consumables, and time). Furthermore, microfluidic devices offer the unrivaled ability to precisely control and perturb the environment of single cells while capturing their behavior using high resolution microscopy. In this report, we will concentrate on how to design, build, operate, and analyze data from single cells growing in the chambers of high-throughput microfluidic devices. We will focus primarily on a device built to monitor the growth of *Saccharomyces cerevisiae* (yeast) in a dynamically changing environment as a case study. This device is known in our lab as the MDAW or Multiple Dial-A-Wave device.

In our lab we strongly believe in the importance of acquiring single cell trajectories from our experimental runs. This requires the ability to track single cells over the course of an experiment, which generally lasts 24–72 h. Indeed, of all technologies available in molecular biology, microfluidics alone offers the ability to track the behavior of a large number of individual cells over the course of an experiment. While other technologies, such as flow cytometry, allow the acquisition of single cell data, the experimenter cannot track each individual cell in time. This leads to "snap shots" of how the population as a whole changes in time, but does not capture how individual cells progress over the course of an experiment.

The difference between the techniques can be illuminated easily if one thinks of a population of cells containing a desynchronized genetic oscillator. In this case much depends on the waveform of the oscillator. For oscillators with sinusoidal output, the population will appear bimodal with a large portion of the cells spread between the two modes. However, for an oscillator with output similar to a triangle wave, the cells will be uniformly distributed between all phases of oscillation and therefore the population will have a fairly evenly distributed set of fluorescent values. Of course the behavior of a real oscillator can be somewhere between these extremes, but the point is that looking at the progression of a population as a whole does not tell you everything about its dynamics. For example, in each of the cases mentioned above, other explanations are possible, such as the transient of a bistable switch, or even a genetically mixed population of cells. In contrast, using a microfluidic device to follow the temporal dynamics of single cells in such a population would allow one to easily see if any cells were oscillating.

While microfluidics is powerful, flow cytometry has the ability to capture a large amount of data quickly, much more quickly than it can be done in traditional microfluidics. For this reason, microfluidic and flow cytometry should be thought of as complimentary, instead of competing, technologies. We often find it useful to first characterize our genetic circuits using flow cytometry, testing as many media or inducer concentrations as possible, to look for behavior indicative of interesting dynamics. Once these conditions are determined we follow up with the more powerful but involved microfluidic experiments.

Thus in the context of this report we will be talking about microfluidic chips designed to capture single cell data over the 1–3 days of the experiment. Unfortunately this limits the architecture of such a chip due to the difficulty of tracking cells. Regrettably cells such as yeast or especially *Escherichia coli* have few unique features which can be used to distinguish them from their brethren. The full details of this will be discussed in a later section describing cell tracking, but suffice it to say, the only truly unique characteristic all cells possess visible by phase contrast microscopy is their position in time. As an added complexity, cells such as yeast or *E. coli* are so fast growing they can quickly fill both a trap and the camera's field of view. Once the trap is full, the colony of cells will begin to move in flows resembling particulate flows (Mather *et al.*, 2010). These flows are due to pressure exerted by the colony on the walls of the trap. Due to this movement, phase contrast images of a colony's growth must be taken often, usually every 30 s to a minute, to prevent excessive movement between images.

Unfortunately, this requirement of frequent imaging imposes a physical limit to the size of the chip, usually determined by the speed of the microscope hardware. Even state of the art, fully automated microscope hardware such as the Nikon TI system, cannot autofocus, acquire phase contrast plus 3–4 fluorescent images, and then move to a new stage location in less than 4–5 s and sometimes as many as 7–10 s depending on the acquisition parameters. This limits the number of chambers and hence the number of independent experiments to at most 8–14, if the 1-min interval between phase images is followed. Of course one also has to worry about overexposing cells to fluorescent excitation light, which can easily kill even the hardiest of cells rather quickly. Thus while phase contrast images are acquired every minute, we normally only capture fluorescent images every 5 min. Since 4 out of 5 acquisitions will not contain fluorescence capture (usually the longest step) this decreases the overall acquisition time somewhat. However, even if the phase contrast interval is lengthened the scope hardware will end up being the limiting factor in determining how large a chip can become. Of course microfluidic chips have been created with thousands of chambers (Taylor *et al.*, 2009); however, these devices cannot capture the type of single cell trajectory data that smaller devices can, at least with current microscope technology.

The types of microfluidic experiments we will discuss here pretty much require the latest in microscope hardware for reasons mentioned above. Automation of most microscope tasks is critical, such as stage movement, phase ring and fluorescent cube changing, and shutter control. Moreover taking images every minute for days on end requires an automated focus routine, which luckily most microscope manufactures can readily provide. This also requires large amounts of hard disk space and equally important a rigorous method for space management, with backup procedures in place to prevent catastrophic data loss. Moreover the sensitivity of the camera used is extremely important. While the background fluorescence (a bound for the minimum detectable signal) of yeast and *E. coli* cells is easily observed using CCD cameras even a decade old, one should always use the most sensitive camera available to minimize the exposure time and hence phototoxicity caused by the fluorescent excitation lamp. The overall idea is that while older hardware may allow you to capture some data like that we discuss here, newer hardware will allow you to capture more data with a higher quality and with less damage to your cells.

1.1. The design of a microfluidic chip

To design a microchemostat chip useful for the type of experiments described in the introduction, one has to know a small amount about fluid mechanics at the microscale. We will briefly describe the physics behind microfluidics here, but the reader is directed to more complete texts if desired (Beebe *et al.*, 2002; Brody *et al.*, 1996; Nguyen and Wereley, 2002; Whitesides *et al.*, 2001a). Those that have not studied fluid mechanics in depth do not have to worry because making a functional microchemostat is not too difficult. The first thing to understand is how fluid flows at the microscale of a microfluidic device. From fluid mechanics we know that there are essentially two major flow regimes: laminar and turbulent flow. Laminar flows contain highly predictable, parallel flow streams resulting in fairly easy to model profiles. In contrast, turbulent flows are unpredictable, difficult to model computationally, and contain complicated flow patterns such as eddies and vortices (there is also a transition regime between these two flow types). For microchemostat devices, the flow will be exclusively laminar as explained below. However, to determine the flow type in a arbitrary system, the most important parameters are the type of fluid used, the dimensions of the fluid channels and the fluid's velocity in these channels. The relationship between these parameters can be expressed as the Reynolds number (*Re*), which is a dimensionless quantity useful for determining the dominant profile in a flow system. The Reynolds number is defined by

$$Re = \frac{\rho v D_{\mathrm{h}}}{\mu}, \tag{14.1}$$

where ρ is the density of the fluid, υ is the mean fluid velocity, D_h is the hydraulic diameter of the channel (a value which depends on the channels dimensions; see Nguyen and Wereley, 2002), and μ is the fluid's viscosity (Beebe *et al.*, 2002). The Reynold's number represents a ratio between the inertial forces and the viscous forces of a fluid's flow. Empirically it has been determined that flows with a high Reynold's number ($Re > 10^3$), indicating the dominance of inertial forces, will be turbulent while low Reynolds number flows ($Re < 1$) will be exclusively laminar (Brody *et al.*, 1996). Typical parameter values for microchemostats with an aqueous fluid are given in Table 14.1. Due to the low Reynolds number in these chips flow is laminar.

1.1.1. Mixing in microchemostat devices

A major consequence of laminar flow is that mixing will only occur due to diffusion, since bulk mixing relies on some type of turbulent flow. An important way to view the effect of diffusion in a microchemostat is to

Table 14.1 Typical physical parameter values for microchemostat devices used in synthetic biology

Parameter	Variable	Value	Units
Density of water	ρ	1×10^3	$kg\ m^{-3}$
Viscosity of water (dynamic)	μ	1×10^{-3}	$kg\ m^{-1}\ s^{-1}$
Hydraulic diameter	D_{h}	1×10^{-4}–1×10^{-6}	m
Mean fluid velocity	υ	1×10^{-4}–1×10^{-6}	$m\ s^{-1}$
Reynolds number	*Re*	1×10^{-2}–1×10^{-6}	N/A

consider the diffusion length scale, which describes the one dimensional distance a molecule can be expected to travel in a given amount of time. The relationship is given as (Beebe *et al.*, 2002)

$$d^2 = 2Dt, \tag{14.2}$$

where d is the distance a molecule travels, D is the molecule's diffusion coefficient, and t is the elapsed time. Since the distance traveled by a molecule is proportional to the square root of the elapsed time, diffusion will become more important at smaller length scales. For a specific example consider the Atto 655 dye, expected to diffuse 10 μm in 0.1 s, but taking over 1000 s to diffuse 1 mm. Diffusion coefficients for representative molecules often encountered in microchemostats are given in Table 14.2.

As expected the diffusion coefficient tends to increase with increasing molecular weight and this is important to compensate for when using a tracer dye to monitor nutrient transport. For example, as can be seen in Table 14.2, one should be careful using Atto 655 dye as a surrogate for bovine albumin transport, or any high molecular weight protein, due to their order of magnitude difference in diffusion coefficients. Another important concept regarding diffusive transport in microchemostats is the Péclet number, which is another dimensionless quantity given by

$$Pe = \frac{vL}{D}, \tag{14.3}$$

where L is the is the characteristic length scale, which in a microchemostat corresponds to the channel width. The Péclet number represents a ratio between advection and diffusion of a substance. Conceptually it can be thought of as the ratio of how far "downstream" a molecule is carried versus how far it diffuses across the channel in a given unit of time. In microfluidic systems reliant on diffusive mixing, knowledge of the Péclet number is critical for designing functional microchemostats. To determine the length

Table 14.2 Diffusion coefficients for ions and molecules commonly used in microfluidic chemostats

Name	Molecular weight (Da)	Diffusion coefficient ($cm^2\ s^{-1}$)	Reference
Sodium ion (Na^+)	22.98	1.3×10^{-5}	Lide (2004)
Glucose	180.16	6.7×10^{-6}	Lide (2004)
Atto 655 dye	528	4.3×10^{-6}	Dertinger *et al.* (2007)
Bovine albumin	67,000	5.9×10^{-7}	Young *et al.* (1980)

required (Δy_{m}) for effective diffusive mixing of a substance, the following relationship is useful (Stroock *et al.*, 2002):

$$\Delta y_{\mathrm{m}} \approx \frac{vL^2}{D}, \tag{14.4}$$

$$\Delta y_{\mathrm{m}} \approx PeL. \tag{14.5}$$

Thus, Eq. (14.4) indicates that for two channels with equivalent Péclet numbers, the narrower one will require a shorter length for complete mixing. This statement is important because often in the design of microchemostats one wishes to carefully manage the volumetric flow rate to ensure optimal reagent use. As derived in the next section, there is often a combination of parameter values for the dimensions of a channel which result in the same resistance (and hence the same volumetric flow rate for a given pressure gradient). Often many of these parameter values will result in the same Péclet numbers as well. For example, two channels, one with a twofold greater width and a twofold smaller length will have the same resistance and equivalent Péclet numbers. However, the length required for diffusive mixing will differ as described by Eq. (14.4) and this is important to consider in the design.

1.1.2. Calculating flow rates and pressure drops

While there has been some debate as to whether the general Navier-Stokes equation is applicable to the small scale of microfluidic devices, recent work has demonstrated that this is so and suggested that previously observed deviations were due to experimental error (Bao and Harrison, 2006). As a consequence of laminar flow in a microchemostat chip, the Navier-Stokes equations reduce to a simple analog of Ohm's law. This equation is

$$\Delta P = QR, \tag{14.6}$$

where ΔP is the pressure drop across a channel, Q is the volumetric flow rate, and R is the resistance of the channel. This allows the simple calculation of flow rate in a chip as a function of external pressure and channel resistance. To calculate R the dimensions of the channel have to be considered. For cylindrical channels, the resistance is given by the Hagen–Poiseuille equation, equal to

$$R = \frac{8\mu L}{\pi r^4}, \tag{14.7}$$

where μ is the fluids viscosity, L is the length of the channel, and r is the radius of the channel. For the rectangular channels usually encountered in microchemostats, this equation has to be modified somewhat, taking into consideration the ratio between the width of the channel and the height, known as the aspect ratio. For channels with a low aspect ratio (w ≈ h), the equation for channel resistance is given by (Beebe *et al.*, 2002)

$$R = \frac{12\mu L}{wh^3}\left[1 - \frac{h}{w}\left(\frac{192}{\pi^5}\sum_{n=1,3,5}^{\infty}\frac{1}{n^5}\tan h\left(\frac{n\pi w}{2h}\right)\right)\right]^{-1}, \tag{14.8}$$

while Eq. (14.8) appears complicated, in practice it is not too difficult to work with if desired. Note the $1/n^5$ term in the infinite sum. Since this term quickly approaches zero for increasing n, only the first five terms need to be considered to get a reasonable approximation. However, this equation can be further reduced when using a chip with high aspect ratio channels ($w \ggg h$). Usually this is the case, as typical channel heights in a yeast or *E. coli* chip will be in the range of 5–10 μm while the width will range from 60 to 300 μm. In this situation, the bracket term in Eq. (14.8) will tend to zero and the resistance simply becomes

$$R = \frac{12\mu L}{wh^3}. \tag{14.9}$$

Using Eqs. (14.6) and (14.9) the flow rates in a microfluidic chip can be solved for in a straightforward manner, using methods similar to nodal analysis for electrical circuits. First consider a sample microfluidic chip depicted in Fig. 14.1A and B, which is shown diagrammatically in stick form to make analysis easier. For each internal node labeled a–d in the figure, the flow entering must equal the flow exiting due to the conservation of mass. This is analogous to Kirchhoff's first law for electrical circuits. Thus for all nodes in the device

$$\sum_{k=1}^{n} Q_k = 0, \tag{14.10}$$

where n is the number of channels joining at the node. Furthermore, note that the system will be solved once the internal pressures at the nodes are determined, since the flow rates between nodes can be found from Eq. (14.6). We will use the system described in Fig. 14.1 as an example to demonstrate how to solve such a problem. The first step would be to come up with a diagram similar to Fig. 14.1A, with the external ports and internal nodes clearly labeled.

Next label the current flow directions with arrows between nodes as shown in Fig. 14.1B, while making sure to obey the conservation of mass.

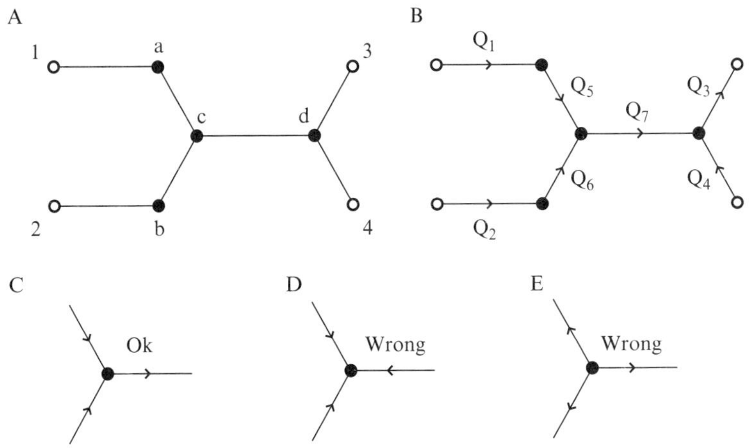

Figure 14.1 Overview of how to conceptually set up microfluidic flow problems. (A) Stick diagram of a conceptual microfluidic device. External ports with specified pressures (open circles) are labeled 1–4. Internal junctions (whose pressures will be solved for, closed circles) are labeled a–d. (B) Same diagram as in part A, except the port and junction numbers are removed for clarity. Volumetric flows to be solved for are given by Q_{1-7}. (C–E) Overview of the correct way to set up flow directions in a microfluidic junction, while obeying the conservation of mass. Part C has the correct setup, containing both inlets and an outlet. Part D is incorrect since there are only inlets. Part E is also incorrect since there are only outlets.

Note that you may not know the direction of flow beforehand (in fact that may be why you are doing this exercise), however, this does not matter initially. As long as the conservation of mass is followed the system can be solved properly. If your initial flow direction guess is incorrect, its solution will be negative, indicating the opposite is the true direction of flow. After this step is complete, develop a system of equations describing the flow in each node. For the example system

$$Q_1 = Q_5, \quad Q_2 = Q_6, \tag{14.11a}$$

$$Q_7 = Q_5 + Q_6, \quad Q_3 = Q_7 + Q_4. \tag{14.11b}$$

Next use Eq. (14.6) to substitute the pressure and resistance for the current

$$\frac{P_1 - P_a}{R_1} = \frac{P_a - P_c}{R_5}, \quad \frac{P_2 - P_b}{R_2} = \frac{P_b - P_c}{R_6}, \tag{14.12}$$

$$\frac{P_c - P_d}{R_7} = \frac{P_a - P_c}{R_5} + \frac{P_b - P_c}{R_6}, \quad \frac{P_d - P_3}{R_3} = \frac{P_c - P_d}{R_7} + \frac{P_4 - P_d}{R_4}. \tag{14.13}$$

Since Eqs. (14.12) and (14.13) contain cumbersome fractions, it is useful to define the conductance G as the inverse of the resistance R,

$$G = \frac{1}{R}. \tag{14.14}$$

By substituting the conductance for the resistance in Eqs. (14.12) and (14.13) we get the following:

$$G_1(P_1 - P_a) = G_5(P_a - P_c), \tag{14.15}$$
$$G_2(P_2 - P_b) = G_6(P_b - P_c), \tag{14.16}$$
$$G_7(P_c - P_d) = G_5(P_a - P_c) + G_6(P_b - P_c), \tag{14.17}$$
$$G_3(P_d - P_3) = G_7(P_c - P_d) + G_4(P_4 - P_d). \tag{14.18}$$

Expanding and rearranging we get

$$G_1P_1 = (G_1 + G_5)P_a - G_5P_c, \tag{14.19}$$
$$G_2P_2 = (G_2 + G_6)P_b - G_6P_c, \tag{14.20}$$
$$0 = G_5P_a + G_6P_b - (G_5 + G_6 + G_7)P_c + G_7P_d, \tag{14.21}$$
$$-G_3P_3 - G_4P_4 = G_7P_c - (G_3 + G_4 + G_7)P_d. \tag{14.22}$$

Or in matrix form

$$\begin{bmatrix} G_1P_1 \\ G_2P_2 \\ 0 \\ -G_3P_3 - G_4P_4 \end{bmatrix} = \begin{bmatrix} G_1 + G_5 & 0 & -G_5 & 0 \\ 0 & G_2 + G_6 & -G_6 & 0 \\ G_5 & G_6 & -G_5 - G_6 - G_7 & G_7 \\ 0 & 0 & G_7 & -G_3 - G_4 - G_7 \end{bmatrix} \begin{bmatrix} P_a \\ P_b \\ P_c \\ P_d \end{bmatrix}. \tag{14.23}$$

Equation (14.23) is a linear system which can be either solved manually or with the aid of a computer program such as Excel or Matlab. Of course the above procedure can become tedious, especially for larger microchemostat chips and a method which lends itself to automation would be preferred. To develop such a system first rearrange Eqs. (14.11a) and (14.11b) to put all currents on the LHS

$$Q_1 - Q_5 = 0, \quad Q_2 - Q_6 = 0, \tag{14.24a}$$
$$-Q_5 - Q_6 + Q_7 = 0, \quad Q_3 - Q_4 - Q_7 = 0. \tag{14.24b}$$

Now arrange Eqs. (14.24a) and (14.24b) into matrix form

$$\begin{bmatrix} 0 \\ 0 \\ 0 \\ 0 \end{bmatrix} = \begin{bmatrix} 1 & 0 & 0 & 0 & -1 & 0 & 0 \\ 0 & 1 & 0 & 0 & 0 & -1 & 0 \\ 0 & 0 & 0 & 0 & -1 & -1 & 1 \\ 0 & 0 & 1 & -1 & 0 & 0 & -1 \end{bmatrix} \begin{bmatrix} Q_1 \\ Q_2 \\ \vdots \\ Q_7 \end{bmatrix}, \tag{14.25}$$

which can be expressed as

$$0 = C\vec{q}, \tag{14.26}$$

where C is an $i \times j$ matrix called the connectivity matrix for a chip with i nodes and j channels. The C matrix is unique for each chip and should be specified from a graph of the chips architecture. The $\vec{q}$ is a vector of length j representing the flows in the chip. Since $\vec{q}$ is unknown we need to use Eqs. (14.6) and (14.14) to substitute flows for pressures and conductivities

$$\begin{bmatrix} Q_1 \\ Q_2 \\ Q_3 \\ Q_4 \\ Q_5 \\ Q_6 \\ Q_7 \end{bmatrix} = \begin{bmatrix} G_1(P_1 - P_a) \\ G_2(P_2 - P_b) \\ G_3(P_d - P_3) \\ G_4(P_4 - P_d) \\ G_5(P_a - P_c) \\ G_6(P_b - P_c) \\ G_7(P_c - P_d) \end{bmatrix} = \begin{bmatrix} G_1P_1 \\ G_2P_2 \\ -G_3P_3 \\ G_4P_4 \\ 0 \\ 0 \\ 0 \end{bmatrix} + \begin{bmatrix} -G_1P_a \\ -G_2P_b \\ G_3P_d \\ -G_4P_d \\ G_5P_a - G_5P_c \\ G_6P_b - G_6P_c \\ G_7P_c - G_7P_d \end{bmatrix}. \tag{14.27}$$

Thus the flow vector can be split into two vectors as shown in the RHS of Eq. (14.27). The first vector contains only known values, being the external pressures and conductances of the channels connected to these ports. The second vector contains known conductances and the unknown internal node pressures which we are interested in solving for. Separating the conductances from the pressures we get

$$\vec{q} = \begin{bmatrix} G_1 & 0 & 0 & 0 \\ 0 & G_2 & 0 & 0 \\ 0 & 0 & -G_3 & 0 \\ 0 & 0 & 0 & G_4 \\ 0 & 0 & 0 & 0 \\ 0 & 0 & 0 & 0 \\ 0 & 0 & 0 & 0 \end{bmatrix} \begin{bmatrix} P_1 \\ P_2 \\ P_3 \\ P_4 \end{bmatrix} + \begin{bmatrix} -G_1 & 0 & 0 & 0 \\ 0 & -G_2 & 0 & 0 \\ 0 & 0 & 0 & G_3 \\ 0 & 0 & 0 & -G_4 \\ G_5 & 0 & -G_5 & 0 \\ 0 & G_6 & -G_6 & 0 \\ 0 & 0 & G_7 & -G_7 \end{bmatrix} \begin{bmatrix} P_a \\ P_b \\ P_c \\ P_d \end{bmatrix} \tag{14.28}$$

or

$$\vec{q} = G\vec{s} + H\vec{p}, \tag{14.29}$$

where G is a $j \times k$ matrix of j channels and k external ports containing conductance values, $\vec{s}$ is a k length vector specifying the known external port pressures, H is a $j \times l$ matrix of j channels and l internal nodes containing conductance values and $\vec{p}$ is a l length vector containing the unknown internal port pressures. Combining Eqs. (14.26) and (14.29) we get

$$0 = C(G\vec{s} + H\vec{p}), \tag{14.30}$$

$$-CG\vec{s} = CH\vec{p}, \tag{14.31}$$

$$\vec{t} = I\vec{p}, \tag{14.32}$$

where $\vec{t} = -CG\vec{s}$ and $I = CH$. Note that Eq. (14.32) is the same as Eq. (14.23) and can be solved in the same ways. To solve the flow profiles for an arbitrary chip, the C, G, and H matrices need to be specified, which can be done once the connectivity and channel geometries are decided upon. To automate this process our lab uses a custom matlab script, written by a former graduate student, called moca. This program has been extended to calculate how the pressure in each external port changes in time as fluid flows from the inlet ports to the outlets.

Alternatives to nodal analysis are commercial software package employing finite element techniques to solve for the flows in a more exact manner. An example of such a software package is the program Comsol, which contains an internal software package explicitly set up to solve microfluidics problems. For the design of microchemostats, this level of computation can be helpful for certain parts of the chip. For example, Comsol, unlike nodal analysis techniques, can model the diffusive transport of nutrients in complicated geometries such as cell traps or junctions. Moreover transient behavior of the chip, including how a cell chamber will respond to pressure surges, can be easily modeled in Comsol but not using nodal analysis techniques. As an additional advantage, Comsol has the ability to create models directly from Autocad files, which can save a considerable amount of time. However, software programs such as Comsol are quite expensive and nodal analysis techniques are generally fine for designing basic microchemostats.

1.1.3. Designing a microchemostat chip

To design a microchemostat device one has to know a little about the overall fabrication process. The complete details will described in the fabrication section, but we will give a brief description here. The general process is known as soft-photolithography, originally developed for the semiconductor industry. When used for microchemostats, soft-photolithography creates reusable master molds with chemicals known as photoresists. Photoresists are viscous chemicals spun on silicon wafers to very precise heights. When exposed to ultraviolet (UV) light, the photoresist cross-links and becomes resistant to developer solvent, while the uncross-linked photoresist remains susceptible. To make a microchemostat, a negative image of the device's features is placed between the photoresist and the UV light source. An example of such a mask is shown in Fig. 14.2. When exposed, the UV light will pass through the clear sections containing the device's features, while the dark regions will prevent the background from being cross-linked. After the uncross-linked photoresist is removed with developer, the process is repeated for the next layer. To align multiple layers, an aptly named mask aligner machine is used. This machine contains a microscopy setup so alignment patterns between the previous photoresist layer and the current mask can be viewed. Once all layers have been completed, the wafer can be used to produce an almost unlimited amount of microchemostat devices.

When designing a device, the first step is to layout the architecture in a vector graphics software program such as Autocad. While it is possible to use other programs, such as Adobe Illustrator, in general Autocad is superior since it is designed for precision fabrication. Furthermore companies offering extremely high resolution mask printing generally require Autocad files. Student versions of Autocad are reasonably priced and offer more capability than is necessary for designing microchemostats. During the design stage, one needs to decide how many different channel heights will be in the device. For example, the cell trap might be 3.5 μm while the channel network is 10 μm, as is often the case for yeast chips. All features with the same height should be on the same layer in the Autocad file to make work easier (see Fig. 14.2).

When designing a chip with multiple layers, care must be taken to provide an accurate method for alignment during fabrication. During the alignment process one will need to look through the mask at the pattern from a previous layer and adjust the controls so the current mask will perfectly overlap. As shown in Fig. 14.3 there are three degrees of freedom which need to be manipulated during the alignment process: xy translation and rotation. To make sure the wafer and mask are in perfect alignment, two locations must be viewed on the wafer to compensate for small errors in rotation. The center of the mask essentially determines the axis of rotation. The further away the two locations are from the center (and each other), the easier small errors in rotation will be to see.

In Fig. 14.2A the alignment locations are in the lower left corner and upper right corner of the mask, the furthest possible from the center. The alignment features shown in Fig. 14.3 are designed to have coarse and fine features to speed the alignment process and work quite well in practice. To align the patterns, one adjusts the mask aligner controls until the points of the squares meet in all locations. Note that a separate alignment pattern will

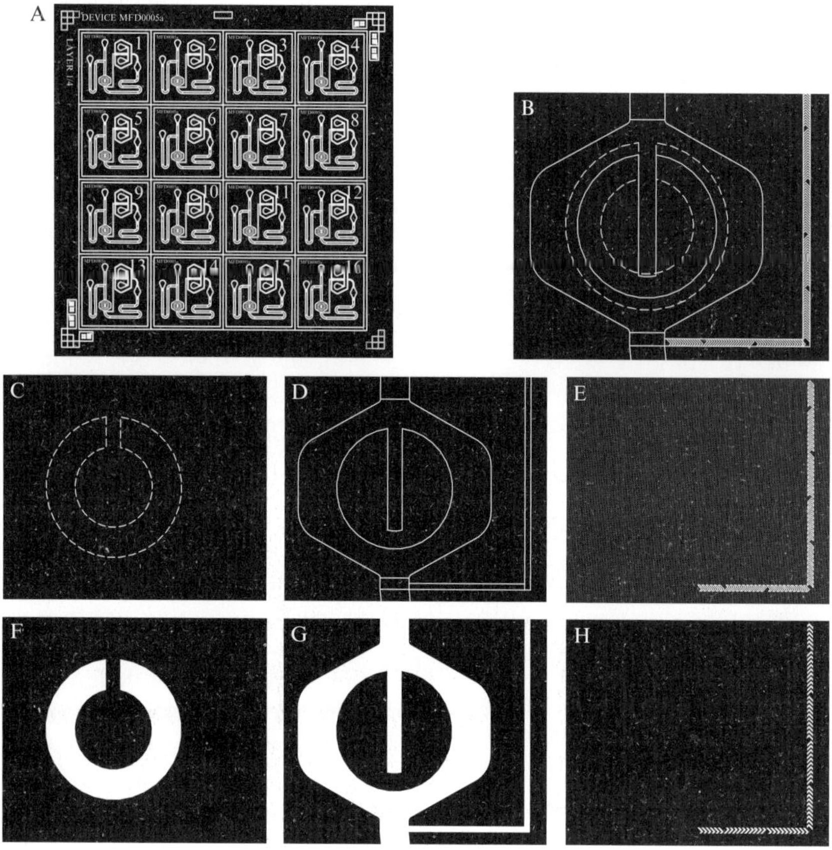

Figure 14.2 Overview of the mask design process for microchemostat devices. (A) Overview of an Autocad file with the features of the microchemostat shown in white. Note the alignment features in the lower left and upper right corners. Each chip is individually numbered so those defective can be tracked. (B) Close-up of the cell trap region from the Autocad file shown in part A. This region contains features of three different heights, which are in different layers of the Autocad file. The cell trap will be of height 3.5 μm and is shown with dashed lines. The central chamber will be 10 μm and is shown with solid lines. The staggered herringbone mixers (SHM) will be of height 3 μm above the 10 μm mixer channel height for a total of 13 μm. Note the overlap between layers. When layers meet there should always be an overlap to compensate for small errors in mask alignment. (C-E) Each layer from part B is shown individually, with the cell trap in part C, the cell chamber in part D, and the SHM features in part E. When sent for printing, the layers should be displayed individually as is shown here. (F–H) Depiction of what the mask will look like after printing. The features of the device will be clear (white in the figure) to allow UV light to pass, while the background is black.

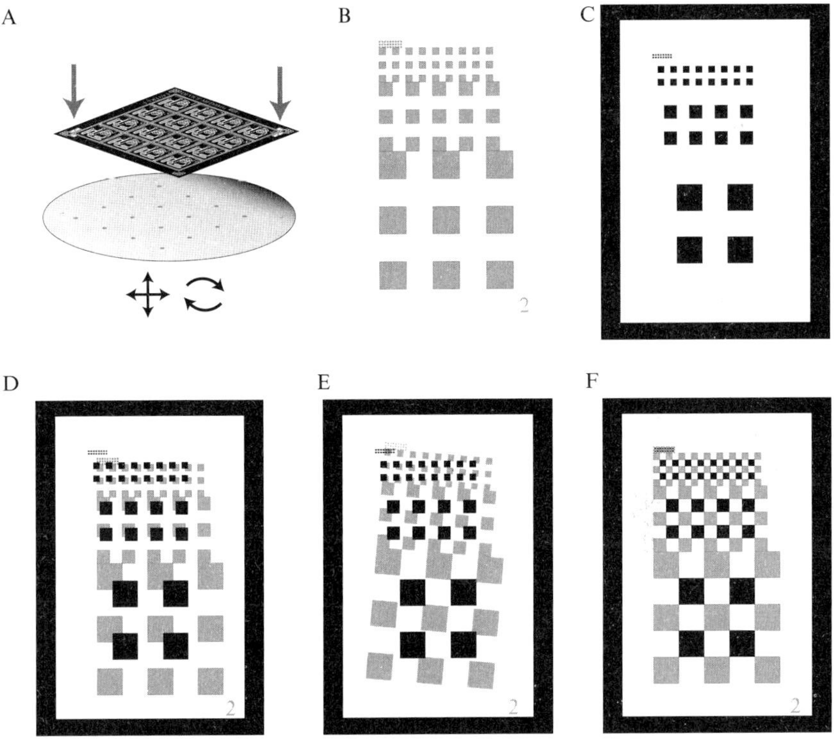

Figure 14.3 Overview of the alignment pattern for microchemostat devices. (A) Overview of the alignment process, with a mask shown above a wafer containing a previously deposited photoresist layer with alignment patterns. The mask aligner will have controls to compensate for both translation and rotation (bottom arrows). The arrows pointing down on the mask show the alignment pattern location. (B) Alignment pattern present on the wafer from the previous photoresist deposition. Each layer will require a separate alignment pattern; the layer number is shown in the lower right. The pattern is composed of sets of squares whose sides are reduced by half in each iteration. (C) Alignment pattern present on the mask. The clear window surrounding the squares allows the fabricator to view the pattern from the previous layer. The objective is to make the points of the squares from the mask and the previous layer touch. (D) Mask and wafer out of alignment by *xy* translation only. (E) Mask and wafer out of alignment by rotation only. (F) Mask and wafer in perfect alignment.

be necessary for each layer other than the first, since the mask's viewing window will cross-link the photoresist (and therefore remove the wafer's alignment pattern for the layer) after each alignment and exposure.

When considering a device design and alignment pattern, it is critical that the thinner layers are fabricated before the thicker ones. For example, a 3.5-μm layer should always be fabricated before a 10-μm layer. We have found that if thicker layers are fabricated first, the later layers will spin unevenly, since the larger features from the previous layer prevent an even coating of the wafer. Furthermore, it is important not to increase the height too greatly between consecutive layers, since this limits the contrast in the mask alignment process. Recall that mask alignment occurs after spinning the current (uncross-linked) photoresist layer, which covers all previous (cross-linked) photoresist layers. Fortunately, the wafer's alignment pattern on the previously cross-linked photoresist layer can usually be seen through the current layer. However, if the height ratio between the two layers is greater than about 5:1, the contrast becomes so poor that it is difficult to see the wafer's alignment pattern. In general, we try and limit the height ratio to 3:1, since mask alignment is generally the most difficult and frustrating part of fabrication.

Once the alignment strategy is settled upon, the device features can be laid out in Autocad. For this purpose simple rectangles are usually sufficient, but arc segments can be used if more complex shapes are desired. We have found that curved sections are superior for cell containing channels, since they prevent clogging. For areas of the chip not expected to contain cells, rectangular segments meeting at sharp corners are fine. When designing channels, all features should be closed objects in Autocad, there should be no open segments. While resulting in lines across channels, these will not be printed since lines are considered to be of infinitesimal thickness by the printer and only closed regions are recognized (compare Fig. 14.2D and G). Ensuring that all regions are closed in Autocad will not only make printing easier, but also facilitates importing into Comsol and Illustrator.

In general when laying out features, one must consider the tradeoff between compacting the device into as small a space as possible and maintaining usability. For example, placing two ports closer than 2 mm is not advisable since it makes it extremely difficult to plug in the port lines upon setup. Furthermore having a channel pass closer than a 1-mm to a port should also be avoided since it can be damaged if the port hole is punched incorrectly. Along these same lines, there should be at least 1 mm between a feature and the edge of the chip, so when the PDMS slab is diced into individual units no features are damaged.

In addition, when two layers are contiguous there should be some overlap between them to compensate for the small errors in alignment that inevitably occur. For example, in Fig. 14.2B the cell trap layer overlaps the cell chamber layer. If the layers were designed with no overlap a small alignment error could create a gap between them resulting in a nonfunctional chip. Even with the alignment patterns described in Fig. 14.3 and a meticulous alignment procedure, small errors will occur and can be compensated for with layer overlap. When layers overlap the total height is usually a smooth transition between the height of the thicker layer alone and the sum of the heights of the overlapping layers. As shown in Fig. 14.2B the cell chamber wall starts out at 10 μm and gradually increases to $\sim$14 μm in the overlapping area. This phenomenon should be remembered when modeling the flow profile of a device in Comsol for example, since a ~40% change in height due to overlap will have a large effect on the channel's resistance.

Another common mistake results from layers unintentionally intersecting due to small alignment errors. This can create fluidic "short circuits" and nonfunctional chips. The solution here is to again make sure an adequate margin is present between nonintersecting layers to compensate for fabrication problems. Most importantly, keep in mind the concept of tolerances. While the feeling for this comes from experience, always assume that some fabrication error is inevitable rather than trying to come up with the most beautiful design in Autocad. The best design will be one that can tolerate some fabrication error and still work properly, even if it is not the most "compact" design. The size of the channels is also affected by these same concepts. We have found that channel widths smaller than 60 μm should be avoided since they are prone to clogging with debris that can enter the chip (often residual PDMS). Moreover long channels should generally be 10 μm or more in height, also to prevent clogging.

Of course the ultimate limitation for microfluidic design is the resolution of the printer making your masks. This limit usually comes into play before that imposed by the UV light source or the photoresist. We use a company named CAD/Art for mask printing which has a 20,000-dpi printer. While this is normally adequate for microchemostats, higher resolution options used in the semiconductor industry are available at far greater expense. Using this process, we have been able to make features separated by as little as 13 μm *as long as they are on the same layer.* However, even this is dependent on the type of photoresist used. For example the spatial resolution of a thinner photoresist, like that used to make a 10-μm layer, is generally greater than that of a thicker resist, used for making a 35-μm layer. General guidelines for recommended channel dimensions are given in Table 14.3. Note it is certainly fine to make channels having dimensions other than those given in the table and for specialized features (like high resistance cell

Table 14.3 General guidelines for channel dimensions in microchemostat chips

Channel type	Organism	Width range	Height range
General flow network (no cells)	Any	60–100 μm	10–15 μm
High flow channel (no cells)	Any	300–400 μm	20–45 μm
General cell channels	*E. coli*	150–300 μm	6–15 μm
	yeast	200–300 μm	10–15 μm
	mammalian	200–300 μm	25–35 μm
Cell trap	*E. coli*	Varies	1 μm
	Yeast	Varies	3.55 μm
	Mammalian	Varies	25 μm

feeding channels) this may be necessary. For the normal fluidic "backbone" of the chip, the channel dimensions listed in Table 14.3 should be fine.

While the general guidelines listed so far should be useful for creating a microchemostat device, as a case study we will describe our design process for an updated dial-a-wave chip. The device, called $MFD005_a$, was designed as an improved version of the chip described in Bennett *et al.* (2008). The chip is designed to grow cells reliably in a monolayer and cope with high growth by flushing excess cells into a waste port. The chip is also designed to generate arbitrary, time varying inducer concentrations, so the cell's response to a dynamic environment can be recorded. Often we use the chip to generate arbitrary waveforms, such as sine waves, square waves, or waves having a random period component. The waves generated by the device have high temporal accuracy and the chip is easy to use. An overview of the device is shown in Fig. 14.4. The chip has five external ports, which is a reduction from eight in the Bennett chip. Reducing ports saves on consumables and eases setup, so finding the minimum number necessary to produce a working chip should always be a design goal. The chip is designed to use hydrostatic pressure and therefore no pumps are required of any kind for operation. We have found that hydrostatic pressure gives the most reliable, steady, and cost effective means of controlling the pressures in a microchemostat device. In a later section, we will describe our use of linear actuators to alter the inlet hydrostatic pressure of our device and why this is advantageous compared to other means such as syringe pumps.

The role of each port of the $MFD005_a$ chip is given in Table 14.4. When an experiment is running, fluid will enter from ports 1 and 2 which meet at the dial-a-wave junction (Fig. 14.4B). The DAW junction has two inlets and three outlets. As described in a later section, the ratio of the inputs from port 1 and 2 leaving the junction to the cell chamber is determined by each port's pressure. Excess fluid is diverted through a shunt network to port 3, which is a waste port. Fluid leaving the central fork of the junction for the cell chamber travels through a long channel where it is mixed into a uniform concentration by staggered herringbone mixers (SHM). The ingenious SHM mixers (as shown in Fig. 14.4C) are designed to induce a corkscrew effect in the fluid stream and increases the surface area available for mixing (Stroock *et al.*, 2002; Williams *et al.*, 2008). Since mixing only occurs due to diffusion in a microchemostat, as mentioned in Section 1.1.1, this increase in surface area will logarithmically reduce the length of a channel necessary for uniform mixing.

Even with the help of SHM features, mixing still requires a length which depends on the log of the Péclet number (Stroock *et al.*, 2002). Thus a central question when designing our device was how long to make the mixing channel from the DAW junction to the cell port. If the channel were too short, the two inputs would not be completely mixed, resulting in a nonuniform and uneven concentration profile over the cell culture.

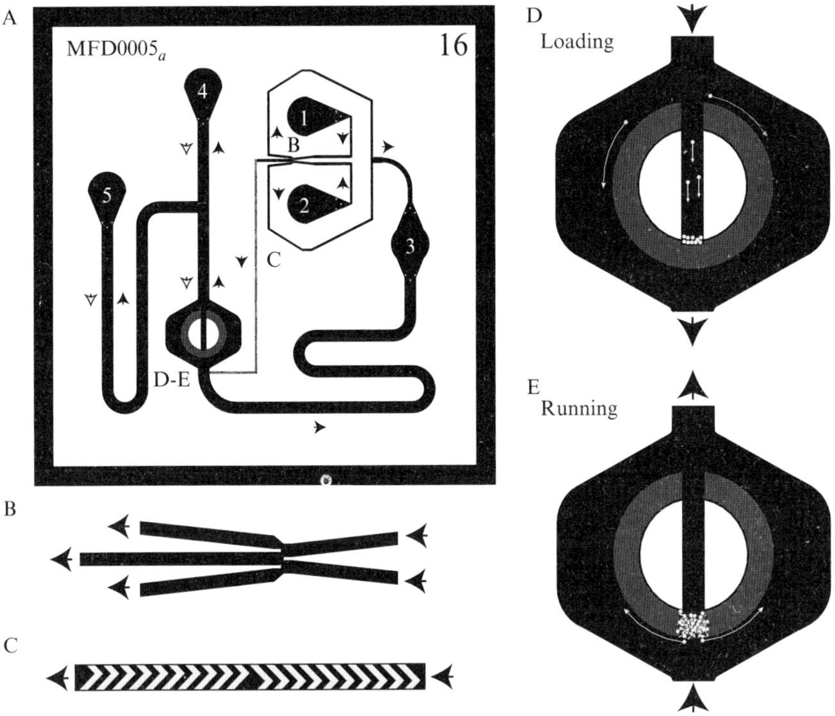

Figure 14.4 Overview of the MFD005$_a$ chip and components. (A) Overview of the MFD005$_a$ chip's architecture. Flow directions in each segment during running conditions are given by black arrows, during loading conditions by white arrows. Note that only flow from ports 4, 5, and across the cell chamber changes direction during loading. Letters represent locations of the features described in other parts of the figure. External ports are numbered 1–5. Each port is described in Table 14.4. (B) Depiction of the DAW junction. Flow direction is indicated by the black arrows. The two inlets on the right come from ports 1 and 2. The flow from the inlets converges in a ratio dependent on the inlet pressures of each. The middle fork of the junction leads to the cell chamber while the two outer forks lead to port 3, the cell and shunt waste port. (C) Depiction of the staggered herringbone mixers (SHM) which reduce the channel length required for mixing. These mixers immediately follow the DAW junction and continue until just before the cell chamber. (D) Overview of trap region of the MFD005$_a$ chip under loading conditions. This trap is known as the yeast doughnut trap. Black region represents the cell chamber with a height of 10 μm. Gray region is the actual cell trap, with a height of 3.525 μm. White circles represent cells entering from the cell port and either passing around the trap to the cell and shunt waste (port 3), or entering the central channel and moving to the trap entry barrier. The yeast cells are slightly too large to move into the trap directly without "flicking" the cell line to assist in their entry. (E) Cell trap upon running of an experiment. Cells begin to grow in the trap and the colony expands (black arrows). Eventually the colony fills up the gray region near where they were loaded. The growth of the cells will force some out of the trap into the outer channel where they will be efficiently carried away to the waste port (white arrows). Over the course of the experiment the cell colony will expand to fill the entire trap.

Table 14.4 Role and pressures for each port in the MFD005$_a$ device

Port	Description	Contents	Run inH_2O	Load inH_2O
1	Inlet 1 for DAW	Media + inducer + tracking dye	25	25
2	Inlet 2 for DAW	Media	25	25
3	Cell and shunt waste	dH_2O	5.5	5.5
4	Alternate outlet	dH_2O	6	17
5	Cell port	Media + cells	6	18

All pressures are given in inH_2O above the height of the microscope stage.

However, making the channel too long is also disadvantageous since it increases the delay time for a signal to propagate the length of the channel. To find the optimum channel length, we require knowledge of the flow velocities as a function of the external port pressures. This is a good example of the usefulness of the modeling techniques mentioned in Section 1.1.2. Using nodal analysis or Comsol it is easy to determine the flow rates and hence the Péclet number for various substances and flow regimes. We performed just such an analysis when designing the MFD005$_a$ device to determine the necessary channel length for efficient mixing, shown in Fig. 14.5.

After mixing, fluid from ports 1 and 2 enters the cell chamber and proceeds to the outlet ports 4 and 5. Fluid also enters a diversion channel and exits at port 3. By controlling the height of port 3 relative to ports 4 and 5, one can set the ratio of fluid passing through the chamber versus exiting through the diversion channel. Modulation of this diversion ratio is important for controlling the flow velocity across the cell chamber. For example, say you wanted to minimize the flow velocity in the cell chamber why still retaining functionality of the DAW junction. Without a diversion channel you could lower the height of the input ports 1 and 2 relative to 4 and 5 and reduce the flow velocity in the cell chamber. However, this would also reduce the flow velocity in the mixing channel between the DAW junction and the cell chamber. This reduction in mixing channel velocity would increase the delay time for fluid transit and negatively impact the chips function. With a diversion channel, an alternative is to maintain the height difference between ports 1–2 and 4–5 and instead lower port 3. This would increase the ratio of fluid entering the diversion channel and hence lower the fluid velocity in the cell chamber. This is another example of flow modeling's usefulness, since the diversion channel's length is critical for determining the amount of fluid diverted for a given height change.

Modeling also allowed us to solve a problem with flow reversal (backflow) in the diversion channel, which would sometimes occur over the course of an experiment in a previous version of this device. The plot in

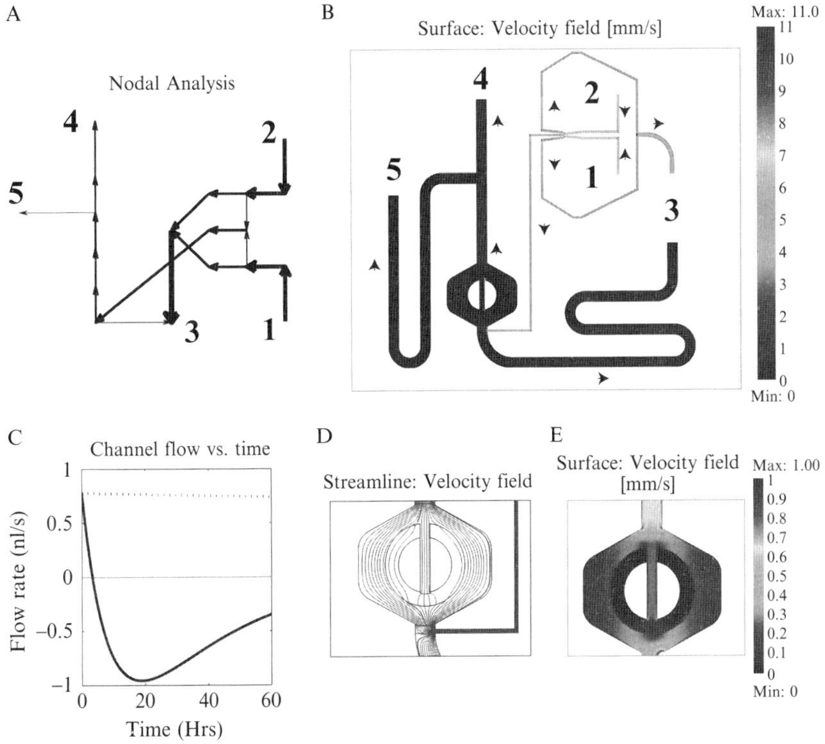

Figure 14.5 Comparison of a nodal analysis tool written in Matlab (moca) and Comsol (finite element analysis package). (A) Graph from the moca matlab script depicting flows for the device pictured in Fig. 14.4A. The arrow thickness and direction represent the volumetric flow in a channel section of the device. Numbers are the external ports of the device. (B) Flow velocity through the same chip modeled using Comsol. The magnitude of the velocity is given by the channel's color, while the direction is indicated by the black arrows. Numbers are again the external ports. The $MFD005_a$ geometry was loaded directly from Autocad, simplifying setup. (C) Flow profile of a channel section over the course of an experiment. In previous designs, we have had problems with backflow problems over the course of an experiment, as the fluid level in the external ports is altered by flow. Modeling an experiment's flow profile using nodal analysis helped to solve these problems, resulting in a redesign of the diversion channel's dimensions and using larger syringes. The blue line represents fluid flow using 1-ml syringes and the red dashed line 60-ml syringes. (D) Streamline plot showing the path fluid particles take upon moving through the cell trap. While this plot was generated under running conditions, the streamlines are very similar for loading conditions (the direction of flow of course is opposite). Note that only about one fourth of the flow enters the central channel, most flow is directed around the trap. Hence, when loading, most cells will not enter the trapping region. (E) Plot of the velocity field inside the trap region. Note that the velocity is lowest inside the trap itself and considerably higher in the outer channel region. This allows nutrients to be continually replenished from the outer channel into the cell trap and helps remove cells once they outgrow the trap. (For interpretation of the references to color in this figure legend, the reader is referred to the Web version of this chapter.)

Fig. 14.5C represents a time dependent solution for the flow profile in the device's diversion channel, compensating for pressure changes due to fluid movement over the course of an experiment. The solid blue line represents the flow rate when small diameter syringes are used for the outlets. The fluid level in these syringes increases in height rapidly for a given volumetric flow. Under certain conditions this height increase can be large enough to change the flow velocity in the chip. When the blue line crosses the zero point of the y-axis, flow reversal has occurred. The red dashed line represents the same initial setup using larger diameter 60 ml syringes. These syringes undergo far less increase in height for a given volumetric flow than the smaller syringes and therefore it takes far longer (much longer than an experiment would last) to reach a flow reversal condition. The solution was reached by redesigning the diversion channel to have a greater resistance and by using larger syringes. While this model was created using nodal analysis, it could also be done in Comsol.

1.1.4. Design of an improved DAW junction

Another opportunity for flow modeling came from designing the DAW junction. As mentioned previously, this junction is designed to combine the inputs from ports 1 and 2 of the MFD005$_a$ device in a precise ratio depending on the input pressures. By controlling the input pressures as a function of time, one can generate precise waves of inducer concentration and hence expose cells to a fluctuating environment. To set the mixing ratio, the pressure of one input is increased and the other decreased by the *same* amount. By changing the input pressures in an opposing manner, the flow rate out of the junction remains constant and hence the downstream flow rates are not altered (this can be easily demonstrated using nodal analysis). Of course, by the conservation of mass, if the total outlet flow does not change, then the total inlet flow must not change either. Instead the ratio between the two inlet flows changes.

Initially one might think that a simple T-junction would suffice to reliably mix the two inputs streams. Indeed, when the output is derived nearly equally from both inputs (near a 50% mixing ratio), a T-junction works fine. However, as depicted in Fig. 14.6, a T-junction does not work well for skewed output ratios, when most of the output is coming from only one of the inputs. As an example of a skewed ratio, consider when 95% of the output is coming from input 1 and the other 5% is coming from input 2, with a total flow rate of 1 nl/s. Under these conditions the input 1 flow rate will be 0.95 nl/s, while the input 2 flow rate will be only 0.05 nl/s. Going further, for a mixing ratio of 100%, the input 1 flow rate will be 1 nl/s and the input 2 flow rate will be 0 nl/s. Of course in practice, even with the most accurate system, an entirely stagnant flow is impossible to achieve. In reality this situation represents an unstable equilibrium, prone to backflow. If attempting this with a real device, either a true 100% ratio will not be

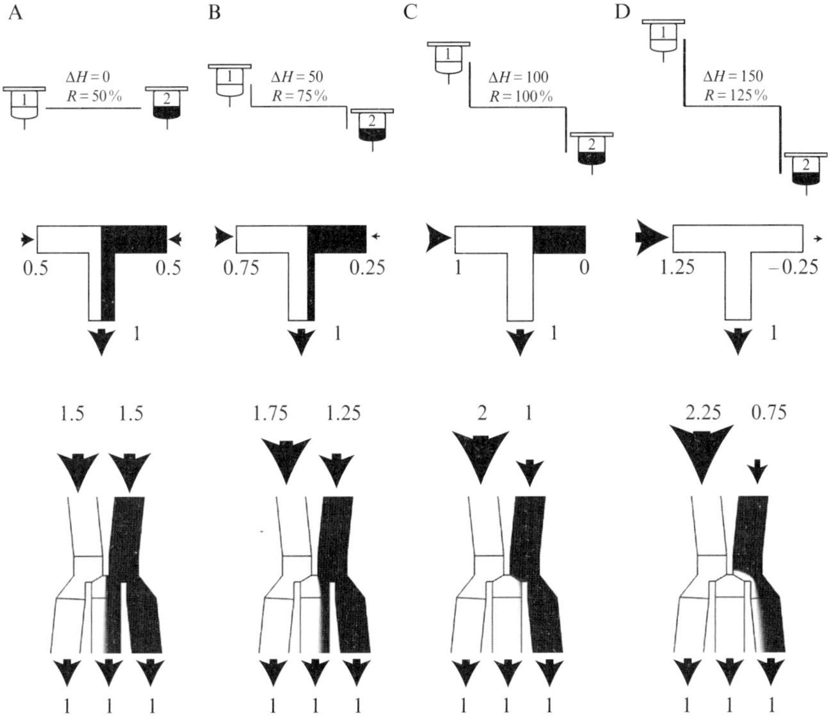

Figure 14.6 Comparison of a T-junction to our improved DAW junction for combining different source fluids in precise ratios. The figure depicts four mixing ratios from 50% to 125% and compares the performance of each junction. Note that since the system is symmetrical, flows for mixing ratios from −25% to 50% will be the reverse of those shown here. Mixing ratios above 100% or below 0% indicate complete diversion of one of the inputs to the shunt. deltaH (please write it as greek lowercase delta and english uppercase H) is an arbitrary unit of distance. (A). Mixing ratio of 50% ($R = 50\%$), corresponding to equal flows from both reservoirs. A fluorescent dye has been added to reservoir 1, displayed in white as it would be seen under the microscope. Top portion of the figure depicts the reservoirs at equal height ($\Delta H = 0$). Middle portion of the figure depicts a T-junction, each input flow is 0.5 nl/s, for a total flow of 1 nl/s. Bottom portion represents the DAW junction. Each inlet has a flow of 1.5 nl/s, for a total inlet flow of 3 nl/s. Note the smooth interface between fluid streams, as diffusion has not yet been able to cause appreciable mixing. (B) Mixing ratio of 75%. The height of the port 1 reservoir has increased while the corresponding port 2 reservoir has decreased by an equivalent amount. Both junctions continue to perform well. Note that the flow rate in inlet 1 has increased in the exact amount it has decreased in inlet 2. (C) Mixing ratio of 100%. The T-junction fails here as the flow rate in input 2 has dropped to zero. In practice, zero flow is unattainable and will likely result in a backflow situation. Note the DAW junction continues to perform well, since all flow from input 2 is directed into a shunt. (D) Mixing ratio of 125%. At this point backflow has occurred in the T-junction, as flow from port 1 begins to enter the input 2 source. In the DAW junction, the excess flow from input 1 is directed into a shunt and flow continues from input 2. Note that the output of the junction directed to the cell chamber will be the same in both C and D (center channel). This is why the output in the cell chamber seems to plateau after increasing ΔH beyond the 100% level.

achieved, or (more likely) fluid from input 1 will begin to flow into input 2. This backflow situation will result in improper mixing of the input 2 source, preventing the system from functioning properly if later switched. For example, consider if backflow had occurred for 1 h and then the system was switched, from a mixing ratio of 100% to 0%. In this situation, the residual flow from input 1 would have to flow back again before fresh input 2 media could again enter the junction. Depending on the residual flow rate from input 1 to 2, this could take a considerable amount of time.

To overcome this difficulty the chip in Bennett *et al.* (2008) contained a shunt network designed to direct some fluid from each input to a waste port at all mixing ratios, in addition to the junction outlet. This system prevents backflow because the inlet flow rates never approach zero, even for skewed outlet ratios. A comparison of a T-junction to the DAW junction used in the MFD005$_a$ device is shown in Fig. 14.6. While the shunt network solved the backflow problem, the response of the junction to input pressures was somewhat different than expected. Ideally the output response of the DAW junction should be linear, but we had found significant deviations from linearity with the Bennett device. These deviations made experimental setups sometimes difficult. To investigate the cause of these deviations we turned to modeling in Comsol.

We determined that diffusive transport between the input streams could cause significant deviations from an ideal response. Diffusion *at the junction* leads to transport of nutrients destined for the output into the shunts, altering the expected response. This deviation was especially pronounced at skewed mixing ratios similar to what we had observed. To correct these problems, we designed a new DAW junction (depicted in Fig. 14.4B) to minimize the contact distance between the two fluid streams and increase the flow velocity. These changes essentially increased the flow's Péclet number in the junction to limit diffusive mixing. Moreover we altered the shunt network compared with the Bennett design so the shunt entrances would be nearly parallel to the outlet. The idea was to minimize any changes in flow direction occurring at the junction. The performance of this new junction is shown in Fig. 14.7.

1.1.5. Calibration of the DAW junction

The junction is designed to be used in conjunction with linear actuators to physically move the input reservoirs up and down thereby altering their hydrostatic pressures. To map the height of the input 1 and 2 reservoirs to a mixing ratio of the DAW junction, we have come up with a simple calibration scheme. First we find two sets of reservoir heights corresponding to mixing ratios beyond 0% and 100%. Each set represents flow from one of the inputs being completely diverted into a shunt. Since the heights do not have to be exact at this step, it is relatively quick and easy to set up (unlike trying to find the exact 0% and 100% heights). Next we program the linear

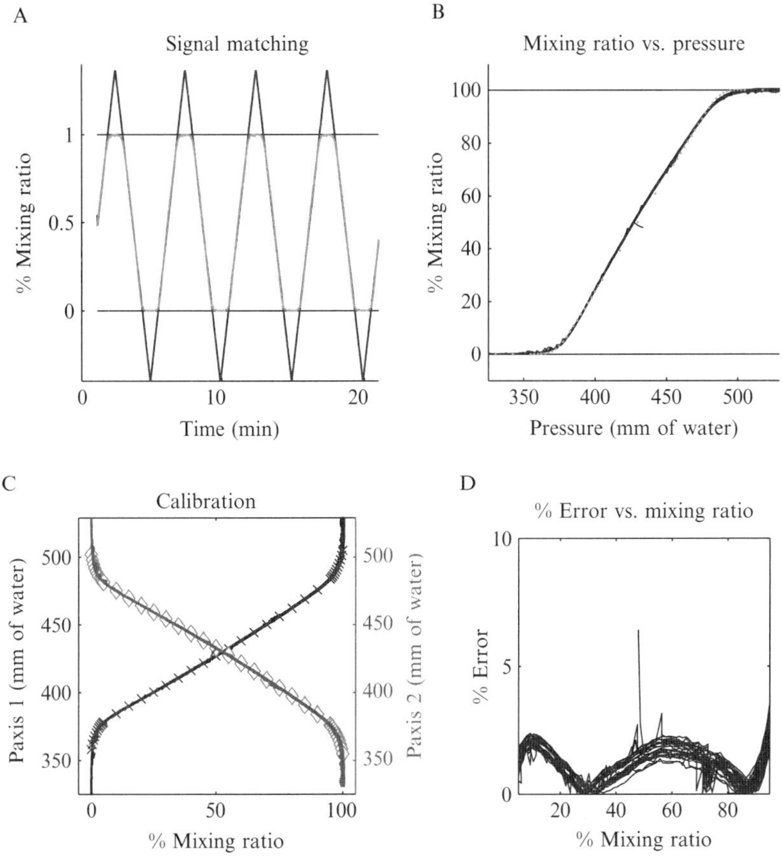

Figure 14.7 Performance of the DAW junction. (A) Calibration signal (red line) overlaid with output signal (green line) after correction for the delay in acquisition. During calibration the system is designed to intentionally overshoot the bounds of the DAW junction. Since the starting and ending points for calibration are not critical, this makes it easier to set up as described in the text. The ideal response would be a closely tracking output signal transitioning to plateaus after the system moves beyond 0% and 100% mixing ratios. As can be seen in the figure, this is what we observe, except for a slight rounding near the plateau region. (B) Compression of the data in part A into a single curve by mapping the input pressure directly to the output mixing ratio. Blue curve is the compressed data, while the green dots are the expected results from Comsol modeling. As can be seen in the figure, the modeling and experimental results are in excellent agreement. (C) Completed calibration for both inputs. Red crosses and pink diamonds represent polynomial fits of inputs 1 and 2, respectively, to the output mixing ratio. These fits can be used to program a linear actuator controller to generate precise inducer waves. (D) Measure of the percent error of the *uncalibrated* output signal, which general is less than 3%. (See Color Insert.)

actuator controllers to generate a triangle input wave and begin to move the reservoirs. Generally we use an input wave with a 5-min period. We then monitor the fluorescence near the cell chamber to record the output signal. The two signals are overlaid and any delay is removed, as shown in Fig. 14.7A. In this figure, the input is shown in red and the output in green. We expect the output to closely track the input signal until a plateau is reached, indicating complete diversion of a inlet into a shunt. As can be seen in the figure, this is essentially what we see, with a slight rounding at skewed mixing ratios.

Once this mapping is complete we compress the data into a single curve, as shown in Fig. 14.7B. This figure depicts one of the external port pressures mapped to a mixing ratio of the DAW junction. An ideal response would be a plateau at 0% leading to a linear ramp until another plateau is reached at 100%. The output of our junction (blue curve) closely approximates this, again with slight rounding near 0% and 100% mixing ratios. As an additional example of Comsol's utility, the green dots represent modeling results generated of the junction's response. As can be seen in the figure, the modeling and experimental results are in excellent agreement. In Fig. 14.7C, the calibration results for each input are shown. A high order polynomial fit is used for each input, which can then be programmed into the linear actuator controller. Figure 14.7D represents the percent error of the output signal as a function of mixing ratio for the *uncalibrated* system. Even without calibration the system is highly accurate, usually having an error of less than 3%.

1.1.6. Design of an improved yeast cell trap

Beyond the flow network or DAW junction, the most important part of a microchemostat chip is the cell trap. Often a successful design will hinge on a properly functioning trap. A microchemostat's cell trap should ideally be easy to load, force the cells to grow in a monolayer so they are all in the same focal plane, allow nutrients to enter the trap even when packed with cells, force cells to grow in well-defined directions to assist with cell tracking and allow cells to exit the trap without clogging the device. For some cell types, specifically mammalian cells, controlling the flow rate in the trap is also extremely important. We have found that even hearty mammalian cell lines, such as 3T3 cells, can be killed by extremely low flow rates (less than 1–5 μm/s). This requires the design of highly specialized traps to prevent any flow from reaching the cells after loading. We have never encountered an issue where yeast or *E. coli* cells seem adversely affected by flow, however the flow rate can be important for intercellular communication by diffusible substances (Danino *et al.*, 2010).

Often the goals mentioned above are difficult to achieve completely, for example a trap with high cell retention is often very difficult to load. This is the case with the TμC chip described in Cookson *et al.* (2005).

To overcome these problems, an improved yeast cell trap, known as the doughnut trap, was designed. Figure 14.4D and E contain an overview of this trap. The salient feature is improved loading while retaining the ability to image cells in a monolayer. Another major issue in the trapping region is clogging of cells from excess growth. Yeast cells grown in glucose can clog a device in several hours if the microchemostat is not properly designed. As shown in Fig. 14.4, the outer channel is designed with a height of 10 μm. This height is large enough that no cells will be able to clog it under normal circumstances. The height of the trap is kept at 3.525 μm for yeast cells of the W303 background. Note that the height of the trap is the most critical parameter of the entire chip as will be stressed in the fabrication section. Even height differences as little as 0.1 μm can make a difference in terms of the effectiveness of the trap. If the trap is too high, yeast cells will flow right through and not be trapped at all. Even those that are trapped may not grow in a monolayer and hence a uniform focal plane will be impossible to achieve. However, if the trap is too low, then it will be impossible to get the cells into the trap. Thus, the height of the trap depends intimately on the cell type and even the cell strain. We have noticed that some larger backgrounds of yeast actually require a slightly higher trap than other common laboratory yeast strains.

Upon loading, when cells flow into the chamber containing the trap, most will actually flow around the trap to the cell and shunt waste (port 3), since this region's flow mostly goes around the trap. This is actually beneficial to the design since it allows growing cells to be quickly whisked away when they overgrow the trap, while minimizing any movements of the cells in the trap due to flow (which can make cell tracking difficult). Furthermore this difference in flow rates is primarily a consequence of the difference in the heights between the two regions. Recall Eq. (14.9) which states that resistance of a channel scales with the cube of the height. Thus while the height difference between the trap and the outer channel is only $\sim$3 fold, the resistance difference will be $\sim$27 fold.

Those cells entering the central channel will move to the base and become stuck at the entrance barrier. Since the trap height is slightly smaller than the diameter of a yeast cell, the cells cannot enter the trap without some assistance from the experimenter. Once enough cells have accumulated behind the entrance barrier, the experimenter will flick the microfluidic line attached to the cell port with his index finger. This perturbation will cause a momentary pressure disturbance which will force some cells under the barrier into the trap. Once in the trap they will be efficiently held between the roof of the trap and the glass cover slip.

During the course of the experiment, cells will divide and enter exponential growth. They will quickly fill up the trap and the colony will come into contact with its walls. The pressure exerted on the trap's walls by the growing colony will generate a flow of cells, which can be modeled as a particulate flow (Mather *et al.*, 2010). This flow will expel some cells from the trap into the outer channel, to be carried away into ports 4 and 5 (note port 5, originally the cell port now functions as a waste port). The design of the cell trap should take cell flow into account so it can be directed in appropriate ways. For example, to track cells often it is useful to direct their movement in a regular direction to limit the difficulty of tracking. With the doughnut trap, cell flow is directed in radial directions which works fairly well. However, we have been considering designing a new trap with internal baffles to limit lateral movements of the cells.

The $MFD005_a$ device has been used successfully to generate many types of input concentration waves for numerous yeast strains and genotypes. In general, the chip takes 1–2 h to set up and can run for several days depending on the conditions. The chip is highly useful for all types of small scale experiments involving dynamic environments. However, upon building this chip we realized that most of the time during an experiment our microscope sat idle between imaging frames. To make better use of our time and resources, we decided to build a parallel version of the $MFD005_a$ device which we have named the MDAW device.

1.2. A parallel DAW device

The parallel version of our $MFD005_a$ device was designed to have eight copies of the smaller device on a single larger device. This parallel architecture greatly increases the throughput of a run by allowing eight independent subexperiments to be conducted at a time. The utility of this design can be seen by comparing the number of ports required to carry out equivalent experiments for the progression of chip designs. With the Bennett chip, 64 ports are required to conduct eight experiments, for the $MFD005_a$ device, 40 ports are required, while the MDAW device requires only 26 ports. Since setup time is directly proportional to the number of ports a chip contains, this reduction represents a significant savings of both time and consumables. Of course designing such a device presents its own challenges, a major one being space. Since we wanted all features to fit entirely on a single 24×40 mm coverslip, space was at even more of a premium than with the $MFD005_a$ device.

To conserve space we compressed the features of the $MFD005_a$ device as much as possible while retaining functionality and maintaining a margin for fabrication errors. We made the device radially symmetric in order to provide equal resistance paths to the ports shared among the subexperiments. To divide the space, we separated the chip into eight circular sectors of equal area, similar to slices of a pizza. While a rectangularly shaped device would have been a better fit for the coverslip, it would have been more difficult to ensure the resistances were equal to the outlets for all subexperiments. Moreover excessive stage movement between locations during acquisition can generate bubbles in the microscopy oil. These bubbles sometimes show up after several hours into an experiment and can cause a severe loss of focus or degradation of image quality. To prevent these problems, the cell chambers were placed as close to each other as possible, which essentially requires radial symmetry. As an added bonus, this lowers the amount of time for stage movement between positions.

An overview of a MDAW subexperiment is shown in Fig. 14.8A. Compare this to the $MFD005_a$ device in Fig. 14.4A. Both contain a DAW junction, SHM features and a cell trap that are essentially identical, although the length of the channel between the DAW junction and the cell chamber has been reduced slightly in order to conserve space. In fact ports 1, 2, and 5 and the channels linking them are essentially equivalent to ports A, B, and C, respectively, in the MDAW device. The major difference is that ports 3 and 4 on $MFD005_a$ have been consolidated in the MDAW device. In the MDAW device, we call the port 3 analog the consolidated shunt port and the port 4 analog, the consolidated alternative outlet.

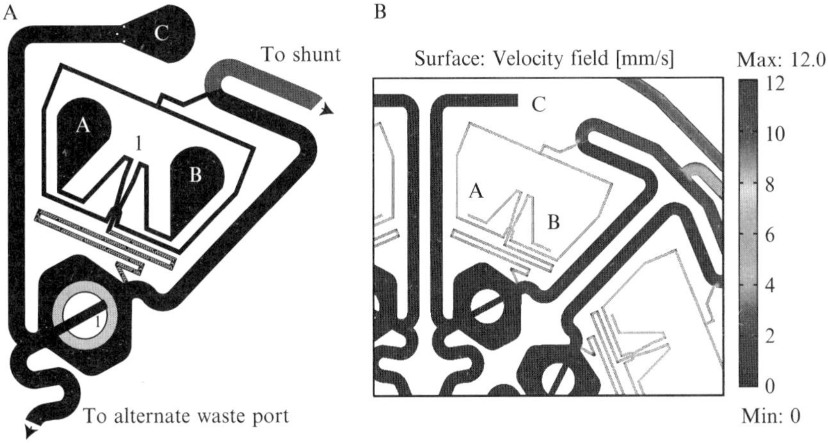

Figure 14.8 Graphic of the individual subexperiments in the MDAW microfluidic device. (A) This is a subexperiment from the MDAW device. It is essentially a compressed version of the $MFD005_a$ device shown in Fig. 14.4A. The ports labeled A, B, and C are equivalent to ports 1, 2, and 5, respectively, in Fig. 14.4A. The equivalents to port 3, the cell and shunt waste and port 4 the alternate outlet port, in the $MFD005_a$ device are shared among all eight subexperiments in this device. The arrows point to these shared ports. This port sharing reduces the number of outlets and eases the setup of such a large device. To make identification easier under the microscope, we have placed the subexperiment number above the DAW junction and near the cell trap. (B) Close-up of a Comsol model of the MDAW device. Comsol modeling was crucial for designing the combined collection network so each subexperiment's shunt would function similar to the $MFD005_a$ device. Since the collection network combines the output of eight subexperiments, the resistance had to be lowered so it would carry the combined flow as efficiently as that in the $MFD005_a$ device.

The consolidated shunt port is connected to each subexperiment by an extensive collection network. This collection network can be seen in Fig. 14.9. To create this collection network, Comsol modeling was essential to ensure that the flows would be equal to their equivalents in the $MFD005_a$

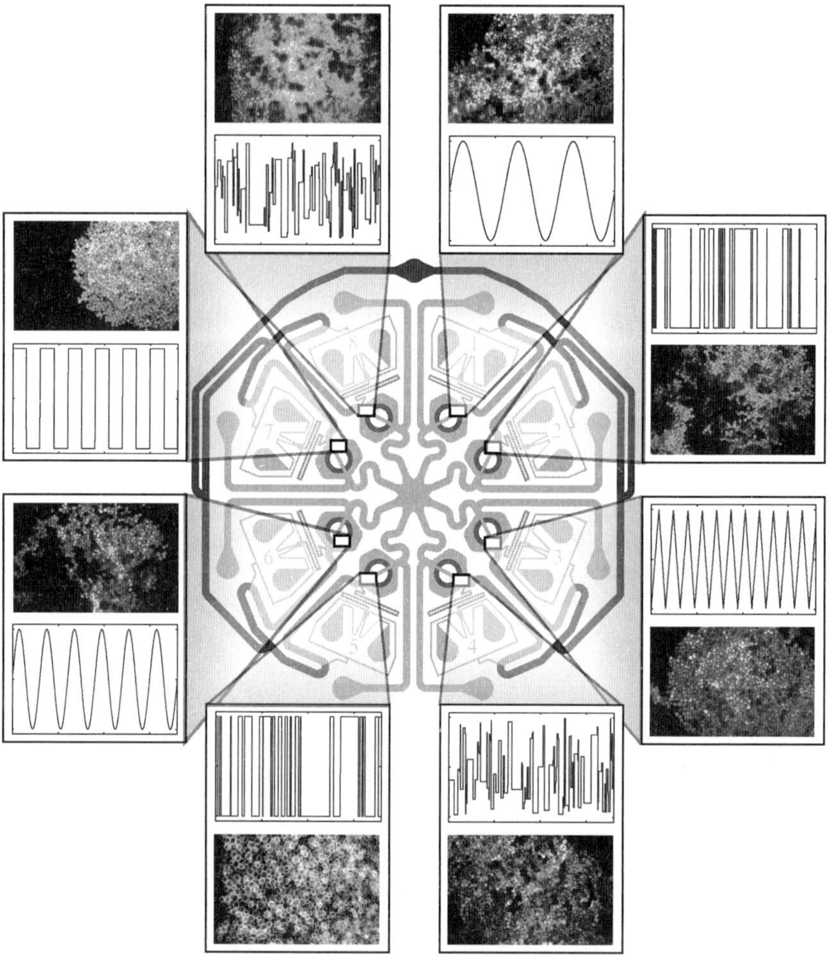

Figure 14.9 Graphic of the MDAW microchemostat device. The MDAW device has eight independent subexperiments. Each subexperiment can generate a separate inducer signal for an independent yeast strain. Examples of each are given in the breakout boxes. The system is capable of generating both periodic and pseudo random waves. The symmetry of the chip is important to ensure that all subexperiments have equal resistance outlet paths to the shared ports: the combined alternate outlet port (center) and the combined cell and shunt waste (top). (For interpretation of the references to color in this figure legend, the reader is referred to the Web version of this chapter.)

device. This modeling indicated that the height of the collection network would have to be increased to 35 μm to sufficiently lower the resistance (shown in dark blue in Fig. 14.9). Moreover the shunt channels from the DAW junction now connect to the diversion channel before it reaches the consolidated shunt port, whereas in MFD005$_a$ they both reach port 3 independently. Comsol modeling indicated that back flow from the shunt into the diversion channel could be a problem if the diversion channel was not long enough. The connection point was extended to ensure this would not happen.

It was easier to consolidate port 4 into the alternate outlet port on the MDAW device since it was in the center of the chip and each subexperiment had an independent path to the port. Thus the height of these channels could remain 10 μm. However, the channel length between cell chamber and the alternate outlet port had to be reduced, which altered the resistance somewhat. Comsol modeling allowed us to determine the port pressures which led to equivalent flow. One might wonder how many ports could be shared among a device of this size. Of course if a multilayer microfluidic device were used then there would be no restriction; however, we believe the time required to manufacture multilayer devices does not justify their added benefits and therefore we avoid their use if possible. For a single layer device, at most two ports can be shared among *all* subexperiments due to geometric constraints. It is possible to share additional ports between adjacent subexperiments, however, with the MDAW device this would have meant sharing the cell ports (port C) and we wished them to remain independent. It is also possible to add *y*-junctions or manifolds to connect multiple outlet ports to a single reservoir. However if this is done, extra care must be taken to ensure no bubbles are introduced in the lines. This is especially a problem with small diameter *y*-junctions.

Even at eight subexperiments you begin to push the limit of what modern microscopes can accomplish. For example, on our current setup using the Nikon TI, the amount of time it takes to autofocus, change filter cubes, acquire a phase contrast image and 2–4 fluorescence channels and move stage positions for eight subexperiments is nearly 1 min. Since phase contrast images must be taken approximately every minute for adequate cell tracking, the microscopy setup becomes limiting before the microfluidics. While laser based focus systems would offer an increase in speed, many, like the Nikon Perfect Focus System, do not work well with PDMS devices. Thus while other microfluidic devices have been produced which offer a far greater number of independent experiments, often they cannot track *individual* cells due to excessive movement between frames (Taylor *et al.*, 2009). This prevents the acquisition of cell trajectories and the device essentially functions similar to a highly parallel flow cytometer. Thus the device chosen should reflect the type of study and data required. For generating large numbers of cell trajectories in a dynamic environment with relative ease of setup, our device works well. For generating population level data using an extremely large set of conditions the device described in Taylor *et al.* (2009) would be superior.

1.3. Cell tracking

For microchemostat experiments cell tracking is essential for capturing high quality data. In fact, one could argue that effective cell tracking is as important as the design of a microfluidic device itself. Like a high powered computer running an early version of DOS, even the best device is not much use if the cells cannot be tracked. Thus most articles making use of microchemostats make a reference to "custom Matlab code" used for cell tracking (Bennett *et al.*, 2008; Hersen *et al.*, 2008; Kurth *et al.*, 2008; Lee *et al.*, 2008; Taylor *et al.*, 2009). Our lab is no different and we have spent much time and effort generating a software package which works quite well but has room for improvement. There is also a program called CellTracer available free online (http://www.stat.duke.edu/research/software/west/celltracer/). It should be stressed that a microchemostat should be designed with cell tracking in mind from the beginning, rather than designing software to track how the cells happen to grow in the device. For example, by making the cell culture expand in defined, regular directions the cell tracking routine becomes less complex and hence works better. An excellent example of this concept is the trap described in Rowat *et al.* (2009) which constrains yeast cells in essentially one dimension and makes lineage tracking quite robust.

The essential problem for tracking all types of cells, and yeast cells are no exception, is that they simply are not unique, at least as viewed under phase contrast microscopy. This can be seen in Fig. 14.10 which compares different parameter values for a population of cells. Ideally each cell would occupy a unique position in some high dimensional space, corresponding to a combination of parameters, such as cell area, eccentricity, and fluorescence, specific for that cell and invariant in time. This would be similar to a bar code or serial number for cells. However, as seen in Fig. 14.10 there is simply no combination of inherent characteristics visible under this type of imaging which can uniquely identify *all* members of the population at once. If there were, there would be no clusters of high density in the histogram. Moreover, since cells grow and divide, there is often a high amount of variability in the geometric properties between frames for the *same* cell. Unfortunately the only parameter which is unique for all cells confined to a monolayer is position. Thus it is of critical importance to keep track of cellular position during a microchemostat experiment and this explains why phase contrast images must be taken frequently. For fast growing cell types such as yeast or *E. coli*, frequent sampling is a necessity. If the cellular movement is greater than one cell diameter between frames, cell tracking becomes next to impossible.

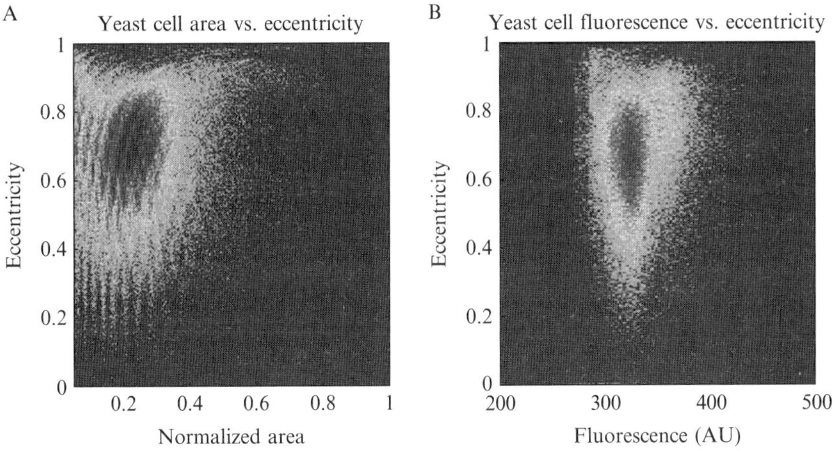

Figure 14.10 Comparison of different cell parameters for a population of yeast cells. (A) Two dimensional histogram of yeast cell eccentricity versus area. Striations in the data are a remnant of the ellipse filter used to segment the cellular boundaries. Notice that most cells have similar values for eccentricity and area. (B) Similar plot as part A, except here eccentricity and mean fluorescence are plotted.

Cell tracking software can be divided into two basic types, segmentation based methods and nonsegmentation based methods (Miura, 2005; Mosig *et al.*, 2009). Segmentation methods are the more common type and will be the focus of this discussion. In a segmentation method, a transmitted light image of the cell population is converted to a binary image containing only the outlines of cells. This is repeated for each image of the experiment and trajectories are formed by linking cellular objects between frames based on shared characteristics. Binary images are preferred since there are a large number of mathematical functions available for processing them. To convert a transmitted light image to a binary image the simplest method to use is a threshold. Essentially anything below the threshold is converted to black and anything above to white. Phase contrast images typically have a light halo around the boundary of cells, which provides high contrast, and thus are perfect for thresholding. A comparison of phase contrast imaging to differential interference contrast imaging, which is less suitable for thresholding, is shown in Fig. 14.11.

Typically a threshold value will be chosen to retain the boundary halo while discarding all other features, thus preserving only the boundary of cells. This procedure works fairly well assuming there are no other "phase objects" present in the cell. Unfortunately, yeast vacuoles are quite prominent under phase contrast microscopy and often are difficult to remove by thresholding alone. This necessitates later postprocessing steps to remove the vacuolar artifacts to prevent errors in segmentation. Some yeast

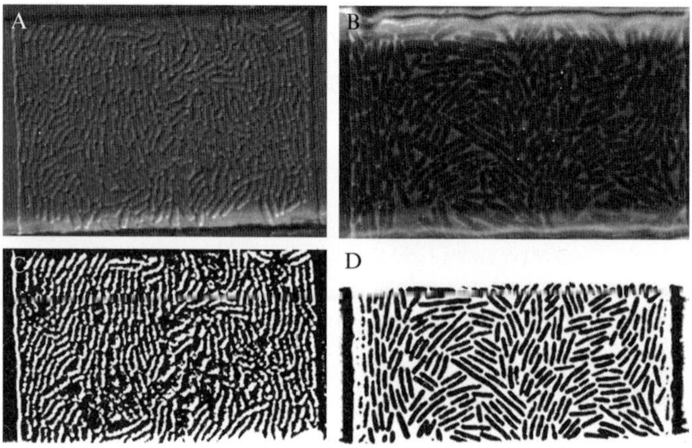

Figure 14.11 Comparison of phase contrast and differential interference contrast (DIC) imaging with regards to cell tracking. (A) DIC image of an *E. coli* colony growing in a microchemostat device. (B) Phase contrast imaging of a similarly grown *E. coli* colony. (C) Binary image created by thresholding the DIC image shown in part A. Notice how difficult it is to distinguish the cellular boundaries. (D) Thresholded version of the phase contrast image in part B. Notice how much more clearly the cellular boundaries are compared to C.

backgrounds or mutants can have especially prominent vacuoles which can be problematic. Moreover, environmental conditions, stress and aging can increase vacuole prominence. Thresholding based segmentation routines will need to cope with vacuoles and this is a downside of the technique for yeast. In spite of these issues, thresholding usually works well enough to be a reliable first step of the tracking procedure when chosen appropriately.

After thresholding the cellular boundaries are generally prominent but incomplete. Due to the aforementioned vacuole problems, often an aggressive threshold value is chosen leaving only the most prominent features of the image. While more successful in removing vacuolar artifacts, this will also remove some the cell's boundaries. For efficient processing of binary images, the image must be composed of only completely closed objects. Thus any cells lacking completely closed boundaries will not be found by the algorithm. Even if vacuoles are not a problem, a morphological closing operation is performed to repair inevitable boundary defects. This closing operation is done using either a structuring element or the watershed algorithm. Structuring elements are small geometrical objects which can reinforce common motifs of the image. We have found them to be very useful for processing *E. coli* cells.

Implemented in ImageJ and Matlab, the watershed algorithm is good at repairing small defects in cellular boundaries. However, if even a small vacuolar remanent remains, the watershed algorithm will bisect the cell through it, causing improper segmentation. An example of how the watershed algorithm performs in segmentation is given in Fig. 14.12. Note that another thresholding tradeoff comes from deciding how much to emphasize cells at the edge versus those in the interior of a microchemostat's colony. Multiple cells in close contact reinforce their boundary halo's, causing the signal from these areas to be greater than from isolated cells at the edge of the colony. When choosing a threshold value that maximizes boundaries and minimizes vacuoles, often the boundaries of cells on the colony edges are removed. This can be seen in Fig. 14.12. These boundary cells are consequently often dropped from the segmented image.

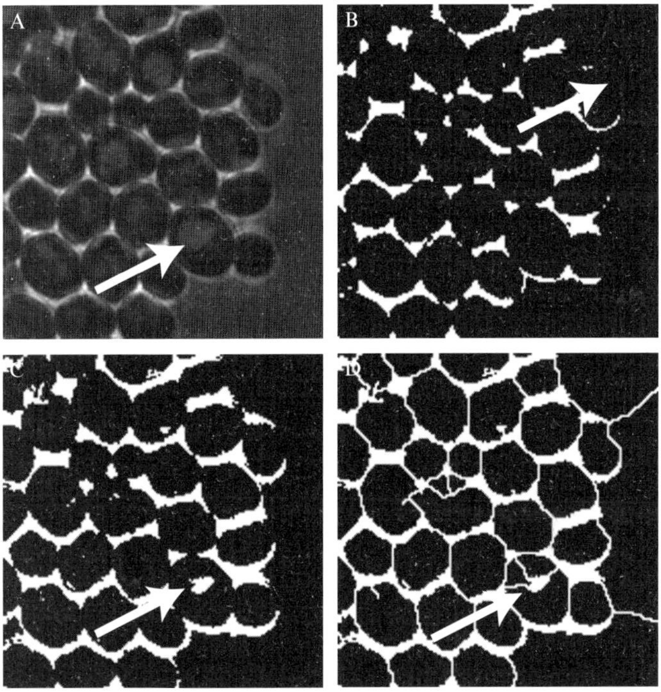

Figure 14.12 Segmentation of yeast phase contrast imagery (A). Phase contrast image of a tightly packed yeast cell colony. The white arrow points to a cell with a prominent vacuole. (B). Binary image created by thresholding the image from part A. The thresholding value was chosen to minimize vacuolar artifacts, but also has the effect of removing boundaries of cells on the colonies edge. The white arrow points to a cell with a deficient boundary. This cell will not be closed by the watershed algorithm and therefore will not be present in the segmented image. (C) Binary image created by thresholding A with a less stringent cutoff value. Notice that the boundaries are thicker and well defined, but that the vacuoles are more prominent than B. White arrow points to a vacuolar artifact. (D) Segmented image made from performing the watershed algorithm on the thresholded image from part C. Note that the vacuolar artifact has caused the segmented cell to be split into three regions. In later processing steps each of these regions will be considered cells, thus potentially causing errors in tracking.

After segmentation the binary image is processed to extract useful data from the contained objects. To assist in this processing, we fit each object to an ellipse since we have found that it generates a good approximation to a yeast cell's shape. After processing each image in an experiment we link cells between images to form trajectories. To accomplish this we have a scoring function which compares two cells and generates a score based on how likely they are to be the same cell. We compare the position, area, eccentricity, and orientation of each pair of cells to be scored. While we could also use the fluorescence values of the cell, we have found that this usually does not improve the score's power and often is not possible since we take phase contrast more often than fluorescence images. The scoring of cells between frames is usually the most computationally intensive part of the entire process. To aid in the computation we only compute scores for cells in the same general location of the two images, since if the cells have moved more than one cell diameter between frames they become virtually impossible to track anyways. This greatly reduces the computational time. Images of the $MFD005_a$ and MDAW traps generally contain ~1000 cells when fully packed; runs sampled every minute for 2–3 days will have thousands of phase contrast images to process. Clearly, any savings in time are important.

After scoring we remove cells which are below a threshold empirically determined to result in a poor match. To match a cell from the current frame to one from a previous frame there are several cases which need to be dealt with. These cases are depicted in Fig. 14.13. The first is the easiest which is a unique match between a cell from frame n and a cell from frame $n - 1$. In this case, the cell from frame n is assigned to the $n - 1$ cell's trajectory. An example of our algorithm's scoring output for the single match case is given in Table 14.5. The next case is when two cells match to the same trajectory. This often happens due to excessive cell movement and the cellular position becomes a less powerful discriminant. In this case, the cell with the highest score will be retained as the trajectory's match and the other cell will be moved to its next highest scoring trajectory. The third case is a skip, where a trajectory was present in frame $n - 2$ but for whatever reason a match was not found in frame $n - 1$. This often happens due to a segmentation error in frame $n - 1$.

If vacuoles are prominent, this type of skipping may happen often and should be corrected for. By keeping trajectories for an extra frame you can match a cell from frame n to a cell from frame $n - 2$. The next case is the start of a trajectory. Here a new cell is formed. The last case is the removal of a trajectory, here the cell either left the field of view or died. One has to be careful that the algorithm is not too "greedy" by always finding a match for a cell in the previous frame. Cells are born and cells die, these events will happen and if an algorithm is too greedy it will end up making improper trajectories. For example, often a greedy algorithm will cause a trajectory, which should have ended due to a cell leaving the trap, to jump to an

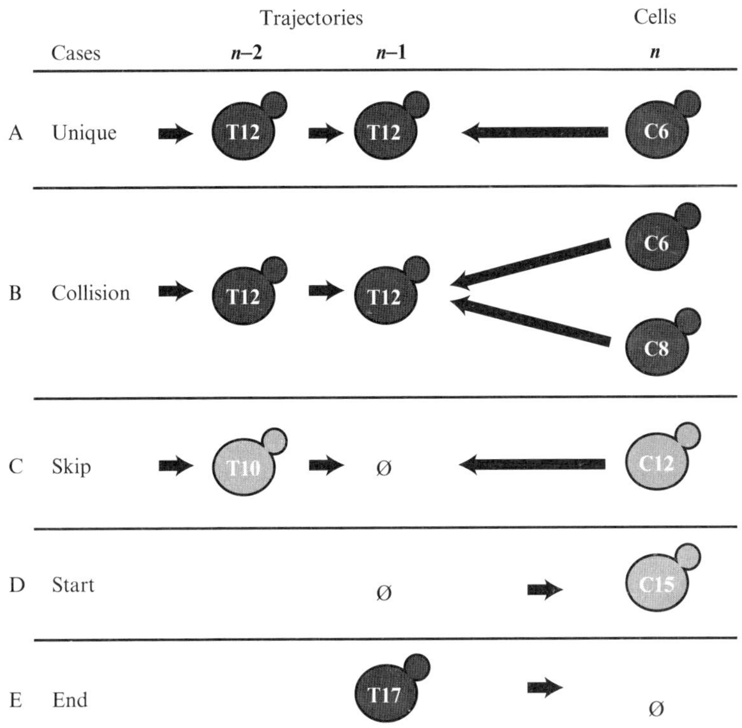

Figure 14.13 Different cases which need to be handled in cell tracking. Each case is given on the left hand side of the figure. The cells representing trajectories present in frames $n - 2$ and $n - 1$ are given in the middle portion of the figure. These cells are labeled with their trajectory number (e.g., T12). Cells in the current frame (n) are shown on the right hand side of the figure. They are labeled with their cell number (e.g., C6). (A) Unique match. A trajectory present in the previous two frames matches a single cell in the current frame. (B) Collision. Two cells have the same trajectory as their best match. Normally the highest scoring cell is chosen as the match; however, this is a symptom of poorly acquired data (cells have moved too much between frames) and will likely lead to mistakes. (C) Frame skipping. A trajectory present in frame $n - 2$ did not find a match in frame $n - 1$ but does find a match in frame n. This often is caused by segmentation errors in the $n - 1$ frame, especially vacuolar splitting of cells (see Fig. 14.12D). If this case is handled, longer trajectories can be generated, however there is a potential for the algorithm to become overly greedy. (D) Start of a trajectory. A cell is either born or moves into the frame. (E) End of a trajectory. A cell either dies of moves out of the frame.

adjacent trajectory. This is sometimes worse than ending a trajectory prematurely because it can be difficult to detect unless one goes through the data very carefully. Thus to obtain long, reliable trajectories one needs above all else good data and an algorithm which is balanced among all cases.

To reliably link cells into trajectories the number of cells uniquely matching a trajectory should be maximized. As stated earlier, the largest

Table 14.5 Sample yeast cell tracking output

Trajectory 36	Base	Predicted	Cells			
Score			0.1	2.1	2.5	2.5
Area	871.0	871.0	856.0	615.0	608.0	560.0
CentroidX	592.4	591.1	591.3	605.1	575.3	572.2
CentroidY	596.2	602.6	599.1	630.9	566.9	631.0
Eccentricity	0.5	0.5	0.5	0.5	0.8	0.5
Orientation	63.0°	63.0°	50.8°	71.7°	51.2°	−53.3°
Object	610	NaN	621	640	603	601
Fluor mean	NaN	NaN	392.6	331.2	390.3	326.4
Fluor std.	NaN	NaN	603.4	316.3	538.5	277.2

Comparison of a cell from a given trajectory and the nearest cells in the next frame. The data for the trajectory is taken from its matched cell in the previous frame. This is called the base cell. The predicted column refers to the algorithm's prediction of how the cell's properties should have changed in the current frame based on its previous behavior. This prediction is usually generated from MatPIV data of the colonies movement. While not shown here, it is also possible to predict a change in area from previous growth data. Note that a lower score is better and all scores above 1 are considered to be below the scoring threshold and thusly discarded.

impediment to unique matching is movement of cells as the colony expands. This can be severe for *E. coli* or even yeast grown in rich media. In fact, sometimes it is possible to see movement of the colony due to growth in real time under high magnification. To correct for bulk movements of cells a particle image velocity (PIV) program can be invaluable. PIV programs are imaging analysis routines which are able to detect particulate flows in a sequence of images by comparing how the field of view changes in time. This is very useful for tracking bulk movements of cells and can often significantly improve the fidelity of tracking. We use a program called MatPIV, which has been conveniently implemented in Matlab, to track cell flow in our images (Sveen, 2004). Using this data we come up with a predicted position for each trajectory present in frame $n - 1$ for frame n. An example of how this is useful is shown in Fig. 14.14. The MatPIV generated velocity field has been used to adjust the position of the cells resulting in more robust tracking.

In principle, the change in a cell's area and eccentricity could also be predicted from previous data. These changes would be most pronounced for newer, smaller cells. However, we have not done this and it is unlikely to improve tracking appreciably. An overview of the entire procedure is given in Fig. 14.14. The overall sequence of events is presented in the figure. We have done much work to improve the visualization of the trajectory data to ensure high quality. While the linking of trajectories works very well, the biggest improvements can be made in the segmentation steps of the process. Indeed, some cell tracking methods have no segmentation step at all, relying on comparison based methods for

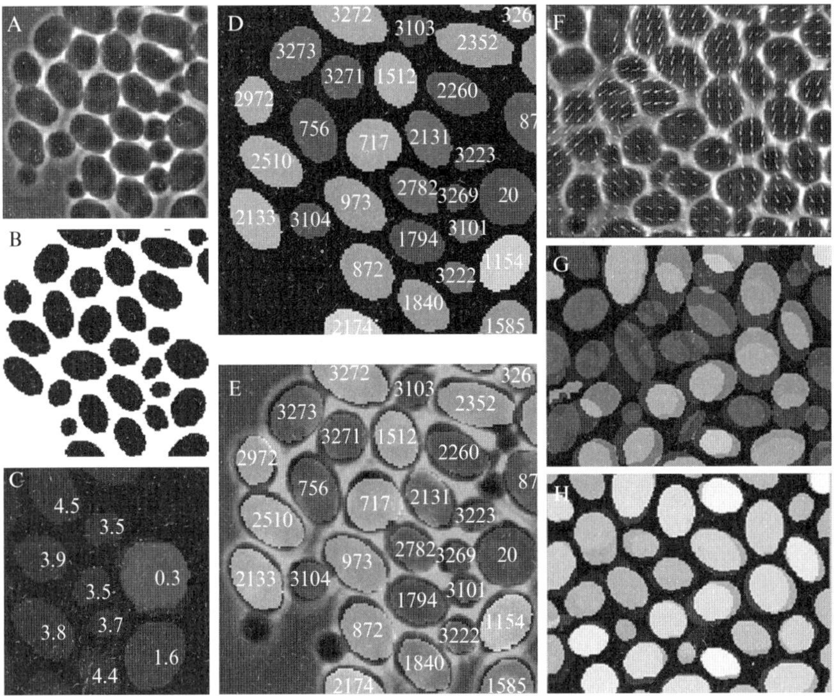

Figure 14.14 Overview of the cell tracking process. (A) Raw data: phase contrast image of yeast cells. Note the high contrast between the boundary of the cell and the exterior. (B) Segmented image after thresholding, application of the watershed algorithm and fitting the resultant objects to ellipses. (C) Scoring of a cell from frame n (shown in red) to trajectories present in frame $n - 1$. Lower score is better. Notice that the red cell has closely overlapped with a previous trajectory and generates a better score. All other scored cells are above the scoring threshold (which is set at 1). Note that the scoring system here has generated good contrast between the ideal match and the neighbors. This is indicative of a good match. (D) Colored image of the masks after trajectory finding is complete. Colored regions represent trajectories which are numbered. (E) Overlay of the trajectory image from part D with the phase contrast image of part A. Note most cells were assigned trajectories except for smaller cells and cells near the exterior. (F) Example of MatPIV processing for cell flow. White arrows indicating the cell flow velocity are overlaid with a phase contrast image of the colony. (G) Image of cells from frame $n - 1$ (opaquely colored objects) overlaid with cells from frame n (translucent objects). Notice there is an overall movement of cells toward the lower left corner of the image due to cell flow. Also note that the distance traveled here is almost one half cell diameter between frames for some cells. One can see how this movement could generate ambiguous situations for similarly shaped cells without prior knowledge of the cell flow. (H) Same cell field as in part G except MatPIV velocity information has been applied to correct for cell flow. Notice the much better overlap compared to part G. This will lead to more reliable matches since cell position is crucial for reliable matching. (See Color Insert.)

identifying cells in an image field (Miura, 2005). These methods generally rely on comparing a reference library of known cells to the current image using a cross-correlation function. The cross-correlation function will be maximized when the reference image matches a cell in the target image. Indeed, MatPIV works in a very similar way for tracking cell flows.

In principle, comparison methods could get around the vacuole problems mentioned above which are the bane of the segmentation approach for yeast. However, comparison methods can have problems if the cells change markedly between frames, which will happen due to growth, division and rotation. While there are ways to correct for this, in general, they are computationally intensive. In fact, the whole process is much more computationally intensive than the segmentation method. We are currently working on a hybrid method which employs an initial segmentation step that is corrected using a comparison step. Since segmentation works quite well, running it initially will reduce the space to be searched by direct comparison. The subsequent comparison step will correct for any initial errors in segmentation. Fortunately, cross-correlation methods lend themselves to parallel processing, and modern graphics cards can be programmed to greatly speed up computation (Owens *et al.*, 2008). In the future, we expect more use of parallel processing and comparison based methods for cell tracking. However, it should be emphasized that no matter how well designed an algorithm is, the most crucial determinants for success are the quality of the initial data and the regularity of a cell colony's movements.

1.4. DAW hardware and software

1.4.1. Hardware

As mentioned in the chips design section (1.1.3) of this chapter, the DAW junction works by changing the relative pressures at DAW ports, while keeping the total pressure the same. Physically this can be achieved in a number of ways: by pneumatically pressurizing the syringes, using a syringe pump, or changing the hydrostatic pressures of the syringes. Our initial design relied on pneumatically pressurized syringes, but due to problems with flow control we switched to a hydrostatic system. We use two vertically mounted linear actuators to change heights of liquid filled syringes that feed into the DAW junction, Fig. 14.15A. The smooth motion of the linear actuators allows for smooth changes in mixing ratios. Linear actuators are also a better solution in case of a hardware malfunction. If the actuators break down or cannot move to a new position, they will still allow the experiment to continue, since the flow depends only on the position of the syringe. The inability to move the syringes will only result in a constant inducer level, while maintaining a steady flow. In case of a malfunction with syringe pump or pneumatically driven system the flows will change over time and might even result in flow reversal, which would most likely ruin the experiment.

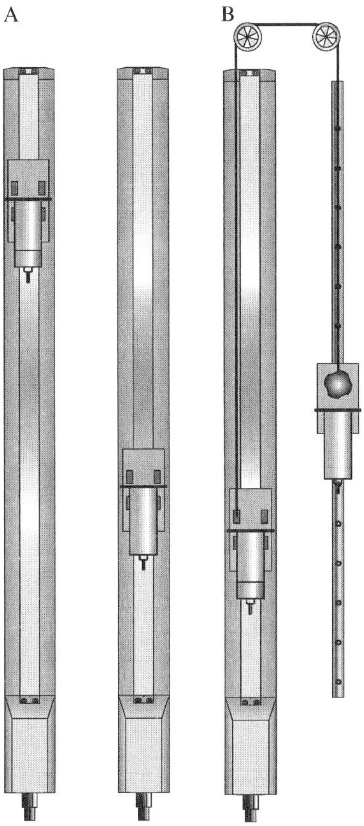

Figure 14.15 Linear actuator setup for DAW. (A) Dual linear actuator setup. Each actuator can be move individually. One of the actuators (left) moves a media syringe with added dye. (B) Alternate design of the DAW system using only a single linear actuator. The actuator controls the position of both syringes simultaneously. To eliminate friction in the system the following components need to be in a single plane in space: both pulleys, line attachment to actuator, and line attachment to linear guide block.

The first version of DAW system had two linear actuators, which could be controlled independently, Fig. 14.15A. By attaching a syringe with media to each actuator and moving them equal distances in opposite directions we were able to change the ratio of pressures at the DAW ports while keeping the total pressure constant. Since we have constrained our total pressure to be constant, the movement of one actuator has to be mirrored by the movement of other actuator in opposite direction. In essence, the second linear actuator could be replaced with a linear guide and a pulley system, as seen in Fig. 14.15B. The linear guide consists of a rail and a guide block that slides along the rail. We found that the guide block does not have enough mass to keep a taught line through the pulley system. A steel block was used to weigh down the guide. Also, the length of the line between the linear actuator cart and the guide block needs to be adjustable. This can be achieved by attaching the line to guide block with a pinch mechanism operated by a screw.

The elimination of the second linear actuator proved to have a major benefit of reduced setup cost per DAW unit. However, when considering the additional parts and labor required to fabricate a pulley system the value of this benefit diminishes. Unless you intend on running a full eight trap MDAW chip, we recommend on installing a dual linear actuator system. During installation the actuators should be securely attached to some sort of a support system. In our case, we attached them to metal struts that are directly connected to the wall studs.

From Table 14.6, which lists all the required parts for a dual linear actuator system, we can see that there are two linear actuator controllers (RPCON) and a single communication gateway (SIO) module. The SIO module is used for communication to a computer, while the RPCON's connect the actuators to the SIO. This setup seems redundant, but it allows for easy expansion. The SIO module can operate up to 16 individual linear actuators, while maintaining only a single connection to the computer. Using this system from the start will allow one to easily expand from 2 to 8 axis DAW system. Also the SIO can be wired to communicate with a computer via USB interface.

1.4.2. Software: *iDAW*

To control the linear actuators we have created a custom software, nicknamed *iDAW*, using the National Instruments LabVIEW environment. Currently there are two major versions of the software, for the 2- and 8-actuator systems. Both versions, manuals and installation guides are freely available by request.

Table 14.6 Hardware required for DAW actuator setup

Equipment	Qty	Part No.	Vendor
Linear actuator: fast speed, 800 mm travel length	2	RCP2-SA7C-I-56P-16-800-P1-M-BE	Valin Corp.
Controller	2	RPCON-56P	Valin Corp.
Communication gateway module	1	RGW-SIO	Valin Corp.
Serial communication cable	1	CB-RCA-SIO-050	Valin Corp.
USB adapter	1	RCB-CV-USB	Valin Corp.
USB cable	1	CB-SEL-USB010	Valin Corp.
24V DC power supply	1	OMRON-S8VS-06024	Valin Corp.
AC power cable	1	70355K34	McMaster-Carr

The graphical user interface presents the user with three main areas: actuator controls, calibration, and experiment setup as seen in Fig. 14.16. During a typical experiment, the actuators first have to be calibrated to the specific chip. This calibration establishes a relationship between relative positions of each actuator and the respective mixing ratios. There are two ways to calibrate the system: manually and automatically. The automatic calibration was already discussed in an earlier section. During manual calibration the actuator positions are changed to create different mixing ratios. Once all the calibration points have been acquired the software creates a calibration function. The software allows up to 11 calibration points, but we have found that a two point calibration performs very well. Also, depending on the number of points, the order of the calibration function can be increased for improved data fit.

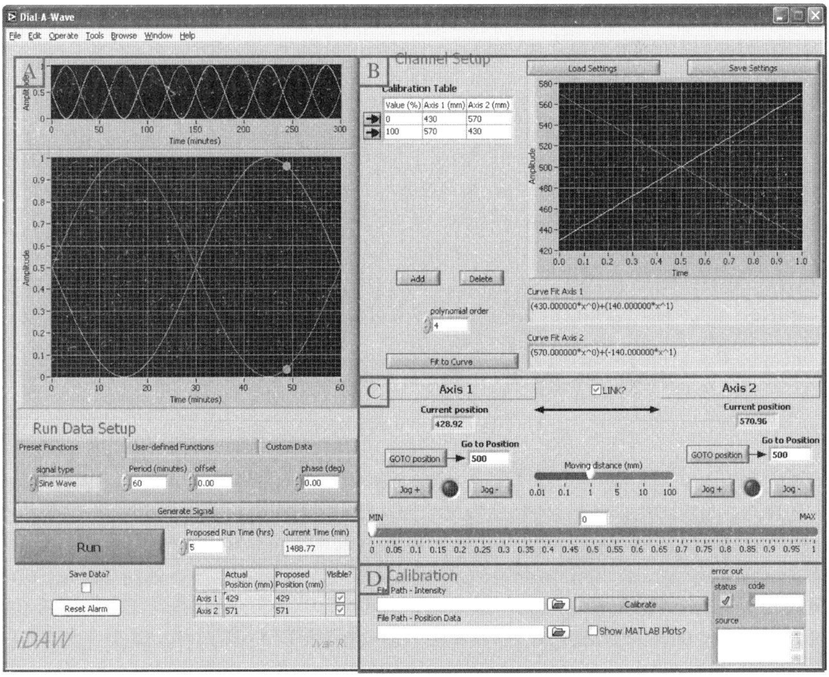

Figure 14.16 Screenshot of *iDAW* software. (A) Experimental parameter setup allows user to set up mixing ratios as a function of time. Most mathematical functions or an arbitrary dataset can be used as templates. (B) Manual calibration. The table records the calibration points and the graph shows the calibration functions. In this example, a two point calibration was used to create a linear calibration profile. (C) Actuator controls allow the actuators to be moved independently or together when in "Linked" mode. (D) Automatic calibration functions take position data from the actuators and fluorescence data from the microscope to determine the calibration profiles.

To start the manual calibration procedure the actuators are moved together to a height that provides the desired flow to the cell trap. Since the pressures at both syringes are the same, this becomes the 50% value for the calibration. Next, the actuators are linked to move equal distances in opposite directions. The positions are adjusted until there is only media with the inducer going through the mixer, this becomes the 100% point. Similarly, the 0% point is recorded. The 0% and the 100% points are used to make a linear calibration function as can be see on the graph in Fig. 14.16B.

The experimental setup area of *iDAW* allows the user to create a profile of induction versus time. The user can choose from a number of built-in functions, such as square or sine waves, or load an arbitrary function. The software automatically adjusts the inducer values to fit between 0% and 100%. The proposed induction profile is plotted for the duration of the whole run and the individual linear actuator positions are constantly updated, Fig. 14.16A. These displays eliminate errors during experimental setup and actual run-time.

2. Part II: Fabrication

With the design of the chip drafted and thoroughly analyzed we begin the fabrication process. An overview of fabrication is shown in Fig. 14.17. The complete fabrication of a microfluidic chip can be broken down into three main phases. In the first phase we create a patterned wafer by photolithography. Next, we use this wafer to create a silicon rubber mold by a process of soft lithography. And finally the silicon is prepped and bonded to a glass coverslip to make a functional microfluidic device.

2.1. Photolithography

Photolithography was initially developed for the semiconductor industry and later applied to a variety of fields, including microfluidics (Xia and Whitesides, 1998). The process relies on transfer of a geometrical pattern from a mask onto a photosensitive layer via light radiation. The first step involves thorough cleaning of the wafer, which will act as the foundation for all the features. It is very important to remove all debris and any chemicals from the surface of the wafer, as they will get incorporated into the final wafer design and will highly affect the adhesion properties of photoresist to the wafer. Next, we deposit a small amount of photoresist onto the wafer and spin the wafer at predetermined speed to create a photoresist film of precise height, this step is called spin coating. The wafer is then soft-baked by gradual heating on a level hot plate, which removes solvent and enhances photoresist adhesion to the wafer. At this point the wafer is exposed to UV

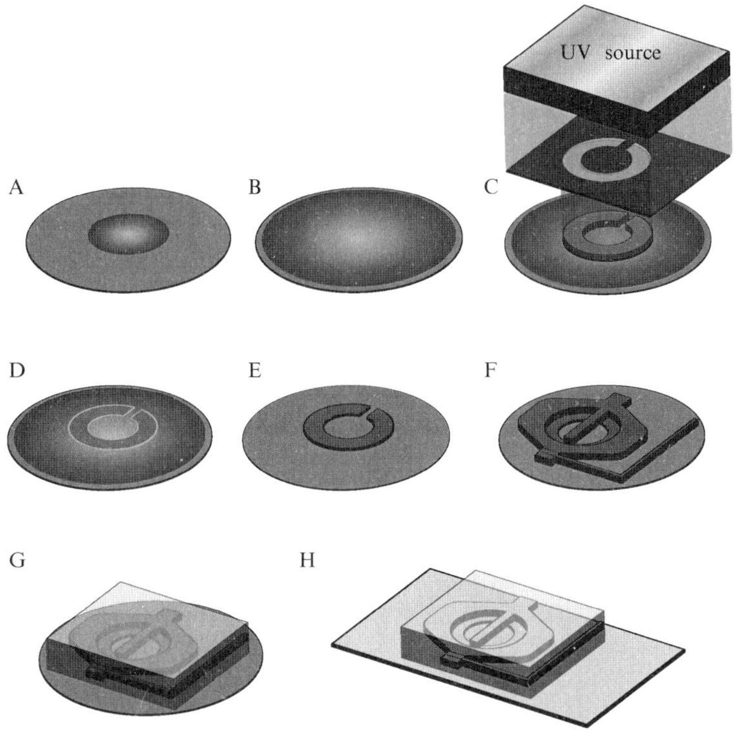

Figure 14.17 Overview of the fabrication process. Photolithography (A–F), soft lithography (G), and PDMS processing (H). (A) Photoresist deposition. (B) Spin coating: the deposited photoresist is spun at a specific speed to create a uniformly thick layer. (C) UV exposure cross-links the photoresist creating a pattern identical to the photomask. (D) Postexposure baking joins the silicon wafer and the cross-linked photoresist. (E) Developing removes the uncross-linked photoresist, revealing the features. (F) Repeating steps A–E creates additional features. (G) Pouring and curing PDMS over the patterned wafer creates a mold. (H) Bonding the PDMS mold to a glass coverslip finishes a microfludic chip.

light through a photomask, this transfers the pattern from the mask onto the photoresist layer. We use the SU-8 2000 line of photoresist from MicroChem Corporation. SU-8 is a negative photoresist, which means that areas of the film exposed to UV radiation will form solid structures, while unexposed areas will be washed away during the developing step. The wafer is then baked again, in the postexposure bake (PEB), to increase the level of cross-linking. And finally to complete a single photolithographic cycle, the wafer is developed by immersion in solvent which removes uncross-linked photoresist leaving only the desired pattern on the wafer. Since all of the chip designs we use require wafers with multiple heights this cycle is repeated a number of times.

2.1.1. Photoresist

Manufacturers, such as MicroChem, make a variety of photoresist formulations. The SU-8 line of resists alone has three subcategories with a total of 18 different formulations, specific for heights ranging from 1.5 to 550 μm (MicroChem, 2010). We use the SU-8 2000 photoresists which have great adhesion to silicon wafers and are able to make high aspect ratio structures (del Campo and Greiner, 2007). The "negative" denomination of a photoresist means that areas of the film exposed to UV radiation will form solid structures, while unexposed areas will be washed away during the developing step. Specifically exposure to UV radiation changes the chemistry of the resist by generating a very strong acid within the film, which starts the cross-linking reaction of the SU-8 epoxy. The main difference between the various SU-8 2000 formulations is the epoxy solids content that directly relates to the viscosity of the liquid as can be seen in Table 14.7.

Commonly there is a need for a nonstandard formulation. It is possible to make new, less viscous, formulations by adding SU-8 thinner to the initial, more viscous, stock of photoresist. It should be noted that the manufacturer's naming scheme loosely relates to the height of the photoresist film when it is spun at 3000 rpm. Thus, for 2002 and 2005 photoresists, spin coating at 3000 rpm would in theory produce 2 and 5 μm film heights, respectively. Using this information we can plot these theoretical heights against the percentage of solids for each formulation, as seen in Fig. 14.18. By making a curve fit function of percent solids (s) as a function of height (h), as written in Eq. (14.33), we are able to estimate the required solids for any new formulation. For example, to make a new formulation, which would produce 3 μm height at 3000 rpm, we use Eq. (14.33) to determine that it requires 35% solids

Table 14.7 SU-8 2000 photoresists formulations

SU-8 2000	% Solids	Viscosity (cSt)
2000.5	14.3	2.49
2002	29.00	7.5
2005	45.00	45
2007	52.50	140
2010	58.00	380
2015	63.45	1250
2025	68.55	4500
2035	69.95	7000
2050	71.65	12,900
2075	73.45	22,000
2100	75.00	45,000
2150	76.75	80,000

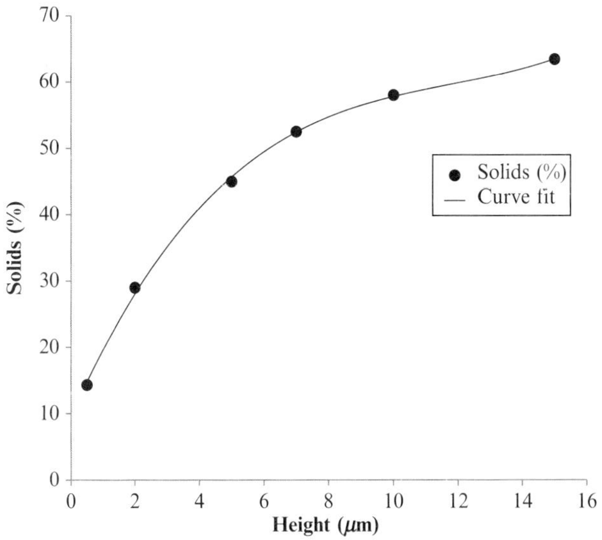

Figure 14.18 Graph of SU-8 formulation versus percent solids. Relationship between estimated height of SU-8 formulations when spun at 3000 rpm and their solids content.

$$s = 0.0235h^3 - 0.834h^2 + 10.807h + 9.5781. \tag{14.33}$$

Next, the amount of thinner required for the new formulation can be calculated using the relationship described in Eqs. (14.34a) and (14.34b), where $\text{mass}_{\text{total}}$ is the desired mass of the new formulation, $\text{mass}_{\text{thinner}}$ is the required mass of thinner, $\text{mass}_{\text{initial}}$ is the required mass of original photoresist, s_{initial} is the percentage of solids in the original photoresist, and s_{final} is the percentage of solids in the desired photoresist formulation.

To make the formulation measure out and deposit the predetermined amounts of photoresist and thinner into a clean amber glass bottle, make sure to do this in a fume hood. Drop a clean stir bar into the bottle and place on a magnetic stirrer, until it is thoroughly mixed. Due to the viscosity of photoresists removing the stir bar could be difficult, so we leave it in the bottle until the photoresist runs out

$$\text{mass}_{\text{thinner}} = \text{mass}_{\text{total}}\left(1 - \frac{s_{\text{final}}}{s_{\text{initial}}}\right), \tag{14.34a}$$

$$\text{mass}_{\text{initial}} = \text{mass}_{\text{total}} - \text{mass}_{\text{thinner}}. \tag{14.34b}$$

Finally, to complete the process it is necessary to characterize the new photoresist formulation by making a spin speed curve. This step should also be performed for any standard formulations that have not been previously characterized by your lab. To create a spin speed curve for a particular photoresist the photolithographic cycle, described later on, should be repeated 3–6 times with various spin-coating speeds. For each speed, measure and record the feature heights using a surface profilometer. Plotting and curve fitting the data will produce enough data to reliably estimate spin speeds for specific heights. As mentioned earlier, the functionality of a cell trap is dependent on its height. Thus, it is critical to manufacture the exact height required by the design. The spin curves allow us to estimate only a rough range of speeds required to achieve a height. Using this range as a starting point, we perform as many spin test as necessary to get the desired height. An example of an actual spin curve for 2003 formulation can be seen in Fig. 14.19. Examining the figure it becomes evident, that our 2003 formulation produces 2.6 and not 3 μm height at 3000 rpm, this fact reinforces the need for photoresist characterization.

2.1.2. Equipment and environment

Due to sensitivity of photolithography to contamination it is usually performed in a cleanroom environment. A number of universities and research centers have shared facilities that house equipment necessary for photolithography and other dust-sensitive processes. We have made wafers in various environments from a Class 100 cleanroom to a basic HEPA filtered room with no rating. The latter type of noncleanroom manufacturing environment is achieved by creating a dedicated fabrication space, installing HEPA filters over the air ducts and changing the ceiling panels to nonparticulate releasing tiles. Also to prevent uncontrolled photoresist cross-linking, the

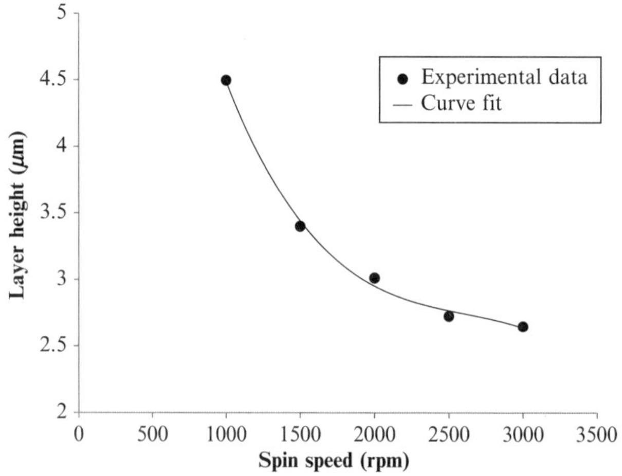

Figure 14.19 Example spin curve for an SU-8 2003 formulation. As spin speed increases the the spun photoresist thickness decreases. The graph levels off at high speeds, with further increases in spin speed having little effect on layer thickness.

lights should be fitted with UV absorbing filters. This can be easily done by placing thin filter sleeves over the fluorescent lights bulbs. Although it is beneficial to carry out the whole manufacturing process in the cleanest possible environment, in our case only photolithography is performed in the cleanroom. While soft lithography and PDMS processing are carried out in regular lab space.

It is important to point out that the chemical safety precautions are more important than the cleanliness of the facility. Some of the chemicals used in photolithography are potentially carcinogenic, labs should use a properly functioning fume hood when working with photoresists and developers at all times. A standard fume hood convects air from the environment past the user, into the hood and out a ventilation shaft. Since users are generally the largest source of particulates in a clean environment, use of a standard fume hood can increase the local concentration of particulates over the work surface in the hood, even if the surrounding environment is clean. In contrast, biosafety cabinets contain a laminar air stream between the interior and the user, preventing the transfer of particulates into the hood. However, unless specially made and calibrated, biosafety cabinets can potentially allow chemical fumes to escape into the work area. Purpose built hoods, protecting both the user from chemical fumes and the interior from particulates, do exist but are expensive. For microchemostat fabrication, we have found that a standard chemical fume hood is sufficient; however, electrical engineering facilities will often contain specialty hoods. Since hot plates and spin processors are used with uncured photoresists, it is essential that they be placed in the fume hood. However, the process of soft-baking removes the solvent from photoresist, allowing one to work with a mask aligner outside of the fume hood.

2.1.3. Photomasks

Conventional photolithography requires expensive chrome photomasks, we use the cheaper photomasks printed on a transparency-like material as described in Whitesides *et al.* (2001b). As mentioned earlier, all of our masks are made by CAD/Art Services, Inc. (Bandon, Oregon). They use a photographic process to print the design on a 0.007″ polyester mylar sheet coated with photographic silver. Since our masks are designed in AutoCAD software, we just provide them with a ⋆.dxf file. However, they do accept a variety of other CAD files, listed in order of preference: ⋆.dwg, ⋆gds, ⋆.cif, ⋆.gerber, and ⋆.eps. Due to the limits of their photoplotting process, the minimum feature size is defined by a circle with 10 μm diameter. Although, it is possible to print various size masks, we usually order an 8 × 10-in. sheet. This gives us enough room to fit up to six individual layer masks and since most of our design require less than six layers we can have a whole chip printed on a single sheet. When ordering the mask, it is important to specify the polarity of the mask, considering that we are working with a negative photoresist, our masks need to have clear features on a black background. Once the masks have been printed, they are cut out and individually glued at the corners to a 3 × 3-in. glass square using clear instant adhesive. It is important to have the emulsion side of the mask facing away from the glass, since it needs to be in contact with the photoresist later on. Also, when gluing the photomask to the glass make sure to keep the glue away from any transparent parts of the mask. For storage and transport we keep the masks in individual plastic bags, this prevents them from getting dirty and scratched.

2.1.4. Sample fabrication parameters

For each individual wafer we create a table with fabrication parameters, this is an effective way of condensing all of the necessary information for manufacturing the wafer. Most of the parameters, such as layer heights and number of layers, will be dictated by your design, however some of them have to be calculated after the design is done. For example, the exposure time will depend on the exposure dose required for the photoresist and on the UV lamp power. MicroChem's datasheets provide exposure energy ranges for different heights. For example, 0.4 μm layer requires 60–80 mJ$\cdot$cm^{-2} and 3 μm layer requires 90–105 mJ$\cdot$cm^{-2}. Given that our mask aligner UV lamp has an effective power of 1.4 mJ$\cdot$cm^{-2}, we can calculate the exposure times using Eq. (14.35), see Table 14.8.

$$\text{Exposure time} = \frac{\text{Exposure dose}}{\text{Effective power}} = \frac{\text{mJ/cm}^2}{\text{mW/cm}^2} = \frac{\text{mW}\cdot\text{s/cm}^2}{\text{mW/cm}^2} = \text{seconds.} \tag{14.35}$$

With additional information from photoresist spin curves we can finalize the fabrication parameters into a table, as seen in Table 14.9 and proceed to fabrication.

2.2. Photolithography: Protocol

All of the necessary equipment, supplies, and chemicals for this protocols are listed in Table 14.10 at the end of this section.

Table 14.8 Calculated and experimental exposure times

Layer height (μm)	Exposure energy (mJ/cm^2)	Calculated exposure time (s)	Experimental exposure time (s)
0.4	60–80	43–57	60
3	90–105	64–75	80

Table 14.9 Sample table of wafer fabrication parameters

Layer number	1	2	3	4
Layer height (μm)	0.4	1	3	10
SU-8 formulation	2000.5	2000.5	2002	2005
Spin speed (rpm)	3750	700	1000	660
Soft-bake at 95 °C (s)	120	120	150	240
Exposure time (s)	60	60	80	100
Postexposure bake at 95 °C (s)	160	160	180	240

Table 14.10 Photolithography Equipment, chemicals and supplies

Equipment	Model No.	Manufacturer
Mask aligner	Model 200	OAI
Spin processor	WS-400BZ-NPP-Lite	Laurell Technologies Corporation
Surface profilometer	Dektak 150	Veeco
Infrared thermometer	62	Fluke
Hot plate		
Fume hood		
Chemicals and supplies	**Part No.**	**Supplier**
SU-8 Photoresists 2000. 5-2050 (500 ml)	Varies	MicroChem
SU-8 Developer (4 L)	Y020100-4000L1PE	MicroChem
SU-8 2000 Thinner (4 L)	G010100-4000L1PE	MicroChem
AlphaLite Polyster swab	18-375	Fisher Scientific
Glass bottle (amber)	41265T31	McMaster-Carr
Instant Adhensive	495045	Loctite
Borosilicate glass square, 3′ × 3′, 1/8″ thick	8476K131	McMaster-Carr
Silicon Wafer	100MM/CZ/1-0-0/Boron/P Type/ Resis-10-20/Thick 500-550/Oxy 9-21/ SLBACK: ETCH ACID	WaferNet, Inc.
Wafer tray	H20-3000-01-1415	Entegris, Inc.
Wafer cover	H20-3000-02-1216	Entegris, Inc.
Wafer tweezers (125 mm)	S3WF	SPI Supplies
Crystallizing Dish (740 ml)	08-741E	Fisher Scientific
Wash bottles (500 ml)	08-647-707	Fisher Scientific
Acetone		
DI Water	Milli-Q or better	
Isopropanol	HPLC grade	
Methanol	HPLC grade	

2.2.1. Cleaning the wafer

Place the wafer inside the spin processor (spinner) with reflective surface facing up, this is your working surface. Try to align the center of the wafer with the center of the vacuum chuck of the spinner, this eliminates uneven rotation. If you have cleanroom paper, line the inside of the spinner with it to help with the clean up process. Set the rotational speed to 3000 rpm and start the spinner. At this point it is recommended to turn on the mask aligner and UV source, as the lamp needs time to warm up.

2.2.2. Applying the cleaning agents

Thoroughly clean the wafer by applying chemicals in the following order: acetone, isopropanol, methanol, and DI water, while gently applying pressure with a polyester swab. Make sure not to press too hard, but rather smoothly move the tip across the spinning surface of the wafer.

2.2.3. Drying the wafer

Place the clean wafer on a hot plate set at 200 °C and let dry for 5 min. Once done with the drying cycle set the temperature to 95 °C, as it will take some time to cool down. By the time you are done with step 6 your hot plate should be at the right temperature.

2.2.4. Centering the wafer on the spinner

Pick up the wafer from the hot plate with wafer tweezers and let it cool prior to positioning on the spinner chuck. Once cool, position the wafer on the chuck, making sure it is centered with respect to the chuck. To check if the wafer is centered, spin it at 500 rpm, if the wafer is centered correctly when spinning it will look like a circle. However, when off center, it will spin creating an oval shape. For best results it is recommended to center the wafer as much as possible. It is helpful to use a wafer alignment tool, although we have made a custom one, there are plenty of commercially available options.

2.2.5. Dispensing photoresist

Dispense 5–10 ml of photoresist in the center of the wafer. The total amount of photoresist depends highly on it's viscosity, with higher volumes needed for more viscous formulations. When working with photoresists make sure to never dispense directly from the main stock. Constantly opening the stock bottle will cause solvent evaporation and build-up of dry photoresist on the mouth of the bottle. This leads to change in the viscosity of the resist and to contamination with solid particles. The best practice is to have a working stock 30 ml amber glass bottle, which you refill from the main stock. The dark glass will limit the amount of UV entering and reacting with the photoresist. Make sure to label the bottles as all photoresist look the same.

2.2.6. Spin coating

Depending on desired layer thickness the spin speed during the second step will vary. Program the spinner for a two step cycle. Step 1: 500 rpm for 15 s, acceleration of 100 rpm/s; Step 2: desired spin speed for 30 s, acceleration of 300 rpm/s. For example, to achieve a layer thickness of 0.4 μm with SU-8 2000.5 we spin for 30 s at 3750 rpm; 3 μm with SU-8 2002 we spin for 30 s at 1000 rpm. These numbers are true for our formulations but might not be correct for your formulations, since the age of photoresist will have an effect. As mentioned earlier, it is absolutely crucial to create spin-curves for each photoresist prior to final wafer fabrication.

2.2.7. Soft-baking at 95 °C

Previously it was recommended to have a pre-bake step at 65 °C prior to soft-baking 95 °C. According to MicroChem and our own experience pre-baking step is not really necessary. We have eliminated it from our protocols and have not noticed any significant effects.

Using an infrared thermometer check the temperature of the hot plate, it should be 95 °C. Place the wafer on the center of the hot plate and be careful as the wafer may sometimes slide off the hot plate. Keep the wafer on the hot plate for 1–3 min, depending on the layer thickness. MicroChem's material datasheets can act as a guide in selecting the baking time, however the exact time can only be determined empirically. A good way of optimizing baking time is to remove the wafer from the hot plate and let it cool. Once cool, place the wafer back on the hot plate. If the photoresist film "wrinkles" keep it on the hot plate for another 30 s. Repeat this process until the film no longer "wrinkles" (MicroChem, 2010).

For example for 0.4 μm layer the soft-bake time is 120 s and for 3 μm layer it is 150 s.

2.2.8. Alignment of photomask and UV exposure

Turn on the mask aligner UV source, if this has not been done in Step 1. Place the wafer on top of the vacuum chuck in the mask aligner. Turn on the vacuum, to secure the wafer on the chuck. Position the photomask in the mask holder on the aligner, with the transparency side facing the wafer, turn on the vacuum to secure the mask. During exposure, the light path should be as follows: glass, printed mask, photoresist film, wafer. Make sure the *z*-axis of the wafer is all the way down, then move the mask into horizontal position. If the wafer is too high it can come in contact with the mask and smear the photoresist film. Move the wafer up slowly until in makes contact with the mask. Usually this creates a number of light diffraction patterns on the mask, which can be observed by looking at the mask at an angle. For alignment, the best distance is usually right after the diffraction patterns appear. This distance allows for independent movement of the wafer and the mask, while keeping them close enough to each other to see the features on the wafer through the mask. For an alignment methodology see Section 2.3 at the end of the protocol.

Once the wafer and the mask have been aligned, bring the wafer in complete contact with the mask without forcing or overextending the *z*-axis. Expose the wafer for a predetermined time. Move the *z*-axis down, lift the mask, turn off the vacuum to the wafer chuck and remove the wafer.

2.2.9. PEB (Post Exposure Bake) at 95 °C

Bake the wafer on the 95 °C hot plate for a specified time. Once again this time will depend on the thickness of the layer, with some rough estimates present by MicroChem's datasheets. For example, for 0.4 μm layer our PEB is 160 s and for 3 μm layer it is 180 s. If the exposure times are correct you should be able to see the pattern within the photoresist film within 15 s of baking.

2.2.10. Developing

Fill up a crystallizing dish with enough SU-8 Developer to cover the wafer. Make sure the wafer has cooled down to room temperature, before immersing it in the developer. Next, while keeping the bottom of the dish on the surface of the fume hood, move the dish in a circular fashion. This technique improves removal of uncross-linked photoresist. Continue this process for 1–2 min. MicroChem suggests other methods, such as ultrasonic or megasonic baths, but we have not needed them in the past.

2.2.11. Cleaning

Pick up the wafer from the dish using tweezers and rinse it with fresh SU-8 Developer, you can let the developer collect in the dish. Follow by a rinse with fresh Isopropanol and air dry using filtered air or nitrogen. At this point you should clearly see the features on the wafer. Sometimes the wafer will have white streaks, this is due to photoresist that has not been removed by development. Clean the wafer with fresh developer, rinse with fresh Isopropanol and dry.

2.2.12. Examining the wafer

Cleaning completes a single photolithographic cycle. At this point it is necessary to examine the wafer under a microscope, if the process was successful then the features will have uniform color and straight, smooth edges.

2.2.13. Measuring feature height

Using a surface profilometer measure a number of height points for each important feature. Since the height of cell traps is absolutely crucial for microfluidic chips, it is necessary to measure the height of the trap in different locations on the wafer and see that it conforms to your design specification.

2.2.14. Hard-baking at 200 °C

If there are no more layers to deposit, place the wafer on 200 °C hot plate for 5 min. If there are any cracks on the surface of the features, this step should remove them. It is beneficial to ramp up the wafer temperature to 200 °C.

2.3. Special notes on alignment

As mentioned in the chip design section, the wafer is made layer-by-layer from the ground up. It is recommended to deposit the smallest height features first and gradually move in increasing order. Although the design of the chip should account for small alignment errors, this sequential approach to wafer manufacturing can result in propagation of errors from one layer to the next. Since the compounded effect of these errors can be significant, it is crucial to have the best possible alignment at each layer. Due to lack of a consistent protocol for alignment, it can be most time consuming and very frustrating step of wafer manufacturing. Here, we propose a simple methodology that should let a minimally experienced person successfully align layers.

Most of the manual mask aligners use micrometers for x, y, z, and θ stage movements. The micrometers are primarily used for very fine axis adjustments, but they also can be used to precisely record the position of the wafer. Also, it is easy to see that if two different alignment elements on the wafer are individually aligned to their respective alignment elements on the photomask, then the whole wafer is completely aligned to the mask. Thus, the positional data should be identical at both alignment elements. By systematically adjusting and recording the x, y, and θ positions we can find a set of values that is identical for both alignment elements.

2.3.1. Protocol

Photomasks presented in Fig. 14.20 will be used as an example. For correct scale it should be noted that alignment elements presented in Fig. 14.20A and C, are located in the center of the mask and are 80% of the width of the mask.

Using the x, y, and θ micrometers on the mask aligner, find the alignment features from layer #1 and roughly position them under alignment elements of photomask for layer #2, Fig. 14.20D. Adjust the magnification of the mask aligner, so that most of your field of view is covered by a single alignment element, Fig. 14.20E.

Next, adjusting only the y-direction, align the top of the features to the top of the photomask alignment box. Record the position of y-direction micrometer, this is the $y1$ point. Repeat this step for the bottom side of the features and record the micrometer position, this is the $y2$ point. In a similar fashion obtain micrometer readings for alignment of left and right sides of the features to the alignment box, $x1$ and $x2$, Fig. 14.20F. Although it would seem that if the edges are aligned then position $y1$ would be equal to $y2$, and $x1$ equal to $x2$, however, this is rarely the case. In reality, the new

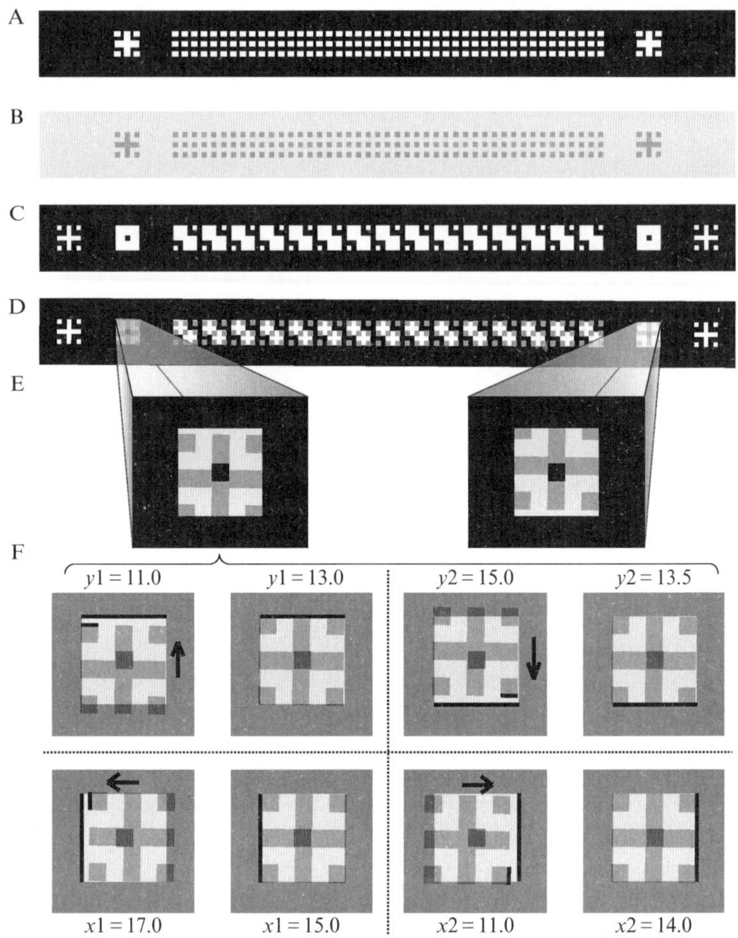

Figure 14.20 Sample layer alignment technique. (A) Photomask of layer #1, features are created by the transparent areas of the mask. (B) Features (green) on wafer (gray) for layer #1. (C) Photomask of layer #2. (D) Alignment of wafer with features from layer #1 to photomask for layer #2, as seen through the microscope of mask aligner. (E) Close-up view of alignment of the outermost left and right features. (F) For each side (left and right), the features on the wafer are aligned to the 4 four sides of the alignment box. The mask aligner micrometer position is averaged for x (15, 14) and y (13, 13.5) directions, to provide a single xy (14.5, 13.25) position. Note that it would seem that $y1$ and $y2$ positions should be identical if the feature sides are aligned to the alignment box. In reality, due to the new photoresist layer the features from previous layer become distorted, resulting in the difference. If the xy positions from the left and the right side are identical the alignment is good, otherwise the θ position needs to be changed and the whole process repeated. The transparency of the photomask has been adjusted for demonstration purposes. (See Color Insert.)

Table 14.11 Sample alignment datasheet

	Left		Right	
θ	X_1, X_2, ($\bar{X}$)	Y_1, Y_2, ($\bar{Y}$)	X_1, X_2, ($\bar{X}$)	Y_1, Y_2, ($\bar{Y}$)
17	15, 14, (14.5)	13, 13.5 (13.25)	16, 17 (16.5)	11, 10 (10.5)
15	14, 13, (13.5)	14, 15 (14.5)	17, 19 (18.0)	9, 10 (9.5)
19	15, 15 (15.0)	12, 12.5 (12.25)	16, 15 (15.5)	14, 12 (13)
20	15, 16 (15.5)	12, 12 (12)	16, 15 (15.5)	12, 12 (12)

photoresist layer makes the features seem somewhat distorted when viewed through the microscope. Though, assuming that the distortion is equal in all directions we can take the average of the two positions to get the actual aligned position, as seen by values in brackets in Table 14.11. Repeat the four measurements for the right side alignment element. Record all the data points into a table, as seen in Table 14.11.

In the first row of the table, the average positions for x and y are different for left and right sides. This would indicate that wafer is not aligned. Change the θ micrometer position by a small amount, in the example case we moved from 17 to 15. Repeating all the measurements it becomes evident that the left side and right sides are diverging from each other. This is probably not the right direction for θ movement. Move the θ from the initial position by the same amount in the opposite direction and repeat the measurements. In our example the θ position changed from 15 to 19. It is clear that the x–y positions are converging, but are not exactly equal yet. In the same direction, change θ position by the smallest possible step, and repeat measurements. If the positions are identical the wafer is aligned, if not, repeat θ movement and measurements. In our example, θ movement from 19 to 20 resulted in identical x–y positions for both sides, successfully terminating alignment procedure.

We have determined through experience that developing a systematic way of placing the wafer and the photomask into the mask aligner greatly reduces the time for alignment. The wafers we use have two flat edges, so when placing the wafer into mask aligner we find a surface on the mask aligner and roughly align the edge to that surface. The same trick is applied for the photomask. This results in relatively consistent placement of wafer and photomask, thus lowering the final alignment adjustments.

For the UV exposure step, set the x and y micrometer positions to the averaged values of x and y, respectively.

2.4. Soft lithography

Soft lithography is a microfabrication technique that relies on the use of a patterned elastomer to create structures, in our case, by cast molding. Although a number of different elastomers can be used, PDMS (polydimethysiloxane) has become the standard choice for microfluidics. PDMS is optically transparent, permeable to biologically important gases, chemically and thermally stable, the surface can be chemically modified and it does not absorb water. The PDMS we use comes as a two part kit: silicone monomer and curing agent. Mixing the components in specific ratio creates PDMS prepolymer that remains liquid for a few hours. The PDMS mold is prepared by pouring liquid prepolymer over a patterned wafer, curing it at elevated temperature and removing from the wafer. Since PDMS is initially in liquid phase, it easily conforms to the geometry of the wafer. Once cured, it remains flexible and allows for easy peel-off from the wafer. Furthermore, treating the wafer with a release agent improves the peel-off process (Duffy *et al.*, 1998; Sia and Whitesides, 2003; Whitesides *et al.*, 2001b; Xia and Whitesides, 1998). All the tools, chemicals and equipment required for soft lithography are listed in Table 14.12.

2.5. Soft lithography: Protocol

2.5.1. Aluminum holder

Cut out a 20-cm circle from aluminum foil. Place the wafer, features up, in the center of the foil. Next, carefully holding the wafer down, start to fold the foil up all the way around the perimeter. This will create 5 cm high walls around the wafer that will hold PDMS in. Make sure that the foil is really tight against the edge of the wafer, this prevents significant leaks of PDMS under the wafer.

2.5.2. Applying release agent (for new wafers only)

It is necessary to perform this step in a fume hood following all safety precautions, as most release agents are toxic. Place the wafer into a dedicated silanizing desiccator. Using a syringe with a needle, draw up the release agent, we use (TRIDECAFLUORO-1,1,2,2-TETRAHYDROOCTYL)-1-TRICHLOROSILANE. Deposit only a single drop ($\sim$30 μl) of the release agent into an open top small container inside the desiccator, see Fig. 14.21. Close the lid of the desiccator and turn on the vacuum. The release agent will vaporize and evenly deposit onto the wafer. Let this reaction happen for about 15 min. Using too much release agent will inhibit PDMS binding to glass coverslip.

2.5.3. Preparing PDMS

In a clean weighing tray measure out, in a 10:1 ratio, and mix 40 grams of silicone elastomer base with 4 grams of silicon curing agent. Continue vigorously mixing with a clean spatula. The consistency of the mixture should start to change from clear to foamy. Mix the components thoroughly for 3 min.

Table 14.12 Soft lithography equipment, chemicals and supplies

Equipment	Qty	Part No.	Vendor
Vacuum pump RV8	1	A65401906	Edwards
Vacuum pump EMF 10 exhaust mist filter	1	A46226000	Edwards
Vacuum pump oil return kit	1	A50523000	Edwards
Vacuum pump inlet connection (NW25 to 3/4″ hose barb)	1	NGT908000	Edwards
Vacuum pump NW25 clamping ring	1	C10514401	Edwards
Desiccators	2	08-642-5	Fisher Scientific
Ceramic desiccator plate	2	08-642-10	Fisher Scientific
Isotemp Oven	1	506G	Fisher Scientific
Vacuum manifold parts	**Qty**	**Part No.**	**Vendor**
1/2″ stainless steel hose clamps	5	6151K51	McMaster-Carr
1″ stainless steel hose clamps	5	6151K53	McMaster-Carr
1′ 3/4″ ID, 1″ OD wire-reinforced tubing	5	5393K45	McMaster-Carr
1′ 1/4″ ID, 1/2″ OD wire-reinforced tubing	10	5393K31	McMaster-Carr
3/4″ MPT to 3/4″ barb adapter	1	5365K23	McMaster-Carr
3/4″ FPT to 3/4″ FPT to 1/4″ FPT tee	1	4429K229	McMaster-Carr
1/4″ MPT to 1/4″ MPT nipple	3	9171K122	McMaster-Carr
1/4″ FPT to 1/4″ to 1/8″ FPT tee	3	4429K223	McMaster-Carr
1/4″ MPT to 1/4″ FPT-handle valve	5	4912K87	McMaster-Carr
−30 in Hg vacuum gauge with 1/8″ MPT at back	3	3935K21	McMaster-Carr
3/4″ MPT to 1/2″ MPT reducing nipple	1	9171K223	McMaster-Carr
1/2″ FPT to 1/2″ FPT to 1/2″ FPT tee	1	4429K253	McMaster-Carr
1/2″ MPT to 1/4″ MPT reducing nipple	2	9171K219	McMaster-Carr
1/4″ FPT to 1/4″ FPT to 1/4″ FPT tee	2	4429K251	McMaster-Carr
1/4″ MPT to 1/4″ barb adapter	2	53505K64	McMaster-Carr
PTFE thread seal tap	1	4591K12	McMaster-Carr
Chemicals and supplies	**Qty**	**Part No.**	**Manufacturer**
Silicone elastomer kit		Sylgard 184	Dow Corning
(TRIDECAFLUORO-1,1,2,2-TETRAHYDROOCTYL)-1-TRICHLOROSILANE		T2492	UCT
Aluminium foil			

2.5.4. Degassing PDMS

Mixing introduces a lot of air bubbles into the PDMS. To degas, place the weighing tray into the dedicated desiccator and turn on the vacuum. As pressure within the desiccator drops, the trapped air bubbles will expand

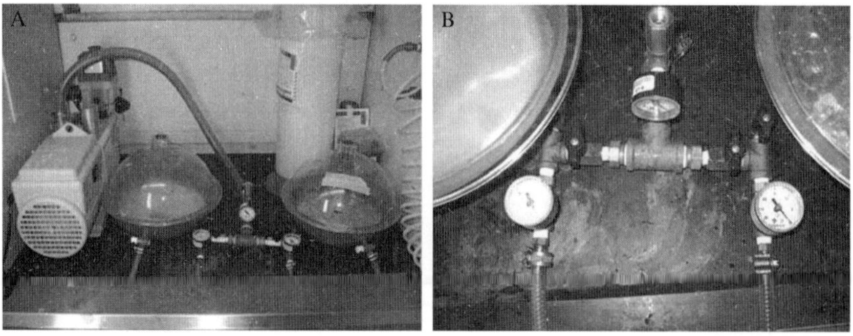

Figure 14.21 Vacuum pump and desiccators. (A) Vacuum pump and desiccators located in the fume hood. Each desiccators for a single purpose: (1) wafer silanizing (left) and (2) PDMS degassing (right). Note the opaqueness of the silanizing desiccator, this is due to silanizing agent vapor deposition over the years. (B) Vacuum manifold connecting the vacuum pump to the desiccators. The manifold allows for individual control of vacuum or atmospheric pressures to each desiccator. See Table 14.12 for parts list.

and PDMS might spill out of the tray. Quickly releasing the vacuum should pop a significant portion of the bubbles. Turn on the vacuum again and repeat this cycle until there are no more bubbles. Depending on the vacuum pressure this should take 10–20 min. Also, it is possible to degas by pouring mixed PDMS into a 50-ml Falcon tube and centrifuging it at ~2700 g for 10 min.

2.5.5. Pouring PDMS

Place the wafer into the degassing desiccator and pour the PDMS over it. Since PDMS is very viscous, you might have to use a spatula to get all of it onto the wafer. This process will introduce new air bubbles into the PDMS. Repeat step 4 until there are no visible bubbles in the PDMS. Sometimes the PDMS will leak under the wafer and you will see bubbles forming around the perimeter of the wafer. You can ignore them when considering to stop degassing.

2.5.6. Curing

Place wafer in 80 °C oven for 1 h.

2.6. PDMS processing

During the final phase of manufacturing the individual chips are cut out, cleaned and bonded to coverslips. Although the processing is performed in regular lab environment it is critical to get the chips and coverslips as clean as possible. This eliminates debris from the chip and improves the overall

Table 14.13 PDMS processing equipment, chemicals and supplies

Equipment	Qty	Part No.	Vendor
Dissecting scope	1		
Fiber optic light source	1	Dynalite 150	A.G. Heinze, Inc.
UVO Cleaner	1	Model No. 42	Jelight Company Inc.
Flowmeter	1	FR4A37	Key Instruments
1/8 Male pipe adapter	2	5454K65	McMaster-Carr
Polyurethane tubing			
Chemicals and supplies	**Qty**	**Part No.**	**Manufacturer**
Leur stub (25 gauge)		75165A686	McMaster-Carr
Biopsy punch Harris Uni-Core 0.5 nm		15071	Tex Pella, Inc.
Razor blades		12-640	Fisher Scientific
10 ml Disposable syringe		14-823-2A	Fisher Scientific
Cover slips No. 1 1/2, size: 24 × 40 mm, thickness: 0.16–0.19 mm		12-530F	Fisher Scientific
Magic Tape		810	Scotch
Kimwipes, Kimberly-Clark No. 34155		06-666A	Fisher Scientific
Compressed O_2		Medical grade	
n-Heptane		HPLC grade	
Methanol		HPLC grade	
DI Water		Milli-Q or better	

quality of the devices. To improve the final bond between PDMS and glass coverslip, it is recommended to complete soft lithography and PDMS processing in the same day. All the required materials and tools for this phase of manufacturing are listed in Table 14.13.

2.7. PDMS processing: Protocol

2.7.1. Removing PDMS layer

Take the wafer out of the oven and let it cool down to room temperature. Carefully peel of the foil from PDMS. Some PDMS may have gotten under the wafer. You need to remove this layer prior to peeling off the top layer of PDMS. Using a razor blade, cut the bottom layer as close to the edge of the wafer as possible. It is also possible to rub the edge of the wafer with your gloved finger. This will break the PDMS on the edge, disconnecting the bottom and top layers of PDMS. Very slowly lift up the top layer of PDMS. Allow the PDMS to lift off from the wafer by itself, this is best done by raising a part of PDMS to a small height, stopping and letting the PDMS catch up. Lift up the PDMS in 3–4 places around the perimeter of the wafer, before peeling it off completely. Wafers are very brittle, so make sure not to twist or apply excessive pressure on it, as it will easily break. For safe storage place the wafer into a labeled wafer holder.

2.7.2. Cutting PDMS

Using the dissecting scope, examine the features on the PDMS. Sometimes the angle of the light source needs to be adjusted to get enough contrast to see the microscopic features. Placing the PDMS on a dark background also improves contrast. Next, using a razor blade carefully cut out individual chips, leaving extra room around the perimeter of the chip. Try to leave at least 3 mm of extra PDMS around each port, it will improve the chip's bonding and prevent port leaks.

2.7.3. Punching ports

Place the chip with feature side up and, using the dissecting scope, locate the outline of the port. Place the tip of 25 gauge leur stub within the outline and, making sure it is as vertical as possible, apply downward pressure. The PDMS should first deform and then break; sometimes a final push is required to completely break through the PDMS on the exit. Next, carefully pick up the PDMS chip and remove the PDMS core using tweezers. Slowly pull out the puncher from the hole, while rotating it back and forth. Continue this for all ports on the chip. Sometimes the punching will tear the PDMS around the port, this is most likely due to a dull punching tip. Simply, swipe the punching tip against an abrasive surface 2–3 times and retry the punching. It is also possible to use a biopsy punch, which combines the leur stub and tweezers in a single tool, to make the holes.

2.7.4. Cleaning ports

Attach a 23-gauge leur stub to a syringe and fill it with DI water. Hold the tip of the leur stub against a port and apply pressure. A stream of water should exit from the other side of the chip. Keep the pressure for 3–5 s. Repeat this process of all ports on both sides of the chip.

2.7.5. Cleaning chips

Spray each chip with 70% Ethanol and gently rub using your gloved finger. Thoroughly rinse the chip with MilliQ quality water and blow dry using clean dry air. Make sure to dry both sides of the chip and all the ports. Place the dry chips in a clean Petri dish. Apply scotch tape to both sides of the chip. The next step is crucial for clean chips. Careful not to tear the PDMS, run your fingernail over the features a few times, covering the area of the whole chip. Repeat the scotch tape cleaning 3–5 times. Once done, use a fresh piece of tape to cover the chip and put the chip in the Petri dish.

2.7.6. Cleaning coverslips

Spray both sides of the coverslips with *n*-Heptane and gently rub the surface using your finger. To prevent the coverslip from breaking, apply pressure using your finger on both surfaces at the same time. Wipe the coverslip completely dry with a Kimwipe. Repeat the process using Methanol. Finally, wash the coverslips with DI water and dry using clean air. Make sure the coverslips are completely free of dust, spot or streaks. If you notice something, redo the DI water wash step. Once done, place the clean coverslips into a Petri dish and cover.

2.7.7. Bonding chips to coverslips

Open the compressed O_2 valve on your tank and make sure the flow through the UVO cleaner is 0.4–0.6 scfm. Warm up the UVO cleaner, by running it for 5 min. Once the warm up is done, open the loading tray, there should be a faint smell of ozone. Place the chips with feature side up and coverslips onto the tray. Close the tray and run the bonder for 3 min. When done, open the tray and place the chip onto the coverslip using tweezers. To improve the bond, using tweezers, gently apply pressure around the perimeter of the chip. Make sure that chip and coverslip come in contact as soon as possible, as the chemistry allowing for bonding changes with time. Place bonded chips in 80 °C oven overnight. If you have a lot of chips, it might be beneficial to break up the bonding step in 2 or more batches.

2.7.8. Troubleshooting

Poor chip bonding: this can be caused by a number of issues

(1) Too much release agent used during wafer preparation. Try lowering the amount of release agent or shortening the coating time.
(2) Check O_2 supply to UVO bonder.
(3) Expose the chips and coverslips exactly for 3 min. Make sure to bond chips and coverslips immediately after exposure to ozone.
(4) Place a weight on top of the chips during overnight baking. Make sure not to break the coverslip.
(5) Make sure all PDMS processing steps are done in one day.

Collapsed features

(1) Usually only the lowest features will collapse, but if enough pressure is applied from the top of the chip even taller features are susceptible. Lower the amount of pressure applied on the top of the chip during bonding.
(2) Try placing the coverslip on top of the chip during bonding. This should prevent features lower than 0.5 μm from collapsing.

3. Part III: Experiments

3.1. Experimental setup for *E. coli*

Although it is possible to perform microfluidic experiments without a lot of specialized equipment, we have found that purpose-built tools, such as our DAW and syringe towers, greatly increase productivity and experimental control. As mentioned in earlier sections, we use linear actuators to control the hydrostatic pressure of syringes, Fig. 14.23B. However, we use special syringe towers, as shown in Fig. 14.22, for controlling the height of our static syringes used for waste, cell and shunt ports. The towers are equipped with rulers, allowing us to record the position of syringes for an experiment. This data is used in subsequent experiments to reliably reproduce flows within the

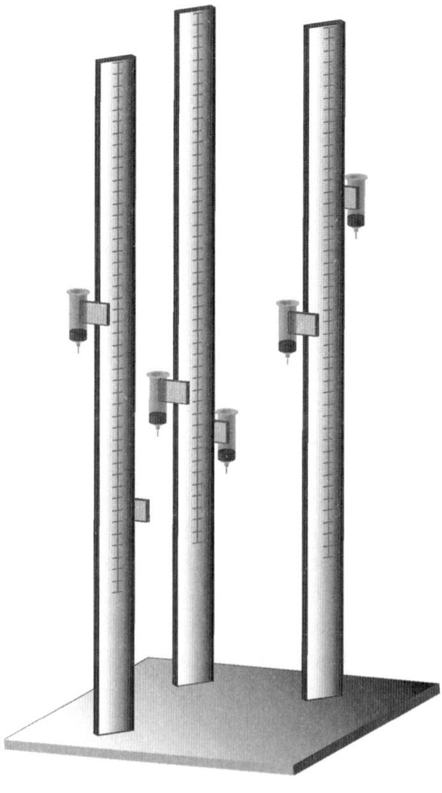

Figure 14.22 Syringe towers. Made from a commercially available erector set, the towers provide support for static syringes. We use a three pillar design with six adjustable platforms, which hold 2–9 syringes each. The ticked lines in the drawing represent rulers that are used for consistent syringe placement. All the parts necessary for constructing the tower are listed in Table 14.14.

chip. Image data acquisition is performed by a Nikon fluorescence microscope, see Appendix for component list. Our complete experimental setup can be seen in Fig. 14.23A, and functional mDAW chip with attached lines in Fig. 14.23C.

For experiments, we modify our standard LB media, by adding 0.075% Tween 80 and filtering it through 0.22 μm filter. Addition of Tween 80 prevents the cells from sticking to chip walls without any noticeable harm to the cells. Depending on your experiment, make sure to add antibiotics and any inducers to the media.

3.1.1. Overnight culture

Grow up an overnight culture of cells from -80 °C stock or from a plate. Grow the cells in 3 ml of LB media with appropriate antibiotics in 37 °C shaker incubator.

3.1.2. Cell growth

Dilute the overnight culture by a factor of 1:1000 into 50 ml of fresh media with appropriate antibiotics and inducers. Let the cells grow up to a culture density of OD600 0.05–1.0, we usually try for OD600 0.1. Depending on the cell type this step should take 2–3 h. During this time perform steps 3–6.

3.1.3. Wetting the chip

Secure the chip in a chip holder, using rubber gaskets for additional contact. We have a custom chip holder, details for which can be provided by request. Basically, it securely holds a 24 $\times$ 40 mm cover slip, while allowing light access from the bottom, physical access from the top, and secure attachment to the microscope's stage. Place the chip under microscope at 4$\times$ magnification. It is important to examine the chip for dirt and collapsed channels. This is best done at lower magnification, as you can see a larger area. Make sure that there is no debris blocking the channels or the imaging areas. Collapsed traps or channels will look darker and generally resemble in shade bonded parts of the chip. If the chip looks good proceed to wetting. Wetting the chip can be done using hydrostatic or manual pressure applied through a syringe. Attach a leur stub, a microfluidic line and a connection pin to a syringe and fill with fresh media. Make sure there are no bubbles in the syringe or the line. Bubbles can be removed by flicking the syringe or the line with your finger. Carefully insert the pin into a port. The color of the channels should start to change as fluid fills them. If using hydrostatic pressure to wet, position the syringe on the towers as high as possible and let the fluid flow through the chip. If using manual pressure, make sure to apply gentle pressure as too much pressure will lift the chip off the coverslip. As media fills the chip it will come up through the open ports and start forming droplets on the surface of the chip. Repeat this process for all the ports and until there are no more air bubbles in the chip. Media removal from the surface of the chip is best accomplished using a kimwipe.

Figure 14.23 Experimental setup. (A) The equipment setup for mDAW experiments. In the background one can see the linear actuators, it is possible to fit all eight actuators and eight linear guides in a compact space behind the microscope. Fluorescent microscope with environmental chamber can be seen in the foreground of the image. (B) Three linear actuators with linear guides and pulley systems. This is a photograph of the system described in Fig. 14.15B. (C) An mDAW chip with all the connection pins and lines attached. U.S. dime coin (17.91 mm) is shown for scale.

3.1.4. Preparing syringes

Attach a sterile 23 gauge leur stub to a clean 60 ml syringe. Take 6 feet of Tygon tubing and gently slide it over the leur stub. Attach a connection pin to the other side of the tubing. A connection pin is basically just the metal part of a 23-gauge leur stub. We make them by pulling out the metals tips from the plastic part of the leur stub using pliers. The phase condensers on some microscope may come in contact with the straight connection pins. To circumvent this issue we make L-shaped pins by bending them around the shank of a 10–32 wood screw using pliers, refer to Fig. 14.24 for exact instructions on making straight and L-shaped pins.

Depending on the intended use for the syringe, remove the plunger and extract 100 μl of media or dH_2O with a P200 pipetman. Insert the tip of the

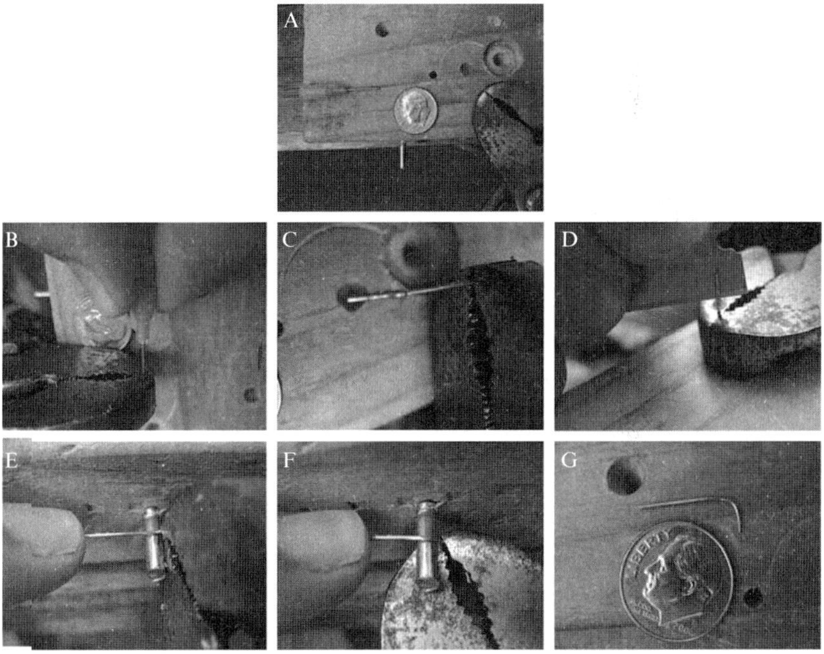

Figure 14.24 Making connection pins. (A) A nonsterile 23 gauge leur stub, dowel pin, and pliers are used to make the connection pins. (B) Using pliers grab onto the metal part of the luer stub, while holding the plastic part with your fingers. Pull them in opposite direction until they separate. (C) The metal pin alone, notice all the sealant and glue on it. (D) Using a razor blade, carefully remove all the glue from the outside of the pin. To make straight connection pins the process is finished at this point and pins just need to be cleaned in a sonicator. (E) Holding the pin with pliers, place it over the dowel pin. (F) While holding one of the ends of the pin with your finger, gently rotate the other end around the dowel pin. (G) Finished L-shaped connection pin. This method preserves the inner radius of the connection pin. Simple bending it will most likely pinch the pin. U.S. dime (17.91 mm) shown for scale.

pipetman into the syringe and make contact with the inside of the leur stub adapter. Slowly expel the fluid into the leur stub adapter which should be enough to fill it, as shown in Fig. 14.25C–E. Adding fluid to the leur stub adapter in this way greatly reduces bubble formation. Tilt the syringe slightly and gently pour the rest of the media or dH_2O into the top of the syringe, letting it run down the side before it reaches the base of the syringe. This also helps in preventing bubbles. Flick the leur stub connector to cause media to flow into the microbore tubing. If difficult bubbles are present, partially unscrew the leur stub adapter about one half turn and then retighten. This can help release bubbles. If fluid still will not enter the microbore tubing, use the syringe plunger to force the fluid in. Note if the plunger is necessary, it usually indicates a severe bubble problem. Make extra sure all bubbles are removed before proceeding. Watch the fluid flow carefully through the microbore tubing line to the exit point at the leur stub. Carefully look over the line to ensure no bubbles are present. If bubbles are present, flick the lines to release them and watch them flow to the end of the microbore tubing. Cover the syringe top using a piece of foil or parafilm,

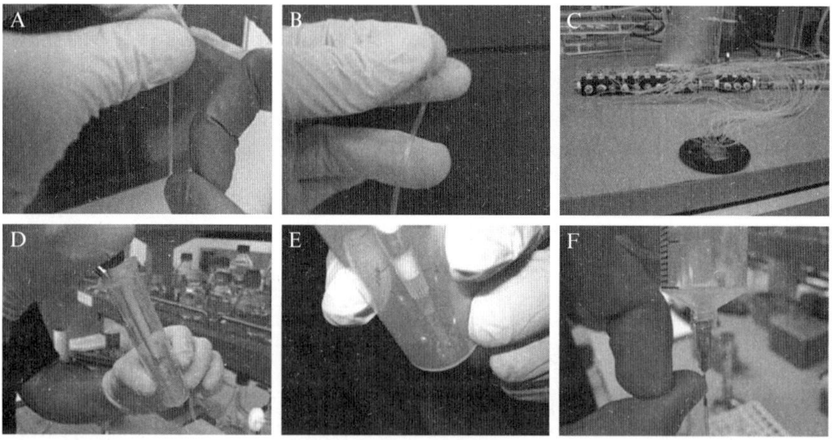

Figure 14.25 Experimental line and syringe techniques. (A) General guide for flicking a microfluidic line. Hold the line between the thumb and index finger of one hand, while flicking the downstream the line with a finger on your other hand. In the figure, the hands are place on the line so that the left hand is closer to the syringe and the right hand is closer to the microfluidic chip. (B) Technique for gentle agitation of fluid within the microfluidic device. Hold the line between thumb and index finger. Gently move the line back and forth using your ring finger. The ring finger is placed on the line toward the microfluidic device. (C) Wetting of the MDAW chip using a specially designed manifold. (D) Technique for minimizing bubbles during syringe preparation. Using a P200 pipetman draw up 100 μl of desired liquid. Remove the plunger from the syringe. Insert the pipetman into the syringe and hold both at a slight angle. (E) Insert the pipette tip all the way into the leur stub adapter and slowly expel the liquid. (F) Flicking the bottom of the syringe to fill the connection pin with liquid.

while leaving a small opening to the atmosphere. Label the syringes appropriately and make sure the connection tips do not touch any surface.

3.1.5. Connecting syringes

Attach all the syringes to the sliders on the microfluidic tower, as seen in Fig. 14.22. Adjust syringe heights appropriately. To prevent contamination of the media source, make sure it is always the highest positioned syringe. One by one, starting with the media, connect each syringe to the chip. Since the cells are not ready yet, connect a syringe filled with DI water to the cell port. Examine that no bubbles were introduced into the chip during this process. If there are bubbles, see if they can slowly disappear on their own. However, it might be necessary flush the chip by disconnecting all the syringes and repeating step 3. Once all the bubbles have been eliminated, using scotch tape secure each line to the chip holder. Tape far enough away from the chip, so that the bending of the tubing is not applying a force on the connection pin.

3.1.6. Setting up DAW software

Using the software, calibrate the syringe heights for correct mixing ratios. Create a desired profile for the syringe movement. Make sure that all the static syringes are in their "running" positions.

3.1.7. Spinning down cells

When the cells are ready, spin them down at 2700 g for 10 min. As a backup, pour the supernatant media back into the flask and place it in incubator. Add 2–3 ml of fresh media to the pelleted cells and gently vortex them, until there are no cell clumps. Load the cells into a prepared syringe. Once again, make sure there are no bubbles.

3.1.8. Loading cells

Move all syringe to their "loading" positions. Disconnect the temporary cell syringe and plug in the actual cell syringe. At this point the flow from the media and the cell port both should be going toward the waste port. At 20× magnification you should be able to see cells flowing through the channels. Adjust the height of the cell syringe so that the cells are slowly moving past the traps. Next, securely hold the cell line between the thumb and index finger of one hand, while flicking the line with a finger on your other hand as seen in Fig. 14.25A. Imagine that the pinching fingers divide the line into two parts: the syringe part and the connection pin part. The flicking should be done on the connection pin part of the line. The cells should rapidly move back and forth within the chip, as the flicking wave propagates down the line. Adjust the flicking strength to have enough force to load the traps. Once enough traps have been loaded, adjust the syringe heights to their "running" positions. Media flow should be 20–200 μm/s.

3.1.9. Starting experiment

Allow the cells to grow in the traps for 3–5 doublings, depending on your cell type this should take 1–2 h. Setup the microscope software for automated image acquisition, this includes auto-focus, various position on the chip and required light channels, i.e. phase-contrast, fluorescence. Start the imaging and the *iDAW* software at the same time.

3.1.10. Checking on cells

During the experiment it might be necessary to remove stuck cells from the channels. Hold the cell line between the thumb and index finger with your middle finger further away from the syringe. Gently move the line back and forth using your ring finger and watch as the cells smoothly mirror the motion, see Fig. 14.25B for a visual representation. This technique is useful for getting rid of stuck cells or controllably reducing cell density within a trap.

3.2. Method to set up a MDAW microfluidic experiment

In this section we will describe how to set up a microfluidics experiment using the MDAW parallel DAW microchemostat chip since this chip presents challenges not seen for smaller chips.

3.2.1. Pre-experiment preparation

See Table 14.14 for the catalog numbers of supplies listed here. The steps described in this section should be performed at least 1 day in advance of the experiment. The microfluidic devices themselves should be prepared as described in Section 2.6. Cut 26 lines using Tygon microbore tubing seven feet in length. Note the line length is dependent on the microscopy setup, there should be some slack to allow for movements of the syringe reservoirs. Obtain 26 sterile 30 ml syringes. Note for the combined ports we use 4 inch stainless steel pipe caps fitted with leur stub adapters. If these pipe caps are used, then only twenty-four 30-ml syringes are necessary. These pipe caps were manufactured by our university's machine shop, details can be provided by request. Due to their large diameter, the liquid height in the pipe caps changes very little for a given volume of accumulated fluid. This is important since the combined ports receive vastly more fluid then the individual ports during an experiment. If a 30-ml syringe were used instead of a pipe cap, it is conceivable that the height increase would affect flows in the chip over the course of a long experiment (2–3 days). See Fig. 14.5 and the accompanying text for an example of these issues.

To clean the metal parts, sonicate twenty-six reusable leur stub adapters, twenty-six 90° curved connection pins (see Figure 14.24), and the pipe caps if they are to be used, for 60 min at 60 °C in a 250-ml beaker containing 1% w/v Alconox. Sonication in 1% w/v Alconox does an excellent job of

Table 14.14 Experimental equipment, Chemicals and supplies

Equipment	Qty	Part No.	Vendor
Inverted fully automated microscope	1	Ti	Nikon
PDMS chip holder	1		Custom
Syringe towers			
Aluminum bread board, 12″ × 12″, 1/4-20 threaded	1	MB12	Thorlabs, Inc.
1″ × 3″ Extrusion 60″ long	3	1030 × 60″	80/20, Inc.
8 Hole inside corner gusset	3	25-4138	80/20, Inc.
Slide-in T-nut	6	3382	80/20, Inc.
Double Slide-in T-nut	6	3280	80/20, Inc.
1/4-20 × 1/2″ Flanged button head socket cap screw	6	3342	80/20, Inc.
1/4-20 × 1/2″ Socket head cap screw	12	3062	80/20, Inc.
1/4-20 × 3/8″ Socket head cap screw	6	3058	80/20, Inc.
1/4″ washer—black zinc	20	3258	80/20, Inc.
Double flange linear bearing brake kit ready	8	6425	80/20, Inc.
Ratcheting L-handle	8	6850	80/20, Inc.
White UHMW Pads w/brake hole	24	6490	80/20, Inc.
#8 × 3/8″ SS standard bearing pad screw	24	3625	80/20, Inc.
48″ stainless steel rule	3	2120A15	McMaster-Carr
1″ adjustable strap	1 Pkg	7565K51	McMaster-Carr
Chip holder rubber gasket		7665K11	McMaster-Carr
Chemicals and supplies	**Qty**	**Part No.**	**Manufacturer**
Connection pins (23 gauge, ID 0.017′, OD 0.025″, 1/2″ Long)		75165A684	McMaster-Carr
Single use, sterile leur stub (23 gauge)		14-826-19E	Fisher Scientific
Reusable leur stub (23 gauge, ID 0.017″, OD 0.025″, 1/2″ Long)		JGM23-0.5D	Jensen Global Inc
Disposable syring 10 ml w/leur lock tip		14-823-2A	Fisher Scientific
Disposable syring 30 ml w/leur lock tip		14-829-48A	Fisher Scientific
Disposable syring 60 ml w/leur lock tip		13-689-8	Fisher Scientific
Tygon flexible microbore tubing (ID 0.020″, OD 0.060″)	1	14-170-15B	Fisher Scientific
Tween 80	1	P8074	Sigma-Aldrich
Alconox 1104	1	04-322-4	Fisher Scientific
Liquid cell media			
DI Water		Milli-Q or better	

removing cell debris and residual media from small metal parts. After sonication rinse the parts in dH_2O. Flush water through the leur stub adapters and connection pins to remove residual Alconox. We use a manifold to flush all metal parts at once, see Fig. 14.25C. In general, we flush 3 l of dH_2O through the entire system for rinsing. After flushing with water, air can be flushed for drying. Autoclave the leur stubs, adapters and, if using them, pipe caps for 30 min on a dry cycle.

Prepare the 26 (24 if using pipe caps) syringe reservoirs as described in Section 3.1.4. Cut 8 sections of red, orange and yellow tape. Write 1–8 on each of the colored tape sets and affix to the syringe bodies. The tape will help to identify the syringe reservoirs later in the setup. Each of the eight subexperiments has two DAW input ports (A and B) and a cell port (C). We use the red tape to refer to A reservoirs, orange for B and yellow for C. Cut another three sets of colored tape and again write 1–8 on each set. Affix the tape near the end of the microbore tubing (just before the connection pin) for the appropriate reservoir. Labeling the end of the tubing helps to identify its connected reservoir, a necessity when many lines are nested together. Use scotch tape to affix the loose microbore tubing end to the syringe.

3.2.2. Cell Growth

Determine how many cell cultures will be needed, a maximum of eight can be used. Inoculate the cell cultures the day before the experiment with the appropriate media and additives. In the next morning check the culture optical density at 600 nm (OD600) using a spectrophotometer. Grow cells at 30 °C for ~4 h, to an OD600 of ~1.0 upon cell loading.

3.2.3. Media preparation

Prepare 4 ml of media for each of the 16 input syringes (ports A and B for each of the eight subexperiments). Generally the media is the same in the two inputs except for the tested component. See Table 14.15 for an example of the media composition. Add dye to one of the two input reservoirs for each subexperiment to use as an inducer tracer. After the media has been prepared, add it to each of the A and B syringe reservoirs as described in Section 3.1.4.

Table 14.15 Experimental equipment, chemicals and supplies

	Ingredients	Stock	Final	Units	Lot	1X (μL)	2.5X (mL)
1	2X SC—met media	2	1	X		2000	5
2	Methionine	50,000	500	μM		40	0.1
3	Galactose	2	0.2	%w/v		400	1
4	Raffinose	20	2	%w/v		400	1
5	ddH20	–	–	–		1160	2.9
					total:	4000	10

Add sterile dH_2O to the shunt and alternate waste reservoirs. If 30 ml syringes are used, add 4 ml of dH_2O. If pipe caps are used add $\sim$100 ml.

3.2.4. Air removal from the chip

Note that this procedure differs from that of a smaller chip (such as $MFD005_{a)}$). While removal of air can also be facilitated using a vacuum, we have found this to interfere with our cell's growth under some conditions. Affix a bonded MDAW chip onto a solid substrate, like a glass plate or a microscopy chip holder, as the one described in Step 3 of *E. coli* experimental section. Fill a 10-ml syringe with a sterile solution of 0.1% v/v Tween 80 and connect it to the central port of the chip (called the alternate waste port). Tween 80 is a surfactant that aids in clearing bubbles. It also acts as a lubricant that prevents cell clogging. Purge air from the chip by applying force to the syringe. Watch for droplets to appear at each port indicating that fluid has propagated through the chip's channel network.

We have a custom built 26 outlet manifold connected to a pressure reservoir filled with dH_2O. Each outlet of the manifold is connected to a half meter of teflon microbore tubing with a connection pin at the end. The manifold is fully autoclavable and by pressurizing all ports at once, achieves better clearance of air. Details of the manifold's construction can be provided upon request. To purge air from the manifold we pressurize it until water flows out of each connection pin. Next we connect each pin to a port on the chip, making sure a "fluidic connection" is made, that is, there is a visible droplet of fluid above each port on the chip and fluid is leaving the connection pin of the manifold. When the manifold is fully connected it is then pressurized to 4 Psi for 5 min to flush all air from the system.

3.2.5. Connecting DAW reservoirs to the device

After the air has been purged from the system, place the chip in a microscopy holder if not done already. Secure both ends of the device with scotch tape. Place the chip above the microscope stage. Our microscope has an acrylic environmental chamber around it, whose top is about 25 cm above the stage height. We place the chip on top of this box. Attach the shunt and alternate waste port reservoirs to the syringe towers. Adjust the height of the shunt reservoir to 30 cm above the stage. Adjust the height of the alternate waste reservoir to 37.5 cm above the stage. Take the connection pin from the shunt reservoir and place it several centimeters below the reservoir's fluid level. Wait for fluid to exit the end of the microbore tubing line and then connect the leur stub to the shunt port at the top of the device. Repeat this procedure to connect the alternate waste reservoir to the device. Connect the DAW input reservoirs for each subexperiment to each set of A and B ports on the device using the same procedure. After connecting the reservoirs attach them to the linear actuators. The linear actuators should be set so each reservoir is 60 cm above the stage height. Once all of the input reservoirs have been connected, bundle the lines together with scotch tape so they do not become unwieldy.

3.2.6. Processing and loading cells

At this time remove the cell cultures from the incubator and record the final OD600 value if desired. Add Tween 80 to each cell culture to a final concentration of 0.1% v/v. Vortex on a medium setting to mix. Add each culture to the appropriate syringe reservoir as described in Section 3.1.4. Be extra careful there are no bubbles in any of the cell reservoirs. While using Tween 80 helps to prevent bubbles, any that remain in the reservoir will make it extremely difficult to load cells later on. The Tween 80 will also prevent clogging of the device by excess cells.

Adjust the height of the shunt port to 11.25 cm above the height of the stage and make sure the cell reservoir holder is set to 32.5 cm above stage height. This adjustment will ensure cells flow into the shunt and not other cell ports. At this time, all cell ports will contain a bead of fluid above them since they are the outlets for the other connected reservoirs. This bead of fluid will essentially function as a small reservoir. Since the device should still be placed at 25 cm above stage height, this will be the pressure of each cell port before its reservoir is connected. Once the shunt port is lowered there will be a net flow between the fluid bead of each cell port and the shunt. When the cell reservoirs are connected their pressure will increase to 32.5 cm. If the shunt port were not lowered some flow would exit at the unconnected cell ports, possibly causing cross-contamination.

Once all cell ports have been connected, place the chip into the microscope and tape down all microbore tubing lines. Adjust the height of the cell ports to 40 cm and observe the cells entering the system at 4× magnification. If cells are not entering from the cell ports it is usually due to residual bubbles in the cell lines. Disconnect if necessary and make sure there are no bubbles. Adjust the height of the alternate waste port to force more cells into the central region of the trap if necessary. Flick the lines for each cell port to load cells into the trapping region. Continue this procedure until an adequate number of cells have been loaded (generally 20–40 yeast cells). Once the cells have been loaded adjust the heights of all reservoirs as follows: Cell ports: 15.5 cm, combined alternate waste: 14 cm, combined shunt: 11.25 cm. All heights above stage height. The level of the DAW inputs should remain at 60 cm above stage height. If desired move to the DAW junction of each subexperiment and record the height positions for 0% and 100% mixing ratios or use the calibration procedure described in Section 1.1.5.

3.2.7. Microscope setup

Record the stage locations for each of the cell traps in the eight subexperiments in the microscopy software. Switch to a 40× or 60× objective and add microscopy oil as necessary. Update the *xy* positions for each trap as they will have changed slightly. Set up the microscopy software for a multiple location experiment, using appropriate exposure settings for phase contrast and any fluorescence wavelengths. Make sure the autofocus routine is properly set up. Since the MDAWchip is quite large, there will likely be a *z*-offset between the cell traps of each subexperiment. This *z* offset needs to be compensated for. Moreover, due to stage drift over the course of an experiment the *z* offset will shift in time. Some microscope software packages cannot cope with this properly and we have written a custom macro for the NI Elements software to compensate for this changing *z* offset. The macro uses the median of the last five focal planes for each cell trap to calculate an updated *z* offset. This *z* offset is used as the best guess for where to start the next iteration's autofocus routine. Taking the median prevents a single poor autofocus result from causing a catastrophic loss of focus, which can happen if a bubble in oil droplet drifts into the field of view. We have had good success with this macro, retaining focus even after almost 72 h of an experiment. Set the linear actuator controller software for the proper input waves as described in Section 1.4.2. Begin image acquisition.

Appendix

Components of Nikon T*i* Nikon T*i* automated fluorescence microscope

Description	Qty	Part No.
Ti-E Inverted Microscope	1	MEA53100
Ti-HUBC/A Hub Controller A	1	MEF55030
Ti-HC/A AC Adapter for HUBC/A	1	MEF51010
Ti-AC120 Power Cord 120 V	1	MEF51200
USB 2.0 Cable A-B 15′, Required for DS-U2 Controller	1	97050
Ti-DH Dia Pillar Illuminator 100 W	1	MEE59905
D-LH/LC Precentered Lamphouse with LC	1	MBE75221
Halogen Lamp 12 V 100 W L.L.	3	84125
Ti-PS100 W Power Supply 100–240 V	1	MEF52250
Ti-100WRC 100 W Lamphouse Remote Cable	1	MEF51001
Power Cord	3	79035
Filter 45 mm GIF	1	MBN11200
Filter 45 mm NCB11	1	MBN11710
45 mm Heat Absorbing Filter	1	MBN11500
Eclipse Microscope Pad	1	92080
Eclipse Large Nylon Cover 14 × 26 × 32	1	92084

(*Continued*)

Appendix (*Continued*)

Description	Qty	Part No.
Package Lens Tissue 50 Sheets, 4 × 6	1	76997
CFI 10× Eyepiece F.N. 22 mm	2	MAK10100
Ti-TD Eyepiece Tube D	1	MEB52320
Ti-T-B Eyepiece Tube D	1	MEB55800
Ti-S-ER Motorized Stage With Encoders	1	MEC56100
Ti-SH Universal Holder for Motor Stage	1	MEC59110
Ti-S-C Motorized Stage Controller	1	MEF55710
Ti-S-EYOU Joystick for Motorized Stage	1	MEF55700
Ti-CT-E Motorized Condenser Turret	1	MEL51910
Ti-C-LWD LWD Lens Unit for System Condenser Turret	1	MEL56200
System LWD Ph L Annuli	1	MEH31040
TE-C LWD Ph1 Module	1	MEH41100
TE-C LWD Ph2 Module	1	MEH41200
TE-C LWD Ph3 Module	1	MEH41300
Ti-ND6-E Sextuple Motor Dic Nosepiece	1	MEP59310
CFI Plan Fluor DL 4× NA 0.13 WD 17.1 mm	1	MRH20041
CFI Plan Fluor DL 10× NA 0.3 WD 16.0 mm	1	MRH20101
CFI Plan Fluor DLL 20× NA 0.5 WD2.1 mm Sprg	1	MRH10201
CFI Plan FLUOR DLL40× OIL NA 1.3/WD 0.2MM	1	MRY10018
CFI Plan APO DM 60× Oil	1	MRD31602
CFI Plan APO DM 100× Oil	1	MRD31901
50 cc Immersion Oil Each	2	MXA20234
C-FL GFP HC HISN Zero Shift	1	96362
C-FL TRITC HYQ	1	96321
C-FL YFP HC HISN Zero Shift	1	96363
C-FL CFP HC HISN Zero Shift	1	96361
C-FL Texas Red HC HISN Zero Shift	1	96365
Ti-FL Epi-Fl Iilluminator for Ti-Series	1	MEE54100
Ti-FLC-E Motorized Epi-Fl Filter Turret for Ti-Series	1	MEV51110
Lumen 200 Illumination System	1	77011315
SmartShutter controller	1	77016099
35 mm SmartShutter w/stand-alone housing	2	77016096
Excitation Adapter for SmartShutter	1	77016169
Transmitted Light Adapter for Ti, 35 mm	1	77016168

(*Continued*)

Appendix *(Continued)*

Description	Qty	Part No.
Ti Shutter Trigger Cable for Sutter	2	MXA22088
Ti Emission Adapter	1	77016182
NIS-Elements software	1	MQS31000
NIS-Elements: Module 6D imaging	1	MQS42560
NIS-Elements: Hardware Module	1	MQS41220
C-Mount/ISO Adapter, 1×	1	MQD42000
CoolSNAP HQ2 Monochrome Camera	1	77018219
Nikon Environment-Chamber	1	77065000

ACKNOWLEDGMENTS

This project was supported by Grant Number P50GM085764 from the National Institutes of Health.

REFERENCES

Bao, J. B., and Harrison, D. J. (2006). Measurement of flow in microfluidic networks with micrometer-sized flow restrictors. *AIChE J.* **52**(1), 75–85.

Beebe, D. J., Mensing, G. A., and Walker, G. M. (2002). Physics and applications of microfluidics in biology. *Annu. Rev. Biomed. Eng.* **4,** 261–286.

Bennett, M. R., Pang, W. L., Ostroff, N. A., Baumgartner, B. L., Nayak, S., Tsimring, L. S., and Hasty, J. (2008). Metabolic gene regulation in a dynamically changing environment. *Nature* **454**(7208), 1119–1122.

Brody, J. P., Yager, P., Goldstein, R. E., and Austin, R. H. (1996). Biotechnology at low reynolds numbers. *Biophys. J.* **71**(6), 3430–3441.

Cookson, S., Ostroff, N., Pang, W. L., Volfson, D., and Hasty, J. (2005). Monitoring dynamics of single-cell gene expression over multiple cell cycles. *Mol. Syst. Biol.* **1** (2005), 0024.

Danino, T., Mondragn-Palomino, O., Tsimring, L., and Hasty, J. (2010). A synchronized quorum of genetic clocks. *Nature* **463**(7279), 326–330.

del Campo, A., and Greiner, C. (2007). Su-8: A photoresist for high-aspect-ratio and 3d submicron lithography. *J. Micromech. Microeng.* **17**(6), R81–R95.

Dertinger, T., Pacheco, V., von der Hocht, I., Hartmann, R., Gregor, I., and Enderlein, J. (2007). Two-focus fluorescence correlation spectroscopy: A new tool for accurate and absolute diffusion measurements. *Chemphyschem* **8**(3), 433–443.

Duffy, D. C., McDonald, J. C., Schueller, O. J. A., and Whitesides, G. M. (1998). Rapid prototyping of microfluidic systems in poly(dimethylsiloxane). *Anal. Chem.* **70,** 4974–4984.

Hersen, P., McClean, M. N., Mahadevan, L., and Ramanathan, S. (2008). Signal processing by the hog map kinase pathway. *Proc. Natl. Acad. Sci. USA* **105**(20), 7165–7170.

Hong, J. W., Studer, V., Hang, G., Anderson, W. F., and Quake, S. R. (2004). A nanoliter-scale nucleic acid processor with parallel architecture. *Nat. Biotechnol.* **22**(4), 435–439.

Kurth, F., Schumann, C. A., Blank, L. M., Schmid, A., Manz, A., and Dittrich, P. S. (2008). Bilayer microfluidic chip for diffusion-controlled activation of yeast species. *J. Chromatogr. A* **1206**(1), 77–82.

Lee, P. J., Helman, N. C., Lim, W. A., and Hung, P. J. (2008). A microfluidic system for dynamic yeast cell imaging. *Biotechniques* **44**(1), 91–95.

Lide, D. (2004). CRC handbook of chemistry and physics: a ready-reference book of chemical and physical data. CRC Pr I Llc.

Mather, W., Mondragon-Palomino, O., Danino, T., Hasty, J., and Tsimring, L. S. (2010). Streaming instability in growing cell populations. *Phys. Rev. Lett.* **104**(20), 208101-1–208101-4.

MicroChem (2010). Microchem corporation. su-8 photoresist product line. http://www.microchem.com/products/su_eight.htm.

Miura, K. (2005). Tracking movement in cell biology. *Adv. Biochem. Eng. Biotechnol.* **95,** 267–295.

Mosig, A., Jager, S., Wang, C., Nath, S., Ersoy, I., Palaniappan, K., and Chen, S. (2009). Tracking cells in life cell imaging videos using topological alignments. *Algorithms Mol. Biol.* **4**(1), 10.

Nguyen, N., and Wereley, S. (2002). Fundamentals and Applications of Microfluidics. Artech House Publishers.

Owens, J., Houston, M., Luebke, D., Green, S., Stone, J., and Phillips, J. (2008). Gpu computing. *Proc. IEEE* **96**(5), 879–899.

Rowat, A. C., Bird, J. C., Agresti, J. J., Rando, O. J., and Weitz, D. A. (2009). Tracking lineages of single cells in lines using a microfluidic device. *Proc. Natl. Acad. Sci. USA* **106** (43), 18149–18154.

Sia, S. K., and Whitesides, G. M. (2003). Microfluidic devices fabricated in poly(dimethylsiloxane) for biological studies. *Electrophoresis* **24**(21), 3563–3576.

Stroock, A. D., Dertinger, S. K., Ajdari, A., Mezic, I., Stone, H. A., and Whitesides, G. M. (2002). Chaotic mixer for microchannels. *Science* **295**(5555), 647–651.

Sveen, J. (2004). An introduction to matpiv v. 1.6.1. Eprint series (Department of Mathematics, University of Oslo, Oslo, Norway, 2004). Author Affiliations Bruno Moulia Biomechanics Group, Institut National de la Recherche Agronomique 234.

Taylor, R. J., Falconnet, D., Niemisto, A., Ramsey, S. A., Prinz, S., Shmulevich, I., Galitski, T., and Hansen, C. L. (2009). Dynamic analysis of mapk signaling using a high-throughput microfluidic single-cell imaging platform. *Proc. Natl. Acad. Sci. USA* **106**(10), 3758–3763.

Thorsen, T., Maerkl, S. J., and Quake, S. R. (2002). Microfluidic large-scale integration. *Science* **298**(5593), 580–584.

Whitesides, G. M., Ostuni, E., Takayama, S., Jiang, X., and Ingber, D. E. (2001a). Soft lithography in biology and biochemistry. *Annu. Rev. Biomed. Eng.* **3,** 335–373.

Whitesides, G. M., Otuni, E., Takayama, S., Jiang, X., and Ingber, D. E. (2001b). Soft lithography in biology and biochemistry. *Annu. Rev. Biomed. Eng.* **3,** 335–373.

Williams, M. S., Longmuir, K. J., and Yager, P. (2008). A practical guide to the staggered herringbone mixer. *Lab Chip* **8**(7), 1121–1129.

Xia, Y., and Whitesides, G. M. (1998). Soft lithography. *Annu. Rev. Mater. Sci.* **28,** 153–184.

Young, M. E., Carroad, P. A., and Bell, R. L. (1980). Estimation of diffusion-coefficients of proteins. *Biotechnology and Bioengineering* **22**(5), 947–955.

CHAPTER FIFTEEN

Plate-Based Assays for Light-Regulated Gene Expression Systems

Jeffrey J. Tabor

Contents

Abstract

Light sensing proteins can be used to control living cells with exquisite precision. We have recently constructed a set of bacterial light sensors and used them to pattern gene expression across lawns of *Escherichia coli* with images of green and red light. The sensors can be expressed in a single cell and controlled independently by applying different light wavelengths. Both sensors also demonstrate continuous input–output behavior, where the magnitude of gene expression is proportional to the intensity of light applied. This combination of features

Department of Bioengineering, Rice University, Houston, Texas, USA

Methods in Enzymology, Volume 497
ISSN 0076-6879, DOI: 10.1016/B978-0-12-385075-1.00015-9

allows complex patterns of gene expression to be programmed across an otherwise homogeneous cell population. The red light sensor has also been connected to a cell–cell communication system and several genetic logic circuits in order to program the bacterial lawn to behave as a distributed computer that performs the image-processing task of edge detection. Here, we will describe protocols for working with these systems in the laboratory.

1. Bacterial Photography Protocol

Equipment required

- 37 °C shaking incubator
- Stirplate
- 42 °C heat block with 50 mL conical adapter
- 37 °C still incubator with hole in top
- Kodak Ektagraphic projector (or other broad wavelength light source)
- 532 or 650 nm bandpass filter (Edmund Optics catalogue #NT43-174 or NT43-189, respectively)
- 34 × 24 mm slide with printed image
- Focusing lens

Media required

- LB medium (Lennox formulation)
- LB + 0.3 mg/mL S-gal (Sigma #S9811) + 0.5 mg/mL ferric ammonium citrate (Sigma #15-1030) + 1% Seaplaque agarose (Cambrex #CLNZR006)

Light patterned gene expression, or "bacterial photography" assays have been developed using *Escherichia coli* strains that express β-galactosidase under the control of green or red light regulated promoters (Fig. 15.1; Tabor *et al.*, 2011). Gene expression patterns are visualized in agarose slabs containing the engineered *E. coli* using the chromogenic substrate S-gal. When expressed, β-galactosidase hydrolyzes S-gal, and a product of this reaction conjugates ferric iron (also supplied in the media) to form a visible black pigment (Fig. 15.5). The pigment does not diffuse away from the cell where it was produced, resulting in crisp, stable images that remain long after the cells die.

1.1. Generation of light-sensing strains

To generate a red light-inactivated bacterial photography strain, transform RU1012 (Utsumi *et al.*, 1989) with pCph8 and pPLPCB(A) (Levskaya *et al.*, 2005; Tables 15.1 and 15.2). For a red light activated strain, transform JT2 with pCph8, pPLPCB(S), and pJT106b (Tabor *et al.*, 2011). To generate a

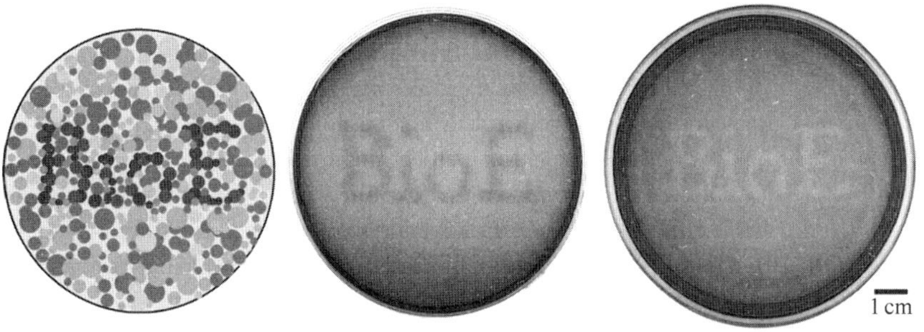

Figure 15.1 Bacterial photography. The green/red pattern shown on the left was used to produce a 34 × 24 mm slide mask. The mask was then used to project the two-color image onto agarose embedded slabs of *E. coli* expressing the green light ON (center) or red light ON (right) systems. In both cases, the light sensitive signaling pathways control the expression of β-galactosidase, which is visualized as patterns of black pigment in the media. The plates were developed in the projector–incubator for 21 h and photographed as described in the text. (See Color Insert.)

green light activated bacterial photography strain transform JT2 with pJT118 and pPLPCB(S) (Tabor *et al.*, 2011). Plate the transformed cells on LB agar containing appropriate antibiotics (Table 15.2) and recover at 37 °C overnight. The following day, recover a single colony in 3 mL LB + the appropriate antibiotics. Shake at 37 °C, 250 rpm, until the culture reaches mid-log phase (OD_{600} ~0.4–0.7). Make a −80 °C freezer stock by mixing 700 μL of the growing culture with 300 μL of autoclaved 60% glycerol in a cryovial. This frozen stock should last years as a source for experiments. In the event that the −80 °C stock begins to show phenotypic problems, retransform fresh cells and make new stocks.

1.2. Setting up the plate-based assay

To begin a bacterial photography experiment, use the −80 °C stock to inoculate 3 mL LB + the appropriate antibiotics with the relevant strain using a sterile inoculating loop or toothpick. Shake 37 °C, 250 rpm for 12–16 h to saturation (OD_{600} =3–4).

Prepare the assay media by adding 50 mg ferric ammonium citrate, 30 mg S-gal and 1 g Seaplaque low melt agarose to 100 mL LB in a 150-mL screw cap Pyrex bottle with a stir bar. Autoclave this solution and then let the media cool to 50–60 °C while slowly stirring. Transfer 15 mL of the molten agarose solution to a 50-mL sterile conical tube that is being warmed to 42 °C in a heat block. Allow the molten agarose to cool to 42 °C. After cooling, add the appropriate antibiotics and then inoculate the mixture with 30 μL of the overnight bacterial culture. Swirl the mixture vigorously while being careful not to introduce bubbles. Immediately pour the mixture into a 90-mm sterile Petri dish and tilt the dish side to side to ensure that the

Table 15.1 Light sensing *E. coli* plasmids

Plasmid	Size (bp)	Origin	Antibiotic	Properties	Reference
pCph8	4231	ColE1	Cm	Red sensor expression	Levskaya *et al.* (2005)
pJT118	8780	ColE1	Cm	Green sensor-*lacZ* output	Tabor *et al.* (2010)
pJT122	11,112	ColE1	Cm	pJT118 + red sensor expression	Tabor *et al.* (2010)
pJT106b	7728	pSC101*	Amp	Red light inverter—lacZ output	Tabor *et al.* (2010)
pJT106b3	7735	pSC101*	Amp	Red light inverter—lacZ output (weak)	Tabor *et al.* (2010)
pED$_L$ 3	9404	pSC101*	Amp	Edge detector	Tabor *et al.* (2009)
pPLPCB	3997	p15A	Kan	PCB biosynthesis	Gambetta and Lagarias (2001)
pPLPCB(A)	3985	p15A	Amp	PCB biosynthesis	Levskaya *et al.* (2005)
pPLPCB(S)	3946	p15A	Spec	PCB biosynthesis	Tabor *et al.* (2010)

Table 15.2 Light sensing *E. coli* strains

Strain	Plasmids			Antibiotics	Properties	Reference
RU1012	pPLPCB(A)	pCph8		Kan, Amp, Cm	Red light OFF	Levskaya *et al.* (2005)
JT2	pPLPCB(S)	pJT118		Kan, Spec, Cm	Green light ON	Tabor *et al.* (2010)
JT2	pPLPCB(S)	pCph8	pJT106b	Kan, Spec, Cm, Amp	Red light ON	Tabor *et al.* (2010)
JT2	pPLPCB(S)	pJT122	pJT106b	Kan, Spec, Cm, Amp	Green OR red ON	Tabor *et al.* (2010)
JT2	pPLPCB(S)	pJT122	pJT106b3	Kan, Spec, Cm, Amp	Green OR red ON (reduced red output)	Tabor *et al.* (2010)
JW3367c	pPLPCB	pCph8	pED$_L$3	Kan, Cm, Amp	Edge detector	Tabor *et al.* (2009)
JT2	pPLPCB(S)	pCph8	pED$_L$3	Kan, Spec, Cm, Amp	Edge detector (larger dynamic range)	

agarose slab will be even. Allow the mixture to solidify on the benchtop for 1 h with the lid on. In the meantime set up the projector-incubator system (Section 3). The unused LB/S-gal/agarose media can be stored solid at room temperature for approximately 1 month and can be melted in the microwave for reuse.

After the inoculated media has solidified, remove the lid and cover the top of the Petri dish with cellophane that is drawn taut. Use a razor blade to cut three narrow slits in the cellophane. This step is important: if no lid is used the slab will dry out overnight, and if the lid cannot breathe condensation will accumulate and drip onto the surface of the agarose slab, distorting the pigment pattern.

Place the agarose slab in the projector–incubator, centered under the image. For red-sensing strains use a 650-nm filtered image with light intensity less than or equal to 0.15 W/m^2. For green sensing-strains use a 532-nm filtered image at the same intensity. Both sensors respond continuously to their cognate light wavelengths at intensities between 0.002 and 0.01 W/m^2. Gently close the incubator door and grow the plate at 37 °C for 12–72 h. The longer the incubation time, the higher the black pigment intensity and contrast of the resulting image. The pigment pattern does not blur nor is it otherwise compromised during longer exposure experiments. The agarose slabs sometimes begin to crack around the edges after ~24 h, but these cracks do not affect the region of interest on the plate.

The red light OFF strain RU1012/pPLPCB(A)/pCph8 produces relatively high levels of β-galactosidase and generates well-contrasted pigment patterns in about 14 h. The β-galactosidase output of the red ON and green ON strains is lower (Tabor *et al.*, 2011). We therefore recommend exposing these strains for longer periods of time.

After exposure, remove the plate from the incubator. Carefully pull away the cellophane, being careful not to touch it to the surface of the agarose slab. Any contact with the slab surface will disrupt the pattern of pigment. Replace the Petri dish lid, wrap the perimeter of the dish in parafilm to reduce drying and place upside-down at 4 °C to stop bacterial growth and gene expression. Images do not change at 4 °C and remain permanently, even after the slabs dry out.

1.3. Two-color bacterial photography

The red and green light sensors can be coexpressed in a single cell and activated independently by cognate light wavelengths (Fig. 15.1). To generate a two-color bacterial photograph use strain JT2/pPLPCB(S)/pJT122/pJT106b3 (Table 15.2). β-galactosidase expression in this strain is induced from P_{cpcG2} in response to green light and from the red activated promoter J64067 in response to red light. Specific illumination considerations must be taken when conducting two-color bacterial photography experiments (described in Section 3).

2. Bacterial Edge Detection Protocol

Equipment required

- 37 °C shaking incubator
- Benchtop Microcentrifuge
- Stirplate
- 42 °C heat block with 50 mL conical adapter
- 37 °C still incubator with hole in top
- Kodak Ektagraphic projector (or other broad wavelength light source)
- 650 nm bandpass filter
- 34 × 24 mm slide with printed image
- Focusing lens

Solutions

- LB + 0.1 *M* HEPES pH = 6.6
- LB + 0.1 *M* HEPES pH = 8.0
- LB + 0.1 *M* HEPES pH = 6.6 + 0.3 mg/mL S-gal + 0.5 mg/mL ferric ammonium citrate + 1% Seaplaque agarose

In bacterial strains engineered to detect the edges of images, the red light responsive promoter P_{ompC} does not control *lacZ* transcription. Rather it controls transcription of *luxI*, whose product synthesizes a diffusible acyl-homoserine lactone (AHL), and *cI*, encoding the transcriptional repressor from phage λ (Tabor *et al.*, 2009). These two coexpressed signals are integrated at a single downstream promoter, R0065, that drives the expression of *lacZ*. R0065 is maximally active in the illuminated areas near dark boundaries, thus allowing the bacterial community to report the edges of the projected image (Fig. 15.2).

2.1. Media preparation

Prior to your experiment, prepare two solutions of LB supplemented with 0.1 *M* HEPES buffer. pH one batch to 6.6 and one to 8.0 and autoclave.

2.2. Generation of edge detector strains

Transform *E. coli* JW3367c with pCph8, pPLPCB, and pED_L3. Select on LB agar plates with the appropriate antibiotics (Table 15.2). Alternatively *E. coli* strain JT2 can be used for edge detection. Transform JT2 with the

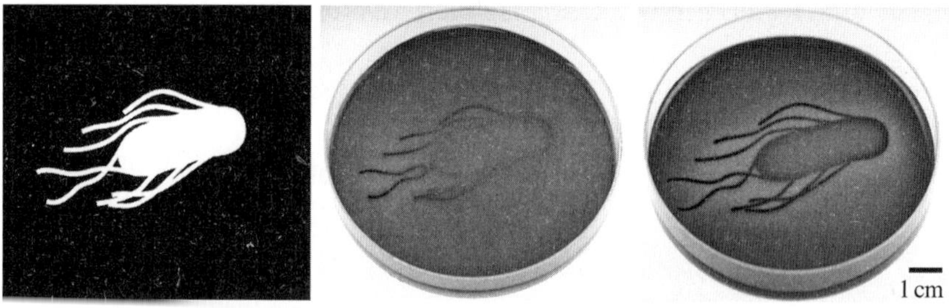

Figure 15.2 Bacterial edge detection. A mask with the pattern of an *E. coli* cell (left) was used to project 650 nm filtered light onto lawns of JW3367c (center) and JT2 (right) carrying the edge detector plasmids (Table 15.2). The red light responsive ompC promoter has a lower dynamic range in JW3367c than in JT2, resulting in thinner, lower contrast edges.

same plasmids, substituting pPLPCB(S) for pPLPCB. Recover a single transformant in 3 mL LB + 0.1 *M* HEPES pH = 8.0 and make a −80 °C stock as in Section 1.

In the edge detector plasmid pED_L3, transcription of the reporter gene *lacZ* is induced by the membrane-diffusible acylated homoserine lactone $3OC_6HSL$ (Tabor *et al.*, 2009). $3OC_6HSL$ is synthesized by LuxI, which is expressed under control of the red light-inactivated *ompC* promoter. Because P_{ompC} is leaky even under saturating light (Levskaya *et al.*, 2005), the edge detector strain should be maintained in LB media pH buffered to 8.0 in order to destabilize background $3OC_6HSL$ (Yates *et al.*, 2002). This in turn reduces the amount of background β-galactosidase, which would otherwise accumulate in cells before the start of a given experiment, reducing the contrast of the edges.

2.3. Setting up the plate-based assay

To begin the edge detection experiment, inoculate 3 mL LB + 0.1 *M* HEPES pH = 8.0 with the freezer stock of the edge detector strain. Shake at 250 rpm at 37 °C overnight until the culture reaches saturation (OD_{600} =3–4; 12–16 h).

While the culture is growing add 50 mg ferric ammonium citrate, 30 mg S-gal and 1 g Seaplaque low melt agarose to 100 mL of the previously autoclaved LB + 0.1 M HEPES pH = 6.6 solution in a 150-mL screw cap Pyrex bottle with a stir bar. Autoclave the solution. After autoclaving, cool 15 mL of the media to 42 °C in a heat block as in Section 1.

Add 500 μL of the saturated overnight culture to a microcentrifuge tube (1.7 mL). Spin at 13,000 rpm in a benchtop centrifuge for 1 min to pellet the cells. Remove the supernatant media. Resuspend the pellet in 500 μL fresh LB media (unbuffered or pH = 8.0). Pellet and resuspend twice more.

Add the appropriate antibiotics to the 42 °C molten agarose. Inoculate the molten agarose with 150 μL of the triple washed bacterial suspension. Swirl, pour, allow to solidify and cover with cellophane as in Section 1.

The intensity of edges on the plate is set by the difference in AHL production between the dark and light areas. This in turn is set by the difference in LuxI abundance in the two areas. Because the rate of transcription of *luxI* from P_{ompC} is inversely proportional to light intensity, the intensity of the edge will be greatest when dark areas are completely dark (0.000 W/m^2) and illuminated areas are at minimal P_{ompC} expression levels (>0.02 W/m^2). In our experience, black inks printed onto standard slide masks are capable of generating 0.000 W/m^2 650 nm in dark areas is if the light flux in the clear areas of the mask is set to 0.08–0.15 W/m^2. As light intensity is increased above this level, light passing through the black areas begins to increase to a level sufficient for detection by Cph8. It is also easy to "blow out" the bacterial slabs with too much light, eliminating all light response (even in the bacterial photography strains). It is unclear whether this is a result of general phototoxicity, disruption of the normal behavior of the light sensors or some combination of effects. We therefore recommend that light intensities be capped at 0.15 W/m^2.

Incubate at 36 °C for 24 h. The patterns are visible after about 14 h, but 24 h or more results in higher contrast. Remove the plate and place at 4 °C to stop growth and expression as above. The patterns appear to be at steady state as soon as they are visible. There are no significant changes in the shape of the pattern or relative intensities at the different locations on the plate if it is allowed to expose for up to 72 h. We have not quantified absolute signal intensity or edge width overtime; however, these are values that could potentially change after the initial appearance of edges.

3. Setting up a Projector–Incubator

The projector–incubator used to shine light on bacterial-embedded agarose slabs can be constructed in a variety of ways. We use a system that is relatively modular and allows for simple control of light intensity at the surface of the bacterial plate (Fig. 15.3).

Start by setting a light source on an adjustable height stand. We have had success with the Kodak Ektagraphic III AMT projector equipped with an 82 V, 300 W Philips FocusLine quartz bulb. This projector has an adjustable telescopic lens that can be used to vary the light intensity at the surface of the agarose slab without moving the projector itself. Next, fix a 5-cm square bandpass filter (e.g., NT43-189, Edmunds Optics, Barrington, NJ) in the projector itself (in the location intended to hold slides) or immediately below the telescopic lens.

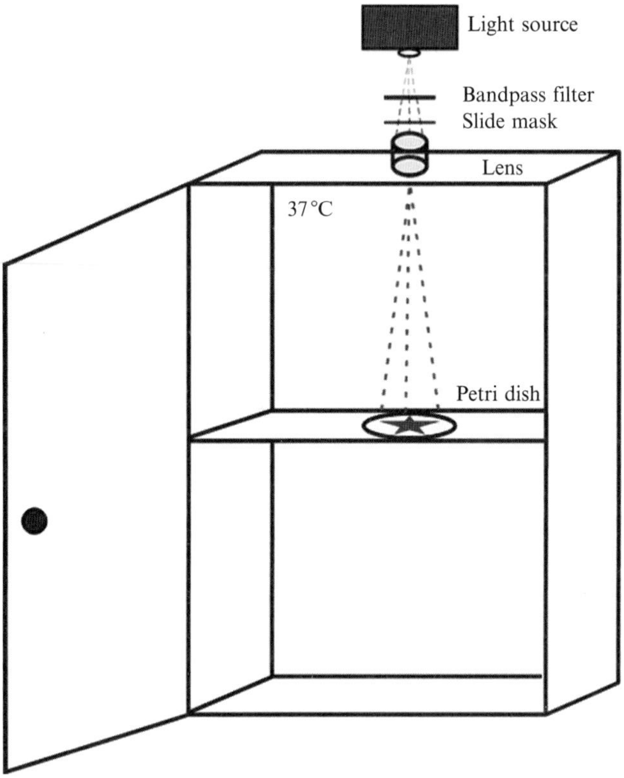

Figure 15.3 Projector–incubator apparatus. A broad wavelength light source is placed on an adjustable height platform on top of a standard laboratory incubator with a 2–3″ hole in the top. The light emitted from the source is filtered through an appropriate bandpass filter and then patterned through a 34 × 24 mm slide mask. The image is focused through a lens onto the growing lawn of bacteria embedded within an agarose slab in a Petri dish inside the incubator.

Place the 34 × 24 mm slide mask (Fig. 15.4) on a second adjustable height platform in the light path. 34 × 24 mm slides can be produced from digital files (.jpg, .tiff, etc.) at many photo laboratories. While it is best to create your source files at the printing resolution of the slide printer you will be using, we generate our source images files at a resolution of 2032 dpi. Adjusting slide height allows focusing of the image on the agarose slab. Set a focusing lens underneath the 35-mm slide, atop the hole of the incubator. The thermometer hole on many incubators works well, otherwise the hole can be drilled. While a variety of focusing lenses can be used, those with wide apertures and adjustable focal lengths are optimal. Finally, set a light absorbing black material such as construction paper on the incubator rack where the Petri dish will be placed. This serves to limit reflection of light that passes through or around the sample.

Figure 15.4 Production of a two-color slide mask. A source image with significant RGB components in most pixels (left) is used as a starting point to make the mask. In Adobe Photoshop, the image is shrunk to ~10 mm in height (2032 dpi) and placed in the center of a 34 × 24 mm black background. Most of the red and blue signal is removed from pixels in the green areas, and most of the green and blue signal is removed from red areas. A .tif file of the enhanced image is then exported from Photoshop and used to print the slide. (See Color Insert.)

Turn on the light source, center, and focus the image on a white piece of paper and then quantify the intensity of the image. The intensity of a given range of light wavelengths at the surface of the plate can be quantified using a calibrated spectroradiometer such as the EPP2000 models from Stellarnet, Inc. (Tampa, FL, USA). These models have a fiber optic probe that can be held in your hand or immobilized at the position of measurement using a tripod. The intensity of light can be changed by moving the projector further from or closer to the focusing lens, adjusting the telescopic lens, placing neutral density filters in the light path or moving the bacterial agar plate further from or closer to the top of the incubator.

When taking a two color bacterial photograph the intensities of the different light wavelengths at the surface of the plate must be carefully controlled. While the green sensor is not activated by red light, the red sensor has a long blue tail and is activated linearly as a function of green light intensity (Tabor *et al.*, 2011). Because the green sensor saturates above 0.1 W/m^2 532 nm light, green light above this level only serves to increase unwanted expression from the red sensor.

The simplest method for projecting a color image is to use a broad wavelength light source masked only by a color slide (Fig. 15.4). When generating the color slide it is important to carefully prescribe the red, green, and blue (or CMYK) intensity values of each pixel in the source file. If starting with a color photograph, for example, many pixels which appear to be predominantly one color will have significant components of other color channels. Multiple light wavelengths passing through a mask can in turn trigger unwanted responses. In this case, the RGB or CMYK values of each pixel can be modified using software such as MATLAB or Adobe Photoshop (Fig. 15.4).

4. The β-Galactosidase/S-Gal Reporter System

In the agarose slab experiments, we have found patterns of fluorescent protein expression difficult to detect. We have, however, had success with the highly sensitive β-galactosidase/S-gal reporter system. Unlike fluorescent proteins, which have quantum yields significantly less than 1, β-galactosidase is an enzyme that turns over many substrates per minute. Substrate-turnover results in signal amplification, which enables more sensitive detection. Indeed, a single β-galactosidase enzyme per cell (1 Miller Unit) can be readily detected by the standard colorimetric Miller assay, and in our plate-based assays (Tabor *et al.*, 2011). By contrast, the lower limit of detection of GFP-derived fluorescent proteins per cell in high sensitivity microscopy experiments is on the order of 100 molecules (Golding *et al.*, 2005).

S-gal (3,4-cyclohexenoesculetin-ß-D-galactopyranoside) allows the sensitive detection of β-galactosidase in solid phase assays. S-gal is an engineered sugar analog that is hydrolyzed by β-galactosidase at a glycosidic bond (James *et al.*, 1996). The products of this hydrolysis are galactose and esculetin (Fig. 15.5). Two esculetin groups coordinate a ferric iron, which is added to the medium in the form of ferric ammonium citrate. Esculetin-coordinated iron is an insoluble black precipitate that does not diffuse outside the cell

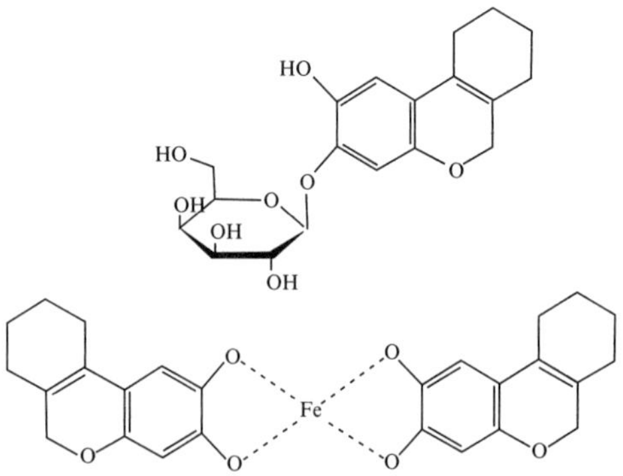

Figure 15.5 S-gal (top) is cleaved by β-galactosidase at the glycosidic bond, liberating galactose from an esculetin group. Two esculetins then coordinate a ferric iron (bottom), which is supplied in the agarose slab in the form of ferric ammonium citrate. Esculetin-coordinated iron forms an insoluble black precipitate which is visualized as pigment in the bacterial photographs.

membrane (James *et al.*, 1997). The high visibility and lack of diffusion allow sharp, crisp patterns of gene expression to be visualized by eye. Moreover, the esculetin-iron precipitate is very stable, making bacterial photographs effectively permanent.

The classic blue chromogenic β-galactosidase substrate X-gal (5-bromo-4-chloro-3-indolyl-beta-D-galactopyranoside) is inappropriate for experiments involving projected light, as X-gal is sensitive to light. S-gal, by contrast, is insensitive to light and has the added benefit of being heat resistant making it microwaveable and autoclavable during media preparation. The color of the S-gal product is also higher contrast than that of the X-gal product against the bacterial growth media (Heuermann and Cosgrove, 2001).

5. Quantifying Signal Intensity on the Plates

In our experiments, black pigment intensity is proportional to the concentration of β-galactosidase at a given location on the plate. Signal intensity at a given position on the plate can be quantified by taking a digital photograph of the plate and analyzing the intensity profile in image analysis software.

To image the plates, place them face-up on a white-light emitting fluorescent light box in a darkened room. If the plate was previously stored at 4 °C, first warm the plate in a 37 °C incubator for 1 h to remove condensation on the plastic. Remove the Petri dish lid and take a photograph with the surface of the lens parallel to the surface of the agarose slab. We find that a digital SLR camera with a macro lens and lens hood is superior to point and shoot models, which can produce photographs with distorted edges. Signal intensity across the plate can be quantified using image analysis software such as MATLAB (The Mathworks Inc.) or ImageJ (NIH). Total luminous intensity will vary significantly across a plate in a given photograph, distorting pixel intensity profiles. To correct for this, each data point must be corrected against a local background value.

6. Microscopic Imaging of Agarose Slabs

The cells inside an agarose slab can be imaged using fluorescence microscopy. Figure 15.6 shows a bacterial photograph taken by a red light OFF strain that constitutively expresses green fluorescent protein. The macroscopic pigment pattern was imaged as above and the bacteria within the slab were visualized using a fluorescent dissecting scope (Lumar V12 Stereo microscope, Zeiss) and GFP filters.

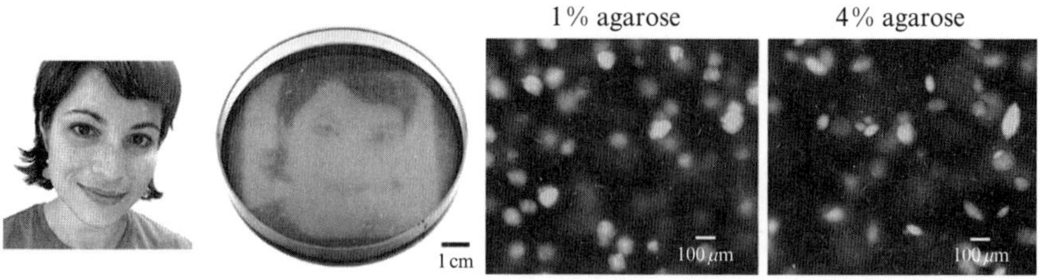

Figure 15.6 Microscopic view of a bacterial photograph. The mask on the left was used to project a 650-nm image onto a lawn of RU1012/pPLPCB(A)/pCph8 which were made to constitutively express green fluorescent protein. The macroscopic view of the plate in visible light (2nd from left) shows a standard bacterial photograph. The bacteria were then imaged at 60× magnification on a fluorescent dissecting microscope (2nd from right). The result when the experiment is conducted using 4% agarose is shown on the right. (See Color Insert.)

The bacteria mainly occupy the top 1 mm of the slab (not shown) and form ~100 μm diameter clumps ~200–300 μm apart in a given X–Y plane. Assuming the clumps are spherical, they have a volume of ~5 × 10^5 μm^3. As *E. coli* 2 μm by 1 μm cylinders, these clumps contain approximately 3 × 10^5 cells. Clump size and shape appears to be dependent on the concentration of agarose but not the genotype of the *E. coli*. For example, under the standard growth conditions of 1% agarose, the photography strains (RU1012 and JT2) and the edge detector strain (JW3367c) form morphologically similar clumps (not shown). However, at 4% agarose the clumps are on average smaller and more elliptical (Fig. 15.6).

It is unclear how the bacterial clumps originate. To generate ~3 × 10^5 descendants, a single cell would need to divide 18 times over the course of a 14-h experiment (~45 min division time). Given the relatively low oxygen environment, the fact that the cells carry several multicopy plasmids, overexpress multiple heterologous proteins and are challenged with 3–4 antibiotics, it seems unlikely that they would achieve this rate of division in these experiments. Another possibility is that that numerous bacteria aggregate during the agarose solidification process (at time 0), become immobilized by the agarose matrix, and then grow together to the form the clumps.

7. Properties of Relevant Strains

The red light sensor Cph8 is a hybrid protein constructed from the photosensory domain of the Synechocystis PCC6803 phytochrome Cph1 and the histidine kinase domain of the *E. coli* protein EnvZ. To avoid interference, Cph8 must be used in a strain lacking *envZ*. A number of *envZ* knockout strains exist and four have been used in plate-based assays with Cph8.

The original bacterial photographs were taken in *E. coli* RU1012 (Utsumi *et al.*, 1989) and its derivative CP919 (Baumgartner *et al.*, 1994). RU1012 is a derivative of *E. coli* MC4100 that is commonly used to study signaling through the EnvZ/OmpR pathway. MC4100 has undergone significant genetic alteration in the laboratory and is estimated to contain at least 123 gene disruptions relative to its ancestor *E. coli* K12 (Peters *et al.*, 2003). RU1012 was engineered to carry a genomic fusion between the *ompC* promoter (P_{ompC}) and *lacZ* making it a natural reporter strain for Cph8. CP919 is a descendent of RU1012 with the genes encoding ribose binding protein (*rbsB*) and ribokinase (*rbsK*) disrupted by a transposon insertion (Baumgartner *et al.*, 1994) and it performs identically to RU1012 in our experiments.

RU1012 and CP919 show an approximate 10-fold reduction in transcription from P_{ompC} as a result of exposure to red light and produce very smooth and high contrast bacterial photographs (Levskaya *et al.*, 2005). We recently constructed a derivative of RU1012, named JT2, with the $P_{ompC-lacZ}$ fragment of the genome knocked out (Tabor *et al.*, 2011). This allows *lacZ* to be used as a reporter for other promoters. As expected, this strain functions qualitatively identically to RU1012 in plate-based assays when the $P_{ompC-lacZ}$ fusion is reintroduced on plasmids.

Plate-based assays with Cph8 have also been conducted in the *E. coli* K12 W3110 *envZ* knockout strain JW3367c (Baba *et al.*, 2006; Tabor *et al.*, 2009). The genome of JW3367c has undergone far less laboratory manipulation than that of RU1012/CP919 and it lacks *lacZ*. The dynamic range of the P_{ompC} promoter under control of Cph8 is lower in JW3367c than in RU1012. Moreover, the dynamic range of the biobrick P_{ompC} promoter R0082, which contains the 108-bp upstream of the transcription start site of the *ompC* gene, is lower than the version of the *ompC* promoter in the RU1012 genome (termed $P_{ompC1157}$) $P_{ompC1157}$ carries the same 108 bp of upstream sequence as well as the first 789 bp of the *ompC* open reading frame.

The green light sensor, its response regulator and promoter are absent from the genome of *E. coli*. Assuming no interference with other two component signaling systems, the green senor should therefore be functional in any *E. coli* strain background.

8. Properties of Relevant Plasmids

8.1. pCph8

The red light sensing protein Cph8 is expressed from the plasmid pCph8 (Levskaya *et al.*, 2005). Cph8 is a fusion between the N-terminal 515 amino acids of the phytochrome Cph1 from Synechocystis PCCC6803 and the C-terminal 229 amino acids of *E. coli* EnvZ. This fragment contains the photosensory core of Cph1 and the histidine kinase domain of EnvZ. Transcription of *cph8* is driven by the TetR-repressible $P_{LTetO-1}$ promoter (Lutz and Bujard, 1997). All experiments using the red light sensor have to this point been conducted in *E. coli* strains lacking the TetR protein, which makes Cph8 expression constitutive.

8.2. pJT106 and derivatives

Cph8 is produced in an active ground state, and it is inactivated by red light. For many applications, a sensor that is produced in an inactive ground state and that can be activated by light is preferred. We have therefore constructed several plasmids that invert the transcriptional output of the red light sensor. Among these plasmids are pJT106 (Tabor *et al.*, 2009) and its derivatives pJT106b and pJT106b3 (Tabor *et al.*, 2011). They are based on the biobrick vector pSB4A3 (Shetty *et al.*, 2008) which contains a 2- to 3-copy pSC101★ origin of replication and an ampicillin resistance marker (Table 15.1, Fig. 15.7). The CI is repressor protein from phage λ is expressed from P_{ompC} and the *lacZ* gene is in turn expressed from a CI-repressible promoter. CI is translated under control of the weak ribosome binding site "RBS 3" (Weiss, 2001) and carries a C-terminal LVA degradation tag that reduces protein half-life to approximately 30 min (Andersen *et al.*, 1998). These two features result in relatively low abundance CI in

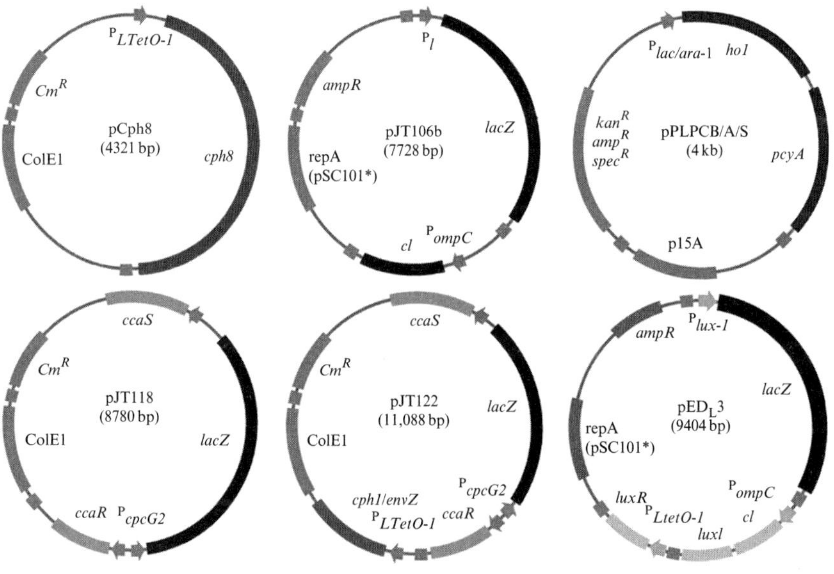

Figure 15.7 Maps of relevant plasmids.

both the high and low expression states. This is important as even small amounts of CI expressed from multicopy plasmids can result in strong repression of a CI-binding promoter (Yokobayashi *et al.*, 2002). On pJT106, *lacZ* is expressed from the hybrid LuxR–CI-binding promoter R0065, while on pJT106b and pJT106b3 it is expressed from J64067, an engineered variant of R0065 which lacks the LuxR-binding site and has higher levels of transcription. pJT106b3 differs from pJT106b only in that it has a much weaker RBS upstream of the *lacZ* output gene, resulting in lighter pigment patterns on plates (Tabor *et al.*, 2011).

8.3. pPLPCB and derivatives

The phycocyanobilin (PCB) biosynthetic plasmid pPLPCB was constructed by Gambetta and Lagarias (2001). This plasmid carries the ~15 copy p15A origin of replication, kanamycin resistance marker and the *ho1* and *pcyA* genes from Synechocystis PCC6803 under the control of the $P_{lac/ara\text{-}1}$ promoter. Ampicillin and Spectinomycin resistant versions of pPLPCB (pPLPCB(A) and pPLPCB(S)) have been constructed for compatibility with other plasmids (Table 15.1).

$P_{lac/ara\text{-}1}$ is induced by L(+)-arabinose in *E. coli* strains containing AraC and dominantly repressed by LacI (Lutz and Bujard, 1997). RU1012, CP919, JT2, and JW3367c all contain AraC and only JW3367c contains LacI. It is not necessary to induce PCB biosynthesis with IPTG in strain JW3367c, however, as a single genomic copy of *lacI* is insufficient to repress lac-repressible promoters on most multicopy plasmids. We have also found that arabinose is not required for light responsivity in any of the strains above.

pPLPCB has been reported to be unstable in *E. coli* (J.C. Lagarias, personal communication) likely owing to the selective disadvantage imparted by the wasteful conversion of heme to PCB. Once pPLPCB is transformed, it is therefore recommended that the strain be propagated for as few generations as possible prior to making −80 °C stocks and conducting experiments. For many light responsive experiments including edge detection, the transformation of three or more plasmids is required. Triple transformation is inefficient (particularly in RU1012 and its derivatives) and it is often useful to transform the plasmids iteratively. In this case, we recommend transforming pPLPCB last in order to avoid unnecessary propagation and unwanted mutations. Alternatively pPLPCB could be carried in a *lacI* overexpressing background to keep PCB biosynthesis repressed until needed.

8.4. Green sensor plasmids

pJT118 is the green light sensor expression and reporter plasmid (Table 15.1, Fig. 15.7; Tabor *et al.*, 2011). It constitutively expresses the green/red photoswitchable histidine kinase *ccaS* and its response regulator *ccaR* under control of their native cyanobacterial promoters and *lacZ* from the phospho–CcaR-binding *cpcG2* promoter. It is based on the same vector backbone as pCph8 (pProTetE333, Clontech), though $P_{LtetO\text{-}1}$ promoter is removed. pJT122 is the green + red sensor expression plasmid which was constructed by adding the *cph8* expression cassette from pCph8 to pJT118 (Fig. 15.7).

REFERENCES

Andersen, J. B., Sternberg, C., Poulsen, L. K., Bjorn, S. P., Givskov, M., and Molin, S. (1998). New unstable variants of green fluorescent protein for studies of transient gene expression in bacteria. *Appl. Environ. Microbiol.* **64,** 2240–2246.

Baba, T., Ara, T., Hasegawa, M., Takai, Y., Okumura, Y., Baba, M., Datsenko, K. A., Tomita, M., Wanner, B. L., and Mori, H. (2006). Construction of *Escherichia coli* K-12 in-frame, single-gene knockout mutants: The Keio collection. *Mol. Syst. Biol.* **2,** 2006.0008.

Baumgartner, J. W., Kim, C., Brissette, R. E., Inouye, M., Park, C., and Hazelbauer, G. L. (1994). Transmembrane signalling by a hybrid protein: Communication from the domain of chemoreceptor Trg that recognizes sugar-binding proteins to the kinase/phosphatase domain of osmosensor EnvZ. *J. Bacteriol.* **176,** 1157–1163.

Gambetta, G. A., and Lagarias, J. C. (2001). Genetic engineering of phytochrome biosynthesis in bacteria. *Proc. Natl. Acad. Sci. USA* **98,** 10566–10571.

Golding, I., Paulsson, J., Zawilski, S. M., and Cox, E. C. (2005). Real-time kinetics of gene activity in individual bacteria. *Cell* **123,** 1025–1036.

Heuermann, K., and Cosgrove, J. (2001). S-Gal: An autoclavable dye for color selection of cloned DNA inserts. *Biotechniques* **30,** 1142–1147.

James, A. L., Perry, J. D., Ford, M., Armstrong, L., and Gould, F. K. (1996). Evaluation of cyclohexenoesculetin-beta-D-galactoside and 8-hydroxyquinoline-beta-D-galactoside as substrates for the detection of beta-galactosidase. *Appl. Environ. Microbiol.* **62,** 3868–3870.

James, A. L., Perry, J. D., Ford, M., Armstrong, L., and Gould, F. K. (1997). Note: Cyclohexenoesculetin-beta-D-glucoside: A new substrate for the detection of bacterial beta-D-glucosidase. *J. Appl. Microbiol.* **82,** 532–536.

Levskaya, A., Chevalier, A. A., Tabor, J. J., Simpson, Z. B., Lavery, L. A., Levy, M., Davidson, E. A., Scouras, A., Ellington, A. D., Marcotte, E. M., and Voigt, C. A. (2005). Synthetic biology: Engineering *Escherichia coli* to see light. *Nature* **438,** 441–442.

Lutz, R., and Bujard, H. (1997). Independent and tight regulation of transcriptional units in *Escherichia coli* via the LacR/O, the TetR/O and AraC/I1-I2 regulatory elements. *Nucleic Acids Res.* **25,** 1203–1210.

Peters, J. E., Thate, T. E., and Craig, N. L. (2003). Definition of the *Escherichia coli* MC4100 genome by use of a DNA array. *J. Bacteriol.* **185,** 2017–2021.

Shetty, R. P., Endy, D., and Knight, T. F., Jr. (2008). Engineering BioBrick vectors from BioBrick parts. *J. Biol. Eng.* **2,** 5.

Tabor, J. J., Salis, H. M., Simpson, Z. B., Chevalier, A. A., Levskaya, A., Marcotte, E. M., Voigt, C. A., and Ellington, A. D. (2009). A synthetic genetic edge detection program. *Cell* **137,** 1272–1281.

Tabor, J. J., Levskaya, A., and Voigt, C. A. (2011). Multichromatic control of gene expression in *Escherichia coli*. *J. Mol. Biol.* **405,** 2, 315–324.

Utsumi, R., Brissette, R. E., Rampersaud, A., Forst, S. A., Oosawa, K., and Inouye, M. (1989). Activation of bacterial porin gene expression by a chimeric signal transducer in response to aspartate. *Science* **245,** 1246–1249.

Weiss, R. (2001). Cellular Computation and Communications Using Engineered Genetic Regulatory Networks. Massachussets Institute of Technology, Cambridge, MA, USA.

Yates, E. A., Philipp, B., Buckley, C., Atkinson, S., Chhabra, S. R., Sockett, R. E., Goldner, M., Dessaux, Y., Camara, M., Smith, H., and Williams, P. (2002). N-acylhomoserine lactones undergo lactonolysis in a pH-, temperature-, and acyl chain length-dependent manner during growth of *Yersinia pseudotuberculosis* and *Pseudomonas aeruginosa. Infect. Immun.* **70,** 5635–5646.

Yokobayashi, Y., Weiss, R., and Arnold, F. H. (2002). Directed evolution of a genetic circuit. *Proc. Natl. Acad. Sci. USA* **99,** 16587–16591.

CHAPTER SIXTEEN

Spatiotemporal Control of Small GTPases with Light Using the LOV Domain

Yi I. Wu,*[,1] Xiaobo Wang,[†] Li He,[†] Denise Montell,[†] *and* Klaus M. Hahn*

Contents

Abstract

Signaling networks in living systems are coordinated through subcellular compartmentalization and precise timing of activation. These spatiotemporal aspects ensure the fidelity of signaling while contributing to the diversity and

* Department of Pharmacology and Lineberger Cancer Center, University of North Carolina, Chapel Hill, North Carolina, USA

† Department of Biological Chemistry, Center of Cell Dynamics, Johns Hopkins School of Medicine, Baltimore, Maryland, USA

1 Present address: Robert D. Berlin Center for Cell Analysis and Modeling, and Department of Genetics and Developmental Biology, University of Connecticut Health Center, Farmington, Connecticut, USA

Methods in Enzymology, Volume 497

ISSN 0076-6879, DOI: 10.1016/B978-0-12-385075-1.00016-0

specificity of downstream events. This is studied through development of molecular tools that generate localized and precisely timed protein activity in living systems. To study the molecular events responsible for cytoskeletal changes in real time, we generated versions of Rho family GTPases whose interactions with downstream effectors is controlled by light. GTPases were grafted to the phototropin LOV (light, oxygen, or voltage) domain (Huala, E., Oeller, P. W., Liscum, E., Han, I., Larsen, E., and Briggs, W. R. (1997). Arabidopsis NPH1: A protein kinase with a putative redox-sensing domain. *Science* **278,** 2120–2123.) via an alpha helix on the LOV C-terminus (Wu, Y. I., Frey, D., Lungu, O. I., Jaehrig, A., Schlichting, I., Kuhlman, B., and Hahn, K. M. (2009). A genetically encoded photoactivatable Rac controls the motility of living cells. *Nature* **461,** 104–108.). The LOV domain sterically blocked the GTPase active site until it was irradiated. Exposure to 400–500 nm light caused unwinding of the helix linking the LOV domain to the GTPase, relieving steric inhibition. The change was reversible and repeatable, and the protein could be returned to its inactive state simply by turning off the light. The LOV domain incorporates a flavin as the active chromophore. This naturally occurring molecule is incorporated simply upon expression of the LOV fusion in cells or animals, permitting ready control of GTPase function in different systems. In cultured single cells, light-activated Rac leads to membrane ruffling, protrusion, and migration. In collectively migrating border cells in the *Drosophila* ovary, focal activation of photoactivatable Rac (PA-Rac) in a single cell is sufficient to redirect the entire group. PA-Rac in a single cell also rescues the phenotype caused by loss of endogenous guidance receptor signaling in the whole group. These findings demonstrate that cells within the border cell cluster communicate and are guided collectively. Here, we describe optimization and application of PA-Rac using detailed examples that we hope will help others apply the approach to different proteins and in a variety of different cells, tissues, and organisms.

1. Introduction

In living cells, transient protein interactions, conformational changes, and posttranslational modifications are controlled with precise kinetics and localization. To better understand how location and timing impact signaling, and to harness local signaling to produce cell behaviors, approaches have been developed to modulate protein activity in living cells and animals. Modulating activity with light was initially accomplished through covalent attachment of photolabile protecting groups to specific residues of proteins, blocking activity until irradiation broke the covalent bond and freed the "caged" residue. These techniques require high-energy light to break covalent bonds and lead to irreversible protein activation. They are technically challenging in that specific residues must be modified, and the resulting covalent protein adduct must be loaded into cells via microinjection or other techniques that perturb the cell membrane. In contrast, the methods described here and elsewhere in this volume enable reversible photoactivation of modified proteins that are entirely genetically encoded, using less harmful wavelengths and naturally occurring cofactors that permit ready application in living animals. Here, we use photoactivatable Rac (PA-Rac) as an example and provide detailed procedures for generating and applying a PA protein.

2. The LOV Domain as a Tool for Protein Caging

The LOV (light, oxygen, or voltage) domain (Huala *et al.*, 1997) denotes a subgroup of the Per-ARNT-Sim (PAS) domains, named for homologous domains that occur within signaling proteins initially identified in the *Drosophila* period circadian rhythm (Per) and single-minded (Sim) proteins and in the vertebrate aryl hydrocarbon receptor nuclear transporter (ARNT). The LOV domain has a conserved central antiparallel β sheet with five strands, and several α helices, within which resides a flavin analogue, typically a flavin mononucleotide (FMN) (Möglich *et al.*, 2009). Upon 400–500 nm illumination, excitation of the flavin molecule leads to the formation of a covalent linkage between the C4(a) atom in the flavin and a thiol from a conserved Cys residue in the LOV domain. This reaction is reversible and undergoes recovery in the dark over seconds to minutes depending on the protein (Swartz *et al.*, 2001). Attached to the LOV domain in phototropin is a helix (named Jα) that is docked on the β-sheet of the LOV domain (Harper *et al.*, 2003). Upon illumination, conformational changes occur throughout the LOV domain, leading to dissociation and unwinding of the Jα helix.

As shown in Fig. 16.1A, this light-induced helix unwinding can be the basis of a generalizable approach to protein caging. When the helix unwinds, the LOV domain is no longer held tightly against the GTPase, but is now on the end of a long tether. This relieves steric inhibition and "activates" the protein. Advantages of the LOV domain include its small size (~15 kDa), well-characterized light-induced conformational change, and the prevalence of the flavin cofactor, which occurs in cultured cells and *in vivo*. For some applications, the ability to reverse activation simply by turning off the light will be valuable, although the current time constants for dark recovery are on the order of seconds to minutes.

3. Design and Structure Optimization of PA-Rac

We were interested in grafting the LOV-Ja domain onto Rac such that activity would be inhibited in the dark, but restored upon illumination. The interconversion of Rho GTPases between active (GTP-bound) and

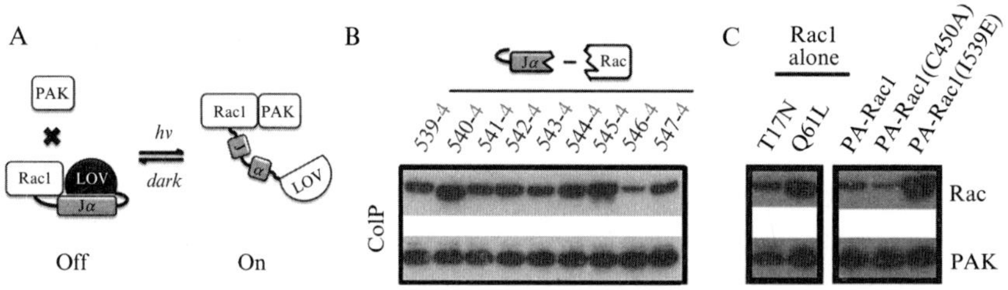

Figure 16.1 Design and characterization of photoactivatable Rac (PA-Rac). (A) cartoon diagram of PA-Rac light-induced conformational change and the resultant binding of PA-Rac to effector PAK. (B) pull down results of different linkages between Jα and Rac using PAK as bait. (C) light control of the binding of PA-Rac to PAK exceeded the difference in binding between inactive (T17N) and active (Q61L) Rac mutants (comparing PA-Rac in the dark, the dark state mutant C450A, and the lit state mutant I539E).

inactive (GDP-bound) states is catalyzed by guanine nucleotide exchange factors (GEFs) and GTPase-activating proteins (GAPs). We wanted control of Rac activity to be solely a function of our cell irradiation, and not subject to normal upstream regulatory pathways. We therefore locked PA-Rac in its GTP-loaded state by using a hydrolysis-deficient mutant (Q61L in Rac1), which also blocks its interaction with GEFs. We introduced mutations at the interface between GTPases and GAP proteins (N91H/E92H in Rac) based on known crystal structures, leading to decreased binding of p50GAP.

Because the C-terminus of Jα was well packed against the LOV domain in the dark, we reasoned that the globular LOV domain could establish a steric block, occluding effector binding, if the C-terminus of Jα were tethered near Rac's effector binding interface. Irradiation and unwinding of the helix should relieve this block. The steric block was achieved by attaching the LOV-Jα to the N-terminus of Rac. We sampled different linkages by varying the junctional residues at the C-terminus of Jα and the N-terminus of Rac, leading to a working linkage (L546-I4) that significantly decreased the interaction of active Rac with downstream effectors. To optimize GTPase caging, we developed a simple pull down assay taking advantage of the efficient FLAG epitope-M2 monoclonal antibody system (Sigma). Different linkers and Rac/Lov truncations were tested to find those best able to inhibit Rac interactions with effectors in the dark, and to restore them in the light (Fig. 16.1B). HEK293 cells were transfected with both FLAG-tagged p21-activated kinase (PAK), a Rac effector, and different LOV domain-Rac fusions. Pull down of Rac by Pak was used to gauge the effect of structural modification.

To avoid unintended Rac activation by light, the cells were maintained in an incubator in a dark room and all steps were performed under red light. The cells were lysed on ice in prechilled lysis buffer (50 m*M* Tris, pH 7.5,

150 mM NaCl, and 1% Triton X-100 with an EDTA-free protease inhibitor cocktail). EDTA was avoided because GTP binding to Rac requires Mg^{2+} ions, which are chelated by EDTA. Two wells (ϕ = 3.5 cm) from a 6-well plate with cells at 50–75% confluency, lysed in 400 μL lysis buffer, provided enough protein at an appropriate concentration for this assay. Incubation on ice for 20 min was sufficient for lysis. The cell lysates were carefully collected and cleared of insoluble debris by centrifuging at 10,000×g for 5 min. Small portions of the supernatants (30–40 μL) were sampled at this time for western blot analysis of protein expression. Forty microliters (50% slurry) of anti-FLAG antibody-conjugated agarose (Sigma) was mixed with the rest of the samples and incubated for 1 h in the dark at 4 °C, followed by three washes with 500 μL of lysis buffer. Handee spin columns (Pierce) made sample handling in the dark less challenging and also minimized the variation between experiments. The bound proteins were eluted with 40 μL of 200 μg/mL 3× FLAG peptide after 5 min of incubation. The pull down samples were subjected to western blot analysis using antibodies against fluorescent protein JL-8 (Clontech) and PAK N-20 (Santa Cruz) or rabbit polyclonal anti-FLAG antibody (Sigma).

To examine enhancement of Pak interaction with light, we first attempted to carry out these assays in room light, but this produced inconsistent results, potentially due to the rapid dark recovery of the LOV domain during sample preparation. Instead, we compared point mutations that mimic the closed/dark (C450A) and open/lit (I539E) conformations of the LOV domain (Fig. 16.1C). These screens led to the optimized PA-Rac described in the original publication (Wu *et al.*, 2009).

4. Activation of PA-Rac in Living Cells

4.1. Cell handling

Constructs derived from the pTriEx vector (Novagen) carry a CMV promoter. These were used, typically at 0.2–0.3 μg DNA per well of a 6-well plate ($\phi = \sim$ 2.2 cm) for transient transfection of mouse embryo fibroblasts with Fugene 6 reagents. The cells were imaged between 12–24 h after transfection. Alternatively, lower but uniform expression of PA-Rac can be achieved in retrovirus-infected MEF cells. We chose a retroviral vector that carries a TetCMV promoter. The expression of PA-Rac was suppressed until the time of imaging using retroviral constructs in which the expression of PA-Rac was driven by a TetCMV promoter. Using this approach, the expression level could be controlled through titration of doxycycline. We used the fluorescent proteins attached to the PA-Rac to gauge expression level, determining optimum expression for light-induced Rac activation with minimal effects from low-level PA-Rac activity in the dark.

Because of the light sensitivity of the LOV domain, cells expressing PA-Rac should not be exposed to light within the LOV action spectrum (<500 nm) immediately before imaging experiments. It is safe to prepare live samples under yellow or red light, available in most dark rooms. If DIC or phase-contrast images are needed, caution must be taken to avoid photoactivation from the transmitted light source. A red filter, such as a Schott RG610 Glass filter (610 nm long pass), can be placed in the transmitted light path (most commercial microscopes have a transmitted light filter holder). Even light emitted from bright computer monitors can be of concern, but this is less consequential at a distance. We transferred the cells immediately after transfection into incubators located in isolated dark rooms to avoid unintended light exposure. In some experiments, extensive manipulations were required prior to imaging. When this was difficult in a dark room, cells could be incubated in the dark for an hour to reduce the effect of prior light exposure and restore responsiveness to photoactivation. It can be useful to place a black paper cone over the objective and live cell chamber of the microscope to eliminate small amounts of ambient light in some rooms.

We constructed PA-Rac constructs with various fluorescent protein tags including mCerulean, mVenus, and mCherry, to facilitate the identification of cells expressing PA-Rac. Based on the excitation wavelengths of these fluorescent proteins, mCherry (590 nm max) can be monitored without concern for excitation of the LOV domain. The common excitation filters for mVenus or YFP do fall in the 480–500 nm range that can activate the LOV domain. We recommend using excitation filters that cut off between 515–520 nm, together with minimal intensity and exposure time, to avoid unintended photoconversion of the LOV domain. Because the excitation wavelengths of mCerulean or CFP (433 nm max) overlap with the photoactivation spectrum, fluorescence from these proteins is best used for confirmation of expression at the end of experiments.

Our preparation of cells for live cell imaging has been previously described (Hodgson *et al.*, 2010). Briefly, cells were seeded onto ϕ25 mm coverslips coated with 1–10 μg/mL fibronectin overnight. The imaging medium was Ham's F-12K medium without phenol red, containing 2% fetal bovine serum. Coverslips were mounted in an Attofluor live cell chamber (Invitrogen) and placed in a microscope stage with a heated stage adaptor (Warner).

4.2. Irradiation in living cells—Light sources, dosage, and spatial control

The action spectrum of plant phototropin is in the UV-A and blue light range (360–500 nm). We tested several common laser lines for their ability to induce membrane ruffles in MEF cells expressing PA-Rac. The wavelengths 405, 458, 473, and 488 nm all proved to be effective. The power dosage of the PA-Rac to 458 nm line was measured in stable MEF cell lines, where expression levels could be well controlled and the areas of induced protrusions readily measured. A light dose of 6.2 μJ over a 10 μm spot at 458 nm induced a cellular response with a single exposure. This was the lowest power setting (0.1% of total power on the mW scale) of our Fluoview 1000 confocal microscope at very fast scan rate (10 μs/pixel). Following this exact photoactivation regime but with increasing laser power or scan duration, we determined that the cellular response (protrusion area) stopped increasing when we reached 1000-fold higher dose. A 100 W Mercury arc lamp used for fluorescence imaging was also an effective source for photoactivation. We can qualitatively state that we found no difficulties in activating PA-Rac through global cell irradiation using a 100 W mercury source filtered through a ND 2.0 (1.0% transmission) filter and a CFP excitation filter (ET430/24 nm). This was sufficient to induce membrane ruffles with a 500 ms exposure. One of the main peaks of emission from the mercury light source falls at these wavelengths, and these observations were made with cells expressing low levels of PA-Rac (empirically optimized for optimal light response without induction of a Rac phenotype in the dark).

Spatial control is perhaps the most valuable feature of light-mediated protein activation. The ability to target a defined cell area enables study of localized signaling milieus. With Rac, localized irradiation can be used to control cell motility. Most conventional wide-field microscopes can be modified to incorporate a field stop/diaphragm or pinhole in a conjugate image plane. The light source used for fluorescence excitation can be used to illuminate a small region of the cell. By modulating illumination intensity, low intensity can be used to target the protein uncaging light, followed by high intensity radiation to activate protein. Alternatively, various scan modes on laser scanning confocal microscopes can be used. Commercial vendors have devised a variety of solutions for laser irradiation of small portions of the field of view, usually for FRAP studies. A laser beam can be coupled into the light path and focused on the focal plane to a diffraction-limited spot or dilated to bigger areas through *z* offset. These spots can also be mobilized to scan across different shapes either manually or using galvanometer-driven mirrors. Complex patterns can be achieved using digital micromirror-based devices such as the commercially available Mosaic digital illumination system (Levskaya *et al.*, 2009), or more advanced methods based on liquid-crystal spatial light modulators, realized in applications involving laser tweezers (Curtis *et al.*, 2002) and adaptive optics (Girkin *et al.*, 2009).

4.3. Detection and quantitation of Rac activation in cells

For any given protein, it will be important to devise an assay that can define successful activation in living cells. For Rac, overexpression in many cell types induces membrane ruffles and lamellipodial protrusions, both phenotypes readily identifiable with DIC or phase-contrast imaging. These phenotypes were scored after global illumination to examine the effects of different mutations on PA-Rac. Induction was much clearer when cells were serum-starved to minimize background activity prior to irradiation. Effects were easier to observe in MEF, HEK293, and HeLa cells, but less apparent in COS-7 cells. This was perhaps due to differences in basal Rac activity and/or the abundance of downstream effectors. Similar differences were seen with dominant-negative mutants of PA-Rac that inhibited endogenous Rac activity through sequestration of Rac GEFs. The production of protrusions was quantified from line scans of the cell border in time-lapse images, or by monitoring protrusion area.

To analyze Rac initiation of downstream signaling, we also tracked translocation of downstream effectors. For wide-field microscopy, artifacts of changing cell volume were avoided using ratiometric imaging. The fluorescence intensity of the tagged target protein was divided by the fluorescence intensity of a separately expressed volume indicator, that is, YFP or mCherry. As demonstrated previously (Wu *et al.*, 2009), PA-Rac redistributed slowly out of sites of illumination with a diffusion coefficient of 0.55 $\mu m^2 s^{-1}$. This led to detectable accumulation of downstream effectors where Rac was activated.

Phosphorylation of endogenous PAK was also traced using immunofluorescence of fixed cells after local activation of PA-Rac. MEF cells stably expressing PA-Rac were plated onto coverslips with etched grids (Bellco), used to locate the cells that had been irradiated. Cells were irradiated at 473 nm through a 20× phase-contrast objective. Immediately after protrusions were induced, the cells were fixed in 3.7% formalin (Sigma), permeabilized in 0.2% Triton X-100, incubated with anti-phospho-PAK antibody (Cell Signaling), and finally incubated with Alexa Fluor 594-conjugated secondary antibody (Molecular Probes). Imaging was performed as described above.

5. Application of PA-Rac in *Drosophila* Ovarian Border Cell Migration

5.1. What is border cell migration?

To test whether PA-Rac would be useful in an intact three-dimensional tissue and in a different organism, we chose to study border cell migration in the *Drosophila* ovary, a well-characterized example of collective cell movement. Border cells are a group of 6–8 cells that arise from a monolayer of ~650 epithelial follicle cells that surround 15 nurse cells and one oocyte in a structure called an egg chamber (Fig. 16.2). Border cells migrate ~175 μm in between the nurse cells, as an interconnected group of two distinct cell

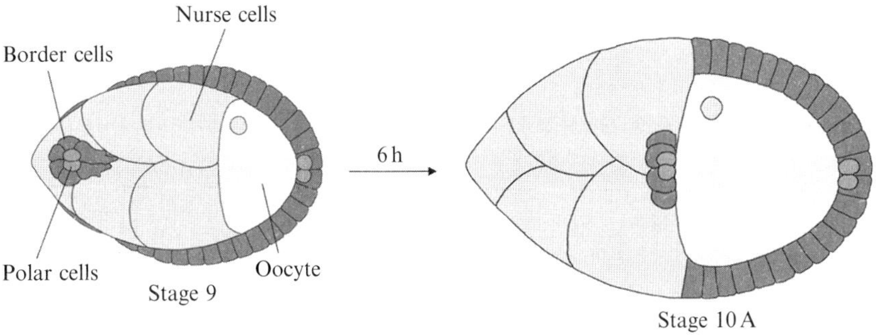

Figure 16.2 Schematic drawing of border cell migration. A pair of polar cells secretes a cytokine that activates the JAK/STAT pathway in immediately adjacent cells to specify the border cells at early stage 9. Then, border cells detach from the epithelium and migrate in between the germ-line cells during stage 9 and arrive at the boundary between nurse cells and oocyte at stage 10.

types: 4–8 migratory cells surrounding two central polar cells. Polar cells cannot migrate but secrete a cytokine that activates the JAK/STAT pathway rendering the outer cells motile. The outer cells carry the polar cells and lose the ability to move in the absence of continuous JAK/STAT activation. Thus each cell type requires the other. Border cells also require steroid hormone, receptor tyrosine kinase, Notch, and other signaling cascades. Thus border cells experience a rich and complex signaling environment, as do most cells *in vivo*.

There are several prominent differences between the migration of single cultured cells and border cells. Firstly, single cells *in vitro* typically migrate on stiff 2-dimensional surfaces coated with fibronectin or collagen substrates, while border cells migrate in a more pliable environment surrounded on all sides by nurse cells. Secondly, border cells experience a complex signaling environment that includes many nondirectional signals that could potentially activate small GTPases uniformly, in addition to directional signals, which could possibly create a high background and interfere with the effect of light-induced Rac activation. Thirdly, single cells in culture have unlimited time and space in which to move, whereas border cells are normally limited to a straight path between the nurse cells, starting from the anterior tip and ending at the oocyte and only during developmental stage 9. Lastly and most importantly, border cells migrate as a coherent cluster. Therefore, it was unclear whether activation of Rac in one cell would have an effect on the other cells.

Previous studies had shown that constitutive activation or inhibition of Rac led to complete inhibition of migration. However, one could only speculate as to why activation or inactivation produced the same effect. The ability to control Rac activity both temporally and spatially with PA-Rac allowed us to address these critical open questions.

5.2. Genetics

Considering the different excitation wavelengths of mCerulean, mVenus, and mCherry and PA-Rac activation, we chose to insert N-terminal-mCherry-tagged PA-RacQ61L, PA-RacT17N, and the light-insensitive control C450M-PA-RacQ61L into pUASt *Drosophila* expression vector, using the Invitrogen Gateway recombination system. Transgenic flies were generated by Bestgene Inc., and confirmed by the detection of mCherry fluorescence. We used the mammalian Rac rather than substituting *Drosophila* Rac for two reasons. *Drosophila* and mammalian Rac proteins share more than 90% similarity and are known to function similarly (Luo *et al.* 1994). We were concerned, however, that the 6 amino-acid difference between the carboxy-terminal tails of *Drosophila* and mammalian Rac might alter the physical interaction with the Lov-Jα domain. We are currently testing PA-*Drosophila* Rac and comparing it with the mammalian protein.

Initially, all fly stocks and crosses were maintained at room temperature in normal light. We used the Gal4/UAS system (Brand and Perrimon, 1993) to drive cell-type specific expression of PA-Rac. Specifically, we used transgenic flies expressing Gal4 under the control of the *slow border cells* (*slbo*) gene, to drive expression of pUASt-PA-Rac. During stage 9 of oogenesis, Slbo-Gal4 drives expression of PA-Rac and other UAS transgenes primarily in border cells. Expression of PA-RacQ61L or PA-RacT17N driven by Slbo-Gal4 had no detectable deleterious effect on border cell migration in the absence of laser illumination. Another Gal4 commonly used to drive border cell expression of UAS transgenes, c306-Gal4, was lethal in combination with UAS-PA-RacQ61L. c306 is not as specific as *slbo*-Gal4 and the lethality was probably due to expression of the transgene earlier in development and inappropriate activation of Rac due to ambient light. UAS-PA-RacT17N driven by C306-Gal4 caused a moderate delay in detachment of border cells from the epithelium in early stage 9. C306-Gal4 drives UAS transgene expression earlier and to higher levels than slbo-Gal4, which starts weakly in a few anterior and posterior follicle cells around stage 7 and gradually becomes stronger during stages 8–10, when border cells complete their migration. Thus, the restricted spatial and temporal expression of Slbo-Gal4 has important advantages in alleviating side effects of "leaky" Rac activity. Newer PA-Rac mutations currently being characterized have greatly reduced residual Rac activity in the dark. To further prevent the effects of residual Rac activity, we maintained progeny flies carrying both Gal4 and UAS-PA-Rac constructs in a culture incubator with lower temperature (18 °C), which decreases Gal4-mediated gene expression due to temperature sensitivity of the Gal4 transcription factor. Prior to dissection, flies were transferred to foil-wrapped vials and incubated at 29 °C overnight to induce UAS transgene expression. This incubation had no negative effect on border cell migration in the absence of laser illumination.

5.3. Illumination of border cells—*In vitro* culture, live imaging, and photomanipulation

Drosophila egg chambers were dissected in Schneider's insect medium supplemented with 20% fetal bovine serum and 0.10 mg/mL insulin, and then mounted in the same medium on a 50 mm Greiner Lumox culture hydrophilic dish (also known as a petriperm plate) and coved with a 22 mm coverslip, as described (Prasad *et al.* 2007). Figure 16.3 shows a schematic drawing of the culture chamber mounting. The following points are critical for successful live imaging of border cells: The final culture medium pH was adjusted to 6.85–6.95, which is essential for normal border cell migration (Prasad *et al.* 2007). Secondly, the surface of the petriperm plate was kept clean to allow O_2/CO_2 exchange. Thirdly, the volume of culture medium

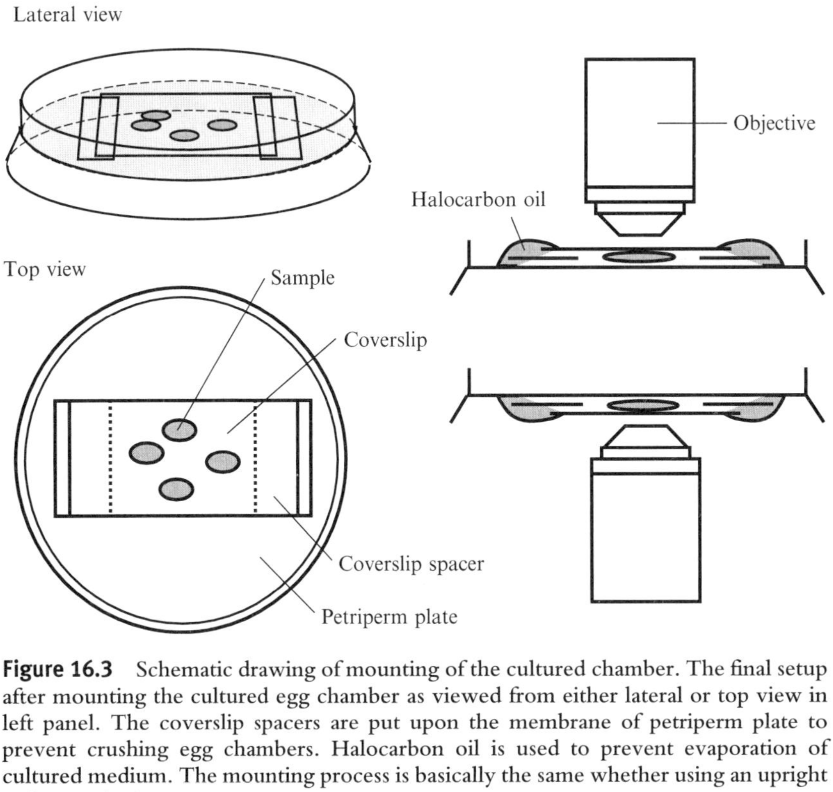

Figure 16.3 Schematic drawing of mounting of the cultured chamber. The final setup after mounting the cultured egg chamber as viewed from either lateral or top view in left panel. The coverslip spacers are put upon the membrane of petriperm plate to prevent crushing egg chambers. Halocarbon oil is used to prevent evaporation of cultured medium. The mounting process is basically the same whether using an upright or inverted microscope.

between the petriperm plate and coverslip was adjusted carefully. Too little medium can crush egg chambers and prevent normal border cell migration. Excess medium allows the egg chamber to float and change position, making subsequent manipulations impossible. This problem is particularly noticeable when the mounted petriperm plate needs to be inverted (as shown in Fig. 16.3). The mounting volume is appropriate when the stages 9 and 10 egg chambers have normal morphology (not too flat) and their position remains stable even when one lightly taps the edge of the petriperm plate with a finger. The dissection and mounting process should be complete within 30 min to prevent general activation of PA-Rac, since these steps are carried out in ambient light.

Photoactivation, time-lapse-imaging, and 3-dimensional morphological reconstruction were carried out using a Zeiss 510-Meta or 710-Meta confocal microscope using a 63×, 1.4 numerical aperture oil lens with 2× zoom. To photoactivate PA-Rac in border cells, a 458 nm laser was set at 10% power for 0.1 ms scan per pixel in a 7 μm spot, and the photoactivation scan took ~25 s at a scan speed of 2× and scan number of 8. Laser scanning with 477 and 488 nm wavelengths produced a much weaker effect. Lower scan energy, less scan time, or smaller scan regions also produced weaker effects. Optimal photomanipulation conditions will likely depend on both the specific tissue under study and the particular microscope and laser setup. Approximately 20–30 s after photoactivation, border cells were imaged by capturing the mCherry-tagged PA-Rac signal using a 568 nm wavelength scan at the minimally detectable laser power, lasting 6–7 s at scan speed of 8× and scan number of 8. This series of photoactivation and live imaging steps was repeated for a few minutes or up to 5 h depending on the experiment. Egg chambers survived this repeated light exposure for up to 5 h before showing signs of phototoxicity. However, the effect of light-induced Rac activation diminished after 2.5–3.5 h of repeated treatment. The 3-dimensional reconstruction of border cell morphology was captured before and after repeated phototreatment of border cells, using 20 focal planes 1.5 μm apart in the *z*-axis, using 568 nm light. The laser energy at 568 nm was adjusted to higher levels to show the detailed filopodia.

5.4. Detection and quantification of effects of Rac activation on border cells

Analysis of the effect of light-induced Rac activation in border cells requires careful quantification and comparison to controls to distinguish induced effects from endogenous cell behavior. As mentioned above, border cells experience a rich cocktail of endogenous signals that lead to substantial Rac activation, protrusion formation, and directional migration. So, for example, when membrane ruffles or lamellipodial protrusions are observed following photo-activation of Rac, how do we distinguish whether PA-Rac caused that effect or whether the cell might have done the same thing without light-activated Rac due to endogenous signals? To clearly identify effects of light-induced Rac activity, we measured multiple parameters including migration speed, protrusion area, number, and density, as well as a directionality index (DI). We compared these features in wild-type and various genetic backgrounds in the presence or absence of light. The light-insensitive versions of PA-Rac proteins were also essential to control for nonspecific effects of the phototreatment itself. Substantial variation in border cell responses were observed, which might be caused by slight variations in the condition of the egg chambers, PA-Rac expression levels, starting position of the cells along the travel path, etc. Statistical analyses were therefore essential.

Migration speed is the simplest parameter to measure. The distance of the center of the border cell cluster between the first and final time points in a time-lapse series was measured using Imaris software. This distance divided by the elapsed time gave the speed. Imaris software can also measure speeds during shorter intervals within the overall experiment. We typically measured the migration speed beginning $\sim$1 h after initiating the light treatment, and we observed border cell migration speeds ranging from 0.1 to 0.5 μm per minute in response to light-induced Rac.

Protrusion area was measured as follows: a confocal image of the cluster edge was captured before light treatment as shown in Fig. 16.4 (left panel). After repeated light treatments, another image was captured, as shown in Fig. 16.4 (middle panel). The difference in position of the cell boundary between the two images, as shown in Fig. 16.4 (right panel), was used to estimate the protrusion area using ImageJ software.

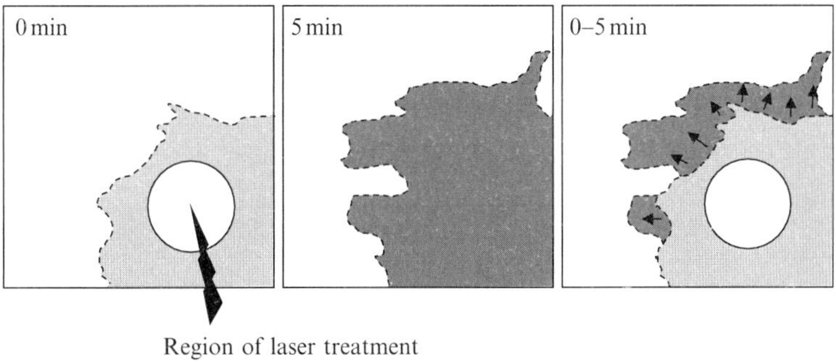

Figure 16.4 Schematic drawing of change in protrusion area in response to light-induced Rac. The light or dark gray region shows the protrusion region at the beginning or end of laser treatment. Black arrows represent the direction of expansion of the protruding region in response to light-induced Rac.

Cell protrusions were defined and counted automatically using MatLab as follows: a circle corresponding to the average cluster diameter was drawn, and major protrusions were defined as extensions at least 2 μm beyond the circle and broader than 2 μm. The number of major protrusions at any given time varied between 2 and 5 in different genetic backgrounds. The DI represents the fraction of forward-directed protrusions relative to the total number of protrusions. And DI was calculated using the following equation:

$$\mathrm{DI} = \frac{\sum_{i=1}^{N} \vec{p_i} \cdot \vec{d}}{\sum_{i=1}^{N} \|\vec{p_i}\|}$$

where N is the total number of major protrusions, $\vec{p_i}$ is the ith protrusion vector, and $\vec{d}$ is the unit vector of migration direction. The protrusion vector is calculated by fitting the major protrusion by a parabola whose peak together with the cluster center gives the vector's direction and length. Protrusion density was generated by dividing the number of all the recognizable membrane protrusions by the measured cell perimeter in micron.

5.5. What do we learn from PA-Rac application to border cell migration?

The studies in cultured cells showed that PA-Rac could induce local membrane ruffling and protrusion and even cell movement. The effects were stronger in some cell types than others and were most pronounced in low-serum or serum-free conditions. It was not obvious therefore how effective the treatment would be *in vivo*, where cells are exposed to myriad endogenous signals both directional and nondirectional. Our results indicated that despite the intense and complex signaling environment, local activation of Rac is sufficient to cause local ruffling and protrusion and to set the direction of movement for cells, forward, backward, or sideways, *in vivo*. Secondly, we found that there were regions of the egg chamber into which the cells could not move. Thus PA-Rac could be used to "map" the permissive and nonpermissive regions of the tissue. Thirdly, border cells migrate collectively as an interconnected group, as do a variety of other cell types including vertebrate neural crest cells and invading carcinomas (Gaggioli *et al.* 2007; Theveneau *et al.* 2010). We found that local activation of Rac in a single cell was sufficient to redirect the entire cluster. Most surprisingly, light-induced Rac activity in one border cell altered the morphology of the other cells. Photo-inactivation of Rac in the leading cell "confused" all of the cells. These findings demonstrate that a group of cells can sense direction collectively according to which cell has the highest level of Rac activity, a finding that would have been difficult or impossible to make with traditional genetic methods. Moreover, these studies demonstrate that PA proteins should be generally useful in a variety of cell types, tissues, and organisms.

REFERENCES

Brand, A. H., and Perrimon, N. (1993). Targeted gene expression as a means of altering cell fates and generating dominant phenotypes. *Development* **118**(2), 401–415.

Curtis, J. E., Koss, B. A., and Grier, D. G. (2002). Dynamic holographic optical tweezers. *Opt. Commun.* **207,** 169–175.

Gaggioli, C., Hooper, S., Hidalgo-Carcedo, C., Grosse, R., Marshall, J. F., Harrington, K., and Sahai, E. (2007). Fibroblast-led collective invasion of carcinoma cells with differing roles for RhoGTPases in leading and following cells. *Nat. Cell Biol.* **9**(12), 1392–1400.

Girkin, J. M., Poland, S., and Wright, A. J. (2009). Adaptive optics for deeper imaging of biological samples. *Curr. Opin. Biotechnol.* **20,** 106–110.

Harper, S. M., Neil, L. C., and Gardner, K. H. (2003). Structural basis of a phototropin light switch. *Science* **301,** 1541–1544.

Hodgson, L., Shen, F., and Hahn, K. (2010). Biosensors for characterizing the dynamics of rho family GTPases in living cells. *Curr. Protoc. Cell Biol.* 1–26, Chapter 14, Unit 14.11.

Huala, E., Oeller, P. W., Liscum, E., Han, I., Larsen, E., and Briggs, W. R. (1997). Arabidopsis NPH1: A protein kinase with a putative redox-sensing domain. *Science* **278,** 2120–2123.

Levskaya, A., Weiner, O. D., Lim, W. A., and Voigt, C. A. (2009). Spatiotemporal control of cell signalling using a light-switchable protein interaction. *Nature* **461,** 997–1001.

Luo, L., Liao, Y. J., Jan, L. Y., and Jan, Y. N. (1994). Distinct morphogenetic functions of similar small GTPases: Drosophila Drac1 is involved in axonal outgrowth and myoblast fusion. *Genes Dev.* **8**(15), 1787–1802.

Möglich, A., Ayers, R. A., and Moffat, K. (2009). Structure and signaling mechanism of Per-ARNT-Sim domains. *Structure* **17,** 1282–1294.

Prasad, M., Jang, A. C., Starz-Gaiano, M., Melani, M., and Montell, D. J. (2007). A protocol for culturing Drosophila melanogaster stage 9 egg chambers for live imaging. *Nat. Protoc.* **2**(10), 2467–2473.

Swartz, T. E., Corchnoy, S. B., Christie, J. M., Lewis, J. W., Szundi, I., Briggs, W. R., and Bogomolni, R. A. (2001). The photocycle of a flavin-binding domain of the blue light photoreceptor phototropin. *J. Biol. Chem.* **276,** 36493–36500.

Theveneau, E., Marchant, L., Kuriyama, S., Gull, M., Moepps, B., Parsons, M., and Mayor, R. (2010). Collective chemotaxis requires contact-dependent cell polarity. *Dev. Cell* **19**(1), 39–53.

Wu, Y. I., Frey, D., Lungu, O. I., Jaehrig, A., Schlichting, I., Kuhlman, B., and Hahn, K. M. (2009). A genetically encoded photoactivatable Rac controls the motility of living cells. *Nature* **461,** 104–108.

CHAPTER SEVENTEEN

Light Control of Plasma Membrane Recruitment Using the Phy–PIF System

Jared E. Toettcher,[*,†] Delquin Gong,[*] Wendell A. Lim,[†,‡] *and* Orion D. Weiner[*]

Contents

Abstract

The ability to control the activity of intracellular signaling processes in live cells would be an extraordinarily powerful tool. Ideally, such an intracellular input would be (i) genetically encoded, (ii) able to be turned on and off in defined temporal or spatial patterns, (iii) fast to switch between on and off states, and (iv) orthogonal to other cellular processes. The light-gated interaction between fragments of two plant proteins—termed Phy and PIF—satisfies each of these constraints. In this system, Phy can be switched between two conformations using red and infrared light, while PIF only binds one of these states. This chapter describes known constraints for designing genetic constructs using Phy and PIF and provides protocols for expressing these constructs in mammalian cells,

[*] Cardiovascular Research Institute and Department of Biochemistry, University of California San Francisco, San Francisco, California

[†] Department of Cellular and Molecular Pharmacology, University of California San Francisco, San Francisco, California

[‡] Howard Hughes Medical Institute, University of California San Francisco, San Francisco, California

Methods in Enzymology, Volume 497
ISSN 0076-6879, DOI: 10.1016/B978-0-12-385075-1.00017-2

purifying the small molecule chromophore required for the system's light responsivity, and measuring light-gated binding by microscopy.

1. Introduction

In recent years, tremendous strides have been made in developing quantitative readouts that report on live cell activity at the molecular scale. Time-lapse microscopy has been combined with fluorescent detection of protein concentration (Heim and Tsien, 1996; Michalet *et al.*, 2005) and protein–protein association (Truong and Ikura, 2001), enabling studies of the temporal dynamics (Lahav *et al.*, 2004; Nelson *et al.*, 2004) and spatial organization (Ilani *et al.*, 2009) of complex signaling pathways. These techniques have also been instrumental in characterizing complex emergent properties such as perfect adaptation (Cohen-Saidon *et al.*, 2009; Shimizu *et al.*, 2010). All of these advances rely on the ability to quantitatively measure signaling *outputs*—the concentration or activity of various pathway components—with precision in living cells.

The ability to quantitatively vary intracellular signaling inputs in time and space, in a user controlled way, would be equally revolutionary, allowing researchers to better manipulate and probe the cellular processes they study. However, comparatively few technologies are available to achieve the goal of manipulation as compared to measurement in living cells. Microfluidic devices are limited to controlling the spatial and temporal pattern of extracellular inputs (Paliwal *et al.*, 2007; Tay *et al.*, 2010). For intracellular inputs, the rapamycin-inducible FRB/FKBP protein–protein interaction (Spencer *et al.*, 1993) offers the opportunity to activate the association of two intracellular species. However, this high-affinity interaction is slow to dissociate, and thus the resulting control is poorly reversible (Terrillon and Bouvier, 2004). More recently, inputs have been designed that utilize light-induced conformational changes in naturally light-responsive proteins such as channelrhodopsin (Gunaydin *et al.*, 2010) LOV domains (Strickland *et al.*, 2008), or proteins incorporating photocaged amino acids (Gautier *et al.*, 2010; Lemke *et al.*, 2007), although each has limitations. Applying light-activatable LOV domains to control additional signaling processes requires considerable protein engineering, and LOV domain inactivation occurs spontaneously but cannot be directly controlled by light. Uncaging of photocaged proteins can be performed quickly and selectively, but is an irreversible modification. Finally, light control of channelrhodopsin is fast and reversible but limited to controlling transmembrane cation flux, a specific signaling currency.

Ideally, a controllable intracellular signaling input would be genetically encoded (easily "wired in" to control a variety of proteins), photoreversible, nontoxic to the cell, and high resolution in both time and space. A recently developed light-dependent binding interaction using plant phytochrome proteins satisfies each of these constraints (Levskaya *et al.*, 2009; Ni *et al.*, 1999). This interaction has already proven useful for applying complex temporal and spatial intracellular inputs to live cells in a variety of species (Levskaya *et al.*, 2009; Shimizu-Sato *et al.*, 2002). In principle, this approach could be used for spatiotemporal control over any cellular process that is dependent on association of two proteins. Because induced protein interaction is such a common mechanism for controlling molecular activity, this method has the potential to be highly generalizable.

Despite these advantages, it can be challenging to prepare light-gated genetic constructs, purify the small molecule chromophore required for their function, and assay the function of all these components inside living cells. In this chapter, we address these challenges by presenting a detailed and optimized methodology based on that introduced in Levskaya *et al.* (2009) for visualizing light-induced binding in mammalian cells. After a brief introduction to the system (Section 2) we present methods for genetically encoding the light-responsive protein domains (Section 3), purifying the small-molecule chromophore that is required for their interaction (Section 4), and validating the light-mediated protein translocation in mammalian cells (Sections 5 and 6).

2. Light-Controlled Phy–PIF Interaction

The Phy–PIF system takes advantage of a light-controllable binding interaction between two genetically encoded components: a fragment of *Arabidopsis thaliana* phytochrome B, referred to here as Phy; and a fragment of phytochrome interaction factor 6, referred to here as PIF. Phy becomes light-responsive following conjugation to the membrane-permeable small molecule chromophore, phycocyanobilin (PCB). Exposure to 650 nm induces association of PIF and Phy, while exposure to 750 nm light induces dissociation of PIF from Phy (Fig. 17.1A). Phy can be reversibly switched between PIF-interacting and -noninteracting states using light within seconds, and switching can be performed for hundreds of cycles without toxicity to the cell or any measurable degradation of the system's performance (Levskaya *et al.*, 2009).

How can this system be used to enable light control of a range of cellular activities? Because protein association and dissociation is such a general currency of cell signaling, this light-gated heterodimerization scheme has been applied to a broad range of signaling processes such as transcription, splicing, plasma membrane signaling, and modulating actin assembly *in vitro* (Leung *et al.*, 2008). It was first used outside of its native context in a two-hybrid approach, in which the split DNA binding and transcriptional

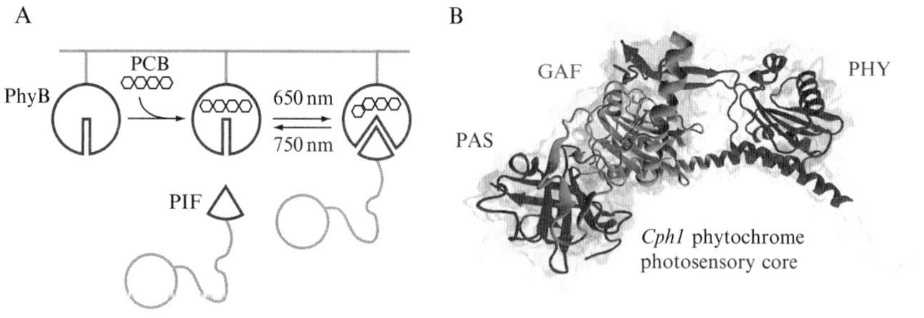

Figure 17.1 Schematic of the Phy–PIF interaction. (A) After incorporation of the chromophore PCB, the conformation of Phy can be controlled by exposure to two wavelengths of light (650 and 750 nm). The PIF domain binds only one of these domains with high affinity. By controlling the ratio of 650:750 nm light, the fraction of Phy in a state permissive for binding can be tuned, modulating the total amount of PIF recruitment. (B) Crystal structure of a fragment of the cyanobacterial phytochrome protein Cph1 bound to the small molecule chromophore PCB (shown as licorice) (PDB ID: 2VEA) (Essen *et al.*, 2008). The PAS, GAF, and PHY domains, each required for Phy–PIF interaction, are shown in purple, orange, and red, respectively, while the N terminal 26 amino acids are shown in green. (See Color Insert.)

activation domains of Gal4 were fused to Phy and PIF to enable light-gated transcriptional control in yeast (Shimizu-Sato *et al.*, 2002). A similar concept using a split protein (intein) was used to activate splicing in yeast using light (Tyszkiewicz and Muir, 2008). The first application of this system to mammalian cell signaling used Phy and PIF to activate GTPase signaling in mouse fibroblasts by recruiting a constitutively active GEF to its G protein target on the plasma membrane (Levskaya *et al.*, 2009). In this chapter, we focus on a simple, direct assay of light-controllable Phy–PIF binding, in which a fluorescently tagged Phy directed to the plasma membrane where it can act as a light-controlled binding site for cytoplasmic fluorescently tagged PIF. To observe this localization change, we will focus on a typical experimental context for using the Phy–PIF system, in which Phy and PIF constructs are introduced into mammalian cells by retroviral infection and imaged by fluorescence microscopy.

3. Genetic Constructs Encoding Phy and PIF Components

The Phy and PIF components of the light system can be expressed as fusions to other proteins of interest to elicit binding and activation of a variety of intracellular signaling processes. However, care must be taken to validate Phy and PIF expression and localization for each fusion construct. In this section, we describe some known constraints of the system and elucidate details for establishing successful light-controllable fusion constructs.

The first of these components is Phy, consisting of residues 1–908 of the *A. thaliana* PhyB protein (Entrez Gene ID: 816394; see Fig. 17.1B for the structure of a related phytochrome, Cph1) (Essen *et al.*, 2008; Levskaya *et al.*, 2009). Phy has been expressed successfully without codon optimization in *Saccharomyces cerevisiae* and NIH-3T3 cells; however, we have found codon optimization to facilitate strong Phy expression in some contexts (as for HL-60 cells and *Dictyostelium*). Whereas PIF tolerates a large range of fusion orientations, we have found that Phy fusion protein expression and function is particularly sensitive to linker lengths and component orientation. Phy appears to work most robustly as an N-terminal fusion component (Fig. 17.2A). While a useful guideline, this is not a strict rule; PIF recruitment has also been observed from some C-terminal and internal Phy fusion constructs. We have had success using a 15 amino acid linker (linker L1: EFDSAGSAGSAGGSS) between the C-terminus of Phy and the N-terminus of downstream fusion constructs; this linker performed better than shorter linkers in some applications.

How can Phy function be validated in different contexts? One useful, previously characterized Phy single mutant (Y276H) fluoresces at far-red

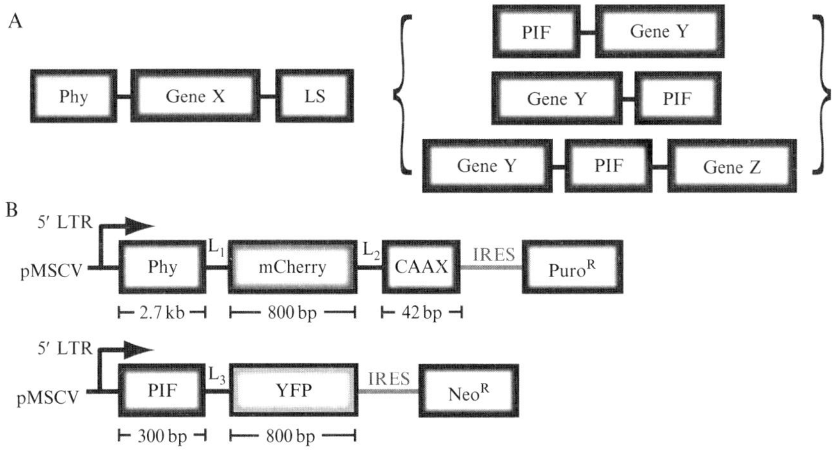

Figure 17.2 Genetic Phy and PIF constructs for use in recruitment assays. (A) Schematic diagram showing typical Phy and PIF fusion constructs. The larger Phy protein is usually expressed as a membrane-localized component, fused to a gene of interest (*Gene X*) as well as a C-terminal membrane localization tag (LS). Phy best tolerates fusion when it encodes the N-terminal component. PIF is typically fused to a freely diffusing cytoplasmic component, and tolerates fusion at either or both termini. (B) Schematic of the specific constructs discussed in this article. Phy is fused to mCherry and a plasma membrane localization tag, the KRas CAAX tail, while PIF is fused to YFP. Both are driven using the MSCV retroviral vector system and are expressed along with puromycin or neomycin selection markers. Gly-Ser linkers L_1, L_2, and L_3 are as described in the text.

frequencies only in the chromophore-bound state (Su and Lagarias, 2007). We previously demonstrated that NIH-3T3 cells expressing this mutant exhibit bright fluorescence after 30 min of incubation with PCB (Levskaya *et al.*, 2009). Thus, the ability of a Phy fusion construct to bind PCB can be tested directly, independently of the recruitment assay described in Section 6. Phy-Y276H can also be used to test the quality of purified PCB (Section 4).

Phy interacts in a light-dependent fashion with PIF, a second component consisting of residues 1–100 of *A. thaliana* PIF6 protein. PIF does not exhibit any preference toward N or C terminal fusions and also tolerates fusions on both termini simultaneously (Fig. 17.2A). We have not observed any dependence of PIF–Phy binding on linker length within PIF fusion constructs. Typically, Gly-Ser spacers of 10 amino acids are placed between PIF and its fusion partners.

When using the Phy–PIF system to induce binding between membrane-tethered and cytoplasmic proteins, we typically attach Phy to the membrane component and PIF to the cytoplasmic component (Fig. 17.2A), as the smaller PIF domain is less likely to significantly affect diffusion of its fusion partner. Phy can be tethered to the plasma membrane using a linker followed by the KRas "CAAX tail" plasma membrane localization signal (linker L2: SAGSAGKASG; CAAX tag: KKKKKKSKTKCVIM). This tag has been validated in both yeast and mammalian cells (Clarke *et al.*, 1988). Although this CAAX sequence can be used to induce robust membrane localization in NIH-3T3 cells, phosphorylation at its serine residue by PKC causes dissociation from the membrane in some cell types (Bivona *et al.*, 2006); incorporating a Ser-Ala mutation (unphosphorylatable CAAX tag: KKKKKKAKTKCVIM) can be used to stabilize membrane association in these contexts.

We have demonstrated light-controlled Phy–PIF interaction in a number cell lines, and NIH-3T3 cells (ATCC catalog number CRL-1658) continue to be our gold standard for these studies. Transient transfection of fluorescently tagged Phy and PIF constructs works for some applications, but we find that generating stable cell lines from retroviral constructs greatly enriches the population of cells expressing both Phy and PIF and also facilitates robust recruitment. Furthermore, these stable cell lines maintain expression in long-term culture and thus cells can be sorted by expression level to hone in on the optimal recruiting population for each application.

To establish stable cell lines, we typically clone Phy and PIF constructs into the pMSCV retroviral vector system (Clontech catalogue number 634401), in which each construct is driven from the viral LTR promoter. These constructs also contain an internal ribosomal entry site (IRES), followed by neomycin/G418 or puromycin antibiotic resistance to allow selection for the stably infected population (Fig. 17.2B). To produce virus, we transfect each construct into 293-GPG cells, a standard retrovirus packaging cell line (Ory *et al.*, 1996) using the TransIT 293 transfection kit (Mirus catalog number 2700). NIH-3T3 cells should be sequentially infected with Phy and PIF viral constructs at a high enough multiplicity of infection (MOI) to lead to $>70\%$ infected cells. There is no need to select using antibiotics before assaying recruitment, although this selection can be performed to enrich for doubly infected cells.

Phy–PIF recruitment is easiest to observe if Phy expression levels are high (see Section 6). For more effective viral transduction, it can be helpful to concentrate Phy-containing retrovirus (Kanbe and Zhang, 2004). Briefly, the collected retrovirus is spun at max speed in 1.5 mL centrifuge tubes in a table top microcentrifuge (Eppendorf Centrifuge model 5415D or similar) at 4 °C for 1 h. After centrifugation, discard all but 100 μL or so of the supernatant, and pool together the "invisible" pellets for infection.

4. Purification of PCB from *Spirulina*

The responsiveness of Phy domains to light depends on their covalent attachment to a small molecule chromophore, PCB. PCB is not synthesized naturally in most nonphotosynthetic cells, but it is easily taken up by all cell types we have tested, penetrating yeast cell walls and freely diffusing through mammalian cell membranes. Thus, free PCB must be obtained and added to cell cultures before light-responsive experiments are conducted. While addition of the PCB chromophore is another required manipulation, it is also an advantage—cells can be handled freely with regard to light exposure until one is ready to perform an experiment, at which point the PCB is added under controlled light conditions. In this section, we provide a detailed step-by-step protocol for PCB purification from *Spirulina* algae based on the procedure described in Smith and Holmes (1984), adapted from Levskaya *et al.* (2009) (see Fig. 17.3). This procedure relies on the fact that PCB is the most prevalent protein-bound tetrapyrrole found in *Spirulina*, so a generic tetrapyrrole purification protocol leads to high enrichment for this compound. Briefly, protein is purified from resuspended *Spirulina*, an 8 h methanolysis separates tetrapyrroles from their protein binding partners, and chloroform extraction is used to isolate these unbound tetrapyrroles and discard the protein fraction. The resulting purified PCB is stable at −20 °C for months.

4.1. Protocol

1. Resuspend 75 g Spirulina powder (Seltzer Chemical) in 2 L doubly distilled water (~30 mL/g). Stir for 10 min in a 4 L plastic beaker, transfer to 1 L screw-top plastic bottles, then spin at 8000 rpm at 4 °C

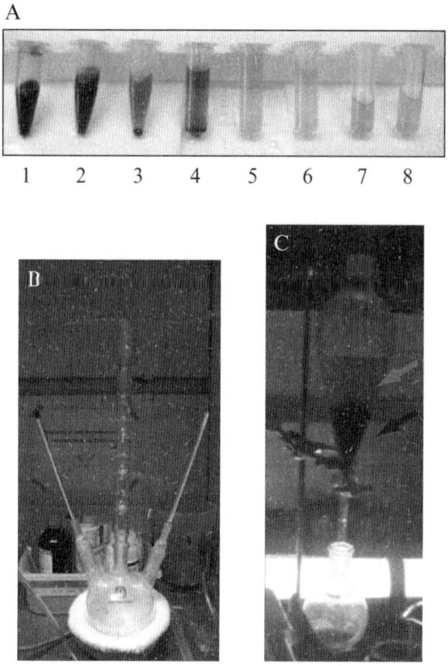

Figure 17.3 Key steps involved in the purification of PCB. (A) Samples collected at different stages of the PCB preparation procedure. Tubes contain the *Spirulina*-water mixture (1); the green supernatant after the first spin (2); the TCA protein precipitation, containing supernatant and pellet (3); the supernatant after each of three methanol washes (4–6); and the supernatant after the first and second 8 h methanolysis (7–8). (B) The assembled methanol reflux apparatus. The two thermometers should be placed in the vapor and fluid phases of the PCB-containing solution, respectively. Adjust heating until the temperature in both phases is the same, and the methanol solution is at a low boil. To prevent methanol loss through evaporation, flow cold water through the condenser (middle column of the apparatus). (C) Chloroform separation of PCB from the protein phase, photographed under a green safelight. Separation should result in a white, cloudy aqueous phase (red arrow) above a dark green chloroform phase (black arrow). If the aqueous phase retains a green color, add hydrochloric acid dropwise to acidify the solution, shake vigorously, and allow separation to occur again. (See Color Insert.)

for 1 h (we use a Sorvall RC5C Plus centrifuge with a Fiberlite F9S-4x1000y rotor).

2. Collect the supernatant, discarding the dark green pellet. Precipitate soluble protein from the supernatant by adding 20 g TCA (final solution: 1%, w/v, TCA). Wrap the beaker in foil to protect from light and stir solution at 4 °C for 1 h. Centrifuge in 1 L screw-top plastic bottles at 3000 rpm for 10 min (Sorvall RC5C Plus centrifuge; Fiberlite F9S-4x1000y rotor).

3. Resuspend and wash pellets with 100% methanol. Centrifuge in 250 mL screw-top plastic bottles at 3000 rpm for 10 min at 4 °C (Beckman J2-21M centrifuge; Beckman JA-14 rotor), and discard supernatant containing free tetrapyrroles. Repeat washes and centrifugation until supernatant runs clear or is only lightly green; this typically takes three washes. During washes, the pellet will change color, finally achieving a bright cyan color when washing is complete (Fig. 17.3A). These washes can be used to consolidate the material into a smaller volume. We typically perform the first wash in eight 250 mL bottles, the second in four bottles, and the third in two bottles for a final volume of ~500 mL.

   Note: during methanolysis and all subsequent steps, you will be working with PCB that is no longer protein-bound. In this state, PCB is light sensitive and the material should be shielded from light using aluminum foil, or by illuminating with a green safelight (Sylvania F40G fluorescent tube wrapped once with a Roscolene 874 sheet and once with a Roscolene 877 sheet to provide illumination at 550 nm) in an otherwise dark room.
4. Collect the washed cyan pellet and transfer to a 1 L three-neck round-bottomed distillation flask for methanolysis. Add 500 mL methanol; the pellet will not fully resuspend. Add boiling chips to prevent bumping of the methanol during heating. Connect the reflux apparatus, including cold water to recirculate through the condenser, thermometers to measure both the fluid and vapor temperatures, and a heat bath for heating the reaction. Reflux by maintaining at a slow boil with the solution and vapor phase temperatures held at 64.7 °C for 8 h (Fig. 17.3B). Be patient and do not overheat—it may take some time to establish a stable temperature for the reaction.
5. Transfer the liquid phase of the methanolysis reaction to 500 mL single-necked round-bottomed flask and connect to a rotary evaporator (but do not discard the remaining pellet—see Step 8). Evaporate the methanol to a final volume of 50 mL.
6. Add 50 mL chloroform and 100 mL water to a separatory funnel, followed by the concentrated PCB solution. Stopper the top of the funnel, and shake vigorously to emulsify the chloroform and aqueous phases. Wait for ~1 min to allow the phases to separate. The aqueous phase should be cloudy and white, while the PCB-containing chloroform phase should be dark green (Fig. 17.3C). If the aqueous phase is green or there is not a distinct color separation between the two phases, it should be taken as an indication that PCB is not well confined to the chloroform phase. In this case, add hydrochloric acid dropwise to acidify the solution. Shake vigorously and repeat until the aqueous phase is colorless.
7. Separate the chloroform phase (bottom liquid) into a 500 mL single-necked round-bottomed flask, discarding the remaining aqueous phase.

Evaporate the chloroform solution using a rotary evaporator to obtain dry PCB. Resuspend the dry PCB in 3 mL DMSO. Aliquot between 20 and 100 μL into dark 0.2 mL tubes. PCB can be frozen and thawed tens of times without degradation, but care must be taken not to expose aliquots to excess light.

8. To obtain the final PCB concentration, dilute the final preparation 1:100 into 1 mL 95:5% MeOH:HCl (37.5%) solution and reading the absorbance at 680 nm. The concentration in m*M* is calculated as $A_{680} \times 2.64$; typical final concentrations from this procedure are between 3 and 15 m*M*.
9. To increase the yield from this preparation, steps 4–8 can be performed a second time on the pellet remaining after the first 8 h methanolysis reaction. Typically, we perform the two 8-h methanolyses (Step 4) overnight on two consecutive nights, and perform steps 5–8 once on each preparation beginning the next morning, keeping each preparation separate and testing each for its final PCB concentration.
10. *Optional step*: The quality of the PCB preparation can be verified independently of Phy–PIF recruitment using cells expressing a Phy mutant (Y276H) that fluoresces upon incorporation of PCB (Levskaya *et al.*, 2009; Su and Lagarias, 2007). Expose cells expressing Phy Y276H to PCB as described in Section 5, and image cells using Cy5 excitation and emission optical configurations.

5. Cell Culture Preparation for Phy–PIF Translocation

After building Phy and PIF recruitment constructs and purifying PCB, it is important to validate the functionality of all components in the experiment of interest. Testing translocation requires two steps: incubating cells with PCB and preparing them for imaging on fibronectin-coated coverslips (discussed here), and imaging these cells using confocal microscopy (Section 6). Both of these steps require some measure of precision and care, as many variables can affect the quality of observed recruitment.

5.1. Protocol

1. Coat the glass coverslip on the bottom of a 35 mm MatTek dish (MatTek catalog number P35G-1.5-14C) with 100 μL of 0.08 mg/mL fibronectin diluted in PBS (Sigma catalog number F4759) for at least 1 h at room temperature.

2. Wash dish twice with 3 mL of Dulbecco's PBS (D-PBS). Plate cells immediately, or store dishes in PBS at 4 °C for no longer than 2 days prior to use.
3. Trypsinize and count NIH-3T3 cells expressing Phy and PIF constructs (Section 3) from an existing culture (maintained as described by ATCC). Plate 150,000 cells in 2 mL media on the fibronectin-coated MatTek dish. Place dish in incubator for 30 min to allow cells to adhere.
4. *Note*: this step is light-sensitive and care should be taken to minimize PCB's exposure to light. Perform under low light conditions or under a green safelight (see Step 3). In this step, media and PCB are premixed before adding to cell culture to ensure that cells are not exposed to high concentrations of DMSO. Transfer 100 μL media from the MatTek dish to a 1.5 mL tube. Pipet 4 nmol of PCB (about 1 μL of the 4 m*M* stock from Step 3) into the tube, and mix well. Add the PCB–media mixture back to dish; swirl to mix. Wrap dish in aluminum foil and place in incubator for at least 30 min. Plates can be maintained in PCB-containing media for a few hours, so multiple plates can be prepared simultaneously and imaged sequentially.
5. Before imaging, exchange the PCB-containing media for an imaging solution containing 3 mL modified Hank's balanced salt solution (mHBSS) supplemented with 2% FBS. NIH-3T3 cells should remain healthy at room temperature without supplemental CO_2 for at least 6 h under these conditions. For prolonged imaging, replace imaging solution to combat evaporation.

6. Imaging PIF Translocation Using Spinning Disk Confocal Microscopy

After preparing cells with the desired Phy- and PIF-containing constructs and purifying PCB, it is important to verify that binding between Phy and PIF are controllable by light. The following procedure relies on a localization change in one of the components (here, a fluorescent PIF construct) upon binding to the other (membrane-localized Phy). To perform this experiment, the cytoplasmic concentration of PIF is measured by fluorescent imaging using confocal microscopy. Here, we outline the protocol for imaging PIF–YFP and Phy-mCherry-CAAX (mCherry is a fluorescent protein with RFP-like excitation and emission).

A confocal microscope can be easily used to supply wavelengths required for association and dissociation of Phy–PIF complexes. A red laser emitting at 561 nm (or alternatively, a white light source and 561 nm filter) can be used to maximally activate Phy–PIF association, and these should be available on fluorescent microscopes capable of imaging RFP or similar proteins.

However, because the transition to the PIF-binding state is so sensitive, unfiltered brightfield light is sufficient to generate measurable Phy–PIF recruitment, as described in the protocol below. To elicit Phy–PIF dissociation, place a 750 nm long-pass filter (Newport, model FSQ-RG9) on top of the microscope's condenser (in the brightfield imaging light path) and turn on brightfield illumination (Fig. 17.4A). It is important to ensure that no filters are present in the light path that could interfere with infrared light transmission. It is possible to use a 750 nm filter in conjunction with other light sources (e.g., mercury halide arc lamp; DG4 light source). However, care must be taken to ensure that all optical components are capable of transmitting infrared light (e.g., mirrors, filters, liquid light guides), and that no infrared-blocking filters are present in the light path.

A

B

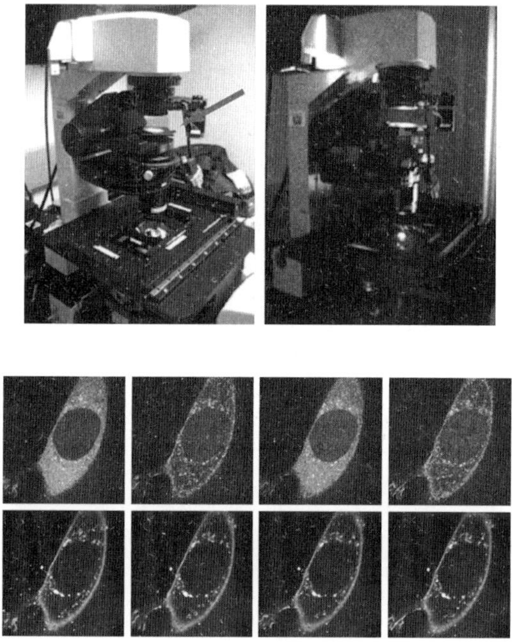

Figure 17.4 Confocal microscopy to image Phy–PIF translocation. (A) A 750 nm filter can be placed in the brightfield light path to elicit Phy–PIF dissociation. With this filter in place, illuminating with brightfield light leads to Phy–PIF dissociation. Simply removing the filter provides enough activating light to induce Phy–PIF translocation. Alternatively, RFP excitation light (650 nm) can be used to induce association. (B) A montage of confocal images of a NIH-3T3 cell showing PIF–YFP translocation in response to light. Cells were prepared harboring the genetic constructs described in Section 3. The upper panel shows PIF–YFP fluorescence after sequential 30 s exposures of activating (brightfield) and inactivating (750 nm filtered) light. Phy-mCherry levels in corresponding timepoints are shown in the lower panels.

Observing high-quality light-dependent recruitment depends strongly on the expression levels of both Phy and PIF fluorescent fusion proteins. Because this assay relies on observing PIF changing localization from cytoplasm to cell membrane, there must be enough Phy on the membrane to appreciably deplete cytoplasmic PIF levels during exposure to activating light. Thus, it is crucial to select cells with high membrane expression of Phy and low to moderate PIF levels. Cell geometry can also play a crucial role: because small cells have a higher surface area to volume ratio than large cells, smaller cells are further enriched for an excess of Phy molecules compared to PIF, leading to better PIF depletion upon activation.

6.1. Protocol

1. Select a cell for imaging based on the criteria described above (high Phy-mCherry membrane expression, moderate PIF–YFP expression, and small cell volume).
2. For the best contrast between recruitment and release, choose an imaging plane in the lower half of the cell, just above the coverslip. Such a focal plane should be close to a large pool of membrane-bound Phy, where PIF cytoplasmic depletion should be maximal. If available, initiate the microscope's autofocus system to prevent focal plane drift.
3. Place a 750 nm square filter on top of the microscope condenser, in the brightfield imaging light path (Fig. 17.4A).
4. Alternate 30 s exposures of brightfield light with and without the 750 nm filter. After each 30 s brightfield exposure, take single PIF–YFP translocation images using the confocal microscope's YFP imaging mode. A typical series of NIH-3T3 cell images showing recruitment by this technique is shown in Fig. 17.4B.

ACKNOWLEDGMENTS

We thank Anselm Levskaya for advice on PCB purification and light system manipulation. This work was partially supported by the National Institutes of Health grant GM084040, the Cancer Research Institute Postdoctoral Fellowship (J. E. T.), the American Cancer Society Fellowship (D. G.).

REFERENCES

Bivona, T. G., Quatela, S. E., Bodemann, B. O., Ahearn, I. M., Soskis, M. J., Mor, A., Miura, J., Wiener, H. H., Wright, L., Saba, S. G., Yim, D., Fein, A., *et al.* (2006). PKC regulates a farnesyl-electrostatic switch on K-Ras that promotes its association with Bcl-XL on mitochondria and induces apoptosis. *Mol. Cell* **21,** 481–493.

Clarke, S., Vogel, J. P., Deschenes, R. J., and Stock, J. (1988). Posttranslational modification of the Ha-ras oncogene protein: evidence for a third class of protein carboxyl methyltransferases. *Proc. Natl. Acad. Sci. USA* **85,** 4643–4647.

Cohen-Saidon, C., Cohen, A. A., Sigal, A., Liron, Y., and Alon, U. (2009). Dynamics and Variability of ERK2 Response to EGF in Individual Living Cells. *Mol. Cell* **36,** 885–893.

Essen, L. O., Mailliet, J., and Hughes, J. (2008). The structure of a complete phytochrome sensory module in the Pr ground state. *Proc. Natl. Acad. Sci. USA* **105,** 14709–14714.

Gautier, A., Nguyen, D. P., Lusic, H., An, W., Deiters, A., and Chin, J. W. (2010). Genetically encoded photocontrol of protein localization in mammalian cells. *J. Am. Chem. Soc.* **132,** 4086–4088.

Gunaydin, L. A., Yizhar, O., Berndt, A., Sohal, V. S., Deisseroth, K., and Hegemann, P. (2010). Ultrafast optogenetic control. *Nat. Neurosci.* **13,** 387–392.

Heim, R., and Tsien, R. Y. (1996). Engineering green fluorescent protein for improved brightness, longer wavelengths and fluorescence resonance energy transfer. *Curr. Biol.* **6,** 178–182.

Ilani, T., Vasiliver-Shamis, G., Vardhana, S., Bretscher, A., and Dustin, M. L. (2009). T cell antigen receptor signaling and immunological synapse stability require myosin IIA. *Nat. Immunol.* **10,** 531–539.

Kanbe, E., and Zhang, D. E. (2004). A simple and quick method to concentrate MSCV retrovirus. *Blood Cells Mol. Dis.* **33,** 64–67.

Lahav, G., Rosenfeld, N., Sigal, A., Geva-Zatorsky, N., Levine, A. J., Elowitz, M. B., and Alon, U. (2004). Dynamics of the p53-Mdm2 feedback loop in individual cells. *Nat. Genet.* **36,** 147–150.

Lemke, E. A., Summerer, D., Geierstanger, B. H., Brittain, S. M., and Schultz, P. G. (2007). Control of protein phosphorylation with a genetically encoded photocaged amino acid. *Nat. Chem. Biol.* **3,** 769–772.

Leung, D. W., Otomo, C., Chory, J., and Rosen, M. K. (2008). Genetically encoded photoswitching of actin assembly through the Cdc42-WASP-Arp2/3 complex pathway. *Proc. Natl. Acad. Sci. USA* **105**(2008), 12797–12802.

Levskaya, A., Weiner, O. D., Lim, W. A., and Voigt, C. A. (2009). Spatiotemporal control of cell signalling using a light-switchable protein interaction. *Nature* **461,** 997–1001.

Michalet, X., Pinaud, F. F., Bentolila, L. A., Tsay, J. M., Doose, S., Li, J. J., Sundaresan, G., Wu, A. M., Gambhir, S. S., and Weiss, S. (2005). Quantum dots for live cells, in vivo imaging, and diagnostics. *Science* **307,** 538–544.

Nelson, D. E., Ihekwaba, A. E., Elliott, M., Johnson, J. R., Gibney, C. A., Foreman, B. E., Nelson, G., See, V., Horton, C. A., Spiller, D. G., Edwards, S. W., McDowell, H. P., *et al.* (2004). Oscillations in NF-kappaB signaling control the dynamics of gene expression. *Science* **306,** 704–708.

Ni, M., Tepperman, J. M., and Quail, P. H. (1999). Binding of phytochrome B to its nuclear signalling partner PIF3 is reversibly induced by light. *Nature* **400,** 781–784.

Ory, D. S., Neugeboren, B. A., and Mulligan, R. C. (1996). A stable human-derived packaging cell line for production of high titer retrovirus/vesicular stomatitis virus G pseudotypes. *Proc. Natl. Acad. Sci. USA* **93,** 11400–11406.

Paliwal, S., Iglesias, P. A., Campbell, K., Hilioti, Z., Groisman, A., and Levchenko, A. (2007). MAPK-mediated bimodal gene expression and adaptive gradient sensing in yeast. *Nature* **446,** 46–51.

Shimizu, T. S., Tu, Y., and Berg, H. C. (2010). A modular gradient-sensing network for chemotaxis in Escherichia coli revealed by responses to time-varying stimuli. *Molecular Syst. Biol.* **6**(382), 1–14.

Shimizu-Sato, S., Huq, E., Tepperman, J. M., and Quail, P. H. (2002). A light-switchable gene promoter system. *Nat. Biotech.* **20,** 1041–1044.

Smith, H., and Holmes, M. G. (1984). Techniques in Photomorphogenesis. Academic Press, Orlando, FL.

Spencer, D. M., Wandless, T. J., Schreiber, S. L., and Crabtree, G. R. (1993). Controlling signal transduction with synthetic ligands. *Science* **262,** 1019–1024.

Strickland, D., Moffat, K., and Sosnick, T. R. (2008). Light-activated DNA binding in a designed allosteric protein. *Proc. Natl. Acad. Sci. USA* **105,** 10709–10714.

Su, Y. S., and Lagarias, J. C. (2007). Light-independent phytochrome signaling mediated by dominant GAF domain tyrosine mutants of Arabidopsis phytochromes in transgenic plants. *Plant Cell* **19,** 2124–2139.

Tay, S., Hughey, J. J., Lee, T. K., Lipniacki, T., Quake, S. R., and Covert, M. W. (2010). Single-cell NF-kappaB dynamics reveal digital activation and analogue information processing. *Nature* **466,** 267–271.

Terrillon, S., and Bouvier, M. (2004). Receptor activity-independent recruitment of betaarrestin2 reveals specific signalling modes. *EMBO J.* **23,** 3950–3961.

Truong, K., and Ikura, M. (2001). The use of FRET imaging microscopy to detect protein-protein interactions and protein conformational changes in vivo. *Curr. Opin. Struct. Biol.* **11,** 573–578.

Tyszkiewicz, A. B., and Muir, T. W. (2008). Activation of protein splicing with light in yeast. *Nat. Methods* **5,** 303–305.

CHAPTER EIGHTEEN

Synthetic Physiology: Strategies for Adapting Tools from Nature for Genetically Targeted Control of Fast Biological Processes

Brian Y. Chow,[1,2] Amy S. Chuong,[1] Nathan C. Klapoetke,[1] *and* Edward S. Boyden

Contents

Abstract

The life and operation of cells involve many physiological processes that take place over fast timescales of milliseconds to minutes. Genetically encoded technologies for driving or suppressing specific fast physiological processes in intact cells, perhaps embedded within intact tissues in living organisms, are critical for the ability to understand how these physiological processes contribute to emergent cellular and organismal functions and behaviors. Such "synthetic physiology" tools are often incredibly complex molecular machines, in part because they must operate at high speeds, without causing side effects. We here explore how synthetic physiology molecules can be identified and deployed in cells, and how the physiology of these molecules in cellular contexts can be assessed and optimized. For concreteness, we discuss these

Synthetic Neurobiology Group, The Media Laboratory and McGovern Institute, Departments of Biological Engineering and Brain and Cognitive Sciences, Massachusetts Institute of Technology, Cambridge, Massachusetts, USA
[1] Authors contributed equally
[2] Future location: Department of Bioengineering, University of Pennsylvania, Philadelphia, Pennsylvania, USA

Methods in Enzymology, Volume 497
ISSN 0076-6879, DOI: 10.1016/B978-0-12-385075-1.00018-4

methods in the context of the "optogenetic" light-gated ion channels and pumps that we have developed over the past few years as synthetic physiology tools and widely disseminated for use in neuroscience for probing the role of specific brain cell types in neural computations, behaviors, and pathologies. We anticipate that some of the insights revealed here may be of general value for the field of synthetic physiology, as they raise issues that will be of importance for the development and use of high-performance, high-speed, side-effect free physiological control tools in heterologous expression systems.

1. Introduction

The life and operation of cells involve many physiological processes that take place over fast timescales of milliseconds to minutes. These physiological changes include variations in cell membrane potential and cellular ionic composition; changes in protein conformation; posttranslational modification, localization, and interaction; and other biochemical and mechanical processes, all occurring at length scales ranging from nanometers to meters. Technologies for driving or suppressing specific fast physiological processes in intact cells, perhaps embedded within intact tissues in living organisms, are critical for the ability to understand how those physiological processes contribute to emergent cellular and organismal functions and behaviors. For example, the ability to drive a specific physiological process can reveal precisely which functions that process is sufficient to initiate or sustain, whereas the ability to suppress a specific physiological process can reveal the set of functions for which the process is necessary. Such precision physiological control technologies may, of course, also serve therapeutic purposes if they offer the ability to remedy a pathway thrown into disarray in a disease context, ideally while leaving other pathways unperturbed.

A diversity of molecular tools have been developed that allow the precision control of physiological processes—including high-specificity pharmacological compounds, caged chemicals that can be activated by pulses of light, and tools whose physiological impact is unleashed by administration of heat or radiofrequency energy. This ongoing effort has led to a number of physiological control tools that are partly or entirely genetically encoded, and therefore easy to use in genetic model organisms in conjunction with commonly available transgenic strategies, for example, viruses for delivery to specific mammalian cells embedded within intact organ systems. One might call the set of capabilities opened up by these tools "synthetic physiology," because these tools enable a synthetic approach to studying physiological pathways, with an emphasis on perturbation of specific pathways, to see what their influence is on other pathways. Many of the synthetic physiology tools in widespread use have come to be known as "optogenetic," because they enable specific physiological processes to be controlled by light, and thus enable temporally and spatially precise control of physiology with microscopes, lasers, and other common laboratory optical equipment, often without the need for exogenous chemical delivery (helpful for use *in vivo*). Such light-driven tools, or prototypes of tools, exist for applications including driving of protein–protein interactions (Kennedy *et al.*, 2010; Levskaya *et al.*, 2009; Yazawa *et al.*, 2009), enzyme activity (Wu *et al.*, 2009), intracellular signaling (Schroder-Lang *et al.*, 2007), and many other fast changes (Moglich and Moffat, 2010). A widely used set of optogenetic tools are the microbial rhodopsins, molecules that respond to light by translocating ions from one side of the plasma membrane to the other, thus enabling electrical activation or silencing of electrically excitable cells such as neurons, in response to pulses of light (Boyden *et al.*, 2005; Chow *et al.*, 2010; Gradinaru *et al.*, 2008, 2010; Han and Boyden, 2007; Zhang *et al.*, 2007a). For example, channelrhodopsins, microbial opsins from algae, admit cations into cells in response to light, depolarizing the cells; halorhodopsins, opsins from archaea, pump in chloride in response to light, resulting in cellular hyperpolarization; archaerhodopsins and bacteriorhodopsins, also isolated from archaea (and other kingdoms), pump out protons, also resulting in cellular hyperpolarization. In the mammalian nervous system, these molecules do not require any exogenous chemical supplementation for their operation, and thus can be treated as fully genetically encoded. The hyperpolarization opsins are used to enable optical silencing of genetically targeted neurons in order to see what neural dynamics, behaviors, and pathologies they are necessary for, whereas the depolarizing opsins are used to drive neural activity in genetically targeted neurons, to assess which downstream neural computations and behaviors causally result. Both sets of tools are in widespread use for investigating the roles that specific cells play within the nervous systems of species ranging from *Caenorhabditis elegans* to nonhuman primate (see the following references for some early papers in the field; Adamantidis *et al.*, 2007; Alilain *et al.*, 2008; Aravanis *et al.*, 2007; Arenkiel *et al.*, 2007; Atasoy *et al.*, 2008; Bi *et al.*, 2006; Douglass *et al.*, 2008; Farah *et al.*, 2007; Han *et al.*, 2009a; Huber *et al.*, 2008; Ishizuka *et al.*, 2006; Lagali *et al.*, 2008; Li *et al.*, 2005; Liewald *et al.*, 2008; Mahoney *et al.*, 2008; Nagel *et al.*, 2005; Petreanu *et al.*, 2007; Schroll *et al.*, 2006; Toni *et al.*, 2008; Wang *et al.*, 2007; Zhang and Oertner, 2007; Zhang *et al.*, 2007b, 2008).

Although one of the goals of synthetic biology is to be able to regard such tools as "black box parts" (Canton *et al.*, 2008; Carr and Church, 2009; Endy and Brent, 2001), whose internal workings can be hidden beneath an abstraction layer, the genetically encoded tools in use for synthetic physiology are incredibly complex machines, in part because they must operate at high speeds. To be useful, their inner workings must be sophisticated enough to enable these tools to accomplish their precision functions when activated by an external stimulus such as light, while avoiding undesired side effects. Understanding and engineering these tools require thinking and working at the level of protein structure and dynamics, which means that the ability to systematically engineer synthetic physiology tools is primitive, compared to, say, DNA synthesis or gene engineering, where design principles emerge from systematic application of straightforward considerations of the structure and chemistry of nucleic acids.

As just one example, halorhodopsins are just a few hundred amino acids long, yet these molecules contain an anchor site for a vitamin A-derived chromophore, and upon illumination undergo structural rearrangements through at least seven coupled photointermediates during the process of translocating a chloride ion from one side of the cell membrane to the other. Halorhodopsins also possess alternate photocycles that involve shifts in the spectrum of light responsivity, as well as secondary transport of protons (Bamberg *et al.*, 1993; Han and Boyden, 2007; Hegemann *et al.*, 1985; Oesterhelt *et al.*, 1985). A great many individual amino acids in halorhodopsins, when mutated, result in impairments or alterations of opsin function, implying that the operation of these proteins relies upon a great many of the residues remaining intact for structural or protein dynamics reasons (Otomo, 1996; Rudiger and Oesterhelt, 1997; Rudiger *et al.*, 1995; Sato *et al.*, 2003a,b). As one might therefore guess, methodologies for the discovery, characterization, and optimization of synthetic physiology parts represent at this time something of an art form.

The purpose of this chapter is to present how synthetic physiology molecules can be identified and implemented (Section 2), how these molecules can be expressed in cells (Section 3), and how the physiology of these molecules in cellular contexts can be assessed (Section 4). For concreteness, we discuss these methods in the context of the "optogenetic" light-gated ion channels and pumps that we have worked on, and widely disseminated, over the last several years. During this time, we have characterized dozens of gene products of microbial rhodopsin sequence homologs from across the tree of life, and we have begun to understand the principles governing how to create, express, and analyze them. However, we anticipate that many of the insights revealed here may be of general use for the field of synthetic physiology, as they raise issues that will be of importance for the development and use of high-performance, high-speed, side-effect free physiological control tools. Because of the scope of methodologies involved in the research, the information here provides the reasoning behind our current best practices, as opposed to the laying out of step-by-step protocols. Detailed protocols will be posted as white papers on our Web site (http://syntheticneurobiology.org) and updated regularly; the goal for this chapter is to lay out the principles that guide these protocols.

Synthetic physiology tools, at this point in protein engineering history, heavily rely upon naturally occuring genetically encoded proteins as the effectors that perform the actual modulation of physiological functions, although one might imagine that in the future entirely artificial designs might be realized. In the case of microbial opsins, which transport ions into or out of cells in response to light, the light sensor is built into the ion translocation machinery, embedded within the middle of the seven-transmembrane domain protein. (In most other classes of optogenetic synthetic physiology tool, genetically encoded light sensors are fused to genetically encoded effectors so as to couple the conformational change of the light sensor under illumination to the physiological function downstream, as reviewed in Moglich and Moffat, 2010.) To date, many opsins have been derived from organisms such as archaea, algae, bacteria, and fungi: there is considerable molecular diversity in such organisms, which provides vast genetically encoded wealth from which one can repurpose proteins as novel molecular tools or building blocks of tools, and the proliferation of publically available sequence information, coupled to the rapidly decreasing costs of *de novo* gene synthesis (Carlson, 2003; Carr and Church, 2009), makes it increasingly easier to mine molecular wealth. Thus, the pipeline for developing opsins as tools begins with isolation of gene sequences from genomes, followed by *de novo* gene synthesis, then mutagenesis and/or appending of useful sequences for visualization and improvement of trafficking of opsins, and then finally embedding of the sequence in a transgenic vector (e.g., a viral vector) for heterologous expression in the cells of a target organism.

2. Molecular Design and Construction

Synthetic physiology tools in the opsin space have been identified from genomic databases by searching for proteins with similar amino acid sequence homology to previously characterized opsins. Microbial opsins were first discovered by biologists around four decades ago, and many members of this class have been identified at a genomic level over the intervening time, although only a subset of these molecules have been characterized at a physiological level. *De novo* gene synthesis has proven important for rapid construction of opsin DNA from sequences derived from genomic and transcriptomic databases (Chow *et al.*, 2010; Han and Boyden, 2007). One can obtain a gene that is codon-optimized to the target organism (Richardson *et al.*, 2010; Welch *et al.*, 2009; Wu *et al.*, 2006), important for proper protein expression in cells of the target organism, within a few days to weeks of sequence identification, from a gene synthesis vendor. Codon optimization is useful for expression of these genes, which are isolated from algae, bacteria, and other nonanimal species, in heterologous systems (e.g., animal cells in an organism of interest). During *de novo* gene synthesis, it is possible to eliminate restriction sites within the gene to ease later molecular cloning steps, so that they can be easily engineered to facilitate opsin function, for example, by enabling fusion of a fluorescent protein tag to the molecule, concatenation of trafficking sequences to the molecule, or addition of a cell type-specific promoter to the gene to delimit the expression to specific cells within the target organism.

After gene synthesis, the next step is to alter the gene as needed, or to append extra sequences, to optimize its function toward a directed physiological control goal. A few opsin crystal structures have been obtained (e.g., Enami *et al.*, 2006; Kolbe *et al.*, 2000; Luecke *et al.*, 1999a,b, 2001; Yoshimura and Kouyama, 2008), and decades of studies have been performed in which specific residues within opsins were mutated, followed by spectroscopic or physiological characterization of the resultant mutated opsins (e.g., Gilles-Gonzalez *et al.*, 1991; Greenhalgh *et al.*, 1993; Hackett *et al.*, 1987; Marinetti *et al.*, 1989; Marti *et al.*, 1991; Mogi *et al.*, 1987, 1988, 1989a,b; Otto *et al.*, 1989; Stern and Khorana, 1989; Subramaniam *et al.*, 1992). These datasets have proven influential in guiding the strategic engineering of these molecules through site-directed mutagenesis (Berndt *et al.*, 2009; Chow *et al.*, 2010; Gunaydin *et al.*, 2010; Lin *et al.*, 2009; Wang *et al.*, 2009), enabling molecules with improved trafficking, or faster or slower kinetics, to be created. In part because many new opsins are being discovered at a rapid rate, antibodies for localizing them in an immunocytochemical fashion are not commonly available. Thus, tagging the proteins with a fluorophore, or with a small epitope to which antibodies already exist, may be helpful in order to determine efficiently which cells within the target organism are expressing the opsin. In addition, such tagging yields critical information about the membrane trafficking and localization of opsins within cells—indeed, fluorophore localization to the plasma membrane of opsin–fluorophore fusions has been used to predict photocurrent magnitude, as measured through electrophysiology characterization (Chow *et al.*, 2010; Wang *et al.*, 2009). We have previously reported a method for quantifying membrane localized proteins in neurons (Chow *et al.*, 2010), based on a Gaussian-blur-based technique developed for the same purpose in HEK293 cells (Wang *et al.*, 2009). Of course, this method is useful for quantifying protein localization in the cells, but it does not provide information on whether the protein is properly folded and functional, within the membrane. Ultimately it is the number of proteins in the membrane that are functional which determines their overall efficacy in physiological control. Since these molecules, when expressed in neurons or other animal cells, are often in a very different lipid environment than the one they evolved to function in, even a properly folded molecule in a lipid membrane may not be fully functional. As a concrete example, photocurrent enhancement of an opsin by appending the flanking sequences of the KiR2.1 protein (as done in Gradinaru *et al.*, 2010) boosts the membrane expression of opsins as observed through microscopy, but may boost the photocurrent even more than might be expected from the cellular appearance alone (unpublished observations); this appearance–current discrepancy may vary from opsin to opsin. Indeed, KiR2.1 sequences may even decrease overall cellular expression for some opsins (unpublished observations), even as they might be increasing the amount of properly folded membrane-embedded protein that is functional. Thus, quantitative confocal microscopy must be supplemented by a functional, physiological assay. This theme, that there are few proxies for function in the assessment of synthetic physiology tools, is partly why they are hard to find, engineer, and assess.

It is important to realize that the complexity of these molecules means that even an innocuous change like creating a fusion protein between an opsin and a fluorophore may modulate the function of the opsin. For example, an observation that requires further investigation is that fluorophore fusions with a target molecule can alter expression or performance of the target molecule, for example reducing viral titer important when viral delivery is the route for transgenically engineering the target organism (Weber *et al.*, 2008). Appending different fluorophores (e.g., EGFP vs. mCherry vs. ECFP) to an opsin can result in different opsin localization patterns (e.g., due to mCherry's greater tendency to aggregate than EGFP or ECFP) and potentially different levels of photo current, for a given cell type. In the event that fusion of a fluorophore to a given opsin is undesirable, alternatives exist to directly fusing fluorophores to opsins, while still enabling identification of cells expressing the opsin, including interposing IRES (internal ribosome entry sites) and 2A sequences ("self-cleaving" linkers first identified in foot-and-mouth-disease virus) in between opsins and fluorophores (Han *et al.*, 2009b; Tang *et al.*, 2009). Protein expression levels for the gene that appears after the IRES is often a small fraction of that of the gene before the IRES (Hennecke *et al.*, 2001; Mizuguchi *et al.*, 2000; Osti *et al.*, 2006; Yu *et al.*, 2003). 2A sequences in principle yield highly stoichiometric amounts of translated protein, but in reality, different functional levels may be observed for the pre- and post-2A proteins, due to alterations in protein trafficking or function that result from the residual amino acids of the 2A sequence left behind after protein translation.

The use of trafficking sequences, export motifs, and other signal sequences, both natural and designed, may be generally useful for improving the heterologous expression of opsins in the cells of target organisms. Opsins come from organisms whose membrane structure and overall cellular architecture is different from neurons. For example, *Natronobacterium pharaonis* halorhodopsin photocurrents can be enhanced several fold in mammalian cells by appending to this molecule the N- and C-terminal sequences of the human KiR2.1 potassium channel protein, which are responsible for endoplasmic reticulum-export and Golgi-export (although, see alternative explanations of the role that these KiR2.1 sequences play in boosting cellular expression, above; Gradinaru *et al.*, 2010; Hofherr *et al.*, 2005; Ma *et al.*, 2001; Stockklausner and Klocker, 2003). The enhancement offered by a given exogenous trafficking sequence is opsin-dependent—for example, appending a trafficking sequence that boosts *N. pharaonis* halorhodopsin expression levels (the ER2 sequence; Gradinaru *et al.*, 2008) has no effect on augmenting the currents of the *Halorubrum sodomense* archaerhodopsin-3, although adding a different sequence (the Prl sequence, derived from the prolactin secretion targeting sequence) does augment archaerhodopsin-3 photocurrent (Chow *et al.*, 2010). We have found, through experiments with combinatorial addition of N- and C-terminal signal sequences, that adding multiple signal sequences does not necessarily improve expression in a linear way, perhaps owing to interactions between the multiple trafficking mechanisms at play. It should be noted that opsins may also possess intrinsic, even covert, sequences that enable them to be expressed very well on the plasma membrane. For example, the light-driven outward proton pump archaerhodopsin-3 from *H. sodomense* (and, in general, members of the archaerhodopsin class of opsins) naturally expresses efficiently and well on plasma membranes (Chow *et al.*, 2010). Opsin mutagenesis and chimeragenesis has pointed toward candidate amino acids that may play a critical role in opsin trafficking and expression on the plasma membrane (Lin *et al.*, 2009; Wang *et al.*, 2009).

3. Transduction of Microbial Opsins into Cells for Heterologous Expression

The analysis of the potential power of a given microbial opsin to control the voltage or ionic composition of a target cell type (e.g., in a given organism under study), should be performed ideally in the target cell type itself, or in a testbed cell type that is as similar as possible to the target cell type. For example, the trafficking-enhancement and protein folding enhancement sequences described above are derived from specific species and were optimized in cells from specific species; accordingly, they may not work equally well in species different from the source species, or in cell types greatly different from the cell types used to assess and optimize the sequences. Similarly, the covert trafficking sequences found within opsins may not function equally well in all cell types. As a concrete example, the *Halobacterium salinarum* bacteriorhodopsin has long been considered a difficult protein to express in *Escherichia coli* (e.g., Dunn *et al.*, 1987), but it expresses readily in mammalian neurons, and can mediate biologically meaningful photocurrents (Chow *et al.*, 2010). Similarly, channelrhodopsin-2 does not express well in

E. coli, but expresses well in mammalian neurons. Conversely, proteorhodopsins from uncultured marine gamma-proteobacteria express and function well in *E. coli*, but do not generate photocurrents in mammalian cells (HEK293 cells or mouse neurons; Chow *et al.*, 2010), despite a rudimentary degree of expression of the proteorhodopsin protein in these mammalian cells. Thus, reliance on just a single heterologous expression cell type (e.g., *E. coli*, yeast, *Xenopus* oocytes, HEK cells) as the sole testbed for characterizing the physiological function of opsins, may lead to a partial picture of how well the opsins assessed will perform across the broad set of cell targets confronted in biology. Similarly, screening for enhancing mutations, trafficking sequences, or other beneficial modifications, using a single heterologous expression cell type, may lead to unintentional optimization of the opsin for function in that particular cell type, and potential deoptimization of expression, trafficking, or function in other cell types of interest within the ultimate spectrum of usage of the tool.

If mammalian neurons in the living mouse or rat are the target, then mammalian neurons in primary culture should be at some point used to assess the function of a given opsin (Boyden *et al.*, 2005; Chow *et al.*, 2010), although ideally *in vivo* assessment should be performed as well, given the different state of neurons *in vivo* versus *in vitro*. It is important to note that different types of neurons, at different ages, may differ in their level and timecourse of opsin expression and function. We typically utilize mouse hippocampal and cortical primary cultures because they contain representatives of different neuron classes (Boyden *et al.*, 2005; Chow *et al.*, 2010; Han and Boyden, 2007). However, primary neuron cultures are laborious to prepare and maintain, and so we and others use HEK293 cell lines as well to perform electrophysiological characterization of opsins (Chow *et al.*, 2010; Lin *et al.*, 2009; Nagel *et al.*, 2003; Wang *et al.*, 2009). HEK cells are more robust, and easier to work with, than neurons, and can be grown for multiple cell division cycles in culture, unlike neuron cultures which do not replicate after plating and differentiation. In addition, HEK cells possess cellular shapes and molecular phenotypes that are somewhat less variable than those of neurons, and possess fewer active conductances than do neurons; both of these features help reduce variability of opsin characterization measurements. Conversely, HEK cells may yield smaller photocurrents than do neurons due to their smaller surface area, and may have limited utility in fully predicting how well a protein will traffic in neurons (and thus, how they will perform as optical modulators of neural physiology) due to their differences from neurons. As a simple example of this latter point, HEK cells do not possess axons or dendrites; some findings have been published claiming that certain opsins preferentially traffic to the synaptic processes of neurons (Li *et al.*, 2005), and of course, any such effects would not be observable in a HEK cell. However, HEK cells are still extremely useful for performing fast screening assays of whether there is any physiological effect of illuminating a given opsin, and may be particularly useful for characterization of amplitude-normalized features of opsins such as the action spectrum, the plot of the relative photocurrent observed upon delivery of light of different colors.

Transfection is the simplest and fastest way to get DNA that encodes for opsins into cells, for rapid characterization of opsins in a cellular context. For HEK cells, transfection can increase the likelihood of delamination from the substrate; the use of Matrigel to promote cell adhesion to a glass coverslip when plating, as opposed to polylysine, is suggested. Well-dissociated HEK cells that are spatially separated from one another are critical for high-quality electrophysiological assays, as HEK cells that grow together can form gap junction-connected syncytia that can preclude accurate electrophysiological analysis of expressed opsins, by compromising voltage-clamp fidelity. In order to improve the quality of HEK cells for physiological assessment, passage the cells for their final plating when they reach medium levels of confluence ($\sim$50%); then, during the final plating step, trypsinize the HEK cells; resuspend the cells in serum-free media; pipette the cells against the sidewalls of the dish or flask to break up clumps of cells, perhaps triturating the cells with a fine-gauged sterile needle (e.g., less than five times to avoid excessive mechanical force on cells, through a $\sim$31 gauge needle); and then add serum-containing media (to halt the trypsinization) before plating the final mixture on glass coverslips. For neuron culture, mouse or rat hippocampal or cortical neurons should be cultured from P0 pups or E18 embryos at moderate densities, using standard protocols (Boyden *et al.*, 2005; Chow *et al.*, 2010; Han and Boyden, 2007). Multiple experimenters in our laboratory have found that the lowest-variability recordings from opsin-expressing cells are often from ones in areas of sparse cell density, often at the edge of the area occupied by cells. The preferred method for HEK and neuron culture transfection is calcium phosphate precipitation of DNA, for example, using commercially available kits. The calcium phosphate precipitation-based process can be harsh on neurons; accordingly, precautions should be taken, if needed, to limit neuronal excitotoxicity, for example, adding AP5, a NMDA receptor antagonist, to the medium. The best transfection rates in neurons, in our hands, are achieved when neurons are transfected 3–4 days *in vitro*, with diminishing efficiency beyond then (although the genes encoding for well-expressed proteins, like Arch, can be delivered at 5 days *in vitro*).

Viral vectors, such as lentiviral vectors, can also be useful for assessing opsin function, because they can result in a high yield of opsin-expressing cells in a cultured cell environment, and they can also be used to insure a precise gene dosage into a cell of interest. For neurons, they also present lower toxicity, at a higher cellular yield, than achieved commonly with calcium phosphate transfection. Many lentiviral preparation protocols, involving the transfection of opsin-containing and helper plasmids into carefully cultivated and healthy HEK cells, exist that work well with opsins (e.g., Boyden *et al.*, 2005; Chow *et al.*, 2010; Han *et al.*, 2009a). One key consideration is that recombination can be an issue when preparing lentiviral vectors (and other viral vectors), due to the presence of repetitive sequences within the genomic vector of the virus, that is, the payload-encoding plasmid. In theory, any *E. coli* with loss of function mutation in *rec* gene(s) should be suitable for growing up such plasmids. In our experience working with lentiviral plasmids, Stbl3 (*recA13-*) *E. coli* have a lower rate of recombination compared to other *rec*-cells such as XL1-Blue (*recA*1-). XL10-Gold *E. coli* may work well, with AAV plasmids. It is recommended to try out different types of *rec*-cells to find the optimal one for a particular viral vector, as recombination events can cause loss of vectors, and require time-consuming plasmid reconstruction. It is also important to check if any special considerations are needed for utilizing these specialized viral plasmid-compatible competent cell lines. For example, Stbl3 is endA+, and thus the endA endonuclease will need to be removed with appropriate washing when purifying the DNA, to prevent DNA degradation. To check for recombination, viral plasmids should regularly be verified in both sequence and topology, using DNA sequencing and restriction digestion, respectively. Both methods are recommended because sequencing short regions, such as the cloned insert, will only inform you whether the sequence is locally correct, but recombination can also occur between unpredictable locations, so that the cloned sequence is largely locally correct but different in global topology. Therefore, it is highly recommended to perform multiple restriction digests to verify that the global sequence topology has not deviated from the designed plasmid. When cloning payloads into viral vectors, it is important to use only parent vectors that have also been tested for recombination, and it is important to perform both sequencing and restriction digests periodically as a viral plasmid stock is generated and propagated.

4. Physiological Assays

Once a molecule is chosen, and expressed in a target cell type for characterization, it must be physiologically characterized by an observation method (e.g., patch clamp, dye imaging)—in the case of opsins, using illumination. Below, we discuss illumination hardware, solutions in which to perform experiments, strategies for selecting cells to be analyzed, and methods for cellular readout.

The millisecond-scale resolution of optogenetic tools enables the remote control of cellular physiology with unprecedented resolution, but also requires illumination sources with increased temporal resolution than what is achievable with conventional fluorescence illuminators. A commonly used programmable excitation source is the Sutter DG-4, which uses a galvanometer mirror to direct light from the lamp into one of four filter slots within the lamp, to determine the excitation wavelength (i.e., no excitation filter should be placed within the actual fluorescence cube in the microscope, if the excitation light is filtered within the DG4 itself); a second mirror is used to shutter and/or adjust the intensity of light by modulating how much is directed to the light collection optics for delivery from the DG4 into the microscope. The output of the DG4 can be fed into an illuminator port of most microscopes used for fluorescence imaging or electrophysiology. Laser-based systems can also be useful; since the action spectra of microbial rhodopsins are quite broad, typically with 100–150 nm bandwidths (full-width at half-maximum), suboptimal excitation at a given wavelength of illumination can easily be compensated for by increased illumination power. For example, a 532 nm solid-state green laser is an order of magnitude cheaper than a 593 nm solid-state yellow laser but will still excite the yellow light-sensitive *N. pharaonis* halorhodopsin quite effectively, just by increasing the delivered power slightly over the amount that would be required if a 593 nm laser was used. Action spectra (but not true absorbance spectra, which requires flash photolysis) can be measured during electrophysiological recording, by scanning through the spectrum with a Till Photonics Polychrome V or analogous color-programmable light source, coupled to a microscope through a standard fiber optic cable. In this particular illuminator, broadband light from a xenon lamp is passed through a programmable monochromator, so as to emit light with narrow-band (~10 nm bandwidths) properties, centered at various wavelengths. Light-emitting diodes (LEDs) have become increasingly popular due to their cheapness and fast switching times; a recent report (Albeanu *et al.*, 2008) offers excellent instructions for constructing a high-power and fast illuminator with two LEDs coaligned for dual-spectral excitation, and commercial systems from Thorlabs and other vendors are also available. Most LEDs can be switched on and off with fast (e.g., nanosecond) resolution, so the temporal resolution of various LED systems is largely limited by the drivers or power sources. LEDs are particularly useful for ultraviolet, orange, red, and infrared wavelengths, since many lamps are only weakly irradiant in these spectral bands.

Solutions used in electrophysiological characterization of mammalian cells (e.g., during patch clamp or imaging) are in some ways more complicated than typical solutions used in molecular biology. We highly suggest preparing electrophysiology solutions from scratch, instead of purchasing premade solutions. These solutions must be osmotically balanced to prevent cell death, ideally within 1–5 mOsm, and pH balanced, ideally within 0.1 pH units; extreme precaution must be taken to avoid contamination of reagents (as even a small change in a low-concentration ion, like calcium, can greatly change the health or electrical properties of a cell under electrophysiological study). For example, our laboratory avoids insertion of spatulas into stock containers to dispense solids for preparing electrophysiological solutions; chemicals are instead poured from stock containers whenever possible. Solutions should be sterile filtered immediately following preparation, to maximize cell health and available recording time. Our bath solution of choice for *in vitro* experiments using both HEK cells and neurons is Tyrode's solution, a HEPES buffer-based saline solution (Boyden *et al.*, 2005; Chow *et al.*, 2010; Han and Boyden, 2007). Artificial cerebrospinal fluid (ACSF), which is bicarbonate-buffered, may improve cell health in certain circumstances over that obtained from use of Tyrode's Solution, but requires fluid manifolds to perfuse CO_2-saturated solutions in order to maintain physiological pH levels, and the added inconvenience is often not justified. For the use of dyes that indicate the levels of ions such as H+ or Ca+2 (e.g., SNARF, fura-2, Oregon Green BAPTA), the manufacturers' instructions provide a good starting point for deriving protocols for the loading and imaging of the dyes, although some optimization of loading conditions and imaging conditions may be required for given cell types and given conditions of joint photostimulation and imaging (Chow *et al.*, 2010; Lin *et al.*, 2009; Prigge *et al.*, 2010).

Even within a culture of a single cell type, different cells will vary in their levels of opsin expression, appearance, and sustained photocurrents, potentially to great degrees. Excessive overexpression of an opsin can lead to poor cell health, so simply picking the brightest cells to record electrophysiologically may yield unrepresentative data, and accurate characterization of the performance of an opsin should be performed with unbiased selection of opsin-expressing cells. Beginning experimenters may have slightly different, even unconscious, biases in cell choice strategy–, for example, choosing the biggest cells in the field of view. To address some of these problems, we often normalize observed photocurrents by cell capacitance, thus obtaining the photocurrent density, which helps compensate for the varying size of the cells being recorded. We often have multiple experimenters in our lab validate key results when photocurrent magnitude is the question, to additionally address this issue (e.g., there were no statistically significant differences in photocurrents measured between the two co-first author experimenters in Chow *et al.*, 2010).

Expression of opsins in a cell can increase over time, as the process of protein expression and trafficking can be slow. For full characterization of an opsin, it is recommended to assess opsin function at various times after transfection, for example, between a few days and a few weeks, to understand the timecourse of expression and trafficking. Importantly, different opsins, and opsins expressed using different gene delivery mechanisms, will present with different timecourses of functional expression. In neurons, we have noted a trend for some microbial rhodopsins from archaea to express and traffic to the membrane more quickly than those from fungi and plants, although specific opsins within these families can violate this trend.

Typically, the photocurrents measured in neurons from archaeal rhodopsins (both bacteriorhodopsins and halorhodopsins) using the protocols in Section 4 do not change after 5–6 days posttransfection or 10 days postviral infection; photocurrents of channelrhodopsin-2 take a few extra days to plateau, compared to the archaeal opsins. These are the times that it takes for currents to saturate; fluorescence levels may saturate earlier, perhaps because although opsins are rapidly expressed at the level of protein, it may take some time for them to traffic, and assemble within the membrane in functional form (perhaps because some may potentially require multimerization within the membrane to attain full function). It is possible that adding trafficking sequences, or inducing mutations, can result in slowed down or sped up functional protein expression versus the wild type form of the opsin.

Similar trends in expression and membrane localization rates as a function of kingdom of origin are also observed in transfected HEK cells, with faithful expression of rhodopsins from archaea requiring 2 days and ones from fungi and plants requiring up to 3 days (although again, individual opsins may violate these rules of thumb). These multiday expression times may present difficulties because HEK cells will divide a few times during this period, and this partly counteracts the goal of performing reliable recordings on isolated cells (or on cells with minimal shared membrane with other cells, as described in Section 3). The addition of sodium butyrate (Dunlop *et al.*, 2008) or lowering the cell culture incubator temperature (32 °C instead of 37 °C) can extend the time between cell divisions and may allow for more time for membrane expression (Wang *et al.*, 2009).

5. Conclusion

The process of assessing the physiological function of a heterologously expressed protein in a target cell is complex. Such explorations are critical for understanding the potential uses of a given synthetic physiology tool, for evaluating potential side effects or toxicity of a candidate tool, or for screening for novel or optimized tools. As a closing example, it was originally believed that channelrhodopsin-1 (ChR1) was a light-gated proton channel, but multiple reports since then have demonstrated that it is indeed a nonspecific cation channel like ChR2 when evaluated at neutral pH and expressed at sufficiently high levels (Berthold *et al.*, 2008; Nagel *et al.*, 2002; Wang *et al.*, 2009). Thus, considerations of the cellular environment in which a protein is evaluated, for example, pH and expression level, are key for understanding the physiological power of a given molecular tool. In summary, assessing the function of a given physiological driver is complex because of the many variables that can modulate the expression and performance of physiological drivers, and the quantitative, high-speed nature of the signal being driven. The creation of new model systems that can replicate key features of targeted physiological systems, in a fashion that could support high-throughput tool assessment or tool optimization, may greatly enhance the ability to generate novel and impactful synthetic physiology tools.

ACKNOWLEDGMENTS

E. S. B. acknowledges funding by the NIH Director's New Innovator Award (DP2OD002002) as well as NIH Grants 1R01DA029639, 1RC1MH088182, 1RC2DE020919, 1R01NS067199, and 1R43NS070453; the NSF CAREER award as well as NSF Grants EFRI 0835878, DMS 0848804, and DMS 1042134; Benesse Foundation, Jerry and Marge Burnett, Department of Defense CDMRP Post-Traumatic Stress Disorder Program, Google, Harvard/MIT Joint Grants Program in Basic Neuroscience, Human Frontiers Science Program, MIT Alumni Class Funds, MIT Intelligence Initiative, MIT McGovern Institute and the McGovern Institute Neurotechnology Award Program, MIT Media Lab, MIT Mind-Machine Project, MIT Neurotechnology Fund, NARSAD, Paul Allen Distinguished Investigator Award, Alfred P. Sloan Foundation, SFN Research Award for Innovation in Neuroscience, and the Wallace H. Coulter Foundation. The reported methods were developed with additional contributions from Xue Han (currently Assistant Professor, Boston University), Xiaofeng Qian, and Aimei Yang of the Synthetic Neurobiology research group at MIT, and with the helpful discussion of Dr. John Y. Lin (UCSD). Thanks also to Xue Han for critically reading the chapter.

REFERENCES

Adamantidis, A. R., Zhang, F., *et al.* (2007). Neural substrates of awakening probed with optogenetic control of hypocretin neurons. *Nature* **450**(7168), 420–424.

Albeanu, D. F., Soucy, E., *et al.* (2008). LED arrays as cost effective and efficient light sources for widefield microscopy. *PLoS ONE* **3**(5), e2146.

Alilain, W. J., Li, X., *et al.* (2008). Light-induced rescue of breathing after spinal cord injury. *J. Neurosci.* **28**(46), 11862–11870.

Aravanis, A. M., Wang, L. P., *et al.* (2007). An optical neural interface: In vivo control of rodent motor cortex with integrated fiberoptic and optogenetic technology. *J. Neural Eng.* **4**(3), S143–S156.

Arenkiel, B. R., Peca, J., *et al.* (2007). In vivo light-induced activation of neural circuitry in transgenic mice expressing channelrhodopsin-2. *Neuron* **54**(2), 205–218.

Atasoy, D., Aponte, Y., *et al.* (2008). A FLEX switch targets Channelrhodopsin-2 to multiple cell types for imaging and long-range circuit mapping. *J. Neurosci.* **28**(28), 7025–7030.

Bamberg, E., Tittor, J., *et al.* (1993). Light-driven proton or chloride pumping by halorhodopsin. *Proc. Natl. Acad. Sci. USA* **90**(2), 639–643.

Berndt, A., Yizhar, O., *et al.* (2009). Bi-stable neural state switches. *Nat. Neurosci.* **12**(2), 229–234.

Berthold, P., Tsunoda, S. P., *et al.* (2008). Channelrhodopsin-1 initiates phototaxis and photophobic responses in chlamydomonas by immediate light-induced depolarization. *Plant Cell* **20**(6), 1665–1677.

Bi, A., Cui, J., *et al.* (2006). Ectopic expression of a microbial-type rhodopsin restores visual responses in mice with photoreceptor degeneration. *Neuron* **50**(1), 23–33.

Boyden, E. S., Zhang, F., *et al.* (2005). Millisecond-timescale, genetically targeted optical control of neural activity. *Nat. Neurosci.* **8**(9), 1263–1268.

Canton, B., Labno, A., *et al.* (2008). Refinement and standardization of synthetic biological parts and devices. *Nat. Biotechnol.* **26**(7), 787–793.

Carlson, R. (2003). The pace and proliferation of biological technologies. *Biosecur. Bioterror.* **1**(3), 203–214.

Carr, P. A., and Church, G. M. (2009). Genome engineering. *Nat. Biotechnol.* **27**(12), 1151–1162.

Chow, B. Y., Han, X., *et al.* (2010). High-performance genetically targetable optical neural silencing by light-driven proton pumps. *Nature* **463**(7277), 98–102.

Douglass, A. D., Kraves, S., *et al.* (2008). Escape behavior elicited by single, channelrhodopsin-2-evoked spikes in zebrafish somatosensory neurons. *Curr. Biol.* **18**(15), 1133–1137.

Dunlop, J., Bowlby, M., *et al.* (2008). High-throughput electrophysiology: An emerging paradigm for ion-channel screening and physiology. *Nat. Rev. Drug Discov.* **7**(4), 358–368.

Dunn, R. J., Hackett, N. R., *et al.* (1987). Structure-function studies on bacteriorhodopsin. I. Expression of the bacterio-opsin gene in *Escherichia coli*. *J. Biol. Chem.* **262**(19), 9246–9254.

Enami, N., Yoshimura, K., *et al.* (2006). Crystal structures of archaerhodopsin-1 and −2: Common structural motif in archaeal light-driven proton pumps. *J. Mol. Biol.* **358**(3), 675–685.

Endy, D., and Brent, R. (2001). Modelling cellular behaviour. *Nature* **409**(6818), 391–395.

Farah, N., Reutsky, I., *et al.* (2007). Patterned optical activation of retinal ganglion cells. *Conf. Proc. IEEE Eng. Med. Biol. Soc.* **1,** 6368–6370.

Gilles-Gonzalez, M. A., Engelman, D. M., *et al.* (1991). Structure-function studies of bacteriorhodopsin XV. Effects of deletions in loops B-C and E-F on bacteriorhodopsin chromophore and structure. *J. Biol. Chem.* **266**(13), 8545–8550.

Gradinaru, V., Thompson, K. R., *et al.* (2008). eNpHR: A Natronomonas halorhodopsin enhanced for optogenetic applications. *Brain Cell Biol.* **36**(1–4), 129–139.

Gradinaru, V., Zhang, F., *et al.* (2010). Molecular and cellular approaches for diversifying and extending optogenetics. *Cell* **141**(1), 154–165.

Greenhalgh, D. A., Farrens, D. L., *et al.* (1993). Hydrophobic amino acids in the retinal-binding pocket of bacteriorhodopsin. *J. Biol. Chem.* **268**(27), 20305–20311.

Gunaydin, L. A., Yizhar, O., *et al.* (2010). Ultrafast optogenetic control. *Nat. Neurosci.* **13**(3), 387–392.

Hackett, N. R., Stern, L. J., *et al.* (1987). Structure-function studies on bacteriorhodopsin. V. Effects of amino acid substitutions in the putative helix F. *J. Biol. Chem.* **262**(19), 9277–9284.

Han, X., and Boyden, E. S. (2007). Multiple-color optical activation, silencing, and desynchronization of neural activity, with single-spike temporal resolution. *PLoS ONE* **2**(3), e299.

Han, X., Qian, X., *et al.* (2009a). Millisecond-timescale optical control of neural dynamics in the nonhuman primate brain. *Neuron* **62**(2), 191–198.

Han, X., Qian, X., *et al.* (2009b). Informational lesions: Optical perturbation of spike timing and neural synchrony via microbial opsin gene fusions. *Front. Mol. Neurosci.* 10.3389/neuro.02.012.2009.

Hegemann, P., Oesterbelt, D., *et al.* (1985). The photocycle of the chloride pump halorhodopsin. I: Azide-catalyzed deprotonation of the chromophore is a side reaction of photocycle intermediates inactivating the pump. *EMBO J.* **4**(9), 2347–2350.

Hennecke, M., Kwissa, M., *et al.* (2001). Composition and arrangement of genes define the strength of IRES-driven translation in bicistronic mRNAs. *Nucleic Acids Res.* **29**(16), 3327–3334.

Hofherr, A., Fakler, B., *et al.* (2005). Selective Golgi export of Kir2.1 controls the stoichiometry of functional Kir2.x channel heteromers. *J. Cell Sci.* **118**(Pt 9), 1935–1943.

Huber, D., Petreanu, L., *et al.* (2008). Sparse optical microstimulation in barrel cortex drives learned behaviour in freely moving mice. *Nature* **451**(7174), 61–64.

Ishizuka, T., Kakuda, M., *et al.* (2006). Kinetic evaluation of photosensitivity in genetically engineered neurons expressing green algae light-gated channels. *Neurosci. Res.* **54**(2), 85–94.

Kennedy, M. J., Hughes, R. M., *et al.* (2010). Rapid blue-light-mediated induction of protein interactions in living cells. *Nat Methods* **7**(12), 973–975.

Kolbe, M., Besir, H., *et al.* (2000). Structure of the light-driven chloride pump halorhodopsin at 1.8 A resolution. *Science* **288**(5470), 1390–1396.

Lagali, P. S., Balya, D., *et al.* (2008). Light-activated channels targeted to ON bipolar cells restore visual function in retinal degeneration. *Nat. Neurosci.* **11**(6), 667–675.

Levskaya, A., Weiner, O. D., *et al.* (2009). Spatiotemporal control of cell signalling using a light-switchable protein interaction. *Nature* **461**(7266), 997–1001.

Li, X., Gutierrez, D. V., *et al.* (2005). Fast noninvasive activation and inhibition of neural and network activity by vertebrate rhodopsin and green algae channel rhodopsin. *Proc. Natl. Acad. Sci. USA* **102**(49), 17816–17821.

Liewald, J. F., Brauner, M., *et al.* (2008). Optogenetic analysis of synaptic function. *Nat Methods* **5**(10), 895–902.

Lin, J. Y., Lin, M. Z., *et al.* (2009). Characterization of engineered channelrhodopsin variants with improved properties and kinetics. *Biophys. J.* **96**(5), 1803–1814.

Luecke, H., Schobert, B., *et al.* (1999a). Structural changes in bacteriorhodopsin during ion transport at 2 angstrom resolution. *Science* **286**(5438), 255–261.

Luecke, H., Schobert, B., *et al.* (1999b). Structure of bacteriorhodopsin at 1.55 A resolution. *J. Mol. Biol.* **291**(4), 899–911.

Luecke, H., Schobert, B., *et al.* (2001). Crystal structure of sensory rhodopsin II at 2.4 angstroms: Insights into color tuning and transducer interaction. *Science* **293**(5534), 1499–1503.

Ma, D., Zerangue, N., *et al.* (2001). Role of ER export signals in controlling surface potassium channel numbers. *Science* **291**(5502), 316–319.

Mahoney, T. R., Luo, S., *et al.* (2008). Intestinal signaling to GABAergic neurons regulates a rhythmic behavior in *Caenorhabditis elegans*. *Proc. Natl. Acad. Sci. USA* **105**(42), 16350–16355.

Marinetti, T., Subramaniam, S., *et al.* (1989). Replacement of aspartic residues 85, 96, 115, or 212 affects the quantum yield and kinetics of proton release and uptake by bacteriorhodopsin. *Proc. Natl. Acad. Sci. USA* **86**(2), 529–533.

Marti, T., Otto, H., *et al.* (1991). Bacteriorhodopsin mutants containing single substitutions of serine or threonine residues are all active in proton translocation. *J. Biol. Chem.* **266**(11), 6919–6927.

Mizuguchi, H., Xu, Z., *et al.* (2000). IRES-dependent second gene expression is significantly lower than cap-dependent first gene expression in a bicistronic vector. *Mol. Ther.* **1**(4), 376–382.

Mogi, T., Stern, L. J., *et al.* (1987). Bacteriorhodopsin mutants containing single tyrosine to phenylalanine substitutions are all active in proton translocation. *Proc. Natl. Acad. Sci. USA* **84**(16), 5595–5599.

Mogi, T., Stern, L. J., *et al.* (1988). Aspartic acid substitutions affect proton translocation by bacteriorhodopsin. *Proc. Natl. Acad. Sci. USA* **85**(12), 4148–4152.

Mogi, T., Marti, T., *et al.* (1989a). Structure-function studies on bacteriorhodopsin. IX. Substitutions of tryptophan residues affect protein-retinal interactions in bacteriorhodopsin. *J. Biol. Chem.* **264**(24), 14197–14201.

Mogi, T., Stern, L. J., *et al.* (1989b). Structure-function studies on bacteriorhodopsin. VIII. Substitutions of the membrane-embedded prolines 50, 91, and 186: The effects are determined by the substituting amino acids. *J. Biol. Chem.* **264**(24), 14192–14196.

Moglich, A., and Moffat, K. (2010). Engineered photoreceptors as novel optogenetic tools. *Photochem. Photobiol. Sci.* **9**(10), 1286–1300.

Nagel, G., Ollig, D., *et al.* (2002). Channelrhodopsin-1: A light-gated proton channel in green algae. *Science* **296**(5577), 2395–2398.

Nagel, G., Szellas, T., *et al.* (2003). Channelrhodopsin-2, a directly light-gated cation-selective membrane channel. *Proc. Natl. Acad. Sci. USA* **100**(24), 13940–13945.

Nagel, G., Brauner, M., *et al.* (2005). Light activation of channelrhodopsin-2 in excitable cells of Caenorhabditis elegans triggers rapid behavioral responses. *Curr. Biol.* **15**(24), 2279–2284.

Oesterhelt, D., Hegemann, P., *et al.* (1985). The photocycle of the chloride pump halorhodopsin. II: Quantum yields and a kinetic model. *EMBO J.* **4**(9), 2351–2356.

Osti, D., Marras, E., *et al.* (2006). Comparative analysis of molecular strategies attenuating positional effects in lentiviral vectors carrying multiple genes. *J. Virol. Methods* **136**(1–2), 93–101.

Otomo, J. (1996). Influence exercised by Histidine-95 on Chloride transport and the photocycle in halorhodopsin. *Biochemistry* **35**(21), 6684–6689.

Otto, H., Marti, T., *et al.* (1989). Aspartic acid-96 is the internal proton donor in the reprotonation of the Schiff base of bacteriorhodopsin. *Proc. Natl. Acad. Sci. USA* **86**(23), 9228–9232.

Petreanu, L., Huber, D., *et al.* (2007). Channelrhodopsin-2-assisted circuit mapping of long-range callosal projections. *Nat. Neurosci.* **10**(5), 663–668.

Prigge, M., Rosler, A., *et al.* (2010). Fast, repetitive light-activation of CaV3.2 using Channelrhodopsin 2. *Channels (Austin)* **4**(3), 241–247.

Richardson, S. M., Nunley, P. W., *et al.* (2010). GeneDesign 3.0 is an updated synthetic biology toolkit. *Nucleic Acids Res.* **38**(8), 2603–2606.

Rudiger, M., and Oesterhelt, D. (1997). Specific arginine and threonine residues control anion binding and transport in the light-driven chloride pump halorhodopsin. *EMBO J.* **16**(13), 3813–3821.

Rudiger, M., Haupts, U., *et al.* (1995). Chemical reconstitution of a chloride pump inactivated by a single point mutation. *EMBO J.* **14**(8), 1599–1606.

Sato, M., Kikukawa, T., *et al.* (2003a). Roles of Ser130 and Thr126 in chloride binding and photocycle of pharaonis halorhodopsin. *J. Biochem.* **134**(1), 151–158.

Sato, M., Kikukawa, T., *et al.* (2003b). Ser-130 of Natronobacterium pharaonis halorhodopsin is important for the chloride binding. *Biophys. Chem.* **104**(1), 209–216.

Schroder-Lang, S., Schwarzel, M., *et al.* (2007). Fast manipulation of cellular cAMP level by light in vivo. *Nat. Methods* **4**(1), 39–42.

Schroll, C., Riemensperger, T., *et al.* (2006). Light-induced activation of distinct modulatory neurons triggers appetitive or aversive learning in *Drosophila* larvae. *Curr. Biol.* **16**(17), 1741–1747.

Stern, L. J., and Khorana, H. G. (1989). Structure-function studies on bacteriorhodopsin. X. Individual substitutions of arginine residues by glutamine affect chromophore formation, photocycle, and proton translocation. *J. Biol. Chem.* **264**(24), 14202–14208.

Stockklausner, C., and Klocker, N. (2003). Surface expression of inward rectifier potassium channels is controlled by selective Golgi export. *J. Biol. Chem.* **278**(19), 17000–17005.

Subramaniam, S., Greenhalgh, D. A., *et al.* (1992). Aspartic acid 85 in bacteriorhodopsin functions both as proton acceptor and negative counterion to the Schiff base. *J. Biol. Chem.* **267**(36), 25730–25733.

Tang, W., Ehrlich, I., *et al.* (2009). Faithful expression of multiple proteins via 2A-peptide self-processing: A versatile and reliable method for manipulating brain circuits. *J. Neurosci.* **29**(27), 8621–8629.

Toni, N., Laplagne, D. A., *et al.* (2008). Neurons born in the adult dentate gyrus form functional synapses with target cells. *Nat. Neurosci.* **11**(8), 901–907.

Wang, H., Peca, J., *et al.* (2007). High-speed mapping of synaptic connectivity using photostimulation in Channelrhodopsin-2 transgenic mice. *Proc. Natl. Acad. Sci. USA* **104**(19), 8143–8148.

Wang, H., Sugiyama, Y., *et al.* (2009). Molecular determinants differentiating photocurrent properties of two channelrhodopsins from chlamydomonas. *J. Biol. Chem.* **284**(9), 5685–5696.

Weber, K., Bartsch, U., *et al.* (2008). A multicolor panel of novel lentiviral "Gene Ontology" (LeGO) vectors for functional gene analysis. *Mol. Ther.* **16**(4), 698–706.

Welch, M., Govindarajan, S., *et al.* (2009). Design parameters to control synthetic gene expression in *Escherichia coli. PLoS ONE* **4**(9), e7002.

Wu, G., Bashir-Bello, N., *et al.* (2006). The synthetic gene designer: A flexible web platform to explore sequence manipulation for heterologous expression. *Protein Expr. Purif.* **47**(2), 441–445.

Wu, Y. I., Frey, D., *et al.* (2009). A genetically encoded photoactivatable Rac controls the motility of living cells. *Nature* **461**(7260), 104–108.

Yazawa, M., Sadaghiani, A. M., *et al.* (2009). Induction of protein-protein interactions in live cells using light. *Nat Biotechnol.* **27**(10), 941–945.

Yoshimura, K., and Kouyama, T. (2008). Structural role of bacterioruberin in the trimeric structure of archaerhodopsin-2. *J. Mol. Biol.* **375**(5), 1267–1281.

Yu, X., Zhan, X., *et al.* (2003). Lentiviral vectors with two independent internal promoters transfer high-level expression of multiple transgenes to human hematopoietic stem-progenitor cells. *Mol. Ther.* **7**(6), 827–838.

Zhang, Y. P., and Oertner, T. G. (2007). Optical induction of synaptic plasticity using a light-sensitive channel. *Nat. Methods* **4**(2), 139–141.

Zhang, F., Wang, L. P., *et al.* (2007a). Multimodal fast optical interrogation of neural circuitry. *Nature* **446**(7136), 633–639.

Zhang, W., Ge, W., *et al.* (2007b). A toolbox for light control of *Drosophila* behaviors through Channelrhodopsin 2-mediated photoactivation of targeted neurons. *Eur. J. Neurosci.* **26**(9), 2405–2416.

Zhang, Y. P., Holbro, N., *et al.* (2008). Optical induction of plasticity at single synapses reveals input-specific accumulation of alphaCaMKII. *Proc. Natl. Acad. Sci. USA* **105**(33), 12039–12044.

SECTION FOUR

DEVICES FOR METABOLIC ENGINEERING

CHAPTER NINETEEN

Metabolic Pathway Flux Enhancement by Synthetic Protein Scaffolding

Weston R. Whitaker* *and* John E. Dueber*,†

Contents

Abstract

Spatial control over enzyme organization presents a promising posttranslational strategy for improving metabolic flux. Directly tethering enzyme polypeptides has had inconsistent success. Use of a separate scaffold molecule, built from modular protein–protein interaction domains, provides designable control over enzyme assembly parameters, including stoichiometry, as well as providing scalability for multiple enzymes. Thus, metabolic flux can be optimized by expression of these scaffolds *in vivo*. It is important to note that exploration of the use of synthetic scaffolds for improving metabolic flux is in its early stages. Accordingly, in this chapter, we describe efforts to date, hypotheses for scaffold function, and parameters to consider for application to new pathways.

* Department of Bioengineering, University of California, Berkeley, California, USA

† Energy Biosciences Institute, University of California, Berkeley, California, USA

Methods in Enzymology, Volume 497

ISSN 0076-6879, DOI: 10.1016/B978-0-12-385075-1.00019-6

1. Introduction

Metabolic engineering has the potential to provide environmentally safe and cost-effective routes for synthesizing a range of compounds, from high-value specialty compounds such as therapeutics to bulk commodities including plastics and biofuels. Particularly for the latter class of compounds, a complement of strategies will be needed to achieve the production yields, near theoretical maximum, necessary to achieve industrial viability. These stringent requirements will likely inspire improvements across many technologies: modeling metabolic and cellular behavior (Price *et al.*, 2004), predictable control over gene expression (Pfleger *et al.*, 2006; Salis *et al.*, 2009; Win and Smolke, 2007), and directed evolution approaches for improved enzyme characteristics (Dougherty and Arnold, 2009; Zhang *et al.*, 2008). In this chapter, we focus on ongoing efforts to improve pathway efficiency through engineered enzyme complex formation using synthetic scaffolds; however, all strategies discussed here must eventually be performed in concert with existing proven methodologies to achieve optimal yields. Since our mechanistic understanding of scaffold function is still at an early stage, we describe here the parameters empirically derived thus far to be important and describe a suggested process for applying scaffolding strategies to a new pathway.

There are numerous natural examples of enzymes forming complexes for optimal metabolic pathway performance. For excellent in-depth reviews on this topic, please see those written by Conrado *et al.* (2008) and Miles *et al.* (1999). The most striking examples of improved pathway efficiency via complex formation are those that have evolved structures capable of physically channeling substrates. Tryptophan synthase, carbamoyl phosphate synthase, and glutamine phosphoribosylpyrophosphate amidotransferase are three examples described in detail by Miles *et al.* (1999) whose structures reveal tunnels connecting catalytic sites that are capable of protecting reactive intermediates from the bulk solution. Another mechanism of channeling substrates is through electrostatic channeling. Thymidylate synthase and dihydrofolate reductase are two enzymatic activities found in a single polypeptide in some plants and some protozoa, including *Leishmania major* for which the crystal structure has been solved. The surface of this structure is predominantly positively charged, suggesting a mechanism of an electrostatic "highway" spanning the 40 Å between the two active sites across which the negatively charged dihydrofolate intermediate would travel (Stroud, 1994). Recently, evidence has grown for the dynamic assembly of complexes, perhaps as a feedback mechanism to achieve a precise concentration of metabolite product (An *et al.*, 2008; Narayanaswamy *et al.*, 2009). These dynamic complexes have been difficult to observe biochemically *in vitro*. For example, purine biosynthesis in eukaryotes involves six enzymes. Despite early anticipation of potential interactions between these enzymes, only recently was it understood, by fluorescently tagging these enzymes *in vivo*, that all six proteins coassemble (An *et al.*, 2008). Interestingly, these proteins dynamically assemble and disassemble depending on purine concentration. Narayanaswamy *et al.* (2009) similarly showed that numerous metabolic enzyme complexes dynamically assemble depending on culture conditions, suggesting these phenomena are considerably more common than would be predicted by *in vitro* biochemical experiments. Likely, many of these complexes are not detected due to characterization under conditions incompatible with complex formation.

Drawing inspiration from natural pathways, engineers have begun assembling synthetic enzyme complexes to improve pathway performance. For degradation of cellulose and hemicellulose *in vitro*, various enzyme combinations have been corecruited to cellulose substrate to include synergistic combinations of activities as found in natural cellulosome complexes (Fierobe *et al.*, 2001, 2005). Recently, our lab has expressed scaffolds built from modular protein–protein interaction domains (Table 19.1) to optimize flux of engineered metabolic pathways *in vivo* (Dueber *et al.*, 2009). Enzymes were tagged with peptide ligands specific for these scaffold protein–protein interaction domains. The modular composition of the scaffolds was used to build various architectures that were critical for optimizing flux as discussed later.

The mevalonate biosynthetic pathway presents an interesting model system for synthetic complex engineering in that it suffers from a flux imbalance between HMG-CoA synthase (HMGS) and HMG-CoA reductase (HMGR) that results in the accumulation of the cytotoxic HMG-CoA intermediate (Martin *et al.*, 2003; Pitera *et al.*, 2007). Scaffolding this pathway improves efficiency, producing higher product titers even at considerably lower enzyme inducer concentrations (Dueber *et al.*, 2009). It should be noted that the relationship between scaffold architecture and titer improvement is not predictable for this pathway, as discussed later. The same set of scaffold architectures was applied to a second pathway engineered by Moon *et al.* (2009, 2010) for the biosynthesis of glucaric acid. This pathway presents an interesting test case in that it is a relatively high titer-producing pathway, on the order of 1 g/L, with a flux bottleneck enzyme, MIOX, that appears to be substrate activated (Moon *et al.*, 2009). Varying the number of domains that recruit the enzyme upstream of MIOX, Ino1, resulted in gradually increased product titers to a maximum of almost fivefold improvement with four Ino1 recruitment domains, whereas varying the number of MIOX recruiting domains had little impact (Moon *et al.*, 2010). This observation is consistent with a model in which the local concentration of substrate for the limiting MIOX activity is modulated by upstream enzyme recruitment via scaffold domain stoichiometry within the synthetic complex.

Table 19.1 Protein–protein interaction domain families potentially useful for scaffold construction

Part family	Tightest affinity K_d	Domain/ligand size (AAs)	Source (Accession #)	Features and issues	Confirmed orthogonal pairs
SH3 domain/ peptide	1×10^{-1} u*M* (Posern *et al.*, 1998)	57/11	AAH31149: 196–274 PPPALPPKRRR (Posern *et al.*, 1998)	Relatively context independent and well-characterized. Ideal for internal insertion. Natural peptides tend to have micromolar affinities	Specificity observed within species (Zarrinpar *et al.*, 2003)
PDZ domain/ peptide	1×10^{0} u*M* (Tonikian *et al.*, 2008)	96/6	EDL06069: 77–171 GVKESLV (Harris *et al.*, 2001)	Generally PDZ peptide must be C-terminal. nNOS domain can be used for non-C-terminal ligands	2 Natural (Fuh *et al.*, 2000)
GBD domain/ peptide	1×10^{0} u*M* (Kim *et al.*, 2000)	80/32	BAA21534: 196–274 P42768: 466–497	Less well-characterized than SH3 or PDZ. Longer linker sequence.	1
Leucine Zippers	6.1×10^{-3} u*M* (Acharya *et al.*, 2002) and 8.3×10^{-2} u*M* (Grünberg *et al.*, 2010)	43/43	ITIRAAFLEKENTALRTEIAE LEKEVGRCENIVSKYETRYGPL LEIRAAFLEKENTALRTRAAEL RKRVGRCRNIVSKYETRYGPL	Significant likelihood of homodimerization, particularly important to test for intramolecular pairs	3 Synthetic (Reinke *et al.*, 2010)
PhyB/Pif3 light switchable binding	2×10^{-2}–1×10^{-1} u*M* (Levskaya *et al.*, 2009)	908/91	AAW56577: 1–908 NP_172424: 120–210	Light dependent binding activated at 720 nm and deactivated at 660 nm light	1

FKBP/FRB	1.2×10^{-3} u*M* (Banaszynski *et al.*, 2005)	107/93	AAI19733: 39–145 EAW71681: 1972–2064	Interaction is inducible with the small molecule Rapamycin at $K_d = 0.2$ n*M* (Bierer *et al.*, 1990). A FRB (T2098L) mutation allows use of a nontoxic rapamycin analog for T2098L characterization see (Grünberg *et al.*, 2010)	1
Cohesin/Dockerin	$<1 \times 10^{-5}$ u*M* (Fierobe *et al.*, 2001)	~150/~70	YP_001039466 YP_001038489	Calcium-dependent binding activity is likely not functional at free cellular Ca^{2+} levels Calcium $K_d = 2.5 \times 10^{-7}$ 1/M^2, half binding at 500 u*M* Ca^{2+} (Leibovitz *et al.*, 1997)	5 Natural (Haimovitz *et al.*, 2008)

Even without direct substrate channeling guiding intermediates between active sites as observed in the natural examples discussed previously, product titers may be improved by colocalizing consecutive metabolic enzymes to produce a higher local concentration of metabolite in close proximity to the downstream enzyme (Conrado *et al.*, 2008; Welch, 1977). This has been the subject of debate in early papers in which consecutive enzymes, β-galactosidase and galactose dehydrogenase, were tethered with a translational fusion, generating a higher product titer (Ljungcrantz *et al.*, 1989). The authors suggested the mechanism for increased flux was substrate channeling; however, a subsequent paper performing a detailed kinetic analysis of the fusion protein challenged this conclusion (Pettersson and Pettersson, 2001). Local concentration effects have been modeled for engineered pathways of heterologous enzymes by simulating native and engineered pathway reaction rates within an *Escherichia coli* discretized into subvolumes, localizing the engineered pathway within a single subvolume, and accounting for metabolite diffusion to simulate compartmentalization (Conrado *et al.*, 2007). Though direct measurements of local intermediate concentrations within enzyme complexes have remained elusive, this mechanism seems to be an attractive explanation for some of the successes observed with scaffolding (Dueber *et al.*, 2009; Moon *et al.*, 2010) and other colocalization engineering examples reviewed in Conrado *et al.* (2008). However, local concentration effects may be acting in conjunction with other mechanisms discussed later in this chapter. Enzyme colocalization may allow achievement of a specific local intermediate concentration with a lower concentration of upstream enzyme than would be possible with freely diffusing enzymes, thus retaining high flux while reducing the metabolic load on the cells (Fig. 19.1A). Reduction of intermediate in the bulk of the cell may also be beneficial if the intermediate is toxic or undergoes undesired reactions through competing pathways (Fig. 19.1A).

We employ modular protein–protein interaction machinery tethered with long flexible synthetic linkers to colocalize enzymes in engineered metabolic pathways. While this approach currently lacks the ability to precisely control the three-dimensional positioning of recruited enzymes, it has the advantage that each protein–protein interaction domain should be capable of targeting its interaction partner in a manner independent of composition context, provided neighboring targeted enzymes do not sterically block a physical interaction. When a binding domain is incorporated into a scaffold using long linkers, it generally will retain the ability to bind its target ligand regardless of where it is located on the scaffold and what domains are encoded up or downstream. This provides a highly designable platform where matrices of scaffolds can be generated in which key parameters are varied while interaction functionality is maintained (Fig. 19.1B). Additionally, once interaction tags have been successfully added to pathway enzymes, the pathway can be used with a variety of scaffold architectures.

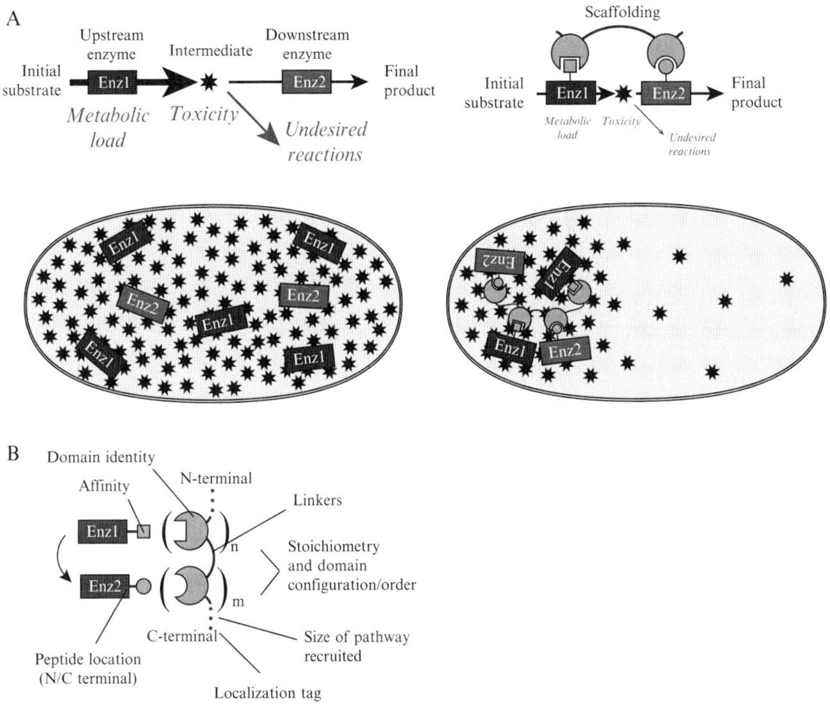

Figure 19.1 (A) High local substrate concentrations may be achieved with low enzyme expression, thereby reducing the cellular burden by using scaffolding to colocalize enzymes. (B) Scaffolding based on modular protein–protein interaction domains provides a highly designable control point with multiple parameters available for optimization.

However, as discussed later in the chapter, applying tags to enzymes can have difficult-to-predict effects on enzyme activity and concentration, both of which must be assessed. To scale recruitment to include additional enzymes, additional binding domains can be fused to existing scaffolds or a second scaffold molecule can be cotargeted to the original scaffold. For interaction domains with a set of ligands ranging in recruitment affinity, binding strength can usually be modulated by point mutations to the ligand without the need to redesign scaffold libraries. Further, different enzyme packing structures can be explored by shuffling the configurations of the scaffold domains.

In this chapter, we describe methodologies by which synthetic complexes can be engineered from metabolic pathways using modular protein scaffolds. Although the synthetic scaffolds built to date, and certainly for the foreseeable future, do not rival the elegance of natural systems, they offer designable control over several assembly parameters, most notably enzyme stoichiometry, incorporation of heterologous enzymes, and potential scalability for increased numbers of enzymes. We hope to provide a practical guide for how we would approach scaffolding a pathway *de novo* together with a discussion of our considerations and experiences thus far. We conclude this chapter with a discussion of several mechanisms that may play a role in observed titer improvements for pathways tested thus far and of systems that may benefit from these possible effects.

2. Method—How to Build Modular Protein Scaffolded Systems for Metabolic Engineering Applications

2.1. Selecting protein–protein interaction domains and ligands for scaffold construction

The first decision to be made for scaffolding a metabolic pathway is the choice of colocation components. Each enzyme is translationally fused to a ligand specific for a protein–protein interaction domain. A translational fusion of these domains will compose a scaffold capable of colocalizing the ligand-fused enzymes. The structural modularity of the protein–protein interaction domains is of primary importance, as they will need to retain binding activity in the nonnative context of the translational fusions. A number of modular protein–protein interaction domains have been characterized and employed in various applications, a partial list of which is compiled in Table 19.1. In our experience, the members of SH3, PDZ, GBD, and leucine zipper families tend to retain binding activity as N-, C-terminal, or internal fusions and, given sufficient linker lengths, often do not require linker optimization to achieve binding activity. However, as discussed later, despite robustness of binding activity, overall flux improvement is likely also influenced by scaffold architecture including parameters such as linker length/composition and number/arrangement of protein–protein interaction domains.

Another parameter of importance is association affinity, particularly under low expression regimes. We generally design our scaffolds to target enzymes in the low micromolar range or tighter. To date, all targetings have been executed with the tightest affinity ligands available; however, many of these domains, as listed in Table 19.1, include lower affinity ligands that could be employed if transient interactions are desired.

Protein–protein interaction domains belonging to families with many members are particularly attractive choices for use in scaffolding, as they may offer a set of domains that potentially recognize specific ligands orthogonally (i.e., minimal cross talk with other ligands used as enzyme tags), yet have conserved folds and can more likely be used interchangeably.

For example, individual SH3 domains appear to have undergone negative selection such that they do not measurably interact with other SH3 domain family ligands within that organism (Zarrinpar *et al.*, 2003). Zarrinpar *et al.* showed that a peptide ligand was highly specific for a single SH3 domain within its native host, *Saccharomyces cerevisiae*, whereas this same peptide ligand interacted with a high percentage of non-*S. cerevisiae* SH3 domains. This selection for reduced cross talk should considerably increase the number of orthogonal domain/ligand pairs available for simultaneous use, particularly of domains recognizing small ligands such as the SH3 and PDZ domain family. Additionally, these interaction domain families often appear to have evolved physical and functional modularity, including characteristics such as robust-independent domain folding and surface-exposed N- and C-termini that are located close together to permit domain functioning as either terminal or internal fusions.

Leucine zipper and synthetic coiled-coil domain folds share many of these characteristics and are attractive targets for expanding the available number of orthogonal interaction partners. Works such as those by Havranek and Harbury (2003), where eight residues between leucine zipper pairs were altered based on computational prediction to create new pairs of either homodimers or heterodimers, show promise for rationally engineering new domains. More recently, Reinke *et al.* (2010) investigated the interaction specificities of a large set of synthetic coiled-coils that do not exhibit measurable self-association providing up to three orthogonal pairs that do not cross talk. These large libraries of structurally similar but orthogonally binding pairs provide excellent candidates for scaffold parts, as presumably they may be interchanged to switch specificity with minimal perturbation. However, as generating very large libraries of orthogonal parts has proven challenging, limited to sets with only several experimentally verified orthogonal pairs, taking parts from different families to minimize likelihood of cross talk is still likely to be fruitful for producing larger numbers of orthogonal protein–protein interaction pairs.

Cohesin–dockerin interaction modules have been successfully used to scaffold multienzyme complexes to function as synthetic cellulosomes *in vitro*. Up to three cellulose degrading enzymes were translationally fused to dockerins that localize to specific cohesins on a synthetic scaffold, which itself localizes to cellulose substrate via a carbohydrate-binding module. The resultant complex enhanced cellulose degradation in the complex substrate of straw sixfold over free enzyme (Fierobe *et al.*, 2005). An in-depth review of a number of applications that have taken advantage of cohesin–dockerin domains to provide controlled extracellular binding has recently been published (Nordon *et al.*, 2009). A study of cohesin–dockerin specificities has demonstrated up to five cohesin–dockerin pairs exhibiting orthogonal-binding specificity, providing a set of modules for further application (Haimovitz *et al.*, 2008). A unique feature of cohesin–dockerin interactions is that they bind with a very tight affinity in a calcium ion dependent manner. This makes them ideal candidates for extracellular scaffolding but likely limits their application *in vivo* due to the low concentration of free calcium in the cytoplasm.

2.2. Assembling scaffolds from domains and tagging enzymes for corecruitment

Domain/ligand choice is particularly important for proteins whose activity, stability, and/or solubility are sensitive to translational fusion. Particular peptide sequences and fusion locations may decrease the flux through these sensitive enzymes beyond the capability of the scaffolding effect to surpass. We attempt to minimize the perturbation to the enzyme of interest by selecting the smaller member of the binding pair to use as a tag. For proteins known to be problematic, we often use either an 11-amino acid peptide with a $K_d = 0.1\ \mu M$ for the Crk SH3 domain on the N- or C-terminus or a six-amino acid peptide with a $K_d = 8\ \mu M$ for the syntrophin PDZ domain as a C-terminal fusion. PDZ peptides must be used as C-terminal fusions since the carboxyl group is critical for binding. In the case of enzymes that are already experiencing solubility problems, it is possible that adding a larger, well-folded, binding domain may increase solubility similar to the oft-used strategy of tethering folding problematic proteins to maltose-binding protein or other highly soluble motifs (Di Guan *et al.*, 1988). The coiled-coil motifs, due to their high solubility, may be good candidates to try for this purpose, although all of these efforts are protein-dependent and currently unpredictable.

Linkers connecting domains of the scaffold are likely to provide another parameter that could be explored for flux optimization. To date, we have limited experience in the effect of different linker types. To connect domains, we have been using linkers expected to be of sufficient length (nine or more amino acids) to avoid sterically obstructing neighboring domains from binding and with compositions predicted to be unstructured and flexible (glycine–serine repeats). We have observed a small improvement in performance when linkers separating blocks of domains (i.e., the linkers in the scaffold GBD_1_linker_$(SH3)_2$_linker_$(PDZ)_2$) were further increased from 9 to 25 total amino acids of Gly-Ser repeats (Dueber, unpublished observation). However, we have not investigated varying linker lengths or the composition of these linkers beyond this initial characterization. Robinson and Sauer (1998) investigated the effect of linker length and composition on the stability of single-chain Arc repressor. For this protein, both linker length and the amino acid composition had a large impact on stability. Initial work on a simple tethering of two enzymes in the mevalonate biosynthetic pathway (Fig. 19.2) also suggested that a linker of adequate length must be used to

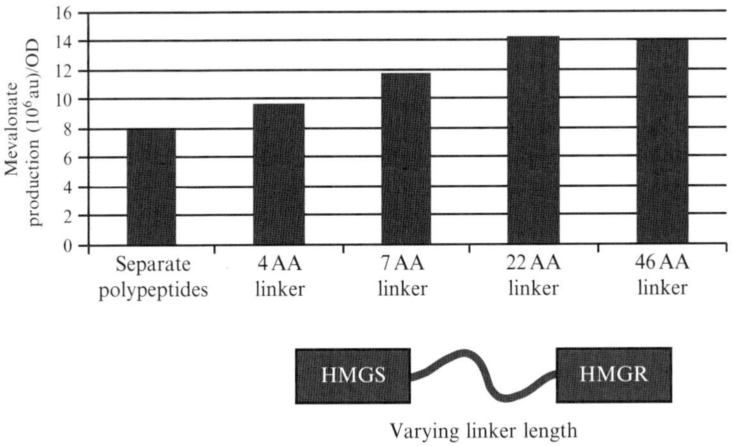

Figure 19.2 Two consecutive enzymes in a synthetic mevalonate pathway were tethered, C- to N-termini, and the length of a simple synthetic linker of alternating glycine and serine residues was varied. Tethering with a short linker length showed small improvements, which were increased to a higher titer with increased linker lengths. Mevalonate production was measured as GC/MS peak area as described in Dueber *et al.* (2009).

achieve highest flux improvement and, in this case, that the improvement is maintained through linkers of increasing lengths.

When making translational fusions of binding domains/ligands and enzymes, it is important to assay enzyme activity and interaction domain functionality. A GST pull-down assay can be employed to ensure binding activity remains functional. When domains from the same family are being employed with the intention of orthogonal function, GST pull-down assays often are very useful for avoiding unexpected intermolecular interactions, though low affinity and high effective concentration intramolecular interactions may be missed. For *in vivo* assays of enzyme activity, untagged enzyme function can be tested against tagged enzymes in absence of scaffolding. However, care must be taken to ensure expression rates are not being altered by domain addition, particularly for N-terminal tags or polycistronic systems which, in prokaryotes, should be expected to alter expression rates through RNA secondary structure interactions with the ribosome-binding site (Mathews *et al.*, 1999). For example, the addition of a C-terminal peptide targeting an enzyme might affect the expression level of an immediately downstream enzyme in a polycistronic message. This effect can be estimated or corrected for with the RBS calculator (Salis *et al.*, 2009). To confirm that expression levels are not changing, enzyme concentration should be carefully measured or expressed on independent transcripts.

2.3. Balancing the scaffold and enzyme concentrations

Balancing concentrations of enzymes has been shown to be an important consideration for optimizing metabolic pathways. Balancing scaffold concentrations is also important, as there is a theoretical optimal concentration for maximizing fully occupied scaffold molecules. This effect was modeled by Levchenko *et al.* (2000) for scaffolding in the MAPK-signaling pathway, where low concentrations of scaffold result in insufficient scaffold to colocalize the targets, while concentrations of scaffold considerably higher than enzymes result in segregation of components and a high percentage of scaffold molecules with low occupancy. We believe this biphasic trend also exists for synthetic scaffolding of multienzyme pathways. We simulated a simple mathematical model of equilibrium-binding reactions for a varying number of different enzymes that bind to different single sites on a scaffolding protein (Fig. 19.3). Differential equations for simple binding kinetics were generated with code written and simulated in MATLAB (The MathWorks, Natick, MA), code available upon request. Enzyme concentrations were held constant at 10 μM with each binding to a single site on the scaffold with a 100 nM K_d, while scaffold concentrations were varied. As expected, the optimal scaffold performance occurs when scaffold concentrations are approximately equal to enzyme concentrations, and concentration optimization becomes increasingly important as pathway size increases. In agreement with these modeled predictions, we observed a strong dependence of production titers on the relative expression levels of both scaffold and metabolic enzymes (Dueber *et al.*, 2009; Moon *et al.*, 2010). Thus, it may be helpful to drive expression of pathway enzymes and scaffold with independent promoters to independently tune expression to the optimal levels.

In addition to optimizing scaffolding levels, it may be beneficial to simultaneously adjust the pathway enzyme concentrations. High induction was found to be the optimal expression level for mevalonate biosynthetic pathway enzymes in the absence of scaffolding. However, when the most effective scaffold was present, low enzyme induction produced optimal production titers, giving higher titers than the maximum achievable in the absence of scaffold, even at the uninduced background expression level of the promoter (Fig. 19.4). Although it would be interesting to independently optimize the expression of the three enzymes scaffolded in this pathway, optimization beyond polycistronic expression level has not yet been carried out. Scaffold architecture, expression level, and enzyme expression levels are all interconnected variables that must be optimized.

2.4. Varying scaffold stoichiometry

Often when metabolic pathways are engineered, one pathway enzyme exhibiting relatively low activity creates a bottleneck in the pathway. In many cases, this limitation can be alleviated by increasing expression of

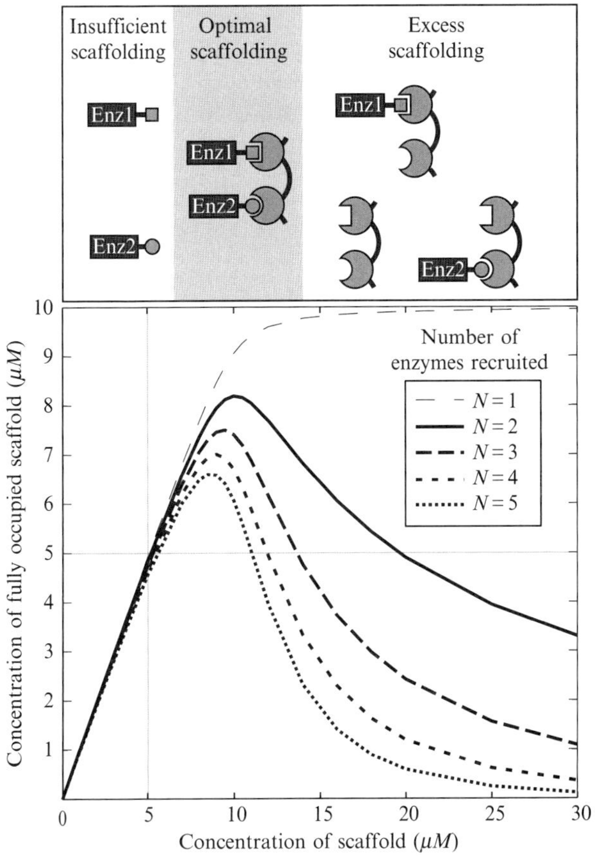

Figure 19.3 A mathematical model of equilibrium binding of scaffolding recruiting enzymes shows an optimal scaffold concentration for maximizing full occupancy. Five scaffolds consisting of a varying number of enzyme recruitment domains are independently simulated. Enzyme levels are held constant at 10 μ*M* each, and each scaffold molecule recruits a specified number of different enzymes each with a dissociation constant of 100 n*M*. Initially, as scaffold concentration is increased, excess enzymes are recruited to fully occupy the scaffold. As enzymes become limiting, scaffold competition for enzyme recruitment leads to low occupancy of scaffold.

that enzyme (Pitera *et al.*, 2007). Pathway scaffolding presents another strategy for addressing this problem at lower enzyme expression regimes. Ability to achieve high product titers with low concentration of enzymes should prove particularly advantageous for systems with enzymes prone to aggregate. Additionally, improved pathway efficiency is likely to become increasingly important as the number of enzymes in the pathway is increased, although this remains to be empirically tested.

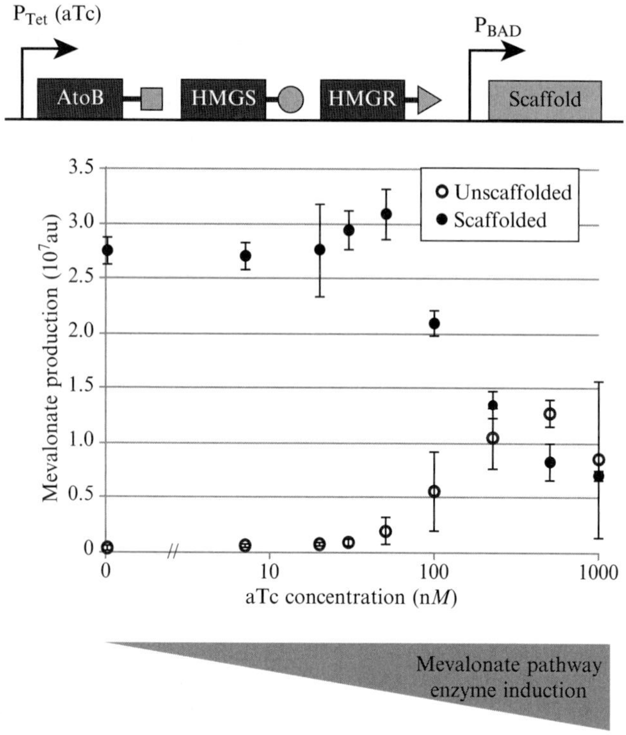

Figure 19.4 Mevalonate titers are measured at varying aTc concentrations, corresponding to induction of P_{Tet} driving peptide-tagged AtoB, HMGS, and HMGR polycistronically. The P_{BAD} promoter either drives the optimal scaffold, GBD_1–$SH3_2$–PDZ_2 (filled circles), or GFP, representing unscaffolded pathway (unfilled circled). Mevalonate production was measured as GC/MS peak area. Error bars represent one standard deviation from three separate experiments (figure adapted from Dueber *et al.*, 2009).

Due to the modular nature of the described scaffold strategy, the relative stoichiometry of enzymes cocomplexed can be controlled by varying the number of repeats of each protein–protein interaction domain. For the mevalonate biosynthetic pathway, a matrix of nine scaffold architectures was assembled with one, two, or four protein–protein interaction domains recruiting the enzymes for the bottleneck intermediate transfer, HMGS and HMGR (Fig. 19.5). Within this matrix, the optimal architecture produced a 77-fold improvement in product titers relative to the unscaffolded pathways. Importantly, although all scaffolds improved titers relative to the unscaffolded pathway, the results were difficult to explain based on stoichiometry alone.

We reapplied the same matrix of scaffolds used to improve flux of the mevalonate biosynthetic pathway to a second pathway, the glucaric acid

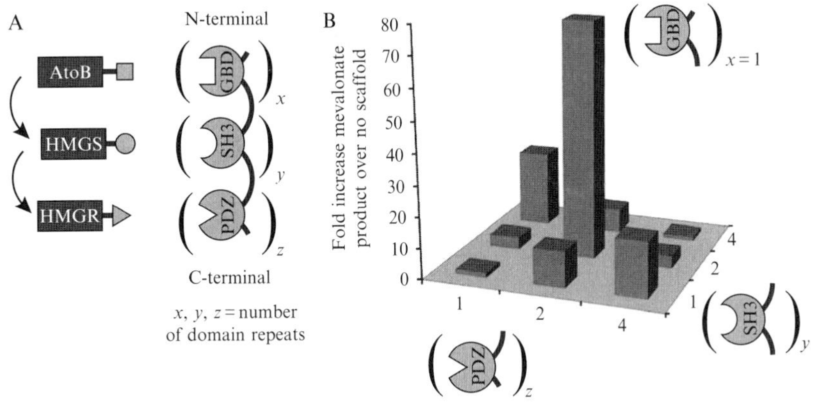

Figure 19.5 Three enzymes in the synthetic mevalonate pathway were tagged with binding peptides and recruited to a synthetic scaffolding protein with varying stoichiometry of binding domains. (A) Schematic of scaffolding with varying stoichiometry, where the number of GBD domains is held constant at one, while the number of SH3 and PDZ domains is varied to be one, two, or four. (B) The different scaffold stoichiometries gave very different results, with the best performing scaffold resulting in a 77-fold increase in mevalonate production compared to when no scaffold was expressed (figure adapted from Dueber *et al.*, 2009).

biosynthetic pathway. This pathway was previously engineered by Moon and Prather from three heterologous enzymes, Ino1, MIOX, and Udh (Moon *et al.*, 2009) and expressed at high levels under the T7 promoter to maximize glucaric acid titers. Interestingly, the Prather group measured higher activities of the limiting enzymatic activity, MIOX, in the presence of high concentrations of *myo*-inositol substrate (Moon *et al.*, 2009, 2010). The three heterologous enzymes were tagged with recruitment peptides and expressed from the P_{Lac} promoter. Similar to our findings with the mevalonate biosynthetic pathway, the scaffolds showed various degrees of titer improvements dependent on architecture. Consistent with a hypothesis of increasing the local concentration of *myo*-inositol at the resultant synthetic complex, titer improvements were dependent on the number of Ino1-recruiting domains producing *myo*-inositol, whereas there was no strong dependence on the number of domains recruiting MIOX enzyme. Titers of 2.3 g/L glucaric acid were produced with the optimal scaffold, giving an almost fivefold increase over the control lacking scaffold expression, a 50% improvement over highest titers previously reached (Fig. 19.6).

There are a few considerations to be made when constructing sequence for scaffolds with domains repeated multiple times. Cloning strategies relying on PCR or homologous overhangs such as recombination-based methods (Shao *et al.*, 2009), sequence and ligation-independent cloning (SLIC; Li and Elledge, 2007), or isothermal enzymatic assembly

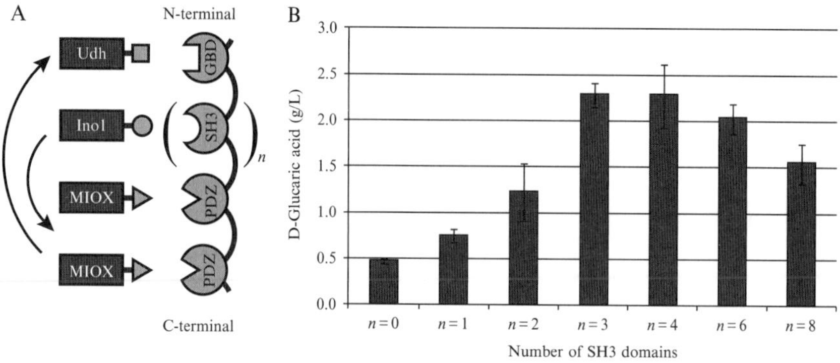

Figure 19.6 A pathway engineered to produce glucaric acid was scaffolded with optimized stoichiometry to improve production. (A) Schematic of scaffolded pathway where the number of SH3 domains corresponding to Ino1 recruitment is varied. Ino1 produces the substrate for MIOX, which MIOX converts to the substrate for Udh. (B) Glucaric acid concentrations were measured under conditions where only the number of SH3 domains on the scaffolding is varied, showing around three or four SH3 domains provided optimal production (figure adapted from Moon *et al.*, 2010).

(Gibson *et al.*, 2009) may result in misannealing for products containing repeated sequences. One construction strategy that deals particularly well with repeated domains is the BioBrick-based cloning strategies (Anderson *et al.*, 2010; Shetty *et al.*, 2008), particularly the *Bgl*II/*Bam*HI-based strategy (Anderson *et al.*, 2010) that leaves generally innocuous and often useful glycine–serine scars that can be used as part of the linker sequence. Interestingly, we have observed problems coming from recombination arising from greater than four identical repeats for both domains (~200 bases) and peptide ligands (~60 bases). A solution to this problem is to design multiple domain "parts" with degenerate codon usage such that repeated domains, linkers, and ligands are sufficiently different to prevent recombination. For making multiple SH3 domain repeats, using six degenerate SH3 parts, we were able to make constructs with 10 repeats without observing a significant number of incorrect products due to recombination (Dueber, unpublished observation; Moon *et al.*, 2010).

2.5. Scaffold composition effects

The three-dimensional structure of the scaffolded complex will determine the efficiency of improving flux. Although these structures are determined by the domain architecture, these architecture/structure relationships are not currently predictable. This is highlighted by the importance of not only the total number of each protein–protein interaction domain but also by the arrangement of these domains. The number and identity of SH3 and PDZ interaction domains in a scaffold was held constant but the order of these domains was rearranged (Dueber *et al.*, 2009). GBD–$(SH3)_2$–$(PDZ)_2$, GBD–$(SH3)_1$–$(PDZ)_2$–$(SH3)_1$, and GBD–$(SH3)_1$–$(PDZ)_1$–$(SH3)_1$–$(PDZ)_1$ scaffolds showed dramatically varied abilities to improve mevalonate titers despite each having the same number of recruitment domains (Fig. 19.7).

Work on synthetic cellulosomes has also supported the importance of scaffold composition (Mingardon *et al.*, 2007). In this study, scaffolds were designed to recruit other scaffolds, creating complexes containing up to four different scaffolds, each in turn recruiting, or directly fused to, a cellulose degrading enzyme. One issue that arose was the importance of enzyme mobility, as redundant binding that was likely to limit flexibility decreased degradation efficiency (Mingardon *et al.*, 2007). Natural cellulosome protein sequences suggest that they are physically flexible complexes, since they generally have long linkers (tens up to 550 residues) predicted to result in highly mobile enzymes (Xu *et al.*, 2003). Another issue was the importance

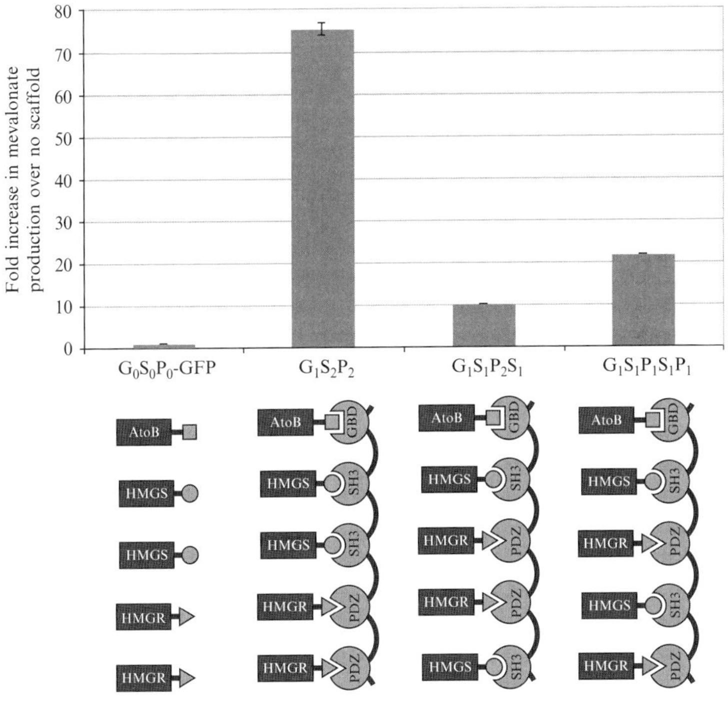

Figure 19.7 Mevalonate titers are measured for ligand tagged enzymes with expression of scaffolds of differing architecture and compared to the titers in the absence of scaffold expression. Rearranging domain order, while retaining the one GBD, two SH3, and two PDZ stoichiometry gives significant changes in titer improvement (figure adapted from Dueber *et al.*, 2009).

of scaffolding complex mobility, suggested from lowered activity with the addition of more than one carbohydrate-binding module. These issues of mobility and flexibility are likely to be particularly relevant to the catalysis of cellulose, an immobilized substrate, and may have limited pertinence to readily diffusing metabolic intermediates.

Interestingly, it has been suggested that when individual enzymes are tethered together, if these enzymes exist in oligomeric form, they may multimerize, forming even larger complexes (Bülow and Mosbach, 1991; Conrado *et al.*, 2008). A scaffold with repeated domains that recruits oligomeric enzymes, as is the case in several applications thus far (Dueber *et al.*, 2009; Moon *et al.*, 2010), may also form large multimeric complexes (Fig. 19.8). These multimeric complexes may improve titers by further increasing local concentrations beyond those achievable with individual scaffolds. This potential phenomenon is another reason we recommend taking a library approach to optimize scaffolded pathway flux, varying as many parameters as practical to empirically determine the optimal combination of architectural parameters.

Scaffold configuration may become an increasingly important variable as pathways are scaled to consist of larger numbers of enzymes. One practical concern is the increasing scaffold protein size to target increasing numbers of enzymes while also achieving stoichiometry control. As explored in the development of synthetic cellulosomes (Mingardon *et al.*, 2007), a potential solution to this problem is building multiple scaffolds that can coassemble either directly or through a separate adaptor molecule. This approach might also prove to be a convenient method for modularizing various sections of a pathway as well as increasing the combinatorial architecture possibilities.

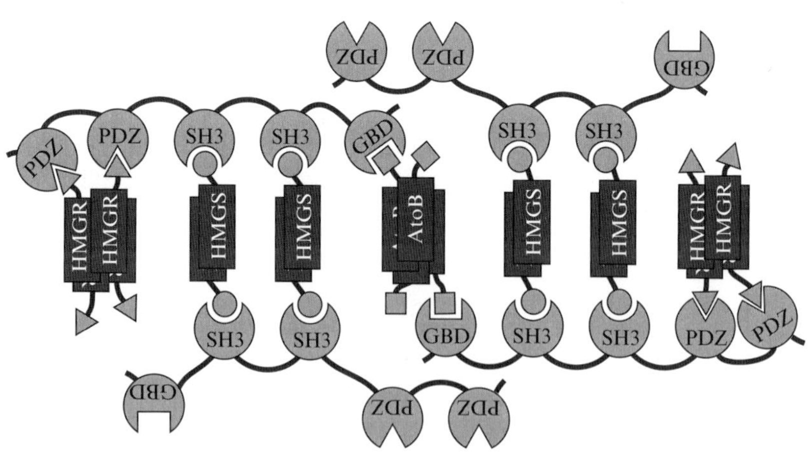

Figure 19.8 Schematic of a multimeric enzyme complex. Enzymes with oligomeric structures could potentially bind multiple scaffolds, resulting in large complexes and difficult to predict positioning in three-dimensional space.

3. Systems that May Benefit from Scaffolding

There are a number of potential mechanisms that may contribute to the increased titers shown in the scaffolded systems discussed. We have thus far focused on the increased local concentration effect by which scaffolding may reduce toxic intermediates, reduce the load on the cell by reducing the necessary enzyme expression levels, and increase pathway efficiency. Another potential benefit of scaffolding is the prevention of enzyme aggregation by lowering the necessary protein expression levels in addition to sequestering individual enzyme molecules. It is also possible that scaffold could be used to increase enzyme stability or activity, though this was shown to not be the case for the glucaric acid pathway (Moon *et al.*, 2010). Substrate-activated enzymes, such as MIOX in the glucaric acid pathway discussed earlier, are also particularly good candidates for scaffolding. Scaffold complexes with higher numbers of protein–protein interaction domains can be used to recruit substrate-producing upstream enzyme (Moon *et al.*, 2010).

Scaffolding a complex of efficient enzymes may also provide a means of limiting intermediate loss. Increased local concentration, or corresponding reduction in bulk cytoplasmic concentrations, would reduce the rate of intermediate loss to competing pathways. Additionally, the reduced enzyme expression also reduces the rate of those enzymes metabolizing unintended substrates. Insulation from unintended interactions is clearly a desired engineering characteristic as it is difficult to predict interactions with the natural cellular metabolism. These undesired interactions may be reduced with coassembled enzyme complexes. This may be increasingly important as engineered metabolic pathways continue to scale in size and complexity. These larger pathways may necessitate building and characterizing different modules of a pathway independently, which when assembled would benefit from reduction of unintended interactions between shared intermediates. Another potential mechanism for reducing unwanted interactions is the localization to different subsections or compartments of the cell. The ability to target the scaffold to a particular location may facilitate pathway localization strategies.

4. Concluding Remarks

In conclusion, scaffolding provides a posttranslational tool that may help increase production yields and deal with problematic enzymes as well as reduce the cellular burden and unintended interactions that may become an issue when scaling to engineer the next generation of biosynthetic metabolic pathways. The methods we describe in this chapter have the advantage of being highly designable and easily adaptable to library approaches. Future research must be done to determine the mechanisms by which these scaffolded systems function such that they can be more rationally applied to other pathways in a predictable manner.

ACKNOWLEDGMENTS

We thank R. Conrado and M. DeLisa as well as K. Tipton and other members of the Dueber lab for comments and discussion regarding the preparation of the chapter. This work was supported by funding from NSF grant no. CBET-0756801 (W. R. W., J. E. D.).

REFERENCES

Acharya, A., Ruvinov, S. B., Gal, J., Moll, J. R., and Vinson, C. (2002). A heterodimerizing leucine zipper coiled coil system for examining the specificity of a position interactions: Amino acids I, V, L, N, A, and K. *Biochemistry* **41,** 14122–14131.

An, S., Kumar, R., Sheets, E., and Benkovic, S. (2008). Reversible compartmentalization of *de novo* purine biosynthetic complexes in living cells. *Science* **320,** 103–106.

Anderson, J. C., Dueber, J. E., Leguia, M., Wu, G. C., Goler, J. A., Arkin, A. P., and Keasling, J. D. (2010). BglBricks: A flexible standard for biological part assembly. *J. Biol. Eng.* **4,** 1.

Banaszynski, L. A., Liu, C. W., and Wandless, T. J. (2005). Characterization of the FKBP–Rapamycin–FRB ternary complex. *J. Am. Chem. Soc.* **127,** 4715–4721.

Bierer, B. E., Mattila, P. S., Standaert, R. F., Herzenberg, L. A., Burakoff, S. J., Crabtree, G., and Schreiber, S. L. (1990). Two distinct signal transmission pathways in T lymphocytes are inhibited by complexes formed between an immunophilin and either FK506 or rapamycin. *Proc. Natl. Acad. Sci. USA* **87,** 9231–9235.

Bülow, L., and Mosbach, K. (1991). Multienzyme systems obtained by gene fusion. *Trends Biotechnol.* **9,** 226–231.

Conrado, R. J., Mansell, T. J., Varner, J. D., and DeLisa, M. P. (2007). Stochastic reaction–diffusion simulation of enzyme compartmentalization reveals improved catalytic efficiency for a synthetic metabolic pathway. *Metab. Eng.* **9,** 355–363.

Conrado, R. J., Varner, J. D., and DeLisa, M. P. (2008). Engineering the spatial organization of metabolic enzymes: Mimicking nature's synergy. *Curr. Opin. Biotechnol.* **19,** 492–499.

Di Guan, C., Li, P., Riggs, P. D., and Inouye, H. (1988). Vectors that facilitate the expression and purification of foreign peptides in *Escherichia coli* by fusion to maltose-binding protein. *Gene* **67,** 21–30.

Dougherty, M. J., and Arnold, F. H. (2009). Directed evolution: New parts and optimized function. *Curr. Opin. Biotechnol.* **20,** 486–491.

Dueber, J. E., Wu, G. C., Malmirchegini, G. R., Moon, T. S., Petzold, C. J., Ullal, A. V., Prather, K. L. J., and Keasling, J. D. (2009). Synthetic protein scaffolds provide modular control over metabolic flux. *Nat. Biotechnol.* **27,** 753–761.

Fierobe, H., Mechaly, A., Tardif, C., Bélaïch, A., Lamedi, R., Shoham, Y., Bélaïch, J., and Bayer, E. A. (2001). Design and production of active cellulosome chimeras: Selective incorporation of dockerin-containing enzymes into defined functional complexes. *J. Biol. Chem.* **276,** 21257–21261.

Fierobe, H., Mingardon, F., Mechaly, A., Bélaïch, A., Rincon, M. T., Pagés, S., Lamed, R., Tardif, C., Bélaïch, J., and Bayer, E. A. (2005). Action of designer cellulosomes on homogeneous versus complex substrates: Controlled incorporation of three distinct enzymes into a defined trifunctional scaffoldin. *J. Biol. Chem.* **280,** 16325–16334.

Fuh, G., Pisabarro, M. T., Li, Y., Quan, C., Lasky, L. A., and Sidhu, S. S. (2000). Analysis of PDZ domain–ligand interactions using carboxyl-terminal phage display. *J. Biol. Chem.* **275,** 21486–21491.

Gibson, D. G., Young, L., Chuang, R.-Y., Venter, J. C., Hutchison, C. A., and Smith, H. O. (2009). Enzymatic assembly of DNA molecules up to several hundred kilobases. *Nat. Methods* **6,** 343–345.

Grünberg, R., Ferrar, T. S., Van der Sloot, A. M., Constante, M., and Serrano, L. (2010). Building blocks for protein interaction devices. *Nucleic Acids Res.* **38,** 2645–2662.

Haimovitz, R., Barak, Y., Morag, E., Voronov-Goldman, M., Shoham, Y., Lamed, R., and Bayer, E. A. (2008). Cohesin–dockerin microarray: Diverse specificities between two complementary families of interacting protein modules. *Proteomics* **8,** 968–979.

Harris, B. Z., Hillier, B. J., and Lim, W. A. (2001). Energetic determinants of internal motif recognition by PDZ domains. *Biochemistry* **40,** 5921–5930.

Havranek, J. J., and Harbury, P. B. (2003). Automated design of specificity in molecular recognition. *Nat. Struct. Biol.* **10,** 45–52.

Kim, A. S., Kakalis, L. T., Abdul-Manan, N., Liu, G. A., and Rosen, M. K. (2000). Autoinhibition and activation mechanisms of the Wiskott–Aldrich syndrome protein. *Nature* **404,** 151–158.

Leibovitz, E., Ohayon, H., Gounon, P., and Béguin, P. (1997). Characterization and subcellular localization of the *Clostridium* thermocellum scaffoldin dockerin binding protein SdbA. *J. Bacteriol.* **179,** 2519–2523.

Levchenko, A., Bruck, J., and Sternberg, P. W. (2000). Scaffold proteins may biphasically affect the levels of mitogen-activated protein kinase signaling and reduce its threshold properties. *Proc. Natl. Acad. Sci. USA* **97,** 5818–5823.

Levskaya, A., Weiner, O. D., Lim, W. A., and Voigt, C. A. (2009). Spatiotemporal control of cell signalling using a light-switchable protein interaction. *Nature* **461,** 997–1001.

Li, M. Z., and Elledge, S. J. (2007). Harnessing homologous recombination *in vitro* to generate recombinant DNA via SLIC. *Nat. Methods* **4,** 251–256.

Ljungcrantz, P., Carlsson, H., Minsson, M., Buckel, P., Mosbach, K., and Biilow, L. (1989). Construction of an artificial bifunctional enzyme, β-galactosidasel/galactose dehydrogenase, exhibiting efficient galactose channeling. *Biochemistry* **28,** 8786–8792.

Martin, V. J., Pitera, D. J., Withers, S. T., Newman, J. D., and Keasling, J. D. (2003). Engineering a mevalonate pathway in *Escherichia coli* for production of terpenoids. *Nat. Biotechnol.* **21,** 796–802.

Mathews, D. H., Sabina, J., Zuker, M., and Turner, D. H. (1999). Expanded sequence dependence of thermodynamic parameters improves prediction of RNA secondary structure. *J. Mol. Biol.* **288,** 911–940.

Miles, E. W., Rhee, S., and Davies, D. R. (1999). The molecular basis of substrate channeling. *J. Biol. Chem.* **274,** 12193–12196.

Mingardon, F., Chanal, A., Tardif, C., Bayer, E. A., and Fierobe, H. (2007). Exploration of new geometries in cellulosome-like chimeras. *Appl. Environ. Microbiol.* **73,** 7138–7149.

Moon, T. S., Yoon, S., Lanza, A. M., Roy-Mayhew, J. D., and Prather, K. L. J. (2009). Production of glucaric acid from a synthetic pathway in recombinant *Escherichia coli*. *Appl. Environ. Microbiol.* **75,** 589–595.

Moon, T. S., Dueber, J. E., Shiuea, E., and Prather, K. L. J. (2010). Use of modular, synthetic scaffolds for improved production of glucaric acid in engineered *E. coli*. *Metab. Eng.* **12,** 298–305.

Narayanaswamy, R., Levy, M., Tsechansky, M., Stovall, G. M., O'connell, J. D., Mirrielees, J., Ellington, A. D., and Marcotte, E. M. (2009). Widespread reorganization of metabolic enzymes into reversible assemblies upon nutrient starvation. *Proc. Natl. Acad. Sci. USA* **106,** 10147–10152.

Nordon, R. E., Craig, S. J., and Foong, F. C. (2009). Molecular engineering of the cellulosome complex for affinity and bioenergy applications. *Biotechnol. Lett.* **31,** 465–476.

Pettersson, H., and Pettersson, G. (2001). Kinetics of the coupled reaction catalysed by a fusion protein of L-galactosidase and galactose dehydrogenase. *Biochim. Biophys. Acta* **1549,** 155–160.

Pfleger, B. F., Pitera, D. J., Smolke, C. D., and Keasling, J. D. (2006). Combinatorial engineering of intergenic regions in operons tunes expression of multiple genes. *Nat. Biotechnol.* **24,** 1027–1032.

Pitera, D. J., Paddon, C. J., Newman, J. D., and Keasling, J. D. (2007). Balancing a heterologous mevalonate pathway for improved isoprenoid production in *Escherichia coli*. *Metab. Eng.* **9,** 193–207.

Posern, G., Zheng, J., Knudsen, B. S., Kardinal, C., Müller, K. B., Voss, J., Shishido, T., Cowburn, D., Cheng, G., Wang, B., Kruh, G. D., Burrell, S. K., *et al.* (1998). Development of highly selective SH3 binding peptides for Crk and CRKL which disrupt Crk-complexes with DOCK180, SoS and C3G. *Oncogene* **16,** 1903–1912.

Price, N., Reed, J., and Palsson, B. (2004). Genome-scale models of microbial cells: Evaluating the consequences of constraints. *Nat. Rev. Microbiol.* **2,** 886–897.

Reinke, A. W., Grant, R. A., and Keating, A. E. (2010). A synthetic coiled-coil interactome provides heterospecific modules for molecular engineering. *J. Am. Chem. Soc.* **132,** 6025–6031.

Robinson, C. R., and Sauer, R. T. (1998). Optimizing the stability of single-chain proteins by linker length and composition mutagenesis. *Proc. Natl. Acad. Sci. USA* **95,** 5929–5934.

Salis, H. M., Mirsky, E. A., and Voigt, Christopher A. (2009). Automated design of synthetic ribosome binding sites to control protein expression. *Nat. Biotechnol.* **27,** 946–952.

Shao, Z., Zhao, H., and Zhao, H. (2009). DNA assembler, an *in vivo* genetic method for rapid construction of biochemical pathways. *Nucleic Acids Res.* **37,** e16.

Shetty, R. P., Endy, D., and Knight, T. F. (2008). Engineering BioBrick vectors from BioBrick parts. *J. Biol. Eng.* **2,** 5.

Stroud, R. (1994). An electrostatic highway. *Nat. Struct. Biol.* **1,** 131–134.

Tonikian, R., Zhang, Y., Sazinsky, S. L., Currell, B., Yeh, J., Reva, B., Held, H. A., Appleton, B. A., Evangelista, M., Wu, Y., Xin, X., Chan, A. C., *et al.* (2008). A specificity map for the PDZ domain family. *PLoS Biol.* **6,** 2043–2059.

Welch, G. R. (1977). On the role of organized multienzyme systems in cellular metabolism: A general synthesis. *Prog. Biophys. Mol. Biol.* **32,** 103–191.

Win, M., and Smolke, C. (2007). A modular and extensible RNA-based gene-regulatory platform for engineering cellular function. *Proc. Natl. Acad. Sci. USA* **104,** 14283.

Xu, Q., Gao, W., Ding, S., Kenig, R., Shoham, Y., Bayer, E. A., and Lamed, R. (2003). The cellulosome system of *Acetivibrio cellulolyticus* includes a novel type of adaptor protein and a cell surface anchoring protein. *J. Bacteriol.* **185,** 4548–4557.

Zarrinpar, A., Park, S., and Lim, W. A. (2003). Optimization of specificity in a cellular protein interaction network by negative selection. *Nature* **426,** 676–680.

Zhang, K., Sawaya, M. R., Eisenberg, D. S., and Liao, J. C. (2008). Expanding metabolism for biosynthesis of nonnatural alcohols. *Proc. Natl. Acad. Sci. USA* **105,** 20653–20658.

CHAPTER TWENTY

A Synthetic Iterative Pathway for Ketoacid Elongation

C. R. Shen *and* J. C. Liao

Contents

Abstract

Iterative formation of nonpolymeric carbon–carbon bonds has been employed by organisms to synthesize fatty acids, polyketides, and isoprenoids. In these biosynthetic schemes, same reaction cycles are used iteratively for functional

Department of Chemical and Biomolecular Engineering, University of California, Los Angeles, Los Angeles, California, USA

Methods in Enzymology, Volume 497
ISSN 0076-6879, DOI: 10.1016/B978-0-12-385075-1.00020-2

modifications that result in the increase in carbon-chain length. This principle has been used in the design of a synthetic module for 2-ketoacid elongation. The system utilizes the *Escherichia coli* enzymes LeuABCD, which were engineered to accept bulkier nonnatural substrates, and was able to extend the chain length iteratively. The success in achieving a diverse range of 2-ketoacids and alcohols from this module via engineering of the 2-isopropylmalate synthase and ketoacid decarboxylase demonstrates the plasticity of LeuABCD and its feasibility for iterative carbon-chain elongations. In addition, this strategy illustrates a principle of designing novel metabolic modules for nonpolymeric carbon-chain elongation, which is essential in the synthesis of nonnative metabolites in microorganisms.

1. Introduction

Synthetic biology typically deals with the design of gene circuits that perform nonnative tasks (Voigt, 2006). Such circuits may interact with metabolism to control the flux in a desirable way (Farmer and Liao, 2000; Fung *et al.*, 2005). In an area overlapping metabolic engineering, synthetic biology also involves transfer of heterologous genes and pathways from foreign organisms to a nonnative host for biosynthetic purposes (Atsumi *et al.*, 2008a,b; Cann and Liao, 2008; Connor and Liao, 2008; Farmer and Liao, 2001; Ro *et al.*, 2006; Rohlin *et al.*, 2001; Shen and Liao, 2008; Yan *et al.*, 2005). In contrast to the design of regulatory circuits or transfer of foreign metabolic pathways, this chapter focuses on the construction of a nonnative module that functions iteratively to increase the carbon–carbon-chain length of 2-ketoacids. The basic unit of this module, LeuABCD, is used in leucine and norvaline biosynthesis in a noniterative manner. Here, we discuss the efforts to engineer this module such that the reactions occur recursively to extend the chain length of 2-ketoacids.

In nature, nonpolymeric carbon-chain elongation schemes are used in the biosynthesis of fatty acids, polyketides, and isoprenoids. In these chain extension systems, a starting substrate with a given functional group undergoes a series of reactions, including condensation, reduction, and dehydration. The end product of one cycle increases the chain length by one unit (two or five carbons) but retains the same functional group as the starting substrate (Fig. 20.1), which allows the reactions to iterate and elongate the carbon chain in each cycle. Fatty acid and polyketide syntheses utilize malonyl-CoA as the repetitive C_2 addition unit and each elongation cycle assembles one acetyl group to the growing acyl chain (Hopwood and Sherman, 1990; Staunton and Weissman, 2001). The extent of processing, which results in either fully reduced β-carbon or other functional groups at the β position, distinguishes fatty acid formation from polyketide synthesis (Katz and Donadio, 1993). Isoprenoid chain elongation, on the other hand, incorporates

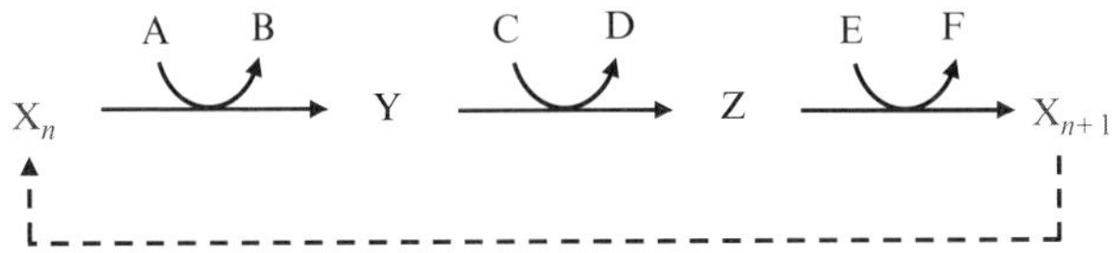

Figure 20.1 Schematic illustration of nonpolymeric iterative chain elongation. X retains the same functional group at the end of each cycle, which is essential for the recursive addition of compound A.

isopentenyl diphosphate (IPP) as the repetitive C_5 unit to its isomer dimethylallyl diphosphate (DMAPP) and extends the growing chain consecutively by five carbons in each cycle (Barkovich *et al.*, 2001; Li *et al.*, 2010).

Ketoacid pathways found in the branched-chain amino acid biosynthesis is another class of reactions that occur in nature for carbon-chain elongations (Atsumi *et al.*, 2008a,b; Connor and Liao, 2009). However, unlike the fatty acid and isoprenoid syntheses, ketoacid chain extension reactions are not naturally recursive. The two major enzymes involved are 2-isopropylmalate synthase (IPMS) in the leucine pathway and acetohydroxy acid synthase (AHAS) in the valine pathway, each responsible for the extension of specific ketoacids to make the corresponding amino acid precursors (Fig. 20.2). While IPMS catalyzes the addition of acetyl group to 2-ketoisovalerate and results in one net carbon gain at the end of chain elongation cycle (Umbarger, 1978; Shen and Liao, 2008), AHAS condenses two pyruvates together and increases the carbon number by two with a branch in the chain (McCourt and Duggleby, 2006). In this review, we will focus on the 2-ketoacid chain elongation initiated by IPMS and discuss its potential for expanding cell metabolites with the initial success achieved on recursive carbon-chain extension upon protein engineering.

2. Natural Pathways Involving Ketoacid Chain Elongations Catalyzed by the LeuABCD-Dependent Mechanisms

2.1. Synthesis of leucine ($C_5 \rightarrow C_6$) and norvaline ($C_4 \rightarrow C_5$) in *Escherichia coli* by LeuABCD

The enzyme IPMS encoded by *leuA* is found in the biosynthetic pathway of leucine (Umbarger, 1978; Connor and Liao, 2008). Naturally LeuA catalyzes the Claisen condensation of acetyl-CoA and the five-carbon branched ketoacid 2-ketoisovalerate to initiate a series of chain elongation reactions that eventually lead to the synthesis of 2-ketoisocaproate via LeuBCD with

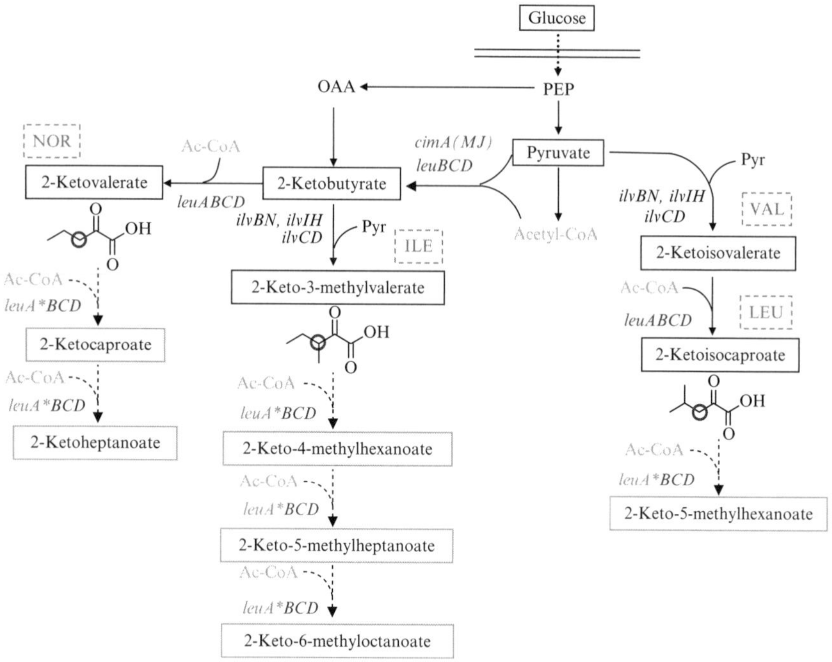

Figure 20.2 Various pathways leading to the synthesis of natural and nonnatural 2-ketoacids by recursive chain elongation catalyzed by *Escherichia coli* enzymes LeuABCD. Native ketoacids are boxed in black while nonnative ones are boxed in blue. Dashed arrows indicate the introduction of new pathways in *E. coli* via protein engineering of LeuA* shown by the blue font. Red circles point out the position of carbon insertion. Biosyntheses of the specific amino acids are indicated with dashed grey boxes above their corresponding ketoacid precursors. NOR, norvaline; ILE, isoleucine; VAL, valine; LEU, leucine. *MJ: Methanococcus jannaschii* (See Color Insert.)

one net carbon gain at the β position (Fig. 20.2). In addition to the formation of leucine precursor 2-ketoisocaproate, native *E. coli* LeuA also catalyzes the condensation of acetyl-CoA and 2-ketobutyrate, a four-carbon straight-chained ketoacid commonly derived from threonine for isoleucine biosynthesis, to make 2-ketovalerate via ketoacid chain elongation and result in production of the nonnatural amino acid norvaline as shown in Fig. 20.2 (Kisumi *et al.*, 1976; Shen and Liao, 2008). The promiscuity of LeuA is revealed by its activity toward the alternative substrate 2-ketobutyrate, which differs from its natural substrate 2-ketoisovalerate by a methyl group at the β position. The broad substrate specificity of LeuABCD suggests the potential evolvability of the pathway for iterative carbon-chain elongations.

2.2. Alternative citramalate pathway ($C_3 \rightarrow C_4$) for isoleucine synthesis by CimA-LeuBCD

Most microorganisms synthesize isoleucine via the common intermediate 2-ketobutyrate derived from threonine. The citramalate pathway is an alternative pathway present in some methanogenic archaea such as *Methanococcus jannaschii* and *Leptospira interrogans* for isoleucine production that bypasses the threonine biosynthesis route for making the essential precursor 2-ketobutyrate (Fig. 20.4). Production of 2-ketobutyrate is important because of its participation in the synthesis of various higher chain alcohols such as 1-propanol, 1-butanol, and 2-methyl-1-butanol when the promiscuous 2-ketoacid decarboxylase (Kivd) from *Lactococcus lactis* and alcohol dehydrogenase 2 (ADH2) from *Saccharomyces cerevisiae* are introduced (Cann and Liao, 2008; Shen and Liao, 2008). Parallel to the reaction schemes found in the production of 2-ketoisocaproate in *E. coli* for leucine biosynthesis, citramalate synthase CimA, a structural homologue of LeuA, catalyzes the condensation of acetyl-CoA and the three-carbon ketoacid pyruvate into (*R*)-citramalate, which subsequently leads to the formation of 2-ketobutyrate with reactions catalyzed by LeuBCD (Fig. 20.4). Similarities between the leucine and citramalate pathways suggest the common usage of ketoacid chain elongations for metabolite biosynthesis and the interchangeability of essential enzymes for substrates with different chain lengths and branch position.

3. IPMS and Similar Enzymes

3.1. Reaction mechanisms of carbon-chain elongation catalyzed by LeuABCD

Ketoacid elongation catalyzed by LeuABCD initiates with the Claisen condensation of acetyl-CoA to 2-ketoisovalerate (Fig. 20.3) by deprotonating the methyl group of acetyl-CoA via concerted acid–base chemistry with adjacent side chains of LeuA to generate an enolate intermediate, which then attacks the electrophilic carbonyl group of the ketoacid (Koon *et al.*, 2004). The resulting 2-isopropylmalate is then converted to 3-isopropylmalate by the isopropylmalate isomerase (consisting of subunits LeuC and LeuD), which transfers the hydroxyl group between adjacent carbons. Finally, oxidation and decarboxylation of 3-isopropylmalate are catalyzed by the metal-dependent 3-isopropylmalate dehydrogenase (LeuB) using NAD^+ as the electron acceptor to yield 2-ketoisocaproate, NADH and CO_2 (Fig. 20.3). Similar catalytic mechanisms are also proposed for CimA (Ma *et al.*, 2008), where the conversion of acetyl-CoA and pyruvate into citramalate follows the three basic steps of enolization, condensation, and

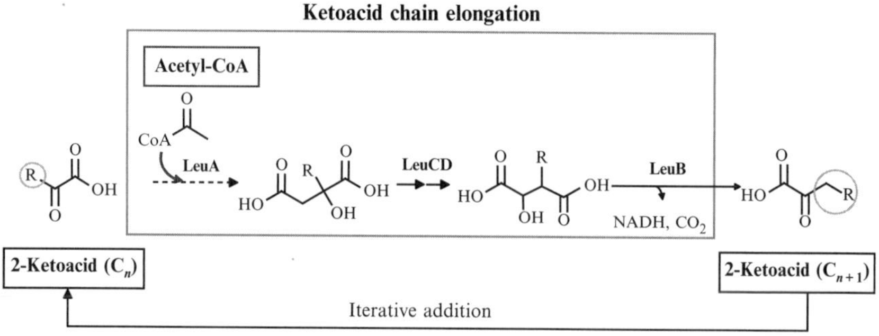

Figure 20.3 Reaction schematics of the recursive 2-ketoacid chain elongation catalyzed by *E. coli* enzymes LeuABCD. R, isopropyl for the native substrate 2-ketoisovalerate. LeuA appears to be the major bottleneck in the 2-ketoacid elongation pathway of various chain lengths (illustrated by the dashed line). One additional carbon is inserted at the β position of ketoacids at the end of each cycle as indicated by the yellow circle. (For interpretation of the references to color in this figure legend, the reader is referred to the Web version of this chapter.)

hydrolysis as described in LeuA-dependent reactions. Promiscuity of LeuBCD then allows the subsequent reactions to proceed with the corresponding substrates. Using 2-ketoacid as the starting material and acetyl-CoA as the recursive addition unit, these are the series of reactions catalyzed by LeuABCD in ketoacid chain elongations that enable the extension of chain length by one carbon with no alteration to the branch position (Fig. 20.3).

3.2. Structural and mechanistic homology among LeuA, CimA, and GlcB

To gain insight into the preferential substrate selectivity of the first committed step of ketoacid chain elongation catalyzed by LeuA and other analogous enzymes, it is interesting to compare the structures and reaction mechanisms among crystallized enzymes that catalyze similar reactions with *E. coli* LeuA, in particular, malate synthase G (*Ec*MSG) from *E. coli* encoded by *glcB*, citramalate synthase (*Li*CMS) from *L. interrogans* encoded by *cimA*, and IPMS from *Mycobacterium tuberculosis* (*Mt*IPMS) encoded by *leuA*. CMS (Ma *et al.*, 2008) and IPMS (Koon *et al.*, 2004) are of special interest to us because of their essential role in catalyzing ketoacid chain elongation reactions involved in the synthesis of alcohol precursors (Fig. 20.2). All three enzymes catalyze the typical acetyl-CoA-dependent Claisen condensation of 2-ketoacids (glyoxylate, pyruvate, and 2-ketoisovalerate, respectively) to yield malate (MSG), (*R*)-citramalate (CMS), and 2-isopropylmalate (IPMS) by deprotonating the methyl group of acetyl-CoA and generate an enolate intermediate which then attacks the electrophilic carbonyl group of the 2-ketoacids. These enzymes, although appear to have no significant sequence identity, seem to have evolved convergently toward a common chemistry by using very similar functional groups acting on similar substrates. *Li*CMS and *Mt*IPMS share decent structural homology with *Ec*MSG in that their catalytic domains also locate near the C-terminal end of the triosephosphate isomerase (TIM) barrel with the key side chains of Glu146/Arg16 (*Li*CMS) and Glu218/Arg80 (*Mt*IPMS) carry essentially the same acid–base function as Asp631/Arg338 found in *Ec*MSG for enolization and intermediate stabilization (Howard *et al.*, 2000; Anstrom *et al.*, 2003; Koon *et al.*, 2004; Ma *et al.*, 2008). Both *Li*CMS and *Mt*IPMS are dimers of individual β8/α8 TIM barrel monomer with catalytic activities dependent on bivalent metals (Mn^{2+} exhibits the highest activity for *Li*CMS and Zn^{2+} for *Mt*IPMS). Nevertheless, some structural difference with *Ec*MSG also exists such as the location of the coordinating metal ion around the barrel and the binding site of acetyl-CoA relative to the substrate (Howard *et al.*, 2000; Anstrom *et al.*, 2003). It is thus interesting to study the catalytic domains of *Li*CMS, *Mt*IPMS, and *Ec*MSG to elucidate possible structural changes that would potentially lead to alteration of their substrate specificity to accommodate iterative 2-ketoacid chain elongations with various chain lengths.

4. Expansion to Nonnatural Pathways

4.1. Iterative ketoacid chain elongation

With the promiscuity of LeuA demonstrated from its intrinsic activity in the leucine and norvaline pathway, it is of great interest to investigate the possibility of recursive ketoacid chain elongations upon protein engineering of the essential enzymes. In order for the enzymes LeuABCD to extend 2-ketoacids in a repetitive fashion, their substrate specificities need to be broadened to take longer 2-ketoacids with decent catalytic activities (Fig. 20.2). Since nonnatural 2-ketoacids have no direct involvement in the synthesis of essential metabolites, random mutagenesis was not employed as the evolution strategy of LeuABCD due to the lack of a selection pressure. As a result, site-directed mutagenesis with rational design was used for expanding the substrate capacity of LeuA. Minimization of steric clash between the nonnatural 2-ketoacids and residues at the active site broadened the substrate specificity of LeuA while maintained the important catalytic functions (Zhang *et al.*, 2008). Fine tailoring of the substrate binding pocket according to the branch position and length of 2-ketoacids allowed the achievement of six different nonnatural longer-chained ketoacids and alcohols as a result of one to three recursive chain elongation events, each with 2-ketoisocaproate, 2-keto-3-methylvalerate, or 2-ketovalerate as the starting material (Fig. 20.2). Interestingly, the first step catalyzed by LeuA appeared to be the only significant bottleneck in the expansion of nonnatural 2-ketoacids; once the LeuA* is optimized for the condensation of acetyl-CoA and a particular 2-ketoacid, subsequent reactions catalyzed by the wild-type LeuBCD proceeds without major limitations (Zhang *et al.*, 2008). It appears that LeuBCD never encountered larger substrates and did not have the evolutionary pressure to become more selective. This may shed light on how enzymes evolved to regulate pathway specificity. The wide range of natural and nonnatural 2-ketoacids and the corresponding long chain alcohols achieved upon iterative chain elongations catalyzed by LeuABCD is listed in Table 20.1.

4.2. Alteration of LeuA selectivity for longer substrates by rational mutagenesis

LeuA belongs to a class of enzymes that catalyze the Claisen condensation of acetyl-CoA and 2-ketoisovalerate to yield the essential leucine precursor and is commonly found in all spectrums of organisms. However, crystal structures of LeuA are quite limited throughout literature and the only one available up to date was reported from the pathogenic bacteria *M. tuberculosis* (Koon *et al.*, 2004). To explore the substrate promiscuity of LeuA in *E. coli* and to extend the repertoire of nonnatural 2-ketoacids with the iterative mechanism, it was necessary to identify the active sites of LeuA and redesign the binding pocket so that it could recursively react with the extending 2-ketoacid chain (Figs. 20.2 and 20.3). Upon examination of the *M. tuberculosis* LeuA crystal structure complexed with its native substrate 2-ketoisovalerate, it was revealed that three residues His-167, Ser-216, and Asn-250 are situated within 4 Å of the γ-methyl group of the bound 2-ketoisovalerate and would result in steric hindrance if the substrate is to be expanded (Zhang *et al.*, 2008). Although protein sequence alignment showed a low sequence identity of only 21% between the two LeuA from *M. tuberculosis* and *E. coli*, essential residues that catalyze the reaction and constitute the binding pocket are well conserved. The three corresponding key residues His-97, Ser-139, and Asn-167 surrounding the active site of *E. coli* LeuA suspected to cause steric clash with longer 2-ketoacids were subjected to site-specific mutagenesis.

To minimize the effect of feedback inhibition of LeuA by its end-product isoleucine, the feedback resistant mutant G462D (Gusyatiner *et al.*, 2002) was used as the starting point for all the additional mutations tested for expanding the possibility of nonnatural 2-ketoacids. In addition to *in vitro* enzymatic assays, *in vivo* production of long chain alcohols was used to assess the variation in substrate specificity of LeuA upon each mutation by overexpression of the entire ketoacid network for alcohol production coupled with the *L. lactis* 2-ketoacid decarboxylase Kivd variant F381L/V461A and alcohol dehydrogenase ADH6 from *S. cerevisiae*. It was found

Table 20.1 List of all the 2-ketoacids, naturally and nonnaturally, synthesized by the iterative ketoacid chain elongation pathway catalyzed by *Escherichia coli* enzymes LeuABCD

2-Ketoacid	Natural?	Corresponding alcohol	LeuA* (G462D)	Alcohol titers (mg/L)
2-Ketoisocaproate	Yes	3-Methyl-1-butanol (C5)	–	963
2-Keto-5-methylhexanoate	No	4-Methyl-1-pentanol (C6)	S139G/H97L	202
2-Ketovalerate	Yes	1-Butanol (C4)	–	382
2-Ketocaproate	No	1-Pentanol (C5)	–	445
2-Ketoheptanoate	No	1-Hexanol (C6)	S139G/H97A	38
2-Keto-3-methylvalerate	Yes	2-Methyl-1-butanol (C5)	S139G/N167L	82
2-Keto-4-methylhexanoate	No	3-Methyl-1-pentanol (C6)	S139G	794
2-Keto-5-methylheptanoate	No	4-Methyl-1-hexanol (C7)	S139G/H97A/N167A	57
2-Keto-6-methyloctanoate	No	5-Methyl-1-heptanol (C8)	S139G/H97A/N167A	22

The corresponding long chain alcohols, titers achieved, and mutations on LeuA that enabled the highest production of that particular ketoacid are also tabulated in the same row. *Lactococcus lactis* ketoacid decarboxylase Kivd variant F381L/V461A and alcohol dehydrogenase ADH6 from *Saccharomyces cerevisiae* were used in all alcohol production experiments. The isoleucine feedback resistant LeuA G462D was utilized as the basis for all mutation constructions. Different shadings are used to group ketoacids by their common initial substrates prior to recursive chain elongation.

that the replacement of Ser-139 with the smallest amino acid glycine successfully doubled the amount of 3-methyl-1-pentanol and resulted in fivefold higher 1-hexanol. In addition, combination with other specific mutations at His-97 and Asn-167 improved the synthesis of four other nonnatural 2-ketoacids and the corresponding alcohols. To summarize, H97L coupled with S139G significantly broaden the binding pocket for 2-keto-5-methylhexanoate while H97A and N167A were required for even bigger substrates 2-keto-5-methylheptanoate and 2-keto-6-methyloctanoate (Table 20.1). Combination of S139G, H97A, and N167A enabled production of the C8 alcohol 5-methyl-1-heptanol, which is three carbons longer than the natural product of LeuA (Zhang *et al.*, 2008). It is noteworthy that all the nonnatural alcohols with different branch positions were achieved with only mutagenesis on LeuA coupled with wild-type LeuBCD, indicating that the first committed step of 2-ketoacid condensation catalyzed by LeuA is the major bottleneck in the synthetic recursive chain elongation. The initial success of expanding LeuA substrate capacity shown by the production of various straight-chained and branch-chained nonnatural alcohols demonstrates the possibility of iterative ketoacid chain elongation using LeuABCD and the potential repertoire of biosynthetic fuels and chemicals that can be generated with this module.

5. Transfer of Citramalate Pathway to *E. coli* for Ketoacid Chain Elongation

5.1. Directed evolution of 2-ketoacid pathways using amino acid auxotrophs

For pathways that are coupled to essential metabolite synthesis that is required for cell growth, auxotrophic strains of such nutrient can be constructed and used as a host for evolution of specific enzymes along that pathway. This is generally applicable to all 20 amino acid biosynthesis and most of the essential cofactors and vitamins. Excitingly, production of higher alcohols via the ketoacid intermediates found in microorganism's native amino acid pathways has recently shown promising results (Atsumi *et al.*, 2008a). Because 2-ketoacids are both precursors for the alcohols and the branched chain amino acids (Fig. 20.2), this production system can adopt enzyme evolution with growth selection scheme using the specific amino acid auxotrophs based on 2-ketoacid deficiency. A demonstration of such an approach is the directed evolution of citramalate synthase (CimA) from *M. jannaschii* for 1-propanol and 1-butanol production in *E. coli* (Atsumi and Liao, 2008) based on the growth phenotype associated with cell's inability to generate 2-ketobutyrate.

5.2. Evolution of citramalate synthase (CimA) for the elongation of pyruvate to 2-ketobutyrate

With both the threonine deaminase (IlvA) and the catabolic threonine dehydratase (TdcB) deleted from the wild-type *E. coli*, 2-keotbutyrate can no longer be produced from the threonine pathway, which leads to the isoleucine auxotrophic phenotype (Fig. 20.4). Such strain can no longer grow without the supplementation of isoleucine, and the growth can only be rescued if the citramalate pathway catalyzed by the *M. jannaschii* CimA and *E. coli* wild-type LeuBCD is sufficiently active to provide the precursors for 2-ketobutyrate synthesis (Fig. 20.4). The citramalate pathway is the shortest route to make 2-ketobutyrate, which bypasses threonine biosynthesis thus is not limited by the availability of NAD(P)H. In addition, it does not involve transamination and the subsequent deamination steps present in the threonine pathway. Once (*R*)-citramalate is synthesized from pyruvate and acetyl-CoA by CimA, it is converted to 2-ketobutyrate via LeuCD (isopropylmalate isomerase) and LeuB (3-isopropylmalate dehydrogenase) mediated reactions, parallel to the ones found in leucine and norvaline biosynthesis. To improve the expression and catalytic activity of the heterologous CimA derived from the thermophilic microorganism, random mutagenesis of the protein was performed, and potential positive mutants were selected based on the growth rate of the isoleucine auxotroph. The evolved CimA mutants from growth selection exhibited activities over a wide range of temperatures (30–70 °C), and the best variant (I47V/E114V/H126Q/T204A/L238S/V373STOP) had a nearly threefold increase in its catalytic efficiency (k_{cat} improved from 0.36 to 0.84 s^{-1}) along with

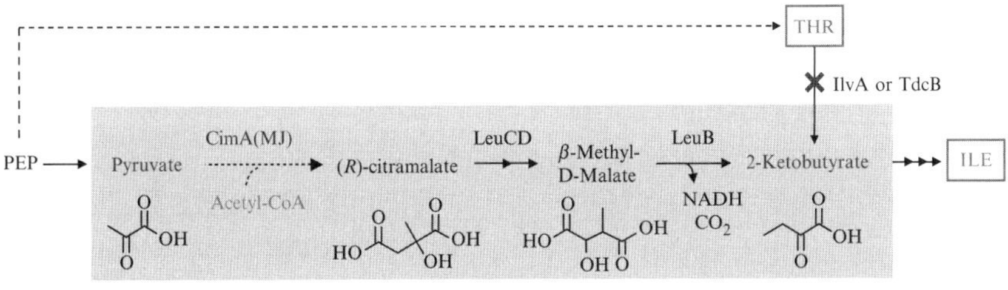

Figure 20.4 Heterologous citramalate pathway (shaded with light blue) bypasses the threonine pathway to make 2-ketobutyrate, an essential isoleucine precursor. When both *ilvA* and *tdcB* are deleted in *Escherichia coli*, the only route available to synthesize isoleucine and enable cell growth becomes the engineered citramalate pathway. The resulting strain BW Δ*ilvA* Δ*tdcB* can be used to evolve the foreign CimA in *E. coli* using growth rescue as the selection system. Citramalate synthase (CimA) from *Methanococcus jannaschii* (MJ) naturally catalyzes the condensation of pyruvate and acetyl-CoA and its activity in *E. coli* was greatly improved upon random mutagenesis and growth selections. (For interpretation of the references to color in this figure legend, the reader is referred to the Web version of this chapter.)

feedback resistance to isoleucine (Atsumi and Liao, 2008). Production of 1-propanol and 1-butanol was increased 9- and 22-fold, respectively, compared to wild-type CimA, reaching a final titer of 3.5 g/L of 1-propanol and 0.5 g/L of 1-butanol in 72 h. The citramalate pathway is another demonstration of ketoacid chain extensions that can be used to replace existing metabolic pathways for the production of essential compounds. This result highlights the plasticity of IPMS-related enzymes. It also illustrates the effectiveness of growth selection platform based on amino acid auxotroph for evolving homologous enzymes along the ketoacid elongation pathway to introduce novel functions or to improve their existing catalytic efficiency.

6. Conclusion Remarks

Ketoacid chain elongations catalyzed by LeuABCD has shown great potential to be utilized as another set of reaction mechanisms for iterative carbon–carbon bond formation. Initial success in expanding the substrate specificity of LeuA by rational design and random mutagenesis has provided us important insights into its ketoacid selectivity and future direction for protein engineering. The high robustness of LeuABCD for 2-ketoacids with a wide range of chain lengths and branch positions could potentially be used for recursive carbon-chain elongation in various industrial applications.

REFERENCES

Anstrom, D. M., Kallio, K., and Remington, S. J. (2003). Structure of the *Escherichia coli* malate synthase G:pyruvate:acetyl-coenzyme A abortive ternary complex at 1.95 A resolution. *Protein Sci.* **12,** 1822–1832.

Atsumi, S., and Liao, J. C. (2008). Directed evolution of *Methanococcus jannaschii* citramalate synthase for biosynthesis of 1-propanol and 1-butanol by *Escherichia coli*. *Appl. Environ. Microbiol.* **74,** 7802–7808.

Atsumi, S., Hanai, T., and Liao, J. C. (2008a). Engineering synthetic non-fermentative pathways for production of branched-chain higher alcohols as biofuels. *Nature* **451,** 86–89.

Atsumi, S., Cann, A. F., Connor, M., Shen, C. R., Smith, K. M., Brynildsen, M. P., Chou, K. J., Hanai, T., and Liao, J. C. (2008b). Metabolic engineering of *Escherichia coli* for 1-butanol production. *Metab. Eng.* **10,** 305–311.

Barkovich, R., and Liao, J. C. (2001). Metabolic engineering of isoprenoids. *Metab. Eng.* **3,** 27–39.

Cann, A. F., and Liao, J. C. (2008). Production of 2-methyl-1-butanol in engineered *Escherichia coli*. *Appl. Microbiol. Biotechnol.* **81,** 89–98.

Connor, M. R., and Liao, J. C. (2008). Engineering *Escherichia coli* for the production of 3-methyl-1-butanol. *Appl. Environ. Microbiol.* **74,** 5769–5775.

Connor, M. R., and Liao, J. C. (2009). Microbial production of advanced transportation fuels in non-natural hosts. *Curr. Opin. Biotechnol.* **20,** 307–315.
Farmer, W. R., and Liao, J. C. (2000). Improving lycopene production in *Escherichia coli* by engineering metabolic control. *Nat. Biotechnol.* **18,** 533–537.
Farmer, W. R., and Liao, J. C. (2001). Precursor balancing for metabolic engineering of lycopene production in *Escherichia coli. Biotechnol. Prog.* **17,** 57–61.
Fung, E., Wong, W. W., Suen, J. K., Bulter, T., Lee, S. G., and Liao, J. C. (2005). A synthetic gene-metabolic oscillator. *Nature* **435,** 118–122.
Gusyatiner, M. M., Lunts, M. G., Kozlov, Y. I., Ivanovskaya, L. V., and Voroshilova, E. B. (2002). DNA coding for mutant isopropylmalate synthase L-leucine producing microorganism and method for producing L-leucine. *US Patent 6403342.*
Hopwood, D. A., and Sherman, D. H. (1990). Molecular genetics of polyketide and its comparison to fatty acid biosynthesis. *Annu. Rev. Genet.* **24,** 37–66.
Howard, B. R., Endrizzi, J. A., and Remington, S. J. (2000). Crystal structure of *Escherichia coli* malate synthase G complexed with magnesium and glyoxylate at 2.0 A resolution: Mechanistic implications. *Biochem. J.* **39,** 3156–3168.
Katz, L., and Donadio, S. (1993). Polyketide synthesis: Prospects for hybrid antibiotics. *Annu. Rev. Microbiol.* **47,** 875–912.
Kisumi, M., Sugiura, M., Kato, J., and Chibata, I. (1976). L-Norvaline and L-homoisoleucine formation by *Serratia marcescens. J. Biochem.* **79,** 1021–1028.
Koon, N., Squire, C. J., and Baker, E. N. (2004). Crystal structure of LeuA from *Mycobacterium tuberculosis*, a key enzyme in leucine biosynthesis. *Proc. Natl. Acad. Sci. USA* **101,** 8295–8300.
Li, H., Cann, C. F., and Liao, J. C. (2010). Biofuels: Biomolecular engineering fundamentals and advances. *Annu. Rev. Chem. Biomol. Eng.* **1,** 19–36.
Ma, J., Zhang, P., Zhang, Z., Zha, M., Xu, H., Zhao, G., and Ding, J. (2008). Molecular basis of the substrate specificity and the catalytic mechanism of citramalate synthase from *Leptospira interrogans. Biochem. J.* **415,** 45–56.
McCourt, J. A., and Duggleby, R. G. (2006). Acetohydroxyacid synthase and its role in the biosynthetic pathway for branched-chain amino acids. *Amino Acids* **31,** 173–210.
Ro, D. K., Paradise, E. M., Ouellet, M., Fisher, K. J., Newman, K. L., Ndungu, J. M., Ho, K. A., Eachus, R. A., Ham, T. S., Kirby, J., Chang, M. C. Y., Whiters, S. T., *et al.* (2006). Production of the antimalarial drug precursor artemisinic acid in engineered yeast. *Nature* **440,** 940–943.
Rohlin, L., Oh, M. K., and Liao, J. C. (2001). Microbial pathway engineering for industrial processes: Evolution, combinatorial biosynthesis and rational design. *Curr. Opin. Microbiol.* **4,** 330–335.
Shen, C. R., and Liao, J. C. (2008). Metabolic engineering of *Escherichia coli* for 1-butanol and 1-propanol production via the keto-acid pathways. *Metab. Eng.* **10,** 312–320.
Staunton, J., and Weissman, K. J. (2001). Polyketide biosynthesis: A millennium review. *Nat. Prod. Rep.* **18,** 380–416.
Umbarger, H. E. (1978). Amino acid biosynthesis and its regulation. *Annu. Rev. Biochem.* **47,** 532–606.
Voigt, C. A. (2006). Genetic parts to program bacteria. *Curr. Opin. Biotechnol.* **17,** 548–557.
Yan, Y., Kohli, A., and Koffas, M. A. G. (2005). Biosynthesis of natural flavanones in *Saccharomyces cerevisiae. Appl. Environ. Microbiol.* **71,** 5610–5613.
Zhang, K., Sawaya, M. R., Eisenberg, D. S., and Liao, J. C. (2008). Expanding metabolism for biosynthesis of nonnatural alcohols. *Proc. Natl. Acad. Sci. USA* **105,** 20653–20658.

SECTION FIVE

EXPANDING CHASSIS

CHAPTER TWENTY-ONE

Synthetic Biology in *Streptomyces* Bacteria

Marnix H. Medema,*,† Rainer Breitling,†,‡ *and* Eriko Takano*

Contents

Abstract

Actinomycete bacteria of the genus *Streptomyces* are major producers of bioactive compounds for the biotechnology industry. They are the source of most clinically used antibiotics, as well as of several widely used drugs against common diseases, including cancer . Genome sequencing has revealed that the potential of *Streptomyces* species for the production of valuable secondary metabolites is even larger than previously realized. Accessing this rich genomic

* Department of Microbial Physiology, Groningen Biomolecular Sciences and Biotechnology Institute, University of Groningen, Groningen, The Netherlands
† Groningen Bioinformatics Centre, Groningen Biomolecular Sciences and Biotechnology Institute, University of Groningen, Groningen, The Netherlands
‡ Institute of Molecular, Cell and Systems Biology, College of Medical, Veterinary and Life Sciences, Joseph Black Building, University of Glasgow, Glasgow, United Kingdom

Methods in Enzymology, Volume 497
ISSN 0076-6879, DOI: 10.1016/B978-0-12-385075-1.00021-4

resource to discover new compounds by activating "cryptic" pathways is an interesting challenge for synthetic biology. This approach is facilitated by the inherent natural modularity of secondary metabolite biosynthetic pathways, at the level of individual enzymes (such as modular polyketide synthases), but also of gene cassettes/operons and entire biosynthetic gene clusters. It also benefits from a long tradition of molecular biology in *Streptomyces*, which provides a number of specific tools, ranging from cloning vectors to inducible promoters and translational control elements. In this chapter, we first provide an overview of the synthetic biology challenges in *Streptomyces* and then present the existing toolbox of molecular methods that can be employed in this organism.

1. Synthetic Biology for Novel Compound Discovery in *Streptomyces*

The rapidly decreasing costs of genome sequencing have made genome mining the most promising source of raw material for drug discovery: a great number of putative secondary metabolite biosynthesis pathways have been discovered *in silico* (Medema *et al.*, 2010). In genome-sequenced *Streptomyces* species (Dyson, 2011; Hopwood, 2007), the number of gene clusters encoding pathways for secondary metabolite biosynthesis has been found to range from about 20 to 50 per genome (Bentley *et al.*, 2002; Ikeda *et al.*, 2003; Medema *et al.*, 2010; Ohnishi *et al.*, 2008; Wang *et al.*, 2010). Intriguingly, a large number of these pathways are cryptic: they are not expressed under standard laboratory conditions, and their products are therefore unknown. Exciting proof-of-principle successes have already been achieved in awakening cryptic secondary metabolites. Untargeted approaches have been used to randomly awaken some clusters (Tala *et al.*, 2009), and recently a targeted approach has resulted in the specific awakening of a cryptic gene cluster in *Streptomyces coelicolor* by inactivating a pathway-specific repressor within the cluster (Gottelt *et al.*, 2010). However, a key limitation to these approaches is the difficulty of implementing a high-throughput screening of the hundreds of cryptic gene clusters that are available in the databases. Synthetic biology may offer the possibility of achieving the large-scale reengineering of the native regulation of these clusters, as well as the directed modification of their metabolic network context that is necessary for successful novel compound identification at a rapid rate (Medema *et al.*, 2011).

One advantage of a synthetic biology approach would be the option to completely remove the native regulation of the cryptic clusters, replacing it by a completely synthetic regulation that is predictable, easy to manipulate, as specifically tuned to the function of the pathway. To achieve this, the cryptic gene clusters need to be completely redesigned *in silico* in terms of promoters, composition of the transcriptional units, ribosome-binding sites (RBSs), and possibly even their codon usage (Bayer *et al.*, 2009; Salis *et al.*, 2009; Widmaier *et al.*, 2009). *Streptomyces* biology is in a good position for achieving this ambitious aim, given the existing body of research into regulatory mechanisms (Martin and Liras, 2010), as will be further discussed in the methods section below.

As the gene clusters encoding secondary metabolite biosynthetic pathways of many compound classes are often very similar to one another and largely consist of the same biosynthetic modules combined in different ways (Fischbach *et al.*, 2008), a synthetic version of one optimized model gene cluster could serve as a chassis for an entire class of cryptic gene clusters, which could be inserted into it in their entirety instead of being modified piece-by-piece. This would enable the rapid and versatile activation of new biosynthetic pathways from a variety of sources. It has the advantage that it achieves high throughput, while at the same time being focused on evolutionarily optimized designs present in naturally occurring cryptic gene clusters. As these have been selected by evolution to exhibit a specific and stable bioactivity, the probability of finding novel antibacterial activities in the resulting compound libraries is much higher than in those produced by more random approaches (Zerikly and Challis, 2009).

This approach can be complemented by other forms of (semi-)synthetic biology in *Streptomyces*, which aim at exploiting the highly modular assembly line mechanism of secondary metabolite biosynthetic pathways (Fischbach and Walsh, 2006) to create new chemical diversity from well-characterized gene clusters (Gokhale *et al.*, 1999; Khosla and Zawada, 1996; Menzella *et al.*, 2005, 2007) or to modify the chemistry of the end product for increased efficacy or novel functionality (Baltz, 2008; Caffrey *et al.*, 2008; Donadio and Sosio, 2008; Heide *et al.*, 2008). These "combinatorial biosynthesis" and "biosynthetic engineering" methods can be used quite independently from the ambitious redesign (and synthesis) of entire gene clusters described in this chapter. They have been reviewed before (Luzhetska *et al.*, 2010; Menzella and Reeves, 2007; Walsh, 2002; Zhang and Tang, 2008) and are not further discussed here.

A sensible target for pioneering this approach is a class of antibiotic compounds called polyketides that are synthesized by so-called type II polyketide synthases (PKSs) (Hertweck *et al.*, 2007). Among the polyketides produced by type II PKSs are chemically very diverse bioactive compounds, including tetracycline antibiotics, anthracycline chemotherapeutics (e.g., daunomycin and doxorubicin), angucyclines with a wide range of antibiotic and antitumor activities (e.g., landomycin), and the benzoisochromanequinones, which include the widely studied antibiotic actinorhodin from *S. coelicolor* (Fig. 21.1) (Hertweck *et al.*, 2007).

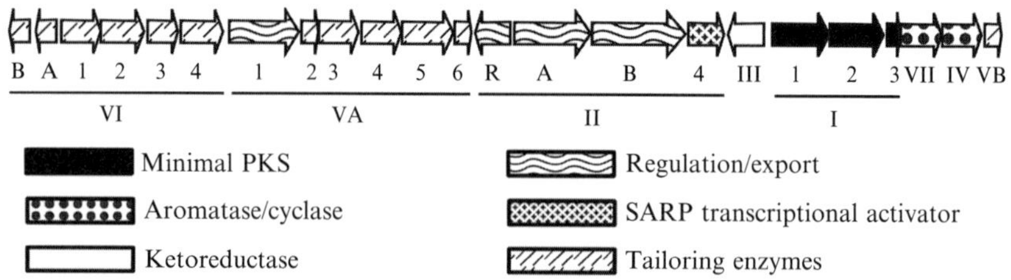

Figure 21.1 The actinorhodin gene cluster, an example of a well-studied type II PKS polyketide biosynthetic gene cluster. Genes are annotated using different patterns as indicated.

The exclusively bacterial type II PKSs are single-domain proteins that form a complex that acts in an iterative fashion to produce the polyketide scaffold that can afterward be modified by a variety of accessory enzymes (Hertweck *et al.*, 2007). The gene clusters containing type II PKSs are smaller than most other secondary metabolite biosynthetic clusters, making a synthetic approach as outlined above particularly feasible. Moreover, the genetic parts of type II PKS systems, including the post-PKS tailoring reactions, have been shown to be generally interchangeable (Ichinose *et al.*, 2001) and combinable (Hopwood *et al.*, 1985; Khosla and Zawada, 1996; McDaniel *et al.*, 1995) to produce functional compounds. Finally, more than one hundred cryptic gene clusters of this type are currently present in the databases, and this number is still increasing rapidly. Consequently, these gene clusters offer great potential for drug discovery.

2. Practical Considerations for Synthetic Biology in *Streptomyces*

2.1. Choice of host organism

Certain biosynthetic enzymes, among which bacterial type II PKSs, are not functional in *Escherichia coli* (Zhang *et al.*, 2008), while the metabolism of *Streptomyces* bacteria is already optimized for the production of secondary metabolites at high rates. Therefore, the host organism of choice for most synthetic biology projects on genome-based drug discovery would be a *Streptomyces* species, such as the model streptomycete *S. coelicolor*, or a closely related actinomycete bacterium used in biotechnological applications. Useful host strains for the heterologous expression of different types of secondary metabolite gene clusters have been reviewed in detail by Baltz (2010).

Starting from the model organism *S. coelicolor* not only has the advantage of being able to use a wide range of existing molecular tools, but additionally a strain of this species is available in which all four highly active ("non-cryptic") antibiotic biosynthesis gene clusters (actinorhodin: *act*; undecylprodigiosin: *red*; calcium dependent antibiotic: *cda*; coelicolorpolyketide: *cpk*) have been deleted from the chromosome, freeing metabolic resources for the synthetic pathways that are inserted (Gottelt *et al.*, 2010; Gomez-Escribano *et al.*, 2011). Alternatively, comprehensively genome-minimized strains would be interesting hosts, such as the recently published genome-minimized *Streptomyces avermitilis* strains (Komatsu *et al.*, 2010) or a plasmid-cured strain of *Streptomyces clavuligerus*, as was recently suggested (Medema *et al.*, 2010).

3. Iterative Reengineering of Secondary Metabolite Gene Clusters

As our understanding of gene regulation in *Streptomyces* is not yet detailed enough to perfectly predict the functioning of all components in an integrated pathway, synthesizing and inserting a whole gene cluster at once is very likely to result in problems that cannot easily be traced to a particular gene. It is, therefore, more promising to use an iterative strategy in which the target gene cluster is subdivided into independent transcriptional units that are individually optimized. After the first design and synthesis of a transcriptional unit, it can be tested for complementation of a deletion mutant of the same genes in the native pathway, by inserting it into the *S. coelicolor* chromosome at the phiC31 or phiBT1 sites (see below). If this step is unsuccessful, the problem can be traced and the design adapted until a successful complementation of pathway functionality is achieved.

Initial proof-of-concept studies can focus on pigmented compounds, such as the actinorhodin antibiotic of *S. coelicolor*. In that way, the "debugging" or "troubleshooting" process can be aided by photospectrometry to rapidly identify eventual blocks in the biosynthetic pathway, as many intermediates or shunt products produced by knock-out mutants of tailoring genes show absorption spectra distinct from the end product. For instance, in the case of actinorhodin, which has a dark blue color (Fig. 21.2), *actVI-orf1* or *ActVI-orf2* mutants produce a brown pigment (Taguchi *et al.*, 2000), *actVI-orf3* mutants produce a reddish pigment (Taguchi *et al.*, 2000), and *actVA-orf5,6* mutants produce a yellowish brown pigment (Okamoto *et al.*, 2009). A second blue pigment, gamma-actinorhodin, is also known to be produced by the same gene cluster (Bystrykh *et al.*, 1996).

When photospectrometry is not informative, as will be the case for most bioactive compounds targeted in high-throughput genome mining approaches, it will still be possible to predict the structures of most pathway intermediates based on genome annotation (Aoki-Kinoshita and Kanehisa,

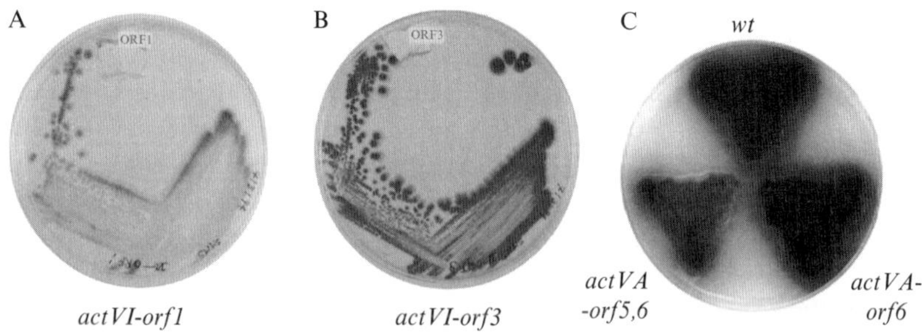

Figure 21.2 (A) Brown pigment produced by *actVI-orf1* mutant (K. Ichinose, personal communication). (B) Red pigment produced by *actVI-orf3* mutant (K. Ichinose personal communication). (C) Yellowish brown pigment produced by the *actVA-orf5,6* mutant compared to the blue actinorhodin produced by wt and the *actVA-orf6* mutant (taken from Okamoto *et al.*, 2009 with permission from Elsevier Limited). (See Color Insert.)

2009). In these cases, high-accuracy liquid chromatography mass-spectrometry can be a promising tool for identifying metabolic signatures that characterize bottlenecks at specific steps in the biosynthetic pathway (Kol *et al.*, 2010).

3.1. Transcriptional control engineering in *Streptomyces*

While the individual transcriptional units are optimized independently, the synthetic operons can all be controlled by thiostrepton-inducible *tipA* promoters (Takano *et al.*, 1995) (see below), and phage fd bidirectional terminators which are functional both in *E. coli* and *Streptomyces* can be used for transcription termination (Ward *et al.*, 1986) (see below). However, once the complete synthetic gene cluster is assembled, one would most likely need to control the timing and expression rate of some operons separately. For instance, in our example of type II PKS engineering, we might want to control the transcriptional unit encoding the core PKS proteins independently of those encoding the tailoring steps. This would especially be advantageous when genes from cryptic pathways will be inserted in these units later on, which may well need a different mRNA expression stoichiometry compared to the initial model cluster. In this case, at least two promoters with different timing and strength would be required. The use of inducible promoters would have the advantage that one can start at low induction rates to avoid build-up of toxic intermediates, and increase induction later.

Concomitant expression of the cluster-specific secondary metabolite transporters will also be required for toxic compounds. In the wild-type gene cluster, expression of the transporter genes is often governed by an intricate system: in our example of the actinorhodin gene cluster, repression of actinorhodin transporter expression by ActR is abolished by binding of ActR to intermediates in the biosynthetic pathway. In this way the transporters will be produced just in time to avoid bacterial suicide (Tahlan *et al.*, 2007, 2008; Willems *et al.*, 2008). It is expected that by simultaneously expressing the tailoring genes and the transporters, toxic effects to the cell can be avoided in a similar fashion. Yet, if toxicity problems arise due to lack of transport capacity, it can also be appropriate to insert a strong constitutive promoter, such as *ermE* (see below), in front of the transporter genes.

3.2. Translational control engineering in *Streptomyces*

As the translational efficiency of redesigned synthetic genes will be different from those of the wild-type genes if the wild-type RBSs are used—RBS functionality is context-dependent (Salis *et al.*, 2009) and codon usage also affects translational efficiency—new synthetic RBSs have to be designed to restore the wild-type stoichiometry of the enzymes. This requires an accurate estimate of the relative wild-type translation rates in each operon. This can be obtained, for example, by fusing the relevant RBS-containing sequence of the wild-type gene to a GFP or RFP reporter in a high-copy number plasmid, such as pTONA5 (Hatanaka *et al.*, 2008) or pIJ8630 (de Jong *et al.*, 2009), with a constitutive promoter in front of it; screening for activity is then easily done by measuring the resulting fluorescence. The necessary synthetic RBSs can be identified in the same way, using a library of *Streptomyces* RBSs in the context of each synthetic gene based on oligonucleotides randomized around known RBS sequences. Subsequently, the translation rates of all proteins in the synthetic cluster can be balanced by inserting RBSs from this library that closely match the wild-type RBS strength. Using this methodology, RBSs can even be constructed to match translation efficiencies of those unconventional genes in *Streptomyces* that do not have RBSs/UTRs and for which translation starts at the far 5′ end of the transcript at the transcription start site itself (Fernandez-Moreno *et al.*, 1994; Strohl, 1992).

4. The Molecular Toolbox for *Streptomyces* Synthetic Biology

Fifty years of genetics and molecular biology in *Streptomyces* have yielded a large and versatile collection of molecular tools, many of which will be useful for synthetic biology applications. The list below provides concise descriptions of some important sets of tools, including the various components discussed in the text above. For detailed protocols on the state-of-art of *Streptomyces* molecular biology and secondary metabolite biosynthesis see the books by Kieser *et al.* (2000), Dyson (2011), and Hopwood (2009a,b).

5. Transcriptional Control

The transcriptional control of gene expression in *Streptomyces* is different from most organisms. Consequently, commercially available tools and kits cannot be utilized readily. Particularly, the high genomic GC content of streptomycetes and the presence of multiple unusual enzymes hinder their use. For example, *lacZ* promoters cannot be used efficiently for screening, although a modified *lacZ*-based system was developed and used for some time (King and Chater, 1986). The reason for the incompatibility is that there are multiple endogenous beta-galactosidase homologues (five in total) encoded in the genome of *S. coelicolor*. Furthermore, IPTG was not transported into the cell; instead, methylumbelliferyl B galactoside had to be used (King and Chater, 1986).

Recently a modified T7 expression system in *Streptomyces* has been reported. However, this system requires a T7 expression strain harbouring the T7 RNA polymerase under the control of a *tipA* promoter (Lussier *et al.*, 2010) (see below). This expression strain is not favorable for use due to the T7 polymerase induction using thiostrepton, but a derivative may be useful in the future. Several—inducible and constitutive—promoters are available and some are listed below.

5.1. Inducible promoters

tipA: induced by thiostrepton. Most commonly used promoter. Very strong promoter, induced with minimum of 5 μg/ml of thiostrepton (Takano *et al.*, 1995). It has several disadvantages: (1) Thiostrepton is only soluble in DMSO, which can result in inaccurate concentration of thiostrepton in the medium, but thiostrepton can potentially be replaced by water soluble derivates (Schoof *et al.*, 2010). (2) The repression of the promoter is weak and it can therefore be expressed without any induction (due to the induction mechanism of the *tipA* protein; Chiu *et al.*, 1999). The strain used needs to contain a thiostrepton resistance gene.

cpkO: Induced by *Streptomyces coelicolor* gamma-butyrolactones (SCBs). Tightly controlled and relatively strong promoter. Promoter activity responds well to the inducer concentration (Fig. 21.3). Only nanomolar concentrations of inducer needed for induction. Inducer diffuses into the cell and is non-toxic. Hosts mutated in the inducer biosynthesis pathway are available. No need to introduce extra resistant gene cassettes (Takano *et al.*, 2005).

tetR: Induced by tetracycline (Tc) and anhydrotetracycline (aTc). Promoter was synthesized by combining the −10 and −35 regions of the strong *ermEp1* promoter (Bibb *et al.*, 1985) with the *tetO1* and *tetO2* operator sequences from the *E. coli* transposon Tn*10*. Promoter activity

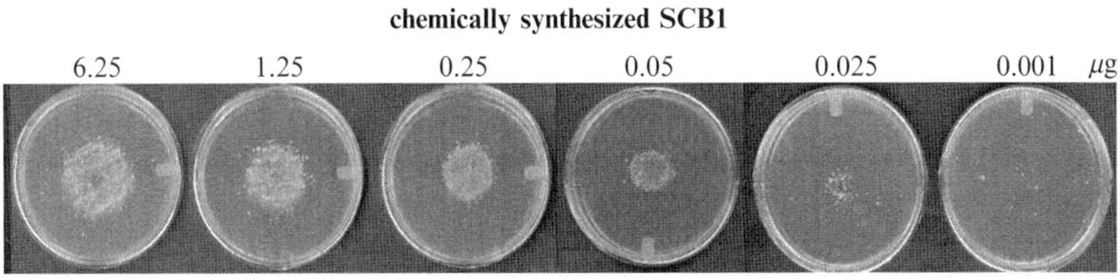

Figure 21.3 The strength of the *cpkO* promoter shown by resistance to kanamycin resulting in the growth of the indicator strain (adapted from Hsiao *et al.*, 2009 with permission from Elsevier Limited).

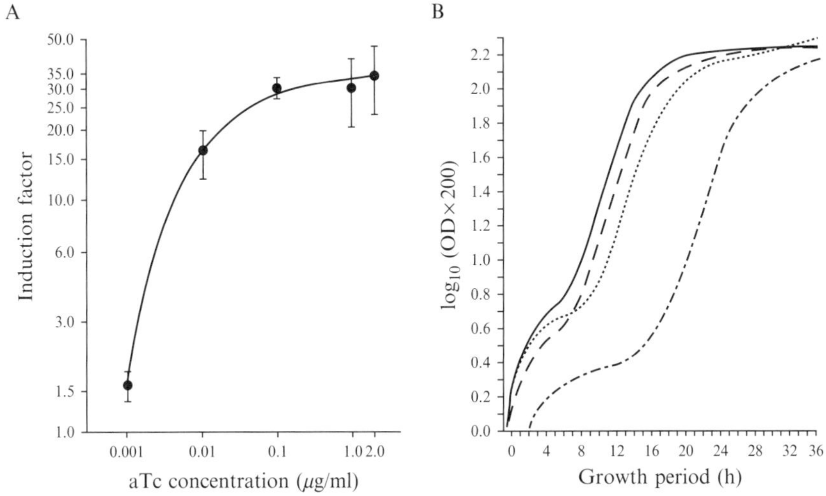

Figure 21.4 (A) Increase of promoter strength in relation to the concentration of aTc. (B) *Streptomyces* growth in the presence of different amounts of inducer. Inducer was added at $OD_{492} = 0.04$. Solid line, cultures without inducer; long dashed line, 1 μg/ml aTc; dotted line, 1 μg/ml Tc; dot and dashed line 1 μg/ml doxycyline (adapted from Rodríguez-García *et al.* (2005), with permission from Oxford University Press).

has a good response to the concentration of the inducer (Fig. 21.4). aTc is not toxic to the cell (Dangel *et al.*, 2010; Rodriguez-Garcia *et al.*, 2005).

5.2. Constitutive promoters

*ermE**: The promoter of the erythromycin resistance gene (*ermE*) which is reported to be constitutive. Commonly used promoter for strong expression. One base pair mutation was introduced which enhanced promoter activity (Bibb *et al.*, 1985).

5.3. Terminators

Terminators from other organisms are used often in *Streptomyces* and work efficiently. The following are the most commonly used:

Fd: the major terminator from *E. coli* phage fd, bidirectional (Ward *et al.*, 1986), and
lambda phage T0 terminator (Scholtissek and Grosse, 1987).

6. Translational Control

6.1. Positive translational control

Not much engineering of RBSs has been done in *Streptomyces* compared to the extensive redesigning employed in *E. coli* (Salis *et al.*, 2009). However, RBSs from proteins that are known to be highly expressed, for example, ribosomal proteins, are used (Takano *et al.*, 1995). A synthetic RBS with the sequence AAGGAGG has also been implemented successfully (Horinouchi *et al.*, 1987).

6.2. Negative translational control

Antisense RNA (asRNA) is a recently developed addition to the *Streptomyces* regulatory toolbox. Applications use 50–100 bp of sequence antisense to coding sequence of the gene which needs to be regulated. The antisense sequence can be cloned into a multicopy plasmid under the control of an inducible promoter, for example, *tipA*, for regulated repression (D'Alia *et al.*, 2010).

7. Vectors

The most widely used vectors in *Streptomyces* molecular biology are self-replicating plasmids, but integrating vectors are also available and can be particularly suitable for synthetic biology applications.

The major advantage of the integrating vectors is their stability, especially when introducing large inserts. There are two attachment sites that can be used for the integration, either the phiC31 or the phiBT1 attachment site. These sites have been shown to be at different locations of the chromosome, phiC31 inside SCO3798 (putative chromosome condensation protein) and phiBT1 inside SCO4848 (putative membrane protein) (Combes *et al.*, 2002; Gregory *et al.*, 2003). This is convenient when it is desired to introduce multiple genetic constructs by integration into *Streptomyces*. Another important technique is the introduction of plasmids via conjugation. Earlier, chemical transformation using protoplasts was the main method of introducing DNA. However, the transformation frequency was often very low (hindering cloning of very large inserts), and the process often encouraged recombination within the chromosome causing unwanted mutations to the host.

The conjugation method developed by Flett *et al.* (1997) allows easy cloning in *E. coli*. Finished constructs can then be transferred in a *dam−/dcm− E. coli* strain (i.e., ET12567) that lacks a restriction modification system and carries a plasmid, pUZ8002, for conjugal transfer to *Streptomyces* by just mixing the *E. coli* with spores of *Streptomyces*. For cloning of very large fragments, especially the antibiotic biosynthetic clusters, which are often more than 100 kb in length, cosmids or artificial bacterial chromosomes (BACs) are used. These large vectors are nowadays all introduced via conjugation. Some vectors introduced to *Streptomyces* by both methods are listed below.

7.1. Low copy number vectors

Self-replicating vectors: Derivatives of SCP2*, the low copy number vector isolated from *S. coelicolor*. Inserts can be very large (> 10 kb), for example, pRM5, an *E. coli* bifunctional vector, and can be introduced into *Streptomyces* by conjugation from *E. coli* (McDaniel *et al.*, 1993).

Integrative vectors: In theory, the insertion of the plasmid is thought to be as a single copy; however, often multiple copies can integrate into the same site (Takano *et al.*, unpublished observations). To reduce multiple integration, the conjugation protocol has been revised so that the *E. coli* is grown with the *Streptomyces* for only 6–8 h.

The following integrative vectors are used for medium insert sizes (< 8 kb) (Fig. 21.5).

pSET152: Integrates into phiC31 attachment site (*attP*), apramycin resistant (aac(3)IV), has *E. coli ori* (Bierman *et al.*, 1992). GenBank: AJ414670.

pMS82: Integrates into phiBT1 attachment site (*attP*), uses hygromycin resistance (hyg) (Gregory *et al.*, 2003).

pIJ10257: Integrates into phiBT1 attachment site (*attP*), constitutive *ermE** promoter in front of the cloning site (Hong *et al.*, 2005). Based on pMS82 (Gregory *et al.*, 2003).

pIJ6902: Integrates into phiC31 attachment site, apramycin resistant (aac(3)IV) and thiostrepton resistant (thio), *tipA* inducible promoter in front of

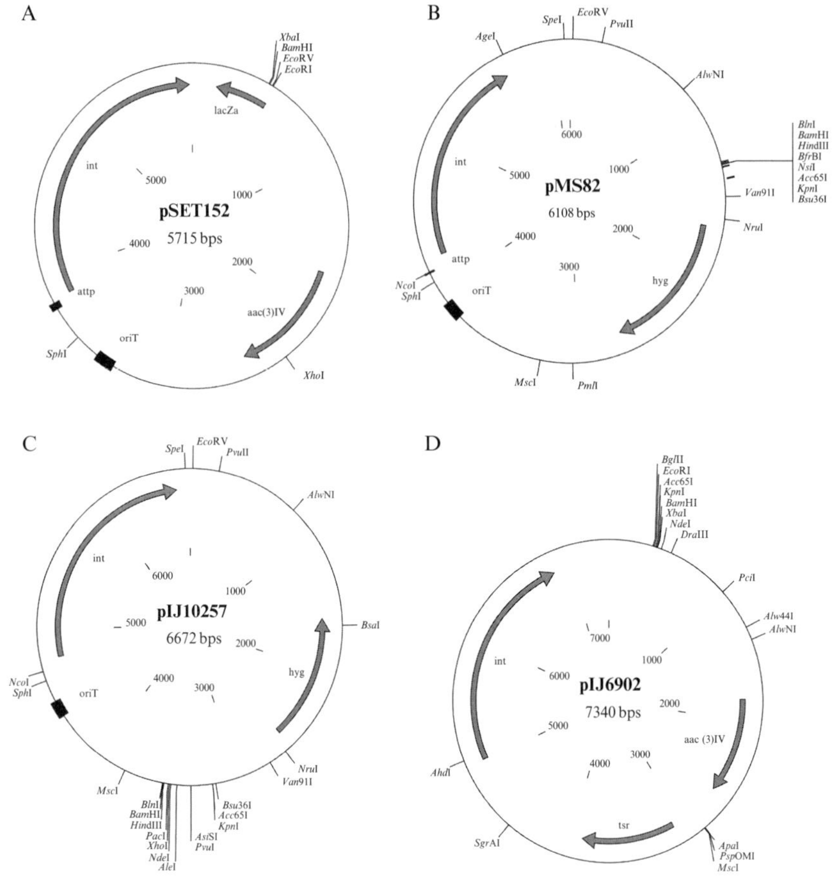

Figure 21.5 Map of (A) pSET152 (GenBank: AJ414670), (B) pMS82 (M. C. Smith personal communication), (C) pIJ10257 (H.J. Hong personal communication), and (D) pIJ6902 (GenBank: AJ937361).

cloning site flanked with *to/fd* termintors (Huang *et al.*, 2005). GenBank: AJ937361.

The following vectors are suitable for large insertion sizes (< 100 kb) (Fig. 21.6):

BACs: pSBAC, integrates into phiBT1 attachment site (*attP-int*). Has the *ori2* to replicate in *E. coli* in a single copy to maintain stability of the large insert and the copy number can be induced by L-arabinose for isolation of DNA. This plasmid has been used to clone the 90 kb meridamycin biosynthetic gene cluster (Liu *et al.*, 2009).

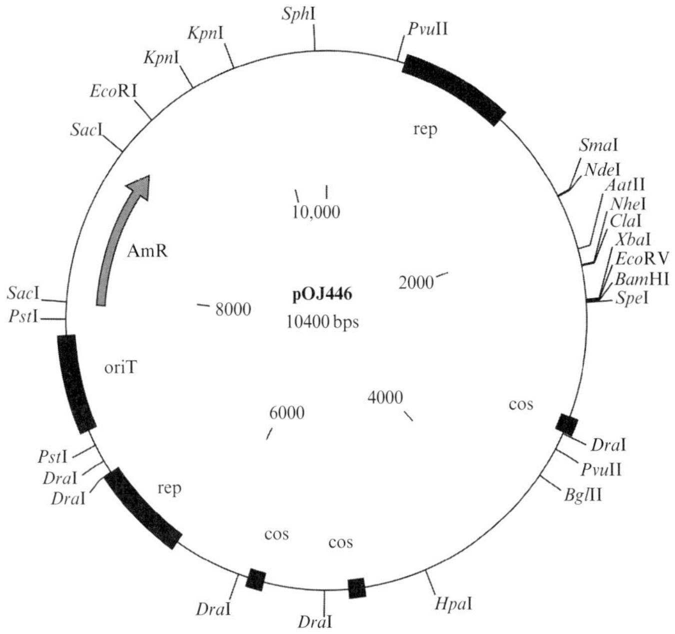

Figure 21.6 Map of pOJ446 (adapted from Bierman *et al.*, 1992).

Cosmids: pOJ446, can be used in both *E. coli* and *Streptomyces* (Bierman *et al.*, 1992). There are many examples of cloning large fragments of 40–70 kb (Kharel *et al.*, 2010).

pKC505, a useful cosmid-based shuttle vector for cloning of large constructs (Richardson *et al.*, 1987), which can be conjugally transferred between Streptomyces strains.

ACKNOWLEDGMENTS

We thank D. Hopwood for critical reading of a draft of the manuscript and K. Ichinose, M.C.M. Smith, and H.J. Hong for providing figures and sequence information. This work was supported by the Dutch Technology Foundation STW, which is the applied science division of NWO, and the Technology Programme of the Ministry of Economic Affairs [STW 10463]. R. B. is supported by an NWO-Vidi fellowship, and E. T. by a Rosalind Franklin Fellowship, University of Groningen.

REFERENCES

Aoki-Kinoshita, K. F., and Kanehisa, M. (2009). Using KEGG in the transition from genomics to chemical genomics. *In* "Bioinformatics for Systems Biology," (S. Krawetz, ed.), pp. 437–452. Humana Press, Heidelberg.

Baltz, R. H. (2008). Biosynthesis and genetic engineering of lipopeptide antibiotics related to daptomycin. *Curr. Top. Med. Chem.* **8,** 618–638.

Baltz, R. H. (2010). *Streptomyces* and *Saccharopolyspora* hosts for heterologous expression of secondary metabolite gene clusters. *J. Ind. Microbiol. Biotechnol.* **37,** 759–772.

Bayer, T. S., Widmaier, D. M., Temme, K., Mirsky, E. A., Santi, D. V., and Voigt, C. A. (2009). Synthesis of methyl halides from biomass using engineered microbes. *J. Am. Chem. Soc.* **131,** 6508–6515.

Bentley, S. D., Chater, K. F., Cerdeno-Tarraga, A. M., Challis, G. L., Thomson, N. R., James, K. D., Harris, D. E., Quail, M. A., Kieser, H., and Harper, D. (2002). Complete genome sequence of the model actinomycete *Streptomyces coelicolor* A3 (2). *Nature* **417,** 141–147.

Bibb, M. J., Janssen, G. R., and Ward, J. M. (1985). Cloning and analysis of the promoter region of the erythromycin resistance gene (*ermE*) of *Streptomyces erythraeus*. *Gene* **38,** 215–226.

Bierman, M., Logan, R., O'Brien, K., Seno, E. T., Rao, R. N., and Schoner, B. E. (1992). Plasmid cloning vectors for the conjugal transfer of DNA from *Escherichia coli* to *Streptomyces* spp.. *Gene* **116,** 43–49.

Bystrykh, L. V., Fernandez-Moreno, M. A., Herrema, J. K., Malpartida, F., Hopwood, D. A., and Dijkhuizen, L. (1996). Production of actinorhodin-related "blue pigments" by *Streptomyces coelicolor* A3(2). *J. Bacteriol.* **178,** 2238–2244.

Caffrey, P., Aparicio, J. F., Malpartida, F., and Zotchev, S. B. (2008). Biosynthetic engineering of polyene macrolides towards generation of improved antifungal and antiparasitic agents. *Curr. Top. Med. Chem.* **8,** 639–653.

Chiu, M. L., Folcher, M., Katoh, T., Puglia, A. M., Vohradsky, J., Yun, B. S., Seto, H., and Thompson, C. J. (1999). Broad spectrum thiopeptide recognition specificity of the *Streptomyces lividans* TipAL protein and its role in regulating gene expression. *J. Biol. Chem.* **274,** 20578–20586.

Combes, P., Till, R., Bee, S., and Smith, M. C. (2002). The *Streptomyces* genome contains multiple pseudo-*attB* sites for the (phi)C31-encoded site-specific recombination system. *J. Bacteriol.* **184,** 5746–5752.

D'Alia, D., Nieselt, K., Steigele, S., Muller, J., Verburg, I., and Takano, E. (2010). Noncoding RNA of glutamine synthetase I modulates antibiotic production in *Streptomyces coelicolor* A3(2). *J. Bacteriol.* **192,** 1160–1164.

Dangel, V., Westrich, L., Smith, M. C., Heide, L., and Gust, B. (2010). Use of an inducible promoter for antibiotic production in a heterologous host. *Appl. Microbiol. Biotechnol.* **87,** 261–269.

de Jong, W., Manteca, A., Sanchez, J., Bucca, G., Smith, C. P., Dijkhuizen, L., Claessen, D., and Wosten, H. A. (2009). NepA is a structural cell wall protein involved in maintenance of spore dormancy in *Streptomyces coelicolor*. *Mol. Microbiol.* **71,** 1591–1603.

Donadio, S., and Sosio, M. (2008). Biosynthesis of glycopeptides: Prospects for improved antibacterials. *Curr. Top. Med. Chem.* **8,** 654–666.

Dyson, P, ed. (2011). *Streptomyces*: Molecular Biology and Biotechnology. Caister Academic Press, Norwich.

Fernandez-Moreno, M. A., Martinez, E., Caballero, J. L., Ichinose, K., Hopwood, D. A., and Malpartida, F. (1994). DNA sequence and functions of the *actVI* region of the actinorhodin biosynthetic gene cluster of *Streptomyces coelicolor* A3(2). *J. Biol. Chem.* **269,** 24854–24863.

Fischbach, M. A., and Walsh, C. T. (2006). Assembly-line enzymology for polyketide and nonribosomal peptide antibiotics: Logic, machinery, and mechanisms. *Chem. Rev.* **106,** 3468–3496.

Fischbach, M. A., Walsh, C. T., and Clardy, J. (2008). The evolution of gene collectives: How natural selection drives chemical innovation. *Proc. Natl. Acad. Sci. USA* **105,** 4601–4608.

Flett, F., Mersinias, V., and Smith, C. P. (1997). High efficiency intergeneric conjugal transfer of plasmid DNA from *Escherichia coli* to methyl DNA-restricting streptomycetes. *FEMS Microbiol. Lett.* **155,** 223–229.

Gokhale, R. S., Tsuji, S. Y., Cane, D. E., and Khosla, C. (1999). Dissecting and exploiting intermodular communication in polyketide synthases. *Science* **284,** 482–485.

Gomez-Escribano, J. P., and Bibb, M. J. (2011). Engineering *Streptomyces coelicolor* for heterologous expression of secondary metabolite gene clusters. *Microb. Biotechnol.* **4,** 207–215.

Gottelt, M., Kol, S., Gomez-Escribano, J. P., Bibb, M., and Takano, E. (2010). Deletion of a regulatory gene within the *cpk* gene cluster reveals novel antibacterial activity in *Streptomyces coelicolor* A3(2). *Microbiology* **156,** 2343–2353.

Gregory, M. A., Till, R., and Smith, M. C. (2003). Integration site for *Streptomyces* phage phiBT1 and development of site-specific integrating vectors. *J. Bacteriol.* **185,** 5320–5323.

Hatanaka, T., Onaka, H., Arima, J., Uraji, M., Uesugi, Y., Usuki, H., Nishimoto, Y., and Iwabuchi, M. (2008). pTONA5: A hyperexpression vector in streptomycetes. *Protein Expr. Purif.* **62,** 244–248.

Heide, L., Gust, B., Anderle, C., and Li, S. M. (2008). Combinatorial biosynthesis, metabolic engineering and mutasynthesis for the generation of new aminocoumarin antibiotics. *Curr. Top. Med. Chem.* **8,** 667–679.

Hertweck, C., Luzhetskyy, A., Rebets, Y., and Bechthold, A. (2007). Type II polyketide synthases: Gaining a deeper insight into enzymatic teamwork. *Nat. Prod. Rep.* **24,** 162–190.

Hong, H. J., Hutchings, M. I., Hill, L. M., and Buttner, M. J. (2005). The role of the novel Fem protein VanK in vancomycin resistance in *Streptomyces coelicolor. J. Biol. Chem.* **280,** 13055–13061.

Hopwood, D. A. (2007). *Streptomyces* in Nature and Medicine: The Antibiotic Makers. Oxford University Press, New York.

Hopwood, D. A, ed. (2009a). Complex enzymes in microbial natural product biosynthesis, part A: Overview Articles and Peptides. *Methods Enzymol.* **458.**

Hopwood, D. A, ed. (2009b). Complex enzymes in microbial natural product biosynthesis, part B: Polyketides, aminocoumarins and carbohydrates. *Methods Enzymol.* **459.**

Hopwood, D. A., Malpartida, F., Kieser, H. M., Ikeda, H., Duncan, J., Fujii, I., Rudd, B. A., Floss, H. G., and Omura, S. (1985). Production of 'hybrid' antibiotics by genetic engineering. *Nature* **314,** 642–644.

Horinouchi, S., Furuya, K., Nishiyama, M., Suzuki, H., and Beppu, T. (1987). Nucleotide sequence of the streptothricin acetyltransferase gene from *Streptomyces lavendulae* and its expression in heterologous hosts. *J. Bacteriol.* **169,** 1929–1937.

Hsiao, N. H., Nakayama, S., Merlo, M. E., de Vries, M., Bunet, R., Kitani, S., Nihira, T., and Takano, E. (2009). Analysis of two additional signalling molecules in *Streptomyces coelicolor* and development of a butyrolactone-specific reporter system. *Chem. Biol.* **16,** 951–960.

Huang, J., Shi, J., Molle, V., Sohlberg, B., Weaver, D., Bibb, M. J., Karoonuthaisiri, N., Lih, C. J., Kao, C. M., Buttner, M. J., and Cohen, S. N. (2005). Cross-regulation among disparate antibiotic biosynthetic pathways of *Streptomyces coelicolor. Mol. Microbiol.* **58,** 1276–1287.

Ichinose, K., Taguchi, T., Bedford, D. J., Ebizuka, Y., and Hopwood, D. A. (2001). Functional complementation of pyran ring formation in actinorhodin biosynthesis in *Streptomyces coelicolor* A3(2) by ketoreductase genes for granaticin biosynthesis. *J. Bacteriol.* **183,** 3247–3250.

Ikeda, H., Ishikawa, J., Hanamoto, A., Shinose, M., Kikuchi, H., Shiba, T., Sakaki, Y., Hattori, M., and Omura, S. (2003). Complete genome sequence and comparative analysis of the industrial microorganism *Streptomyces avermitilis. Nat. Biotechnol.* **21,** 526–531.

Kharel, M. K., Nybo, S. E., Shepherd, M. D., and Rohr, J. (2010). Cloning and characterization of the ravidomycin and chrysomycin biosynthetic gene clusters. *Chembiochem.* **11,** 523–532.

Khosla, C., and Zawada, R. J. (1996). Generation of polyketide libraries via combinatorial biosynthesis. *Trends Biotechnol.* **14,** 335–341.

Kieser, T., Bibb, M. J., Buttner, M. J., Chater, K. F., and Hopwood, D. A. (2000). Practical *Streptomyces* Genetics. The John Innes Foundation, Norwich.

King, A. A., and Chater, K. F. (1986). The expression of the *Escherichia coli lacZ* gene in *Streptomyces. J. Gen. Microbiol.* **132,** 1739–1752.

Kol, S., Merlo, M. E., Scheltema, R. A., De, V. M., Vonk, R. J., Kikkert, N. A., Dijkhuizen, L., Breitling, R., and Takano, E. (2010). Metabolomic characterization of the salt stress response in *Streptomyces coelicolor. Appl. Environ. Microbiol.* **76,** 2574–2581.

Komatsu, M., Uchiyama, T., Omura, S., Cane, D. E., and Ikeda, H. (2010). Genome-minimized *Streptomyces* host for the heterologous expression of secondary metabolism. *Proc. Natl. Acad. Sci. USA* **107,** 2646–2651.

Liu, H., Jiang, H., Haltli, B., Kulowski, K., Muszynska, E., Feng, X., Summers, M., Young, M., Graziani, E., Koehn, F., Carter, G. T., and He, M. (2009). Rapid cloning and heterologous expression of the meridamycin biosynthetic gene cluster using a versatile *Escherichia coli-Streptomyces* artificial chromosome vector, pSBAC. *J. Nat. Prod.* **72,** 389–395.

Lussier, F. X., Denis, F., and Shareck, F. (2010). Adaptation of the highly productive T7 expression system to *Streptomyces lividans. Appl. Environ. Microbiol.* **76,** 967–970.

Luzhetska, M., Harle, J., and Bechthold, A. (2010). Combinatorial and synthetic biosynthesis in actinomycetes. *Fortschr. Chem. Org. Naturst.* **93,** 211–237.

Martin, J. F., and Liras, P. (2010). Engineering of regulatory cascades and networks controlling antibiotic biosynthesis in *Streptomyces. Curr. Opin. Microbiol.* **13,** 263–273.

McDaniel, R., Ebert-Khosla, S., Hopwood, D. A., and Khosla, C. (1993). Engineered biosynthesis of novel polyketides. *Science* **262,** 1546–1550.

McDaniel, R., Ebert-Khosla, S., Hopwood, D. A., and Khosla, C. (1995). Rational design of aromatic polyketide natural products by recombinant assembly of enzymatic subunits. *Nature* **375,** 549–554.

Medema, M. H., Trefzer, A., Kovalchuk, A., van den Berg, M., Müller, U., Heijne, W., Wu, L., Alam, M. T., Ronning, C. M., Nierman, W. C., Bovenberg, R. A. L., Breitling, R., *et al.* (2010). The sequence of a 1.8-Mb bacterial linear plasmid reveals a rich evolutionary reservoir of secondary metabolic pathways. *Genome Biol. Evol.* **2,** 212–224.

Medema, M. H., Breitling, R., Bovenberg, R., and Takano, E. (2011). Exploiting plug-and-play synthetic biology for drug discovery and production in microorganisms. *Nat. Rev. Microbiol.* **9,** 131–137.

Menzella, H. G., and Reeves, C. D. (2007). Combinatorial biosynthesis for drug development. *Curr. Opin. Microbiol.* **10,** 238–245.

Menzella, H. G., Reid, R., Carney, J. R., Chandran, S. S., Reisinger, S. J., Patel, K. G., Hopwood, D. A., and Santi, D. V. (2005). Combinatorial polyketide biosynthesis by de novo design and rearrangement of modular polyketide synthase genes. *Nat. Biotechnol.* **23,** 1171–1176.

Menzella, H. G., Carney, J. R., and Santi, D. V. (2007). Rational design and assembly of synthetic trimodular polyketide synthases. *Chem. Biol.* **14,** 143–151.

Ohnishi, Y., Ishikawa, J., Hara, H., Suzuki, H., Ikenoya, M., Ikeda, H., Yamashita, A., Hattori, M., and Horinouchi, S. (2008). Genome sequence of the streptomycin-producing microorganism *Streptomyces griseus* IFO 13350. *J. Bacteriol.* **190,** 4050–4060.

Okamoto, S., Taguchi, T., Ochi, K., and Ichinose, K. (2009). Biosynthesis of actinorhodin and related antibiotics: Discovery of alternative routes for quinone formation encoded in the *act* gene cluster. *Chem. Biol.* **16,** 226–236.

Richardson, M. A., Kuhstoss, S., Solenberg, P., Schaus, N. A., and Rao, R. N. (1987). A new shuttle cosmid vector, pKC505, for streptomycetes: Its use in the cloning of three different spiramycin-resistance genes from a Streptomyces ambofaciens library. *Gene* **61,** 231–241.

Rodriguez-Garcia, A., Combes, P., Perez-Redondo, R., Smith, M. C., and Smith, M. C. (2005). Natural and synthetic tetracycline-inducible promoters for use in the antibiotic-producing bacteria *Streptomyces*. *Nucleic Acids Res.* **33,** e87.

Salis, H. M., Mirsky, E. A., and Voigt, C. A. (2009). Automated design of synthetic ribosome binding sites to control protein expression. *Nat. Biotechnol.* **27,** 946–950.

Scholtissek, S., and Grosse, F. (1987). A cloning cartridge of lambda t(o) terminator. *Nucleic Acids Res.* **15,** 3185.

Schoof, S., Pradel, G., Aminake, M. N., Ellinger, B., Baumann, S., Potowski, M., Najajreh, Y., Kirschner, M., and Arndt, H. D. (2010). Antiplasmodial thiostrepton derivatives: Proteasome inhibitors with a dual mode of action. *Angew. Chem. Int. Ed. Engl.* **49,** 3317–3321.

Strohl, W. R. (1992). Compilation and analysis of DNA sequences associated with apparent streptomycete promoters. *Nucleic Acids Res.* **20,** 961–974.

Taguchi, T., Itou, K., Ebizuka, Y., Malpartida, F., Hopwood, D. A., Surti, C. M., Booker-Milburn, K. I., Stephenson, G. R., and Ichinose, K. (2000). Chemical characterisation of disruptants of the *Streptomyces coelicolor* A3(2) *actVI* genes involved in actinorhodin biosynthesis. *J. Antibiot. (Tokyo)* **53,** 144–152.

Tahlan, K., Ahn, S. K., Sing, A., Bodnaruk, T. D., Willems, A. R., Davidson, A. R., and Nodwell, J. R. (2007). Initiation of actinorhodin export in *Streptomyces coelicolor*. *Mol. Microbiol.* **63,** 951–961.

Tahlan, K., Yu, Z., Xu, Y., Davidson, A. R., and Nodwell, J. R. (2008). Ligand recognition by ActR, a TetR-like regulator of actinorhodin export. *J. Mol. Biol.* **383,** 753–761.

Takano, E., White, J., Thompson, C. J., and Bibb, M. J. (1995). Construction of thiostrepton-inducible, high-copy-number expression vectors for use in *Streptomyces* spp. *Gene* **166,** 133–137.

Takano, E., Kinoshita, H., Mersinias, V., Bucca, G., Hotchkiss, G., Nihira, T., Smith, C. P., Bibb, M., Wohlleben, W., and Chater, K. (2005). A bacterial hormone (the SCB1) directly controls the expression of a pathway-specific regulatory gene in the cryptic type I polyketide biosynthetic gene cluster of *Streptomyces coelicolor*. *Mol. Microbiol.* **56,** 465–479.

Tala, A., Wang, G., Zemanova, M., Okamoto, S., Ochi, K., and Alifano, P. (2009). Activation of dormant bacterial genes by *Nonomuraea* sp. strain ATCC 39727 mutant-type RNA polymerase. *J. Bacteriol.* **191,** 805–814.

Walsh, C. T. (2002). Combinatorial biosynthesis of antibiotics: Challenges and opportunities. *Chembiochem.* **3,** 125–134.

Wang, X. J., Yan, Y. J., Zhang, B., An, J., Wang, J. J., Tian, J., Jiang, L., Chen, Y. H., Huang, S. X., Yin, M., Zhang, J., Gao, A. L., *et al.* (2010). Genome sequence of the milbemycin-producing bacterium *Streptomyces bingchenggensis*. *J. Bacteriol.* **192,** 4526–4527.

Ward, J. M., Janssen, G. R., Kieser, T., Bibb, M. J., Buttner, M. J., and Bibb, M. J. (1986). Construction and characterisation of a series of multi-copy promoter-probe plasmid vectors for *Streptomyces* using the aminoglycoside phosphotransferase gene from Tn5 as indicator. *Mol. Gen. Genet.* **203,** 468–478.

Widmaier, D. M., Tullman-Ercek, D., Mirsky, E. A., Hill, R., Govindarajan, S., Minshull, J., and Voigt, C. A. (2009). Engineering the *Salmonella* type III secretion system to export spider silk monomers. *Mol. Syst. Biol.* **5,** 309.

Willems, A. R., Tahlan, K., Taguchi, T., Zhang, K., Lee, Z. Z., Ichinose, K., Junop, M. S., and Nodwell, J. R. (2008). Crystal structures of the *Streptomyces coelicolor* TetR-like protein ActR alone and in complex with actinorhodin or the actinorhodin biosynthetic precursor (S)-DNPA. *J. Mol. Biol.* **376,** 1377–1387.

Zerikly, M., and Challis, G. L. (2009). Strategies for the discovery of new natural products by genome mining. *Chembiochem.* **10,** 625–633.

Zhang, W., and Tang, Y. (2008). Combinatorial biosynthesis of natural products. *J. Med. Chem.* **51,** 2629–2633.

Zhang, W., Li, Y., and Tang, Y. (2008). Engineered biosynthesis of bacterial aromatic polyketides in *Escherichia coli*. *Proc. Natl. Acad. Sci. USA* **105,** 20683–20688.

CHAPTER TWENTY-TWO

Methods for Engineering Sulfate Reducing Bacteria of the Genus *Desulfovibrio*

Kimberly L. Keller,*,† Judy D. Wall,*,† *and* Swapnil Chhabra†,‡,§

Contents

Abstract

Sulfate reducing bacteria (SRB) are physiologically important given their nearly ubiquitous presence and have important applications in the areas of bioremediation and bioenergy. This chapter provides details on the steps used for homologous-recombination mediated chromosomal manipulation of *Desulfovibrio vulgaris* Hildenborough, a well-studied sulfate reducer. More specifically,

* University of Missouri, Columbia, Missouri, USA
† VIMSS (Virtual Institute of Microbial Stress and Survival), Berkeley, California, USA
‡ Physical Biosciences Division, Lawrence Berkeley National Laboratory, Berkeley, California, USA
§ Joint BioEnergy Institute, Emeryville, California, USA

Methods in Enzymology, Volume 497
ISSN 0076-6879, DOI: 10.1016/B978-0-12-385075-1.00022-6

we focus on the implementation of a "parts" based approach for suicide vector assembly, important aspects of anaerobic culturing, choices for antibiotic selection, electroporation-based DNA transformation, as well as tools for screening and verifying genetically modified constructs. These methods, which in principle may be extended to other SRB, are applicable for functional genomics investigations, as well as metabolic engineering manipulations.

1. Introduction

Sulfate reducing bacteria (SRB) use sulfate as the terminal electron acceptor during growth under anoxic conditions. Some of these microorganisms, however, can also grow in the presence of other electron acceptors such as nitrate, and indeed, ferment substrates in the absence of any inorganic electron acceptor (Postgate, 1984). SRB play important roles in the global sulfur and carbon cycles and, not surprisingly, inhabit widely diverse natural and man-made environments ranging from high-temperature hydrothermal vents and hypersaline microbial mats to Arctic marine sediments and highly toxic waste-water treatment facilities (Ensley and Suflita, 1995; Fauque, 1995). Biotechnological interest in the SRB stems from their potential applications in bioremediation (Lovley and Phillips, 1992) and bioenergy (Gieg *et al.*, 2008). Over the past decade, genomes of several SRB and archaea have been sequenced and can be accessed on genome sites such as MicrobesOnline (www.microbesonline.org), the Integrated Microbial Genomes (IMG; http://img.jgi.doe.gov/cgi-bin/pub/main.cgi) and the Genome database from the National Center for Biotechnology Information (NCBI; http://www.ncbi.nlm.nih.gov/sites/genome). As additional genomic sequences become available for the numerous SRB, the ability to genetically manipulate those strains becomes increasingly important to further our knowledge. Extensive research has been done in the past few years on the genus *Desulfovibrio*, a member of the δ-proteobacteria. This chapter discusses methods and techniques associated with genetic manipulation of the most widely studied member of the genus, *Desulfovibrio vulgaris* Hildenborough. We describe strategies that we have successfully employed for homologous-recombination mediated targeted chromosomal insertions and deletions; tagged insertions for elucidating protein–protein interactions (PPIs) by tandem-affinity purification and mass spectrometry-based identification of interacting partners; plasmid introduction and replication; as well as heterologous protein expression. These methods have been utilized in *D. vulgaris* Hildenborough and can provide a starting point for developing genetic systems in other SRB.

2. Chromosomal Modifications Through Homologous Recombination

Homologous recombination mediated deletions and gene tagging require the generation of a suicide vector for *D. vulgaris* that can be propagated in *Escherichia coli*, and then transferred to *D. vulgaris* by electroporation. One method for introducing genetic modifications in the *D. vulgaris* chromosome is through single recombination with a suicide vector carrying a homologous DNA segment resulting in chromosomal integration of the complete vector in the target region of the homologous segment of DNA. We originally employed this method for generating tagged mutants of *D. vulgaris* for PPI studies (Chhabra *et al.*, 2011b). The main concern with single recombination modifications results from the continued presence of the entire target sequence within the mutant strain that provides the possibility for recombination functions to restore the wild-type gene (Rousset *et al.*, 1998). In addition, release of selection for vector encoded antibiotic resistance will permit Campbell recombination and removal of the inserted plasmid. Plasmid integration modifications need to remain under constant antibiotic selection and monitoring.

In contrast, the marker exchange approach is devoid of such problems when implemented correctly and is the focus of further discussions in this chapter. This approach results in *no* undesired components of the suicide vector being integrated into the chromosome. We are using a "parts" strategy to implement double homologous recombination in *D. vulgaris* for high throughput mutagenic vector generation (Fig. 22.1; Chhabra

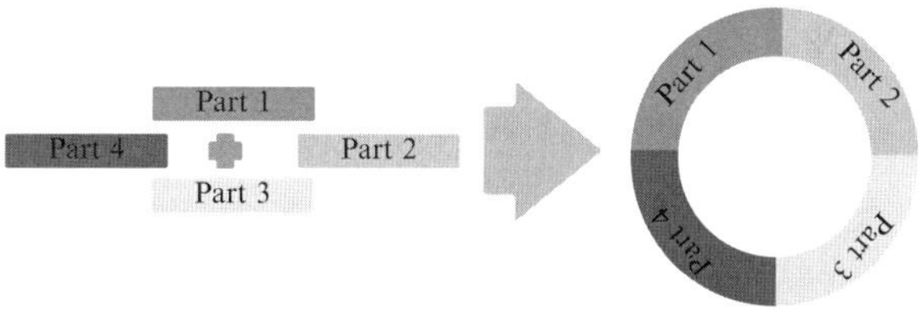

Parts based approach for *D. vulgaris* chromosomal modifications			
Suicide vector structure		Sample applications	
Parts list	Property	Gene deletion	Tag insertion at 3′ end
Part 1	Variable (homology region)	Sequence upstream to gene (750 bp)	Gene sequence – *stop codon removed* (750 bp)
Part 2	Variable (homology region)	Sequence downstream to gene (750 bp)	Sequence downstream to gene (750 bp)
Part 3	Constant (application specific)	AB1	Tag sequence + AB1
Part 4	Constant (application specific + vector backbone)	Replication origin + AB2	Replication origin + AB2

AB = antibiotic resistance marker gene; replication origin (such as pUC) is recognized only in *E. coli.*

Figure 22.1 Simplified suicide vector construction approach for enabling high throughput chromosomal modifications of *D. vulgaris* using double homologous recombination. "Parts" are assembled using ligation independent techniques such as SLIC (Li and Elledge, 2007).

et al., 2011a). Four-parts are typically necessary for mutagenic plasmid assembly in *E. coli* facilitated by sequence and ligation independent cloning (SLIC; Li and Elledge, 2007). For a typical application common to multiple genes, only the regions homologous (typically 750 bp in length) to the sequences flanking the specific target loci are varied. The rest of the plasmid components (those not varied) are specifically designed for the particular application, and are utilized for the construction of all the mutagenic plasmids for that application. Once the nonvariable regions are created and a "library" of the required homologous DNA regions is produced, a large number of mutagenic plasmids required for the specific chromosomal modifications could be rapidly assembled by mixing and matching of the "parts."

We have tested this approach for enabling marker exchange modifications, which utilize two antibiotic resistance genes (kanamycin and spectinomycin), to differentiate between single and double recombinants. An example of this approach is shown in Fig. 22.2. The deletion of DVU0890, putatively encoding homoserine dehydrogenase, would require generation of the following parts for mutagenic plasmid assembly: upstream sequence of DVU0890 (750 bp of DVU0889), downstream sequence of DVU0890 (750 bp of DVU0891), kanamycin resistance cassette, and a spectinomycin resistance cassette coupled to the pUC origin of replication. Similarly modification of the chromosome for the production of a tag fused to the carboxy-terminus of DVU0890 would require generation of the following parts: the 750 bp of DVU0890 (lacking the stop codon), sequence the downstream of DVU0890 (750 bp of DVU0891), Strep-TEV-FLAG tag (Chhabra *et al.*, 2011a) plus kanamycin resistance cassette, and the spectinomycin resistance cassette coupled to the pUC origin of replication. Thus both applications have two parts in common, (1) the spectinomycin resistance cassette coupled to the pUC origin of replication, and (2) the downstream sequence of DVU0890 (750 bp of DVU0891). To further the high throughput capabilities of this process, the Strep-TEV-FLAG tag plus the kanamycin resistance cassette may be switched between different functional tags, such as those required for protein localization, versus tandem-affinity purification of the protein. The homologous recombination for marker replacement can be visualized beginning with integration of the entire suicide construct in the chromosome (Fig. 22.2A-ii and B-ii) followed by a second recombination step (Fig. 22.2A-i and B-i), resulting in the desired outcome. We have specifically included two selectable markers in our mutagenic plasmids to distinguish between single versus double recombination events, the methods of which are discussed in detail below. Successful implementation of the "parts" approach requires careful consideration of the choice of growth conditions, antibiotic resistance markers, transformation methods, and tools to confirm the genetic modification. The following sections present a detailed discussion of these parameters applicable to chromosomal manipulations of *D. vulgaris* Hildenborough.

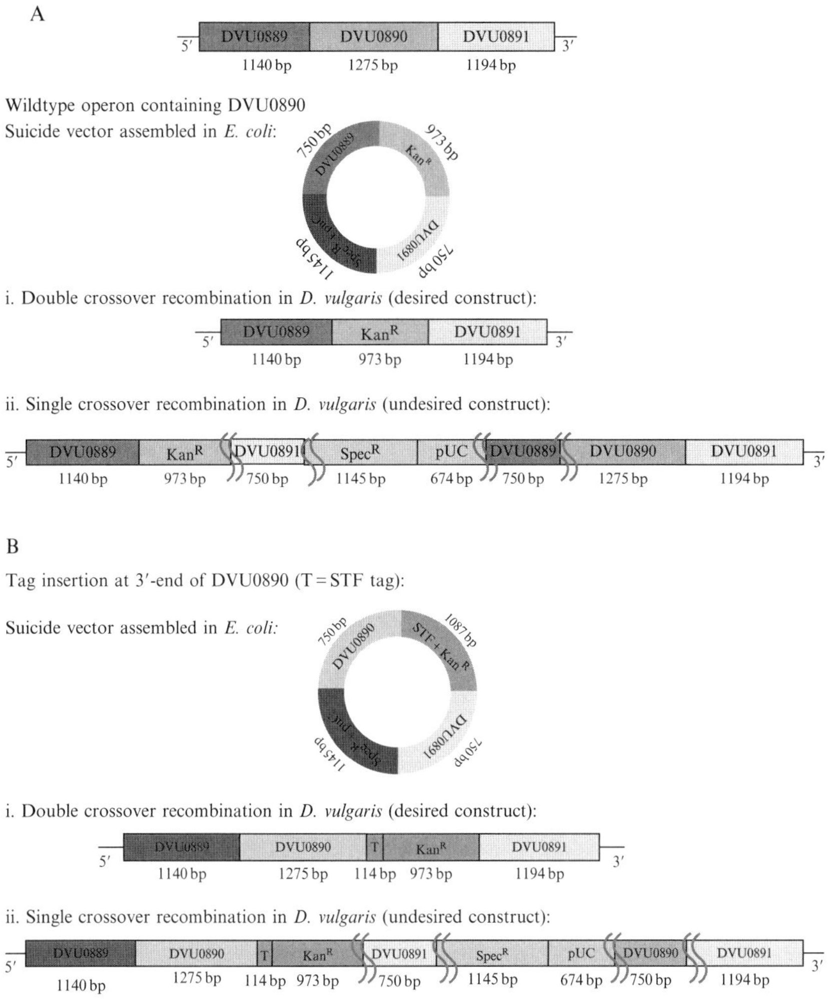

Figure 22.2 Chromosomal modifications of DVU0890 using suicide constructs described in Fig. 22.1 for (A) a marker exchange deletion or (B) a tagged gene mutant and potential outcomes resulting from single and double crossover recombinations.

3. Culturing Conditions and Antibiotic Selection

3.1. Anaerobiosis

Most *Desulfovibrio* can tolerate small amounts of exposure to air; however, oxygen in plating medium, can delay or inhibit growth of colonies all together. Therefore, manipulation and growth of cultures should be performed in an anaerobic growth chamber (Coy Laboratory Product, Inc., Grass Lake, MI) with an atmosphere of ~95% N_2 and ~5% H_2 at ~32 °C, unless indicated otherwise. It is important to remember that plastic items (petri dishes, eppendorf tubes, pipette tips, 50-ml conical tubes, etc.) contain oxygen, which can retard the growth of *D. vulgaris* and other *Desulfovibrio* strains. Therefore, plastic items should be allowed to "degas" inside the anaerobic chamber for at least 7 days prior to use. The use of glass items (test tubes and bottles) can provide more consistent growth.

3.2. Growth medium

When genetically manipulating *D. vulgaris*, it is most convenient to optimize growth by providing rich medium with appropriate electron donor(s) and acceptor(s). Because mutants can have a growth requirement different from the wild-type cells, rich medium provides an excess of components that are often limiting in minimal medium; and therefore, rich medium is generally permissive for mutant growth. *D. vulgaris* cultures are grown in medium adapted from Postgate (1984) named MOYLS4 medium (Zane *et al.*, 2010) [60 m*M* sodium lactate, 30 m*M* Na_2SO_4, 8 m*M* $MgCl_2$, 20 m*M* NH_4Cl, 0.6 m*M* $CaCl_2$, 2 m*M* phosphate (K_2HPO_4/NaH_2PO4), 60 μ*M* $FeCl_2$, 120 m*M* EDTA, 30 m*M* Tris (pH 7.4), 0.1% (w/v) yeast extract, 1 ml Thauers vitamin solution per liter (Brandis and Thauer, 1981), and 6 ml trace elements solution per liter, with pH adjusted to 7.2]. The trace elements solution contains 2.5 m*M* $MnCl_2$, 1.26 m*M* $CoCl_2$, 1.47 m*M* $ZnCl_2$, 210 μ*M* Na_2MoO_4, 320 μ*M* H_3BO_3, 380 μ*M* $NiSO_4$, 11.7 μ*M* $CuCl_2$, 35 μ*M* Na_2SeO_3, and 24 μ*M* Na_2WO_4. For plating, MOYLS4 medium is solidified with 1.5% (w/v) agar and two reductants are added: sodium thioglycolate (1.2 m*M*, added aerobically, presterilization) and titanium citrate (380 μ*M*, added anaerobically, poststerilization). The redox potential indicator, rezasurin, is added to 0.0016% (w/v) to medium such that a pink color develops when the redox potential exceeds 110 mV.

3.3. Culture maintenance

Freezer stocks of *D. vulgaris* are generated by growing a liquid culture to mid- to late-log phase and adding sterile glycerol to a final concentration of 10% (v/v). Within 15 min of the addition of glycerol, ~1 ml portions of the mixture are stored in cryovials at −80 °C. It is important to ensure that freezer stocks and working cultures are free of aerobic contaminants. For this purpose, *D. vulgaris* cultures are routinely streaked on LC plates (components per liter of medium: 10 g tryptone, 5 g NaCl, 5 g yeast extract, and 15 g agar) containing 40 m*M* glucose and incubated in air to detect potential aerobic contaminants.

3.4. Antibiotic sensitivity

A limitation in developing a genetic system in SRB strains is the fact that many exhibit natural resistance to many antimicrobials (Postgate, 1984). In order to determine which, if any, antibiotic resistances can be used for genetic manipulation in a particular bacterium, antibiotic sensitivity studies need to be performed. Once a sensitivity range is established, the introduction of genes conferring antibiotic resistance into the SRB must be successful before confirmation that increased antibiotic resistance can be achieved. Kanamycin sensitivity and selection works well in most SRB studied to date. However, sensitivity studies have revealed that G418 (400 mg/ml) is more effective for kanamycin resistance selection in *D. vulgaris* than kanamycin itself (Ringbauer *et al.*, 2004). *Desulfovibrio* G20 is more sensitive to kanamycin, although the concentration for selection is rather high (800 mg/ml). A list of antibiotic resistances and sensitivities currently used for genetic manipulation of *D. vulgaris* and *Desulfovibrio G20* is compared in Table 22.1.

3.5. Varying the electron donor/acceptor

The most commonly used electron donor:acceptor for genetic manipulation of *D. vulgaris* is lactate:sulfate medium (60 m*M*:30 m*M*). Alternative electron donors or acceptors provide the opportunity of obtaining conditional lethal mutants in other pathways (Zane *et al.*, 2010). Commonly used electron donors:acceptors for *D. vulgaris* and *Desulfovibrio G20* are found in Table 22.2.

Warning: To obtain and test mutants of genes in various metabolic pathways, it may be necessary to grow *Desulfovibrio* in fermenting conditions (pyruvate only) or dismutating fumarate (*Desulfovibrio* G20). Caution needs to be used in growing mutants in these conditions while maintaining selective pressure, because many antibiotics (including kanamycin and G418) are supplied only as sulfate salts that could supply enough sulfate to interfere with establishing growth capabilities.

Table 22.1 Concentration (mg/ml) of different antibiotics currently being used for genetic manipulation and selection

Antibiotic	*D. vulgaris* Hildenborough	*Desulfovibrio* G20
Kanamycin	NU[a]	800
Geneticin (G418)	400	NU
Spectinomycin	100	800
Chloramphenicol	10	NU
Gentamycin	ND[b]	75
Tetracycline	20	NU

[a] NU, not utilized in strain for antibiotic selection.
[b] ND, not determined.

Table 22.2 Concentrations (m*M*) of different electron donors and electron acceptors currently being used for growth of *Desulfovibrio* strains

Electron donor:electron acceptor	*D. vulgaris* Hildenborough	*Desulfovibrio* G20
Lactate:sulfate	60:30	60:30
Lactate:sulfite	30:20	15:10
Pyruvate:sulfate	60:15	60:15
Pyruvate:sulfite	30:10	30:10
Pyruvate	60	60
Fumarate	NG[a]	60

[a] NG, no growth observed.

4. DNA Transformation

Foreign DNA (plasmid or linear) may be introduced into *Desulfovibrio* using conjugation or electroporation. Conjugation has been successfully used in both *D. vulgaris* (Ringbauer *et al.*, 2004) and *Desulfovibrio* G20 (Li *et al.*, 2009) for generating transposon libraries (random chromosomal mutagenesis) as well as site-directed mutants. Protocols related to conjugal transfer have been described extensively elsewhere (van Dongen *et al.*, 1994). In contrast, the introduction of DNA into the SRB with methods other than conjugation has been minimally utilized or described (Bender *et al.*, 2007; Keller *et al.*, 2009; Zane *et al.*, 2010). We have successfully employed electroporation with *D. vulgaris* to generate plasmid insertion mutants (single recombinational events), to tag proteins (single and double recombinational events), and to generate marker exchange deletion mutants (double recombinational events).

4.1. Method for electroporation

For genetic manipulation, *D. vulgaris* are grown anaerobically in MOYLS4 medium and plated in this medium solidified with 1.5% (w/v) agar as described above.

1. To prepare competent *D. vulgaris* for electroporation, thaw a 1 ml *D. vulgaris* freezer stock, introduce it into the anaerobic chamber, and immediately add 4 ml of MOYLS4 medium in a glass test tube and allow to grow for ~16–20 h.
2. The entire 5 ml cells is subcultured into 45 ml of fresh MOYLS4 medium (in a glass bottle), and grown to an optimal OD_{600} of 0.4–0.7.
3. Transfer the culture anaerobically to a 50-ml plastic conical tube and the culture at ~22,000×*g* for 12 min at 4 °C in a refrigerated centrifuge. Be sure to keep cells on ice from this point forward.

4. Wash the cells by resuspending in 50 ml of chilled, sterile electroporation wash buffer (30 m*M* Tris–HCl buffer, pH 7.2, not anaerobic). Centrifuge the resuspended cells as in step 3.
5. Resuspend the resulting pellet in 0.5 ml of chilled wash buffer and keep on ice as competent cells for electroporation.
6. Place 50 μl of prepared cells in a chilled 0.5 ml eppendorf tube and add up to 5 μl of a plasmid (between 0.5 and 1 μg) and mix gently by flicking.
7. Transfer the entire mixture to a 1-mm gapped electroporation cuvette (Molecular BioProducts, San Diego, CA) that has been prechilled on ice.

   Note: There are differences in transformation efficiencies among commercially available cuvettes for unidentified reasons.
8. To determine proper transformation and recombination efficiencies, electroporation controls are carried out with prepared cells without DNA, and cells with a stable plasmid know to be transformable (for *D. vulgaris*, pSC27 [Fig. 22.3] is used as a positive control for the kanamycin markers described in these protocols).

   Note: If spectinomycin is used as an exchange marker, pMO719 (Fig. 22.3) should be used as a positive control.
9. Electroporations in *D. vulgaris* are performed with an ECM 630 electroporator, BTX (Genetronix, San Jose, CA). To insure proper current transfer, be sure to remove any water from the outside of the cuvette prior to placing it in the safety stand. We have typically used the parameters 1750 V, 250 Ω, and 25 μF under anaerobic conditions inside the chamber. Typical voltage and time constants were 1650 V and 200 ms for any given electroporation, respectively and a visible arc was observed. There seemed to be a correlation between this arcing during electroporation and transformation efficiency during marker exchange mutagenesis.
10. After pulsing the cells/DNA mixture, transfer the cells from the cuvettes to a 1.5 ml eppendorf tube with 1 ml of MOYLS4 medium and let recover anaerobically overnight at ~32 °C.
11. Unlike previously described procedures for plating that use a soft agar overlay of solidified medium, the procedure here involves plating cells directly in the molten medium. All plating steps are performed anaerobically, inside the growth chamber. For plating, dispense different amounts of cells (i.e., 50, 250, and 700 ml) into separate sterile, empty petri dishes. Molten MOYLS4 (≤50 °C containing 400 mg of G418/ml) is poured over the cells and the plates swirled in a figure eight motion to distribute the cells within the medium. Once solidified, the plates are inverted, placed in an airtight rectangular jar (Mitubishi Gas Chemical Co., Inc., Japan) and incubated at ~32 °C for 4–7 days until individual colonies appear.
12. Further screening of colonies is necessary to determine if the appropriate strain has been constructed.

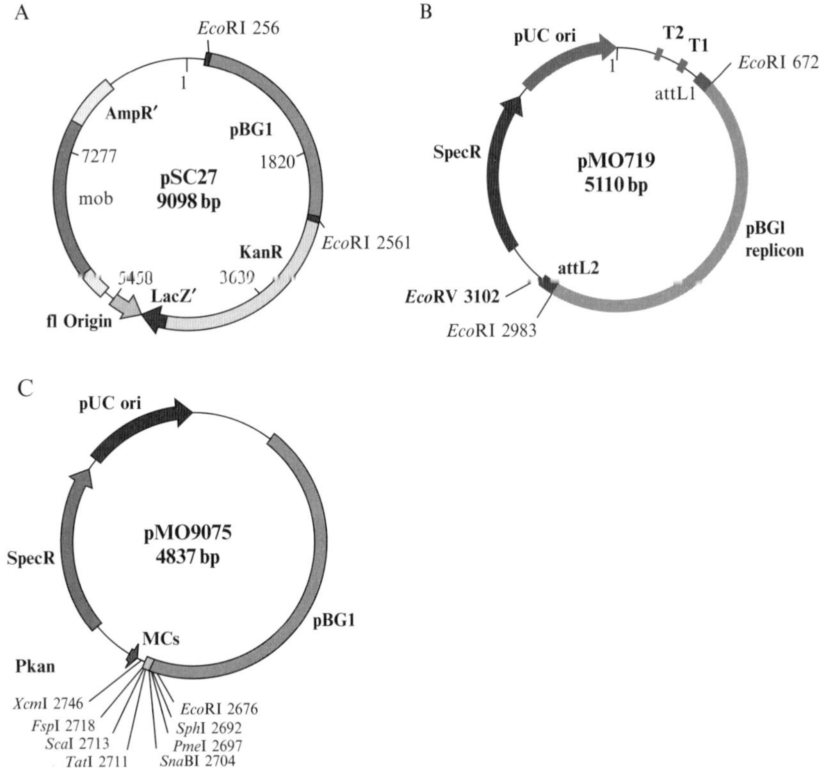

Figure 22.3 Stable plasmids currently being utilized in *Desulfovibrio vulgaris*. (A) Stable kanamycin plasmid used as positive control for kanamycin marker exchange transformations; (B) Stable spectinomycin plasmid, used as positive control for spectinomycin marker transformations; and (C) Vector plasmid used to generate complementation plasmid for marker exchange deletion mutants.

Recent variations to the protocol described above have included lowering the voltage to 1500 V (250 Ω and 25 μF), performing the electroporation aerobically on the benchtop, and adding less plasmid DNA (between 0.25 and 0.5 μg). The new parameters result in typical voltage and time constants of 1420 V and >1 ms for a given electroporation, and arcing at this voltage is reduced or does not occur. To increase cell recovery following aerobic electroporation on the benchtop, the transformed cells are transferred to eppendorf tubes that have been conditioned for ~ 1 week inside the anaerobic chamber and contain 1 ml of anaerobic MOYLS4. These prepared tubes are removed from the anaerobic chamber just prior to electroporation. Following electroporation, the cells are immediately transferred to these tubes and taken back inside the chamber. The lids of the eppendorf tubes are opened briefly to allow gas exchange with the chamber atmosphere (~30–60 s), the lids are then closed, and the cells allowed to recover overnight anaerobically at ~32 °C as above.

5. Screening Colonies for Proper Integration

5.1. Secondary antibiotic screening

Transformation of SLIC generated suicide constructs designed for marked modifications results in single or double homologous recombination events. Testing individual colonies with a secondary antibiotic screen can distinguish between the two events.

1. In the anaerobic chamber, pour and let solidify MOYLS4 plates modified with each of the following: spectinomycin (100 mg/ml); G418 (400 mg/ml); and no antibiotics.
2. With sterile tweezers, grasp a sterile toothpick at one end. Insert the toothpick into the colony and swirl to cover tip of toothpick with cells.
3. Then sequentially insert the toothpick into the spectinomycin-containing plate, then the G418-containing plate, and then into the MOYLS4 plate without antibiotic.
4. Isolates that are resistant to both antibiotics and grow on all three plates are most likely single recombinants and probably not the desired mutants that require a double recombinational event (Fig. 22.2A-ii and B-ii).
5. Isolates that are sensitive to spectinomycin, resistant to G418 and also grow on the MOYLS4 lacking antibiotics are potential candidates for being double recombinants (Fig. 22.2A-i and B-i).
6. Two to four of the isolates with phenotype identified in Step 5 are each toothpicked from the G418 plate into a separate 1.5 ml eppendorf tube containing 1 ml of fresh MOYLS4 medium amended with G418 (400 mg/ml) and the isolated cultures are allowed to grow overnight in the anaerobic chamber at 34 °C.
7. The 1 ml cultures are added into 4 ml of MOYLS4 medium containing G418 (400 mg/ml) and again allowed to grow to amplify the isolates. Cells from this culture are used to make genomic DNA for Southern blot analysis and freezer stocks for further use.

5.2. Southern blot analysis

Because these genetic transformations result in a chromosome modification, a Southern blot should be used to verify the appropriate changes. To confirm a marker exchange mutant, the general scheme includes indentifying a restriction endonuclease with the following three features:

1. Cuts outside the upstream and downstream DNA regions (Parts 1 and 3, Fig. 22.1) used in the suicide vector;
2. Does not cut within the gene being deleted; and
3. Cuts once inside the kanamycin gene marker.

The DNA region upstream of the target gene/site (Part 1, Fig. 22.1) is generally used as the probe for the Southern blot, and therefore it is important to choose restriction endonucleases that produce DNA fragments containing the upstream region of different theoretical sizes between wild-type and the expected mutant. To date, there are three enzymes (*Nae*I, *Pvu*I, and *Bss*HII) that have been found to be problematic for digesting genomic DNA from *D. vulgaris* and should be avoided.

6. Complementing Gene Deletions

6.1. Requirements for complementing plasmid

Once a gene deletion is verified, growth studies in minimal medium are required to determine the effect of the deletion. If a strain with a deletion has a phenotype different than that of the wild type, the missing gene must be complemented to verify that the lack of function of this gene alone caused the change. Complementation becomes even more important when the deleted gene lies promoter proximal to other genes within an operon. The construct could have a polar effect on downstream gene such that the phenotype is caused not only by the deletion of the gene of interest but also because of the loss of the function of downstream genes. Currently, gene deletions are being complemented in *D. vulgaris* by introducing the gene of interest back into the marker exchange mutant on a stable plasmid. Therefore, a second antibiotic sensitivity in the *Desulfovibrio* strain must be available for complementation. We have found spectinomycin to work well in *D. vulgaris* and there is no interference with independent kanamycin selection. It has been found that the pBG1 replicon (Rousset *et al.*, 1998) provides stability and affective replication of plasmids in *D. vulgaris*. An inducible reporter has yet to be properly identified in *D. vulgaris*; however, studies have shown that the constitutive kanamycin promoter is sufficient to get expression in cultures (Keller *et al.*, 2009; Zane *et al.*, 2010). Therefore, complementation plasmids can readily be generated with the vector plasmid pMO9075 (Fig. 22.3C), which contains the Km^R gene-*aph(3′)-II* promoter, pBG1, Sp^R-determinant, and a convenient restriction endonuclease site. To insure proper translation of the gene, a 21 bp ribosomal-binding site (TGC AGT CCC AGG AGG TAC CAT) is added between the start codon of the gene and the kanamycin promoter. As a control, the empty vector pMO9075 is also transformed into the mutant strain.

6.2. Electroporation of stable plasmids

The electroporation protocol for introducing stable plasmids into *D. vulgaris* is similar to the protocol described above, and can be performed aerobically on the benchtop or anaerobically in the chamber. However, lowering the voltage to 1500 or even 1250 V (250 Ω, and 25 μF) and using less plasmid DNA (between 0.25 and 0.5 μg) does appear to increase the transformation efficiency of the stable plasmids when compared to electroporation with higher voltages. Recovery of electroporated cells still occurs in MOYLS4 medium; however, plating is performed with MOYLS4 medium containing both spectinomycin (100 mg/ml) and G418 (400 mg/ml).

Once individual isolates have been amplified, plasmid is purified from 1.5 ml of a grown *D. vulgaris* culture. Since plasmid yields from *D. vulgaris* are often ≤20 ng/ml, appropriate DNA concentrations cannot be achieved in the limited sequencing volume requirements. Therefore, plasmids purified from *D. vulgaris* are routinely transformed back into competent *E. coli* cells to obtain enough plasmid for sequencing. Plasmids purified from such spectinomycin resistant isolates of *E. coli* are sent for DNA sequencing, and sequence comparisons made to insure the plasmid originally isolated from *D. vulgaris* matches the original sequence of the gene in the complementing plasmid.

7. Concluding Remarks

We have successfully applied the methods described in this chapter for chromosomal manipulations for the deletion and tagging of several genes in *D. vulgaris* (Table 22.3). These methods result in chromosomal incorporation of one of the antibiotic selection markers (present in the suicide construct) and work best for singular modifications. Further chromosomal manipulations on the same strain requires a multistep approach, taking advantage of antibiotic selection and counter-selection measures that ultimately generates an in-frame deletion void of any antibiotic markers.

Table 22.3 Gene targets chromosomally manipulated in *D. vulgaris*

DVU ID	Size (bp)	Operon size (no. genes)	Position in operon	Strain ID—deletion	Strain ID—STF tag insertion
DVU1585	2415	6	1	CAD400198	CAT400249
DVU3371	2358	1	1	CAD400164	CAT400256
DVU0890	1275	3	2	CAD400243	CAT400211
DVU1913	1227	2	1	CAD400244	CAT400250
DVU0171	1182	1	1	CAD400242	CAT400151

We have recently demonstrated an unmarked approach for *D. vulgaris* strains lacking the gene encoding for uracil phosphoribosyltransferase (*upp*, DVU1025) and details for that approach are described in Keller *et al.* (2009). We are currently in the process of developing a "parts" based approach for enabling high throughput applications using the markerless strategy.

ACKNOWLEDGMENTS

This work was part of the U.S. Department of Energy Genomics Sciences program: ENIGMA is a Scientific Focus Area Program supported by the U.S. Department of Energy, Office of Science, Office of Biological and Environmental Research, Genomics: GTL Foundational Science through contract DE-AC02-05CH11231 between Lawrence Berkeley National Laboratory and the U.S. Department of Energy. This work conducted by the Joint BioEnergy Institute was supported by the Office of Science, Office of Biological and Environmental Research, of the U.S. Department of Energy under Contract No. DE-AC02-05CH11231.

REFERENCES

Bender, K. S., Yen, H. C., Hemme, C. L., Yang, Z., He, Z., He, Q., Zhou, J., Huang, K. H., Alm, E. J., Hazen, T. C., Arkin, A. P., and Wall, J. D. (2007). Analysis of a ferric uptake regulator (Fur) mutant of *Desulfovibrio vulgaris* Hildenborough. *Appl. Environ. Microbiol.* **73,** 5389–5400.

Brandis, A., and Thauer, R. K. (1981). Growth of *Desulfovibrio* species on hydrogen and sulphate as sole energy source. *J. Gen. Microbiol.* **126,** 249–252.

Chhabra, S. R., Butland, G., Elias, D., Chandonia, J. M., Fok, O. Y., Juba, T., Gorur, A., Allen, S., Leung, C. M., Keller, K., Reveco, S., Zane, G., *et al.* (2011a). Generalized schemes for high throughput manipulation of bacterial genomes. In review.

Chhabra, S. R., Joachimiak, M. P., Petzold, C. J., Zane, G. M., Price, M. N., Gaucher, S., Reveco, S. A., Fok, O.-Y., Johanson, A. R., Batth, T. S., Singer, M., Chandonia, J. M., *et al.* (2011b). A network of protein–protein interactions of the model sulfate reducer Desulfovibrio vulgaris Hildenborough. In review.

Ensley, B. D., and Suflita, J. M. (1995). Metabolism of environmental contaminants by mixed and pure cultures of sulfate-reducing bacteria. *In* "Biotechnology Handbooks," (L. L. Barton, T. Atkinson, and R. F. Sherwood, eds.), Vol. 8. Plenum Press, New York.

Fauque, G. D. (1995). Ecology of sulfate-reducing bacteria. *In* "Biotechnology Handbooks," (L. L. Barton, T. Atkinson, and R. F. Sherwood, eds.), Vol. 8. Plenum Press, New York.

Gieg, L. M., Duncan, K. E., and Suflita, J. M. (2008). Bioenergy production via microbial conversion of residual oil to natural gas. *Appl. Environ. Microbiol.* **74,** 3022–3029.

Keller, K. L., Bender, K. S., and Wall, J. D. (2009). Development of a markerless genetic exchange system for *Desulfovibrio vulgaris* Hildenborough and its use in generating a strain with increased transformation efficiency. *Appl. Environ. Microbiol.* **75,** 7682–7691.

Li, M. Z., and Elledge, S. J. (2007). Harnessing homologous recombination *in vitro* to generate recombinant DNA via SLIC. *Nat. Methods* **4,** 251–256.

Li, X. Z., Luo, Q. W., Wofford, N. Q., Keller, K. L., McInerney, M. J., Wall, J. D., and Krumholz, L. R. (2009). A molybdopterin oxidoreductase is involved in H-2 oxidation in *Desulfovibrio desulfuricans* G20. *J. Bacteriol.* **191,** 2675–2682.

Lovley, D. R., and Phillips, E. J. (1992). Reduction of uranium by *Desulfovibrio desulfuricans*. *Appl. Environ. Microbiol.* **58,** 850–856.

Postgate, J. R. (1984). The Sulphate-Reducing Bacteria. Cambridge University Press, London, UK.

Ringbauer, J. A., Zane, G. M., Emo, B. M., and Wall, J. D. (2004). Efficiencies of various transformation methods for the mutagenesis of *Desulfovibrio vulgaris* Hildenborough. American Society for Microbiology, New Orleans.

Rousset, M., Casalot, L., Rapp-Giles, B. J., Dermoun, Z., de Philip, P., Belaich, J. P., and Wall, J. D. (1998). New shuttle vectors for the introduction of cloned DNA in *Desulfovibrio*. *Plasmid* **39,** 114–122.

van Dongen, W. A. M., Stokkermans, J. P. W. G., and van den Berg, W. A. M. (eds.), (1994). Genetic Manipulation of *Desulfovibrio*, Academic Press, New York.

Zane, G. M., Yen, H. C. B., and Wall, J. D. (2010). Effect of the deletion of qmoABC and the promoter-distal gene encoding a hypothetical protein on sulfate reduction in *Desulfovibrio vulgaris* Hildenborough. *Appl. Environ. Microbiol.* **76,** 5500–5509.

CHAPTER TWENTY-THREE

Modification of the Genome of *Rhodobacter sphaeroides* and Construction of Synthetic Operons

Paul R. Jaschke,* Rafael G. Saer,* Stephan Noll,[†] *and* J. Thomas Beatty*

Contents

Abstract

The α-proteobacterium *Rhodobacter sphaeroides* is an exemplary model organism for the creation and study of novel protein expression systems, especially membrane protein complexes that harvest light energy to yield electrical energy. Advantages of this organism include a sequenced genome, tools for genetic engineering, a well-characterized metabolism, and a large membrane surface area when grown under hypoxic or anoxic conditions. This chapter provides a framework for the utilization of *R. sphaeroides* as a model organism for membrane protein expression, highlighting key advantages and shortcomings. Procedures covered in this chapter include the creation of chromosomal gene deletions, disruptions, and replacements, as well as the construction of a

* Department of Microbiology and Immunology, University of British Columbia, Life Sciences Centre, Vancouver, British Columbia, Canada
† Gene Bridges GmbH, Im Neuenheimer Feld 584, Heidelberg, Germany

Methods in Enzymology, Volume 497
ISSN 0076-6879, DOI: 10.1016/B978-0-12-385075-1.00023-8

synthetic operon using a model promoter to induce expression of modified photosynthetic reaction center proteins for structural and functional analysis.

1. Introduction

The defining trait of purple photosynthetic bacteria, such as in the genus *Rhodobacter*, is the ability to perform anoxygenic photosynthesis, an ancient form of photosynthesis that does not generate oxygen (Xiong and Bauer, 2002). Both *Rhodobacter sphaeroides* (formerly *Rhodopseudomonas sphaeroides*) and *Rhodobacter capsulatus* have been studied since the 1940s, but herein we focus on *R. sphaeroides*. *R. sphaeroides* is a member of the α-proteobacteria, and descended from bacteria that eventually became mitochondria in eukaryotic cells (Yang *et al.*, 1985). Interestingly, most genome-sequenced strains of *R. sphaeroides* contain two circular chromosomes, as well as a variable number of plasmids (NCBI accession numbers CP000143-CP000147; DQ232586-DQ232587). Other characteristics of *R. sphaeroides* include an aerobic metabolism that functions in O_2 concentrations from atmospheric down to microaerophilic, in addition to anaerobic respiration using substances such as dimethyl sulfoxide (DMSO) as a terminal electron acceptor (Zannoni *et al.*, 2009). *R. sphaeroides* is able to fix N_2 as well as CO_2 (Calvin–Benson–Bassham cycle), although organic compounds are preferred over CO_2 (Madigan, 1995; Romagnoli and Tabita, 2009).

The key environmental signal that controls the bioenergetic properties of *R. sphaeroides* is the concentration of O_2, and cultures shift from aerobic to photosynthetic metabolism in response to a reduction in oxygen tension. During adaptation, the cytoplasmic membrane expands and folds inward to create a highly invaginated intracytoplasmic membrane system, which houses the photosynthetic apparatus (Chory *et al.*, 1984). Photosynthetic membrane protein complexes execute cyclic electron transfer, and pump protons from the cytoplasm to the periplasmic space. This electrochemical gradient is utilized by the cell to generate ATP. In autotrophic growth, electrons may be bled out of the electron transport chain for use in synthesis of NAD(P)H, while electrons enter by oxidation of H_2 (Herter *et al.*, 1997) or reduced sulfur compounds such as H_2S (Brune, 1995).

The core of the photosynthetic apparatus is a dimer of the reaction center (RC) complexes surrounded by the light-harvesting 1 (LH1) complex and the PufX protein (Qian *et al.*, 2005; Scheuring *et al.*, 2004, 2005). The LH1 complex consists of α/β heterodimer subunits that bind two coupled bacteriochlorophyll (BChl) pigments that absorb light (Kohler, 2006). About 24–28 LH1 heterodimers form an S-shaped structure surrounding two RCs (Fig. 23.1A). Each RC contains three proteins called L, M, and H; the structurally similar (33% sequence identity) RC L and M proteins consist largely of five transmembrane helices with pseudo-twofold symmetry, whereas the RC H protein has only one transmembrane helix

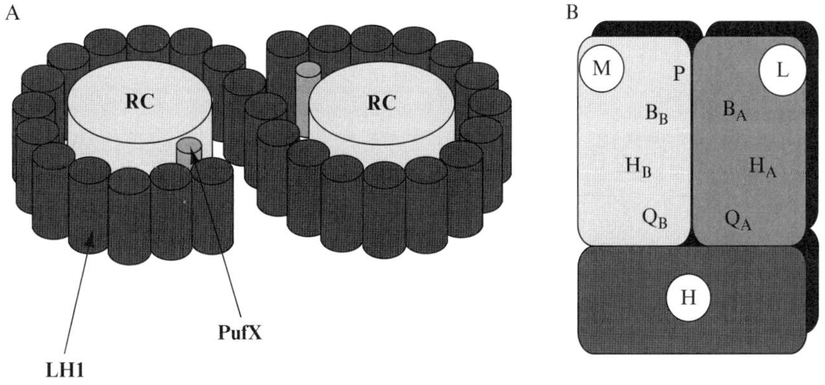

Figure 23.1 Representation of the *Rhodobacter sphaeroides* photosynthetic reaction center and light-harvesting 1 complex. (A) the RC/LH1/PufX supercomplex dimer: RC, reaction center complex; LH1, light-harvesting 1 complex subunit, an α/β protein heterodimer; PufX, the PufX protein (needed for quinone exchange). (B) The RC artificially in isolation from the LH1 to show the organization of the RC proteins and cofactors: M, the RC M protein; L, the RC L protein; H, the RC H protein; P, the "special pair" or "primary donor" dimer of BChls; B_B and B_A, accessory BChls; H_B and H_A, bacteriopheophytins; Q_B and Q_A, quinones.

and a large cytoplasmic domain (Fig. 23.1B; Yeates *et al.*, 1988). The overlapping *pufL* and *pufM* genes encoding the RC L and M proteins are located within a 66.7 kb region of chromosome 1 called the photosynthesis gene cluster, or PSGC (Fig. 23.2). The *pufLM* genes are flanked by the *pufBA* genes (encoding LH1 α/β proteins) upstream and *pufX* downstream, all of which are transcribed as a polycistronic mRNA. In contrast, the RC H gene *puhA* is transcribed from an operon 38 kb distant from the *puf* genes, but still within the PSGC (Chen *et al.*, 1998; Donohue *et al.*, 1986). The *pucBAC* operon, a further 22 kb separated from *puhA* (Fig. 23.2), codes for the LH2 complex which acts as a variable-sized antenna, funneling photons toward the RC–LH1 core complex (Gabrielsen *et al.*, 2009).

A large catalog of research has accumulated on *R. sphaeroides*, and many of the basic parameters of metabolism have been examined and quantified, allowing the construction of models of the electron transport chain (Klamt *et al.*, 2008). Further, transcriptomic and proteomic characterization of cells in various growth modes have been published (Arai *et al.*, 2008; Callister *et al.*, 2006; Zeng *et al.*, 2007), along with an understanding of oxygen and redox signal transduction pathways and the mechanism of regulation of some promoters (Bauer *et al.*, 2009; Eraso and Kaplan, 2009; Eraso *et al.*, 2008; Moskvin *et al.*, 2007; Oh and Kaplan, 2000; Roh *et al.*, 2004).

R. sphaeroides has garnered considerable interest for biotechnology applications. Recently, a large multidisciplinary project has been initiated to look into biological hydrogen production of this organism (Curtis *et al.*, 2010). Additionally, *R. sphaeroides* has been targeted in a structural genomics initiative

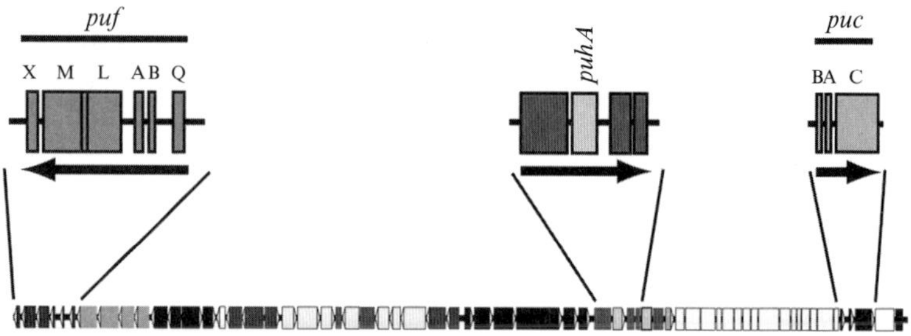

Figure 23.2 The *R. sphaeroides* photosynthesis gene cluster. A 66.7 kb section of chromosome 1 from coordinates 1980460–2047208. This region of the chromosome contains genes for the formation of BChl and carotenoid pigments, some regulatory proteins, and the structural proteins of photosynthetic complexes. The *pufQBALMX* operon (coding for PufQ, a protein involved in BChl synthesis; LH1 β and α proteins; RC L and M proteins; PufX), a segment of the *puhA* (RC H) operon, and *pucBAC* (LH2) operons are expanded to show relative gene sizes and direction of transcription (arrows). Chromosome sequence is available from NCBI Entrez Genome database (accession number NC_007493.1).

(Laible *et al.*, 2004, 2009) for use as a high-throughput membrane protein expression system, because of its inducible promoters and extensive and well characterized membrane system. Additional potential applications include: metal nanoparticle synthesis (Narayanan and Sakthivel, 2010), reduction of odors in large-scale farming waste (Kobayashi, 1995; Schweizer, 2003), heavy metal bioremediation (Italiano *et al.*, 2009; Van Fleet-Stalder *et al.*, 2000), production of plant hormones (Rajasekhar *et al.*, 1999a,b), and photovoltaics (Lebedev *et al.*, 2008; Takshi *et al.*, 2009). Thus, *R. sphaeroides* has much to offer synthetic biologists who are willing to leave the more familiar model organisms behind. In fact, several teams (Utah State and Washington University) in the 2009 International Genetically Engineered Machine (iGEM) competition used *R. sphaeroides* as a chassis for their projects.

This chapter is intended as an introduction to a bacterial chassis that has potential to create devices and study phenomena outside of the realm of possibility of the dominant model organisms. We will outline methods to: (1) delete or disrupt *R. sphaeroides* genes and (2) construct synthetic operons expressed *in trans* from broad-host range plasmids.

2. Gene Disruption and Deletion

2.1. General scheme

This section outlines how to create a null mutation (knockout) in a *R. sphaeroides* gene, using strain 2.4.1 as an example. The two circular chromosomes of 3.2 Mb (RefSeq NC_007493) and 0.9 Mb (NC_007494) in length average 69% G+C. There are also five plasmids ranging from 37 to 114 kb in length. Essential functions are shared by both chromosomes, with the majority of the genes that encode the photosynthetic apparatus found grouped in the PSGC on chromosome 1 (Fig. 23.2; Choudhary and Kaplan, 2000).

Unlike *Escherichia coli*, there is not an efficient transformation or electroporation method for introduction of DNA into *R. sphaeroides*, perhaps because of endogenous restriction enzymes (D. Jun and J. T. Beatty, unpublished). Therefore, the directed genetic manipulation of the genome of *R. sphaeroides* requires the construction of circular gene replacement suicide vectors that encode for the desired gene modification(s).

As outlined in Fig. 23.3, the general scheme for generation of a directed *R. sphaeroides* gene knockout consists of several phases: (1) cloning the gene

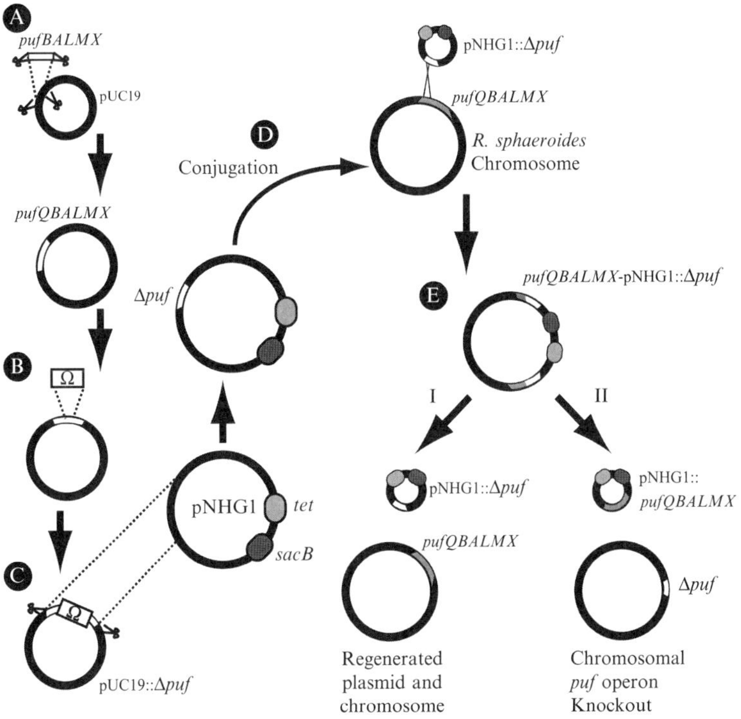

Figure 23.3 Generation of knockout in the *R. sphaeroides* genome. (A) Cloning of the *pufQBALMX* operon and flanking sequences in plasmid pUC19. (B) Insertion of an antibiotic-resistance gene (Ω) in place of the *pufQBALMX* genes. (C) Transfer of Ω and *puf* operon flanking sequences (*Δpuf*), to suicide plasmid pNHG1 (*tet*, tetracycline resistance; *sacB*, levansucrase gene). (D) Conjugation of recombinant suicide plasmid and integration into a *pufQBALMX* flanking region by homologous recombination. (E) Resolution of the cointegrate leading to a mixed wild type (I) and mutant (II) cell population.

of interest in *E. coli*, (2) disruption of the plasmid-borne gene sequence by deletion and/or insertion of an antibiotic-resistance marker, (3) conjugation of a disrupted copy of the gene into *R. sphaeroides*, (4) selection and screening for an initial single-recombination, followed by counter-selection for a double-recombination event.

The gene of interest is first amplified from the chromosome of *R. sphaeroides*, using PCR protocols modified for high GC-content DNA. We have found that Platinum *Pfx* (Invitrogen) and Vent (NEB) work well. DMSO is routinely used at 3–10% (v/v) final concentration in the PCR mixes, to lower melting temperature and reduce secondary structure of the chromosomal DNA template. In our hands, enhancing buffers provided in PCR enzyme kits that are designed for use with high-GC templates do not work as well as DMSO added to the kit's standard buffer.

The PCR product is subcloned by classical methods into a high copy number *E. coli*-compatible vector (Fig. 23.3A), commonly pUC19 (Messing, 1983). Alternatively the commercially available TA (Zhou *et al.*, 1995) or TOPO cloning systems (Shuman, 1994) may be used. Upon plasmid purification from *E. coli*, the gene of interest is then cut with appropriate restriction enzyme(s) and either ligated to yield an unmarked deletion or, as shown in Fig. 23.3B, ligated with an antibiotic-resistance marker such as the spectinomycin resistance gene on the Ω cartridge (Prentki and Krisch, 1984), or the *neo* gene on the KIXX cartridge (Barany, 1985). Both resistance markers are functional in *E. coli* and *R. sphaeroides*, which simplifies subsequent selection and screening steps. The disrupted gene should have >0.4 kb of continuous flanking sequence identity to the chromosomal target locus on each end, to allow for efficient homologous recombination into the genome and subsequent recovery of the desired mutant.

Tangentially, we note that disruption of a gene 5′ of other genes in an operon may interfere with transcription of 3′ genes (a polar effect). The chance of a polar effect cannot be eliminated, but it can be reduced best by use of a translationally inframe deletion, or to a lesser likelihood by use of the KIXX cartridge in the same transcriptional orientation as the disrupted gene. However, to rule out polar effect(s), complementation in *trans* by a plasmid borne wild-type gene should be performed in the final *R. sphaeroides* mutant. The Ω cartridge was designed to halt translation and transcription (Prentki and Krisch, 1984), and is almost guaranteed to have a polar effect when inserted 3′ of a single promoter that drives transcription of multiple genes in an operon. The different phenotypic effects of KIXX (usually nonpolar) and Ω cartridge (polar) disruption of genes may be used to experimentally define and dissect operons of uncertain composition, as described for *R. capsulatus* (Aklujkar *et al.*, 2000) and *R. sphaeroides* (Chen *et al.*, 1998).

As shown in Fig. 23.3C, the mutant gene is transferred to an appropriate suicide plasmid that encodes (i) an origin of replication (usually from ColE1 or p15A) that allows maintenance in *E. coli* but not in *R. sphaeroides*, and (ii) an origin of transfer sequence (*oriT*) needed for conjugation from the appropriate *E. coli* strain to *R. sphaeroides*. Examples of suicide vectors in common use are the pLO-series, the pSUP-series, and pNHG1 (Jeffke *et al.*, 1999; Lenz *et al.*, 1994; Simon *et al.*, 1983).

The preferred diparental conjugation method utilizes *E. coli* donor strain S17-1, which contains key genes that facilitate the transfer of *oriT*-containing plasmids (such as the suicide plasmid) into *R. sphaeroides* cells (Simon *et al.*, 1983). Triparental mating is also efficient, utilizing a DH10B or other auxotrophic donor strain harboring the plasmid of interest along with an HB101(pRK2013) helper strain (Ditta *et al.*, 1985).

After transfer to *R. sphaeroides*, the suicide plasmid cannot replicate, and so selection for the antibiotic-resistance marker on the plasmid ensures that cells in colonies that arise on an agar medium containing the appropriate antibiotic have the plasmid integrated into the chromosome by homologous recombination (Fig. 23.3D). Using the methodology described above, the frequency of RecA-dependent homologous recombination is on the order of $\sim 10^{-4}$ per potential plasmid recipient for a single event (a crossover on one or the other side of the disrupted gene).

After isolation of strains where a single crossover has occurred, growth in liquid culture *without selection* well into stationary phase (5–10 generations) allows time for a second homologous recombination to occur. As shown in Fig. 23.3E, there are two possibilities for this event: one is that the suicide vector will reform and leave the genome, thereby restoring the state prior to the first recombination (Fig. 23.3E(I)); alternatively, the disrupted copy of the gene of interest may be left in the chromosome, while the suicide plasmid backbone leaves the chromosome with the wild-type copy of the gene (Fig. 23.3E(II)).

In addition to a selectable marker, the backbone of a suicide plasmid may contain a counter-selectable marker, which under appropriate growth conditions, allows for selection of colonies of cells that have undergone plasmid loss. A frequently used system is the *sacB* gene from the Gram-positive *Bacillus subtilis*, which allows for counter-selection by growing cells on an agar medium containing a high concentration of sucrose (Gay *et al.*, 1985). In the presence of sucrose, the *sacB*-encoded levansucrase polymerizes fructose from the degradation of sucrose that, in Gram-negative species, inhibits colony formation (Gay *et al.*, 1983; Steinmetz *et al.*, 1983). Apparently, the native promoter of *sacB* functions in *R. sphaeroides*, but *sacB* has also been put under the control of the *R. sphaeroides puc* promoter in pJE2864 (Eraso and Kaplan, 2002), to improve the efficiency of the selective process.

Thus, by using a *sacB*-containing plasmid, plating recipient cells on an agar medium that contains sucrose (10–15%) results in a great enrichment of cells that have lost the plasmid. In the case where an antibiotic-resistance marker has been inserted into the gene of interest, the relevant antibiotic is included in the agar medium to inhibit the growth of cells that have retained the native gene. Colonies are screened for the presence of the disrupted or deleted gene, as indicated by a change in the size of PCR product, using primers that flank the gene of interest.

2.2. Construction of the ΔRCLH mutant

The ΔRCLH mutant (Tehrani and Beatty, 2004) serves as a good example of the gene disruption and deletion techniques described above. This mutant contains deletions of the *puhA* gene (encodes the RC H protein), the *pucBA* genes (encode the LH2 proteins), and the *pufBALMX* operon (coding for LH1, RCL, RC M, and PufX proteins). These modifications resulted in a mutant that does not contain any structural protein of the photosynthetic complexes, which was created to serve as a null background in which modified photosynthetic complexes could be expressed (see Section 3).

The pathway to the ΔRCLH mutant began with the creation of a translationally inframe deletion of the *puhA* gene (Chen *et al.*, 1998). The deletion was obtained using a plasmid-borne copy of *puhA* in *E. coli* and "loop-out" oligonucleotide mutagenesis, to replace a 561 bp segment of the coding region with an *Eco*R V site. This technique removes a section of DNA by using oligonucleotide primers to bridge two separate parts of the gene, causing the intervening sequence to loop-out and be lost upon amplification. The modified *puhA* gene was then inserted into the suicide vector pSUP203 and conjugated into the *R. sphaeroides* strain PUH1 (Chen *et al.*, 1998). After selection for tetracycline resistance resulting from a single homologous recombination event, the resultant strain was grown in liquid medium and plated onto solid medium in the absence of selection. Colonies were replica-plated to identify colonies that had lost the tetracycline resistance marker on the suicide plasmid, because the pSUP203 vector lacks the counter-selection marker *sacB*. Tetracycline sensitive colonies were screened for a decrease in size of the *puhA* sequence by Southern blot hybridization and a clone was named ΔPUHA (Chen *et al.*, 1998).

The method outlined in Fig. 23.3 was used to delete the *pufBALMX* operon from the ΔPUHA strain, using the pNHG1::PUFDEL suicide plasmid. This suicide plasmid was constructed in several steps, starting with a modified pUC19 plasmid (pAli2) at the subcloning stage (Tehrani and Beatty, 2004). A 4.6 kb chromosomal DNA fragment containing the *puf* operon was cloned into pAli2 and modified by replacing the *Bsp*EI to *Bcl*I sequence (from *pufB* to *pufX*, inclusive) with a linker (Tehrani and Beatty, 2004). This markerless deletion was transferred as an *Eco*RI fragment into the suicide plasmid pNHG1 (Jeffke *et al.*, 1999) to generate pNHG1::

PUFDEL, which was conjugated into *R. sphaeroides* ΔPUHA, followed by selection for tetracycline resistance (integration by homologous recombination), and followed by counter-selection on a sucrose-containing medium, and screening for the desired double-crossover event. To delete the *pucBA* genes in the resultant strain, the construction and deployment of the pNHG1::DELPUC suicide plasmid utilized similar principles (Tehrani and Beatty, 2004).

With the creation of the mutant ΔRCLH, which lacks all photosynthetic complexes, we had a blank slate that allows expression of a wide variety of engineered photosynthetic complexes. We describe below how the ΔRCLH strain was used as a key ingredient in the expression of plasmid-borne synthetic operons of RC genes.

3. Construction of Synthetic Operons

The aim of this section is to describe how we initially created and expressed synthetic operons in *R. sphaeroides*. The main principles of design and implementation are similar to principles guiding work on *E. coli*, but several differences between these systems are highlighted.

We first turned to the design of synthetic operons to aid in the study of mutant RC proteins within the native host. Several general considerations that must be kept in mind when designing synthetic expression systems will be explored within the context of the *R. sphaeroides* host system: (1) operon objectives and composition, (2) utilization of a suitable background strain, (3) use or design of an appropriate expression vector, (4) choice of genetic control elements.

3.1. Operon objectives and composition

This example focuses on expressing site-directed mutants of endogenous genes within a synthetic operon in *R. sphaeroides*; see Laible *et al.* (2009) for a review of foreign gene expression in this host. Our general goal was to create a system for expression of variants of the RC genes to further our work on fundamental and applied aspects of RC structure and function (Lin *et al.*, 2009; Takshi *et al.*, 2009).

3.2. Host strain

We chose the ΔRCLH strain (Tehrani and Beatty, 2004) for this purpose, because it contains precise deletions of the genes encoding the photosynthetic complexes, as outlined in Section 2.2. Additionally, because this strain cannot grow photosynthetically without a functional RC complex, the photosynthetic growth phenotype served as a simple test of RC electron transfer efficiency. Although a deletion was made within the chromosomal *puhA* operon, the expression of downstream chromosomal genes was needed for maximal production of photosynthetic complexes (Aklujkar *et al.*, 2005; Chen *et al.*,1998). The determination of the capability for photosynthetic growth and measurement of photosynthetic culture growth rate, coupled with absorption spectroscopy of cells, are rapid and simple ways to evaluate the functional properties of RC variants.

Using the methods described in Section 2, it should be feasible to create many different types of *R. sphaeroides* host strains, depending on the process that is to be engineered. This methodology may be used to deliver novel genes and operons to the genome, as well as creating knockouts.

3.3. Expression vector

A well-designed vector backbone can simplify synthetic operon design and facilitate the genetic manipulations necessary for its construction. The approach is to modify a preexisting vector by tailoring of the backbone to the nature of their work. This may include adding or removing restriction sites, and other key sequences.

Plasmids are introduced into *R. sphaeroides* from *E. coli* by conjugation using shuttle vectors that are stably replicated within both organisms. Some examples of broad host-range plasmids currently used in *R. sphaeroides* research include: pRK415, pBBR1, pJRD215, and pATP19P (Davison *et al.*, 1987; Keen *et al.*, 1988; Kovach *et al.*, 1994; Tehrani and Beatty, 2004).

Typically, a synthetic operon would first be created in a small, high-copy *E. coli* vector such as pUC19 (Messing, 1983), and subsequently transferred to a broad host-range plasmid as a cluster of genes on a single DNA fragment. This is because most broad host-range plasmids are large, low copy number, and lack a wide variety of unique restriction sites—therefore, there are practical reasons for why it is easier to create a synthetic operon in an *E. coli* cloning vector before moving the operon into a plasmid capable of replication in *R. sphaeroides*.

To create a plasmid backbone for *R. sphaeroides* mutant RC expression, the hypoxia-inducible *puc* promoter (Lee and Kaplan, 1995) was inserted into the broad host-range plasmid pRK415 as a 0.75 kb *Hin*dIII fragment, along with part of the multiple cloning site of pUC19, such that the resultant plasmid pATP19P (Tehrani and Beatty, 2004) now had seven unique restriction sites for insertion of genes downstream of the promoter. A copy of the *puhA* gene was inserted downstream of the *puc* promoter as a 1.3 kb *Bam*HI fragment yielding plasmid pATSHR. The two additional (native or mutant) RC genes, *pufL* and *pufM*, were added by inserting a 4.5 kb *Eco*RI fragment that contains the *pufQBALMX* cluster (Tehrani and Beatty, 2004). Another derivative of pATP19P was created by adding a *puhA* gene modified by the addition of six histidine codons on the 3′-end of the gene (Abresch *et al.*, 2005). This plasmid, p6His-C, was found to yield amounts of the His-tagged RC H protein sufficient for purification of the RC or RC/LH1/PufX complex using Ni/NTA affinity chromatography (Abresch *et al.*, 2005; Jaschke and Beatty, 2007; Lin *et al.*, 2009). The C-terminal 6× His tag was used because an N-terminal 6× His tag disrupted RC formation (unpublished).

The plasmid pATSHR described above was used to express deletions of RC genes in *R. sphaeroides* strain ΔRCLH, to investigate protein–protein interactions and membrane-insertion (Tehrani and Beatty, 2004). The method we have typically used to create the desired RC gene mutants is to first create the desired modification to RC genes in an *E. coli* high copy number plasmid, then to transfer the mutant gene as either a *Bam*HI to *Sac*I fragment (for *puhA* mutants), or a *Sac*I to *Eco*RI fragment (*puf* mutants).

The synthetic RC expression operon was created using native coding sequences, but several specific modifications were necessary to achieve the desired results. Our first design of the synthetic operon was found to express RC in insufficient quantities to enable photosynthetic growth of the host strain (Fig. 23.4). A search for potential mRNA stem-loop structures indicated a sequence shortly after the 3′ end of the *puhA* gene that might attenuate transcription into the downstream *pufQBALMX* genes. Replacement of 83 bp, starting 3 bp downstream of the His-tagged *puhA* stop codon, with a *Sac*I restriction site yielded plasmid pRS1. It was found that *R. sphaeroides* strain ΔRCLH(pRS1) was capable of photosynthetic growth (Fig. 23.4). This synthetic RC expression system using the ΔRCLH host strain and plasmid pRS1 has been used by our group to rapidly create, express, and purify a large number of RCs with modifications of all three subunits.

We also considered several factors prior to the creation of this synthetic operon, including (1) the characteristics of the novel mRNA made from this synthetic operon, and whether it would be resistant to nucleases and allow an appropriate level of translation; (2) whether the RC genes could be expressed in *trans* or whether the assembly of the RC complex required *cis*-active factors not present on our plasmid. The successful generation of fully functional RC/LH1/PufX core complexes from our synthetic operon indicates that no essential information was encoded in the relative genome locations of the RC genes within the PSGC.

3.4. Regulation of synthetic operon expression

Regulation is one of the most important features of a synthetic operon, and regardless of whether genes are expressed in the native or heterologous host, cryptic regulatory sequences may be present in the coding or intergenic regions as was seen in the first iteration of our synthetic RC operon (Fig. 23.4A).

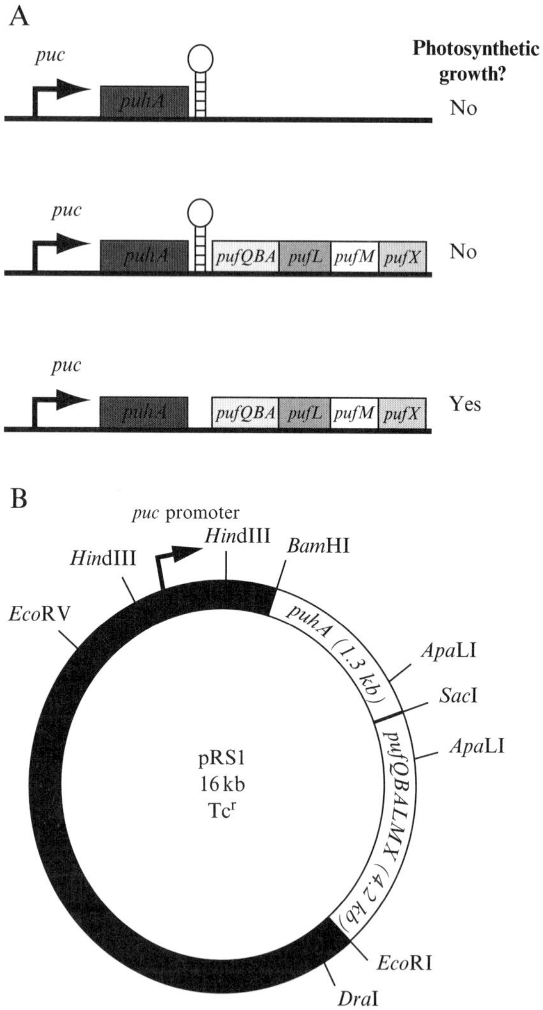

Figure 23.4 Design and construction of the synthetic RC/LH1 gene cluster. (A) The *puc* promoter and *puhA* gene, including *puhA* downstream sequences that later were found to attenuate transcription, were inserted into a plasmid pRK415 backbone; the *pufQBALMX* genes were inserted downstream of *puhA* sequences, but the expression of RC genes was found to be insufficient for photosynthetic growth of strain ΔRCLH. Sequences immediately following the *puhA* stop codon were removed, and it was found that RC gene expression was sufficient for photosynthetic growth of ΔRCLH. (B) Map of plasmid pRS1, which contains the final synthetic operon; bent arrow indicates the *puc* promoter; key unique restriction sites are indicated; Tc^r, plasmid encodes resistance to tetracycline.

Genes within the operon should be coupled to appropriate transcriptional promoters and attenuators to obtain desired levels of mRNA synthesis, and to appropriate ribosome binding sites (RBSs) for desired levels of mRNA translation. The stability of mRNA also affects the level of gene expression, and it is interesting that the *Rhodobacter puf* operon was an early model system in this area (Chen *et al.*, 1988), although *E. coli* has emerged as the prokaryotic model system (Schuck *et al.*, 2009). Codon sequence composition affects translation efficiency, and codon usage varies significantly between species, approximately as a function of genome GC-content (Kane, 1995; Lee *et al.*, 2009).

If construction of a synthetic operon requires isolation of genes of interest from the native host, as opposed to gene synthesis, it may or may not be desirable to include the native regulatory sequences. This decision is colored by how well defined these elements are in the host strain.

Little is known about fundamental properties of *Rhodobacter* transcription promoters, except that they often differ from well-understood *E. coli* promoters in −10 and −35 sequence composition (Leung, 2010; Swem *et al.*, 2001), and hence may not be recognized in a heterologous host. We routinely use the *R. sphaeroides puc* promoter because it is thought to be a strong promoter, and can be regulated by control of culture aeration. A fructose-inducible promoter from *R. capsulatus* was reported to have a high-dynamic range (Duport *et al.*, 1994), and presumably the *R. sphaeroides* homologue would function similarly. Recently, a description was published of the *puc* promoter fused to *lacO* under the control of *lacI*^{q}, so that the promoter is induced under low concentrations of O_2 only when IPTG is present (Hu *et al.*, 2010).

Rhodobacter rho-independent transcription terminators appear to be similar to *E. coli* terminators (Chen *et al.*, 1988). Several algorithms to detect transcriptional terminators, such as TransTermHP (Kingsford *et al.*, 2007) (http://transterm.cbcb.umd.edu/) or RNAfold (Hofacker *et al.*, 1994; McCaskill, 1990; Zuker and Stiegler, 1981) (http://rna.tbi.univie.ac.at/cgi-bin/RNAfold.cgi) can be used to scan the 3′ end of ORFs to find potential terminators that may be present.

The RBS provides another level of regulation (Nakamoto, 2009). If a gene lacks an obvious RBS, it is possible to introduce an RBS sequence to ensure that translation initiation does not pose a bottleneck in protein expression. An analysis of RBS effects on protein expression in *E. coli* was published (Salis *et al.*, 2009), and an online RBS calculation tool for genome-sequenced prokaryotes may be found at https://salis.psu.edu/software/.

Genes may be synthesized for expression in another species, and the coding sequences modified to match the codon usage frequency of the heterologous host (Villalobos *et al.*, 2006). Issues of translation efficiency and relationships to tRNA abundance in *R. sphaeroides* are somewhat unclear, despite recent advances (Cannarozzi *et al.*, 2010; Tuller *et al.*, 2010).

4. Future Directions

4.1. Genome modification

Although the genetics of *R. sphaeroides* allows modification of the genome in a reasonable time-frame, work on this organism is still far from the speed and ease of *E. coli* genetics. Some of the difficulty with working with this organism is due to the high GC-content of the genome, and the longer doubling-time of *R. sphaeroides* (3–5 h) compared to *E. coli* (20–30 min). Neither of these factors can be remedied in the near future, but additional tools could aid the *in vitro* and *in vivo* manipulation of the construct prior to the conjugation step.

Red/ET cloning ("recombineering") has come to be recognized as superior technology for the size- and sequence-independent manipulation of DNA in *E. coli* and related *Enterobacteriacea*. Cells that express λ phage-derived *red* genes, or their functional prophage *rac* equivalents, promote base-precise exchange of linear single- or double-stranded donor DNA into the bacterial chromosome. Therefore, only short flanking homology arms are required; see Sawitzke *et al.* (2007) and references therein.

Recombineering methods for direct genome targeting in non-*Enterobacteriaceae* have been recently developed for *Mycobacterium tuberculosis* (van Kessel and Hatfull, 2007) and *Pseudomonas syringae* (Swingle *et al.*, 2010), but no such system exists for *R. sphaeroides*. However, methods for plasmid recombineering in *E. coli* (Noll *et al.*, 2009; Thomason *et al.*, 2007) facilitate the engineering of gene replacement vectors for *Rhodobacter* and other species as outlined below.

A drug marker suitable for *E. coli* and *Rhodobacter* has to be flanked by ~50-base tails homologous to the subcloned target region. This can easily be achieved by PCR. Thereby, the primer design determines exactly where the cassette recombines into the plasmid because no specific recombination sites are required. The recombination step takes place *in vivo*. To minimize unwanted side effects of plasmid recombineering, that is, multimer formation and mixtures of mutated and parental plasmids, low amounts of substrate plasmid (~10 ng) and linear marker (~100 ng) should be coelectroporated into Red/ET proficient *E. coli* cells. Nevertheless, the isolation of monomeric recombinant plasmids requires careful monitoring of the plasmid topology and a retransformation step (Noll *et al.*, 2009; Thomason *et al.*, 2007).

Recombineering approaches allow freedom from the need for restriction cleavage sites, and an antibiotic-resistance marker can be recombined into an appropriate suicide plasmid to obtain a base-precise disruption or deletion of the subclone target gene. However, as outlined in Section 2.1, a modification of a cloned gene does not necessarily need to be marked by a resistance cassette. Interestingly, all kinds of markerless plasmid modifications (deletion, insertion, replacement) can be introduced in a two-step "hit and fix" approach (Noll *et al.*, 2009).

Therefore, the PCR-amplified marker used in the first Red/ET step has to introduce a unique restriction site into the target plasmid. This can easily be achieved by oligo design. Following drug selection and isolation of recombinant plasmids, nonselectable DNA coding for all kinds of modification(s) is used to replaces the cassette and the unique restriction site in the second step. Upon selective digestion of parental plasmids (i.e., unique restriction site elimination) and retransformation, recombined plasmids are obtained with reasonable efficiency. Given its flexibility, plasmid recombineering should prove to be a welcome alternative for the construction of gene replacement vectors in *R. sphaeroides*.

Additionally, use of the Flp/FRT system from *Saccharomyces cerevisiae* (Sadowski, 1995; Schweizer, 2003) and the Cre/LoxP system from bacteriophage P1 (Abremski *et al.*, 1986; Sternberg *et al.*, 1986), which are commonly used in *E. coli* for marker removal, would facilitate the generation of markerless *R. sphaeroides* mutants in a fraction of the time of the traditional methods. To our knowledge, no group is actively working on adapting these systems for use in *R. sphaeroides*.

4.2. Synthetic operons

It is often desirable when generating synthetic operons to utilize proteins stemming from a wide variety of different species. This "mix and match" approach may prove useful in the creation of novel protein systems with function not found in nature. In such cases, there may be difficulty in finding a good host strain, because no one strain may be able to express at appropriate levels all the heterologous genes in a gene system. In order to overcome such a barrier, it may be necessary to refactor coding sequences such that they are better suited for a particular organism, as well as change the regulatory elements to match the heterologous host. For example, Widmaier *et al.* (2009) changed both the codons and regulatory sequences of spider silk genes to obtain high-level synthesis and secretion of spider silk in *Salmonella* SPI-1 T3SS.

REFERENCES

Abremski, K., Wierzbicki, A., Frommer, B., and Hoess, R. H. (1986). Bacteriophage P1 Cre-LoxP site-specific recombination—Site-specific DNA topoisomerase activity of the Cre recombination protein. *J. Biol. Chem.* **261,** 391–396.

Abresch, E. C., Axelrod, H. L. A., Beatty, J. T., Johnson, J. A., Nechushtai, R., and Paddock, M. L. (2005). Characterization of a highly purified, fully active, crystallizable RC/LH1/PufX core complex from *Rhodobacter sphaeroides*. *Photosynth. Res.* **86,** 61–70.

Aklujkar, M., Harmer, A. L., Prince, R. C., and Beatty, J. T. (2000). The *orf162b* sequence of *Rhodobacter capsulatus* encodes a protein required for optimal levels of photosynthetic pigment–protein complexes. *J. Bacteriol.* **182,** 5440–5447.

Aklujkar, M., Prince, R. C., and Beatty, J. T. (2005). The PuhB protein of *Rhodobacter capsulatus* functions in photosynthetic reaction center assembly with a secondary effect on light-harvesting complex 1. *J. Bacteriol.* **187,** 1334–1343.

Arai, H., Roh, J. H., and Kaplan, S. (2008). Transcriptome dynamics during the transition from anaerobic photosynthesis to aerobic respiration in *Rhodobacter sphaeroides* 2.4.1. *J. Bacteriol.* **190,** 286–299.

Barany, F. (1985). Single-stranded hexameric linkers—A system for in-phase insertion mutagenesis and protein engineering. *Gene* **37,** 111–123.

Bauer, C. E., Setterdahl, A., Wu, J., and Robinson, B. R. (2009). Regulation of gene expression in response to oxygen tension. *In* "The Purple Phototrophic Bacteria," (C. N. Hunter, F. Daldal, M. C. Thurnauer, and J. T. Beatty, eds.), pp. 707–725. Springer Science, Dordrecht, The Netherlands.

Brune, D. C. (1995). Sulfur compounds as photosynthetic electron donors. *In* "Anoxygenic Photosynthetic Bacteria," (R. E. Blankenship, M. T. Madigan, and C. E. Bauer, eds.), pp. 847–870. Kluwer Academic, Dordrecht, The Netherlands.

Callister, S. J., Nicora, C. D., Zeng, X., Roh, J. H., Dominguez, M. A., Tavano, C. L., Monroe, M. E., Kaplan, S., Donohue, T. J., Smith, R. D., and Lipton, M. S. (2006). Comparison of aerobic and photosynthetic *Rhodobacter sphaeroides* 2.4.1 proteomes. *J. Microbiol. Methods* **67,** 424–436.

Cannarozzi, G., Schraudolph, N. N., Faty, M., von Rohr, P., Friberg, M. T., Roth, A. C., Gonnet, P., Gonnet, G., and Barral, Y. (2010). A role for codon order in translation dynamics. *Cell* **141,** 355–367.

Chen, C.-Y. A., Beatty, J. T., Cohen, S. N., and Belasco, J. G. (1988). An intercistronic stem-loop structure functions as an mRNA decay terminator necessary but insufficient for *puf* mRNA stability. *Cell* **52,** 609–619.

Chen, X. Y., Yurkov, V., Paddock, M. L., Okamura, M. Y., and Beatty, J. T. (1998). A *puhA* gene deletion and plasmid complementation system for site directed mutagenesis studies of the reaction center H protein of *Rhodobacter sphaeroides*. *Photosynth. Res.* **55,** 369–373.

Chory, J., Donohue, T. J., Varga, A. R., Staehelin, L. A., and Kaplan, S. (1984). Induction of the photosynthetic membranes of *Rhodopseudomonas sphaeroides*: Biochemical and morphological studies. *J. Bacteriol.* **159,** 540–554.

Choudhary, M., and Kaplan, S. (2000). DNA sequence analysis of the photosynthesis region of *Rhodobacter sphaeroides* 2.4.1. *Nucleic Acids Res.* **28,** 862–867.

Curtis, W., Chapelle, J., Logan, B. E., and Salis, H. (2010). Development of *Rhodobacter* as a versatile platform for fuels production. (PI: W. Curtis; with J. Chapelle, B.E. Logan, H. Salis). ArpaE, June 2010–May 2013. The Pennsylvania State University.

Davison, J., Heusterspreute, M., Chevalier, N., Hathi, V., and Brunel, F. (1987). Vectors with restriction site banks. pJRD215, a wide-host-range cosmid vector with multiple cloning sites. *Gene* **51,** 275–280.

Ditta, G., Schmidhauser, T., Yakobson, E., Lu, P., Liang, X. W., Finlay, D. R., Guiney, D., and Helinski, D. R. (1985). Plasmids related to the broad host range vector, pRK290, useful for gene cloning and for monitoring gene expression. *Plasmid* **13,** 149–153.

Donohue, T. J., Hoger, J. H., and Kaplan, S. (1986). Cloning and expression of the *Rhodobacter sphaeroides* reaction center H gene. *J. Bacteriol.* **168,** 953–961.

Duport, C., Meyer, C., Naud, I., and Jouanneau, Y. (1994). A new gene-expression system based on a fructose-dependent promoter from *Rhodobacter capsulatus*. *Gene* **145,** 103–108.

Eraso, J. M., and Kaplan, S. (2002). Redox flow as an instrument of gene regulation. *Methods Enzymol.* **348,** 216–229.

Eraso, J. M., and Kaplan, S. (2009). Regulation of gene expression by PrrA in *Rhodobacter sphaeroides* 2.4.1: Role of polyamines and DNA topology. *J. Bacteriol.* **191,** 4341–4352.

Eraso, J. M., Roh, J. H., Zeng, X., Callister, S. J., Lipton, M. S., and Kaplan, S. (2008). Role of the global transcriptional regulator PrrA in *Rhodobacter sphaeroides* 2.4.1: Combined transcriptome and proteome analysis. *J. Bacteriol.* **190,** 4831–4848.

Gabrielsen, M., Gardiner, A. T., and Cogdell, R. J. (2009). Peripheral complexes of purple bacteria. *In* "The Purple Phototrophic Bacteria," (C. N. Hunter, F. Daldal, M. C. Thurnauer, and J. T. Beatty, eds.), pp. 135–153. Springer Science.

Gay, P., Le Coq, D., Steinmetz, M., Ferrari, E., and Hoch, J. A. (1983). Cloning structural gene *sacB*, which codes for exoenzyme levansucrase of *Bacillus subtilis*: Expression of the gene in *Escherichia coli*. *J. Bacteriol.* **153,** 1424–1431.

Gay, P., Lecoq, D., Steinmetz, M., Berkelman, T., and Kado, C. I. (1985). Positive selection procedure for entrapment of insertion-sequence elements in Gram-negative bacteria. *J. Bacteriol.* **164,** 918–921.

Herter, S. M., Kortlüke, C. M., and Drew, G. (1997). Complex I of *Rhodobacter capsulatus* and its role in reverted electron transport. *Arch. Microbiol.* **169,** 98–105.

Hofacker, I. L., Fontana, W., Stadler, P. F., Bonhoeffer, L. S., Tacker, M., and Schuster, P. (1994). Fast folding and comparison of RNA secondary structures. *Monatsh. Chem.* **125,** 167–188.

Hu, Z. L., Zhao, Z. P., Pan, Y., Tu, Y., and Chen, G. P. (2010). A powerful hybrid *puc* operon promoter tightly regulated by both IPTG and low oxygen level. *Biochemistry (Moscow)* **75,** 519–525.

Italiano, F., Buccolieri, A., Giotta, L., Agostiano, A., Valli, L., Milano, F., and Trotta, M. (2009). Response of the carotenoidless mutant *Rhodobacter sphaeroides* growing cells to cobalt and nickel exposure. *Int. Biodeter. Biodegrad.* **63,** 948–957.

Jaschke, P. R., and Beatty, J. T. (2007). The photosystem of *Rhodobacter sphaeroides* assembles with zinc-bacteriochlorophyll in a *bchD* (magnesium-chelatase) mutant. *Biochemistry* **46,** 12491–12500.

Jeffke, T., Gropp, N. H., Kaiser, C., Grzeszik, C., Kusian, B., and Bowien, B. (1999). Mutational analysis of the *cbb*$_3$ operon (CO_2 assimilation) promoter of *Ralstonia eutropha*. *J. Bacteriol.* **181,** 4374–4380.

Kane, J. F. (1995). Effects of rare codon clusters on high-level expression of heterologous proteins in *Escherichia coli*. *Curr. Opin. Biotech.* **6,** 494–500.

Keen, N. T., Tamaki, S., Kobayashi, D., and Trollinger, D. (1988). Improved broad-host-range plasmids for DNA cloning in gram-negative bacteria. *Gene* **70,** 191–197.

Kingsford, C. L., Ayanbule, K., and Salzberg, S. L. (2007). Rapid, accurate, computational discovery of Rho-independent transcription terminators illuminates their relationship to DNA uptake. *Genome Biol.* **8,** R22.

Klamt, S., Grammel, H., Straube, R., Ghosh, R., and Gilles, E. D. (2008). Modeling the electron transport chain of purple non-sulfur bacteria. *Mol. Syst. Biol.* **4,** 156.

Kobayashi, M. (1995). Waste remediation and treatment using anoxygenic phototrophic bacteria. *In* "Anoxygenic Photosynthetic Bacteria," (R. E. M. Blankenship, T. Michael, and C. E. Bauer, eds.), pp. 1269–1282. Springer, Dordrecht, The Netherlands.

Kohler, J. (2006). Single molecule spectroscopy of pigment protein complexes from purple bacteria. *In* "Chlorophylls and Bacteriochlorophylls: Biochemistry, Biophysics, Functions and Applications," (B. Grimm, R. J. Porra, W. Rudiger, and H. Scheer, eds.), Vol. 25, pp. 309–321. Springer, Dordrecht.

Kovach, M. E., Phillips, R. W., Elzer, P. H., Roop, R. M., 2nd, and Peterson, K. M. (1994). pBBR1MCS: A broad-host-range cloning vector. *Biotechniques* **16,** 800–802.

Laible, P. D., Scott, H. N., Henry, L., and Hanson, D. K. (2004). Towards higher-throughput membrane protein production for structural genomics initiatives. *J. Struct. Funct. Genomics* **5,** 167–172.

Laible, P. D., Mielke, D. L., and Hanson, D. K. (2009). Foreign gene expression in photosynthetic bacteria. *In* "The Purple Phototrophic Bacteria," (C. N. Hunter, F. Daldal, M. C. Thurnauer, and J. T. Beatty, eds.), pp. 839–860. Springer Science, Dordrecht, The Netherlands.

Lebedev, N., Trammell, S. A., Tsoi, S., Spano, A., Kim, J. H., Xu, J., Twigg, M. E., and Schnur, J. M. (2008). Increasing efficiency of photoelectronic conversion by encapsulation of photosynthetic reaction center proteins in arrayed carbon nanotube electrode. *Langmuir* **24,** 8871–8876.

Lee, J. K., and Kaplan, S. (1995). Transcriptional regulation of *puc* operon expression in *Rhodobacter sphaeroides*. Analysis of the *cis*-acting downstream regulatory sequence. *J. Biol. Chem.* **270,** 20453–20458.

Lee, S. F., Li, Y. J., and Halperin, S. A. (2009). Overcoming codon-usage bias in heterologous protein expression in *Streptococcus gordonii*. *Microbiology* **155,** 3581–3588.

Lenz, O., Schwartz, E., Dernedde, J., Eitinger, M., and Friedrich, B. (1994). The *Alcaligenes eutrophus* H16 *hoxX* gene participates in hydrogenase regulation. *J. Bacteriol.* **176,** 4385–4393.

Leung, M. M.-Y. (2010). CtrA and GtaR: Two systems that regulate the gene transfer agent in *Rhodobacter capsulatus*. Microbiology and Immunology. University of British Columbia, Vancouver.

Lin, S., Jaschke, P. R., Wang, H. Y., Paddock, M., Tufts, A., Allen, J. P., Rosell, F. I., Mauk, A. G., Woodbury, N. W., and Beatty, J. T. (2009). Electron transfer in the *Rhodobacter sphaeroides* reaction center assembled with zinc bacteriochlorophyll. *Proc. Natl. Acad. Sci. USA* **106,** 8537–8542.

Madigan, M. T. (1995). Microbiology of nitrogen fixation by anoxygenic photosynthetic bacteria. *In* "Anoxygenic Photosynthetic Bacteria," (R. E. Blankenship, M. T. Madigan, and C. E. Bauer, eds.), pp. 915–928. Kluwer Academic, Dordrecht, The Netherlands.

McCaskill, J. S. (1990). The equilibrium partition-function and base pair binding probabilities for RNA secondary structure. *Biopolymers* **29,** 1105–1119.

Messing, J. (1983). New M13 vectors for cloning. *Methods Enzymol.* **101,** 20–78.

Moskvin, O. V., Kaplan, S., Gilles-Gonzalez, M. A., and Gomelsky, M. (2007). Novel heme-based oxygen sensor with a revealing evolutionary history. *J. Biol. Chem.* **282,** 28740–28748.

Nakamoto, T. (2009). Evolution and the universality of the mechanism of initiation of protein synthesis. *Gene* **432,** 1–6.

Narayanan, K. B., and Sakthivel, N. (2010). Biological synthesis of metal nanoparticles by microbes. *Adv. Coll. Interface Sci.* **156,** 1–13.

Noll, S., Hampp, G., Bausbacher, H., Pellegata, N. S., and Kranz, H. (2009). Site-directed mutagenesis of multi-copy-number plasmids: Red/ET recombination and unique restriction site elimination. *Biotechniques* **46,** 527–533.

Oh, J. I., and Kaplan, S. (2000). Redox signaling: Globalization of gene expression. *EMBO J.* **19,** 4237–4247.

Prentki, P., and Krisch, H. M. (1984). *In vitro* insertional mutagenesis with a selectable DNA fragment. *Gene* **29,** 303–313.

Qian, P., Hunter, C. N., and Bullough, P. A. (2005). The 8.5 angstrom projection structure of the core RC-LH1-PufX dimer of *Rhodobacter sphaeroides*. *J. Mol. Biol.* **349,** 948–960.

Rajasekhar, N., Sasikala, C., and Ramana, C. V. (1999a). Photoproduction of indole 3-acetic acid by *Rhodobacter sphaeroides* from indole and glycine. *Biotech. Lett.* **21,** 543–545.

Rajasekhar, N., Sasikala, C., and Ramana, C. V. (1999b). Photoproduction of L-tryptophan from indole and glycine by *Rhodobacter sphaeroides* OU5. *Biotech. Appl. Biochem.* **30,** 209–212.

Roh, J. H., Smith, W. E., and Kaplan, S. (2004). Effects of oxygen and light intensity on transcriptome expression in *Rhodobacter sphaeroides* 2.4.1. *J. Biol. Chem.* **279,** 9146–9155.

Romagnoli, S., and Tabita, F. R. (2009). Carbon dioxide metabolism and its regulation in nonsulfur purple photosynthetic bacteria. *In* "The Purple Phototrophic Bacteria," (C. N. Hunter, F. Daldal, M. C. Thurnauer, and J. T. Beatty, eds.), pp. 563–576. Springer Science, Dordrecht, The Netherlands.

Sadowski, P. D. (1995). The Flp Recombinase of the 2-Mu-M Plasmid of *Saccharomyces cerevisiae*. *Prog. Nucleic Acid Res. Mol. Biol.* **51**(51), 53–91.

Salis, H. M., Mirsky, E. A., and Voigt, C. A. (2009). Automated design of synthetic ribosome binding sites to control protein expression. *Nat. Biotech.* **27,** 946–950.

Sawitzke, J. A., Thomason, L. C., Costantino, N., Bubunenko, M., Datta, S., and Court, D. L. (2007). Recombineering: *In vivo* genetic engineering in *E. coli, S. enterica,* and beyond. *Methods Enzymol.* **421,** 171–199.

Scheuring, S., Francia, F., Busselez, J., Melandri, B. A., Rigaud, J. L., and Levy, D. (2004). Structural role of PufX in the dimerization of the photosynthetic core complex of *Rhodobacter sphaeroides*. *J. Biol. Chem.* **279,** 3620–3626.

Scheuring, S., Busselez, J., and Levy, D. (2005). Structure of the dimeric PufX-containing core complex of *Rhodobacter blasticus* by *in situ* atomic force microscopy. *J. Biol. Chem.* **280,** 1426–1431.

Schuck, A., Diwa, A., and Belasco, J. G. (2009). RNase E autoregulates its synthesis in *Escherichia coli* by binding directly to a stem-loop in the *rne* 5′ untranslated region. *Mol. Microbiol.* **72,** 470–478.

Schweizer, H. P. (2003). Applications of the *Saccharomyces cerevisiae* Flp-FRT system in bacterial genetics. *J. Mol. Microbiol. Biotech.* **5,** 67–77.

Shuman, S. (1994). Novel approach to molecular cloning and polynucleotide synthesis using vaccinia DNA topoisomerase. *J. Biol. Chem.* **269,** 32678–32684.

Simon, R., Priefer, U., and Puhler, A. (1983). A broad host range mobilization system for *in vivo* genetic engineering transposon mutagenesis in gram-negative bacteria. *Bio/Technology* **1,** 784–791.

Steinmetz, M., Le Coq, D., Djemia, H. B., and Gay, P. (1983). Genetic analysis of *sacB*, the structural gene of a secreted enzyme, levansucrase of *Bacillus subtilis* Marburg. *Mol. Gen. Genet.* **191,** 138–144.

Sternberg, N., Sauer, B., Hoess, R., and Abremski, K. (1986). Bacteriophage P1 *cre* gene and its regulatory region—Evidence for multiple promoters and for regulation by DNA methylation. *J. Mol. Biol.* **187,** 197–212.

Swem, L. R., Elsen, S., Bird, T. H., Swem, D. L., Koch, H., Myllykallio, H., Daldal, F., and Bauer, C. E. (2001). The RegB/RegA two-component regulatory system controls synthesis of photosynthesis and respiratory electron transfer components in *Rhodobacter capsulatus*. *J. Mol. Biol.* **309,** 121–138.

Swingle, B., Bao, Z. M., Markel, E., Chambers, A., and Cartinhour, S. (2010). Recombineering Using RecTE from *Pseudomonas syringae*. *Appl. Environ. Microbiol.* **76,** 4960–4968.

Takshi, A., Madden, J. D., and Beatty, J. T. (2009). Diffusion model for charge transfer from a photosynthetic reaction center to an electrode in a photovoltaic device. *Electrochim. Acta* **54,** 3806–3811.

Tehrani, A., and Beatty, J. T. (2004). Effects of precise deletions in *Rhodobacter sphaeroides* reaction center genes on steady-state levels of reaction center proteins: A revised model for reaction center assembly. *Photosynth. Res.* **79,** 101–108.

Thomason, L. C., Costantino, N., Shaw, D. V., and Court, D. L. (2007). Multicopy plasmid modification with phage lambda Red recombineering. *Plasmid* **58,** 148–158.

Tuller, T., Carmi, A., Vestsigian, K., Navon, S., Dorfan, Y., Zaborske, J., Pan, T., Dahan, O., Furman, I., and Pilpel, Y. (2010). An evolutionarily conserved mechanism for controlling the efficiency of protein translation. *Cell* **141,** 344–354.

Van Fleet-Stalder, V., Chasteen, T. G., Pickering, I. J., George, G. N., and Prince, R. C. (2000). Fate of selenate and selenite metabolized by *Rhodobacter sphaeroides*. *Appl. Environ. Microbiol.* **66,** 4849–4853.

van Kessel, J. C., and Hatfull, G. F. (2007). Recombineering in *Mycobacterium tuberculosis*. *Nat. Methods* **4,** 147–152.

Villalobos, A., Ness, J. E., Gustafsson, C., Minshull, J., and Govindarajan, S. (2006). Gene designer: A synthetic biology tool for constructing artificial DNA segments. *BMC Bioinform.* **7,** 285.

Widmaier, D. M., Tullman-Ercek, D., Mirsky, E. A., Hill, R., Govindarajan, S., Minshull, J., and Voigt, C. A. (2009). Engineering the *Salmonella* type III secretion system to export spider silk monomers. *Mol. Syst. Biol.* **5,** 1–9.

Xiong, J., and Bauer, C. E. (2002). Complex evolution of photosynthesis. *Annu. Rev. Plant Biol.* **53,** 503–521.

Yang, D., Oyaizu, Y., Oyaizu, H., Olsen, G. J., and Woese, C. R. (1985). Mitochondrial origins. *Proc. Natl. Acad. Sci. USA* **82,** 4443–4447.

Yeates, T. O., Komiya, H., Chirino, A., Rees, D. C., Allen, J. P., and Feher, G. (1988). Structure of the reaction center from *Rhodobacter sphaeroides* R-26 and 2.4.1: Protein–cofactor (bacteriochlorophyll, bacteriopheophytin, and carotenoid) interactions. *Proc. Natl. Acad. Sci. USA* **85,** 7993–7997.

Zannoni, D., Schoepp-Cothenet, B., and Hosler, J. (2009). Respiration and respiratory complexes. *In* "The Purple Phototrophic Bacteria," (C. N. Hunter, F. Daldal, M. C. Thurnauer, and J. T. Beatty, eds.), pp. 537–561. Springer Science, Dordrecht, The Netherlands.

Zeng, X., Roh, J. H., Callister, S. J., Tavano, C. L., Donohue, T. J., Lipton, M. S., and Kaplan, S. (2007). Proteomic characterization of the *Rhodobacter sphaeroides* 2.4.1 photosynthetic membrane: Identification of new proteins. *J. Bacteriol.* **189,** 7464–7474.

Zhou, M. Y., Clark, S. E., and Gomezsanchez, C. E. (1995). Universal cloning method by TA strategy. *Biotechniques* **19,** 34–35.

Zuker, M., and Stiegler, P. (1981). Optimal computer folding of large RNA sequences using thermodynamics and auxiliary information. *Nucleic Acids Res.* **9,** 133–148.

CHAPTER TWENTY-FOUR

Synthetic Biology in Cyanobacteria: Engineering and Analyzing Novel Functions

Thorsten Heidorn, Daniel Camsund, Hsin-Ho Huang, Pia Lindberg, Paulo Oliveira, Karin Stensjö, *and* Peter Lindblad

Contents

Department of Photochemistry and Molecular Science, Ångström Laboratories, Uppsala University, Uppsala, Sweden

Methods in Enzymology, Volume 497
ISSN 0076-6879, DOI: 10.1016/B978-0-12-385075-1.00024-X

Abstract

Cyanobacteria are the only prokaryotes capable of using sunlight as their energy, water as an electron donor, and air as a source of carbon and, for some nitrogen-fixing strains, nitrogen. Compared to algae and plants, cyanobacteria are much easier to genetically engineer, and many of the standard biological parts available for Synthetic Biology applications in *Escherichia coli* can also be used in cyanobacteria. However, characterization of such parts in cyanobacteria reveals differences in performance when compared to *E. coli*, emphasizing the importance of detailed characterization in the cellular context of a biological chassis. Furthermore, cyanobacteria possess special characteristics (e.g., multiple copies of their chromosomes, high content of photosynthetically active proteins in the thylakoids, the presence of exopolysaccharides and extracellular glycolipids, and the existence of a circadian rhythm) that have to be taken into account when genetically engineering them.

With this chapter, the synthetic biologist is given an overview of existing biological parts, tools and protocols for the genetic engineering, and molecular analysis of cyanobacteria for Synthetic Biology applications.

1. Introduction

Cyanobacteria form a morphologically and developmentally diverse group of Gram-negative prokaryotes, consisting of unicellular, colonial, and filamentous strains, some with differentiated cell types (Stanier and Cohen-Bazire, 1977; Waterbury, 2006). They are believed to have been the first organisms to develop oxygenic photosynthesis more than 2 billion years ago, which lead to a dramatic change for life on Earth from anaerobic to aerobic conditions (Olson, 2006). According to the endosymbiotic theory the chloroplasts of plants and algae are derived from cyanobacteria (Martin and Kowallik, 1999). Today, cyanobacteria are important primary producers in the oceans responsible for at least 20–30% of the overall photosynthetic productivity (Hall and Rao, 1999). They are found in highly diverse and extreme environments, concerning temperature (Antarctica and hot springs), salinity (marine and freshwater environments), pH (from acidic hot springs to highly alkaline lakes), and water availability (tropical/Antarctic deserts). Although many cyanobacteria can tolerate these extreme conditions, only relatively few characterized strains are obligate extremophiles (Waterbury, 2006). Furthermore, some strains may live in symbiosis with plants or fungi. Cyanobacteria possess chlorophyll *a* as the primary photosynthetic pigment, phycobiliproteins as auxiliary light-harvesting pigments, and they carry out oxygenic, CO_2-fixing photosynthesis, similar to higher plants. They mainly grow photoautotrophically, but some can also grow photoheterotrophically or chemoheterotrophically in the dark (Waterbury, 2006).

Another feature of cyanobacteria is the capacity of some strains to fix molecular nitrogen (Waterbury, 2006), which makes them independent of any combined carbon and nitrogen source. These strains grow on water, air, sunlight, and minimal mineral nutrients.

In the last decades, molecular tools for genetic investigation and modification of cyanobacteria have been developed: exogenous DNA can be transferred into several unicellular and filamentous strains by natural transformation, electroporation, and conjugation with autonomous or genome-integratable vectors. Selection markers for negative and positive selection and reporter constructs for promoter activity measurements or localization studies are available, and possible restriction/modification systems are known in many strains (Cohen and Gurevitz, 2006; Flores *et al.*, 2008; Koksharova and Wolk, 2002; Thiel, 1994). Currently, at least 39 cyanobacterial genomes of unicellular and filamentous species are sequenced and annotated (Cyanobase, http://genome.kazusa.or.jp/cyanobase, accessed March 2011), which facilitates genetic modifications and transcriptomic and proteomic analyses.

The ability of cyanobacteria to use sunlight and CO_2 as energy and carbon sources, respectively, together with faster growth rates (compared to plants) and the relative ease with which they can be genetically engineered (compared to algae), make cyanobacteria stand out from all other organisms so far used in biotechnological applications. Most promising is the use of cyanobacteria for the sustainable production of biomass, biofuels (e.g., ethanol, butanol, biodiesel, and hydrogen), and bioplastics; furthermore, they can be employed in bioremediation, biofertilization, aquaculture, and the production of biologically active compounds or of high-value products, such as vitamins and pharmaceuticals (Abed *et al.*, 2009; Angermayr *et al.*, 2009; Tamagnini *et al.*, 2007).

Synthetic Biology aims for the genetic engineering of organisms from an engineer's perspective, which includes the use of standardized, well-characterized biological parts and modeling of the genetic and metabolic networks within cells, based on systems biology approaches. These aspects have not yet been given great attention in cyanobacterial biotechnology, but preliminary analyses show that certain widely used promoters in *Escherichia coli* behave very differently in the cyanobacterium *Synechocystis* PCC 6803 (Huang *et al.*, 2010), which reveals the necessity for part characterization in cyanobacteria and the need of new biological parts for cyanobacterial strains.

In addition, cyanobacteria possess some special characteristics that must be taken into account when genetically engineering them: many cyanobacteria contain multiple copies of their chromosomes (e.g., about 12 genomic copies per cell in *Synechocystis* PCC 6803; Labarre *et al.*, 1989), making several rounds of segregation necessary if a particular gene is to be completely knocked out. They possess an efficient mechanism of recombination (RecA; Murphy *et al.*, 1987), which, on the one hand, makes integration of DNA into the genome by homologous recombination possible, but, on the other hand, could also integrate replicative vectors that contain DNA fragments homologous to the host genome, making the behavior of these mutants unpredictable. The existence of an additional membrane structure, the thylakoids, and high amounts of antenna pigments containing the phycobiliproteins (up to 60% of the soluble protein content; Bogorad, 1975; Colyer *et al.*, 2005; Viskari *et al.*, 2001), as well as the presence of mucilaginous sheaths of varying thicknesses and composition on the outside of cells (Stanier and Cohen-Bazire, 1977), make special protocols for the preparation of DNA, RNA, and proteins necessary. Finally, the circadian rhythm, a cell-wide regulatory system, which enhances the fitness of the cells in rhythmic environments (Dong and Golden, 2008), may drive time-dependent behavior in introduced genetic circuits.

In this chapter, we give the synthetic biologist an overview of existing biological parts, tools and protocols from our laboratory for the genetic engineering, and molecular analysis of cyanobacteria. For further information, the reader is advised to consult several book chapters (Cohen and Gurevitz, 2006; Flores *et al.*, 2008; Thiel, 1994), review articles (Koksharova and Wolk, 2002; Vioque, 2007), and internet databases (Wackett, 2010).

2. Cyanobacterial Chassis

For the choice of a cyanobacterial chassis, characteristic properties, like growth temperature, substrate utilization (e.g., mixotrophic, nitrogen fixing), salinity requirements, or doubling times, but also the existence of tools for genetic engineering, have to be taken into consideration. In Table 24.1 we list some widely used cyanobacterial model strains, which are all fully sequenced and susceptible to genetic engineering.

These strains are mainly cultivated under photoautotrophic conditions on either agar plates or in liquid medium. BG11 is widely used as a culture medium, and it can either be supplemented with additional salts (see http://www.crbip.pasteur.fr) for marine strains, or it can be used without sodium nitrate for nitrogen-fixing strains. Typical cultivation parameters are as follows: temperature 25–30 °C, pH 7–8.5, light intensity of ~50 μmol photons m^{-2} s^{-1}. Bubbling with CO_2 enriched air (1–2% CO_2) can increase the growth rate, but the addition of a suitable buffer (e.g., HEPES, MOPS, TES) into the medium is recommended.

A few parameters can be used to normalize measurements to biomass content: The optical density (usually at a wavelength of 700–750 nm), total protein content, and chlorophyll *a* content are the most used ones. For filamentous strains, using optical density is not suitable due to the

Strain	Genome size[a] (Mbp)	Plasmids size[a] (kbp)	Unicellular (u)/ filamentous (f)[b]	Marine (m)/ freshwater (f)	Culture medium[c]	Typical doubling times [h][d]	Methods for gene transfer[e]
Synechocystis PCC 6803	3.6	120, 106, 103, 44, 5.2, 2.4, 2.3	u	f	BG11	8–12[f]	t, c, e
Synechococcus elongatus PCC 7942	2.7	46	u	f	BG11	5[g]	t, c
Synechococcus PCC 7002	3.0	186, 124, 39, 32, 16, 4.8	u	m	BG11 + salts	5[h]	t, c
Thermosynechococcus elongatus BP-1	2.6	–	u	m	BG11 + salts, at 45 °C	15[i]	t
Anabaena variabilis ATCC 29413	6.4	366, 301, 36	f: het	f	BG11 or $BG11_0$	16[j,k]	c
Nostoc PCC 7120	6.4	408, 187, 102, 55, 40, 5.6	f: het	f	BG11 or $BG11_0$	20[l]	c, e
Nostoc punctiforme ATCC 29133	8.2	355, 255, 123, 66, 26	f: het, akin, horm	f	BG11 or $BG11_0$	16[m]/ 26[j,m]	c, e

[a] http://genome.kazusa.or.jp/cyanobase.
[b] het, heterocysts; akin, akinetes; horm, hormogonia.
[c] See Pasteur Culture Collection's Web site, http://www.crbip.pasteur.fr, for recipes.
[d] All values are typical values for orientation, but strongly dependent on cultivation conditions.
[e] t, natural transformation; c, conjugation; e, electroporation (see Section 4.2).
[f] Williams (1988).
[g] Lehmann and Wöber (1978).
[h] Xu *et al.* (2006).
[i] Boussac *et al.* (2004).
[j] Nitrogen-fixing conditions.
[k] Haury and Spiller (1981).
[l] Lluisma *et al.* (2001).
[m] Summers *et al.* (1995).

nonhomogenous growth mode of the cells. To determine chlorophyll *a* content in a sample, the most common method is to extract the cells with 90% methanol in the dark and measure the absorbance of the extract at 665 nm. The chlorophyll *a* content can be calculated using an extinction coefficient of 78.74 L g^{-1} cm^{-1} at 665 nm (Meeks and Castenholz, 1971). The proper selection of standardization parameter depends on the organism used and the process being studied.

3. Biological Parts in Cyanobacteria

So far, not many standardized biological parts are available for cyanobacteria, and very few have been characterized in these organisms. As most work in Synthetic Biology has been done in *E. coli*, there is already a substantial amount of parts available for this organism (see Registry of Standard Biological Parts, http://partsregistry.org), which may also be used in cyanobacteria. But our own experiments showed that the performance of those parts in cyanobacteria can be significantly different (Huang *et al.*, 2010), which underlines the necessity of part characterization in dependence on the specific environmental context. In this section, we will give a list of parts that have been shown to function in cyanobacteria. Most of the parts are endogenous to cyanobacteria; only a small selection is orthogonal.

3.1. Transcriptional control

3.1.1. Promoters

In bacteria, transcription is driven by the RNA polymerase (RNAP) holoenzyme, which is composed of a sigma (σ) subunit (σ factor) and a RNAP core enzyme. The recognition of a promoter sequence by the σ factor is important for an efficient initiation of the transcription. In dependence of environmental or cell internal conditions, the core enzyme is combined with different σ factors ("σ switching"), and because different groups of σ factors specifically recognize different types of promoters, the set of transcribed genes can be controlled by this mechanism. In cyanobacteria, both the structure of the RNAP core enzyme and the set of available σ factors differ from other bacteria. Most bacterial RNAP holoenzymes consist of a core enzyme with the subunits $\alpha_2\beta\beta'$ (RpoA × 2/RpoB/RpoC), while in cyanobacteria, the β′ subunit is split into two subunits RpoC1 (γ) and RpoC2 (β′), and the C-terminal of RpoC2 has a large insertion domain, which is suggested to be a DNA binding domain (Imashimizu *et al.*, 2003). The σ factors of cyanobacteria belong to the σ^{70} family, while the σ^{54} family, which most bacteria possess as a second distinct type of sigma factors, is missing. For more detailed information, the reader is referred to a recent review about σ factors in cyanobacteria (Imamura and Asayama, 2009).

Because of these differences in the central proteins of the transcription machinery in cyanobacteria, it is likely that the common promoters used in *E. coli* may perform differently in cyanobacteria. And indeed, when we characterized the frequently used promoters P_{trc}, P_{lac}, P_{tet}, and λP_R in *Synechocystis* PCC 6803, the activities of the promoters P_{lac}, P_{tet}, and λPR under our standardized experimental conditions were very low or nondetectable, which we discussed in Huang *et al.* (2010).

Only a limited number of promoters have been used to drive the expression of a gene of interest (GOI) in cyanobacteria, promoters that confer a different regulation than the GOI's promoter permits. In addition, most of the promoters used so far are native to the organism of implementation, which also creates several limitations, such as potential crosstalk. These aspects derive from the fact that the characterization of heterologous promoters in cyanobacteria has been limited, and also because the understanding of the overall promoter structure of cyanobacterial promoters is far from being complete. Consequently, the chosen promoter will either be a relatively well-characterized cyanobacterial native promoter, for which an environmental signal is known to trigger a clear up- or downregulation, or a heterologous promoter, which has most likely not yet been fully studied in cyanobacteria, but that can be insulated from the regulatory networks of the organism (Table 24.2). Therefore, we believe that new promoters have to be designed in order to introduce the possibility of using different regulation modules and of exploring multiple functions for Synthetic Biology applications.

3.1.2. Transcription factors

Besides promoter sequences, transcription factors (TFs) are important components for the transcriptional control of protein expression. In Wu *et al.* (2007), 1288 putative TFs were identified from 21 fully sequenced cyanobacterial genomes and compiled in a freely accessible internet database (http://cegwz.com). This information could be valuable if certain metabolic pathways are to be modified, but, on the other hand, the change of the expression of an endogenous TF could have unpredictable effects, and the introduction of orthogonal TFs should be preferred. One example of an orthogonal TF is the introduction of the *lac* repressor LacI (Elhai, 1993; Geerts *et al.*, 1995; Huang *et al.*, 2010), which represses the activity of the promoters P_{tac} and P_{trc}, respectively.

3.1.3. Terminators

Examples of terminators that have been used in cyanobacteria are the early transcription terminator of bacteriophage T7 (Haselkorn, 1991), the terminator of the gene *rrnB* of *E. coli* (Geerts *et al.*, 1995), or the combination of both (BioBrick BBa_B0015; Huang *et al.*, 2010). But to our knowledge, no terminators have been characterized in cyanobacteria.

3.2. Translational control

3.2.1. Ribosome binding sites

In most studies involving heterologous gene expression in cyanobacteria, the ribosome binding site (RBS) of the respective wild-type promoters have been used, and not much is known about the translation efficiency in dependence on the RBS sequence. In prokaryotes in general, the translation is initiated by the binding of a ribosome to the mRNA at the RBS, which contains a core Shine-Dalgarno (SD) sequence (5′-GGAGG-3′). The effectiveness of an RBS sequence depends on both the base-pairing potential

Table 24.2 Selected native and orthogonal promoters

Promoter		Host strain	Comments
Native	*petE*[a,b]	*Nostoc* PCC 7120	Active if sufficient copper is present
	nirA[c,d,e]	*Nostoc* PCC 7120	Active if sufficient nitrate is present
	patS[f,g]	*Nostoc* PCC 7120	Active in heterocysts (nitrogen-fixing conditions), but not in vegetative cells
	patB[h]	*Nostoc* PCC 7120	Active in heterocysts (nitrogen-fixing conditions), but not in vegetative cells
	psbA2[i]	*Synechococcus elongatus* PCC 7942	Active in light, but not in darkness
	psbA2[j,k,l,m,n]	*Synechocystis* PCC 6803	Active in light, but not in darkness
	psbA3[j,k,l,m,o]	*Synechocystis* PCC 6803	Active in light, but not in darkness
	petE[p,q]	*Synechocystis* PCC 6803	Active if sufficient copper is present
	nirA[r]	*Synechocystis* PCC 6803	Active if sufficient nitrate is present
	rnpB[s,t]	*Synechocystis* PCC 6803	Considered as constitutive under standard cultivation conditions
	rbcL[t]	*Synechocystis* PCC 6803	
Orthogonal	T7[u]	*Nostoc* PCC 7120	Constitutive in the presence of T7 polymerase
	tac[v]	*Nostoc* PCC 7120	Repressed by LacI, induced by IPTG[w]
	trc[t]	*Nostoc* PCC 7120	LacI repression/IPTG induction has not been tested

Table 24.2 (*continued*)

Promoter	Host strain	Comments
trc[t]	*Nostoc punctiforme* ATCC 29133	LacI repression/IPTG induction has not been tested
trc[x,y]	*Synechococcus elongatus* PCC 7942	Repressed by LacI, induced by IPTG[w]
trc[t,z]	*Synechocystis* PCC 6803	Repressed by LacI, induced by IPTG[w]

[a] Ghassemian *et al.* (1994).
[b] Buikema and Haselkorn (2001).
[c] Frias *et al.* (1997).
[d] Desplancq *et al.* (2005).
[e] Agervald *et al.* (2010).
[f] Yoon and Golden (1998).
[g] Yoon and Golden (2001).
[h] Jones *et al.* (2003).
[i] Agrawal *et al.* (2001).
[j] Mohamed and Jansson (1989).
[k] Mohamed and Jansson (1991).
[l] Mohamed *et al.* (1993).
[m] Eriksson *et al.* (2000).
[n] Lindberg *et al.* (2010).
[o] He *et al.* (1999).
[p] Briggs *et al.* (1990).
[q] Oliveira and Lindblad (2008).
[r] Ivanikova *et al.* (2005).
[s] Alfonso *et al.* (2000, 2001).
[t] Huang *et al.* (2010).
[u] Wolk *et al.* (1993).
[v] Elhai (1993).
[w] LacI is not endogenously expressed. For repression a LacI cassette was introduced into the strain.
[x] Geerts *et al.* (1995).
[y] Luque *et al.* (2004).
[z] Ng *et al.* (2000).

with the anti-SD sequence (3′ end of the 16S rRNA sequence) as well as on the spacing between the core SD sequence and the start codon. Interestingly, in an analysis of the RBS sequences of all genes (≥100 amino acids) in *Synechocystis* PCC 6803, it was found that only 26% of the genes contained the core SD sequence, while in *E. coli* it was 57% (Ma *et al.*, 2002). Based on that work, we designed an RBS (RBS*: TAGTGGAGGT), which is complementary to the anti-SD sequence (AUCACCUCCUUU; the core anti-SD motif underlined) of *Synechocystis* PCC 6803 and has an optimal spacing of 9 bp between the A of the core SD sequence and the first base of the start codon. We cloned this RBS and three BioBrick RBS (BBa_B0030,

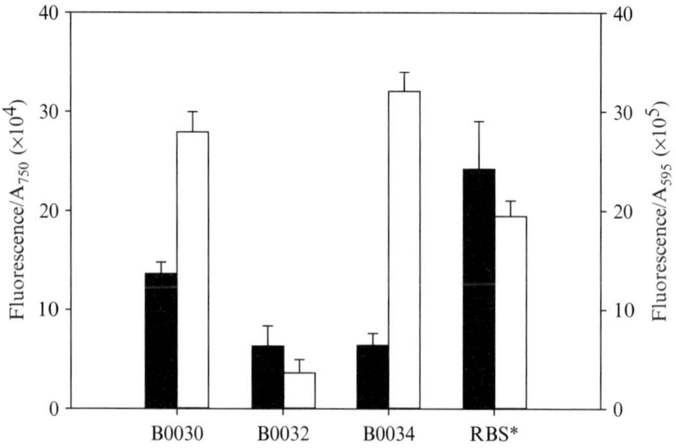

Figure 24.1 Comparison of four different ribosome binding sites BBa_B0030, BBa_B0032, BBa_B0034 (Registry of Standard Biological Parts), and RBS* in *Escherichia coli* DH5α (white bars) and *Synechocystis* PCC 6803 (black bars). The translational efficiencies were measured by means of GFPmut3B fluorescence and divided by the absorbance of the cultures at 595 nm (*E. coli*) or 750 nm (*Synechocystis*), respectively. The data represent mean ± S.D. of 4 measurements each of three independent cultivations.

BBa_B0032, and BBa_B0034) in reporter constructs together with the promoter P_{trc1O}, the fluorescent protein GFPmut3B and a BioBrick terminator (BBa_B0015), and measured the translation efficiency by means of fluorescence intensity (Fig. 24.1). In *Synechocystis* PCC 6803, the fluorescence intensity of the construct with RBS* was five times higher compared to the construct with the BioBrick RBS BBa_B0034, while in *E. coli* the fluorescence intensity of the construct with BBa_B0034 was around one-third stronger than the construct with RBS*. These results emphasize again the importance of the characterization of biological parts in the respective cellular context. The fluorescence intensities of the constructs with the BioBrick RBS BBa_B0032 were the lowest for both organisms, those for BBa_B0030 were in between.

3.2.2. Antisense RNA

It is known that antisense RNA (asRNA) plays an important role in cyanobacterial gene regulation (Dühring *et al.*, 2006; Georg *et al.*, 2009), but to our knowledge just one example of asRNA used for the controlled suppression of a certain gene in the cyanobacterium *Synechococcus elongatus* PCC 7942 has been reported (Holtman *et al.*, 2005). Recently, using differential RNA sequencing, a genome-wide map of 3,527 transcriptional start sites (TSS) of the cyanobacterium *Synechocystis* PCC 6803 was established (Mitschke *et al.*, 2011). Interstingly, one-third of all TSS were located on the reverse complementary strand of 866 genes, suggesting massive antisense transcription. Complementary microarray-based RNA profiling verified a high number of non-coding transcripts and identified strong ncRNA regulations.

3.2.3. Protein coding sequences: Codon usage and gene optimization

Cyanobacterial codon usage is often similar to that of other bacteria, such as *E. coli*, and in many strains there are few or no strongly unfavored codons. Nevertheless, among the model strains, the unicellular strains tend to have more codons that are used with a frequency below 10% for a specific amino acid than do the filamentous strains. Especially strains of *Synechococcus* typically have several codons used at very low frequency (based on codon tables from The Kazusa Codon Usage Database; Nakamura *et al.*, 2000).

To improve expression of a gene in a specific cyanobacterium, codon usage may be adjusted to better fit the codon usage of the intended expression host strain. In a study using *Synechocystis* PCC 6803, heterologous expression of the plant enzyme isoprene synthase, under control of the *psbA2* promoter, was enhanced about 10-fold after adjustment of the entire gene sequence to a *Synechocystis* codon usage (Lindberg *et al.*, 2010).

3.3. Posttranslational control: Degradation tags/proteases

The accumulation of proteins in the cell can be controlled by cytoplasmic proteases, like ATP-dependent Ser-type Clp proteases. In *E. coli*, a single ClpP proteolytic core is flanked with HSP100 chaperone partners. For *Synechocystis* PCC 6803 multiple ClpP paralogs have been found (Sokolenko *et al.*, 2002). As recognition sequences for the proteases, C-terminal oligopeptide extensions, based on the *ssrA* RNA sequence, have been identified, and a change in the last three amino acids can alter the protein stability (Keiler and Sauer, 1996).

In *Synechocystis* PCC 6803, we tested reporter constructs based on the fluorescent protein EYFP, which was tagged with the three different degradation tags LVA, AAV, and ASV (available as BioBricks BBa_E0432, BBa_E0434, BBa_E0436). The experiments showed that the stability of the tagged EYFP was lower than untagged, and that the stability decreased in dependence of the tag from ASV over AAV to LVA (Huang *et al.*, 2010), which resembles the results found in *E. coli* (Andersen *et al.*, 1998). These degradation tags, which have already been used in *E. coli* for Synthetic Biology applications, for example, an oscillator circuit (Elowitz and Leibler, 2000), can apparently also be used in cyanobacteria.

4. Genetic Engineering of Cyanobacteria

Many cyanobacteria are amenable to genetic modification. Of the model strains most commonly used, some can be naturally transformed simply by mixing the appropriate DNA molecules with the cyanobacterial culture. Others require transformation via conjugation with *E. coli*, or by electroporation. DNA can be introduced for integration into the genome, or on replicating plasmids. These techniques are used to eliminate or modify native genes as well as to introduce and express heterologous genes in cyanobacteria. Many papers cover the different techniques in more detail than can be entered here (for reviews, see Elhai and Wolk, 1988; Koksharova and Wolk, 2002; Thiel, 1994).

4.1. Vectors

As vectors for introducing genetic instructions into the cell, plasmids are essential parts to consider when devising a strategy for engineering cyanobacteria. An important decision to be made is whether the construct should be carried on a replicative vector or be integrated into the host genome. Integration of a vector or a construct into the genome of a host cyanobacterium is relevant when high stability, long-term inheritance, or a modification of the host genome is desired. Also, insertion of a construct into the host genome could potentially reduce gene dosage variation caused by copy number variation of replicative plasmids. However, since average copy number and variation will differ between different plasmids, and since cyanobacteria often contain several copies of the chromosome, exemplified by *Synechocystis* PCC 6803 that has been found to contain about 12 genomic copies per cell (Labarre *et al.*, 1989), integrating a construct into the host genome may not always lead to a lower variation in gene dosage. Because of the presence of several copies of the chromosome in cyanobacteria it is important to make sure that an integrated construct is present in all copies, which necessitates an often time-consuming segregation procedure (see Section 4.5). Therefore, it is often easier to introduce and select for a replicative vector than to integrate a construct into the genome of cyanobacteria. Although both replicative and integrative vectors have specific characteristics, they share some important functional features, such as antibiotic resistance cassettes, mobilization elements, and cloning sites. Mobilization genes and origins of transfer (*oriT*) enable conjugative transfer. Cloning sites come in a large variety, but there are examples of standardized cloning sites such as the BioBrick system (Knight, 2003), which is used for the modular assembly of standard BioBrick DNA parts (http://partsregistry.org).

4.1.1. Replicative vectors

As stated above, in most cases the introduction of new functions in cyanobacteria is easier to achieve by a replicative vector compared to genome integration. However, the apparent ease of use comes with several potentially problematic issues, such as the possible spread of the plasmid, its host-range, incompatibility with other plasmids, possible modes of transfer, stability, and copy number. In Table 24.3, we have compiled a list of replicative vectors useful for engineering cyanobacteria.

As molecular cloning is generally carried out in *E. coli*, it is necessary to have shuttle vectors, which are able to replicate both in *E. coli* and in the cyanobacterial strains. In cyanobacteria two types of shuttle vectors have been used: Vectors that are composed of a cyanobacterial specific replicon and an *E. coli* cloning plasmid replicon, and vectors with a broad-host-range replicon that functions both in cyanobacteria and in *E. coli*. An example of a shuttle vector with two different replicons for *E. coli* and cyanobacteria is pSCR119 that contains the *E. coli* pMB1 replicon and the pDC1 plasmid replicon from *Nostoc* MAC, which enables shuttling of DNA constructs between *E. coli* and *Nostoc* (Summers *et al.*, 1995). Depending on the cyanobacterial origin of replication, such plasmids may be more host-specific than single origin broad-host-range plasmids. Examples of broad-host-range plasmids are incompatibility group Q (IncQ) plasmids, such as RSF1010 (Scholz *et al.*, 1989) and plasmids with a replicon derived from RSF1010, exemplified by the BioBrick compatible pPMQAK1 constructed in our lab (Huang *et al.*, 2010). IncQ plasmids are believed to have a broad-host-range in part because the replication initiation is host-independent (Meyer, 2009). Using broad-host-range plasmids, DNA constructs can be shuttled between the construction host and a wide range of final destination cyanobacteria.

Plasmids of the same incompatibility group do not exhibit stable replication when coreplicated in the same host. This is caused by similar replication regulation systems interfering with each other, causing instability. Incompatibility problems can be avoided by using a combination of plasmids from different incompatibility groups, or alternatively, part of the construct can be integrated into the host genome (see integrative vectors below). Also, the plasmid copy number is an important factor to consider, since variation in copy number will affect gene dosage of constructs transcribed from replicative vectors. This is also of importance for constructing accurate computer models of biological systems involving transcription. Higher copy-number plasmids can be anticipated to vary more in copy number between host cells than lower copy-number plasmids, because relying on a larger number for stability leads to a larger variation in the number of plasmids that are transferred to new daughter cells. RSF1010 has been found to have average copy numbers in *Synechocystis* PCC 6803 from 10 (Marraccini *et al.*, 1993) to 30 (Ng *et al.*, 2000) per cell, or 10 in *E. coli* (Frey and Bagdasarian, 1989).

Table 24.3 Selected replicative vectors useful for engineering cyanobacteria

Replicative vector	Replicons	OriT	Selective markers	Host strains	Comments
pPMQAK1[a]	RSF1010	RSF1010	Km and Nm, Amp, CcdB[b]	Broad-host-range[c,d]	BioBrick cloning site: contains BBa_P1010[b]
pSL1211[e]	RSF1010	RSF1010	Gm	Broad-host-range[c,d]	Expression vector with P_{trc}, LacI
pFC1[f]	RSF1010	RSF1010	Sm, Cm	Broad-host-range[c,d]	Expression vector with λP_r, λcI repressor
pSB2A[g]	RSF1010	RSF1010	Km, Sm	Broad-host-range[c,d]	Promoter-probe vector using *cat*[h]
pSCR119/202[i]	pMB1, pDC1	–	Km and Nm, Amp	*E. coli*, *N. punctiforme*[j]	Electroporation vectors
pSUN119/202[k]	pMB1, pDC1	–	Km and Nm, Amp	*E. coli*, *N. punctiforme*[j]	GFP promoter-probe vectors based on pSCR119/202
pRL2697[l]	pMB1, pDU1	pMB1	Cm, Em	*E. coli*, *N.* 7120[m,n]	
pRL25C[n]	pMB1, pDU1	pMB1	Km and Nm	*E. coli*, *N.* 7120[m,n]	Cosmid shuttle vector

pRL1050[o]	pMB1, pDU1	pMB1	Sm, Sp	*E. coli*, *N.* 7120[m,n]	T7 promoter followed by *luxAB*, does not contain T7 polymerase[p]
pSG111M[q,r]	pMB1, pUH24	pMB1	Cm, Amp	*E. coli*, *S. elongatus*	
pPBH201[s]	pMB1, pGL3	pMB1	Cm	*E. coli*, *P. boryanum*, *N.* 7120	

Abbreviations: Amp, ampicillin; Km, kanamycin; Nm, neomycin; Gm, gentamicin; Sm, streptomycin; Cm, chloramphenicol; Em, erythromycin; Sp, spectinomycin. *E. coli*, *Escherichia coli*; *N. punctiforme*, *Nostoc punctiforme* ATCC 29133 (also PCC 73102); *N.* 7120, *Nostoc* PCC 7120 (also *Anabaena* PCC 7120); *S. elongatus*, *Synechococcus elongatus* PCC 7942; *P. boryanum*, *Plectonema boryanum* (also *Leptolyngbya boryana*).

[a] Huang *et al.* (2010).

[b] BBa_P1010 expresses the toxin CcdB, which is useful to avoid false positive colonies when cloning into the BioBrick site.

[c] RSF1010-plasmids belong to the IncQ-group, have a broad host-range in gram-negative bacteria, and have been shown to replicate in several cyanobacteria.

[d] Meyer (2009).

[e] Ng *et al.* (2000).

[f] Mermet-Bouvier and Chauvat (1994).

[g] Marraccini *et al.* (1993).

[h] *cat* provides conditional chloramphenicol resistance through chloramphenicol acetyl transferase.

[i] Summers *et al.* (1995).

[j] As the pDC1 plasmid replicon is derived from *Nostoc* strain Mac, this vector may also replicate in other *Nostoc* strains.

[k] Argueta *et al.* (2004).

[l] Wolk *et al.* (2007).

[m] As the pDU1 plasmid replicon is derived from *Nostoc* PCC 7524, and was found to replicate in several other strains of filamentous cyanobacteria, this vector may also replicate in other *Nostoc* strains.

[n] Wolk *et al.* (1988).

[o] Wolk *et al.* (1993).

[p] Can act as an amplifier for a promoter-probe construct expressing T7 polymerase.

[q] Elhai and Wolk (1988).

[r] Wolk *et al.* (1984).

[s] Walton *et al.* (1993).

4.1.2. Integrative vectors

Integrative plasmids are in most cases suicide vectors, that is, vectors that are unable to replicate in the destination host and therefore must either integrate or disappear, and hence, any plasmid that can be efficiently transferred into the recipient may be used. In Table 24.4, we have compiled a list of integrative vectors useful for engineering cyanobacteria.

Single or double recombination using native cyanobacterial recombinases is commonly used to achieve target-specific insertion of a construct into a cyanobacterial genome, for example, to introduce mutations in a gene or to insert a construct into a neutral site. A single recombination will lead to the integration of the complete plasmid at the target site, whereas a double recombination event will lead to replacement of the wild-type DNA by the construct between the sites of recombination. Recombination sites are determined by the host-homologous DNA sequences flanking the construct and the necessary length of these sequences differ between different cyanobacteria. In *Synechocystis* PCC 6803, the shortest flanking homologous target sequences have been at least 300 bp (Domain *et al.*, 2004), or 200 bp together with 1300 bp (Zang *et al.*, 2007), in *Nostoc punctiforme* ATCC 29133 700 bp (Cohen *et al.*, 1998); in *Nostoc* PCC 7120 about 400 bp (Charaniya *et al.*, 2008), or 200 bp but only for a second recombination step (Cohen *et al.*, 1998); and in *Synechococcus elongatus* PCC 7942 about 300 bp (Clerico *et al.*, 2007). Generally, longer host-homologous target sequences tend to increase the recombination efficiency. Further, frequencies for the occurrence of single and double recombination events vary between cyanobacterial strains.

To remove nondesirable elements, such as antibiotic resistance cassettes, or to select for double recombinants only, negative selection methods such as the use of *sacB* are relevant. The *sacB* gene expresses the enzyme levan sucrase, which leads to the production of potentially toxic products from sucrose in many Gram-negative bacteria (Gay *et al.*, 1985), and has been used in several cyanobacterial strains to select for double recombinants (Andersson *et al.*, 2000; Cai and Wolk, 1990; Summers *et al.*, 1995). For example, in "Hit and run allele replacement" a plasmid containing a GOI, an antibiotic selection marker, and *sacB* is first integrated into the target site by single recombination and selected for using the antibiotic. Then the antibiotic selection marker and *sacB* are removed in a second recombination step by selection using sucrose, while keeping the inserted GOI intact in the genome (Andersson *et al.*, 2000).

Transposons are broad-host-range mobile DNA elements able to insert into new sites, and have been used extensively for random insertion based studies in functional genomics (Hayes, 2003). Placing the GOI inside a transposon carried by a suicide vector, and transferring the vector to the recipient host can facilitate genome integration of a GOI. The size or function of a transposon can be optimized by the deletion of unnecessary

Integrative vector	Integron	OriT	Selective markers	Host/target strains	Comments
pRL1063a[a,b,c]	Tn5-1063	RK2	Bm, Nm, Sm	*E. coli, N.* 7120, *S.* AMC395, *N. punctiforme*[d,e,f]	Promoterless *luxAB* reporter[g] for functional genomics
pRL1058[a]	Tn5-1058	RK2	Sm, Nm	*E. coli, N.* 7120[d,e,f]	As pRL1063a but without *luxAB* reporter
pAM1303[b]	Recombination	pMB1	Sp	*E. coli, S. elongatus*	Neutral site I vector. Potential selection problem for *E. coli* strains with *rpsL* allele
pAM1573[b]	Recombination	pMB1	(Amp), Cm	*E. coli, S. elongatus*	Neutral site II vector. Amp outside sites of recombination
pRL271[h,i,j]	Recombination	pMB1	Cm, Em, *sacB*	*E. coli,* for example, *N.* 7120[k]	For example, "hit and run allele replacement"
pRL278[i,j,l]	Recombination	pMB1	Km and Nm, *sacB*	*E. coli,* for example, *N.* 7120[k]	For example, "hit and run allele replacement"
pSCR104[m]	Recombination	RP4	Nm	*E. coli,* for example, *N. punctiforme*[k]	Can serve as general recombination vehicle

Abbreviations: Bm, bleomycin; Nm, neomycin; Sm, streptomycin; Sp, spectinomycin; Amp, ampicillin; Em, erythromycin; Km, kanamycin. *E. coli*, *Escherichia coli*; *S. elongatus*, *Synechococcus elongatus* PCC 7942; *N. punctiforme*, *Nostoc punctiforme* ATCC 29133 (also PCC 73102); *N.* 7120, *Nostoc* PCC 7120 (also *Anabaena* PCC 7120).

[a] Wolk *et al.* (1991).

[b] Andersson *et al.* (2000).

[c] Cohen *et al.* (1994).

[d] As Tn5 has broad-host-range, transposition is expected also in other cyanobacteria.

[e] Davison (2002).

[f] Hayes (2003).

[g] Requires *n*-decanal substrate, which can also be supplied through *in vivo* expression of, for example, the *Xenorhabdus luminescens luxC-DE.*

[h] GenBank accession number L05081.

[i] Jones *et al.* (2003).

[j] Clerico *et al.* (2007).

[k] Target strain defined by host-homologous sequences of inserted construct.

[l] GenBank accession number L05083.

[m] Summers *et al.* (1995).

or nondesired elements, such as extra antibiotic resistance genes. Also, to increase the stability of acquired mutants, the transposase gene can be moved to a region of the suicide vector outside the transposon itself, limiting secondary transposition (Davison, 2002). Transposons have been used extensively in cyanobacteria, for example, for functional genomics in *N. punctiforme* ATCC 29133 (Cohen *et al.*, 1994), in *Nostoc* PCC 7120 (Ernst *et al.*, 1992; Wolk *et al.*, 1991), and for *S. elongatus* PCC 7942 (Holtman *et al.*, 2005).

4.1.3. Construction of new plasmid vectors

If no suitable vector exists, a new or modified plasmid vector can be constructed using one of several strategies. One strategy is to isolate an endogenous plasmid from the cyanobacterial strain of interest, insert the construct, and use it as a vector in the host of origin. For example, the first plasmid in the pAQE-series was constructed by ligating the *Synechococcus* PCC 7002 plasmid pAQ1 with an *E. coli* plasmid (Buzby *et al.*, 1985). With this approach a stable plasmid can be expected since it has evolved to replicate in the host. On the other hand, the coevolution confers a potential risk for host-interference affecting, for example, the modified plasmid's stability or copy-number regulation. Instead of inserting a construct of interest into an endogenous plasmid, solely its replicon, as excised using restriction enzymes, amplified with PCR or chemically synthesized, may be used to construct a new plasmid. The success of this approach depends on the chosen replicon, and if it is transformed into its wild-type host or into a new host. In the case of replicons from broad-host-range plasmids the chance of successful replication in a new host is obviously much higher, but the risk for host-interference or cross talk is difficult to predict. An example of using a replicon for constructing a new plasmid is the broad-host-range BioBrick vector pPMQAK1, which we constructed using a PCR-amplified RSF1010 replicon. This vector was confirmed to replicate in several commonly used cyanobacteria, and is anticipated to work in many others as well (Huang *et al.*, 2010). Another elegant approach is the use of a flexible system, where a new vector can be assembled by combining replicon(s) and antibiotic resistance cassettes as standard modular parts (Shetty *et al.*, 2008). Finally, DNA synthesis of a complete new vector provides the most options since a new plasmid can be designed to include any desired function and minimize the risk for host cross talk or interference.

4.2. Methods for transfer of DNA

4.2.1. Natural transformation

Some cyanobacteria are able to take up foreign DNA in the form of plasmids or linear DNA from the medium. This ability can be exploited to introduce exogenous DNA into the strains in a quick and simple way.

The DNA to be transferred can be carried on a plasmid prepared using standard procedures, or linear DNA.

The model strains *Synechocystis* PCC 6803 (Grigorieva and Shestakov, 1982), *Synechococcus* PCC 7942 (Shestakov and Khyen, 1970), *Synechococcus* PCC 7002 (Stevens and Porter, 1980), *Thermosynechococcus elongatus* BP-1 (Iwai *et al.*, 2004; Onai *et al.*, 2004) are all naturally transformable. The growth phase of the cells has been shown to be important at least for some strains, where the transformation efficiencies are greatly reduced when the cells enter stationary phase (Grigorieva and Shestakov, 1982; Iwai *et al.*, 2004; Stevens and Porter, 1980).

For *Synechocystis* PCC 6803, the following protocol can be used (adapted from Jansson *et al.*, 1998; Williams, 1988). Grow cultures in BG11 medium for 2–3 days until the cell density reaches about 3×10^7 cells mL^{-1} ($OD_{730} = 0.3$). Harvest the cells by centrifugation and resuspend in fresh BG11 to a density of 10^9 cells mL^{-1}. Add 1 μg of plasmid DNA to 100 μL of cell suspension in a microcentrifuge tube and mix (lower concentrations of DNA may be used with lower numbers of transformants as a result; Kufryk *et al.*, 2002; Zang *et al.*, 2007). Incubate the tube at 25 °C in low light for 4–6 h before spreading on nitrocellulose filters (we use Millipore HAWG filters) on top of BG11 agar plates. After 24 h, move the filters to selective plates containing the appropriate antibiotic for selection. The cells may also be spread directly on agar plates without filters. In this case, antibiotics can be applied underneath the agar by lifting it up using a spatula and pipetting the solution containing antibiotics under the agar. Single colonies can be isolated after about 2 weeks, and grown in liquid culture for analysis.

Natural transformation favors double crossover resulting in gene replacement over single crossover resulting in integration of an entire plasmid into the genome, probably as a result of DNA fragmentation during the transformation process (Elhai and Wolk, 1988; Iwai *et al.*, 2004; Tsinoremas *et al.*, 1994). If single recombination is required, or if the transfer of an intact replicating plasmid is desired, conjugation or electroporation may be the preferred route of DNA transfer.

4.2.2. Conjugation

Conjugation with *E. coli* can be used to transform many cyanobacterial strains. This technique has been widely used for the filamentous type strains *Nostoc* PCC 7120 (Wolk *et al.*, 1984), *N. punctiforme* ATCC 29133 (Cohen *et al.*, 1994; Flores and Wolk, 1985; Lindberg *et al.*, 2002), and *Anabaena variabilis* ATCC 29413 (Murry and Wolk, 1991), as well as for unicellular strains *Synechococcus* PCC 7942, PCC 6301 (Tsinoremas *et al.*, 1994), and *Synechocystis* PCC 6803 (Marraccini *et al.*, 1993). What follows here will be a short introduction to the technique followed by a sample protocol. Many excellent publications describe the available plasmids and protocols in great detail (Cohen *et al.*, 1994, 1998; Koksharova and Wolk, 2002; Thiel and Wolk, 1987).

Briefly, the technique is based on the ability of *E. coli* carrying a conjugal plasmid, such as RP4 or a related plasmid, to transfer DNA via conjugation. When a cyanobacterial strain is mixed with such an *E. coli* strain, the cyanobacteria will be able to receive a plasmid transferred from the *E. coli* cells. The DNA to be transferred, the cargo plasmid, must be either a replicating plasmid, or contain DNA that can be integrated into the genome by recombination or transposition (see further Section 4.1 above). It must also carry a *bom*-site, also called the *oriT*-region, and a *mob*-gene encoding a nickase, which recognizes the *bom*-site and enables transfer of the plasmid. The *mob* gene may also be provided in trans, on a so-called helper plasmid. Furthermore, the DNA must be protected against degradation by restriction enzymes that may be present in the recipient cyanobacterial strain. This is accomplished by methylases that modify the recognition sites of those enzymes, making them inaccessible for restriction. The genes encoding such methylases are usually carried on helper plasmids (see further Section 4.3 below).

Typically, two different *E. coli* strains are used. One is carrying the conjugal plasmid, and the other the helper plasmid(s) and the cargo plasmid to be transferred to the cyanobacterium. Methylases encoded by the helper plasmids may cause DNA to be restricted by the host *E. coli* strain, and thus the strains must be chosen with this in mind. HB101 can be used for the plasmids we work with for conjugations with *Synechocystis* PCC 6803 and *Nostoc* strains (see further Elhai and Wolk, 1988). The *E. coli* strains are mixed with cyanobacteria and spread on plates, where conjugation takes place and exconjugant colonies will grow.

For *Nostoc* PCC 7120 or for *Synechocystis* PCC 6803, the following protocol may be used (from Cohen *et al.*, 1998; Elhai and Wolk, 1988): use as a conjugal strain HB101 carrying the conjugal plasmid pRL443, and as a cargo strain HB101 carrying the cargo plasmid, and, for *Nostoc* PCC 7120, the helper plasmid pRL623. For each conjugation plate that is to be prepared, grow 10 mL each of the cargo and conjugal *E. coli* strains to late exponential phase in LB medium. Mix the two strains and harvest by centrifugation. If the strains are grown with different antibiotics, a washing step is necessary before the cultures are mixed. Resuspend the pellet in a small volume of fresh LB, 200 μL is suitable for one plate. Then, prepare the cyanobacteria: Spin down 50 mL of a healthy, growing (exact growth phase is not important) cyanobacterial culture, resuspend in 2.5 mL of fresh BG11 medium. Determine the chlorophyll *a* concentration (see Section 2 above) and mix a volume of cyanobacteria corresponding to 15 μg of chlorophyll *a* with the previously prepared *E. coli* mixture. Serial dilutions may also be performed where the cyanobacterial suspension is diluted 10- to 1000-fold before mixing with the *E. coli* cells, in order to ensure single colonies on the plates. Spread the mixture on a nitrocellulose filter (we use Millipore HAWG filters) on top of a BG11 agar plate containing 5% LB, without antibiotics. Incubate the plate at low light for 24 h, and then transfer the filter to a fresh BG11 plate, without LB, containing the appropriate antibiotics. Colonies will be visible within 2–3 weeks.

For filamentous cyanobacteria like *Nostoc* PCC 7120, the filaments may be disrupted, down to a size of only a few cells, by ultrasonic treatment before conjugation. This will promote clonal colonies after selection. However, disrupting the filaments is not necessary for conjugation to take place.

The protocol above can be adapted for other cyanobacterial strains, but optimization may be necessary both in the choice of plasmids used and the conditions and amounts of cells used for conjugation. For example, for *N. punctiforme* ATCC 29133 a modified protocol is available (Cohen *et al.*, 1994, 1998).

Frequencies for the occurrence of single and double recombination events as a result of conjugal transfer of homologous DNA vary between cyanobacterial strains. For *Nostoc* PCC 7120, double recombination is rare even when long homologous flanking sequences are used (Cai and Wolk, 1990). However, this can be solved by a second selection step utilizing the conditionally lethal *sacB* gene (Cai and Wolk, 1990). This gene is included on the cargo plasmid and thus incorporated into the genome as the result of a single recombination event. To select for double recombination, isolated exconjugants are spread on a plate containing sucrose, typically to a concentration of 5% (w/v) (Cai and Wolk, 1990), in addition to the appropriate antibiotics. The presence of sucrose will kill all cells still carrying the *sacB* gene and thus promote growth only of those cells, which have undergone a second recombination, resulting in colonies of double recombinants. Since the activity of *sacB* may be lost due to mutations, it is important to test its efficiency in *E. coli* before transfer to the cyanobacterium.

4.2.3. Electroporation

Electroporation may be used to transfer plasmids into cyanobacterial cells. It was first described for a strain of *Anabaena* (Thiel and Poo, 1989), and has subsequently been developed for many other strains (e.g., Marraccini *et al.*, 1993; Mühlenhoff and Chauvat, 1996; Sui *et al.*, 2004; Summers *et al.*, 1995; see further Koksharova and Wolk, 2002; and references therein). The following protocol is used in our laboratory for transfer of replicating plasmids to *N. punctiforme* ATCC 29133 (Holmqvist *et al.*, 2009; adapted from Summers *et al.*, 1995; Thiel and Poo, 1989):

Grow *Nostoc* cells for 2–5 days in medium supplemented with combined nitrogen. Disrupt the filaments by sonication down to a size of 4–6 cells per filament, and let them recover in fresh medium over night. The next day, wash the cells four times in distilled, room temperature water, and resuspend to a concentration of 50–100 μg Chl *a* mL^{-1}. Chill electroporation cuvettes, with a 2 mm gap, on ice. Mix 40 μL of the cyanobacterial suspension with 10 μg of plasmid DNA dissolved in water, and transfer to the cuvette. Perform electroporation with 25 μF, 300 Ω and 1.6 kV.

Immediately rinse the suspension out with fresh growth medium and transfer to a 10 mL volume in a 25 mL flask. Let the cells recover at low light over night. The next day, harvest the cells and plate on selective plates. Colonies will appear within 2–3 weeks.

A similar protocol may be used for other strains. Optimization of the protocol for different strains may be necessary. For *N. punctiforme* ATCC 29133, electroporation is used for transfer of replicating plasmids, but there are no reports of integration into the genome of a construct transferred by electroporation. For *Nostoc* PCC 7120, integration into the genome after transfer of a suicide plasmid by electroporation has been reported (Chaurasia *et al.*, 2008).

4.3. Considerations regarding restriction enzymes present in certain strains

Some cyanobacteria contain restriction enzymes that may degrade the transferred DNA. To improve transfer efficiency to such strains, a recipient strain where endonucleases have been inactivated can be used (Onai *et al.*, 2004), or, more often, specific methylases are employed to protect the exogenous DNA before transfer to the cyanobacteria (Elhai and Wolk, 1988). For *Nostoc* PCC 7120, helper plasmid pRL623 (Elhai *et al.*, 1997) carries methylases that will protect DNA from all identified restriction endonucleases in that strain. In *N. punctiforme* ATCC 29133, no restriction activity has been observed.

4.4. Selection

The selection of transformants or exconjugants is mainly done by the use of an antibiotic resistance encoded on the introduced DNA fragment in combination with the presence of this antibiotic in the culture medium. Typical antibiotic concentrations for liquid cultures are listed in Table 24.5 for a selection of cyanobacterial strains. However, it must be pointed out that for several antibiotics different antibiotic cassettes with different promoters or genes, respectively, are in use, and also the plasmid copy number influences the antibiotic resistance. The appropriate working concentration of the chosen antibiotic has to be identified for each strain and antibiotic cassette used. Ampicillin, commonly used for cloning with *E. coli*, is less useful in cyanobacteria because of its low stability in combination with the long generation times of cyanobacteria. Also the light sensitive antibiotics tetracycline and rifampicin are less useful in cyanobacteria, since the normal growth of the organisms is in the light. An alternative, non-antibiotic dependent selection method was used by Xu *et al.* (2006), where the *psbEF* genes of *Synechococcus* PCC 7002, which are essential for PSII activity, were inactivated on the genome and integrated in a plasmid vector. The selection for transformants

Table 24.5 Typical concentrations[a] of frequently used antibiotics for different cyanobacterial strains

Strain	Ampicillin	Neomycin	Kanamycin	Spectinomycin	Streptomycin	Chloramphenicol	Erythromycin	Gentamicin
Synechocystis PCC 6803			25[b], 100[c]	10[d], 25[b]	10[e]	15[c], 25[b]	25[b]	2[d], 10[f]
Synechococcus elongatus PCC 7942			5– 20[g]	5–20[g]	2[g] + 2 spectinomycin	7.5–10[g]		1–2[g]
Synechococcus PCC 7002	10[h]		100[h,i]	50[h], 100[i]	100[h,i]	20[i]	25[h]	50[i]
Anabaena variabilis ATCC 29413				2[j] + 2 streptomycin	2[j] + 2 spectinomycin			
Nostoc PCC 7120		25[k]		2[l]	2[l]	15[m]	5[l]	
Nostoc punctiforme ATCC 29133	10[n]	25[o]			1[o]	60[o]	15[n]	

[a] Concentrations in μg mL^{-1} for liquid medium. All values are typical values for orientation. Several antibiotic cassettes with different promoters or genes are available, the proper concentration must be determined for each individual case.
[b] Eaton-Rye (2004).
[c] Oliveira and Lindblad (2008).
[d] Thornton *et al.* (2004).
[e] van Thor *et al.* (1999).
[f] Meetam *et al.* (1999).
[g] Mackey *et al.* (2007).
[h] Shen *et al.* (2002).
[i] Zhu *et al.* (2010).
[j] Schmetterer *et al.* (2001).
[k] Agervald *et al.* (2010).
[l] Frias *et al.* (2003).
[m] Scherzinger *et al.* (2006).
[n] Wong and Meeks (2001).
[o] Soule *et al.* (2007).

is done by cultivation under photoautotrophic conditions; the cells can only grow if the vector is present, and antibiotic selection is not necessary.

4.5. Segregation

Since cyanobacteria have multiple copies of the chromosome and plasmids that make up their genome, incorporation of DNA by homologous recombination into the genome initially yields heterozygous cells. Extended periods of selection may be required to segregate the modified allele from all wild-type copies. In cases where the modified locus impairs the viability of the organism, complete segregation may prove impossible to achieve.

4.6. Cryopreservation of cyanobacterial strains

Most cyanobacteria will survive freezing with DMSO as a cryoprotectant (Day, 2007). For *Nostoc* strains, we use the following protocol (http://microbiology.ucdavis.edu/meeks/expromeeks.htm): 100% DMSO is sterilized by filtration and added to a cyanobacterial culture to a concentration of 5%. The culture should be quite dense, in late log to early stationary phase, but still healthy. After mixing with DMSO, the culture is frozen at $-80\,°C$ in Eppendorf tubes. To thaw, remove the tube from the freezer and transfer the cells as soon as possible to fresh growth medium for recovery. The cultures are kept at low light intensity until they have started growing.

For *Synechocystis* PCC 6803, we routinely freeze the cells using glycerol as a cryoprotectant, using standard protocols for *E. coli*. *Synechocystis* PCC 6803 cells frozen with glycerol in this way have been stable for at least several years in our lab.

5. Molecular Analysis of Cyanobacteria

One basic principle of Synthetic Biology is the use of well-characterized biological parts, devices, and whole systems. Most analytical techniques used, for example, in *E. coli* can be employed in cyanobacteria as well. However, some unique features found in cyanobacteria (e.g., the abundance of exopolysaccharides, extracellular glycolipids, and phycobiliproteins) have to be taken into account when designing experiments, and protocols have to be adapted. The blue-green color typical for many cyanobacteria is due to their ability to synthesize chlorophyll *a*, as well as to the presence of pigments called phycobilins. These pigments are associated with proteins arranged in the phycobilisomes, which have the same function as the light-harvesting complexes in higher plants. Phycobiliproteins can constitute up to 60% of the soluble protein content of cyanobacteria (Bogorad, 1975; Colyer *et al.*, 2005; Viskari *et al.*, 2001). This is an important fact to bear in mind, since the identification and quantification of low-abundance proteins may be more difficult in such a background.

In addition, most of the well-studied cyanobacteria possess a complex membrane system, composed of the outer membrane, the plasma membrane, and the thylakoid membranes. The first two are distinctive for Gram-negative bacteria, and between these two membrane systems there is a peptidoglycan layer. Furthermore, outside the outer membrane there is a surface layer, the S-layer, the composition of which is largely unknown. Externally to this S-layer there is a mucilaginous sheath, consisting of polysaccharides and glycolipids, whose thickness varies according to the strain, and even cell-type. These extracellular components tremendously affect the yield and quality of any DNA, RNA, and protein preparation. In our opinion, it is most relevant to bring up these facts since a good understanding of the composition of the cyanobacterial cell wall structure will contribute to designing appropriate protocols for the analysis of cyanobacteria.

5.1. Isolation of heterocysts

Some filamentous nitrogen-fixing cyanobacteria have the ability to differentiate specialized nitrogen-fixing cells, named heterocysts, upon removal of combined nitrogen from the growth medium. These may constitute between 5% and 10% of the biomass in a cyanobacterial culture and are unique in many senses: they perform nitrogen fixation, are terminally differentiated, possess a cell wall that differs in structure and composition from the vegetative cells walls, the pigment composition is different, they lack photosystem II activity and their cytoplasmic environment is microaerobic among other traits. With these features, heterocysts are suitable compartments for the expression of oxygen-sensitive proteins under otherwise aerobic conditions. Therefore, heterocysts have attracted attention for a long time, and methods for separating and isolating them have been developed (Almon and Bohme, 1980; Cardona *et al.*, 2009). These methods exploit the singular chemical and/or mechanical properties of heterocysts, aiming to achieve highly pure and physiologically active preparations. In our laboratory, heterocysts are routinely purified using a combination of lysozyme (which selectively attacks the peptidoglycan layer of the vegetative cells) and sonication treatments (which break the vegetative cells, leaving the heterocysts intact), followed by a series of centrifugation steps (in order to improve the quality of the preparation, by removing any remaining cell debris). Typically, after harvesting the cyanobacteria, the cells are equilibrated for 30 min in a lysis buffer composed of 0.4 *M* sucrose, 50 m*M* HEPES/NaOH, pH 7.2, 10 m*M* NaCl, and 10 mM EDTA, to a final chlorophyll *a* concentration of 150 μg mL^{-1}. A freshly prepared solution of lysozyme is then added to the cell suspension to a final concentration of 1 mg mL^{-1}, and the suspension is incubated at 37 °C for 60 min in the dark. After this step, the cell suspension is sonicated six times with 10 s pulses (27 W in a Sonics Vibracell, VC-130), followed by careful inspection in a light microscope (if possible, it is advisable to use a fluorescence microscope and examine chlorophyll *a* fluorescence to check for remaining intact vegetative cells). If necessary, a new round of sonication may be performed. The suspension is then centrifuged at 1000×*g* for 5 min, the pellet resuspended in lysis buffer (the supernatant is discarded) and subjected to a new centrifugation step: 500×*g* for 5 min. The pellet is again resuspended in the same buffer and the mixture is once more centrifuged at 500×*g* for 5 min. Finally, two more washing steps are carried out as described above, but changing the centrifugation speed and time to 250×*g* and 3 min, respectively. The final heterocyst pellet is resuspended in lysis buffer to a concentration of ca. 0.5 mg chlorophyll *a* mL^{-1}. This protocol results in a highly pure heterocyst preparation, as determined by confocal microscopy, Western blot (assessing the presence of vegetative and/or heterocyst specific peptides), and heterocyst specific staining analyses.

5.2. DNA analysis

For confirmation of the presence of introduced DNA in the transformed cells and for characterization of new biological parts, cyanobacterial DNA may be analyzed by PCR and/or Southern blot. For the first rounds of screening after introducing foreign DNA into the cyanobacteria, whole cells, either from a plate or out of a liquid culture, are usually sufficient for PCR analyses. For more reliable PCR analyses and Southern blotting, the DNA can be extracted from the cyanobacteria according to the protocol below. After DNA preparation, PCR and Southern blot analyses, and other downstream DNA-based applications, can be conducted without special considerations.

5.2.1. Preparation of DNA

A general protocol to isolate cyanobacterial chromosomal DNA is as follows (Tamagnini *et al.*, 1997): Resuspend harvested cells from up to 50 mL of cyanobacterial culture in 500 μL 50 m*M* Tris–HCl, pH 8.0, with 10 m*M* EDTA and add 0.5 g of 0.6-mm-diameter glass beads, 25 μL of 10% sodium dodecyl sulfate (SDS), and 500 μL of phenol-chloroform (1:1, v/v). Disrupt the cells by using a bead homogenizer (typically, three cycles of 5800 rpm for 30 s, with 2 min of rest on ice between the runs is sufficient) or by vortexing. Separate the liquid phases by centrifugation at 14,000×*g* for 15 min, and extract the upper aqueous phase twice with an equal volume of chloroform. Precipitate the DNA with 1/10 volume of 3 *M* sodium acetate (pH 5.2) and 2.5 volumes of 100% ethanol at −20 °C for several hours before washing, drying, and resuspending the DNA in water or a suitable buffer.

5.3. mRNA/transcription analysis

There are two main reasons for analyzing cyanobacterial mRNA, both important for the development of Synthetic Biology applications: First, the transcriptional units can be studied, which includes various aspects like identifying the starting point of a particular transcript, determining how long it is and how many genes it harbors, and characterizing the specific end of the transcript. Second, the transcription level of introduced genes or the global transcriptional changes of the organism in response to introduced genetic circuits may be analyzed. The mRNA can either be analyzed directly or by means of the intensity of an expressed reporter gene, a method often used for the analysis of promoter activities. After RNA preparation, Northern blot analysis of the length and the relative abundance of a certain transcript can be conducted without special considerations. Also the preparation of cDNA and the further analysis by qPCR and 5′ or 3′ RACE (rapid amplification of cDNA ends) can be conducted with common protocols. However, we recommend to prepare cDNA by tag-based reverse transcription (RT) PCR for minimizing the risk of amplification of genomic DNA (Lopes Pinto *et al.*, 2006, 2007), and for 5′ RACE the template-switch protocol (Lopes Pinto and Lindblad, 2010). For expression profiling, cDNA microarrays of several cyanobacterial strains have been produced (e.g., *Synechocystis* PCC 6803 (Hihara *et al.*, 2001, 2003), *Synechococcus* WH8102 (Tetu *et al.*, 2009), *Nostoc* PCC 7120 (Higo *et al.*, 2007), *N. punctiforme* ATCC 29133 (Campbell *et al.*, 2007)). At least for *Synechocystis* PCC 6803, microarray chips are commercially available (IntelliGene® Cyano CHIP, Takara Bio Inc., Japan).

5.3.1. Preparation of RNA

A general protocol to isolate high quality cyanobacterial RNA is as follows (modified from Axelsson and Lindblad, 2002): Resuspend harvested cells from up to 50 mL of cyanobacterial culture in 250 μL 10 m*M* Tris–HCl, pH 8.0, with 1 m*M* EDTA. Add 0.5 g of 0.6-mm-diameter glass beads, 25 μL of 10% SDS, and 250 μL of phenol, equilibrated with 0.1 *M* sodium citrate at pH 4.3, and 250 μL chloroform. Disrupt the cells by using a bead homogenizer (typically, three cycles of 5800 rpm for 30 s, with 2 min of rest on ice between the runs is sufficient) or by vortexing. Separate the liquid phases by centrifugation at 14,000 × *g* for 15 min, and extract the upper aqueous phase twice with an equal volume of chloroform. Precipitate the DNA with 1/10 volume of 10 *M* LiCl and 2.5 volumes of 100% ethanol at −20 °C for several hours, before washing, drying, and resuspending the RNA in water or a suitable buffer.

Alternatively, the RNA can be isolated with TRI® reagent (Sigma-Aldrich, St. Louis, USA) according to the manufacturer's instructions, with the modification that after addition of TRI® reagent, glass beads are added and the cells disrupted using a bead homogenizer as described above.

An analysis of phenol based RNA extraction methods can be found in Lopes Pinto *et al.* (2009).

To remove traces of contaminating genomic DNA for cDNA synthesis, treat the total RNA with RNase free DNaseI at 37 °C for 30 min, followed by phenol-chloroform extraction and precipitation with ethanol. As an alternative to DNase treatment, we recommend the use of tags for the RT PCR (Lopes Pinto *et al.*, 2006, 2007).

5.4. Protein analysis

One important approach to analyze the "synthetic cell" is to use a systems biology approach, which aims to resolve the cellular dynamics on different levels. The concept of systems biology originates from the sequencing of genomes and includes genomics, transcriptomics, and proteomics. The latter is the technique we use as the tool to investigate the dynamics of the cyanobacterial cell focused on the proteome (Ow and Wright, 2009). In synthetic biology, quantitative proteomics is useful to interpret the impact engineering of the cell has on proteome dynamics and, indirectly, on the overall cellular metabolic network. This knowledge could be used to modify existing life forms as well as to validate new designs and organisms.

In this section, quantitative proteomic techniques as well as other protein separation and detection methods that we use for cyanobacteria will be described.

5.4.1. Preparation of soluble/total proteins

The method employed to break the cyanobacterial cells should be chosen based on the downstream application. Soluble or membrane, native or denatured, total or fractionalized proteins are some of the points that must be considered before selecting the technique for cell disruption. Several approaches have been adopted, employing methods that chemically attack the cell wall structure (e.g., lysozyme), mechanically break the cells (e.g., glass beads, French press, Parr cell disruption vessel, sonication), or a combination of both. The methods described below represent the working protocols commonly used in our laboratory.

Extractions of the soluble and total cell protein fractions of *Synechocystis* and vegetative cells of heterocyst-forming cyanobacteria: The harvested cells are resuspended in an appropriate buffer (see below for different downstream applications, that is 2DE, iTRAQ) and an equal volume of acid washed glass beads (425–600 μm) is added. The cells are disrupted mechanically by a bead PRECELLYS®24 lyser/homogenizer (Bertin Technologies, France; four times 30 s intervals of 6000 rpm, resting 2 min. on ice between the intervals). The homogenate, making up the total cell fraction, can then be centrifuged (12,000×*g*, 10 min, 4 °C), to isolate the soluble fraction in the resulting supernatant. In our hands, the mechanical disruption by glass beads works better than disruption by grinding in liquid nitrogen, which has been used for unicellular as well as for filamentous heterocyst-forming cyanobacteria (Barrios-Llerena *et al.*, 2006). For the disruption of vegetative cells of *Nostoc* species, sonication works as well. However, for cell extractions from heterocysts we use either mechanical disruption by glass beads in a bead beater or a Parr cell disruption vessel (Agervald *et al.*, 2010; Cardona *et al.*, 2009).

In our laboratory, we also use a Parr cell disruption vessel to attain the isolation of thylakoid membranes from heterocyst-forming cyanobacteria, breaking the cells by N_2 pressurization and decompression (Cardona *et al.*, 2007, 2009). By applying this method, neither heat damage nor chemical stress is imposed, assuring intact and highly active thylakoid preparations. In short, the cells are concentrated by centrifugation and the pellet is resuspended in an appropriate buffer. The cell suspension is pressurized with N_2 to 150 or 170 bar, for vegetative cells or heterocysts respectively, in a Parr cell disruption bomb (Parr Instrument Company, IL, USA), and maintained in such conditions for at least 5 min. The rapid decompression step follows next by simply opening the sample outlet valve. This pressurization/decompression cycle is repeated three times to maximize cell disruption, before the cell debris is removed by mild centrifugation. All steps described should be performed on ice or at 4 °C. In our hands, the use of this method on *Synechocystis* PCC 6803 gives poor results, since the cells remain intact after running the above described method. Therefore, for *Synechocystis* PCC 6803 we suggest using glass beads, French press or sonication-based methods instead.

5.4.2. Preparation of membrane proteins

As mentioned earlier, cyanobacteria are quite unique prokaryotes in the sense that they possess complex and highly differentiated membrane systems. In addition to having an outer and inner (i.e., plasma) membrane systems, inherent to Gram-negative bacteria, cyanobacteria also possess thylakoid membranes, where the electron transport chains of photosynthesis and respiration are located. When engineering cyanobacteria in the frame of synthetic biology, all these membrane systems are important targets to take in consideration; moreover, any novel functions implemented in cyanobacteria will most likely have an impact on the metabolic capability, composition, and function of these membrane systems. Therefore, it is crucial to have the tools to analyze and characterize the cyanobacterial membrane systems.

For total membrane isolation, we currently use the following protocol: the cyanobacterial cells are disrupted with an appropriate method (see above) in a buffer composed of 10 m*M* MES-NaOH, pH 6.35, 0.8 *M* sucrose, 5 m*M* $CaCl_2$, 5 m*M* $MgCl_2$, 10 m*M* EDTA, 1 m*M* benzamidine, and 1 m*M* phenylmethylsulphonyl fluoride. The cell extract is then centrifuged at 12,000×*g*, at 4 °C for 20 min, and the resulting pellet is discarded. The supernatant is mixed with a washing buffer (10 m*M* MES–NaOH, pH 6.35, 0.8 *M* sucrose, 20 m*M* $CaCl_2$, 20 m*M* $MgCl_2$, 10 m*M* EDTA) at a ratio of 2:1 of supernatant and buffer, and ultracentrifuged at 140 000×*g*, at 4 °C for 40 min. The supernatant is finally discarded, the membranes in the pellet carefully resuspended in the described washing buffer and frozen at −80 °C (Cardona *et al.*, 2007). For isolation of membranes from heterocysts a modified protocol is used (Cardona *et al.*, 2009).

It is worth mentioning that great efforts have been put into the differential separation of the various cyanobacterial membrane systems. We recommend the reader to consult the literature for more detailed protocols, when pure outer, plasma or thylakoid membranes are necessary (Huang *et al.*, 2002, 2004; Srivastava *et al.*, 2005).

To determine the activity of isolated thylakoids, oxygen evolution measurements are usually performed in a Clark-type oxygen electrode (Hansatech, UK), using a light induced DCPIP reduction assay.

5.4.3. Gel-based proteomics

Previous efforts in cyanobacterial proteomics for complete separation of highly complex protein mixtures mostly used gel-based proteomics either in one or two dimensions (2DE). SDS-PAGE is used to separate proteins according to a single parameter, molecular size. It gives a good resolution, is a fast method and immunological detection of specific proteins (Western blots) can be used. The protocols used for 2DE in most part follow the early methods of O'Farrell (1975). The major differences are the extraction procedures used for cyanobacteria. Due to the cell envelope and thylakoid membranes of cyanobacterial cells the extraction methods developed are more similar to what is used for plant cells as compared to the extraction protocols used for bacterial cells. Often they include urea and thiourea and detergents such as Triton X-100 and CHAPS are often used for complete solubilization, also of hydrophobic proteins.

5.4.3.1. 2D gel electrophoresis: Isoelectric focusing (IEF) and SDS-PAGE The cell pellet is extracted in sample buffer containing 8 *M* urea, 2 *M* thiourea, 2% CHAPS (w/v). Add 6.3 μg DTT per mL buffer, and protease inhibitor (Complete, Mini, EDTA-free, Roche Applied Science) just before extraction. The proteins are extracted with acid washed glass beads as described above in preparation for soluble/total proteins. For a good separation of between 200 and 800 proteins on 2DE, 90–200 μg of proteins is applied to an IEF gel. For general protocols of 2DE see (GE Healthcare, 2D Electrophoresis handbook). For details of the IEF and SDS separations and protein detection and analysis modifications for cyanobacteria see Ekman *et al.* (2006). For quantitative analysis of the data, we always use at least three replica gels from three biological replicates. Protein spots with a consistent change in abundance of two times or more are further analyzed.

5.4.3.2. Clear- and Blue-Native Gels Cyanobacterial protein analysis may also include the evaluation of multi-protein complexes, both soluble and membrane bound. In order to determine the size, the relative abundance, and the subunit composition of multi-protein complexes, gel-based techniques can be used. We have successfully employed Clear- and Blue-Native polyacrylamide gel electrophoresis associated with SDS-PAGE to resolve these aspects (Cardona *et al.*, 2007, 2009). The protocols for the execution of these techniques are fairly similar to the ones generally described for proteins extracted from other organisms. However, one point deserves special attention: when working with cyanobacterial thylakoids, it is imperative to perform their isolation and solubilization steps, as well as the electrophoresis run, in the dark or in a room with dim green light.

5.4.4. Quantitative shotgun proteomics: iTRAQ

The technique we use for quantitative shotgun proteomics is based on isobaric tags for relative and absolute quantification (iTRAQ, AB SCIEX). One advantage of iTRAQ is the possibility to analyze eight different protein extracts (8-plex) in one experiment. This allows for comparison of up to four different biological samples while still using biological duplicates, in the same experiment (Ow *et al.*, 2008). Among iTRAQ users it has been recognized that the magnitude of iTRAQ quantification cannot be taken as exact values. However, in a properly designed experiment the iTRAQ trends reliably represent the direction of changes in protein abundance. This issue has been thoroughly experimentally investigated and discussed by Phillip Wright and coworkers (Ow *et al.*, 2009). This is a paper we strongly recommend all new users of iTRAQ to read before designing their first experiment.

For an iTRAQ experiment with the aim to primarily identify the soluble proteome of cyanobacteria the following protocol is used: Cells are extracted in 500 m*M* TEAB, pH 8.5, 0.01% (w/v) SDS, and 0.1% (v/v) Triton X-100 with acid washed glass beads as described above. For the most parts we follow the manufacturer's instructions for solubilization, trypsination, and labeling (AB SCIEX). However, the concentration of the dissolution buffer is increased twofold to enhance solubilization of the acetone-precipitated proteins (1 *M* TEAB, pH 8.5, 0.2%, w/v, SDS, and 1% CHAPS), and the reduction and alkylation process is lengthened by twofold to ensure improved efficiency. Our modifications during labeling with the iTRAQ tags include a twofold increase in concentration of ethanol and increased labeling time (2 h). We have used both SCX (strong cation exchange) as well as mixed mode VAX1 (Dionex, reversed phase-anion-exchange) columns for peptide separations with similar results (Phillips *et al.*, 2010). For identification of differentially regulated proteins, only proteins with at least two distinct peptides are taken into account, and we use either threshold based or statistical modeling methods. The threshold method is based on the variability observed between biological replicates in each specific experiment (Stensjö *et al.*, 2007). In the statistical method, proteins with a *p*-value, as defined in the statistical model used, below 5% are considered differentially regulated.

5.4.5. Protein–DNA interactions

Protein–DNA interactions play an important role when characterizing cyanobacterial parts (e.g., determining which transcription factor binds to which promoter, resolving promoter structure, identifying a transcription factor binding box, etc.) as well as different heterologous parts that can be used in cyanobacteria for Synthetic Biology applications. Assessment of these interactions can be done in relatively simple, rapid, and extremely sensitive *in vitro* methods. Proteins that bind to a specific DNA fragment can be analyzed by electrophoretic mobility shift assays, DNA affinity/chromatography assays, SELEX (systematic evolution of ligands by exponential enrichment) and DNase I footprint analyses. All of these techniques and respective variations have been successfully applied to DNA–protein interactions in cyanobacteria without any special considerations.

5.5. Gene expression analysis based on reporter proteins

Various reporter proteins have been used in cyanobacteria to probe different biological questions, including β-galactosidase, luciferase, and fluorescent proteins (Elhai, 1993; Huang *et al.*, 2010; Scanlan *et al.*, 1990). The latter ones, however, have the clear advantages of being broad-host applicable and do not require cell lysis or the addition of substrates (Ghim *et al.*, 2010). In the following, we will exclusively discuss fluorescent proteins as reporter proteins.

Cyanobacteria contain many photoactive pigments that could potentially disturb fluorescent measurements by, for example, energy transfer or quenching. However, when we expressed the fluorescent proteins Cerulean (Cormack *et al.*, 1996; Rizzo *et al.*, 2004), GFPmut3B (Cormack *et al.*, 1996), or EYFP (Ormo *et al.*, 1996) in *Synechocystis* PCC 6803, the measured fluorescence spectra of the cultures were, after background correction, similar to those reported for the pure proteins (Huang *et al.*, 2010).

The effectiveness of a promoter (for transcription) or an RBS (for translation) can be indirectly quantified by measuring the fluorescence intensity of a fluorescent protein, which is expressed by this promoter or RBS, respectively. By keeping standardized conditions during the measurements, the direct comparison of different promoters or RBS is possible, and by using relative units, the variation of the results can be reduced (Kelly *et al.*, 2009). In cyanobacteria, the promoter P_{rnpB} can serve as a promoter reference standard. The *rnpB* gene has been widely used as a housekeeping gene probe, and it has been shown that the transcript level is not affected by light or redox potential (Alfonso *et al.*, 2000, 2001). We use the following protocol for standardized promoter characterization in *Synechocystis* PCC 6803:

Grow triplicates of *Synechocystis* PCC 6803 cultures (strains containing the empty vector for background correction and the constructs to be characterized, including P_{rnpB}) at 30 °C and 50 μmol photons m^{-2} s^{-1}, in 100 mL E-flasks, containing 25 mL BG11 medium (with appropriate antibiotics), on a rotary shaker at 120 rpm for 4 days. Use this seed culture to inoculate another 100 mL E-flask with 25 mL BG11 medium at an initial optical density at 750 nm (OD_{750}) of 0.01, and grow for another 48 h under the same growth conditions as the seed culture.

For the absorbance and fluorescence measurements we use a Plate Chameleon V Microplate Reader (Hidex, Finland), which is suitable for high-throughput applications, with Microtest 96-well Optilux Black Assay Plates with clear flat bottom (Ref. 353293, BD Biosciences, USA). Determine the linear ranges of the instrument for both absorbance and fluorescence measurements, and conduct all measurements in these linear ranges. The OD_{750} of the samples should be ca. 0.1 for the fluorescence measurements to avoid inner filter effects.

For promoter characterization, determine the optical density and the fluorescence of the background control (empty vector) and all constructs of interest. We use 200 μL of sample per well in the 96-well plate and three technical replicates for each biological replicate. Subtract the absorbance and fluorescence values of the background control from the corresponding values of all constructs, and divide the corrected fluorescence values by the corrected absorbances to normalize for the number of cells per sample. Finally, the promoter activities can be related to P_{rnpB}.

The same procedure is suitable for RBS characterization using the same promoter for all constructs to be compared (see Section 3.2.1).

Besides multiplate readers, other instruments, like fluorescence (confocal) microscopes or fluorescence activated cell sorters (FACS), can be used for promoter characterization based on fluorescent proteins. In contrast to a multiplate reader, which determines whole cell population based data, single cell based measurements using microscopes and FACS give information about cell-to-cell variations within a population (Muller and Nebe-von-Caron, 2010). However, for filamentous cyanobacteria the use of FACS is difficult because of their inhomogeneous growth and the resulting clogging. The development of micro-fabricated, integrated lab-on-a-chip systems make preparation, separation, reaction, detection, and synthesis on a single cell level possible, which can increase data throughput and signal resolution (Gawad *et al.*, 2010).

6. Conclusion and Outlook

As shown in this chapter, the techniques for classical genetic engineering and molecular analysis of cyanobacteria are well developed. For Synthetic Biology approaches, however, new standardized biological parts need to be constructed and characterized in cyanobacteria. This will enable the introduction of novel functions, based on artificial genetic circuits designed *in silico*, in a cyanobacterial chassis, which perform as predicted by a computer model. In the long run, the development of orthogonal systems for cyanobacteria, like T7 RNAP-based transcription and orthogonal ribosome-based translation (An and Chin, 2009), may reduce cross talk with the endogenous metabolism and allow chassis-independent use of biological parts. Our ultimate goal is to obtain a well-developed, photosynthetic cellular platform for the implementation of useful and sustainable applications, like bioremediation and production of biofuels and high-value chemical products.

ACKNOWLEDGMENTS

We thank Thiyagarajan Gnanasekaran for conducting the experimental work for the RBS characterization, Martin Ekman for discussions on gel-based proteomics, and Sean Gibbons for critical reading of the manuscript.
The work was supported by the Swedish Energy Agency, the Knut and Alice Wallenberg Foundation, the Nordic Energy Research Program (project BioH_2), the Royal Norwegian Embassy in New Delhi, India (project BioCO_2), EU/NEST FP 6 project BioModularH2 (contract # 043340), and the EU/Energy FP7 project SOLAR-H2 (contract # 212508).

REFERENCES

Abed, R., *et al.* (2009). Applications of cyanobacteria in biotechnology. *J. Appl. Microbiol.* **106,** 1–12.

Agervald, A., *et al.* (2010). CalA, a cyanobacterial AbrB protein, interacts with the upstream region of *hypC* and acts as a repressor of its transcription in the cyanobacterium *Nostoc* sp. strain PCC 7120. *Appl. Environ. Microbiol.* **76,** 880–890.

Agrawal, G. K., *et al.* (2001). An AU-box motif upstream of the SD sequence of light-dependent *psbA* transcripts confers mRNA instability in darkness in cyanobacteria. *Nucleic Acids Res.* **29,** 1835–1843.

Alfonso, M., *et al.* (2000). Redox control of *psbA* gene expression in the cyanobacterium *Synechocystis* PCC 6803. Involvement of the cytochrome b(6)/f complex. *Plant Physiol.* **122,** 505–516.

Alfonso, M., *et al.* (2001). Redox control of *ntcA* gene expression in *Synechocystis* sp. PCC 6803. Nitrogen availability and electron transport regulate the levels of the NtcA protein. *Plant Physiol.* **125,** 969–981.

Almon, H., and Bohme, H. (1980). Components and activity of the photosynthetic electron transport system of intact heterocysts isolated from the blue-green alga *Nostoc muscorum*. *Biochim. Biophys. Acta* **592,** 113–120.

An, W., and Chin, J. W. (2009). Synthesis of orthogonal transcription-translation networks. *Proc. Natl. Acad. Sci. USA* **106,** 8477–8482.

Andersen, J. B., *et al.* (1998). New unstable variants of green fluorescent protein for studies of transient gene expression in bacteria. *Appl. Environ. Microbiol.* **64,** 2240–2246.

Andersson, C. R., *et al.* (2000). Application of bioluminescence to the study of circadian rhythms in cyanobacteria. *In* "Bioluminescence and Chemiluminescence, Part C," (M. M. Ziegler and T. O. Baldwin, eds.), pp. 527–542. Academic Press Inc, San Diego.

Angermayr, S. A., *et al.* (2009). Energy biotechnology with cyanobacteria. *Curr. Opin. Biotechnol.* **20,** 257–263.

Argueta, C., *et al.* (2004). Construction and use of GFP reporter vectors for analysis of cell-type-specific gene expression in *Nostoc punctiforme*. *J. Microbiol. Methods* **59,** 181–188.

Axelsson, R., and Lindblad, P. (2002). Transcriptional regulation of *Nostoc* hydrogenases: Effects of oxygen, hydrogen, and nickel. *Appl. Environ. Microbiol.* **68,** 444–447.

Barrios-Llerena, M. E., *et al.* (2006). Shotgun proteomics of cyanobacteria—Applications of experimental and data-mining techniques. *Brief. Funct. Genomic. Proteomic.* **5,** 121–132.

Bogorad, L. (1975). Phycobiliproteins and complementary chromatic adaptation. *Annu. Rev. Plant Physiol.* **26,** 369–401.

Boussac, A., *et al.* (2004). Biosynthetic Ca^{2+}/Sr^{2+} exchange in the photosystem II oxygen-evolving enzyme of *Thermosynechococcus elongatus*. *J. Biol. Chem.* **279,** 22809–22819.

Briggs, L. M., *et al.* (1990). Copper-induced expression, cloning, and regulatory studies of the plastocyanin gene from the cyanobacterium *Synechocystis* sp. PCC 6803. *Plant Mol. Biol.* **15,** 633–642.

Buikema, W. J., and Haselkorn, R. (2001). Expression of the *Anabaena hetR* gene from a copper-regulated promoter leads to heterocyst differentiation under repressing conditions. *Proc. Natl. Acad. Sci. USA* **98,** 2729–2734.

Buzby, J. S., *et al.* (1985). Expression of the *Escherichia coli lacZ* gene on a plasmid vector in a cyanobacterium. *Science* **230,** 805–807.

Cai, Y. P., and Wolk, C. P. (1990). Use of a conditionally lethal gene in *Anabaena* sp. strain PCC 7120 to select for double recombinants and to entrap insertion sequences. *J. Bacteriol.* **172,** 3138–3145.

Campbell, E. L., *et al.* (2007). Global gene expression patterns of *Nostoc punctiforme* in steady-state dinitrogen-grown heterocyst-containing cultures and at single time points during the differentiation of akinetes and hormogonia. *J. Bacteriol.* **189,** 5247–5256.

Cardona, T., *et al.* (2007). Isolation and characterization of thylakoid membranes from the filamentous cyanobacterium *Nostoc punctiforme*. *Physiol. Plant.* **131,** 622–634.

Cardona, T., *et al.* (2009). Electron transfer protein complexes in the thylakoid membranes of heterocysts from the cyanobacterium *Nostoc punctiforme*. *Biochim. Biophys. Acta* **1787,** 252–263.

Charaniya, S., *et al.* (2008). Mining bioprocess data: Opportunities and challenges. *Trends Biotechnol.* **26,** 690–699.

Chaurasia, A. K., *et al.* (2008). An integrative expression vector for strain improvement and environmental applications of the nitrogen fixing cyanobacterium, *Anabaena* sp. strain PCC7120. *J. Microbiol. Methods* **73,** 133–141.

Clerico, E. M., *et al.* (2007). Specialized techniques for site-directed mutagenesis in cyanobacteria. *Methods Mol. Biol.* **362,** 155–171.

Cohen, Y., and Gurevitz, M. (2006). The cyanobacteria-ecology, physiology and molecular genetics. *In* "The Prokaryotes," (M. Dworkin, *et al.*, eds.), pp. 1074–1098. Springer, New York.

Cohen, M. F., *et al.* (1994). Transposon mutagenesis of *Nostoc* sp. strain ATCC 29133, a filamentous cyanobacterium with multiple cellular differentiation alternatives. *Microbiology* **140,** 3233–3240.

Cohen, M. F., *et al.* (1998). Transposon mutagenesis of heterocyst-forming filamentous cyanobacteria. *In* "Photosynthesis: Molecular Biology of Energy Capture," (L. McIntosh, ed.), pp. 3–17. Academic Press Inc, San Diego.

Colyer, C. L., *et al.* (2005). Analysis of cyanobacterial pigments and proteins by electrophoretic and chromatographic methods. *Anal. Bioanal. Chem.* **382,** 559–569.

Cormack, B. P., *et al.* (1996). FACS-optimized mutants of the green fluorescent protein (GFP). *Gene* **173,** 33–38.

Davison, J. (2002). Genetic tools for pseudomonads, rhizobia, and other Gram-negative bacteria. *Biotechniques* **32,** 386–401.

Day, J. G. (2007). Cryopreservation of microalgae and cyanobacteria. *Methods Mol. Biol.* **368,** 141–151.

Desplancq, D., *et al.* (2005). Combining inducible protein overexpression with NMR-grade triple isotope labeling in the cyanobacterium *Anabaena* sp. PCC 7120. *BioTechniques* **39,** 405–411.

Domain, F., *et al.* (2004). Function and regulation of the cyanobacterial genes *lexA*, *recA* and *ruvB*. LexA is critical to the survival of cells facing inorganic carbon starvation. *Mol. Microbiol.* **53,** 65–80.

Dong, G., and Golden, S. S. (2008). How a cyanobacterium tells time. *Curr. Opin. Microbiol.* **11,** 541–546.

Dühring, U., *et al.* (2006). An internal antisense RNA regulates expression of the photosynthesis gene *isiA*. *Proc. Natl. Acad. Sci. USA* **103,** 7054–7058.

Eaton-Rye, J. J. (2004). The construction of gene knockouts in the cyanobacterium *Synechocystis* sp. PCC 6803. *Methods Mol. Biol.* **274,** 309–324.

Ekman, M., *et al.* (2006). Protein expression profiles in an endosymbiotic cyanobacterium revealed by a proteomic approach. *Mol. Plant-Microbe Interact.* **19,** 1251–1261.

Elhai, J. (1993). Strong and regulated promoters in the cyanobacterium *Anabaena* PCC 7120. *FEMS Microbiol. Lett.* **114,** 179–184.

Elhai, J., and Wolk, C. P. (1988). Conjugal transfer of DNA to cyanobacteria. *In* "Cyanobacteria," (L. Packer and A. N. Glazer, eds.), pp. 747–754. Academic Press, San Diego.

Elhai, J., *et al.* (1997). Reduction of conjugal transfer efficiency by three restriction activities of *Anabaena* sp. strain PCC 7120. *J. Bacteriol.* **179,** 1998–2005.

Elowitz, M. B., and Leibler, S. (2000). A synthetic oscillatory network of transcriptional regulators. *Nature* **403,** 335–338.

Eriksson, J., *et al.* (2000). Deletion mutagenesis of the 5' *psbA2* region in *Synechocystis* 6803: Identification of a putative cis element involved in photoregulation. *Mol. Cell Biol. Res. Commun.* **3,** 292–298.

Ernst, A., *et al.* (1992). Synthesis of nitrogenase in mutants of the cyanobacterium *Anabaena* Sp. strain PCC 7120 affected in heterocyst development or metabolism. *J. Bacteriol.* **174,** 6025–6032.

Flores, E., and Wolk, C. P. (1985). Identification of facultatively heterotrophic, N_2-fixing cyanobacteria able to receive plasmid vectors from *Escherichia coli* by conjugation. *J. Bacteriol.* **162,** 1339–1341.

Flores, E., *et al.* (2008). Gene transfer to cyanobacteria in the laboratory and in nature. *In* "The Cyanobacteria: Molecular Biology, Genomics and Evolution," (A. Herrero and E. Flores, eds.), pp. 45–57. Caister Academic press, Norfolk.

Frey, J., and Bagdasarian, M. (1989). The molecular biology of IncQ plasmids. *In* "Promiscuous Plasmids of Gram-negative Bacteria," (C. M. Thomas, ed.), pp. 79–94. Academic Press, London.

Frias, J. E., *et al.* (1997). Nitrate assimilation gene cluster from the heterocyst-forming cyanobacterium Anabaena sp. strain PCC 7120. *J. Bacteriol.* **179,** 477–486.

Frias, J. E., *et al.* (2003). Open reading frame all0601 from *Anabaena* sp. strain PCC 7120 represents a novel gene, *cnaT*, required for expression of the nitrate assimilation nir operon. *J. Bacteriol.* **185,** 5037–5044.

Gawad, S., *et al.* (2010). Impedance spectroscopy and optical analysis of single biological cells and organisms in microsystems. *Methods Mol. Biol.* **583,** 149–182.

Gay, P., *et al.* (1985). Positive selection procedure for entrapment of insertion-sequence elements in Gram-negative bacteria. *J. Bacteriol.* **164,** 918–921.
Geerts, D., *et al.* (1995). Inducible expression of heterologous genes targeted to a chromosomal platform in the cyanobacterium *Synechococcus* sp. PCC 7942. *Microbiology* **141,** 831–841.
Georg, J., *et al.* (2009). Evidence for a major role of antisense RNAs in cyanobacterial gene regulation. *Mol. Syst. Biol.* 5.
Ghassemian, M., *et al.* (1994). Cloning, sequencing and transcriptional studies of the genes for cytochrome c-553 and plastocyanin from *Anabaena* sp. PCC 7120. *Microbiology* **140,** 1151–1159.
Ghim, C. M., *et al.* (2010). The art of reporter proteins in science: Past, present and future applications. *BMB Rep.* **43,** 451–460.
Grigorieva, G., and Shestakov, S. (1982). Transformation in the Cyanobacterium *Synechocystis* Sp. PCC 6803. *FEMS Microbiol. Lett.* **13,** 367–370.
Hall, D. O., and Rao, K. K. (1999). Photosynthesis. Cambridge University Press in association with the Institute of Biology, Cambridge.
Haselkorn, R. (1991). Genetic systems in cyanobacteria. *In* "Bacterial Genetic Systems," (J. H. Miller, ed.), pp. 418–430. Academic Press, San Diego.
Haury, J. F., and Spiller, H. (1981). Fructose uptake and influence on growth of and nitrogen-fixation by *Anabaena variabilis*. *J. Bacteriol.* **147,** 227–235.
Hayes, F. (2003). Transposon-based strategies for microbial functional genomics and proteomics. *Annu. Rev. Genet.* **37,** 3–29.
He, Q., *et al.* (1999). Expression of a higher plant light-harvesting chlorophyll a/b-binding protein in *Synechocystis* sp. PCC 6803. *Eur. J. Biochem.* **263,** 561–570.
Higo, A., *et al.* (2007). Dynamic transcriptional changes in response to rehydration in *Anabaena* sp. PCC 7120. *Microbiology* **153,** 3685–3694.
Hihara, Y., *et al.* (2001). DNA microarray analysis of cyanobacterial gene expression during acclimation to high light. *Plant Cell* **13,** 793–806.
Hihara, Y., *et al.* (2003). DNA microarray analysis of redox-responsive genes in the Genome of the cyanobacterium *Synechocystis* sp. strain PCC 6803. *J. Bacteriol.* **185,** 1719–1725.
Holmqvist, M., *et al.* (2009). Characterization of the *hupSL* promoter activity in *Nostoc punctiforme* ATCC 29133. *BMC Microbiol.* **9,** 54.
Holtman, C. K., *et al.* (2005). High-throughput functional analysis of the *Synechococcus elongatus* PCC 7942 genome. *DNA Res.* **12,** 103–115.
Huang, F., *et al.* (2002). Proteomics of *Synechocystis* sp. strain PCC 6803: Identification of plasma membrane proteins. *Mol. Cell. Proteomics* **1,** 956–966.
Huang, F., *et al.* (2004). Isolation of outer membrane of *Synechocystis* sp. PCC 6803 and its proteomic characterization. *Mol. Cell. Proteomics* **3,** 586–595.
Huang, H.-H., *et al.* (2010). Design and characterization of molecular tools for a Synthetic Biology approach towards developing cyanobacterial biotechnology. *Nucleic Acids Res.* **38,** 2577–2593.
Imamura, S., and Asayama, M. (2009). Sigma factors for cyanobacterial transcription. *Gene Regul. Syst. Biol.* **3,** 65–87.
Imashimizu, M., *et al.* (2003). Thymine at -5 is crucial for cpc promoter activity of *Synechocystis* sp. strain PCC 6714. *J. Bacteriol.* **185,** 6477–6480.
Ivanikova, N. V., *et al.* (2005). Construction and characterization of a cyanobacterial bioreporter capable of assessing nitrate assimilatory capacity in freshwaters. *Limnol. Oceanogr. Methods* **3,** 86–93.
Iwai, M., *et al.* (2004). Improved genetic transformation of the thermophilic cyanobacterium, *Thermosynechococcus elongatus* BP-1. *Plant Cell Physiol.* **45,** 171–175.

Jansson, C., *et al.* (1998). Use of *Synechocystis* 6803 to study expression of a *psbA* gene family. *In* "Photosynthesis: Molecular Biology of Energy Capture," (L. McIntosh, ed.), pp. 166–182. Academic Press Inc, San Diego.

Jones, K. M., *et al.* (2003). Heterocyst-specific expression of *patB*, a gene required for nitrogen fixation in *Anabaena* sp. strain PCC 7120. *J. Bacteriol.* **185,** 2306–2314.

Keiler, K. C., and Sauer, R. T. (1996). Sequence determinants of C-terminal substrate recognition by the Tsp protease. *J. Biol. Chem.* **271,** 2589–2593.

Kelly, J. R., *et al.* (2009). Measuring the activity of BioBrick promoters using an in vivo reference standard. *J. Biol. Eng.* **3,** 4.

Knight, T. (2003). Idempotent Vector Design for Standard Assembly of Biobricks. Massachusetts Institute of Technology, Cambridge.

Koksharova, O., and Wolk, C. (2002). Genetic tools for cyanobacteria. *Appl. Microbiol. Biotechnol.* **58,** 123–137.

Kufryk, G. I., *et al.* (2002). Transformation of the cyanobacterium *Synechocystis* sp. PCC 6803 as a tool for genetic mapping: Optimization of efficiency. *FEMS Microbiol. Lett.* **206,** 215–219.

Labarre, J., *et al.* (1989). Insertional mutagenesis by random cloning of antibiotic resistance genes into the genome of the cyanobacterium *Synechocystis* strain PCC 6803. *J. Bacteriol.* **171,** 3449–3457.

Lehmann, M., and Wöber, G. (1978). Continuous cultivation in a chemostat of the phototrophic procaryote, *Anacystis nidulans*, under nitrogen-limiting conditions. *Mol. Cell. Biochem.* **19,** 155–163.

Lindberg, P., *et al.* (2002). A hydrogen-producing, hydrogenase-free mutant strain of *Nostoc punctiforme* ATCC 29133. *Int. J. Hydrogen Energy* **27,** 1291–1296.

Lindberg, P., *et al.* (2010). Engineering a platform for photosynthetic isoprene production in cyanobacteria, using *Synechocystis* as the model organism. *Metab. Eng.* **12,** 70–79.

Lluisma, A. O., *et al.* (2001). Suitability of *Anabaena* PCC7120 expressing mosquitocidal toxin genes from *Bacillus thuringiensis* subsp. *israelensis* for biotechnological application. *Appl. Microbiol. Biotechnol.* **57,** 161–166.

Lopes Pinto, F., and Lindblad, P. (2010). A guide for in-house design of template-switch-based 5' rapid amplification of cDNA ends systems. *Anal. Biochem.* **397,** 227–232.

Lopes Pinto, F., *et al.* (2006). Generation of non-genomic oligonucleotide tag sequences for RNA template-specific PCR. *BMC Biotechnol.* **6,** 31.

Lopes Pinto, F., *et al.* (2007). Webtag: A new web tool providing tags/anchors for RT-PCR experiments with prokaryotes. *BMC Biotechnol.* **7,** 73.

Lopes Pinto, F., *et al.* (2009). Analysis of current and alternative phenol based RNA extraction methodologies for cyanobacteria. *BMC Mol. Biol.* **10,** 79.

Luque, I., *et al.* (2004). In vivo activity of the nitrogen control transcription factor NtcA is subjected to metabolic regulation in *Synechococcus* sp. strain PCC 7942. *FEMS Microbiol. Lett.* **236,** 47–52.

Ma, J., *et al.* (2002). Correlations between Shine-Dalgarno sequences and gene features such as predicted expression levels and operon structures. *J. Bacteriol.* **184,** 5733–5745.

Mackey, S. R., *et al.* (2007). Detection of rhythmic bioluminescence from luciferase reporters in cyanobacteria. *Methods Mol. Biol.* **362,** 115–129.

Marraccini, P., *et al.* (1993). A conjugative plasmid vector for promoter analysis in several cyanobacteria of the genera *Synechococcus* and *Synechocystis*. *Plant Mol. Biol.* **23,** 905–909.

Martin, W., and Kowallik, K. V. (1999). Annotated English translation of Mereschkowsky's 1905 paper 'Uber Natur und Ursprung der Chromatophoren im Pflanzenreiche'. *Eur. J. Phycol.* **34,** 287–295.

Meeks, J. C., and Castenholz, R. W. (1971). Growth and photosynthesis in an extreme thermophile, *Synechococcus lividus* (Cyanophyta). *Arch. Microbiol.* **78,** 25–41.

Meetam, M., *et al.* (1999). The PsbY protein is not essential for oxygenic photosynthesis in the cyanobacterium *Synechocystis* sp. PCC 6803. *Plant Physiol.* **121,** 1267–1272.

Mermet-Bouvier, P., and Chauvat, F. (1994). A conditional expression vector for the cyanobacteria *Synechocystis* sp. strains PCC6803 and PCC6714 or *Synechococcus* sp. strains PCC7942 and PCC6301. *Curr. Microbiol.* **28,** 145–148.

Meyer, R. (2009). Replication and conjugative mobilization of broad host-range IncQ plasmids. *Plasmid* **62,** 57–70.

Mitschke, J., *et al.* (2011). An experimentally anchored map of transcriptional start sites in the model cyanobacterium *Synechocystis* sp. PCC6803. *Pro. Natl. Aca. Sci.* **108,** 2124–2129.

Mohamed, A., and Jansson, C. (1989). Influence of light on accumulation of photosynthesis-specific transcripts in the cyanobacterium *Synechocystis* 6803. *Plant Mol. Biol.* **13,** 693–700.

Mohamed, A., and Jansson, C. (1991). Photosynthetic electron transport controls degradation but not production of *psbA* transcripts in the cyanobacterium *Synechocystis* 6803. *Plant Mol. Biol.* **16,** 891–897.

Mohamed, A., *et al.* (1993). Differential expression of the *psbA* genes in the cyanobacterium *Synechocystis* 6803. *Mol. Gen. Genet* **238,** 161–168.

Mühlenhoff, U., and Chauvat, F. (1996). Gene transfer and manipulation in the thermophilic cyanobacterium *Synechococcus elongatus*. *Mol. Gen. Genet.* **252,** 93–100.

Muller, S., and Nebe-von-Caron, G. (2010). Functional single-cell analyses: Flow cytometry and cell sorting of microbial populations and communities. *FEMS Microbiol. Rev.* **34,** 554–587.

Murphy, R. C., *et al.* (1987). Molecular cloning and characterization of the *recA* gene from the cyanobacterium *Synechococcus* sp. strain PCC 7002. *J. Bacteriol.* **169,** 2739–2747.

Murry, M. A., and Wolk, C. P. (1991). Identification and initial utilization of a portion of the smaller plasmid of *Anabaena variabilis* ATCC 29413 capable of replication in *Anabaena* sp. strain M-131. *Mol. Gen. Genet.* **227,** 113–119.

Nakamura, Y., *et al.* (2000). Codon usage tabulated from international DNA sequence databases: Status for the year 2000. *Nucleic Acids Res.* **28,** 292.

Ng, W. O., *et al.* (2000). PhrA, the major photoreactivating factor in the cyanobacterium *Synechocystis* sp. strain PCC 6803 codes for a cyclobutane-pyrimidine-dimer-specific DNA photolyase. *Arch. Microbiol.* **173,** 412–417.

O'Farrell, P. H. (1975). High resolution two-dimensional electrophoresis of proteins. *J. Biol. Chem.* **250,** 4007–4021.

Oliveira, P., and Lindblad, P. (2008). An AbrB-Like protein regulates the expression of the bidirectional hydrogenase in *Synechocystis* sp. strain PCC 6803. *J. Bacteriol.* **190,** 1011–1019.

Olson, J. (2006). Photosynthesis in the Archean Era. *Photosynth. Res.* **88,** 109–117.

Onai, K., *et al.* (2004). Natural transformation of the thermophilic cyanobacterium *Thermosynechococcus elongatus* BP-1: A simple and efficient method for gene transfer. *Mol. Genet. Genomics* **271,** 50–59.

Ormo, M., *et al.* (1996). Crystal structure of the *Aequorea victoria* green fluorescent protein. *Science* **273,** 1392–1395.

Ow, S. Y., and Wright, P. C. (2009). Current trends in high throughput proteomics in cyanobacteria. *FEBS Lett.* **583,** 1744–1752.

Ow, S. Y., *et al.* (2008). Quantitative shotgun proteomics of enriched heterocysts from *Nostoc* sp. PCC 7120 using 8-plex isobaric peptide tags. *J. Proteome Res.* **7,** 1615–1628.

Ow, S. Y., *et al.* (2009). iTRAQ underestimation in simple and complex mixtures: "The good, the bad and the ugly" *J. Proteome Res.* **8,** 5347–5355.

Phillips, H. L., *et al.* (2010). Shotgun proteome analysis utilising mixed mode (reversed phase-anion exchange chromatography) in conjunction with reversed phase liquid chromatography mass spectrometry analysis. *Proteomics* **10,** 2950–2960.

Rizzo, M. A., *et al.* (2004). An improved cyan fluorescent protein variant useful for FRET. *Nat. Biotechnol.* **22,** 445–449.

Scanlan, D. J., *et al.* (1990). Construction of *lacZ* promoter probe vectors for use in *Synechococcus*: Application to the identification of CO_2-regulated promoters. *Gene* **90,** 43–49.

Scherzinger, D., *et al.* (2006). Retinal is formed from apo-carotenoids in *Nostoc* sp. PCC7120: In vitro characterization of an apo-carotenoid oxygenase. *Biochem. J.* **398,** 361–369.

Schmetterer, G., *et al.* (2001). The coxBAC operon encodes a cytochrome c oxidase required for heterotrophic growth in the cyanobacterium *Anabaena variabilis* strain ATCC 29413. *J. Bacteriol.* **183,** 6429–6434.

Scholz, P., *et al.* (1989). Complete nucleotide sequence and gene organization of the broad-host-range plasmid RSF1010. *Gene* **75,** 271–288.

Shen, G., *et al.* (2002). Assembly of photosystem I. I. Inactivation of the *rubA* gene encoding a membrane-associated rubredoxin in the cyanobacterium *Synechococcus* sp. PCC 7002 causes a loss of photosystem I activity. *J. Biol. Chem.* **277,** 20343–20354.

Shestakov, S. V., and Khyen, N. T. (1970). Evidence for genetic transformation in blue-green alga *Anacystis nidulans*. *Mol. Gen. Genet.* **107,** 372–375.

Shetty, R. P., *et al.* (2008). Engineering BioBrick vectors from BioBrick parts. *J. Biol. Eng.* **2,** 5.

Sokolenko, A., *et al.* (2002). The gene complement for proteolysis in the cyanobacterium *Synechocystis* sp. PCC 6803 and *Arabidopsis thaliana* chloroplasts. *Curr. Genet.* **41,** 291–310.

Soule, T., *et al.* (2007). Molecular genetics and genomic analysis of scytonemin biosynthesis in *Nostoc punctiforme* ATCC 29133. *J. Bacteriol.* **189,** 4465–4472.

Srivastava, R., *et al.* (2005). Proteomic studies of the thylakoid membrane of *Synechocystis* sp. PCC 6803. *Proteomics* **5,** 4905–4916.

Stanier, R. Y., and Cohen-Bazire, G. (1977). Phototrophic prokaryotes: The cyanobacteria. *Annu. Rev. Microbiol.* **31,** 225–274.

Stensjö, K., *et al.* (2007). An iTRAQ-based quantitative analysis to elaborate the proteomic response of *Nostoc* sp. PCC 7120 under N_2 fixing conditions. *J. Proteome Res.* **6,** 621–635.

Stevens, S. E., and Porter, R. D. (1980). Transformation in *Agmenellum quadruplicatum*. *Proc. Natl. Acad. Sci. USA* **77,** 6052–6056.

Sui, Z. H., *et al.* (2004). Cloning and characterization of the phycoerythrin operon upstream sequence of *Gracilaria lemaneiformis* (Rhodophyta). *J. Appl. Phycol.* **16,** 167–174.

Summers, M. L., *et al.* (1995). Genetic-evidence of a major role for glucose-6-phosphate-dehydrogenase in nitrogen-fixation and dark growth of the cyanobacterium *Nostoc* Sp. strain ATCC 29133. *J. Bacteriol.* **177,** 6184–6194.

Tamagnini, P., *et al.* (1997). Hydrogenases in *Nostoc* sp. strain PCC 73102, a strain lacking a bidirectional enzyme. *Appl. Environ. Microbiol.* **63,** 1801–1807.

Tamagnini, P., *et al.* (2007). Cyanobacterial hydrogenases: Diversity, regulation and applications. *FEMS Microbiol. Rev.* **31,** 692–720.

Tetu, S. G., *et al.* (2009). Microarray analysis of phosphate regulation in the marine cyanobacterium *Synechococcus* sp. WH8102. *ISME J.* **3,** 835–849.

Thiel, T. (1994). Genetic analysis of cyanobacteria. *In* "The Molecular Biology of Cyanobacteria," (D. Bryant, ed.), pp. 581–611. Kluwer Academic Publishers, Dordrecht.

Thiel, T., and Poo, H. (1989). Transformation of a filamentous cyanobacterium by electroporation. *J. Bacteriol.* **171,** 5743–5746.

Thiel, T., and Wolk, P. C. (1987). Conjugal transfer of plasmids to cyanobacteria. *In* "Recombinant DNA, Part D," (R. Wu and L. Grossman, eds.), pp. 232–243. Academic Press, San Diego.

Thornton, L. E., *et al.* (2004). Homologs of plant PsbP and PsbQ proteins are necessary for regulation of photosystem ii activity in the cyanobacterium *Synechocystis* 6803. *Plant Cell* **16,** 2164–2175.

Tsinoremas, N. F., *et al.* (1994). Efficient gene transfer in *Synechococcus* sp. strains PCC 7942 and PCC 6301 by interspecies conjugation and chromosomal recombination. *J. Bacteriol.* **176,** 6764–6768.

van Thor, J. J., *et al.* (1999). Localization and function of ferredoxin: $NADP^+$ reductase bound to the phycobilisomes of *Synechocystis*. *EMBO J.* **18,** 4128–4136.

Vioque, A. (2007). Transformation of cyanobacteria. *Adv. Exp. Med. Biol.* **616,** 12–22.

Viskari, P. J., *et al.* (2001). Determination of phycobiliproteins by capillary electrophoresis with laser-induced fluorescence detection. *Electrophoresis* **22,** 2327–2335.

Wackett, L. P. (2010). Phototrophs in biotechnology. *Microb. Biotechnol.* **3,** 487–488.

Walton, D. K., *et al.* (1993). DNA sequence and shuttle vector construction of plasmid pGL3 from *Plectonema boryanum* PCC 6306. *Nucleic Acids Res.* **21,** 746.

Waterbury, J. (2006). The cyanobacteria-Isolation, purification and identification. *In* "The Prokaryotes," (M. Dworkin, *et al.*, eds.), pp. 1053–1073. Springer, New York.

Williams, J. G. K. (1988). Construction of specific mutations in photosystem II photosynthetic reaction center by genetic engineering methods in *Synechocystis* 6803. *In* "Cyanobacteria," (L. Packer and A. N. Glazer, eds.), pp. 766–778. Academic Press, San Diego.

Wolk, C. P., *et al.* (1984). Construction of shuttle vectors capable of conjugative transfer from *Escherichia coli* to nitrogen-fixing filamentous cyanobacteria. *Proc. Natl. Acad. Sci. USA* **81,** 1561–1565.

Wolk, C. P., *et al.* (1988). Isolation and Complementation of Mutants of *Anabaena* Sp Strain PCC 7120 Unable to Grow Aerobically on Dinitrogen. *J. Bacteriol.* **170,** 1239–1244.

Wolk, C. P., *et al.* (1991). Use of a transposon with luciferase as a reporter to identify environmentally responsive genes in a cyanobacterium. *Proc. Natl. Acad. Sci. USA* **88,** 5355–5359.

Wolk, C. P., *et al.* (1993). Amplified expression of a transcriptional pattern formed during development of *Anabaena*. *Mol. Microbiol.* **7,** 441–445.

Wolk, C. P., *et al.* (2007). Paired cloning vectors for complementation of mutations in the cyanobacterium *Anabaena* sp strain PCC 7120. *Arch. Microbiol.* **188,** 551–563.

Wong, F. C., and Meeks, J. C. (2001). The *hetF* gene product is essential to heterocyst differentiation and affects HetR function in the cyanobacterium *Nostoc punctiforme*. *J. Bacteriol.* **183,** 2654–2661.

Wu, J., *et al.* (2007). cTFbase: A database for comparative genomics of transcription factors in cyanobacteria. *BMC Genomics* **8,** 104.

Xu, D., *et al.* (2006). Construction of a non-antibiotic expression system in a marine cyanobacterium *Synechococcus* sp. PCC 7002 and its application in production of oral vaccine against enterotoxin of *Escherichia coli*. *J. Appl. Phycol.* **18,** 127–134.

Yoon, H. S., and Golden, J. W. (1998). Heterocyst pattern formation controlled by a diffusible peptide. *Science* **282,** 935–938.

Yoon, H. S., and Golden, J. W. (2001). PatS and products of nitrogen fixation control heterocyst pattern. *J. Bacteriol.* **183,** 2605–2613.

Zang, X. N., *et al.* (2007). Optimum conditions for transformation of *Synechocystis* sp. PCC 6803. *J. Microbiol.* **45,** 241–245.

Zhu, Y., *et al.* (2010). Roles of xanthophyll carotenoids in protection against photoinhibition and oxidative stress in the cyanobacterium *Synechococcus* sp. strain PCC 7002. *Arch. Biochem. Biophys.* **504,** 86–99.

CHAPTER TWENTY-FIVE

Developing a Synthetic Signal Transduction System in Plants

Kevin J. Morey, Mauricio S. Antunes, Kirk D. Albrecht, Tessa A. Bowen, Jared F. Troupe, Keira L. Havens, *and* June I. Medford

Contents

Abstract

One area of focus in the emerging field of plant synthetic biology is the manipulation of systems involved in sensing and response to environmental signals. Sensing and responding to signals, including ligands, typically involves

Department of Biology, Colorado State University, Fort Collins, Colorado, USA

Methods in Enzymology, Volume 497
ISSN 0076-6879, DOI: 10.1016/B978-0-12-385075-1.00025-1

biological signal transduction. Plants use a wide variety of signaling systems to sense and respond to their environment. One of these systems, a histidine kinase (HK) based signaling system, lends itself to manipulation using the tools of synthetic biology. Both plants and bacteria use HKs to relay signals, which in bacteria can involve as few as two proteins (two-component systems or TCS). HK proteins are evolutionarily conserved between plants and bacteria and plant HK components have been shown to be functional in bacteria. We found that this conservation also applies to bacterial HK components which can function in plants. This conservation of function led us to hypothesize that synthetic HK signaling components can be designed and rapidly tested in bacteria. These novel HK signaling components form the foundation for a synthetic signaling system in plants, but typically require modifications such as codon optimization and proper targeting to allow optimal function. We describe the process and methodology of producing a synthetic signal transduction system in plants. We discovered that the bacterial response regulator (RR) PhoB shows HK-dependent nuclear translocation *in planta*. Using this discovery, we engineered a partial synthetic pathway in which a synthetic promoter (PlantPho) is activated using a plant-adapted PhoB (PhoB–VP64) and the endogenous HK-based cytokinin signaling pathway. Building on this work, we adapted an input or sensing system based on bacterial chemotactic binding proteins and HKs, resulting in a complete eukaryotic signal transduction system. Input to our eukaryotic signal transduction system is provided by a periplasmic binding protein (PBP), ribose-binding protein (RBP). RBP interacts with the membrane-localized chemotactic receptor Trg. PBPs like RBP have been computationally redesigned to bind small ligands, such as the explosive 2,4,6-trinitrotoluene (TNT). A fusion between the chemotactic receptor Trg and the HK, PhoR, enables signal transduction via PhoB, which undergoes nuclear translocation in response to phosphorylation, resulting in transcriptional activation of an output gene under control of a synthetic plant promoter. Collectively, these components produce a novel ligand-responsive signal transduction system in plants and provide a means to engineer a eukaryotic synthetic signaling system.

1. Introduction

1.1. Signal transduction and synthetic biology

A broad goal of synthetic biology is to engineer novel genetic circuits to produce a specific and regulatable behavior (Khalil and Collins, 2010). Endogenous biological circuits that involve signal transduction allow physical and molecular cues from an organism's environment to activate a cellular response (Kiel *et al.*, 2010). In plants, signaling pathways are activated in response to specific molecules, such as hormones and secondary messengers (Jaillais and Chory, 2010; Zhang and McCormick, 2009), and to a variety of environmental stimuli, such as light, temperature, abiotic stress, certain types of bacteria, viruses, and herbivores (De Moraes *et al.*, 2004). One approach to engineer novel synthetic genetic circuits takes advantage of the mechanisms naturally used by endogenous signal transduction pathways. These natural signaling pathways can then be modified by the addition of synthetic components that interact in a similar way.

1.2. Endogenous kinase-based signaling systems in plants

Kinase-based signal transduction pathways represent a major means of information transfer in plants and include: serine–threonine kinases (Rodriguez *et al.*, 2010), histidine kinases (HKs; To and Kieber, 2008), and kinases with dual specificity for threonine and tyrosine (Oh *et al.*, 2009; Sessa *et al.*, 1996). These systems have been studied extensively, and therefore could serve as starting points for building synthetic signal transduction pathways.

1.2.1. Light sensing

Light is essential for every aspect of plant life. Plants regulate growth and development in response to light by modulating seed germination, seedling de-etiolation, phototropism, circadian rhythms, and flowering time. Light is sensed by at least three photoreceptors; phototropins, cryptochromes, and phytochromes, all of which have been shown to have kinase activity in the presence of the appropriate wavelength (Christie, 2007; Osgur and Sancar, 2006; Yeh and Lagarias, 1998). Phototropins respond to UV-A light, while cryptochromes respond to blue light, and phytochromes respond to red and far-red light. By integrating complex signals from different photoreceptors, plants are able to respond as needed to widely differing light conditions.

Phytochromes are the best studied class of light sensors. Phytochromes contain an N-terminal photo-sensory domain and a C-terminal regulatory HK-related domain. There are five known phytochromes, each with both unique and overlapping functions (Castillon *et al.*, 2007). All switch between two reversible states depending on exposure to red or far-red light. Red light switches phytochromes into a biologically active state, whereas "dark-decay" or far-red light reverses this process and inactivates phytochromes. The active phytochromes translocate to the nucleus and directly interact with transcription factors to regulate expression of thousands of light responsive genes (Sharrock, 2008). While light sensing proteins have been previously used in synthetic biology (Levskaya *et al.*, 2009), a plant synthetic signaling pathway that is based on phytochrome signaling would most likely be influenced by light, making it difficult to control such a pathway independent of endogenous processes.

1.2.2. Plant signaling based on MAP kinases

Another endogenous kinase system, mitogen-activated protein kinase (MAPK) signal transduction, mediates responses to environmental and developmental cues in plants. MAPK signaling typically involves a tyrosine or serine–threonine phosphorylation cascade between its membrane receptor and MAP3Ks, MAP2Ks, and MAPKs. MAPKs in turn regulate activity of downstream kinases, other enzymes, and transcription factors, effecting responses to abiotic stress and hormones. In Arabidopsis, 60 MAP3Ks, 10 MAP2Ks, and 20 MAPKs have been identified, although the function of many of these proteins has yet to be determined (Rodriguez *et al.*, 2010). This multitude of MAP kinases requires a multitude of spatial–temporal controls, specific phosphatases, scaffolding proteins, and feedback controls to maintain signal specificity.

One type of MAP kinase signaling involves the plant hormone ethylene. Ethylene is a regulator of a variety of processes including germination, growth, development, fruit ripening, and stress responses. Plant transmembrane receptors for ethylene are localized to the endoplasmic reticulum (ER). When ethylene is bound the receptor complex is inactivated and a MAPK signaling cascade is initiated, activating transcription of primary ethylene response genes. However, the sensitivity of this response is thought to differ among tissues and developmental stages (Stepanova and Alonso, 2009; Yoo *et al.*, 2009), revealing yet another measure of MAP kinase signaling complexity.

The complexity of MAP kinase-based signaling is a major drawback in using these components to build a synthetic signal transduction pathway; a large number of MAPK proteins need to be expressed to obtain a desired response. Furthermore, a synthetic signal pathway built out of parts from this system would likely suffer from significant unwanted crosstalk with endogenous signaling components. This complication would make testing a synthetic system and obtaining the desired specificity with MAP kinase components difficult.

1.2.3. Cytokinin signaling based on histidine kinases

In addition to tyrosine and serine–threonine kinase-based signaling, plants use a HK signal transduction system to respond to the hormone cytokinin. Like other plant hormones, cytokinins affect a number of processes such as branching, chloroplast development, leaf senescence, some stress responses, and pathogen resistance (Heyl and Schmulling, 2003). Early work in this area revealed that plant HKs used in cytokinin perception have homology to the first HK systems discovered, bacterial two-component systems (TCS).

Bacterial TCS range from simple to complex and control a variety of responses that include: chemotaxis (nutrient sensing), osmotic sensing, quorum sensing, regulation of pathogenesis, redox sensing, and sporulation (Parkinson and Kofoid, 1992). The simplest bacterial TCS use only two proteins: a membrane-localized HK, and a soluble, cytoplasm-localized response regulator (RR). Structurally, a typical membrane-localized HK protein consists of an N-terminal extracellular receptor, a transmembrane alpha helix, and a C-terminal cytoplasmic region, which contains a dimerization and histidine phosphorylation (DHP) domain and a hATPase domain. In a simple TCS, a receptor HK binds a ligand or recognizes a stimulus outside the cell. The receptor then undergoes a conformational change that is propagated by a transmembrane alpha helix (transmembrane signaling) leading to activation of the cytoplasmic region. This activation causes autophosphorylation of a histidine residue on the DHP domain. The high energy phosphate is then transferred from the HK to an aspartate residue on its cognate RR (Fig. 25.1A). Upon phosphorylation, many RRs (e.g., OmpR/PhoB family) undergo a conformational change that activates their effector domain. In many cases the effector domain binds DNA and activates transcription, controlling expression of downstream genes involved in the

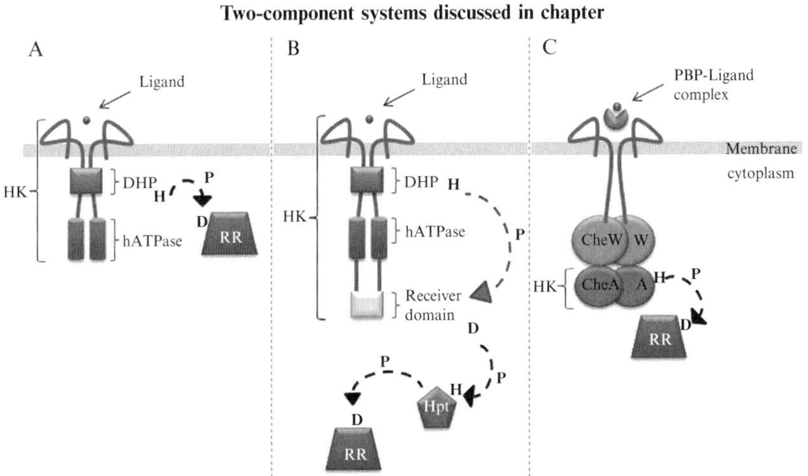

Figure 25.1 HK-based signaling systems in bacteria and plants: (A) "Simple" bacterial TCS—extracellular input is provided by a transmembrane HK. In the HK molecule, a histidine residue in the dimerization and histidine phosphorylation (DHP) domain autophosphorylates in response to ligand binding. The high energy phosphate is then transferred to a cytoplasmically localized response regulator (RR). (B) Hybrid HK (e.g., AHKs, RcsC) signaling found in plants and bacteria—these systems have more complex HKs and signaling components. In response to a ligand binding, a histidine residue in the DHP domain autophosphorylates. The high energy phosphate is then transferred internally to an aspartate in the receiver domain, and subsequently to a histidine in a cytoplasmic Hpt protein (called AHPs in Arabidopsis), which then phosphorylates an RR on its aspartate residue. (C) Bacterial chemotactic TCS—a chemotactic receptor (e.g., Trg) binds a PBP–ligand complex or directly binds a ligand. This binding event initiates a signal that is transmitted to a cytoplasm-localized histidine kinase CheA. CheW acts as an adaptor protein for the chemotactic receptor/CheA complex. CheA then phosphorylates RRs that do not activate transcription.

response to the ligand or stimulus. More elaborate TCS include more complex HK proteins, multiple RRs, and a third component, histidine phosphotransfer (Hpt) proteins (Fig. 25.1B). Regardless of the added complexity, one hallmark of all known TCS is that the high energy phosphate moiety is always relayed between histidine and aspartate residues.

Plant HK systems are similar to the hybrid TCS found in bacteria and involve complex HK receptors with multiple RRs. Three cytokinin-responsive receptor HKs have been identified in Arabidopsis, AHK2, AHK3, and AHK4/CRE1. Upon binding cytokinin, the AHKs autophosphorylate at a histidine residue, then undergo an internal phosphotransfer to an aspartate residue located on the C-terminal receiver domain. This phosphate is transferred to a histidine residue on a Hpt protein (known as AHPs in Arabidopsis) (Fig. 25.1B). AHPs have been shown to relay the phosphate moiety to nuclear-localized RRs (known as ARRs in Arabidopsis), as well as cytoplasm-localized cytokinin response factors (CRFs). Upon activation by aspartate phosphorylation, ARRs control expression of genes involved in cytokinin responses (Argueso *et al.*, 2009; To and Kieber, 2008).

This similarity in the signaling mechanism is corroborated by evolutionary conservation of the components involved in bacterial TCS and plant HK systems. Protein sequence alignment of the respective phosphotransfer and/or phospho-accepting domains of *Escherichia coli* and Arabidopsis HKs, Hpts, and RRs show significant homology, showing these domains are evolutionarily conserved (Fig. 25.2). Moreover, phylogenetic analyses suggest plants obtained these genes from bacteria via lateral gene transfer (Koretke *et al.*, 2000). Collectively, these data and results from the Mizuno laboratory (Section 2.1), lead us to hypothesize that bacteria and plant HK components may have interchangeable functionality.

The evolutionary conservation of plant and bacterial HK signaling systems provides an attractive platform for developing a eukaryotic synthetic signal transduction pathway. First, because bacterial TCS use only two proteins from stimulus to a response, they provide a framework for a synthetic system with a minimal number of components. Second, if the components conserve functionality, it will provide a rapid testing platform (bacteria) where components can be rationally designed and techniques such as directed evolution applied prior to introduction into a eukaryotic host.

2. Foundation for Developing a Molecular Testing Platform for HK Systems

2.1. Early work showing plant HKs function in bacteria

Evidence that the Arabidopsis proteins AHK2, AHK3, and AHK4/CRE1 function as cytokinin receptors was obtained from bacterial studies. These plant HKs in bacterial cells complement a deletion mutant of the *E. coli* HK RcsC

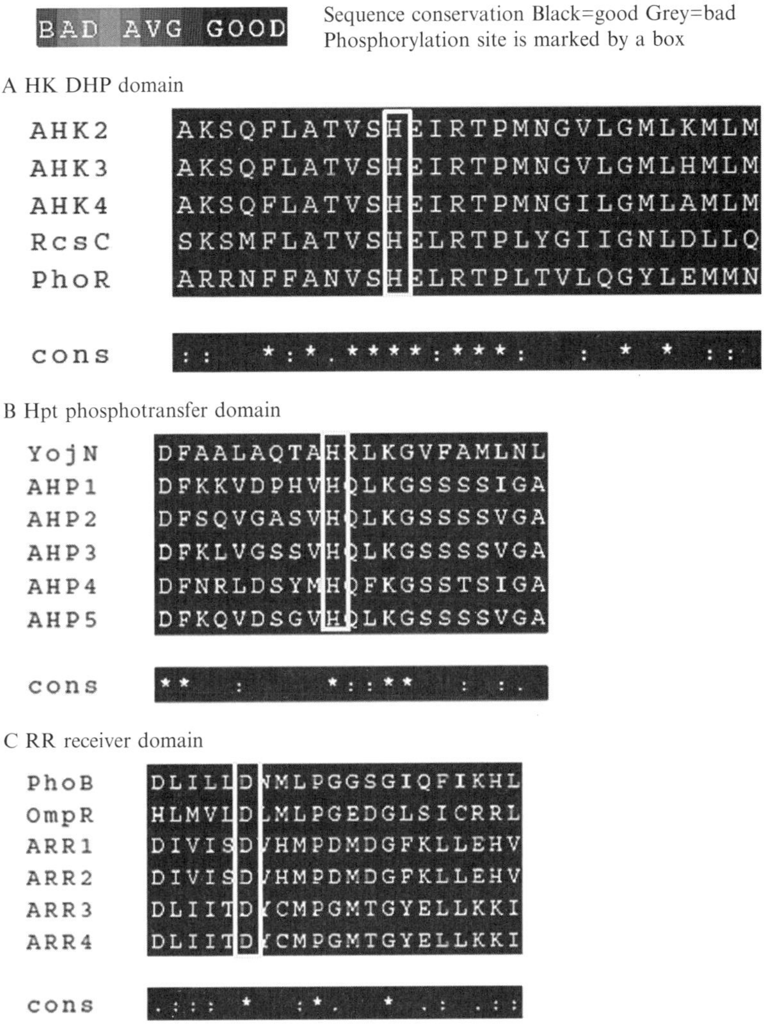

Figure 25.2 T-Coffee (Notredame *et al.*, 2000) alignment of conserved phosphotransfer and phosphoaccepting domains of Arabidopsis and *E. coli* HKs, Hpts, and RRs. These domains are conserved in both prokaryotes and eukaryotes.

and properly initiate a signal in the presence of the plant hormone cytokinin (Suzuki *et al.*, 2001; Yamada *et al.*, 2001). RcsC is a hybrid HK (Fig. 25.1B) that signals, via phosphorelay, to an Hpt factor (YojN) and a RR (RcsB), and activates the capsular polysaccharide synthesis operon (RcsC → YojN → RcsB) (Takeda *et al.*, 2001). The Arabidopsis HKs are able to functionally replace RcsC and interact with the bacterial Hpt YojN to produce the signaling pathway, AHK → YojN → RcsB. In addition, AHK4 has also been coexpressed with the Arabidopsis Hpts (AHP1, 2, 3, and 5) in a signal titration assay which established that AHK4 is capable of interacting with AHPs in bacteria (Suzuki *et al.*, 2001). These results demonstrate that not only is it possible to functionally express heterologous plant HK signaling proteins in bacteria, but also that these heterologous components can interact or "cross talk" with endogenous bacterial components.

These experiments provided a basis for our work, whereby various components from plant or synthetic HK systems can be rapidly tested and improved in a bacterial system prior to functional adaptation for plants. The ability to use bacteria to test synthetic eukaryotic signaling components facilitates the use of synthetic design and other experimental approaches, such as directed evolution. In addition, basic questions about aspects of transmembrane signaling and receptor function can be addressed using this system.

2.2. Development of a synthetic signaling system based on bacterial TCS

The functional expression of plant HK signaling components in bacteria (Suzuki *et al.*, 2001) lead us to hypothesize that the reciprocal might also be possible, that is, bacterial components could function in plants. If plant HK components were used to engineer a plant synthetic signal transduction system there is likely to be unwanted interference from endogenous HK signaling, which in Arabidopsis involves 28 AHPs and ARRs (Dortay *et al.*, 2006).

When designing a synthetic signaling pathway one needs to address how the signal is initiated (input), propagated (transmission), and finally, what is activated by the signal (output). The signal transmission needs to be regulated by an input dedicated to the synthetic signaling pathway to prevent nonspecific activation of the pathway. Periplasmic binding proteins (PBPs), specifically those from the bacterial chemotactic TCS, constitute a framework for synthetic signaling input. PBPs are a family of proteins involved in the scavenging of a wide variety of substances in bacteria. They undergo a distinct conformational change from an open form (without ligand) to a ligand bound closed form able to initiate signal transduction. This conformational change presented us with a relatively "clean" input, that is, an unbound PBP will not initiate signal transduction. In addition, the PBP–ligand binding pocket is highly specific to its particular ligand and the protein itself can be used as a scaffold for computational redesign, providing a broad potential for input (Dwyer and Hellinga, 2004). Computationally redesigned PBPs have been shown to bind novel molecules, including a surrogate for a nerve toxin, the metal zinc, L-lactate, and 2,4,6-trinitrotoluene (TNT) (Allert *et al.*, 2004; Dwyer and Hellinga, 2004; Dwyer *et al.*, 2003, 2004; Looger *et al.*, 2003). Initial attempts to change binding specificity of PBPs focused on computationally redesigning the binding site, but simply modifying the residues in the binding site may introduce unintentional problems into the structural and binding stability of the PBP (Schreier *et al.*, 2009; Vercillo *et al.*, 2007). Despite stability issues, computationally redesigned PBPs have been shown to activate TCS in bacteria (Looger *et al.*, 2003; Rodrigo *et al.*, 2007).

Once the signal is initiated by the PBP outside of the cell, it needs to be transmitted to the interior. Transmembrane signaling for a synthetic pathway can also be based on bacterial TCS. In *E. coli*, a subset of PBPs are capable of interacting with the chemotactic receptors Trg, Tar, Tsr, and Tap, and initiating a TCS signaling cascade that allows the bacteria to respond to the presence of certain sugars (glucose, ribose, maltose, galactose), amino acids (asp, glu, ser, ala, gly), and dipeptides, by directing the cells to environments containing these nutrients (Mowbray and Sandgren, 1998; Fig. 25.1C). However, while the signal is transmitted to the interior of the cell through the chemotactic receptor, the RRs receiving the chemotactic signal do not activate transcription. Hence, to produce a transcriptional output with PBP input, we engineered a fusion protein between the extracellular portion of a chemotactic receptor and the cytoplasmic portion of an HK that is capable of interacting with a RR that can activate transcription. These receptor/HK fusions were developed and tested in the rapid bacterial testing system and further discussed in Section 5.

The third component of a synthetic signaling system is the "output," which typically consists of transcriptional activation of one or more genes, leading to a detectable phenotype. Therefore, for a eukaryotic organism, the synthetic signaling system requires a RR capable of both accepting the signal from a transmembrane HK and subsequently localizing to the nucleus to activate transcription in response to the signal. Work in our laboratory (described in Section 4) established that some bacterial RRs (PhoB and OmpR) are capable of signal-dependent nuclear translocation in Arabidopsis (Antunes *et al.*, 2009). We chose PhoB as the RR for our synthetic signaling system because it shows strong signal-dependent nuclear translocation and it has a simple activation mechanism, allowing us to develop a plant synthetic signal transduction pathway.

In the following sections, we describe considerations for properly expressing synthetic components in bacteria or plants, how the partial signaling system was engineered, how a complete pathway was assembled, and finally how system output was quantified using the reporter gene β-glucuronidase (GUS).

3. Technical Considerations in Developing a Eukaryotic Synthetic Signal Transduction System Based on Bacterial TCS Components

During the process of developing a functional synthetic signal transduction pathway in plants we found it necessary to address a number of technical issues to ensure the bacterial components were properly expressed.

Differences in cell structure, transcription activity, and codon usage between prokaryotes and eukaryotes can all produce problems if not accounted for. The following sections address these concerns.

3.1. Proper membrane and compartment targeting

When developing a eukaryotic synthetic signaling system that uses bacterial components, it is important to consider some fundamental differences between prokaryotic and eukaryotic cells, one of which is their distinct subcellular structure. In membrane-localized bacterial HKs, the transmembrane segments are sufficient for proper targeting to the plasma membrane. Eukaryotic cells have multiple membrane-enclosed organelles; thus, any protein localized in a particular membrane needs to contain a signal peptide that specifically targets it to that membrane. To ensure proper membrane localization in plants, we added the signal peptide from the plasma membrane-localized Arabidopsis flagellin receptor FLS2 (Gomez-Gomez and Boller, 2000) to the N-terminal end of HKs.

We also had to target the aforementioned PBPs as plants do not have a periplasm. Instead, plants have a space outside the cell plasma membrane known as the apoplast, where small proteins like PBPs can freely diffuse and contact exogenous substances (Somerville *et al.*, 2004). We targeted the PBPs to the apoplast using a plant signal peptide that is normally used to secrete proteins to the apoplastic pollen-stigma matrix (Baumberger *et al.*, 2003).

3.2. Bacterial RRs: Signal dependent nuclear translocation and adaptation for transcriptional activation

The evolutionary conservation of TCS genes between bacteria and plants allows some functional interchange of individual proteins from one organism to another. However, a fundamental difference between bacteria and plants is the compartmentalization of transcriptional responses in plants to the nucleus. And, although bacterial cells do not have a nucleus, many bacterial proteins related to pathogenicity show nuclear localization in eukaryotic cells (Kay and Bonas, 2009). In addition, protein sequences that function as nuclear localization signals are commonly short sequences of positively charged amino acids. DNA-binding domains, such as those found in bacterial RRs, are also known to contain positively charged amino acids. This foundational knowledge provided us a framework to investigate bacterial RRs in plants.

In order to build a functional synthetic system in plants, a bacterial RR must receive the transmembrane signal in the cytoplasm and translocate to the nucleus to activate transcription. Previous work in our laboratory (below and Antunes *et al.* 2009) established that some bacterial RRs are capable of signal-dependent nuclear translocation. We then modified the bacterial RR PhoB to interact with eukaryotic RNA polymerase II and activate transcription in plants. A commonly used transcriptional activator is the C-terminal domain of VP16, a herpes simplex virus encoded protein. VP16 has been shown to activate transcription in eukaryotic cells (Triezenberg *et al.*, 1988), and tetrameric repeats (VP64) of amino acids 437–447 of VP16 fused to DNA-binding domains have been shown to produce efficient transcriptional activators in eukaryotes (Seipel *et al.*, 1992). Hence, we fused VP64 to the C-terminus of PhoB to produce PhoB–VP64, a modified or synthetic PhoB that can activate endogenous eukaryotic transcription machinery.

3.3. Consideration of codon bias when designing synthetic signaling systems

One last consideration when engineering a synthetic signaling system is codon usage bias. The presence of codon usage bias among different organisms has been well documented (Batard *et al.*, 2000; Lessard *et al.*, 2002; Suo *et al.*, 2006). The preference of one codon over another by an organism can be a barrier to expressing bacterial genetic circuits in plants or testing plant proteins in bacteria. Codon optimization of heterologously expressed genes can improve expression levels and, in some cases, simply allow a gene to be heterologously expressed (Perlak *et al.*, 1991).

Our system presents a stark example of conflicting codon usage bias. Four of the seven rare codons utilized by *E. coli* (used at a frequency $< 0.5\%$) code for the amino acid arginine (Chen and Texada, 2006). These codons account for over 74% of the arginine codon usage in Arabidopsis (Nakamura *et al.*, 2000), with the extremely rare AGG and AGA codons making up 57% of Arabidopsis usage. Hence, if not addressed when expressing heterologous genes, rare codon clustering can lead to drastically reduced protein levels and mRNA degradation (Li *et al.*, 2006; Sunohara *et al.*, 2004). For instance, the AHK4/CRE1 gene contains 40 of the rare arginine codons, including two arginine codons rare to *E. coli* appearing contiguously at positions 16 and 17, as well as at positions 133 and 134. Our laboratory has encountered gene instability/toxicity issues when expressing this protein in *E. coli*, a result that has also been reported elsewhere (Mizuno and Yamashino, 2010).

The effect of codon usage bias can be complicated during cloning in bacterial cells by the unwanted activity of promoters used to drive plant genes. The *CaMV35S* promoter, commonly used for constitutive expression of genes in plants, has been shown to have activity in *E. coli* (Assaad and Signer, 1990). Therefore, cloning a gene that has been codon-optimized for plants downstream of the *CaMV35S* promoter could potentially lead to instability in bacterial cells if the gene has a large number of rare *E. coli* codons. Hence, codon usage may need to be considered when simply cloning plant genes in *E. coli* as well as when developing a bacterial testing system.

4. A Partial Synthetic Signal Transduction System Using Cytokinin Input

One challenging aspect of developing a eukaryotic signal transduction pathway that is activated by an external input is moving the signal from the membrane to bring about a transcriptional response in the nucleus. Plasma membrane to transcriptional responses in eukaryotes typically involve multiple proteins which themselves interact with numerous other proteins. In contrast bacterial signaling systems can use as few as one protein from membrane to transcriptional response.

Given these considerations, we experimentally tested a simple membrane to nucleus signaling system in eukaryotes. Components from simpler systems (bacteria and yeast) were expressed in plants as N-terminal fusions to smGFP (Davis and Vierstra, 1996). As an initial means to activate HK signaling in plants we used application of exogenous cytokinin. We tested the bacterial RRs OmpR (Mizuno *et al.*, 1982; Wurtzel *et al.*, 1982), PhoB (Makino *et al.*, 1986, 1989), and RcsB (Chen *et al.*, 2001), a bacterial Hpt, YojN (Chen *et al.*, 2001), and the yeast Hpt, Ypd1 (Posas *et al.*, 1996) in transient assays. PhoB and OmpR appeared to show signal-dependent nuclear accumulation in plant cells, whereas the responses of RcsB, YojN, and Ypd1 were equivocal. Therefore, we focused our subsequent work on PhoB and OmpR.

To corroborate these transient assay results, we then generated stable transgenic Arabidopsis plants that constitutively expressed either PhoB-GFP or OmpR-GFP (Antunes *et al.*, 2009). Plants were grown for 6-8 days, and images were collected showing the subcellular distribution of GFP in roots, leaves, and hypocotyls. The same tissues were treated with 1 μM *t*-zeatin (a cytokinin) and re-imaged to determine the effect of cytokinin mediated HK activation on the bacterial RRs. To verify whether the compartment where RR-GFP localized corresponded to nuclei, the roots were stained with 1 ng/μL DAPI for 10 min. Images were collected using either a Nikon Diaphot fluorescence microscope, or a Carl Zeiss LSM 510 META confocal microscope. These results, and others detailed in Antunes *et al.* (2009) indicated that bacterial RRs accumulate in the nucleus in response to a cytokinin-initiated signal and could, therefore, be used as a component of a synthetic signal transduction pathway in plants.

To produce a PhoB responsive promoter in a plant, we designed and constructed a synthetic promoter entitled PlantPho, which consists of four copies of PhoB's DNA-binding element (Pho box) placed upstream of a minimal promoter (−46 region from the *CaMV35S* promoter) (Antunes *et al.*, 2009). To test if the PlantPho promoter was responsive to cytokinin-mediated HK activation, we placed the GUS reporter gene (Jefferson *et al.*, 1987) under its control and generated transgenic Arabidopsis plants that constitutively expressed PhoB–VP64. The partial synthetic signaling system is diagrammed in Fig. 25.3. GUS expression is assayed by incubating developmentally equivalent leaves (leaf 1 and leaf 2) from transgenic plants in the presence or absence of 10 μ*M* *t*-zeatin for 16 h, followed by total protein extraction and GUS activity measurement in the presence of the enzyme's substrate, 4-methylumbelliferyl β-D-glucuronide (MUG). Enzyme activity results in the production of a fluorescent product, 4-methylumbeliferone (4-MU). GUS activity is then normalized to a 4-MU standard curve and the total protein content of the sample, as determined by the Bradford assay. Figure 25.4 is an example of a screening assay of primary transgenic Arabidopsis plants containing the partial synthetic system. Results indicate that cytokinin can activate a partial synthetic signaling system in plants. These experiments provide information on two critical steps in developing a eukaryotic signaling pathway with bacterial components: bacterial RRs are capable of signal-dependent nuclear translocation and, once nuclear localized, they are able to activate a synthetic promoter (provided they contain a eukaryotic transcriptional activation domain).

5. A Eukaryotic Synthetic Signal Transduction Pathway

While the partial synthetic system provides two key aspects of a pathway, transmission and output, it still lacks specific input from an exogenous signal and transmembrane signaling. To produce an entirely synthetic signal transduction pathway, we replaced the input from the native cytokinin sensing components of the partial system described above with synthetic components that are capable of sensing a specific substance of interest and transmitting this information across the membrane. As mentioned previously, when a PBP binds its ligand a conformational change is induced, leading to increased affinity of the PBP–ligand complex for the extracellular domain of a bacterial chemotactic receptor, Trg, with a specific PBP, ribose-binding protein (RBP). We used the computationally redesigned RBP, TNT.R3, that binds the explosive TNT (Looger *et al.*, 2003) to provide a specific input to our synthetic signal transduction pathway. The RR in our partial synthetic signaling system, PhoB–VP64 is capable of receiving a phosphate from PhoB's cognate HK, PhoR (Lamarche *et al.*, 2008). We therefore linked input from the computationally redesigned PBPs to our partial signal transduction system by producing a functional receptor HK fusion between Trg (to bind the ligand-redesigned PBP complex) and

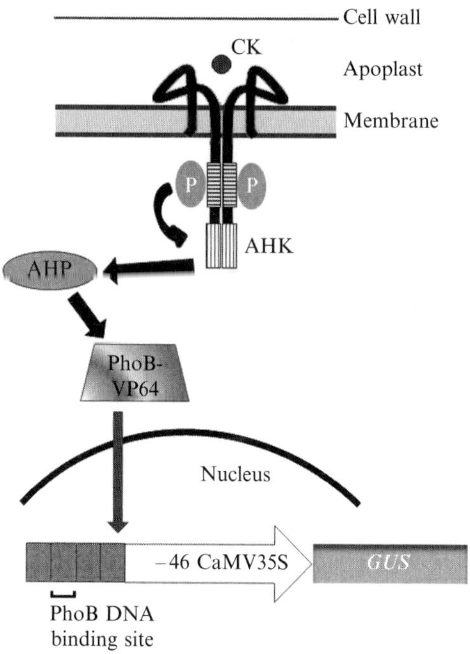

Figure 25.3 Diagram of the partial synthetic signaling system in plants. Endogenous membrane-localized cytokinin receptor HKs (AHK2, AHK3, and AHK4) autophosphorylate upon binding cytokinin resulting in a phosphorelay to endogenous AHPs. We postulate the AHPs then phosphorylate the synthetic RR, PhoB–VP64. PhoB–VP64 translocates to the nucleus and activates a synthetic PlantPho promoter consisting of four PhoB DNA-binding domains upstream of a minimal 35S promoter. CK, cytokinin; P, phosphate moiety.

PhoR to activate PhoB–VP64. A Trg:PhoR fusion that links Trg at its HAMP domain to position M19 of PhoR was produced by experimental analysis in bacteria (Antunes *et al.*, 2011). We re-engineered the Trg–PhoR fusion for plant expression by adding an N-terminal plasma membrane targeting signal peptide from FLS2 (Gomez-Gomez and Boller, 2000) to produce Fls:Trg:PhoR. The genes encoding ssTNT.R3 and Fls:Trg:PhoR were placed downstream of the constitutive *CaMV35S* and nopaline synthase (*PNOS)* promoters, respectively, then combined with the partial synthetic system to generate transgenic Arabidopsis plants expressing the synthetic signal transduction pathway: ssTNT.R3 → Fls:–Trg:–PhoR → PhoB–VP64 → PlantPho promoter::GUS (Fig. 25.5). To test the function of our system in plants, we measured GUS activity in primary transgenic plants by incubating developmentally equivalent leaves (leaf 1 and leaf 2) in the presence or absence of 10 μM TNT for 16 h. Protein extraction and GUS activity were performed as described in Section 7. Figure 25.6 shows an

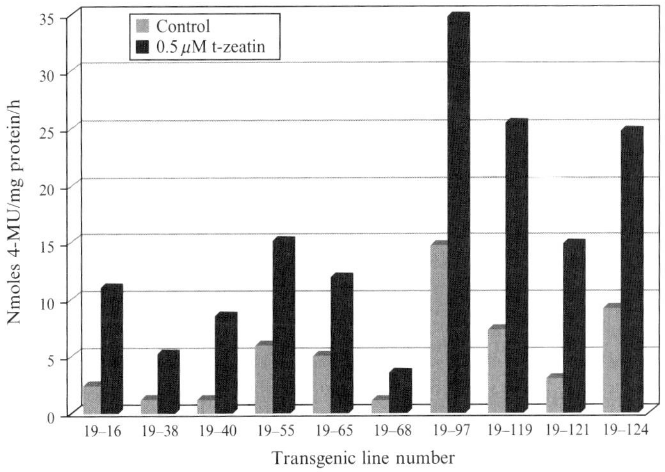

Figure 25.4 GUS activity of primary (T0) transgenic plants expressing the partial synthetic signaling system. Primary transgenic plants expressing the partial synthetic signal transduction system show induction of the GUS reporter in the presence of cytokinin (t-zeatin). Shown are some examples of induction levels of GUS in the presence of 0.5 μ*M* t-zeatin relative to uninduced samples (Control). The partial synthetic system has background GUS activity that is variable among independent transgenic lines. This is a single replicate T0 screen preventing statistical analysis.

example of a screening assay of primary transgenic Arabidopsis plants containing this system. Plants exposed to the TNT ligand showed higher GUS expression than the controls. These results, and details previously described (Antunes *et al.*, 2011) demonstrate that it is possible to produce a complete synthetic signaling pathway in plants using TCS components from bacteria.

6. Conclusions

Using the above-described approaches we have developed the first fully synthetic signal transduction pathway in eukaryotes (Antunes *et al.*, 2011). The design and refinement of this pathway was made possible by our bacterial testing platform. The months to years needed to test signaling components in plants was reduced to only a few weeks with our ability to use bacteria to test signaling components, to identify issues, and to redesign those elements quickly and efficiently. Eukaryotic synthetic signal transduction pathways have extensive applications in basic research, such as in assisting the study of complex endogenous pathways by allowing functional isolation of components. In addition, synthetic signal transduction pathways will create novel tools such as, for example, new signaling protein fusions that could enable better understanding of transmembrane signaling.

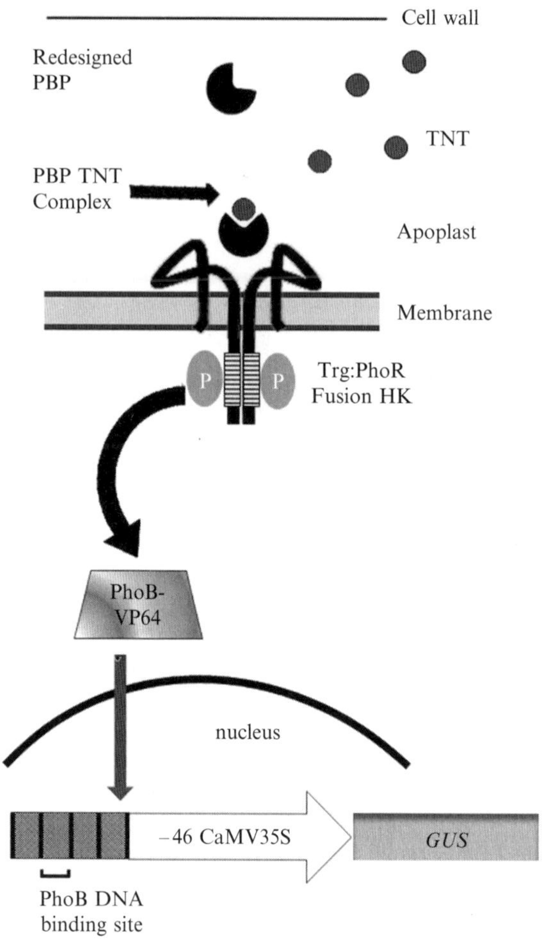

Figure 25.5 Diagram of the complete signal transduction system. In the apoplast, computationally re-designed PBPs bind TNT. The PBP–TNT complex interacts with a membrane-localized Trg-PhoR fusion, causing it to autophosphorylate and transfer the high energy phosphate to PhoB–VP64. PhoB–VP64 translocates to the nucleus and activates the PlantPho promoter.

This synthetic signal transduction pathways will have the practical use of producing systems where a specific biological input reliably results in a desired output. This has implications in many areas of interest, from biosensors capable of responding to the presence of a pollutant, to control of economically important agricultural traits like flowering, ripening, and pathogen defense. The synthetic signaling system we described represents the first step in achieving the goal of a reliable biological input/output system. Ongoing research in our laboratory is seeking to further refine the

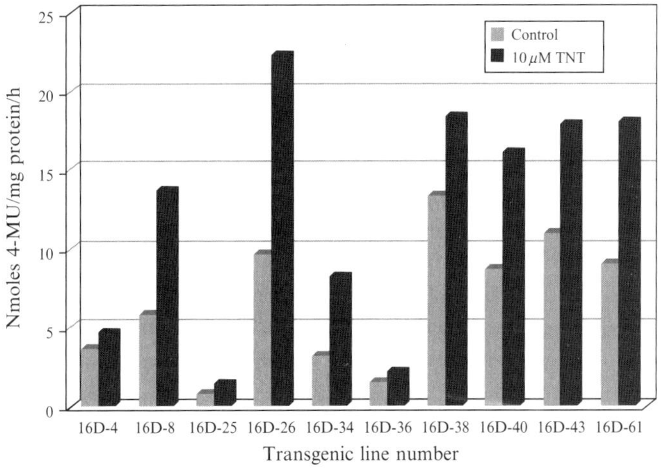

Figure 25.6 GUS activity of primary (T0) transgenic plants expressing the eukaryotic synthetic signaling pathway. Transgenic plants expressing the complete synthetic signal transduction system show induction of the GUS reporter in the presence of 2,4,6-trinitrotoluene (TNT). Shown are some examples of GUS activity seen in (T0) transgenic plants using a paired leaf assay. One leaf was treated with water as a control and the other leaf was exposed for 16 h to 10 μ*M* TNT. The complete synthetic system also has some background GUS activity that is variable among samples. This is a single replicate T0 screen preventing statistical analysis.

synthetic pathway by improving the HK fusion, reducing or eliminating the cross talk between PhoB and the endogenous plant HK systems, developing more complex gene circuits, and improving the response time for the assembly of these components as biosensors.

7. Protocols

7.1. Fluorometric GUS assay (fluorometer)

We modified a widely used GUS assay (Jefferson *et al.*, 1987) to measure the expression in our synthetic signaling pathway. In our system, GUS expression is regulated by the PlantPho promoter described above. A detailed list of solutions is found at the end of this protocol.

(1) After incubation of plants with inducer (e.g., cytokinin or TNT), grind plant tissue in GUS extraction buffer (100 μL for excised leaves, 200 μL for whole plants) in a 1.5-mL microfuge tube, using a drill fitted with autoclaved plastic pestles. Keep tubes on ice.

(2) Centrifuge samples at 10,000 rpm (9,000 g) for 10–15 min at 4 °C.
(3) Transfer the cleared supernatant to a new tube and place it on ice.
(4) In a 0.5-mL microfuge tube combine 50 μL of the protein extract obtained above with 50 μL of GUS assay buffer (2 m*M* MUG in GUS extraction buffer).
(5) Mix well and centrifuge for 15 s.
(6) Incubate reaction at 37 °C for 1 h.
(7) Stop the reaction by mixing 50 μL of the reaction mixture into 1.95 mL carbonate stop buffer (0.2 *M* Na_2CO_3, see Solutions) contained in a 4-mL tube (avoid excessive light exposure of tubes because 4-MU is light sensitive).
(8) Read 4-MU fluorescence units of samples in a DyNA Quant 200 Fluorometer (Hoefer, Inc., San Francisco, CA), excitation 365 nm, emission 460 nm, using a 4-MU standard solution to calibrate the instrument. Calibrate by first adding 1.9 mL of carbonate stop buffer to the cuvette and zero the instrument, then add 100 μL of the 4-MU standard solution to the stop buffer (see Solutions), and set the instrument to 500 units to standardize.
(9) Transfer 20 mL of the remaining protein extract to a new 1.5 mL centrifuge tube and read the protein concentration using the Bradford protein assay at 595 nm. Use bovine serum albumin solution diluted in extraction buffer (described above) as a standard, ranging in concentration from 0 to 0.6 mg/mL.

Solutions:

(1) *GUS extraction buffer*: final concentrations: 50 m*M* $NaHPO_4$, pH 7.0, 10 m*M* β-mercaptoethanol, 10 m*M* Na_2EDTA, pH 8.0, 0.1% sarcosyl, 0.1% Triton X-100.
(2) *GUS assay buffer* (2 m*M* MUG in GUS extraction buffer): Mix 22 mg of MUG with 25 mL of GUS extraction buffer.
(3) *Carbonate stop buffer* (0.2 *M* Na_2CO_3): Weigh out 6.2 g $Na_2CO_3 \cdot H_2O$ and dissolve it in 250 mL of diH_2O.
(4) 4-MU standard solution. First, prepare a concentrated MU calibration stock solution containing 1 μ*M* 7-hydroxy-4-methylcoumarin, MU, in diH_2O. Mix 19.8 mg (0.0198 g) of MU into 100 mL of diH_2O to produce a 1 m*M* solution. Dilute 10 μL of the 1 m*M* MU solution into 10 mL of diH_2O to make the 1 μ*M* MU stock (store this solution at 4 °C, protected from light).
(5) For calibration of the fluorometer, make a 50 n*M* MU solution, by diluting 100 μL of the 1 μ*M* concentrated MU calibration stock solution prepared above into 1.9 mL carbonate stop buffer. This solution must be made fresh just before use.

7.2. Fluorometric GUS assay (microplate reader)

(1) Remove leaves 1 and 2 from 14-day-old Arabidopsis and incubate them overnight in the presence or absence of the inducer of the gene circuit.
(2) Gently blot the leaves dry with a soft paper wipe and transfer the leaves to 1.5 mL centrifuge tubes containing 150 μL extraction buffer.
(3) Grind leaves using a drill fitted with a plastic pestle, keeping samples on ice at all times.
(4) Centrifuge tubes at 10,000 rpm (9,000 g) for 10 min at 4 °C.
(5) Transfer 25 μL of the protein extract to a 96-well microplate containing 25 μL of GUS assay buffer. Cover and incubate at 37 °C for at least an hour.
(6) After the incubation period, add 250 μL of carbonate stop buffer (0.2 *M* Na_2CO_3) to each well to stop the reaction.
(7) Read the fluorescence on a microplate reader; we used a Biotek Synergy HT: excitation 365 nm, emission 460 nm. Use a 1 μ*M* 4-MU standard solution (described above) to prepare a standard curve ranging from 0 to 400 n*M* 4-MU.
(8) Accomplish protein normalization as described in step 9 of 7.1

ACKNOWLEDGMENTS

This work was supported by grants from Defense Advanced Research Project Agency and Office of Naval Research.

REFERENCES

Allert, M., Rizk, S. S., Looger, L. L., and Hellinga, H. W. (2004). Computational design of receptors for an organophosphate surrogate of the nerve agent soman. *Proc. Natl. Acad. Sci. USA* **101,** 7907–7912.

Antunes, M. S., Morey, K. J., Tewari-Singh, N., Bowen, T. A., Smith, J. J., Webb, C. T., Hellinga, H. W., and Medford, J. I. (2009). Engineering key components in a synthetic eukaryotic signal transduction pathway. *Mol. Syst. Biol.* **5,** 270.

Antunes, M. S., Morey, K. J., Smith, J. J., Albrecht, K. D., Bowen, T. A., Zdunek, J. K., Troupe, J. F., Cuneo, M. J., Webb, C. T., Hellinga, H. W., and Medford, J. I. (2011). Programmable ligand detection in plants through a synthetic signal transduction pathway. *PloS ONE* 6(1):e16292.

Argueso, C. T., Ferreira, F. J., and Kieber, J. J. (2009). Environmental perception avenues: The interaction of cytokinin and environmental response pathways. *Plant Cell Environ.* **32,** 1147–1160.

Assaad, F. F., and Signer, E. R. (1990). Cauliflower mosaic-virus P35S promoter activity in *Escherichia coli*. *Mol. Gen. Genet.* **223,** 517–520.

Batard, Y., Hehn, A., Nedelkina, S., Schalk, M., Pallett, K., Schaller, H., and Werck-Reichhart, D. (2000). Increasing expression of P450 and P450-reductase proteins from monocots in heterologous systems. *Arch. Biochem. Biophys.* **379,** 161–169.

Baumberger, N., Doesseger, B., Guyot, R., Diet, A., Parsons, R. L., Clark, M. A., Simmons, M. P., Bedinger, P., Goff, S. A., Ringli, C., and Keller, B. (2003). Whole-genome comparison of leucine-rich repeat extensins in Arabidopsis and rice. A conserved family of cell wall proteins form a vegetative and a reproductive clade. *Plant Physiol.* **131,** 1313–1326.

Castillon, A., Shen, H., and Huq, E. (2007). Phytochrome interacting factors: Central players in phytochrome-mediated light signaling networks. *Trends Plant Sci.* **12,** 514–521.

Chen, D. Q., and Texada, D. E. (2006). Low-usage codons and rare codons of *Escherichia coli*. *Gene Ther. Mol. Biol.* **10A,** 1–12.

Chen, M. H., Takeda, S., Yamada, H., Ishii, Y., Yamashino, T., and Mizuno, T. (2001). Characterization of the RcsC → YojN → RcsB phosphorelay signaling pathway involved in capsular synthesis in *Escherichia coli*. *Biosci. Biotechnol. Biochem.* **65,** 2364–2367.

Christie, J. M. (2007). Phototropin blue-light receptors. *Annu. Rev. Plant Biol.* **58,** 21–45.

Davis, S. J., and Vierstra, R. D. (1996). Soluble derivatives of green fluorescent protein (GFP) for use in *Arabidopsis thaliana*. *Weeds World* **3,** 43–48.

De Moraes, C. M., Schultz, J. C., Mescher, M. C., and Tumlinsoni, J. H. (2004). Induced plant signaling and its implications for environmental sensing. *J. Toxicol. Environ. Health A* **67,** 819–834.

Dortay, H., Mehnert, N., Burkle, L., Schmulling, T., and Heyl, A. (2006). Analysis of protein interactions within the cytokinin-signaling pathway of *Arabidopsis thaliana*. *FEBS J.* **273,** 4631–4644.

Dwyer, M. A., and Hellinga, H. W. (2004). Periplasmic binding proteins: A versatile superfamily for protein engineering. *Curr. Opin. Struct. Biol.* **14,** 495–504.

Dwyer, M. A., Looger, L. L., and Hellinga, H. W. (2003). Computational design of a Zn^{2+} receptor that controls bacterial gene expression. *Proc. Natl. Acad. Sci. USA* **100,** 11255–11260.

Dwyer, M. A., Looger, L. L., and Hellinga, H. W. (2004). Computational design of a biologically active enzyme. *Science* **304,** 1967–1971.

Gomez-Gomez, L., and Boller, T. (2000). FLS2: An LRR receptor-like kinase involved in the perception of the bacterial elicitor flagellin in Arabidopsis. *Mol. Cell* **5,** 1003–1011.

Heyl, A., and Schmulling, T. (2003). Cytokinin signal perception and transduction. *Curr. Opin. Plant Biol.* **6,** 480–488.

Jaillais, Y., and Chory, J. (2010). Unraveling the paradoxes of plant hormone signaling integration. *Nat. Struct. Mol. Biol.* **17,** 642–645.

Jefferson, R. A., Kavanagh, T. A., and Bevan, M. W. (1987). GUS fusions: Beta-glucuronidase as a sensitive and versatile gene fusion marker in higher plants. *EMBO J.* **6,** 3901–3907.

Kay, S., and Bonas, U. (2009). How Xanthomonas type III effectors manipulate the host plant. *Curr. Opin. Microbiol.* **12,** 37–43.

Khalil, A. S., and Collins, J. J. (2010). Synthetic biology: Applications come of age. *Nat. Rev. Genet.* **11,** 367–379.

Kiel, C., Yus, E., and Serrano, L. (2010). Engineering signal transduction pathways. *Cell* **140,** 33–47.

Koretke, K. K., Lupas, A. N., Warren, P. V., Rosenberg, M., and Brown, J. R. (2000). Evolution of two-component signal transduction. *Mol. Biol. Evol.* **17,** 1956–1970.

Lamarche, M. G., Wanner, B. L., Crepin, S., and Harel, J. (2008). The phosphate regulon and bacterial virulence: A regulatory network connecting phosphate homeostasis and pathogenesis. *FEMS Microbiol. Rev.* **32,** 461–473.

Lessard, P. A., Kulaveerasingam, H., York, G. M., Strong, A., and Sinskey, A. J. (2002). Manipulating gene expression for the metabolic engineering of plants. *Metab. Eng.* **4,** 67–79.

Levskaya, A., Weiner, O. D., Lim, W. A., and Voigt, C. A. (2009). Spatiotemporal control of cell signalling using a light-switchable protein interaction. *Nature* **461,** 997–1001.

Li, X., Hirano, R., Tagami, H., and Aiba, H. (2006). Protein tagging at rare codons is caused by tmRNA action at the 3' end of nonstop mRNA generated in response to ribosorne stalling. *RNA* **12,** 248–255.

Looger, L. L., Dwyer, M. A., Smith, J. J., and Hellinga, H. W. (2003). Computational design of receptor and sensor proteins with novel functions. *Nature* **423,** 185–190.

Makino, K., Shinagawa, H., Amemura, M., and Nakata, A. (1986). Nucleotide sequence of the phoR gene, a regulatory gene for the phosphate regulon of *Escherichia coli*. *J. Mol. Biol.* **192,** 549–556.

Makino, K., Shinagawa, H., Amemura, M., Kawamoto, T., Yamada, M., and Nakata, A. (1989). Signal transduction in the phosphate regulon of *Escherichia coli* involves phosphotransfer between PhoR and PhoB proteins. *J. Mol. Biol.* **210,** 551–559.

Mizuno, T., and Yamashino, T. (2010). Biochemical characterization of plant hormone cytokinin-receptor histidine kinases using microorganisms. *Methods Enzymol.* **471,** 335–356.

Mizuno, T., Wurtzel, E. T., and Inouye, M. (1982). Cloning of the regulatory genes (OmpR and EnvZ) for the matrix proteins of the *Escherichia coli* outer-membrane. *J. Bacteriol.* **150,** 1462–1466.

Mowbray, S. L., and Sandgren, M. O. J. (1998). Chemotaxis receptors: A progress report on structure and function. *J. Struct. Biol.* **124,** 257–275.

Nakamura, Y., Gojobori, T., and Ikemura, T. (2000). Codon usage tabulated from international DNA sequence databases: Status for the year 2000. *Nucleic Acids Res.* **28,** 292.

Notredame, C., Higgins, D. G., and Heringa, J. (2000). T-Coffee: A novel method for fast and accurate multiple sequence alignment. *J. Mol. Biol.* **302,** 205–217.

Oh, M. H., Wang, X., Kota, U., Goshe, M. B., Clouse, S. D., and Huber, S. C. (2009). Tyrosine phosphorylation of the BRI1 receptor kinase emerges as a component of brassinosteroid signaling in Arabidopsis. *Proc. Natl. Acad. Sci. USA* **106,** 658–663.

Osgur, S., and Sancar, A. (2006). Analysis of autophosphorylating kinase activities of Arabidopsis and human cryptochromes. *Biochemistry* **45,** 13369–13374.

Parkinson, J. S., and Kofoid, E. C. (1992). Communication modules in bacterial signaling proteins. *Annu. Rev. Genet.* **26,** 71–112.

Perlak, F. J., Fuchs, R. L., Dean, D. A., Mcpherson, S. L., and Fischhoff, D. A. (1991). Modification of the coding sequence enhances plant expression of insect control protein genes. *Proc. Natl. Acad. Sci. USA* **88,** 3324–3328.

Posas, F., WurglerMurphy, S. M., Maeda, T., Witten, E. A., Thai, T. C., and Saito, H. (1996). Yeast HOG1 MAP kinase cascade is regulated by a multistep phosphorelay mechanism in the SLN1-YPD1-SSK1 "two-component" osmosensor. *Cell* **86,** 865–875.

Rodrigo, G., Montagud, A., Aparici, A., Aroca, M. C., Baguena, M., Carrera, J., Edo, C., Fernandez-de-Cordoba, P., Ferrando, A., Fuertes, G., Gimenez, D., Mata, C., *et al.* (2007). Vanillin cell sensor. *IET Synth. Biol.* **1,** 74–78.

Rodriguez, M. C., Petersen, M., and Mundy, J. (2010). Mitogen-activated protein kinase signaling in plants. *Annu. Rev. Plant Biol.* **61,** 621–649.

Schreier, B., Stumpp, C., Wiesner, S., and Hocker, B. (2009). Computational design of ligand binding is not a solved problem. *Proc. Natl. Acad. Sci. USA* **106,** 18491–18496.

Seipel, K., Georgiev, O., and Schaffner, W. (1992). Different activation domains stimulate transcription from remote (enhancer) and proximal (promoter) positions. *EMBO J.* **11,** 4961–4968.

Sessa, G., Raz, V., Savaldi, S., and Fluhr, R. (1996). PK12, a plant dual-specificity protein kinase of the LAMMER family, is regulated by the hormone ethylene. *Plant Cell* **8,** 2223–2234.

Sharrock, R. A. (2008). The phytochrome red/far-red photoreceptor superfamily. *Genome Biol.* **9,** 230.

Somerville, C., Bauer, S., Brininstool, G., Facette, M., Hamann, T., Milne, J., Osborne, E., Paredez, A., Persson, S., Raab, T., Vorwerk, S., and Youngs, H. (2004). Toward a systems approach to understanding plant cell walls. *Science* **306,** 2206–2211.

Stepanova, A. N., and Alonso, J. M. (2009). Ethylene signaling and response: Where different regulatory modules meet. *Curr. Opin. Plant Biol.* **12,** 548–555.

Sunohara, T., Jojima, K., Tagami, H., Inada, T., and Aiba, H. (2004). Ribosome stalling during translation elongation induces cleavage of mRNA being translated in *Escherichia coli*. *J. Biol. Chem.* **279,** 15368–15375.

Suo, G. L., Chen, B., Zhang, J. Y., Duan, Z. Y., He, Z. Q., Yao, W., Yue, C. Y., and Dai, J. W. (2006). Effects of codon modification on human BMP2 gene expression in tobacco plants. *Plant Cell Rep.* **25,** 689–697.

Suzuki, T., Miwa, K., Ishikawa, K., Yamada, H., Aiba, H., and Mizuno, T. (2001). The Arabidopsis sensor His-kinase, AHK4, can respond to cytokinins. *Plant Cell Physiol.* **42,** 107–113.

Takeda, S., Fujisawa, Y., Matsubara, M., Aiba, H., and Mizuno, T. (2001). A novel feature of the multistep phosphorelay in *Escherichia coli*: A revised model of the RcsC → YojN → RcsB signalling pathway implicated in capsular synthesis and swarming behaviour. *Mol. Microbiol.* **40,** 440–450.

To, J. P. C., and Kieber, J. J. (2008). Cytokinin signaling: Two-components and more. *Trends Plant Sci.* **13,** 85–92.

Triezenberg, S. J., Kingsbury, R. C., and Mcknight, S. L. (1988). Functional dissection of Vp16, the trans-activator of herpes-simplex virus immediate early gene-expression. *Genes Dev.* **2,** 718–729.

Vercillo, N. C., Herald, K. J., Fox, J. M., Der, B. S., and Dattelbaum, J. D. (2007). Analysis of ligand binding to a ribose biosensor using site-directed mutagenesis and fluorescence spectroscopy. *Protein Sci.* **16,** 362–368.

Wurtzel, E. T., Chou, M. Y., and Inouye, M. (1982). Osmoregulation of gene-expression. 1. DNA-sequence of the Ompr gene of the Ompb operon of *Escherichia coli* and characterization of its gene-product. *J. Biol. Chem.* **257,** 3685–3691.

Yamada, H., Suzuki, T., Terada, K., Takei, K., Ishikawa, K., Miwa, K., Yamashino, T., and Mizuno, T. (2001). The Arabidopsis AHK4 histidine kinase is a cytokinin-binding receptor that transduces cytokinin signals across the membrane. *Plant Cell Physiol.* **42,** 1017–1023.

Yeh, K.-C., and Lagarias, J. C. (1998). Eukaryotic phytochromes: Light-regulated serine/ threonine protein kinases with histidine kinase ancestry. *Proc. Natl. Acad. Sci. USA* **95,** 13976–13981.

Yoo, S. D., Cho, Y., and Sheen, J. (2009). Emerging connections in the ethylene signaling network. *Trends Plant Sci.* **14,** 270–279.

Zhang, Y., and McCormick, S. (2009). AGCVIII kinases: At the crossroads of cellular signaling. *Trends Plant Sci.* **14,** 689–695.

CHAPTER TWENTY-SIX

Lentiviral Vectors to Study Stochastic Noise in Gene Expression

Kate Franz, Abhyudai Singh, *and* Leor S. Weinberger

Contents

Abstract

Lentiviral vectors are vehicles for gene delivery that were originally derived from the human immunodeficiency virus type-1 (HIV-1) lentivirus. These vectors are defective for replication, and thus considered relatively safe, but are capable of stably integrating into the genomic DNA of a broad range of dividing and nondividing mammalian cell types. The ability to stably integrate at semi-random genomic positions make lentiviral vectors a unique and ideal tool for studying stochastic variation in gene expression. Here, we describe the experimental and mathematical methods for using lentiviral vectors to study stochastic noise in gene expression.

Abbreviations

cPPT	central polypurine tract
CV	coefficient of variation

Department of Chemistry and Biochemistry, University of California, San Diego, La Jolla, California, USA

Methods in Enzymology, Volume 497
ISSN 0076-6879, DOI: 10.1016/B978-0-12-385075-1.00026-3

FACS	fluorescence-activated cell-sorting
FSC	forward scatter
HIV-1	human immunodeficiency virus type-1
LTR	long terminal repeat
RRE	Rev responsive element
SIN vectors	self-inactivating vectors
SSC	side scatter
IU	infectious units

1. Introduction

Fluctuations in the levels of gene products are an inevitable consequence of the inherent stochastic nature of biochemical processes that constitute gene expression (Blake *et al.*, 2006; Kaern *et al.*, 2005; Raj and van Oudenaarden, 2008). These fluctuations are referred to as stochastic 'noise' and in a population of cells the noise is often characterized as either due to 'intrinsic' sources or 'extrinsic' sources (Elowitz *et al.*, 2002; Swain *et al.*, 2002). In isogenic populations (i.e., clonal populations where all cells are derived from a single parent), intrinsic noise manifests as cell-to-cell variation that is uncorrelated, while extrinsic noise manifests as variation that is correlated between cells. The origin of the noise in gene products is biochemical: intracellular processes are driven by reactant molecules randomly diffusing and colliding within the cell and are thus inherently stochastic. Specifically, noise in gene expression can arise from the random timing in individual reactions associated with promoter remodeling, transcription, and translation (Blake *et al.*, 2003; Elowitz *et al.*, 2002; Kaern *et al.*, 2005; Swain *et al.*, 2002). Moreover, intercellular differences in the *amount* of cellular components (e.g., RNA polymerase, transcription factors, and ribosomes) also cause variations in expression levels. Measurements in live, single cells have shown that gene expression noise can lead to large statistical fluctuations in protein and mRNA levels in both prokaryotes and eukaryotes (Bar-Even *et al.*, 2006; Golding *et al.*, 2005; Newman *et al.*, 2006; Raj *et al.*, 2006). These fluctuations (i.e., noise) can have significant effects on biological function and phenotype and noise is now recognized to exert significant influence on probabilistic fate decisions in bacteria (Eldar *et al.*, 2009; Suel *et al.*, 2006), viruses (Singh and Weinberger, 2009), and stem cells (Hanna *et al.*, 2009).

A common method to analyze gene-expression noise is to measure the variation in protein levels across isogenic cells as a function of mean protein levels (Bar-Even *et al.*, 2006; Blake *et al.*, 2003). This method is often referred to as CV^2 versus mean analysis, where CV^2 is a dimensionless statistical measure of intercellular variability in protein levels. More specifically, CV^2 stands for the coefficient of variation (CV) squared and is defined as $CV^2 = \sigma^2/\langle \text{protein} \rangle^2$, where σ^2 is the variance in protein abundance and $\langle \text{protein} \rangle$ is the average number of protein molecules per cell (Paulsson, 2004).

To study how CV^2 varies as a function of $\langle \text{protein} \rangle$, it is essential that one be able to alter mean expression levels by changing the transcriptional rate of the promoter of interest. This modulation can be challenging for many promoters, especially when transcriptional activity cannot be modulated using small-molecule compounds. In such cases, transcriptional efficiency is typically altered through mutations in the promoter sequence (Ozbudak *et al.*, 2002). Studying expression noise using lentiviral vectors provides a unique advantage as it allows one to exploit their known ability to integrate semi-randomly into sites across the human genome (Schroder *et al.*, 2002). Differences in local chromatin microenvironment at each integration site generate differences in promoter strength (Jordan *et al.*, 2001) and this difference provides a natural method to study CV^2 as a function of mean protein levels.

Our laboratory has extensively studied gene-expression noise using lentiviral vectors. We have used lentiviral vectors to study: (i) the role of stochastic noise in viral gene expression (Weinberger and Shenk, 2007; Weinberger *et al.*, 2005), (ii) the influence of stochastic noise upon probabilistic decision making in HIV (Weinberger *et al.*, 2008), and (iii) the molecular source of noise in viral gene expression (Singh *et al.*, 2010). Here, we describe the experimental methods for establishing isogenic populations using lentiviral vectors as well as the quantitative methods used to analyze stochastic noise in these isogenic populations.

The classical method for studying gene-expression noise is to work in an isogenic background—where each cell is grown from a single parent cell—and analyze expression from a specific genetic locus within the isogenic population. Here, we describe how lentiviral-vector technology provides a convenient method for constructing isogenic populations of cells and for minimizing external variation sources in the analysis of gene-expression noise. While usage of isogenic backgrounds does not eliminate all nonexpression sources of cell-to-cell variation (e.g., cell size, cell-cycle state, other extrinsic factors), below we describe how these other sources of variation can be dealt with.

2. The Lentiviral-Vector Approach

The need to deliver foreign genetic information to living cells (i.e., genetically transduce cells with foreign DNA or RNA) has long presented a challenge for molecular and cell biologists. Various methods of gene delivery have been developed including bioballistic 'gene-gun' approaches, electroporation, chemical methods, and viral methods. Lentiviral vectors are one of the most recent vehicles for gene delivery having been derived originally from the human immunodeficiency virus type-1 (HIV-1) lentivirus (Naldini *et al.*, 1996) and subsequently from related lentiviruses (Mitrophanous *et al.*, 1999; Olsen, 1998; Poeschla *et al.*, 1998). These vectors are defective for replication, and thus considered relatively safe, but are capable of stably transducing a broad range of dividing and non-dividing cell types. This provides a significant advantage over other methods of gene transfer, such as gammaretroviral-based vectors (e.g., Murine Leukemia Viruses, MLVs), which can only transduce dividing cells. Lentiviral vectors also carry a number of other recognized benefits over other gene-delivery methods: (i) lentiviral vectors are less susceptible to position-effect variegation as compared to retroviral vectors which are rapidly silenced (Challita and Kohn, 1994); (ii) unlike stable transfections lentiviral vectors do not concatenate or continually change in copy number over multiple cell passages; (iii) lentiviral-integrated constructs are not progressively diluted out with each cell division (as occurs with transient transfections), which makes lentiviral-transduced cell lines easy to maintain. Finally, lentiviral vectors integrate throughout the genome and preferentially integrate, with ~69% preference, in active transcriptional units (Schroder *et al.*, 2002). This semi-random integration allows gene-expression characteristics to be analyzed in various chromatin contexts throughout the genome.

Lentiviral-vector systems have been developed from many nonprimate lentiviruses such as equine infectious anemia virus (EIAV) and feline immunodeficiency virus (FIV). However, our discussion will focus on the generation of HIV-1 derived lentiviral vectors as they are the most widely used vector system.

The HIV-1 genome encodes 15 proteins necessary for infection and pathogenesis of the virus. Additionally, the genome contains *cis*-acting elements that are required for genomic RNA packaging (e.g., the Ψ signal), reverse transcription (e.g., the central polypurine tract or cPPT), and viral gene expression (i.e., the 5′ and 3′ LTRs and the Rev responsive element or RRE) (Fig. 26.1). *Trans*-acting elements required for infection and assembly of viral particles are expressed from constructs lacking the above *cis*-acting sequences. The *cis*-acting sequences are found only on the transfer vector containing the promoter/reporter gene cassette. When both cis-acting and trans-acting constructs are transfected into a cell, infectious viral particles will assemble. However, the particles can only encapsidate the transfer vector RNA which encodes the *cis*-acting elements and no *trans*-acting elements thus, thus limiting the lentiviral vectors to a single round of infection.

Either a three-plasmid or four-plasmid approach can be used to generate lentiviral vectors depending on whether the 2nd or 3rd generation lentiviral system is used. In the 2nd-generation system, one plasmid, the packaging construct, encodes all *trans*-acting elements necessary for assembly of the

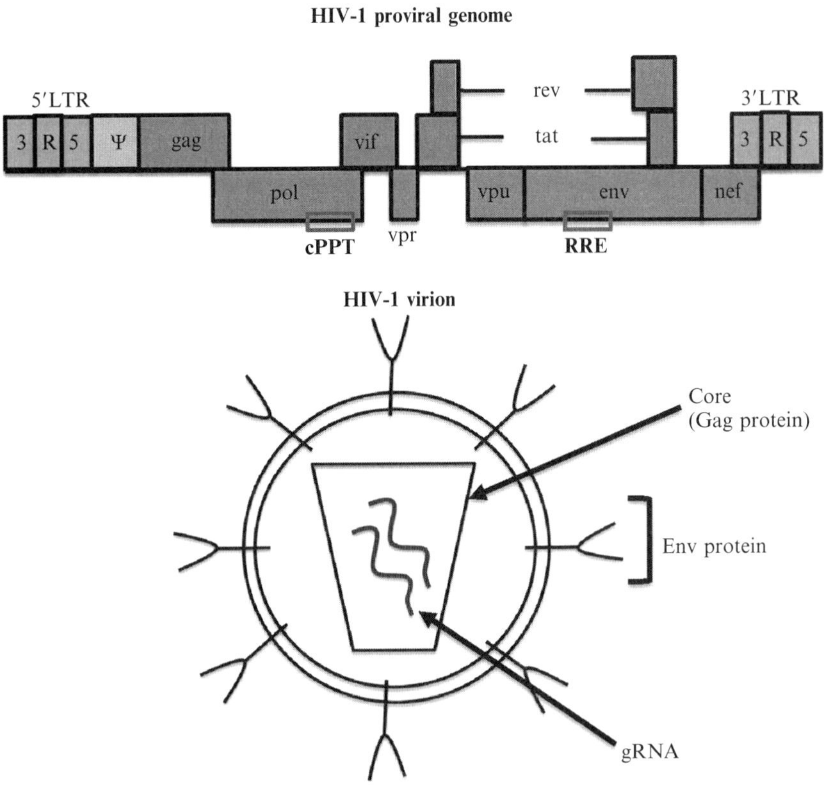

Figure 26.1 HIV-1 genomic organization and virion structure. HIV-1 genome contains nine open reading frames (in red) that code for 15 proteins. *cis*-acting elements important for reverse transcription, gene expression, and packaging are in green. Note that the integrated HIV-1 genome has a full 5′LTR (in blue). (See Color Insert.)

lentiviral core. Accessory genes that act only in the pathogenesis of infection (*nef, vif, vpu*, and *vpr* genes in HIV-1) are eliminated, as well as the trans-activator of transcription (*tat*) and *env* gene (Dull *et al.*, 1998; Kim *et al.*, 1998). As explained above, none of the *cis*-acting elements required for encapsidation or transfer into the target cell are present on this construct. This reduces the risk of recombinant, replication-competent viruses arising. The 3rd-generation packaging system encodes the same *trans*-acting elements, but on two separate plasmid constructs to further minimize the risk of recombinant viruses (Fig. 26.2).

The envelope construct expresses a heterologous envelope protein, which serves to expand the cell-targeting tropism of the viral vector.

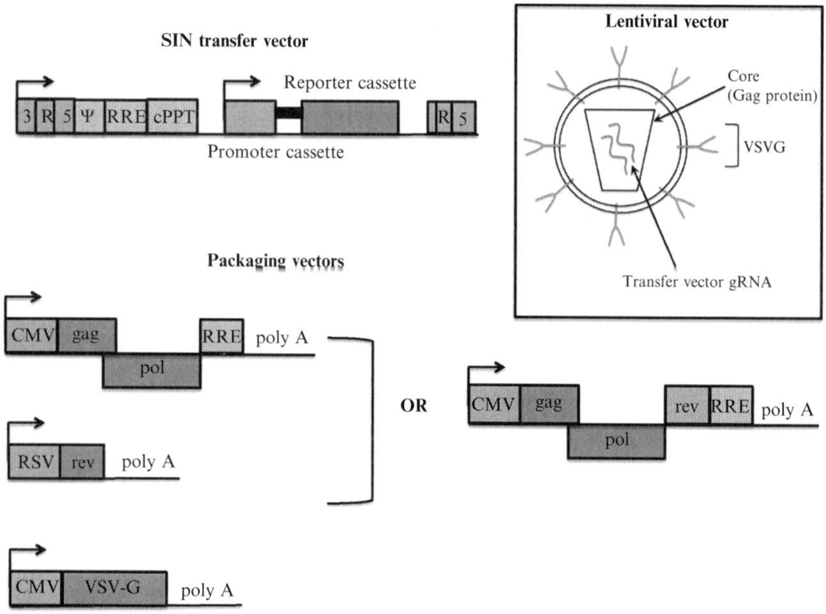

Figure 26.2 Schematic of the 2nd and 3rd generation lentiviral plasmid system. Promoters are in blue, open reading frames are in red, and *cis*-acting elements are in green. The U3 region of the 3′LTR of the transfer vector is mostly deleted. Upon reverse transcription, this deletion will be carried into the 5′LTR of the integrated construct and inactivate it. After transfection of the system into HEK 293FT cells, lentiviral particles are generated that contain transfer vector as the gRNA and VSVG protein for envelope protein (inset). (For interpretation of the references to color in this figure legend, the reader is referred to the Web version of this chapter.)

Exchanging the envelope protein for one of another species is termed 'pseudo-typing.' Most lentiviral vectors utilize the vesicular stomatitis virus glycoprotein (VSVG) due to its broad cell-targeting tropism and its ability to withstand ultracentrifugation and freeze-thaw cycles with only minimal loss of infectivity (Akkina *et al.*, 1996; Reiser *et al.*, 1996). Many other envelope proteins may be used (Cronin *et al.*, 2005), and it is vital to determine whether a given choice of envelope can effectively target the cell type being studied.

All relevant *cis*-acting elements for infection and are located on the transfer vector (a.k.a. the lentiviral "backbone" vector). The transfer vector includes the LTR promoter, Ψ signal, RRE, and cPPT followed by the promoter and reporter cassettes to be studied. The Ψ signal enables the dimerization and encapsidation of only the transfer vector and the cPPT functions in reverse transcription. Since nuclear export is blocked for unspliced mRNAs in eukaryotic cells, inclusion of the RRE is important for efficient expression of the transcribed construct. Rev—coded for within the packaging plasmid—binds the RRE and utilizes the CRM1 nuclear export pathway to transport unspliced RNAs into the cytoplasm (Fischer *et al.*, 1995; Neville *et al.*, 1997).

One advance in transfer-vector design that greatly increased the biosafety of lentiviral vectors has been the creation of *s*elf-*in*activating vectors, or SIN vectors. SIN vectors carry a near complete deletion of the promoter-enhancer sequences in the U3 region in the 3′LTR. Since reverse transcriptase uses the 3′LTR as the template for the proviral copy of the 5′LTR, upon reverse transcription this deletion is transferred to the 5′LTR, rendering the 5′LTR transcriptionally inactive. This deletion prevents future mobilization of the construct, decreases risk of endogenous oncogene activation, and eliminates any interference in gene expression due to transcription from the lentiviral LTR (Miyoshi *et al.*, 1998).

Design of a transfer vector depends significantly on the aims of the study. Often the transferred construct consists of the promoter of interest driving the expression of a fluorescent reporter gene. Fluorescent reporters can be fused (either transcriptionally or translationally) to another gene product. Flanking the promoter and/or reporter gene with restriction enzymes allows for quick swapping of elements within the transfer vector and generation of new vectors that vary with respect to their promoter and reporter gene. Design of transfer vectors can be simplified by using lentivirus expression vectors like those available from Invitrogen (Carlsbad, CA), which use Gateway and TOPO cloning to insert the construct of interest into a transfer-vector backbone. Transfer vectors can also be generated using basic cloning techniques to ligate your construct into a previously described transfer vector.

For example, we have studied gene-expression noise in HIV and its consequences, by utilizing lentiviral vectors expressing fluorescent reporter genes (e.g., green fluorescent protein, GFP) from the HIV-1 LTR promoter. These lentiviral vectors allowed us transduce Jurkat T cells (ATCC # TIB-152), isolate single cells using fluorescence-activated cell-sorting (FACS), and expand these cells into isogenic populations of cells for flow cytometry analysis to obtain data on the CV and mean fluorescence. Below, we describe the methods used to construct these lentiviral vectors and analyze these isogenic populations for stochastic noise (Singh *et al.*, 2010).

3. Production of Lentiviral Vectors and Transduced Cell Lines

All steps should be performed using BSL 2+ containment procedures. Overview of protocol found in Fig. 26.3.

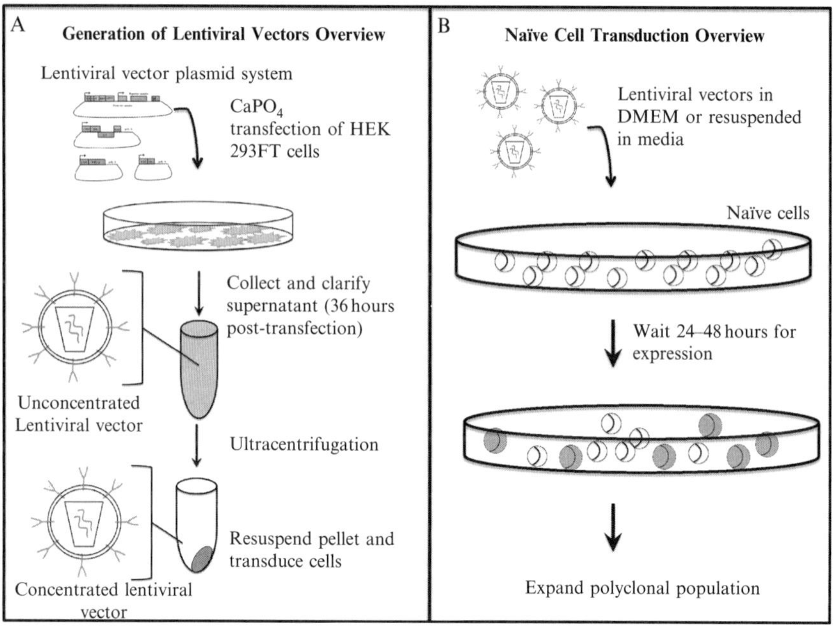

Figure 26.3 Overview of the protocol for production of lentiviral particles (A) and transduction of naïve cells (B). (See Color Insert.)

Reagents and equipment

Human Embryonic Kidney (HEK) 293FT cells (Invitrogen cat. no. R700-07)
Dulbecco's modified eagle medium (DMEM) (Cellgro cat. no. 10-013-CV) supplemented with 10% fetal bovine serum (FBS) (Cellgro 35-016-CV) and 1% penicillin/streptomycin (PS) (Cellgro cat. no. 30-009-CI)
1× phosphate buffered saline (PBS) (Cellgro cat. no. 21-030-CV)
0.05% trypsin (Cellgro cat. no. 25-051-CI)
Plasmids: lentiviral transfer vector, lentiviral packaging vectors (Invitrogen cat. no. K4975-00)
Molecular biology grade water (Cellgro cat. no. 46-000-CM)
2× HEPES-buffered saline (HeBS) (SIGMA cat. no. 51558), filter-sterilized
2.5 *M* $CaCl_2$, autoclaved
10-cm tissue culture dishes
15-mL conical tube
0.45-μm syringe filters
30-mL syringe
Beckman ultracentrifuge tube (Beckman cat. no. 344059)
Beckman SW41 rotor

Beckman ultracentrifuge
Fluorescence microscope
Humidified tissue-culture incubator at 5% CO_2

1. HEK 293FT cell culture
   1.1 Seed low passage-number (<20 passages) HEK 293FT cells on 10-cm dishes in a total of 10 mL DMEM + 10% FBS + 1% PS.
      1.1.1 If seeding from a 10-cm dish of confluent cells, aspirate the media and wash the cells with 4 mL of PBS. Add 0.5 mL of 0.05% trypsin and incubate dish at 37 °C in 5% CO_2 for 5 min. Resuspend cells in 4.5 mL of DMEM + 10% FBS + 1% penicillin and streptomycin. Aliquot resuspended cells in 10-cm dishes. Incubate HEK 293FT cells at 37 °C in 5% CO_2. HEK 293FT cells express the SV40 T antigen which allows plasmids containing the SV40 origin of replication to be retained and replicated during cell division and enhances expression of lentiviral plasmids.
   1.2 Observe the seeded dishes. Cells should be in exponential growth. When cells are 60–80% confluent, they are ready for transfection.

   Important note: Never use 293FT cells that are growing slowly. Switch to lower passage-cells if growth is impaired.
2. Transfection of plasmids by $CaPO_4$ precipitation
   2.1 Prepare one aliquot of DNA/$CaCl_2$ solution per dish of 293FT cells to transfect. In a 15-mL conical tube, add 3.5 μg of pVSVG, 1.5 μg pRev, 5 μg pMDL, and 10 μg transfer vector to 120 μL 2.5 *M* $CaCl_2$. Bring solution to 1 mL with ultrapure water. Mix.
   2.2 Perform the following one aliquot at a time. Add 1 mL 2× HeBS. Pipette entire volume four times. Let solution incubate 1.5 min. Gently add HeBS/DNA/$CaCl_2$ solution drop-wise to HEK 293FT cells. Swirl the dish very gently to evenly distribute the transfection mixture being careful not to detach cells.

   Important note: $CaPO_4$ transfection efficiency is pH-dependent. Titrate 2× HeBS to pH 6.95 before using.

   (alternatively, we use Fugene™ transfection with the 3-plasmid packaging system in the following ratios: 5.6 μg Δ8.9, 2.8 μg of VSVG, and 2.2 μg of transfer vector in 30 μL of Fugene and 300 μL of serum-free DMEM)
   2.3 Incubate the cells overnight at 37 °C, 5% CO_2 in a humidified environment.
   2.4 Visually inspect Observe 12-18 hours later; transfection efficiency can be assessed visually through expression of fluorescent reporter proteins. Transfection efficiency should be >90%. Aspirate the

media and carefully add 10 mL of fresh DMEM + 10% FBS + 1% PS. Incubate 24-28 hours at 37 °C, 5% CO_2 in a humidified environment.

3. Harvest the supernatant
   3.1 Collect the supernatant from the dish. *Use BSL 2+ containment procedures* as all supernatants and cell culture contain infectious vector. If a second harvest of lentiviral vector is desired, carefully add 10 mL of fresh DMEM + 10% FBS + 1% PS to the dish. Incubate cells 24-28 hours at 37 °C, 5% CO_2. Spray exteriors of all tubes containing lentivirus vector with 70% ethanol before removing from hood. Store the collected supernatant in a secondary container at 4 °C.
   3.2 The next morning, collect the second harvest of supernatant from the cells. Pool supernatants from both harvests and centrifuge at 500×*g* for 5 min to pellet cell debris. Clarify the supernatant by filtering through a 0.45-μm filter attached to a 30-mL syringe. Clarified supernatants can be stored up to a year at −80 °C. At this point, the titer of the supernatant can be estimated to be $\sim 10^6$ IU/mL (and up to twofold lower for vectors with transfer constructs over 4kb). If desired, store 1 mL of this filtrate at −80 °C until ready to titer. Supernatants can now be used for transduction, but ultracentrifugation is recommended to concentrate the viral prep and increase transduction efficiency.
4. Ultracentrifugation of clarified lentiviral prep

   For ultracentrifugation, we use a Beckman SW41 swinging-bucket rotor and a Beckman ultracentrifuge, however equivalent equipment will work as well.
   4.1 Pipette filtrate into Beckman ultracentrifuge tubes. It is critical that the tubes opposite from one another in the rotor are precisely balanced. Place tubes into the appropriate rotor buckets, secure the lids on the buckets and weigh opposing tubes to ensure they are balanced.
   4.2 Centrifuge the viral prep at 20,000×*g* for 90 min at 4 °C. Aspirate the supernatant, tilting the tube slightly horizontally to avoid aspirating the viral pellet. The pellet will most likely not be visible, but may appear as a glassy spot on the bottom of the tube.
   4.3 Resuspend the pellet in the amount of PBS or media that allows you to transduce your cells at your target multiplicity of infection (MOI) and at MOIs 1–2 logs above and below target. A good target MOI for noise analysis is MOI = 0.1, which limits double integrations to 10–15% of your transduced population. We usually resuspend the pellets in 500 μL of media. Viral prep can be used directly from this step or aliquoted and stored at −80 °C for up to a year.
5. Transduction of lentiviral vector into cell line of interest
   5.1 Maintain a low passage-number cell line of interest. Use only freshly passaged cells for transduction.

5.1.1 *Suspension cells*: Count cells and resuspend 500,000 cells for each transduction in 1–2 mL of media (depending on your cell line's preferred growth conditions) in a 12-well plate.

5.1.2 *Adherent cells*: Count cells and resuspend 500,000 cells for each transduction in 1.5 mL of media in a 6-well plate. Plating confluency may need to be adjusted to accommodate cell size, growth rate, and optimal culture confluency.

5.2 Add the appropriate amount of viral prep suspended in PBS or media to each well. We resuspend the viral pellet in 500 μL of media and add 100 μL, 10 μL, or 1 μL of a 1:10 dillution of the viral stock to 500,000 cells to obtain target MOIs of 10, 1, 0.1, and 0.01, respectively. Incubate the cells normally. If viral prep is suspended in the same culture media of the cells or PBS, a media change after incubation is not necessary. If viral prep is in a different media, let the cells and virus incubate for 1–4 h. Centrifuge the cells at 300×*g* for 10 min and resuspend in the appropriate media.

5.3 Visually inspect the cells under the microscope the next day. Fluorescent reporters can usually be seen after 24 h. After 48 h, quantify transduction efficiency using flow cytometry. If necessary, small-molecule transcriptional activators can be added to your cells the day before flow cytometry to increase the expression of the fluorescent protein and activate silent promoters. Any activation treatment should also be performed on naive cells to control for any changes in auto-fluorescence induced by the small-molecule activator treatment. Cells can be analyzed for expression of the transgene using flow cytometry.

5.4 Expand cultures using standard tissue culture techniques and freeze down cells for storage. Cells can usually be frozen in a cryovial at concentration of 10^6 cells/mL in media supplemented with 10% DMSO and stored at −80 °C.

3.1. Sorting isoclonal populations

Isoclonal (isogenic) populations are populations that have been derived from a single parent cell. Single cells can be isolated through the use of fluorescence-activated cell-sorting (FACS). Polyclonal populations used for sorting should be between 5% and 10% positive for fluorescent reporter expression; this ensures that the frequency of double integration is low, as calculated using a Poisson distribution.

Reagents and equipment

Live-cell FACS sorting capabilities
Control (naïve) cells
Lentiviral vector-transduced population, expanded

PBS supplemented with 10% FBS
96-well tissue culture plates
24-well tissue culture plates
Culture media supplemented with 10–20% FBS and 1% penicillin/streptomycin
5 mL FACS polypropylene tubes with cell-strainer caps (sterile)
Mylar plate sealers (Thermo Scientific cat. no. 5701ROC)
Humidified tissue-culture incubator at 5% CO_2

1. Preparing polyclonal lentiviral vector-transduced cells for sorting
   1.1 Using standard tissue-culture techniques, passage polyclonal transduced cell population and a naïve untransduced population of the same cell type. Maintain growth in the exponential phase. If necessary, activators can be added to the transduced population to maximize gene expression prior to sorting so that an increased number of transduced cells can be sorted. All activator treatments should also be performed on the naïve population.
   1.2 Two hours before sorting, count the cells in each population and centrifuge 1–2 $\times$ 10^6 cells of each population in 15-mL conical tubes at $300\times g$ for 5 min. Aspirate the media and resuspend the cells in 2 mL of PBS supplemented with 10% FBS.
   1.3 Pipette the population through a sterile cell strainer (to remove debris), into a sterile FACS tube. Place cells on ice.
2. Prepare 96-well plates for collection of sorted cells
   2.1 Each well of the plate will collect one sorted cell, so prepare as many wells as cells you intend to collect. It is a good assumption that only 5–15% of the single-cell sorts will survive. For this reason, we generally collect between 200 and 300 single cells of each population to be sorted to recover 20–30 isoclonal populations.
   2.2 In each well of the plate add 150 μL of culture media supplemented with 10–20% FBS and 1% penicillin and streptomycin. Some cells types are more sensitive to the sorting procedure and may require higher percentages of FBS in the collection media.
3. Gate cells for sorting and collect cells
   3.1 The FACS operator will need the naïve cells in order to distinguish the autofluorescence background of the cells from the fluorescent reporter signal. Gate the live cells according to the forward and side scattering measurements (see Fig. 26.4). From this gate, observe the distribution of fluorescence. We gate three regions of fluorescence for sorting, a DIM region encompassing the lowest third of fluorescence signal, a MID region which consists of the middle third of the fluorescence signal, and a BRIGHT region which consists of the highest signal expressed (see Fig. 26.4).

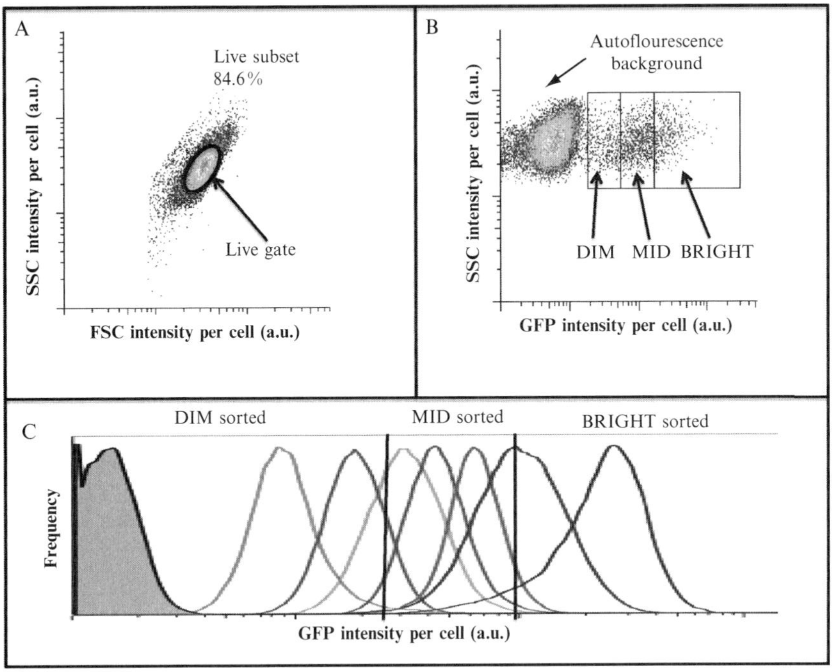

Figure 26.4 (A) Example of live gating off forward-scatter versus side-scatter plots of Jurkat T-cells. (B) Plot shows side scatter versus GFP fluorescence. Gates are drawn to show example sorting gates of DIM, MID, and BRIGHT populations. (C) GFP fluorescence histograms of sorted isoclonal populations.

3.2 Collect 200–300 cells from each gate. After sorting, verify under a microscope that each well has only one cell. Incubate cells under normal conditions. To prevent evaporation of media from the plate, remove the lid and cover the wells with a gas-permeable mylar plate sealer.

4. Expand isoclonal populations.

4.1 Depending on cell type, isoclonal populations may take 1–2 weeks to become confluent in the 96-well plate. During this time, check that the media in the plate has not evaporated below 1/3 the well height and replace media as needed. Keep 100–150 μL of media in each well. Every 3–4 days, check the confluency of the cells in each well under a microscope. Do not check cells more often than every 2–3 days since incandescent light can be toxic to cells when they are at low concentration. Alternatively, if the culture media has a pH indicator, wells can be screened visually by looking for a change in media color in the well indicating that the population is reaching a high confluency.

4.2 Once an isoclonal population has reached 80–90% confluency in the well, transfer the population to 0.5 mL of media in a 24-well tissue culture plate. Continue expanding the population to maintain cells at a healthy confluency using standard tissue-culture techniques. Store aliquots of the expanded populations at −80 °C. Isoclonal populations can now be analyzed by flow cytometry (Fig. 26.4C).

4. Procedure for Constructing a CV^2 Versus Mean Plot

To perform a CV^2 versus mean analysis, the first step is to create a library of isogenic populations each carrying a single integrated copy of the promoter of interest driving a reporter gene, as described previously. These clonal populations will exhibit considerable differences in mean expression levels as each clone corresponds to a different integration site of the lentiviral vector (Fig. 26.4C). Reporter expression in each clone is measured by flow cytometry with data from at least 100,000 single-cells collected per clone. To quantify the cell-to-cell reporter variation within each clonal population, flow cytometry data is analyzed using standard software packages like FlowJo™ (Treestar Inc., Ashland, Oregon). Before quantifying this variation it is important that one minimizes differences in protein levels due to heterogeneity in cell size, cell shape, and cell-cycle state (i.e., extrinsic noise). A standard approach to reduce extrinsic noise is to draw a small gate around the forward scatter (FSC) and side scatter (SSC) medians that contains at least 30,000 cells (Newman *et al.*, 2006). CV and mean protein levels for different clones is computed from this gated population using the statistics toolbox in FlowJo. For computing CV^2, it should be kept in mind that many software packages, like FlowJo, report CV as a percentage. Next, mean protein levels, $\langle \text{protein} \rangle$, which are quantified in terms of fluorescence intensities, are converted into absolute protein molecular counts. For example, EGFP Calibration Beads™ (BD Biosciences, Clontech, San Jose, CA) can be used to convert GFP fluorescence intensities into GFP molecular equivalents of solubilized flourophores (MESF), a measure of GFP molecular abundance. Once CV^2 and $\langle \text{protein} \rangle$ have been quantified for all clones, the final step is to look at correlations between them by making a scatter plot with CV^2 on the *y*-axis and $\langle \text{protein} \rangle$ on the *x*-axis.

5. Inferring Promoter Regulatory Architecture from CV^2 Versus Mean Analysis

To understand how the relationship between CV^2 and $\langle \text{protein} \rangle$ can inform upon promoter architecture, we review mathematical predictions from two different gene-expression models below.

5.1. Constitutive promoter architecture

Constitutive promoter models (Fig. 26.5A), where mRNAs are continuously created from the promoter at exponentially distributed time intervals generate the prediction:

$$CV^2 = \frac{C}{\langle \text{protein} \rangle} \tag{26.1}$$

where C is a proportionality constant (Paulsson, 2004). Equation (26.1) shows that for constitutive gene expression, increasing the mean protein count will decrease CV^2 such that the product $CV^2 \times \langle \text{protein} \rangle$ remains

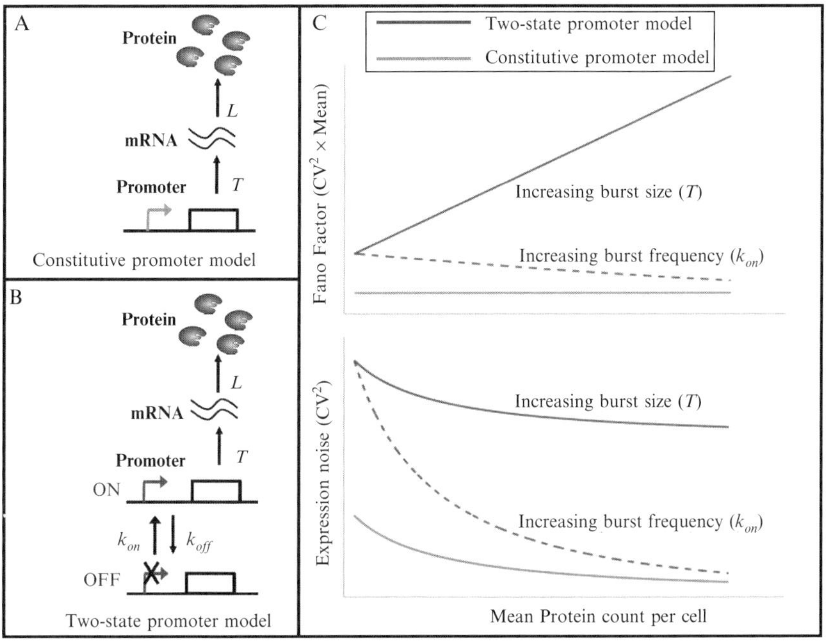

Figure 26.5 (A) Constitutive promoter model where mRNAs are produced continuously from the promoter one at a time. (B) Two-state promoter model where mRNAs are only produced when the promoter transitions to an ON state before returning to an OFF state. (C) Predictions for the scaling of gene-expression noise (CV^2) and Fano factor ($CV^2 \times$ Mean) as a function of mean protein count for different promoter architectures. Two-state promoter architecture is predicted to generate elevated levels of expression noise compared to a constitutive promoter architecture. In a constitutive promoter, CV^2 decreases with mean such that the Fano factor remains fixed (red line). In a two-state promoter model, CV^2 also decreases with mean but the Fano factor can either increase or decrease with mean depending on the mode of transcriptional activation (soild and dashed blue lines). (See Color Insert.)

unchanged. A useful way to detect this inverse correlation between CV^2 and mean protein levels is by constructing a secondary scatter plot of the Fano factor on the y-axis versus the ⟨protein⟩ on the x-axis, where Fano factor is defined as $CV^2 \times \langle \text{protein} \rangle$. The main rationale behind this plot is that since Fano factor is predicted to be a constant for constitutive promoter architecture (Eq. (26.1)), it will appear to be completely uncorrelated with mean protein levels on this plot (Fig. 26.5C). If Fano factor is indeed uncorrelated with mean protein levels, then for constitutive gene expression, the average Fano factor across different clonal populations should be equal to the proportionality constant C given by:

$$C = \frac{L}{\mathrm{d}m + \mathrm{d}p} = \frac{\langle \text{protein} \rangle}{\langle \text{mRNA} \rangle} \frac{\mathrm{d}p}{\mathrm{d}m + \mathrm{d}p} \tag{26.2}$$

where L is mRNA translation rate, ⟨mRNA⟩ is the mean mRNA count per cell and dm and dp represent mRNA and protein degradation rate, respectively (Paulsson, 2004). In essence, to confirm constitutive promoter architecture one needs to independently compute C for the reporter gene in consideration and match it with the experimentally obtained Fano factor. Assuming that the mRNA half-life, protein half-life, and either the mRNA translation rate L or the ratio of protein and mRNA abundance are known for the reporter gene, C can be directly computed from Eq. (26.2). Genome-wide gene-expression studies in yeast have put the average value of C across different genes to be $\sim$1000 molecules; however, this value can vary considerably for different genes (Bar-Even *et al.*, 2006). In summary, an inverse correlation between CV^2 and ⟨protein⟩ such that the Fano factor is uncorrelated with ⟨protein⟩ and equal to the proportionality constant C is a signature of a constitutive promoter architecture. These predictions from stochastic gene-expression models have been useful for analyzing expression noise in different *Escherichia coli* and *Saccharomyces cerevisiae* genes, and have confirmed that many essential genes in these organisms encode constitutive promoters (Bar-Even *et al.*, 2006; Newman *et al.*, 2006; Ozbudak *et al.*, 2002).

5.2. Two-state promoter architecture

Many genes exhibit elevated levels of expression noise that are inconsistent with constitutive gene expression models (Raj *et al.*, 2006). CV^2 or Fano factor values that are much higher than that predicted by Eq. (26.1) are indicative of a two-state promoter architecture, where the promoter fluctuates between an inactive and active state with rates k_{on}, k_{off} and transcription only occurs from the active state at a rate T (Fig. 26.5B and C). In such 'two-state' models, mRNAs are created in bursts during promoter transitions from inactive to active state, with k_{on} and T/k_{off} denoting the burst frequency and the average size of the transcriptional bursts, respectively (Kepler and Elston, 2001; Simpson *et al.*, 2004). Stochastic analysis of two-state promoter models shows a complex relationship between expression noise and mean protein levels. This is both parameter-dependent and also dependent on whether promoter transcriptional efficiency is increased by increasing the burst frequency or burst size (Kaern *et al.*, 2005). At a qualitative level, CV^2 is always predicted to decrease with mean protein levels in a two-state promoter model; however, the product $CV^2 \times \langle\text{protein}\rangle$ or Fano factor can increase (Case I), decrease (Case II), or stay independent of the mean protein levels (Case III). We next discuss how these different noise profiles inform upon the two-state promoter architecture.

Case I. Fano factor increasing with mean protein counts indicates a two-state promoter architecture where higher promoter transcriptional efficiency is obtained by increasing the transcription rate T, that is, higher transcriptional burst size T/k_{off}.

Case II. Fano factor decreasing with mean protein counts indicates a two-state promoter architecture, where promoter infrequently transitions between stable active and inactive states, and higher promoter transcriptional efficiency is obtained by increasing the burst frequency k_{on}.

Case III. Fano factor uncorrelated with mean protein counts indicates a two-state promoter architecture, where promoter infrequently transitions to an unstable active state that rapidly transitions back to the inactive state. Furthermore, higher prompter transcriptional efficiency is obtained by increasing the burst frequency k_{on}.

A recent study of expression noise in HIV-1 LTR promoter using lentiviral vectors shows noise profiles that are a combination of Cases I and III (Singh *et al.*, 2010). More specifically, across some integration sites CV^2 decreases with mean protein counts such that the Fano factor is invariant. However, at many other integration sites Fano factor increases with mean protein counts. Collectively, this data suggests that HIV-1 encodes a two-state promoter architecture, which infrequently transitions to an unstable active promoter state that is rapidly switched OFF after making a burst of few mRNAs. Moreover, the site of integration uses both burst frequency and burst size to modulate viral gene expression.

In summary, CV^2 versus mean analysis has been instrumental in studying expression noise in both prokaryotic and eukaryotic genes. It has shown that while many essential genes encode constitutive promoters to minimize noise levels, other promoters such as HIV-1 LTR use two-state promoter architectures to increase expression noise. The unique property of lentiviral vectors to stably integrate a promoter or genetic circuit at a semi-random position in the human genome makes them an ideal tool for studying gene-expression noise and they will likely find an increasing use in inferring and comparing the regulatory architecture of human promoters.

6. Conclusion

Lentiviral vectors provide several unique experimental advantages for studying gene-expression noise in mammalian systems including the ability to easily introduce transgenes that remain stable as single integrations in host genomic DNA, and the ability to generate isogenic populations of cells where the transgene is genetically stable and often transcriptionally stable. In addition, the semi-random integration pattern of lentiviruses and the natural variation in the chromatin microenvironment inherent at the lentiviral integration site generates differences in lentiviral-encoded promoter strength and this difference provides a natural method to study CV^2 as a function of mean protein levels.

Above, we have focused exclusively on the use of lentiviral vectors as probes to measure magnitude of noise, but we have not touched upon the analysis of temporal correlations in noise or frequency-domain analysis of noise (Austin *et al.*, 2006; Cox *et al.*, 2008). Such frequency-domain analysis of noise can be highly informative about underlying biology and lentiviral vectors are also very effective tools for analyzing the frequency and temporal correlations of noise. We have used lentiviral vectors to study temporal correlations in HIV gene-expression noise and to characterize HIV gene-expression circuitry (Weinberger *et al.*, 2008). Thus, the utility of lentiviral vectors for the study of noise is not limited to measurements of noise magnitude (i.e., CV) and lentiviral vectors are effective tools to probe biological noise in multiple dimensions.

Acknowledgments

This work was supported by the NIH (GM083395), the California HIV/AIDS Research Fund (ID08-SD-01), and UCSD Center for AIDS Research (NIAID 5 P30 AI36214).

References

Akkina, R. K., Walton, R. M., Chen, M. L., Li, Q. X., Planelles, V., and Chen, I. S. (1996). High-efficiency gene transfer into CD34+ cells with a human immunodeficiency virus type 1-based retroviral vector pseudotyped with vesicular stomatitis virus envelope glycoprotein G. *J. Virol.* **70,** 2581–2585.

Austin, D. W., Allen, M. S., McCollum, J. M., Dar, R. D., Wilgus, J. R., Sayler, G. S., Samatova, N. F., Cox, C. D., and Simpson, M. L. (2006). Gene network shaping of inherent noise spectra. *Nature* **439,** 608–611.

Bar-Even, A., Paulsson, J., Maheshri, N., Carmi, M., O'Shea, E., Pilpel, Y., and Barkai, N. (2006). Noise in protein expression scales with natural protein abundance. *Nat. Genet.* **38,** 636–643.

Blake, W. J., Kaern, M., Cantor, C. R., and Collins, J. J. (2003). Noise in eukaryotic gene expression. *Nature* **422,** 633–637.

Blake, W. J., Balazsi, G., Kohanski, M. A., Isaacs, F. J., Murphy, K. F., Kuang, Y., Cantor, C. R., Walt, D. R., and Collins, J. J. (2006). Phenotypic consequences of promoter-mediated transcriptional noise. *Mol. Cell* **24,** 853–865.

Challita, P. M., and Kohn, D. B. (1994). Lack of expression from a retroviral vector after transduction of murine hematopoietic stem cells is associated with methylation in vivo. *Proc. Natl. Acad. Sci. USA* **91,** 2567–2571.

Cox, C. D., McCollum, J. M., Allen, M. S., Dar, R. D., and Simpson, M. L. (2008). Using noise to probe and characterize gene circuits. *Proc. Natl. Acad. Sci. USA* **105,** 10809–10814.

Cronin, J., Zhang, X. Y., and Reiser, J. (2005). Altering the tropism of lentiviral vectors through pseudotyping. *Curr. Gene Ther.* **5,** 387–398.

Dull, T., Zufferey, R., Kelly, M., Mandel, R. J., Nguyen, M., Trono, D., and Naldini, L. (1998). A third-generation lentivirus vector with a conditional packaging system. *J. Virol.* **72,** 8463–8471.

Eldar, A., Chary, V. K., Xenopoulos, P., Fontes, M. E., Loson, O. C., Dworkin, J., Piggot, P. J., and Elowitz, M. B. (2009). Partial penetrance facilitates developmental evolution in bacteria. *Nature* **460,** 510–514.

Elowitz, M. B., Levine, A. J., Siggia, E. D., and Swain, P. S. (2002). Stochastic gene expression in a single cell. *Science* **297,** 1183–1186.

Fischer, U., Huber, J., Boelens, W. C., Mattaj, I. W., and Luhrmann, R. (1995). The HIV-1 Rev activation domain is a nuclear export signal that accesses an export pathway used by specific cellular RNAs. *Cell* **82,** 475–483.

Golding, I., Paulsson, J., Zawilski, S. M., and Cox, E. C. (2005). Real-time kinetics of gene activity in individual bacteria. *Cell* **123,** 1025–1036.

Hanna, J., Saha, K., Pando, B., van Zon, J., Lengner, C. J., Creyghton, M. P., van Oudenaarden, A., and Jaenisch, R. (2009). Direct cell reprogramming is a stochastic process amenable to acceleration. *Nature* **462,** 595–601.

Jordan, A., Defechereux, P., and Verdin, E. (2001). The site of HIV-1 integration in the human genome determines basal transcriptional activity and response to Tat transactivation. *EMBO J.* **20,** 1726–1738.

Kaern, M., Elston, T. C., Blake, W. J., and Collins, J. J. (2005). Stochasticity in gene expression: From theories to phenotypes. *Nat. Rev. Genet.* **6,** 451–464.

Kepler, T. B., and Elston, T. C. (2001). Stochasticity in transcriptional regulation: Origins, consequences, and mathematical representations. *Biophys. J.* **81,** 3116–3136.

Kim, V. N., Mitrophanous, K., Kingsman, S. M., and Kingsman, A. J. (1998). Minimal requirement for a lentivirus vector based on human immunodeficiency virus type 1. *J. Virol.* **72,** 811–816.

Mitrophanous, K., Yoon, S., Rohll, J., Patil, D., Wilkes, F., Kim, V., Kingsman, S., Kingsman, A., and Mazarakis, N. (1999). Stable gene transfer to the nervous system using a non-primate lentiviral vector. *Gene Ther.* **6,** 1808–1818.

Miyoshi, H., Blomer, U., Takahashi, M., Gage, F. H., and Verma, I. M. (1998). Development of a self-inactivating lentivirus vector. *J. Virol.* **72,** 8150–8157.

Naldini, L., Blomer, U., Gallay, P., Ory, D., Mulligan, R., Gage, F. H., Verma, I. M., and Trono, D. (1996). In vivo gene delivery and stable transduction of nondividing cells by a lentiviral vector. *Science* **272,** 263–267.

Neville, M., Stutz, F., Lee, L., Davis, L. I., and Rosbash, M. (1997). The importin-beta family member Crm1p bridges the interaction between Rev and the nuclear pore complex during nuclear export. *Curr. Biol.* **7,** 767–775.

Newman, J. R., Ghaemmaghami, S., Ihmels, J., Breslow, D. K., Noble, M., DeRisi, J. L., and Weissman, J. S. (2006). Single-cell proteomic analysis of *S. cerevisiae* reveals the architecture of biological noise. *Nature* **441,** 840–846.

Olsen, J. C. (1998). Gene transfer vectors derived from equine infectious anemia virus. *Gene Ther.* **5,** 1481–1487.

Ozbudak, E. M., Thattai, M., Kurtser, I., Grossman, A. D., and van Oudenaarden, A. (2002). Regulation of noise in the expression of a single gene. *Nat. Genet.* **31,** 69–73.

Paulsson, J. (2004). Summing up the noise in gene networks. *Nature* **427,** 415–418.

Poeschla, E. M., Wong-Staal, F., and Looney, D. J. (1998). Efficient transduction of nondividing human cells by feline immunodeficiency virus lentiviral vectors. *Nat. Med.* **4,** 354–357.

Raj, A., and van Oudenaarden, A. (2008). Nature, nurture, or chance: Stochastic gene expression and its consequences. *Cell* **135,** 216–226.

Raj, A., Peskin, C. S., Tranchina, D., Vargas, D. Y., and Tyagi, S. (2006). Stochastic mRNA synthesis in mammalian cells. *PLoS Biol.* **4,** e309.

Reiser, J., Harmison, G., Kluepfel-Stahl, S., Brady, R. O., Karlsson, S., and Schubert, M. (1996). Transduction of nondividing cells using pseudotyped defective high-titer HIV type 1 particles. *Proc. Natl. Acad. Sci. USA* **93,** 15266–15271.

Schroder, A. R., Shinn, P., Chen, H., Berry, C., Ecker, J. R., and Bushman, F. (2002). HIV-1 integration in the human genome favors active genes and local hotspots. *Cell* **110,** 521–529.

Simpson, M. L., Cox, C. D., and Sayler, G. S. (2004). Frequency domain chemical Langevin analysis of stochasticity in gene transcriptional regulation. *J. Theor. Biol.* **229,** 383–394.

Singh, A., and Weinberger, L. S. (2009). Stochastic gene expression as a molecular switch for viral latency. *Curr. Opin. Microbiol.* **12,** 460–466.

Singh, A., Razooky, B., Cox, C. D., Simpson, M. L., and Weinberger, L. S. (2010). Transcriptional bursting from the HIV-1 promoter is a significant source of stochastic noise in HIV-1 gene expression. *Biophys. J.* **98,** L32–L34.

Suel, G. M., Garcia-Ojalvo, J., Liberman, L. M., and Elowitz, M. B. (2006). An excitable gene regulatory circuit induces transient cellular differentiation. *Nature* **440,** 545–550.

Swain, P. S., Elowitz, M. B., and Siggia, E. D. (2002). Intrinsic and extrinsic contributions to stochasticity in gene expression. *Proc. Natl. Acad. Sci. USA* **99,** 12795–12800.

Weinberger, L. S., and Shenk, T. (2007). An HIV feedback resistor: Auto-regulatory circuit deactivator and noise buffer. *PLoS Biol.* **5,** e9.

Weinberger, L. S., Burnett, J. C., Toettcher, J. E., Arkin, A. P., and Schaffer, D. V. (2005). Stochastic gene expression in a lentiviral positive-feedback loop: HIV-1 Tat fluctuations drive phenotypic diversity. *Cell* **122,** 169–182.

Weinberger, L. S., Dar, R. D., and Simpson, M. L. (2008). Transient-mediated fate determination in a transcriptional circuit of HIV. *Nat. Genet.* **40,** 466–470.

Author Index

G

I

M

N

O

T

U

V

X

Y

Subject Index

D

T

U

W

Y

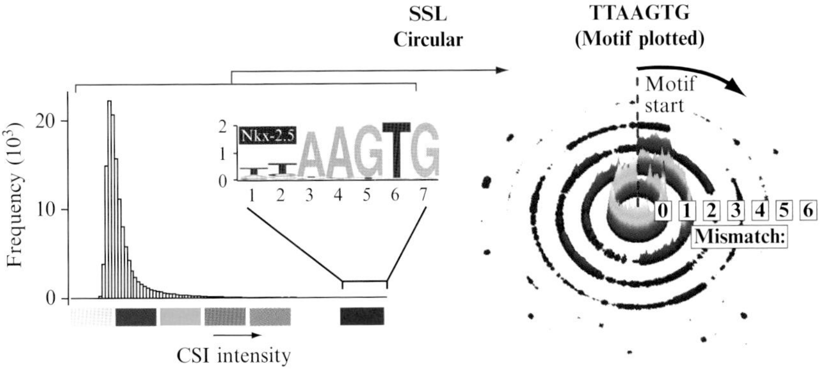

Joshua R. Tietjen *et al.*, Figure 1.4 *Sequence-specificity landscapes (SSLs).* SSL generated from CSI intensity data for the Nkx-2.5 protein. The best bound sequences were used to generate the seed motif (shown in the LOGO and above the landscape). Color scale indicates normalized fluorescence intensity corresponding to how well a given sequence is bound. The sequences are distributed among the rings as described in the text. The distance between each ring and the next concentric ring outward is equivalent to a Hamming distance of 1. (Adapted from Carlson *et al.*, 2010).

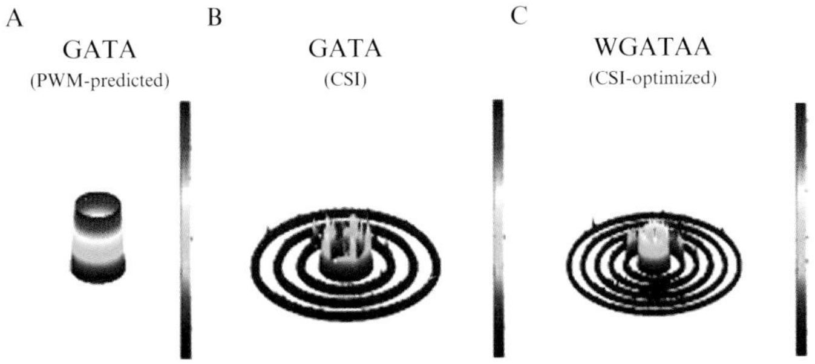

Joshua R. Tietjen *et al.*, Figure 1.5 *SSLs for GATA4.* Panel (A) displays the specificity landscape predicted from the position weight matrix (PWM) generated from the CSI data. The PWM-based landscape predicts uniform binding wherever the exact GATA sequence is found. Panel (B) shows the actual intensity of every sequence on the array that contains the GATA motif. The data clearly indicate that unlike the prediction from the PWM, the binding is not uniform in every instance where this sequence is found. Panel (C) displays the SSL obtained after iterative refinement of the motif permitted by landscapes. Inspection of binding data in this manner reveals that many of the motifs of DNA-binding proteins and molecules compress the data and do not properly account for the influences of flanking sequences. The color scale reflects the fluorescence intensity of each sequence on the *z*-axis and the seed motif is shown above each landscape.

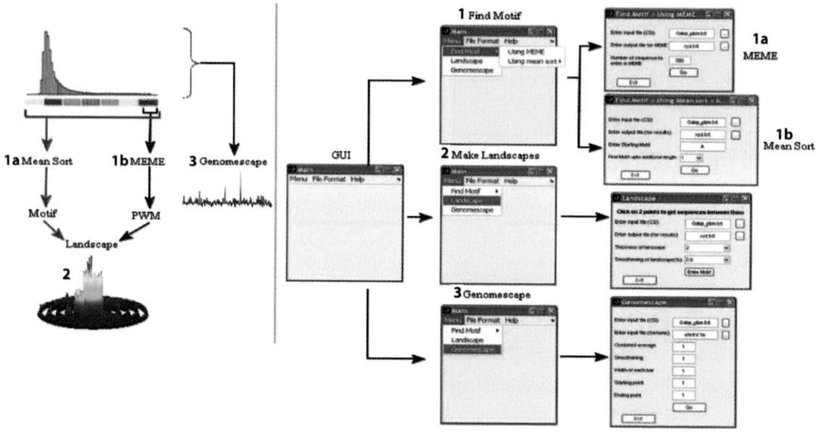

Joshua R. Tietjen *et al.*, Figure 1.6 *SSL and genomescape software process flow diagram.* The left panel displays the experimental process flow for determining the DNA-binding profile of a given protein or small molecule by CSI. The range of fluorescence intensities of the sequences bound is displayed as a histogram, and subsets of these data are used to find the motif that most accurately represents the binding preferences of the molecule. The motif can be found by the program in two ways, as shown in (1a) and (1b) in both left and right panels, using either the Mean Sort algorithm or MEME. Once a motif is identified, it can be used to seed the SSL shown in (2). Finally, the CSI data can be applied to the genome to generate genomescapes, as shown in (3). On the right are windows displayed in the SSL and genomescape software at each step of the workflow of this process.

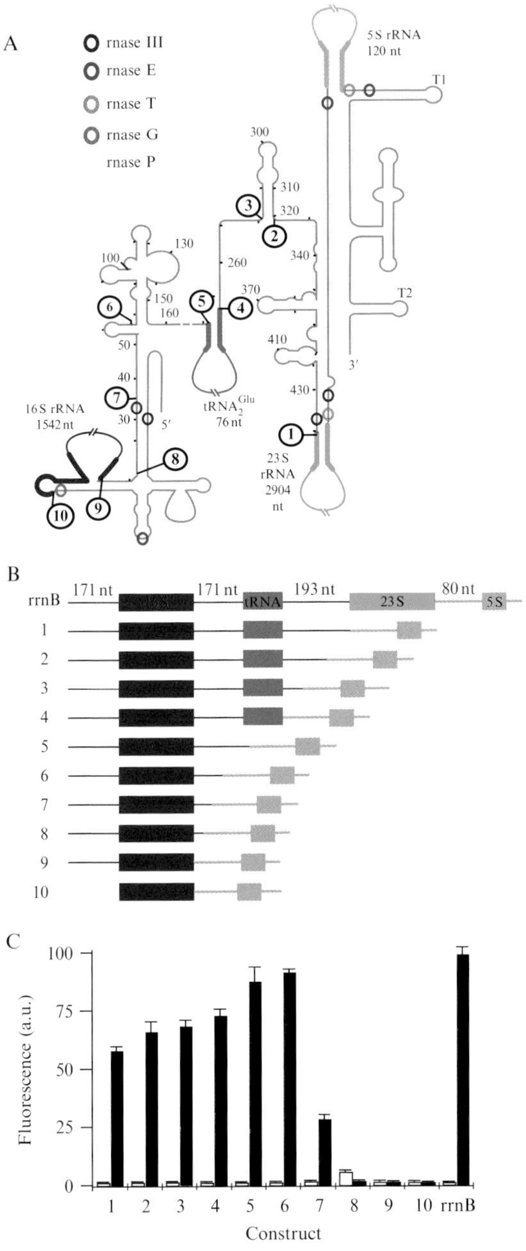

Wenlin An and Jason W. Chin, Figure 5.3 Creation of a minimal O-ribosome. (A) The proposed secondary structure of the rrnB primary transcript. T1 and T2 are terminator sequences. Numbered circles indicate the sites of truncation. Colored circles are sites that being recognized and processed via indicated Ribonuclease (rrnase). (B) Schematic of the truncations examined for the production of active O-ribosomes. (C) The activity of O-ribosomes produced from each truncation construct (constructs 1–10, filled bars) compared with the full-length operon (O-rrnB). Fluorescence was measured in cells containing pXR1 (a tetracycline-resistant p15A plasmid that directs GFP expression from a constitutive promoter and O-rbs). The empty bars show the expression of GFP produced when pXR1 is combined with wild-type ribosomes in the absence of O-ribosomes.

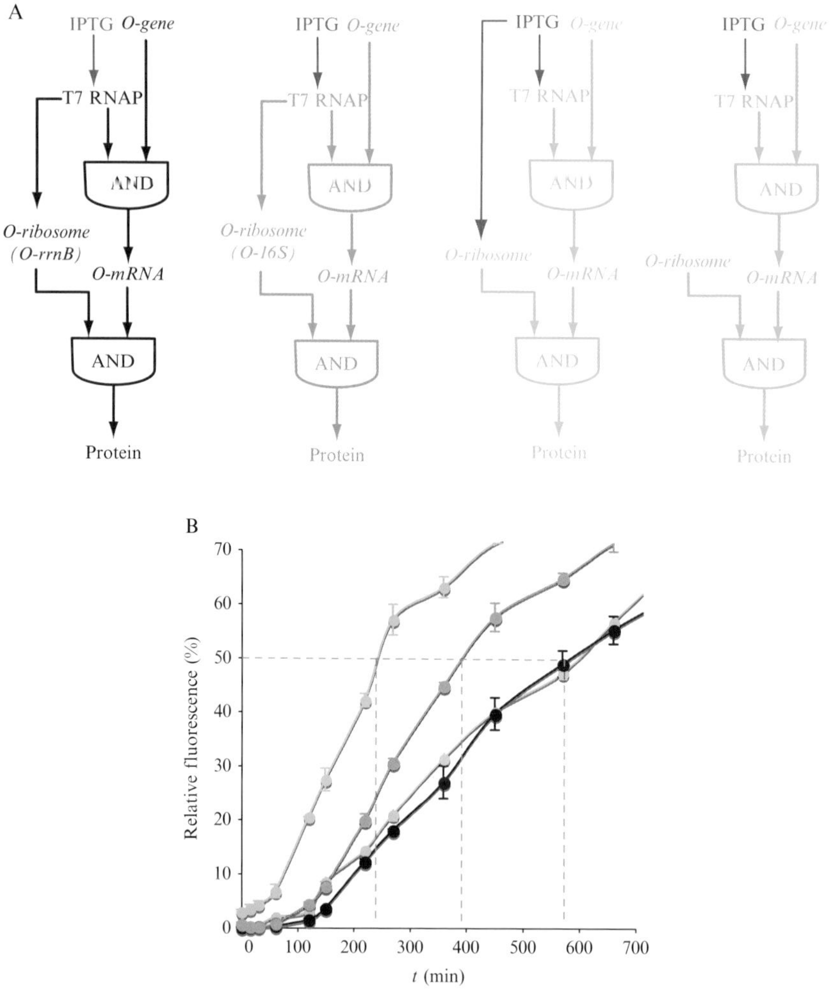

Wenlin An and Jason W. Chin, Figure 5.4 The FFLs with the minimal O-ribosome and the progenitor O-ribosome have identical topologies but mediate distinct delays. (A) The FFL using the progenitor O-ribosome (maroon) and the minimal O-ribosome (orange). (B) The delays in gene expression created by the orthogonal transcription–translation networks. Orange solid circles, BL21 (DE3), pT7 RSF O-16S, pT7 O-rbs GST-GFP. The time taken to reach 50% of maximal expression, used to quantify the delay (Mangan and Alon, 2003), is indicated.

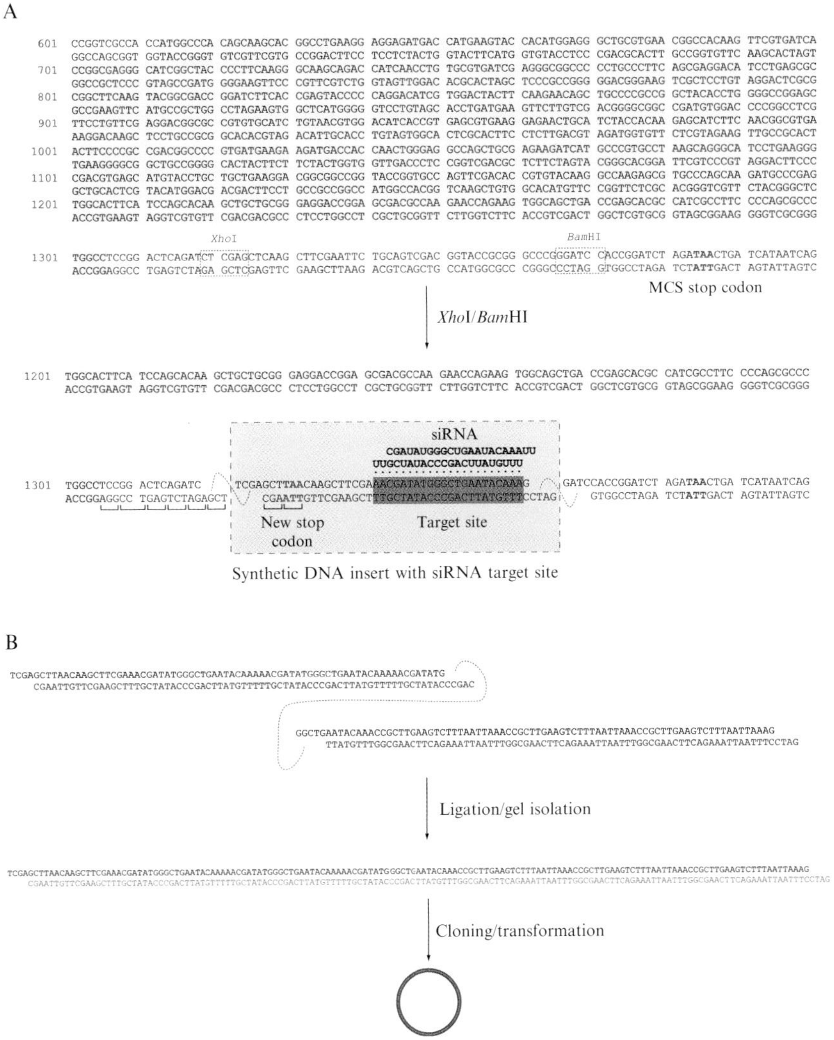

Yaakov Benenson, Figure 8.2 Examples of cloning approaches. (A) Subcloning an siRNA target in 3′-UTR of a fluorescent reporter. A plasmid pZsYellow-C1 (Clontech) is used for illustration. Coding sequence of ZsYellow protein is shaded. The insert is designed to contain a preformed sticky end, a stop codon, a spacer, and an sRNA target site/s. (B) Subcloning a long target sequence. The desired insert containing triple repeats of two different targets is made from two separate dsDNA building blocks by test-tube ligation. The resulting insert is subcloned in the output UTR as in (A).

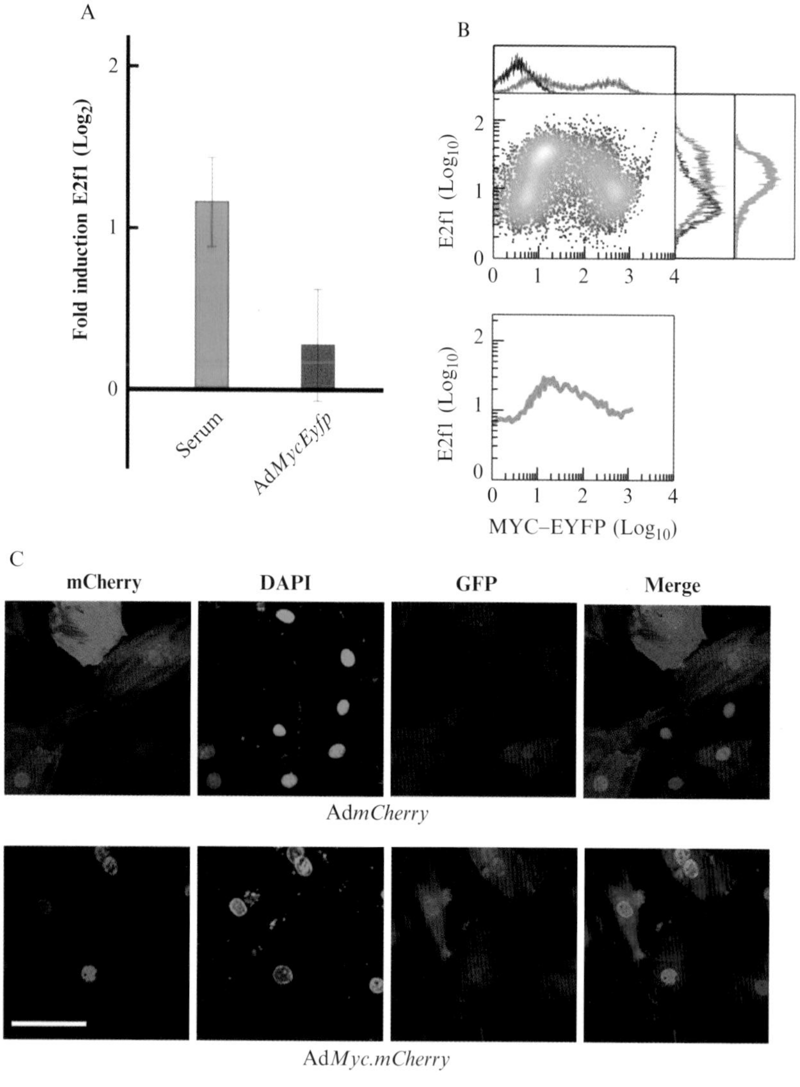

Jeffrey V. Wong *et al.*, Figure 10.2 Cell–cell variability in MYC input levels reveals biphasic E2f1 response. (A) Real-time PCR results for E2f1 responses. Rat embryonic fibroblasts (REF52) cultured in 0.02% bovine growth serum (BGS) were either switched to 10% BGS (Serum; blue bar) or infected with an Ad vector expressing MYC fused to EYFP (AdMycEyfp; red bar) at an MOI of 1000 for 36 h. Endogenous E2f1 mRNA levels in each sample are expressed relative to cells that were starved or staved and infected with AdEyfp control virus, respectively. (B) Typical flow cytometry results. Data from serum starved REF52 cells harboring integrated GFP reporter under control of E2f1 promoter. (Top) Scatter plot shows E2f1 reporter activity (green fluorescence) in individual cells as function of MYC–EYFP input (yellow fluorescence). Histograms summarize fluorescence output for each respective channel in response to 10% serum (blue); control infection with Ad expressing β-galactosidase (black); and infection with AdMycEyfp (red). (Bottom) Moving median values from scatter plot in (B). (C) Fluorescence microscopy. Starved REF52 cells harboring E2f1 reporter (GFP) were infected with Ad vectors expressing mCherry or MYC-mCherry fusion (500 MOI) for 36 h. DNA was subsequently stained with DAPI. Merge represents overlay of mCherry, DAPI, and GFP signals. Scale bar: 100 μm.

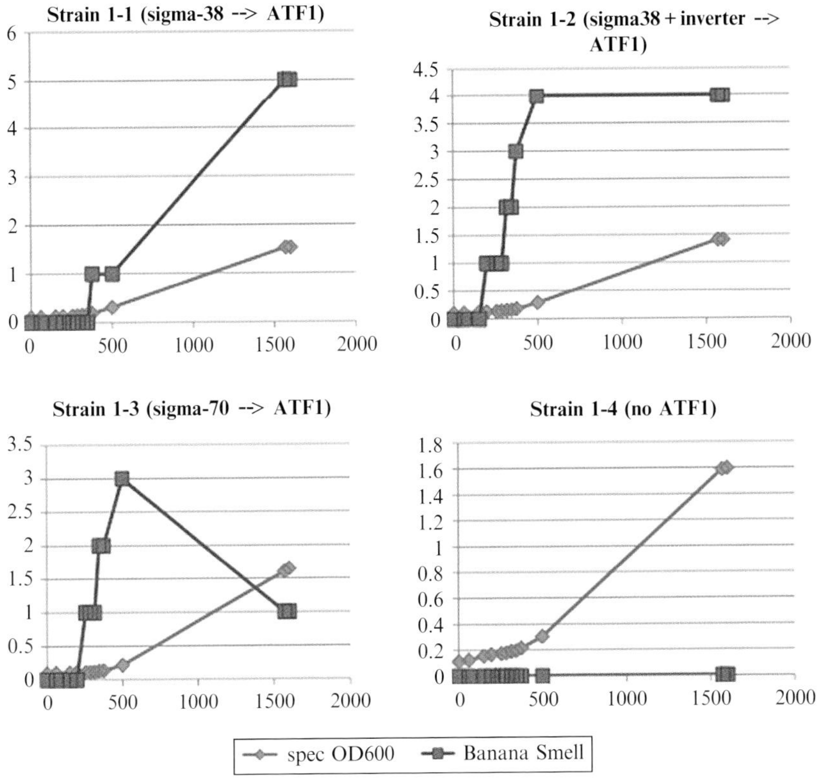

James Dixon and Natalie Kuldell, Figure 12.5 Behavior of "Eau that Smell" experimental and control strains. Growth and banana-smell generated by each experimental strain was assessed. Growth curves, shown with the blue lines, are measured as changes in turbidity over time. Banana smell, shown with red lines, was calibrated to the smell standards and is plotted for each time point on the growth curve. Growth time, in minutes, is shown on the x-axis. Unexpectedly, the log-phase promoter (strain 1-3) generates a less pronounced banana-flavored smell but is more tightly controlled to express only during the log-phase of growth. Source: data collected by J. Dixon.

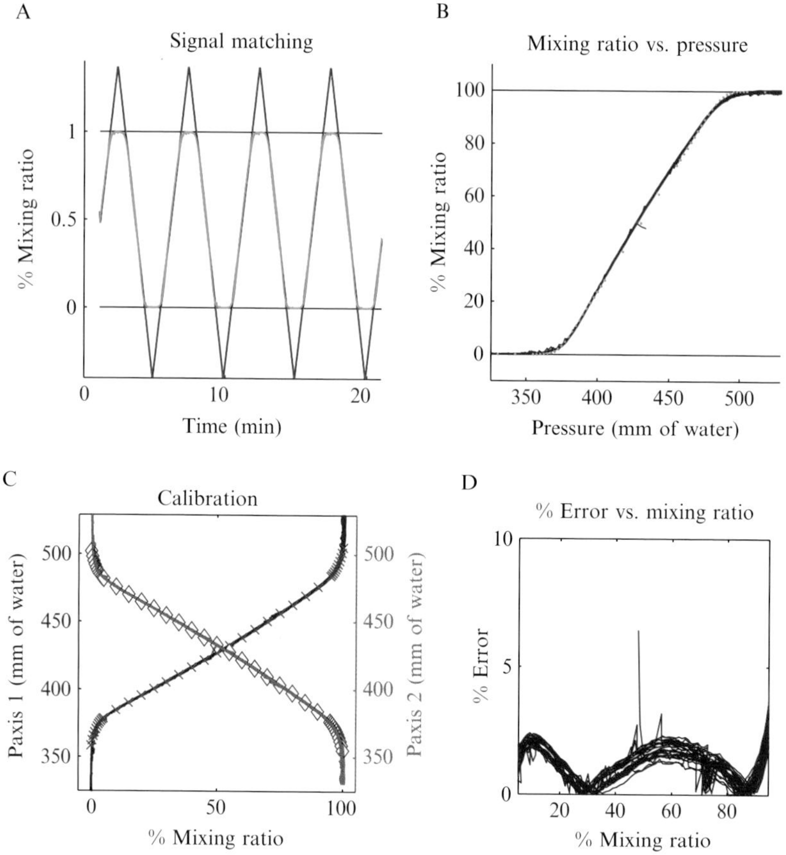

M. S. Ferry *et al.*, Figure 14.7 Performance of the DAW junction. (A) Calibration signal (red line) overlaid with output signal (green line) after correction for the delay in acquisition. During calibration the system is designed to intentionally overshoot the bounds of the DAW junction. Since the starting and ending points for calibration are not critical, this makes it easier to set up as described in the text. The ideal response would be a closely tracking output signal transitioning to plateaus after the system moves beyond 0% and 100% mixing ratios. As can be seen in the figure, this is what we observe, except for a slight rounding near the plateau region. (B) Compression of the data in part A into a single curve by mapping the input pressure directly to the output mixing ratio. Blue curve is the compressed data, while the green dots are the expected results from Comsol modeling. As can be seen in the figure, the modeling and experimental results are in excellent agreement. (C) Completed calibration for both inputs. Red crosses and pink diamonds represent polynomial fits of inputs 1 and 2, respectively, to the output mixing ratio. These fits can be used to program a linear actuator controller to generate precise inducer waves. (D) Measure of the percent error of the *uncalibrated* output signal, which general is less than 3%.

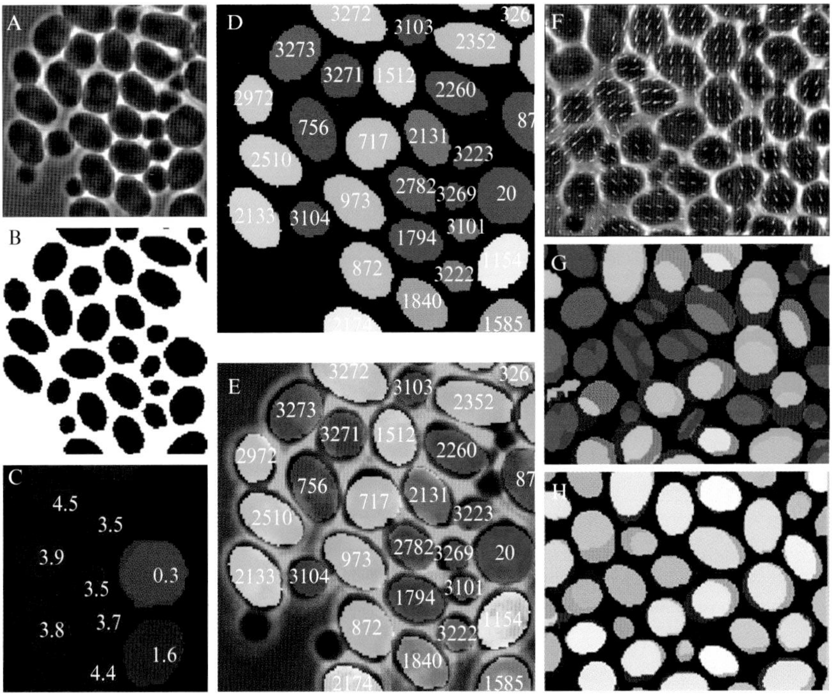

M. S. Ferry *et al.*, Figure 14.14 Overview of the cell tracking process. (A) Raw data: phase contrast image of yeast cells. Note the high contrast between the boundary of the cell and the exterior. (B) Segmented image after thresholding, application of the watershed algorithm and fitting the resultant objects to ellipses. (C) Scoring of a cell from frame n (shown in red) to trajectories present in frame $n - 1$. Lower score is better. Notice that the red cell has closely overlapped with a previous trajectory and generates a better score. All other scored cells are above the scoring threshold (which is set at 1). Note that the scoring system here has generated good contrast between the ideal match and the neighbors. This is indicative of a good match. (D) Colored image of the masks after trajectory finding is complete. Colored regions represent trajectories which are numbered. (E) Overlay of the trajectory image from part D with the phase contrast image of part A. Note most cells were assigned trajectories except for smaller cells and cells near the exterior. (F) Example of MatPIV processing for cell flow. White arrows indicating the cell flow velocity are overlaid with a phase contrast image of the colony. (G) Image of cells from frame $n - 1$ (opaquely colored objects) overlaid with cells from frame n (translucent objects). Notice there is an overall movement of cells toward the lower left corner of the image due to cell flow. Also note that the distance traveled here is almost one half cell diameter between frames for some cells. One can see how this movement could generate ambiguous situations for similarly shaped cells without prior knowledge of the cell flow. (H) Same cell field as in part G except MatPIV velocity information has been applied to correct for cell flow. Notice the much better overlap compared to part G. This will lead to more reliable matches since cell position is crucial for reliable matching.

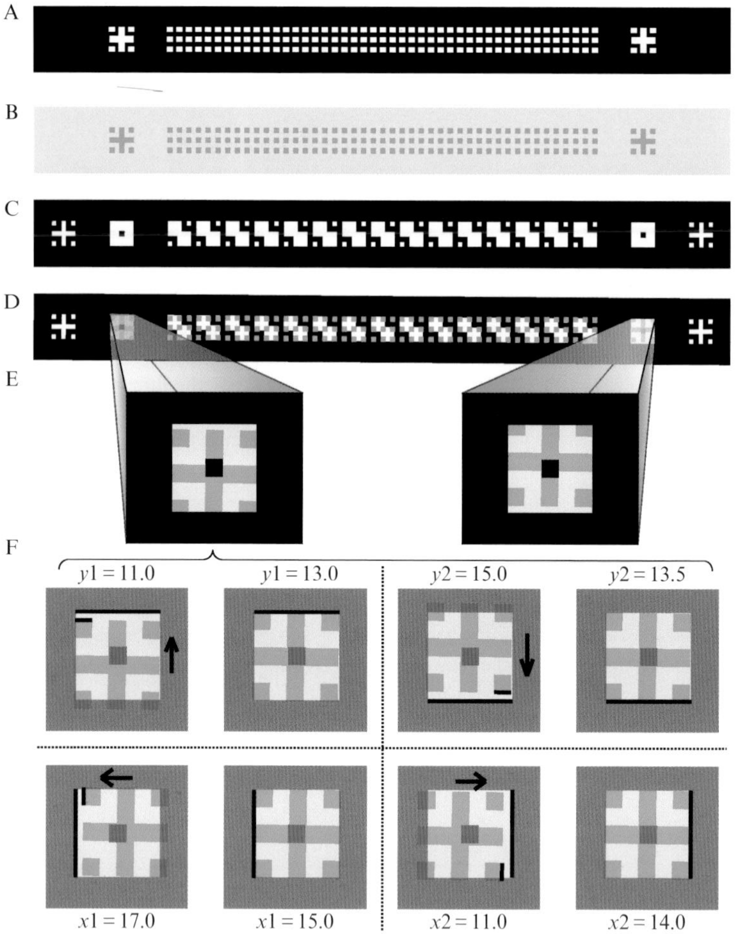

M. S. Ferry *et al.*, Figure 14.20 Sample layer alignment technique. (A) Photomask of layer #1, features are created by the transparent areas of the mask. (B) Features (green) on wafer (gray) for layer #1. (C) Photomask of layer #2. (D) Alignment of wafer with features from layer #1 to photomask for layer #2, as seen through the microscope of mask aligner. (E) Close-up view of alignment of the outermost left and right features. (F) For each side (left and right), the features on the wafer are aligned to the 4 four sides of the alignment box. The mask aligner micrometer position is averaged for x (15, 14) and y (13, 13.5) directions, to provide a single xy (14.5, 13.25) position. Note that it would seem that $y1$ and $y2$ positions should be identical if the feature sides are aligned to the alignment box. In reality, due to the new photoresist layer the features from previous layer become distorted, resulting in the difference. If the xy positions from the left and the right side are identical the alignment is good, otherwise the θ position needs to be changed and the whole process repeated. The transparency of the photomask has been adjusted for demonstration purposes.

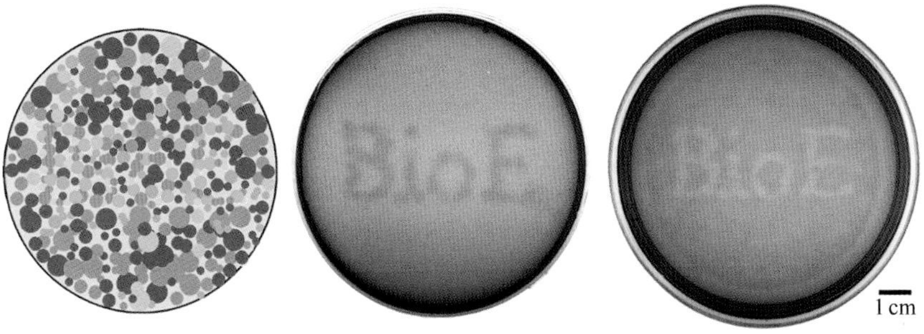

Jeffrey J. Tabor, Figure 15.1 Bacterial photography. The green/red pattern shown on the left was used to produce a 34 × 24 mm slide mask. The mask was then used to project the two-color image onto agarose embedded slabs of *E. coli* expressing the green light ON (center) or red light ON (right) systems. In both cases, the light sensitive signaling pathways control the expression of β-galactosidase, which is visualized as patterns of black pigment in the media. The plates were developed in the projector–incubator for 21 h and photographed as described in the text.

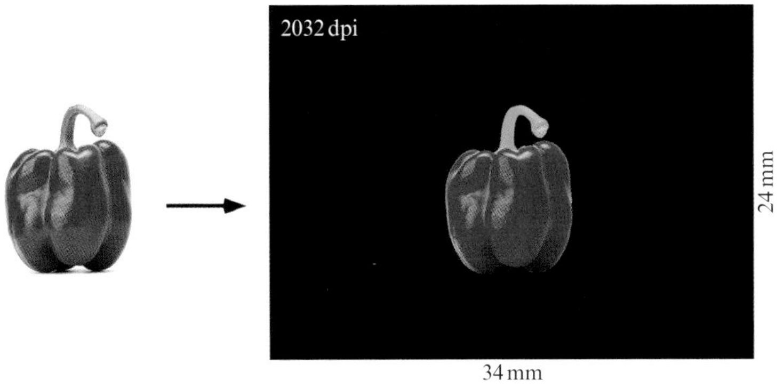

Jeffrey J. Tabor, Figure 15.4 Production of a two-color slide mask. A source image with significant RGB components in most pixels (left) is used as a starting point to make the mask. In Adobe Photoshop, the image is shrunk to ~10 mm in height (2032 dpi) and placed in the center of a 34 × 24 mm black background. Most of the red and blue signal is removed from pixels in the green areas, and most of the green and blue signal is removed from red areas. A .tif file of the enhanced image is then exported from Photoshop and used to print the slide.

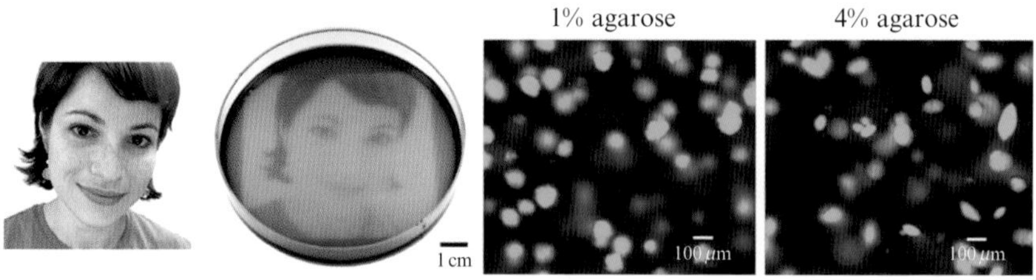

Jeffrey J. Tabor, Figure 15.6 Microscopic view of a bacterial photograph. The mask on the left was used to project a 650-nm image onto a lawn of RU1012/pPLPCB(A)/pCph8 which were made to constitutively express green fluorescent protein. The macroscopic view of the plate in visible light (2nd from left) shows a standard bacterial photograph. The bacteria were then imaged at 60× magnification on a fluorescent dissecting microscope (2nd from right). The result when the experiment is conducted using 4% agarose is shown on the right.

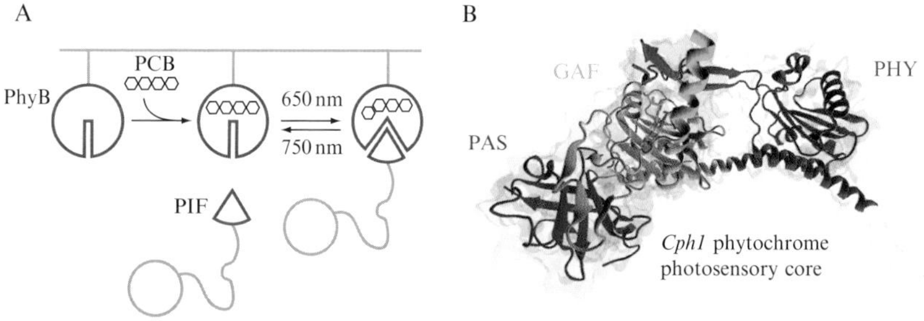

Jared E. Toettcher *et al.*, Figure 17.1 Schematic of the Phy–PIF interaction. (A) After incorporation of the chromophore PCB, the conformation of Phy can be controlled by exposure to two wavelengths of light (650 and 750 nm). The PIF domain binds only one of these domains with high affinity. By controlling the ratio of 650:750 nm light, the fraction of Phy in a state permissive for binding can be tuned, modulating the total amount of PIF recruitment. (B) Crystal structure of a fragment of the cyanobacterial phytochrome protein Cph1 bound to the small molecule chromophore PCB (shown as licorice) (PDB ID: 2VEA) (Essen *et al.*, 2008). The PAS, GAF, and PHY domains, each required for Phy–PIF interaction, are shown in purple, orange, and red, respectively, while the N terminal 26 amino acids are shown in green.

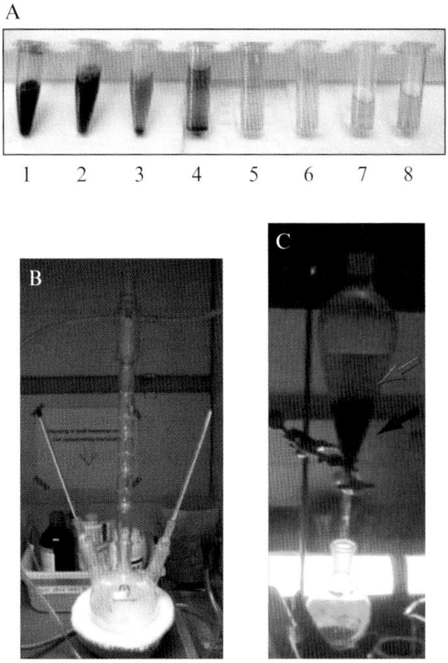

Jared E. Toettcher *et al.*, Figure 17.3 Key steps involved in the purification of PCB. (A) Samples collected at different stages of the PCB preparation procedure. Tubes contain the *Spirulina*-water mixture (1); the green supernatant after the first spin (2); the TCA protein precipitation, containing supernatant and pellet (3); the supernatant after each of three methanol washes (4–6); and the supernatant after the first and second 8 h methanolysis (7–8). (B) The assembled methanol reflux apparatus. The two thermometers should be placed in the vapor and fluid phases of the PCB-containing solution, respectively. Adjust heating until the temperature in both phases is the same, and the methanol solution is at a low boil. To prevent methanol loss through evaporation, flow cold water through the condenser (middle column of the apparatus). (C) Chloroform separation of PCB from the protein phase, photographed under a green safelight. Separation should result in a white, cloudy aqueous phase (red arrow) above a dark green chloroform phase (black arrow). If the aqueous phase retains a green color, add hydrochloric acid dropwise to acidify the solution, shake vigorously, and allow separation to occur again.

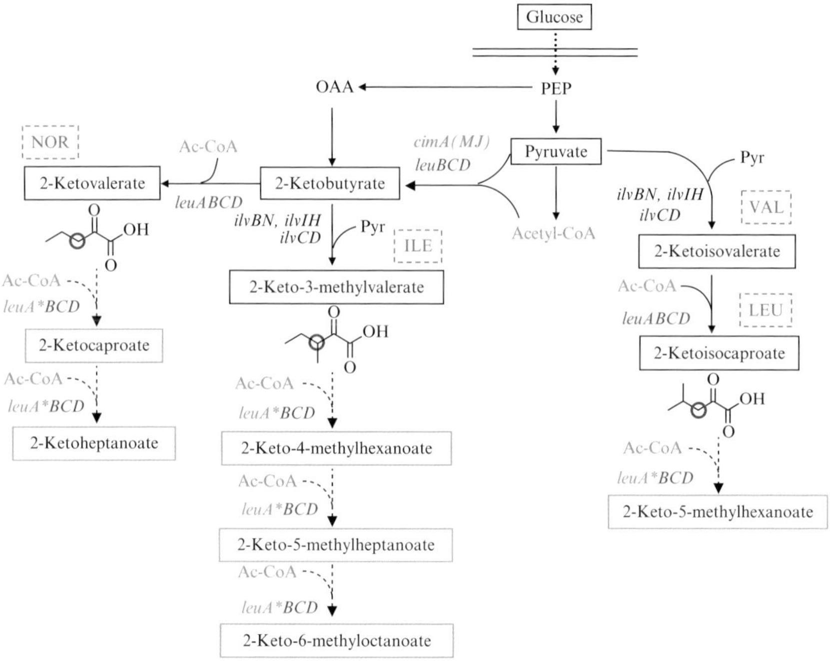

C.R. Shen and J.C. Liao, Figure 20.2 Various pathways leading to the synthesis of natural and nonnatural 2-ketoacids by recursive chain elongation catalyzed by *Escherichia coli* enzymes LeuABCD. Native ketoacids are boxed in black while nonnative ones are boxed in blue. Dashed arrows indicate the introduction of new pathways in *E. coli* via protein engineering of LeuA⋆ shown by the blue font. Red circles point out the position of carbon insertion. Biosyntheses of the specific amino acids are indicated with dashed grey boxes above their corresponding ketoacid precursors. NOR, norvaline; ILE, isoleucine; VAL, valine; LEU, leucine. *MJ: Methanococcus jannaschii.*

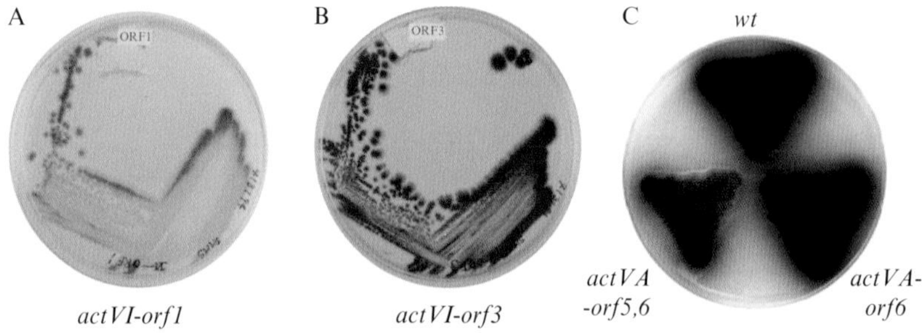

Marnix H. Medema *et al.*, Figure 21.2 (A) Brown pigment produced by *actVI-orf1* mutant (K. Ichinose, personal communication). (B) Red pigment produced by *actVI-orf3* mutant (K. Ichinose personal communication). (C) Yellowish brown pigment produced by the *actVA-orf5,6* mutant compared to the blue actinorhodin produced by wt and the *actVA-orf6* mutant (taken from Okamoto *et al.*, 2009 with permission from Elsevier Limited).

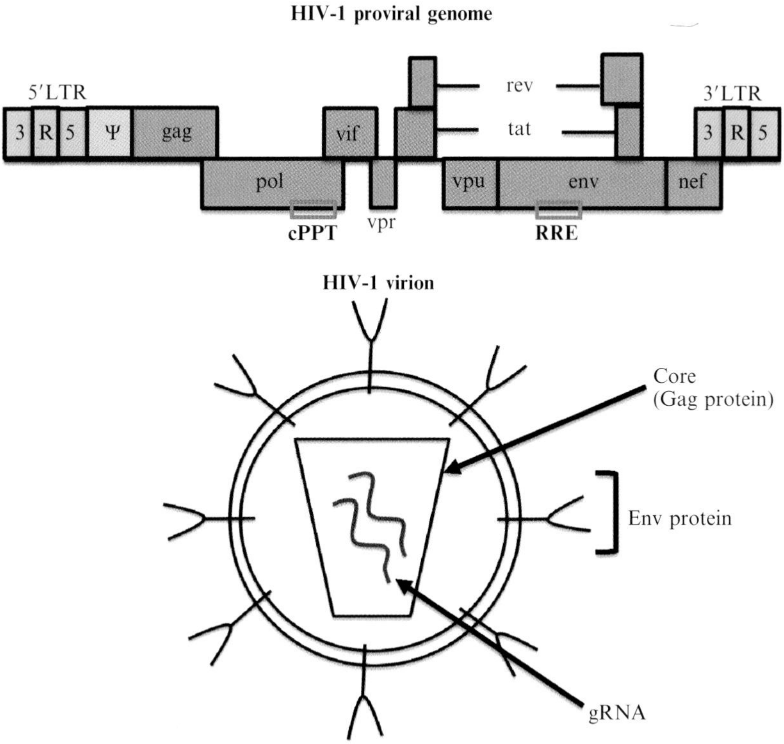

Kate Franz *et al.*, Figure 26.1 HIV-1 genomic organization and virion structure. HIV-1 genome contains nine open reading frames (in red) that code for 15 proteins. *cis*-acting elements important for reverse transcription, gene expression, and packaging are in green. Note that the integrated HIV-1 genome has a full 5′LTR (in blue).

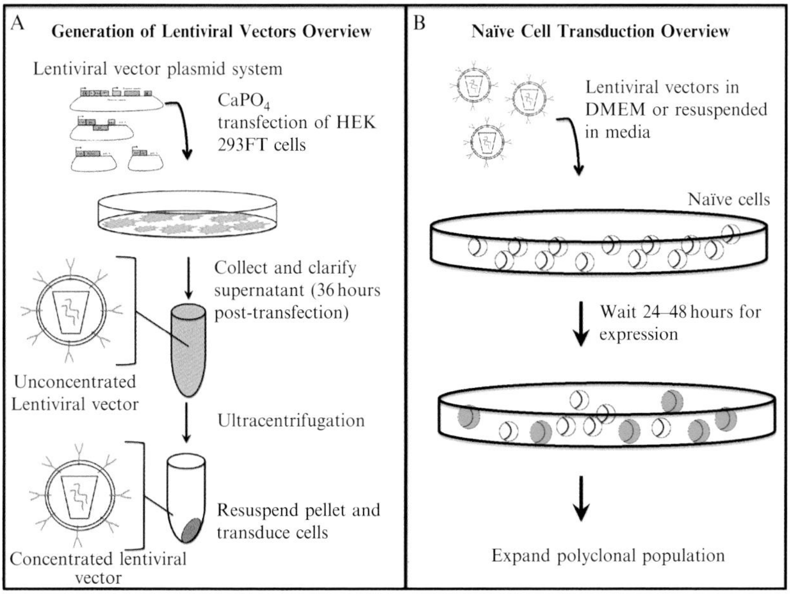

Kate Franz *et al.*, Figure 26.3 Overview of the protocol for production of lentiviral particles (A) and transduction of naïve cells (B).

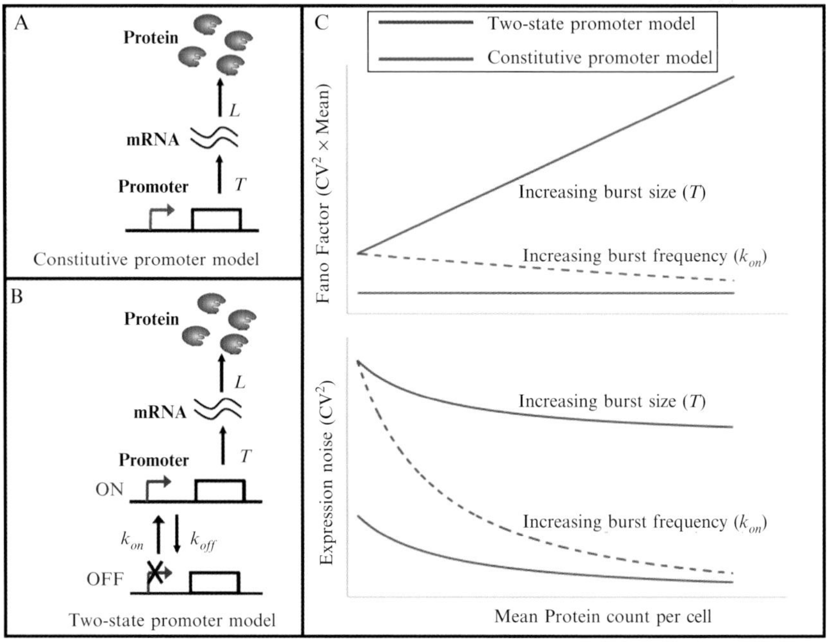

Kate Franz *et al.*, Figure 26.5 (A) Constitutive promoter model where mRNAs are produced continuously from the promoter one at a time. (B) Two-state promoter model where mRNAs are only produced when the promoter transitions to an ON state before returning to an OFF state. (C) Predictions for the scaling of gene-expression noise (CV^2) and Fano factor ($CV^2 \times$ Mean) as a function of mean protein count for different promoter architectures. Two-state promoter architecture is predicted to generate elevated levels of expression noise compared to a constitutive promoter architecture. In a constitutive promoter, CV^2 decreases with mean such that the Fano factor remains fixed (red line). In a two-state promoter model, CV^2 also decreases with mean but the Fano factor can either increase or decrease with mean depending on the mode of transcriptional activation (soild and dashed blue lines).